Neuropeptides in
Respiratory Medicine

CLINICAL ALLERGY AND IMMUNOLOGY

Series Editor

MICHAEL A. KALINER, M.D.

Medical Director
Institute for Asthma and Allergy
Washington, D.C.

ADDITIONAL VOLUMES IN PREPARATION

Neuropeptides in Respiratory Medicine

edited by

Michael A. Kaliner
Institute for Asthma and Allergy
Washington, D.C.

Peter J. Barnes
National Heart and Lung Institute
London, England

Gert H. H. Kunkel
Free University of Berlin
Berlin, Germany

James N. Baraniuk
Georgetown University Medical Center
Washington, D.C.

Marcel Dekker, Inc.   New York•Basel•Hong Kong

Library of Congress Cataloging-in-Publication Data

Neuropeptides in respiratory medicine / edited by Michael A. Kaliner
 ... [et al.].
     p.   cm. — (Clinical allergy and immunology ; 4)
   Based on a conference held in Pottsdam, Germany in May 1993.
   Includes bibliographical references and index.
   ISBN 0-8247-9199-1 (alk. paper)
   1. Respiratory organs—Pathophysiology—Congresses.
 2. Neuropeptides—Pathophysiology—Congresses. 3. Neuropeptides-
 -Physiological effect—Congresses.   I. Kaliner, Michael A.
 II. Series.
   [DNLM: 1. Neuropeptides—physiology—congresses. 2. Respiratory
 System—physiology—congresses.   WL 104 N49429 1994]
 RC711.N48 1994
 612.2—dc20
 DNLM/DLC
 for Library of Congress   94-5869
                                                                     CIP

The publisher offers discounts on this book when ordered in bulk quantities. For more information, write to Special Sales/Professional Marketing at the address below.

This book is printed on acid-free paper.

Marcel Dekker, Inc.
270 Madison Avenue, New York, New York 10016

Current printing (last digit):
10 9 8 7 6 5 4 3 2 1

PRINTED IN THE UNITED STATES OF AMERICA

Series Introduction

The decision to initiate a series of books on clinical allergy and immunology was based on the need to create a library of texts useful for both clinicians and scientists in these rapidly enlarging fields. There already are excellent textbooks providing overviews of the fields of allergy and immunology, and the scientific journals attempt to provide concise reviews of selected topics of interest. However, there is no library of books that takes relevant topics and expands them into texts, with the express purpose of making them of interest to both clinicians and scientists. Thus, this new series.

Clinical Allergy and Immunology will develop into the premier series of texts for our field. The initial book, edited by Howard Druce, was directed at sinusitis and represents the type of amalgamation of pathophysiology with treatment that will be reflected in most of the books yet to be published. The second book was an extremely timely discussion of the eosinophil edited by Gleich and Kay. It too will be a useful addition to the library of many clinicians and scientists. The series will subsequently include books focused on areas of allergy, immunology, specific diseases and important clinical entities, developing areas of research relevant to clinicians, major changes in therapeutic approaches, and research areas that warrant a text. The benefactors of this series should be the patient, because physicians will now have a series to turn to when searching for concise, but authoritative summaries of a field. The other major benefactors will be clinicians. Despite the fact that allergy is the single most prevalent chronic disease suffered by humans, only a small number of medical schools incorporate allergy in their curricula, and medical students are notably deficient in the knowledge of these diseases. As allergy is largely an outpatient specialty, few house officers see allergic patients, other than asthmatics and patients experiencing anaphylaxis. To clinicians, this series offers the chance to develop an extensive knowledge about selected topics, each chosen because of its clinical relevance.

Immunology is the field of the future. Our understanding of immune processes and their relevance to health and disease is advancing at a breakneck pace. As therapeutic approaches become defined, the relevant immunological advances will become incorporated into texts in this series.

The fourth book in the series is the product of many years of labor by the editors. The topic is neuropeptides and the specific focus is the role of neuropeptides in the airways, both in health and in disease. The book derives from a superb conference held at Pottsdam in May 1993 where neuropeptides were discussed for three days. The conference was organized by Gert Kunkel, and the book would not have developed without his organizational skills. He arranged for support for the conference from a number of pharmaceutical firms, for which we are all very appreciative. The meeting was a wonderful opportunity for basic scientists and clinicians from many fields to mingle and discuss neuropeptides from every possible aspect. The book includes discussions of some chapters derived from the original discussions at the meeting, and these discussions were expertly organized by my close colleague, Martha V. White, M.D. The addition of discussions always adds an exciting element to any book that comes from a lively meeting, such as occurred in Pottsdam.

In this work, the newest understanding of the biology of the neuropeptides has been summarized and discussed in a manner that makes the information concisely available to the reader, be he a clinician or scientist. As an appropriate extension of the scientific chapters, there are excellent summaries of the roles of the neuropeptides in respiratory diseases as well. Therefore, this excellent book personifies the type of volume to be included in this series: current, clinically relevant, and of use to clinicians and scientists alike.

Michael A. Kaliner

Preface

In 1986, two of us (P.J.B. and M.A.K.) organized a conference that led to publication of *The Airways: Neural Control in Health and Disease* in 1988 in the series of books edited by Claude Lenfant, Lung Biology in Health and Disease. It was clear at that conference that neuropeptides were an extremely fertile area of investigation in regard to their roles in the airways. As recently as 1986, only vasoactive intestinal peptides had been studied extensively in the airways, and almost no human models of airways had been examined. P.J.B and M.A.K. initiated an extensive array of studies on the roles of neuropeptides, using both animal and human models. We joined a select number of other investigators who were also working in this area to create a reasonable body of information. Thus, by 1990, sufficient data were available to support the publication of a book on the topic.

At this time M.A.K. was approached by two young investigators studying at the National Institutes of Health, Drs. James Baraniuk and Jens Lundgren, who wished to help create such a book. Unfortunately, the idea ran into problems with funding, and we were unable to get sufficient support to hold a conference on this area until Gert Kunkel volunteered to raise the funds and organize the meeting. Thus, in May 1993, 40 scientists and clinicians met in Pottsdam and spent three days discussing neuropeptides in the respiratory tract. The meeting was exciting, current, and stimulating, and the discussions that took place were equally stimulating.

This book, based on the meeting, deals first with the basic biology of neuropeptides in general and then discusses the most interesting of the neuropeptides, one by one. Neuropeptide receptors and antagonists are discussed, espe-

v

cially as they may apply to clinical medicine, as are the enzymes involved in metabolizing neuropeptides. Finally, there are a series of chapters on the various diseases in which neuropeptides may play a role. The book is testimony to the rapid growth of information about neuropeptides in the respiratory tract and the soon to be realized potential for neuropeptide antagonists to become useful therapeutic agents. We are now entering the stage of our understanding where we can apply the data on neuropeptide physiology to a variety of disease states, with the expectation that patients will benefit. Surely, this comprehensive book will assist the interested scientist or clinician in applying the extensive new information contained within to his or her needs and should stimulate more investigation into this rewarding area.

Michael A. Kaliner
Peter J. Barnes
Gert H. H. Kunkel
James N. Baraniuk

Contents

PHYSIOLOGICAL FUNCTIONS OF NEUROPEPTIDES IN THE
AIRWAYS

Contributors

James N. Baraniuk, B.Sc.(Hons), B.Sc.(Med), M.D., F.R.C.P.(C.) Department of Medicine, Georgetown University Medical Center, Washington, D.C.

Peter J. Barnes, D.M., D.Sc., F.R.C.P. Department of Thoracic Medicine, National Heart and Lung Institute, London, England

C. B. Baron, M.D. Department of Physiology, University of Pennsylvania School of Medicine, Philadelphia, Pennsylvania

Maria G. Belvisi, B.Sc., Ph.D. Department of Thoracic Medicine, National Heart and Lung Institute, London, England

Hugh P. J. Bennett, B.A., Ph.D. Endocrine Laboratory, Royal Victoria Hospital, and Department of Medicine, McGill University, Montreal, Quebec, Canada

Melvin Berger, M.D., Ph.D. Immunology and Allergy Division, Department of Pediatrics, Case Western Reserve University School of Medicine, and Rainbow Babies and Children's Hospital, Cleveland, Ohio

John Bienenstock, F.R.S.C., F.R.C.P.C. Faculty of Health Sciences, McMaster University, Hamilton, Ontario, Canada

Richard C. Boucher, Jr., M.D. Division of Pulmonary Diseases, Department of Medicine, University of North Carolina at Chapel Hill, Chapel Hill, North Carolina

R. F. Coburn, M.D. Department of Physiology, University of Pennsylvania School of Medicine, Philadelphia, Pennsylvania

Ervin G. Erdös, M.D.   Departments of Pharmacology and Anesthesiology, University of Illinois College of Medicine, Chicago, Illinois

Axel Fischer, M.D.   Department of Anatomy and Cell Biology, Philipps-University Marburg, Marburg, Germany

Claire M. Fraser, Ph.D.   Department of Molecular and Cellular Biology, The Institute for Genomic Research, Gaithersburg, Maryland

Jens Furkert, B.A., Dr.rer.nat.   Department of Peptide Pharmacology, Research Institute of Molecular Pharmacology, Berlin, Germany

Allan Garland, M.D.   Department of Medicine, UMDNJ–Robert Wood Johnson Medical School, New Brunswick, New Jersey

Edward J. Goetzl, M.D.   Departments of Medicine and Microbiology-Immunology, University of California Medical Center, San Francisco, California

Cristina Goso, M.D.   Department of Pharmacology, Menarini Ricerche Sud, Rome, Italy

Michael A. Kaliner, M.D.   Institute for Asthma and Allergy, Washington, D.C.

H. Benfer Kaltreider, M.D.   Department of Medicine, University of California Medical Center, San Francisco, California

Michael R. Knowles, M.D.   Clinical Services and Adult Cystic Fibrosis Clinic, Department of Medicine, Division of Pulmonary Disease, University of North Carolina at Chapel Hill, Chapel Hill, North Carolina

Doris Koesling, M.D.   Institute for Pharmacology, Free University of Berlin, Berlin, Germany

Marek L. Kowalski, M.D. Ph.D.   Department of Clinical Immunology and Allergy, Faculty of Medicine, Medical Academy, Łódź, Poland

Wolfgang Kummer, M.D.   Department of Anatomy and Cell Biology, Philipps-University Marburg, Marburg, Germany

Gert H. H. Kunkel, M.D.   Department of Clinical Immunology, and Asthma Clinic, Free University of Berlin, Berlin, Germany

Annika Laitinen, M.D., Ph.D.   Department of Anatomy, University of Helsinki, Helsinki, Finland

Lauri A. Laitinen, M.D., Ph.D.   Department of Pulmonary Medicine, University Central Hospital, Helsinki, Finland

Rudolf E. Lang, M.D.   Institute for Physiology and Pathophysiology, Philipps-University Marburg, Marburg, Germany

Eduardo Lazarowski, Ph.D. Division of Pulmonary Diseases, Department of Medicine, University of North Carolina at Chapel Hill, Chapel Hill, North Carolina

Norman H. Lee, Ph.D. Receptor Laboratory, Department of Molecular and Cellular Biology, The Institute for Genomic Research, Gaithersburg, Maryland

Xiongbin Lin, M.D. Institute for Normal and Pathological Physiology, Philipps-University Marburg, Marburg, Germany

R. Ilona Linnoila, M.D. Experimental Pathology Section, Biomarkers and Prevention Research Branch, National Cancer Institute, National Institutes of Health, Rockville, Maryland

Jan M. Lundberg, M.D. Department of Pharmacology, Karolinska Institute, Stockholm, Sweden

Jens D. Lundgren, M.D., D.Sc. Department of Infectious Diseases, Hvidovre Hospital, University of Copenhagen, Copenhagen, Denmark

Carlo A. Maggi, M.D. Department of Pharmacology, A. Menarini Pharmaceuticals, Florence, Italy

Judith Choi Wo Mak, Ph.D. Department of Thoracic Medicine, National Heart and Lung Institute, London, England

Stefano Manzini, Ph.D. Department of Pharmacology, Menarini Ricerche Sud, Rome, Italy

Sarah J. Mason, D.V.M., Ph.D. Division of Pulmonary Diseases, Department of Medicine, University of North Carolina at Chapel Hill, Chapel Hill, North Carolina

H. Matsumoto, M.D. Department of Physiology, University of Pennsylvania School of Medicine, Philadelphia, Pennsylvania

Bernd Mayer, Ph.D. Institute for Pharmacology and Toxicology, Karl-Franzens-University, Graz, Austria

Donald M. McDonald, M.D., Ph.D. Department of Anatomy and Cardiovascular Research Institute, University of California at San Francisco, San Francisco, California

Stefania Meini Department of Pharmacology, A. Menarini Pharmaceuticals, Florence, Italy

Agnes Modin, M.D. Department of Pharmacology, Karolinska Institute, Stockholm, Sweden

Terry W. Moody, Ph.D.   Biomarkers and Prevention Research Branch, National Cancer Institute, National Institutes of Health, Rockville, Maryland

Jay A. Nadel, M.D.   Cardiovascular Research Institute, University of California at San Francisco, San Francisco, California

Karen Nieber, M.D.   Research Institute of Molecular Pharmacology, Berlin, Germany

J. Niehus, M.D.   Department of Clinical Immunology and Asthma Clinic, Free University of Berlin, Berlin, Germany

Regis Olry, M.D.   Department of Chemistry and Biology, University of Quebec at Trois-Rivières, Trois-Rivières, Quebec, Canada

Julia M. Polak, D.Sc., M.D., F.R.C. Path.   Department of Histochemistry, Royal Postgraduate Medical School, London, England

H. Ratti, M.D.   Department of Clinical Immunology and Asthma OPD, Universitätsklinikum Rudolf Virchow (Wedding), Free University of Berlin, Berlin, Germany

Daniel W. Ray, M.D.   University of Chicago, Chicago, and Department of Medicine, Northwestern University and Evanston Hospital, Evanston, Illinois

Sami I. Said, M.D.   Departments of Medicine and Physiology, State University of New York at Stony Brook, Stony Brook, and, Northport VA Medical Center, Northport, New York

James H. Shelhamer, M.D.   Department of Critical Care Medicine, Clinical Center, National Institutes of Health, Bethesda, Maryland

Randal A. Skidgel, Ph.D.   Departments of Pharmacology and Anesthesiology, University of Illinois College of Medicine, Chicago, Illinois

Maria Śliwinska-Kowalska, M.D., Ph.D.   Institute of Occupational Medicine, Łódź, Poland

Julian Solway, M.D.   Section of Pulmonary and Critical Care Medicine, Department of Medicine, University of Chicago, Chicago, and Northwestern University and Evanston Hospital, Evanston, Illinois

David R. Springall, M.Sc., Ph.D.   Department of Histochemistry, Royal Postgraduate Medical School, London, England

Sunil P. Sreedharan, Ph.D.   Division of Allergy and Immunology, Department of Medicine, University of California Medical Center, San Francisco, California

M. Jackson Stutts, Ph.D. Division of Pulmonary Diseases, Department of Medicine, University of North Carolina at Chapel Hill, Chapel Hill, North Carolina

Arpad Szallasi, M.D. Department of Pharmacology, Menarini Ricerche Sud, Rome, Italy

Steven R. White, M.D. Department of Medicine, The University of Chicago, Chicago, Illinois

John Widdicombe, D.M., D.Phil., F.R.C.P. Department of Physiology, St. George's Hospital Medical School, London, England

M. Zhang, M.D. Department of Clinical Immunology and Asthma Clinic, Free University of Berlin, Berlin, Germany

Neuropeptides in Respiratory Medicine

Neuropeptides and Classic Innervation: Neural Structures in Human Airways

Annika Laitinen
University of Helsinki, Helsinki, Finland

Lauri A. Laitinen
University Central Hospital, Helsinki, Finland

I. INTRODUCTION

The earlier concepts of innervation of human airways have been undergoing change owing to expanded research. The autonomic nervous system of the airways is more complex than originally believed. The autonomic nervous system is composed of the sympathetic (adrenergic) and parasympathetic (cholinergic) systems consisting of efferent (motor) and afferent (sensory) pathways. The efferent fibers to the bronchial smooth muscle and submucosal glands arise from ganglia. In the parasympathetic system, airway ganglia receiving preganglionic fibers from the vagal nuclei from the central nervous system (CNS) are mainly located in the airway wall external to the smooth muscle and cartilage. Only a few smaller ganglia are situated in the submucosa. In the sympathetic system in humans, the preganglionic fibers leave the spinal cord and synapse with the prevertebrate ganglia. Postganglionic fibers then enter the lungs. It is generally accepted that virtually all afferent nerve fibers to the CNS from the airways travel in the vagus nerve (Richardson and Ferguson, 1979; Murray, 1986).

Sensory receptors of the airways can be described as those primarily involved in the physiological control of breathing pattern (slowly adapting pulmonary stretch receptors) and those primarily concerned with changes in pathological conditions (rapidly adapting irritant and C-fiber receptors) (Sant'Ambrogio, 1982; Mortola et al., 1975). Neuroepithelial bodies may also be receptors with afferent nerves (Lauweryns and Peuskens, 1972; Hung et al., 1973) responding to changes in the partial pressure of oxygen (Lauweryns and Cocelaere, 1973). Complete understanding of airway receptors is handicapped, however, by the fact that their histological appearance has not been completely analyzed (Widdicombe, 1982). Therefore, studies on structure-function relationships of the airway receptors have been difficult to conduct.

The efferent and afferent neural pathways are linked together by reflex mechanisms, as suggested by physiological studies (Mills et al., 1969; Sant'Ambrogio, 1982; Widdicombe, 1982; Empey et al., 1976). Earlier the reflex mechanisms were mainly considered to be CNS-mediated. Especially during recent years, more attention has been paid to possible local axonal reflexes with antidromic stimulation of the nerves (Jansco et al., 1967; Lundberg et al., 1983; Barnes, 1986b).

In addition to regulating airway smooth muscle tone and secretion from submucosal glands, the autonomic system may influence permeability and blood flow in the bronchial circulation; the function of neuroendocrine-like cells in the epithelium; transport of fluid across the epithelium; and release of mediators from inflammatory cells (Barnes, 1991). In addition to classical adrenergic and cholinergic mechanisms, there exists a third component of the autonomic nervous system, which is mediated by neither catecholamines nor acetylcholine and is thus called the "nonadrenergic noncholinergic nervous system" (NANC) (Campbell, 1971; Coburn and Tomita, 1973; Richardson and Beland, 1976; Irwin et al., 1980; Diamond and O'Donnell, 1980; Richardson, 1981; Doidge and Satchell, 1982; Barnes, 1986a). Current evidence has implicated neuropeptides as possible neurotransmitters for the NANC system in the airways (Matsuzaki et al., 1980; Laitinen, A., 1985b; Barnes, 1987; Springall et al., 1988; Dey et al., 1990).

In this chapter we concentrate on classic understanding of innervation and especially on the role of neuropeptides in humans.

II. INNERVATION OF AIRWAY SMOOTH MUSCLE

A. Cholinergic Innervation

Preganglionic parasympathetic fibers travel to the airways by way of the vagus nerves and synapse in parasympathetic ganglia located within the airway wall (Larsell, 1922). Short postganglionic fibers proceed from the ganglia to airway smooth muscle cells, where they release the excitatory neurotransmitter, acetylcholine, from axon varicosities.

There is abundant physiological evidence that the airway smooth muscle in humans is under cholinergic constrictor control (Richardson and Beland, 1976; Doidge and Satchell, 1982; Widdicombe, 1985; Laitinen et al., 1987a). Stimulation of the parasympathetic nerves causes constriction from the trachea to the small airways. Bronchoconstriction is most pronounced in airways with resting diameters of 1–5 mm and is less significant in airways smaller than 0.5 mm in diameter (Nadel, 1971). Alveolar ducts apparently are not affected by vagal stimulation. Human studies using helium-oxygen mixtures suggest that cholinergic blockade dilates central airways preferentially (McFadden, 1977). Physiological results are supported by findings from histological studies (Laitinen, A., et al., 1985b). If small agranular vesicles (SAGVs) contain acetylcholine, then the majority of motor nerves in the airways are cholinergic. Nerve fibers containing many clear vesicles can be seen using the electron microscope (Laitinen, A., et al., 1985b; Burnstock and Iwayama, 1971). The same nerves also contain large granular vesicles (LGVs) as well as mitochondria in varying proportions. Thus, these nerves may also represent the NANC system (McDonald, 1977). The distribution of cholinergic nerves in the respiratory tract is heterogeneous. Histochemically nerves that stain for acetylcholinesterase (AChE) are found in the human airways. They are abundant in the smooth muscle of the trachea and lower bronchi (Partanen et al., 1982; Laitinen, 1988). There is a decreased density of motor nerve ganglia in the small airways (Larsell, 1922).

B. Muscarinic Receptors

The neurotransmitter, acetylcholine, released from the cholinergic nerves, binds to muscarinic receptors. Recently, muscarinic receptor subtypes have been defined using both pharmacological (M_1, M_2, M_3, and M_4) (Lazareno et al., 1990) and genomic cloning (m1, m2, m3, m4, and m5) criteria (Hulme et al., 1990). In human lung the distribution of muscaric receptor subtype mRNAs has been determined using in situ hybridization; the m1, m2, and m3 mRNAs could be detected in the human lung, whereas m4 and m5 mRNAs could not be detected in any of the structures studied (alveolar wall, airway epithelium, airway smooth muscle, and submucosal glands). The presence of a small amount of m2 mRNA and a relatively large amount of m3 mRNA was demonstrated in human airway smooth muscle (Mak et al., 1992). Second-messenger pathways for M_2 receptors have been reported to mediate an inhibition of adenyl cyclase, whereas M_3 receptors have been shown to be involved in triggering airway smooth muscle contraction and phosphoinositide hydrolysis (Yang et al., 1991).

C. Sympathetic Innervation

It has been considered that the main inhibitory pathway in humans is the NANC pathway, with no detectable adrenergic component (Richardson and Beland, 1976; Davis et al., 1982). However, catecholamine-containing nerve fibers have been found in airway smooth muscle in humans (Partanen et al., 1982; Pack and Richardson, 1984). Although it is evident that adrenergic nerves are present in the smooth muscle layer of adult human airways, they are clearly fewer in number than AChE-containing nerve fibers, representing only a small percentage of the total innervation of smooth muscle in humans (Partanen et al., 1982). One reason may be that the parasympathetic ganglionic transmission is modulated by adrenergic stimuli. Histochemically, adrenergic axon profiles have been demonstrated in parasympathetic ganglia (Jacobowitz et al., 1973). There is evidence in humans and animals that α_1-, α_2-, and β_2-adrenoceptor agonists reduce ganglionic transmission (Grundström and Andersson, 1985; Skoogh and Svedmyr, 1983).

Nerves that are considered adrenergic by ultrastructural criteria have also been found in the proximity of smooth muscle (Laitinen, A., et al., 1985b). In these studies the nerve endings of autonomic axons have been distinguished by their distinct population of vesicles (Richardson, 1966; Hökfelt and Jonsson, 1968). According to these criteria, the nerve fibers (which, after conventional fixation, exhibit small granular vesicles 30–50 nm in diameter) are considered adrenergic.

D. Nonadrenergic, Noncholinergic Innervation: The Role of Neuropeptides

Immunohistochemical studies on neuropeptides have revealed that nerve fibers containing vasoactive intestinal peptide (VIP) are present in the airway smooth muscle layer (Dey et al., 1981; Laitinen, A., et al., 1985a; Carstairs and Barnes, 1986) and that their density exceeds that of adrenergic nerves in humans (Partanen et al., 1982; Laitinen et al., 1985a). Because VIP also exhibits the general features of a neurotransmitter and is a potent relaxant of smooth muscle (Said et al., 1980; Barnes, 1989; Belvisi et al., 1991), it may be that the neuronally mediated smooth muscle relaxation is mainly derived from a VIP-containing nervous system. In a recent study, VIP was not found in the airways ranging from 200 μm to 1.2 cm in diameter in five patients with asthma (Ollerenshaw et al., 1989). These investigators used the avidin-biotin-peroxidase complex technique for immunostaining, but the result was not confirmed by radioimmunoassay and extraction of VIP.

On the basis of light-microscopic observation of VIP-like immunoreactivity (Dey et al., 1981; Laitinen, A., et al., 1985a) in the human airways, immunohistochemistry and electron microscopy have been combined to obtain more information about the localization of VIP. These studies have led to the demonstration of VIP-like immunoreactivity in the vesicles of nerve profiles close to airway smooth muscle (Laitinen, A., et al., 1985a). Under the electron microscope, VIP-like immunoreactivity has been found in LGVs ranging from 90 to 210 nm. The number of LGVs varies between nerve profiles. Nerve profiles containing several LGVs (120–210 nm) are in the majority, but profiles containing only a few smaller, VIP-positive, large granules (90–150 nm) are also present (Laitinen, A., et al., 1985a). This tendency toward size-number correlation has been found to correspond to axon profiles containing SAGVs in addition to LGVs, identified by conventional electron microscopy (Laitinen, A., et al., 1985a,b).

Although the NANC neural pathway is predominantly inhibitory in its action on airway smooth muscle, it has also been shown to have an excitatory action under certain circumstances. It is not known whether this is the same neural pathway that inhibits smooth muscle in the guinea pig, since the possible mediator of the demonstrated excitation is substance P, which may be released from afferent neurons present in the airways of the guinea pig. SP-like immunoreactive nerves occur in the lower respiratory tract of many species, including rat, mouse, human (Lundberg et al., 1984), and rabbit (Laitinen et al., 1983). It has even been claimed that abundant amounts of substance P are seen within nerves in tissue from the lungs of asthma patients (Ollerenshaw et al., 1989). With regard to neuropeptides, the significance of cotransmission has been the subject of debate. Any neural fiber may contain two or more transmitters that may have more than one action on effector tissues. Examples of this have been shown in nonrespiratory tissues: S-hydroxytryptamine, neuropeptide Y, and noradrenaline all occur in the same sympathetic nerves to the heart. In afferent fibers, SP, neurokinin A, and calcitonin gene-related peptide (CGRP) may also occur in the same fibers (Barnes, 1991).

E. Afferent Innervation

In human bronchial smooth muscle, only a few studies of presumed afferent terminals are available (Jabonero and Sabadell, 1972). Physiological evidence indicates that the slowly adapting stretch receptors lie within or are associated with the smooth muscle of the airways (Widdicombe, 1982; Bartlett et al., 1976; Sant'Ambrogio et al., 1977). Afferent nerves have been identified at this location in several species by light and electron microscopy (Krauhs, 1984; Pack et al., 1984). In human bronchial smooth muscle, only a few studies of afferent terminals are available (Larsell and Dow, 1933; Jabonero and Sabadell, 1972).

III. NEURAL PATHWAYS OF BRONCHIAL GLANDS

Physiological studies show that the dominant innervation of submucosal glands is cholinergic and parasympathetic (Phipps, 1981). However, it can also be shown that secretion is induced by sympathetic adrenergic pathways, and probably by the NANC system as well (Peatfield and Richardson, 1983). Histochemical studies have shown the presence of AChE-containing fibers around the glands (Partanen et al., 1982), and electron microscopic studies show the presence of SAGVs in axon profiles close to the gland cells. Histology does not give definite evidence of a cholinergic innervation but is considered to support it. Sympathetic nerve stimulation promotes secretion from submucosal glands, as does the application of β-adrenoceptor agonists to the glands in vitro (Phipps, 1981).

Histological studies support the presence of an adrenergic innervation (Partanen et al., 1982; Laitinen, 1988; Pack and Richardson, 1984). Fluorescence histochemistry shows that nerves containing catecholamines lie between acini as well as in the acini between gland cells. Nerve profiles containing small granular vesicles presumed to be adrenergic are located close to gland cells in humans.

VIP and probably also substance P (Peatfield et al., 1983) can promote glandular secretion. VIP-like immunoreactive nerves have been described in bronchial glands in human airways (Dey et al., 1981; Laitinen, A., et al., 1985a). Ultrastructurally. VIP-like immunoreactivity was localized in LGVs ranging from 90 to 210 nm (Laitinen, A., et al., 1985a).

IV. NEURAL ELEMENTS OF THE AIRWAY VASCULAR BED

A. Classic Innervation

Physiological studies suggest that the main control of the tracheobronchial vascular bed is sympathetic and adrenergic, with alfa-adrenoceptors mediating constrictor effects (Hebb, 1969; Widdicombe, 1993). This conclusion has histological support with both histochemical fluorescence methods and electron microscopy (Partanen et al., 1982). Whether parasympathetic cholinergic fibers to the airway vasculature can cause relaxation is more controversial. In animal studies, parasympathetic nerves dilate the blood vessels (Laitinen et al., 1987b; Martling et al., 1985). This action can be mimicked by neurotransmitters such as acetylcholine and vasoactive intestinal polypeptide or peptide histidine isoleucine or methionine (PHI/PHM) (Laitinen, L. A., et al., 1987a,b,c). However, histological studies do not support a close cholinergic innervation of the vessels. In another study of the parasympathetic nervous control of tracheal vascular resistance in the dog (Laitinen, L. A., et al., 1987b), the vasodilator response to electrical stimulation of the superior laryngeal nerve was halved after atropine, indicating a cholinergic mechanism. The residual vasodilatation was reduced by hexamethonium, which would block orthodromic synaptic pathways to the trachea; this result suggests that an antidromic vasodilator pathway can be activated.

It has been proposed that noncholinergic vasodilatation can be elicited principally in two different ways: by parasympathetic preganglionic nerve stimulation where the vasodilatation is abolished by hexamethonium (Martling et al., 1985) and by antidromic stimulation of sensory nerves (Lundblad, 1984), which is hexamethonium-resistant. The parasympathetic noncholinergic vasodilatation is probably due to the release of peptides such as VIP and PHI from postganglionic neurons (Lundberg, 1981), whereas the second type of stimulation is due to antidromic impulse propagation and the release of neurokinins such as SP (Lundblad, 1984).

B. The Role of Neuropeptides

It has been widely accepted that neuropeptides act as neurotransmitters of the third nervous system, the NANC innervation. There are many examples of NANC innervation of blood vessels (Burnstock, 1983; Barnes, 1986a; Polak and Bloom, 1982). A number of peptides have been found to be related to perivascular nerves (Dey et al., 1981; Laitinen, A., et al., 1985a; McDonald et al., 1988). This observation is in accordance with electron microscopic findings showing nerves containing LGVs that may represent peptidergic nerves. Among peptides abundantly distributed in the vicinity of bronchial and pulmonary vessels are SP, CGRP, neuropeptide tyrosine (NPY), VIP, PHI (McDonald et al., 1988), PHM, and galanin (Polak and Bloom, 1985).

V. NEURAL ELEMENTS OF AIRWAY GANGLIA

The intrapulmonary microganglia are mostly located inside the extrachondrial plexuses, but a few are also present subchondrially as well as being located close to periarterial plexuses. These plexuses contain AChE-positive fibers, and the nerve cell bodies inside the microganglia give a positive reaction. Extrachondrially the microganglia are located inside the plexuses, whereas subchondrially the AChE-positive nerve cell bodies are either inside or beside the nerve bundles emerging in the tissue. Catecholamine-positive nerve fibers have been located around the nerve cell bodies in some microganglia in human airways (Partanen et al., 1982). In the human upper respiratory tract, VIP-immunoreactive nerve cell bodies have been observed in local microganglia (Dey et al., 1981; Uddman and Sundler, 1979). VIP-immunoreactive nerve fibers have also been reported in the human vagus nerve (Lundberg et al., 1979). VIP-immunoreactive nerve cell bodies have not been observed in the microganglia of the lower respiratory tract, but instead a dense network of VIP-immunoreactive nerve fibers has been found to be scattered around the nerve cell bodies of these microganglia.

VI. THE HUMAN AIRWAY EPITHELIAL NEURAL PATHWAYS

A. Classic Innervation

The airway epithelium serves as a primary target organ for exogenous luminal irritants. Several physiological studies, as well as theories of mechanisms behind airway diseases, suggest that excitation of afferent receptors in the epithelium initiates reflexes mediated to bronchial smooth muscle, glands, and bronchial vessel.

Nerve fibers have been found in the bronchial airway epithelium in several species, including humans. The human bronchial epithelium receives terminal fibers from the peribronchial plexus (Larsell and Dow, 1933; Gaylor, 1934; Pessacq, 1971). The earliest observations were made by light microscopy using different staining methods such as methylene blue and silver (Larsell and Dow, 1933; Gaylor, 1934). The presence of intraepithelial axons in both animals and humans has been confirmed at the ultrastructural level using electron microscopy (Richardson, 1979). In human tracheal epithelium, Rhodin (1966), using electron microscopy, described nonmyelinated axons that contained only neurotubuli and no vesicles. Animal studies have proposed the presence of different types of axonal profiles at ultrastructural level (Jeffery and Reid, 1973; King et al., 1974; Das et al., 1978; Lauweryns et al., 1985). In humans, the published reports include illustration of single intraepithelial nerve fibers (Rhodin, 1966; Laitinen, L.A., et al., 1985) and evaluation of their number and distribution (Laitinen, 1985a).

The concentration of intraepithelial nerves at various airway levels in humans has been established (Laitinen, 1985a). In the epithelium, axon profiles seem to have two predominant locations: either close to the airway lumen or at the base of the epithelium close to the basement membrane. Nerve fibers penetrating the basement membrane and traversing the epithelium from the base to the luminal side are also present. Since only a few nerves are seen to penetrate the basement membrane and enter the epithelium, probably the nerve fibers entering the airway epithelium divide further and form several axon profiles. There is a difference between different airway levels in the location of intraepithelial nerves. Nerves near the lumen are mainly found in the larger airways and relatively few are observed in the smaller airways (Laitinen, 1985a). This is in accordance with the findings of previous animal studies (Fillenz and Woods, 1970; Jeffery and Reid, 1973; King et al., 1974; Das et al., 1978). Ultrastructural studies in animals have rarely shown nerves in the intrapulmonary airways (Richardson, 1979).

The possible functional role of nerves cannot be judged on the morphological data only. The only definitive morphological methods of locating and identifying afferent nerves in the airways are the use of intracellular labels (Bower et al., 1978; Lacey, 1978) and degeneration studies (Das et al., 1979; Pack et al., 1984). Since such studies are difficult to perform in humans, the ultrastructure of axon profiles in human airway epithelium has to be compared with more definitive findings in other species, and with nerves in some sensory organs in humans. The axon profiles containing mitochondria and some small agranular vesicles conform well to the classic morphological criteria for afferent axonal fibers (Cauna et al., 1969; King et al., 1974; Das et al., 1978; Hoyes and Barber, 1981). Similar axon profiles have been described by several authors in the airway epithelium of several animal species (Hung et al., 1973; Jeffery and Reid, 1973; King et al., 1974; Das et al., 1978). These nerves may represent a part of a sensory unit arising from the lung and representing the so-called rapidly adapting "irritant receptors."

B. Granule-Containing (APUD Cells)

Granule-containing (GC) cells have been described in adult (Bensch et al., 1965; Lauweryns and Goddeeris, 1975) and fetal (Lauweryns and Peuskens, 1972) human lung. The airway epithelium contains both individual granule-containing cells and also groups of cells defined as neuroepithelial bodies (NEB) (Lauweryns and Peuskens, 1972; Gould, 1983; DiAugustine and Sonstegard, 1984; Laitinen,

1988). Various names have been given to these cells based on morphological and cytochemical characteristics. The single cells have been referred to as Feyrter cells; Kulchitsky cells; argyrophil, fluorescent, and granulated (AFG) cells; and neuroendocrine-like cells (diAugustine and Sonstegard, 1984); and by cytochemical characteristics the GC cells were classified as members of the widespread amine precursor uptake and decarboxylation (APUD) cell system (Pearse, 1969). These cells may have a neurosecretory function. In humans, GC cells contain biogenic monoamine and peptide hormones (Pearse, 1976; DiAugustine and Sonstegard, 1984).

The cells usually are triangular and rest on the basement membrane. The larger basal part of the cells contains many dense-cored vesicles. Often a fenestrated capillary in the lamina propria opposes the basement membrane at the point where the granule-containing cells is located (Laitinen, 1988). Nerve profiles have been observed close to single granule-containing cells (Laitinen, 1988) and NEBs (Lauweryns et al., 1985). The nerve profiles contain clear vesicles of variable size, a few larger vesicles, and mitochondria. When synaptic organizations were found between GC cells and neuronal profiles, the dense projections were observed on each side of the synapses. The function of these nerves has not been established physiologically, but experimental degeneration studies show that the majority of them are afferent (Lauweryns et al., 1985), and the most detailed investigations indicate that the NEBs may be stimulated to release their mediators under the influence of hypoxia.

The apical pole of the cells could contact the airway lumen (Laitinen, 1988). One function of CG cells may be sensory: a GC cell body sends a long, narrow "fingertip" through thick epithelium to reach the lumen, and this fingertip may act as a receptor unit for stimuli coming through the airways. The granule-containing cells, by releasing their dense core vesicles, may produce their effects via either the nervous system or the blood circulation. Thus, either a neural stimulation or a stimulus from the airway lumen to the amine-containing cells may release amines and possibly also neuropeptides into the local bloodstream to be carried into the deeper mucosa and smooth muscle.

C. Neuropeptides

Although it should be safe to assume that intraepidermal endings are afferent, one cannot exclude the possibility that these axons in the respiratory epithelium might be efferent to either ciliated or mucous cells. Nevertheless, histochemical studies have not revealed AChE- (Partanen et al., 1982) or catecholamine-positive (Partanen et al., 1982; Pack and Richardson, 1984) nerve fibers in human airway epithelium. Moreover, the human epithelium of secondary bronchi or of the more distal parts of the bronchial tree does not receive VIP-like immunoreactive nerve fibers (Dey et al., 1981; Laitinen, A., et al., 1985a).

Sensory neuropeptides such as neurokinin A (NKA) and substance P (SP), also called tachykinins, have been identified in mammalian and human airways (Lundberg et al., 1984; Barnes, 1987; Martling et al., 1987; Naline et al., 1989). However, controversial results suggesting lack of SP in the human airway epithelium have also been reported (Laitinen et al., 1983). In addition to tachykinins, peripheral nociceptive sensory neurons contain CGRP (Martling et al., 1987). SP, NKA, and CGRP may be released simultaneously from the same neurons, and their corelease may cause either synergistic or antagonistic effects in target tissue (Saria et al., 1988). In the respiratory tract, stimulation of the sensory nerves results in a series of responses, including vasodilatation, increased vascular permeability, gland secretion, airway smooth muscle contraction, increased neutrophil adherence, and cough. This cascade of events is known as "neurogenic inflammation." The actions are due to the release of substance P and other neuropeptides from the sensory nerves (McDonald, 1988; Nadel, 1991).

VII. NEUROPEPTIDES IN ASTHMA

It has been proposed that the airway inflammation in asthma (Laitinen et al., 1993a) has a neurogenic nature owing to activation of C-fiber afferent nerve endings (Barnes, 1986b). This could lead to axon reflexes with local release of the tachykinins (Barnes, 1987). There is accumulating evidence from animal models in vitro and in vivo that these sensory neuropeptides produce neurogenic inflammation and thereby many of the pathological features of asthma (Laitinen, L.A., and Laitinen, A., 1990; Laitinen, L.A., et al., 1993b), including contraction of the airway smooth muscle, plasma extravasation, mucus hypersecretion, and increased neutrophil adherence (Barnes, 1991; Nadel and Borson, 1991).

Neurogenic inflammatory responses are normally limited by enzymatic degradation. Although substance P can be degraded by various enzymes, most neurogenic inflammatory responses are modulated normally to a large extent by a membrane-bound enzyme, neutral endopeptidase (NEP), also called enkephalinase. This enzyme was described in 1974 by Kerr and Kenny (1974). However, its physiological implications have only recently been found. It cleaves substance P (Skidgel et al., 1984) and bradykinin effectively, but it cleaves CGRP and some other peptides only slowly (Katayama et al., 1991). As the release sites of neuropeptides in the sensory nerves are concentrated in the airway epithelium in close association with the NEP containing basal cells, neuropeptides could be cleaved and inactivated near their sites of release. NEP may be present on the surface of the cells that contain peptide receptors (e.g., glands, smooth muscle, and postcapillary venules) (Ellis and Framer, 1989). It has been suggested that when the released neuropeptides diffuse from the release sites to the receptors, NEP on the surface of the target cell reduces the concentration of the neuropeptide reaching the cell surface receptor. NEP thus may modulate some of the neurogenic inflammatory responses (Nadel, 1992).

VIII. EFFECT OF THERAPY ON INNERVATION

Corticosteroids have been shown to increase the expression of NEP in cultured human airway epithelial cells (Borson and Gruenert, 1991). In humans, inhaled SP and NKA lead to bronchoconstriction in asthmatics but not in normal control subjects (Joos et al., 1987). Recently, it has been shown that inhibition of neutral endopeptidase in humans with inhaled thiorphan enhances bronchoconstriction to NKA in nonasthmatic subjects without affecting metacholine-induced broncho-constriction. The results suggest that NEP can modulate neuropeptide responses in human airways in vivo (Cheung et al., 1992).

In a recent study (Laitinen et al., 1992), the number of intraepithelial axon profiles was expressed as axon profile number/mm^2. The numbers were counted in the airway epithelium of newly defined asthmatic patients, who were randomized to receive either inhaled corticosteroid or inhaled β_2-agonist. After inhaled corticosteroid treatment, the number of intraepithelial nerves increased signifi-cantly from a mean value of 10.6 to 38 nerves/mm^2, whereas after β_2-agonist therapy, no increase in the number of intraepithelial nerves was observed. The increase in the number of the epithelial nerves was associated with structural improvement of the airway epithelium and a decrease in the overall inflammatory reaction. It could be suggested that the normal epithelial structure may be needed for the normal growth of the epithelial nerves.

REFERENCES

Barnes, P.J. (1986a). Neural control of human airways in health and disease. *Am. Rev. Respir. Dis. 134*:1289–1314.

Barnes, P.J. (1986b). Asthma as an axonreflex. *Lancet 1*:242–245.

Barnes, P.J. (1987). Neuropeptides in the lung: localization, function and pathophysiologic implications. *J. Allergy Clin. Immunol. 79*:285–295.

Barnes, P.J. (1989). Vasoactive intestinal peptide and asthma. *N. Engl. J. Med. 321*:1128–1129.

Barnes, P.J. (1991). Neuropeptides and asthma. *Am. Rev. Respir. Dis. 143* (Suppl):28–32.

Bartlett, D., Jeffery, D., Sant'Ambrogio, G., and Wise, J.C.M. (1976). Location of stretch receptors in the trachea and bronchi of the dog. *J. Physiol. 258*:419–420.

Belvisi, M.G., Stretton, D., Verleden, G.M., Yacoub, M.H., and Barnes, P.J. (1991). Inhibitory NANC nerves in human tracheal smooth muscle: involvement of VIP and NO. *Am. Rev. Respir. Dis.*

Bensch, K.G., Gordon, G.B., and Miller, L.R. (1965). Studies on the bronchial counter-part of the Kultschitzky (Argentaffin) cell and innervation of bronchial glands. *J. Ultrastruct. Res. 12*:668–686.

Borson, D.B., and Gruenert, D.C. (1991). Glucocorticoids induce neutral endopeptidase in transformed human tracheal epithelial cells. *Am. J. Physiol. 260*:L83–L89.

Bower, A.J., Parker, S., and Molony, V. (1978). An autoradiographic study of the afferent innervation of the trachea, syrinx and extrapulmonary primary bronchus of *Gallus gallus domesticus*. *J. Anat. 126*:169–180.

Burnstock, G., and Griffith, S.G. (1983). Neurohumoral control of the vasculature. In: *Biology and Pathology of the Vessel Wall. A Modern Appraisal*, (Woolf N., ed.) Praeger, Eastbourne, England, pp. 15–40.

Burnstock, H., and Iwayama, T. (1971). Fine-structural identification of autonomic nerves and their relation to smooth muscle. *Prog. Brain Res. 34*:389–404.

Campbell, G. (1971). Autonomic innervation of the lung musculature of a toad (*Bufo marinus*). *Comp. Gen. Pharmacol. 2*:281–286.

Carstairs, J.R., and Barnes, P.J. (1986). Visualization of vasoactive intestinal peptide receptors in human and guinea pig lung. *J. Pharmacol. Exp. Ther. 239*:249–255.

Cauna, N., Hinderer, K.H., and Wentges, R.T. (1969). Sensory receptor organs of the human nasal respiratory mucosa. *Am. J. Anat. 14*:295–300.

Cheung, D., Bel, E.H., Den Hartig, J., Dijkman, J.H., and Sterk, P.J. (1992). The effect of an inhaled neutral endopeptidase inhibitor, thiorphan, on airway responses to neurokinin A in normal humans in vivo. *Am. Rev. Respir. Dis. 145*:1275–1280.

Coburn, R.F., and Tomita, T. (1973). Evidence for nonadrenergic inhibitory nerves in the guinea pig trachealis muscle. *Am. J. Physiol. 224*:1072–1080.

Das, R.M., Jeffrey, P.K., and Widdicombe, J.G. (1978). The epithelial innervation of the lower respiratory tract of the cat. *J. Anat. 126*:123–131.

Davis, C., Kannan, M.S., Jones, T.R., and Daniel, E.E. (1982). Control of human airway smooth muscle: in vitro studies. *J. Appl. Physiol. 53*:1080–1087.

Dey, R.D., Shannon, W.A., Jr., and Said, S.I. (1981). Localization of VIP-immunoreactive nerves in airways and pulmonary vessels of dogs, cats, and human subjects. *Cell Tissue Res. 220*:231–238.

Dey, R.D., Mitchell, H.W., and Coburn, R.F. (1990). Organization and development of peptide-containing neurons in the airways. *Am. J. Respir. Cell Mol. Biol. 3*:187.

Diamond, L., and O'Donnell, M. (1980). A non-adrenergic vagal inhibitory pathway to feline airways. *Science 208*:185–188.

DiAugustine, R.P., and Sonstegard, K.S. (1984). Neuroendocrine-like (small granule) epithelial cells of the lung. *Environ. Health Perspect. 55*:271–295.

Doidge, J.M., and Satchell, D.G. (1982). Adrenergic and non-adrenergic inhibitory nerves in mammalian airways. *J. Auton. Nerv. Syst. 5*:83–99.

Ellis, J.L., and Framer, S.G. (1989). Effects of peptidases on nonadrenergic, noncholinergic inhibitory responses of tracheal smooth muscle: a comparison with effects on VIP- and PHI-induced relaxation. *Br. J. Pharmacol. 96*:521–526.

Empey, D.W., Laitinen, L.A., Jacobs, L., Gold, W.N., and Nadel, J.A. (1976). Mechanisms of bronchial hyperreactivity in normal subjects after upper respiratory tract infection. *Am. Rev. Respir. Dis. 113*:131–139.

Fillenz, M., and Woods, R.I. (1970). Sensory innervation of the airways. In: *Breathing: Hering–Breuer Centenary Symposium* (Porter, R., ed.) Churchill, London, pp. 101–107.

Gaylor, J.B. (1934). The intrinsic nervous mechanism of the human lung. *Brain 57*: 143–160.

Gould, V.E. (1983). The endocrine lung (editorial). *Lab. Invest. 48*:507–509.

Grundström, N., and Andersson, R.G.G. (1985). Inhibition of the cholinergic neurotransmission in human airways via prejunctional α2-adrenoceptors. *Acta Physiol. Scand. 125*:513.

Hebb, C. (1969). Motor innervation of the pulmonary blood vessels of mammals. In: *The Pulmonary Circulation and Interstitial Space* (Fishman A.P., Hecht, H.H., eds.), University of Chicago Press, Chicago, pp. 195–222.

Hoyes, A.D., and Barber, P. (1981). Morphology and response to vagus nerve section of the intraepithelial axons of the rat trachea. A quantitative ultrastructural study. *J. Anat. 132*: 331–339.

Hulme, E.C., Birdll, N.J.M., and Buckley, N.J. (1990). Muscarinic receptor subtypes. *Annu. Rev. Pharmacol. Toxicol. 30*:633–673.

Hung, K.S., Hertwech, M.S., Hardy, J.D., and Loosli, C.G. (1973). Ultrastructure of nerves and associated cells in bronchiolar epithelium of the mouse lung. *J. Ultrastruct. Res. 43*:426–437.

Irwin, C.G., Boileau, R., Trembley, J., Martin, R.R., and MacKlem, P.T. (1980). Bronchodilatation: noncholinergic mediation demonstrated in vivo in the cat. *Science 207*: 791–792.

Jabonero, V., and Sabadell, J. (1972). Die sensible Innervation der glatten Muskulatur de Luftwege. *Z. Mikrosk. Anat. Forsch. 86*:213–243.

Jacobowitz, D., et al. (1973). Histofluorescent study of catecholamine containing elements in cholinergic ganglion from the calf and dog lung. *Proc. Soc. Exp. Biol. Med. 144*:464.

Jancso, N., Jancso-Gabor, A., and Szolcsanyi, J. (1967). Direct evidence for neurogenic inflammation and its prevention by denervation and by pretreatment with capsaicin. *Br. J. Pharmacol. 31*:131–151.

Jeffrey, P., and Reid, L. (1973). Intra-epithelial nerves in normal rat airways: a quantitative electron microscopic study. *J. Anat. 114*:35–45.

Joos, G., Pauwels, R., and Van Straeten, M. (1987). Effect of inhaled substance P and neurokinin A on the airways of normal and asthmatic subjects. *Thorax 42*:779–783.

Katayama, M., Nadel, J.A., Bunnett, N.W., Di Maria, G.U., Haxhiu, M., and Borson, D.B. (1991). Catabolism of calcitonin gene-related peptide and substance P by neural endopeptidase. *Peptides 12*:563–567.

Kerr, M.A., and Kenny, A.J. (1974). The purification and specificity of a neutral endopeptidase from rabbit kidney brush border. *Biochem. J. 137*:477–488.

King, A.S., McLelland, J., Cook, R.D., King, D.A., and Walsh, C. (1974). The ultrastructure of afferent nerve endings in the avian lung. *Respir. Physiol. 22*:21–40.

Krauhs, J.M. (1984). Morphology of presumptive slowly adapting receptors in dog trachea. *Anat. Rec. 210*:73–85.

Lacey, M. (1978). Studies of pulmonary innervation in the rat using horseradish peroxidase as a neuronal marker. *J. Physiol. 277*:16–17P.

Laitinen, A. (1985a). Ultrastructural organization of intra-epithelial nerves in the human airway tract. *Thorax 40*:488–492.

Laitinen, A. (1985b). Autonomic innervation of the human respiratory tract as revealed by histochemical and ultrastructural methods. *Eur. J. Respir. Dis. 66* (Suppl):140.

Laitinen, L.A., and Laitinen, A. (1987). Innervation of airway smooth muscle. *Am. Rev. Respir. Dis. 136*(4):S38–S42.

Laitinen, L.A., and Laitinen, A. (1990). Overview of the pathology of asthma. In: *Pharmacology of Asthma, Handbook of Experimental Pharmacology* (Page, C.P., and Barnes, P.J., eds.), Springer Verlag, New York.

Laitinen, A., Partanen, M., Hervonen, A., Pelto-Huikko, M., and Laitinen, L.A. (1985a). VIP like immunoreactive nerves in the human lower respiratory tract: light and electron microscopic study. *Histochemistry 82*:313–319.

Laitinen, A., Partanen, M., Hervonen, A., and Laitinen, L.A. (1985b). Electron microscopic study on the innervation of the human lower respiratory tract: evidence of adrenergic nerves. *Eur. J. Respir. Dis. 67*:209–215.

Laitinen, L.A. (1988). Detailed analysis of neural elements in human airways. In: *The Airways: Neural Control in Health and Disease* (Kaliner, M.A., and Barnes, P.J., eds.), Marcel Dekker, New York, pp. 35–56.

Laitinen, L.A., Laitinen, A., Panula, P.A., Partanen, M., Tervo, K., and Tervo, T. (1983). Immunohistochemical demonstration of substance P in the lower respiratory tract of the rabbit and not of man. *Thorax 38*:531–536.

Laitinen, L.A., Heino, M., Laitinen, A., Kava, T., and Haahtela, T. (1985). Damage of the airway epithelium and bronchial reactivity in patients with asthma. *Am. Rev. Respir. Dis. 131*:599–606.

Laitinen, L.A., Robinson, N.P., Laitinen, A., and Widdicombe, J.G. (1986). Relationship between mucosal thickness and vascular resistance in dogs. *J. Appl. Physiol. 61*(6): 2186–2193.

Laitinen, L.A., Laitinen, A., Salonen, R.O., and Widdicombe, J.G. (1987a). Vascular actions of airway neuropeptides. *Am. Rev. Respir. Dis. 136*:S59–S64.

Laitinen, L.A., Laitinen, A., and Widdicombe, J.G. (1987b). Parasympathetic nervous control of tracheal vascular resistance in dogs. *J. Physiol. 385*:135–146.

Laitinen, L.A., Laitinen, A., and Widdicombe, J.G. (1987c). Dose-related effects of pharmacological mediators on tracheal vascular resistance in dogs. *Br. J. Pharmacol. 92*: 703–709.

Laitinen, L.A., Laitinen, A., and Haahtela, T. (1992). A comparative study of the effects of an inhaled corticosteroid, budesonide, and a beta$_2$-agonist, terbutaline on airway inflammation in newly diagnosed asthma: a randomized, double-blind, parallel-group controlled trial. *J. Allergy Clin. Immunol. 90*:32–42.

Laitinen, L.A., Laitinen, A., and Haahtela, T. (1993a). Airway mucosal inflammation even in patients with newly diagnosed asthma. *Am. Rev. Respir. Dis. 147*:697–704.

Laitinen, L.A., Laitinen, A., Haahtela, T., Vilkka, V., Spur, B.W., and Lee, T.H. (1993b). Leukotriene E4 and granulocytic infiltration into asthmatic airways. *Lancet 341*:989–990.

Lammers, J.-W.J., Minnette, P., McCusker, M., Chung, K.F., and Barnes, P.J. (1989). Capsaicin-induced bronchodilatation in mild asthmatic subjects: possible role of nonadrenergic inhibitory system. *J. Appl. Physiol. 66*:856–861.

Larsell, O. (1922). The ganglia, plexuses, and nerve terminations of the mammalian lung and pleura pulmonalis. *J. Comp. Neurol. 35*:97–132.

Larsell, D., and Dow, R.S. (1933). The innervation of the human lung. *Am. J. Anat. 52*: 125–146.

Lauweryns, J.M., and Cocelaere, M. (1973). Intrapulmonary neuro-epithelial bodies: hypoxia sensitive neuro-(chemo) receptors. *Experientia 29*:1384–1386.

Lauweryns, J.M., and Goddeeris, P. (1975). Neuroepithelial bodies in the human child and adult lung. *Am. Rev. Respir. Dis. 111*:469–476.

Lauweryns, J.M., and Peuskens, J.C. (1972). Neuroepithelial bodies (neuroreceptor of secretory organs?) in human infant bronchial and bronchiolar epithelium. *Anat. Rec.* *172*:471–482.

Lauweryns, J.M., van Lommel, A.T., and Dom, R.J. (1985). Innervation of rabbit intra-pulmonary neuroepithelial bodies. Quantitative ultrastructural study after vagotomy. *J. Neurol. Sci.* *67*:81–92.

Lazareno, S., Buckley, N.J., and Roberts, F. (1990). Characterization of muscarinic M4 binding sites in rabbit lung, chicken heart, and NG 108-15 cells. *Mol. Pharmacol. 38*: 805–815.

Lundberg, J.M. (1981). Evidence for coexistence os vasoactive intestinal polypeptide (VIP) and acetylcholine in neurons of cat exocrine glands. Morphological, biochemical and functional studies. *Acta Physiol. Scand. 496* (Suppl):1–57.

Lundberg, J.M., Hokfelt, T., Kewenter, J., et al. (1979). Substance P-, VIP- and enkephalin-like immunoreactivity in the human vagus nerve. *Gastroenterology 77*:468–471.

Lundberg, J.M., Martling, C.R., and Saria, A. (1983). Substance P and capsaicin-induced contraction of human bronchi. *Acta Physiol. Scand. 19*:49–53.

Lundberg, J.M., Hokfelt, T., Martling, C.R., Saria, A., and Cuello, C. (1984). Substance P-immunoreactive sensory nerves in the lower respiratory tract of various mammals including man. *Cell Tissue Res. 235*:251–261.

Lundblad, L. (1984). Protective reflexes and vascular effects in the nasal mucosa elicited by activation of capsaicin-sensitive substance P-immunoreactive trigeminal neurons. *Acta Physiol. Scand. 529* (Suppl):1–42.

Mak, J.C.W., Baraniuk, J.N., and Barnes, P.J. (1992). Localization of muscarinic receptor subtype mRNAs in human lung. *Am. J. Respir. Cell Mol. Biol. 7*:344–348.

Mann, S.P. (1971). The innervation of mammalian bronchial smooth muscle: the localization of catecholamines and cholinesterases. *Histochem. J. 3*:319–331.

Martling, C.-R., Anggard, A., and Lundberg, J.M. (1985). Noncholinergic vasodilatation in the tracheobronchial tree of the cat induced by vagal nerve stimulation. *Acta Physiol. Scand. 125*:343–346.

Martling, C.R., Theodorsson-Norheim, E., and Lundberg, J.M. (1987). Occurrence and effects of multiple tachykinins, substance P, neurokinin A and neuropeptide K in human lower airways. *Life Sci. 40*:1633–1643.

Matsuzaki, Y., Hamasaki, Y., and Said, S.I. (1980). Vasoactive intestinal peptide: a possible transmitter of nonadrenergic relaxation of guinea pig airways. *Science 210*:1252–1253.

McDonald, D. (1977). Structure–function relationships of chemoreceptive nerves in the carotid body. *Am. Rev. Respir. Dis. IIs*:193–207.

McDonald, D.M. (1988). Respiratory tract infections increase susceptibility to neurogenic inflammation in the rat trachea. *Am. Rev. Respir. Dis. 137*:1432–1440.

McDonald, D.M., Mitchell, R.A., Gabella, G., and Haskell, A. (1988). Neurogenic inflammation in the rat trachea. II. Identity and distribution of nerves mediating the increase in vascular permeability. *J. Neurocytol. 17*:605–628.

McFadden, E.R., Jr., et al. (1977). Predominant site of flow limitation and mechanisms of postexertional asthma. *J. Appl. Physiol. 42*:746.

Mills, J., Sellick, H., and Widdicombe, J.G. (1969). The role of lung irritant receptors in respiratory responses to multiple pulmonary embolism, anaphylaxis and histamine-induced bronchoconstriction. *J. Physiol. 203*:337–357.

Mortola, J., Sant'Ambrogio, G., and Clement, M.G. (1975). Localization of irritant receptors in the airways of dogs. *Respir. Physiol. 24*:107–114.

Murray, J.F. (1986). Lymphatic and nervous systems. In: *The Normal Lung*, W.B. Saunders, Philadelphia.

Nadel, J.A. (1991). Neutral endopeptidase modulates neurogenic inflammation. *Eur. Respir. J. 4*:745–754.

Nadel, J.A. (1992). Membrane-bound peptidases: endocrine, paracrine, and autocrine effects. *Am. J. Respir. Cell. Mol. Biol. 7*:469–470.

Nadel, J.A., and Borson, D.B. (1991). Modulation of neurogenetic inflammation by neutral endopeptidase. *Am. Rev. Respir. Dis. 143* (Suppl):33–36.

Nadel, J.A., Cabezas, G.A., and Austin, J.H.M. (1971). In vivo roentgenographic examination of parasympathetic innervation of small airways: use of powdered tantulum and a fine focal spot x-ray. *Invest. Radiol. 6*:9–17.

Naline, E., DeVillier, P., Drapeau, G., et al. (1989). Characterization of neurokinin effects and receptor selectivity in human isolated bronchi. *Am. Rev. Respir. Dis. 140*:679–686.

Ollerenshaw, S., Jarvis, D., Woolcock, A., Sullivan, C., and Scheibner, T. (1989). Absence of immunoreactive intestinal polypeptide in tissue from the lungs of patients with asthma. *N. Engl. J. Med. 320*:1244–1248.

Pack, R.J., and Richardson, P.S. (1984). The aminergic innervation of the human bronchus: a light and electron microscopic study. *J. Anat. 138*:493–502.

Pack, R.J., AlUgaily, L.H., and Widdicombe, J.G. (1984). The innervation of the trachea and extrapulmonary bronchi of the mouse. *Cell. Tissue Res. 238*:61–68.

Partanen, M., Laitinen, A., Hervonen, A., Toivanen, M., and Laitinen, L.A. (1982). Catecholamine and acetylcholinesterase containing nerves in human lower respiratory tract. *Histochemistry 76*:175–188.

Pearse, A.J. (1976). Neurotransmission and the APUD concept. In: *Chromaffin, Enterochromaffin and Related Cells* (Coupland, R.E., and Fujita, T., eds.), Elsevier, New York, pp. 147–154.

Peatfield, A.C., and Richardson, P.S. (1983). Evidence for noncholinergic, non-adrenergic nervous control of mucus secretion into the cat trachea. *J. Physiol. (Lond.) 342*: 335–345.

Peatfield, A.C., Barnes, P.J., Bratcher, C., Nadel, J.A., and Davies, B. (1983). Vasoactive intestinal peptide stimulates tracheal submucosal gland secretion in ferret. *Am. Rev. Respir. Dis. 128*:89–93.

Pessacq, T.P. (1971). The innervation of the lung of newborn children. *Acta Anat. 79*: 93–101.

Phipps, R.F. (1981). The airway mucociliary system. In: *International Review of Physiology. Respiratory Physiology III*. (Widdicombe, J.G., ed.), University Park Press, Baltimore, pp. 213–260.

Polak, J.M., and Bloom, S.R. (1982). Regulatory peptides in the respiratory tract of man and other animals. *Exp. Lung Res. 3*:313–328.

Polak, J.M., and Bloom, S.R. (1985). Occurrence and distribution of regulatory peptides in the respiratory tract. *Recent Results Cancer Res. 99*:1–6.

Richardson, K.C. (1966). Electron microscopic identification of autonomic nerve endings. *Nature 210*:756.

Richardson, J.B. (1979). Nerve supply to the lungs. *Am. Rev. Respir. Dis. 119*:785–802.

Richardson, J.B. (1981). Nonadrenergic inhibitory innervation of the lung. *Lung 159*: 315–322.

Richardson, J., and Beland, J. (1976). Nonadrenergic inhibitory nervous system in human airways. *J. Appl. Physiol. 41*:764–771.

Richardson, J.B., and Ferguson, C.C. (1979). Neuromusculator structure and function in the airways. *Fed. Proc. 38*:292–308.

Rhodin, J.A.G. (1966). Ultrastructure and function of the human tracheal mucosa. *Am. Rev. Respir. Dis. 93*:1–15.

Said, S.I., Giachetti, A., and Nicosia, S. (1980). VIP: possible functions as a neural peptide. In: *Neural Peptides and Neuronal Communication* (Costa, E., and Trabucchi, M., eds.), Raven Press, New York, pp. 75–82.

Sant'Ambrogio, G. (1982). Information arising from the tracheobronchial tree of mammals. *Physiol. Rev. 62*:531–569.

Sant'Ambrogio, G., Bartlett, D., and Mortola, J. (1977). Innervation of stretch receptors in the extrathoracic trachea. *Respir. Physiol. 29*:93–99.

Saria, A., Martling, C.R., Yan, Z., Theodorsson-Norheim, E., Gamse, R., and Lundberg, J.M. (1988). Release of multiple tachykinins from capsaisin-sensitive sensory nerves in the lung by bradykinin, histamine, dimethyl piperazinium, and vagal nerve stimulation. *Am. Rev. Respir. Dis. 137*:1330–1335.

Skidgel, R.A., Engelbrecht, A., Johnson, A.R., and Erdos, E.G. (1984). Hydrolysis of substance P and neurotensin by converting enzyme and neutral endoproteinase. *Peptides 5*:769–776.

Spencer, H., and Leof, D. (1964). The innervation of the human lung. *J. Anat. 98*:599–609.

Springall, D.R., Bloom, S.R., and Polak, J.M. (1988). Distribution, nature, and origin of peptidecontaining nerves in mammalian airways. In: *The Airways* (Kaliner, M.A., and Barnes, P.J., eds.), Marcel Dekker, New York, pp. 299–342.

Uddman, R., and Sundler, F. (1979). Vasoactive intestinal peptide nerves in human upper respiratory tract. *Otorhinolaryngology 41*:221–226.

Widdicombe, J.G. (1982). Pulmonary and respiratory tract receptors. *J. Exp. Biol. 100*: 41–57.

Widdicombe, J.G. (1985). Control of airway caliber. *Am. Rev. Respir. Dis. 131* (Suppl): 533–535.

Widdicombe, J.G. (1993). Why are the airways so vascular? *Thorax 48*:290–295.

Yang, C.M., Chou, S.-P., and Sung, T.-C. (1991). Muscarinic receptor subtypes coupled to generation of different second messengers in isolated tracheal smooth muscle cells. *Br. J. Pharmacol. 104*:613–618.

DISCUSSION

Bienenstock: Have you quantified the surface area of the nerve profiles and expressed these as a percentage of the total surface area you have examined?

Laitinen: As part of the study in which we counted the numbers of inflammatory cells in the epithelium and lamina propria applying in graphical program and expressing the cell numbers/mm^2, we also counted the numbers of the intra-epithelial axonal profiles and nerve bundles in the lamina propria. The total surface area of the nerve bundles in the section was counted and the percentage it formed of the total area of the lamina propria were calculated. The results are currently under analysis.

Polak: You have shown an apparent increase of intraepithelial nerve profiles in asthmatic patients after steroid treatment. Do you have data comparing this with *non*asthmatic control biopsies?

Laitinen: After inhaled corticosteroid treatment, the number of intraepithelial nerves increased to normal levels in the newly diagnosed asthmatics. This increase was associated with the restoration of epithelial structure and a decrease in the overall inflammatory reaction.

Barnes: In sensory nerve profiles of human airway epithelial nerves, can you see electron-dense granules that might indicate the presence of neuropeptides?

Laitinen: The intraepithelial nerves, especially those close to the lumen, contained mainly small agranular vesicles (SAGVs) with a diameter of 30–50 nm. Usually some granular vesicles with a diameter of 40–60 nm were also seen in the epithelial axonal profiles, but their number was clearly less than that of SAGVs.

McDonald: Do you have quantitative data regulating the frequency of intraepi-thelial nerve endings in asthmatics that are directly exposed to the airway lumen? Might the four examples you have found represent cases where epithelial cells were damaged when the biopsy was obtained?

Laitinen: In the airway epithelium of asthmatic patients, the nerves look abnormal, showing features of destruction, in comparison to nerves in normal airways. The epithelium also shows features of damage. However, these changes are not associated with bronchoscopy and biopsy because biopsies from control subjects show normal epithelium.

Barnes: The increased number of epithelial nerve profiles after budesonide treatment in asthma suggests that the sensory nerves grow back. Is anything known about regrowth of sensory nerves after tissue damage?

Laitinen: Because the increase in the number of nerves in the epithelium during inhaled steroid treatment was accompanied by improvement of the epithelial structure and a clear decrease in the inflammatory reaction, we have speculated that normal nerve growth may require interactions with a normal epithelium. In the β_2-agonist-treated group of asthmatic patients, the structure and inflammatory changes did not return to normal, nor was there any increase in the number of nerves. The growth of intraepithelial nerves may also relate to interactions between the extracellular matrix and changes in the basement membrane.

Said: Which neuropeptides are formed in nerve terminals within the airway epithelium? Such presence might imply the localization of these peptides in sensory nerves.

Polak: There are intraepithelial nerve fibers in human airways, as revealed by the panneuronal marker PGP 9.5, but the neurotransmitter remains to be elucidated, since the use of antibodies to the so-called sensory neurotransmitters CGRP and SP shows little or no immunoreactivity. VIP-immunoreactive nerves are seen principally in smooth muscle and around glands.

Solway: Is there definite evidence that the preprotachykinin-1 gene is expressed in human jugular-nodose or thoracic spinal dorsal root ganglia?

Polak: The gene for preprotachykinins have been shown to be expressed in dorsal root ganglion by humans.

2

Functional Autonomic Innervation of the Airways: The Cholinergic and Adrenergic Systems

Steven R. White

The University of Chicago, Chicago, Illinois

I. INTRODUCTION

The function of the respiratory tract, and its constitutive cells and structures, is regulated at many levels by the actions of multiple mediators and several nervous systems present within the airways. Stimuli to airways from the environment or from inflammatory cells present within the airways are converted to electrical impulses and transmitted to the central nervous system by afferent nerves (Widdicombe, 1977; Nadel, 1977). Efferent transmission occurs through one or more of three major autonomic neural systems, each characterized by having a neural axon leading from a cell body within the central nervous system (CNS) to a synapse at a ganglion outside the CNS, and a second axon leading from this ganglion to a final neuroeffector junction with an airway cell. From this junction a neurotransmitter mediator (e.g., acetylcholine, norepinephrine, nitric oxide) is released, whence an effect on airway function may be observed, such as bronchoconstriction or mucous secretion.

Airway function is modulated at all levels of the neural arc from stimulus to response. Neural transmission may be augmented or inhibited at the afferent visceral sensory receptor, within the CNS, at the efferent ganglion, along the efferent axons, or at the neuroeffector junction. An airway structure such as smooth muscle may be innervated by both excitatory and inhibitory pathways (Davis et al., 1982), with the final physiological event (contraction or relaxation) determined by the net contribution of both pathways at a given point in time.

Simultaneous secretion of multiple mediators at the same site within this arc may elicit either augmentatory (Poch and Holzman, 1980) or inhibitory (Leff and Munoz, 1980) functional effects. These mediators may be derived from constitutive cells within airways (e.g., epithelial cells) or from circulating cells that migrate into the airway in disease states (e.g., lymphocytes or eosinophils). Final neural control of airway function, then, is a balance of the contribution of each pathway, each of which may be modified by chemical mediators from other constitutive, inflammatory, or external stimuli.

There are three efferent and three afferent neural pathways in mammalian lungs (Widdicombe, 1981; Richardson, 1988). Two of the efferent pathways, the cholinergic (excitatory) and adrenergic (inhibitory), have been well studied, but the third efferent pathway, the nonadrenergic, noncholinergic (NANC) pathway, until recently was poorly characterized. Recent advances in determining the innervation and neurotransmitters involved in NANC efferent pathways, which include both excitatory and inhibitory pathways, should lead to a better understanding of their role in regulating lung function.

Studies in animals have provided much information on the functional innervation of the respiratory tract, as there are comparatively few studies of the functional innervation of human airways. There is obvious and extensive variability in the autonomic innervation of the respiratory tract among species of animals, both in anatomical structure and in physiological function. For example, in human lungs the principal inhibitory pathway is the NANC inhibitory (i-NANC) pathway, with little detectable adrenergic component (Richardson and Béland, 1976; Davis et al., 1982). In canines, however, the only inhibitory neural pathway is adrenergic (Russell, 1980). Extrapolation of data derived from animals studies to humans, therefore, must be made with caution. In this chapter we will refer to human studies when possible and illustrate potential mechanisms (and the many gaps in our knowledge) with appropriate animal studies.

This chapter will review the functional autonomic innervation of the lung and will focus on the efferent cholinergic and adrenergic systems. Structural autonomic innervation and the emerging role of neuropeptides as neurotransmitters for the NANC pathways are discussed in other chapters.

II. THE PARASYMPATHETIC NERVOUS SYSTEM

A. Parasympathetic Innervation of the Lung

The airways are richly innervated by the parasympathetic nervous system. Cholinergic efferent nerves arise in the vagal nuclei of the brain stem and travel in the vagus nerve to synapse in ganglia located within the airway wall. Many of these ganglia are networked into plexi and receive input from the sympathetic and NANC systems (Skoogh, 1988). Postganglionic fibers innervate smooth muscle, mucosal glands, and mast cells (Richardson, 1979). Cholinergic innervation of smooth muscle is derived from the peribronchial plexi and is greatest in the central airways, decreasing with airway diameter such that relatively few cholinergic fibers are found in the terminal bronchioli (Partenan et al., 1982).

Acetylcholine is the cholinergic neurotransmitter released by both preganglionic and postganglionic fibers. Cholinergic receptors on ganglia are nicotinic and can be stimulated by nicotinic agonists such as dimethylphenylpiperazinium and blocked by the nicotinic antagonist hexamethonium. Cholinergic receptors on constitutive structures within the airways are muscarinic and can be blocked by the muscarinic antagonist atropine. As will be discussed below, recent advances in defining muscarinic receptor subtypes have advanced our understanding of the regulation of cholinergic-mediated airway tone.

B. Parasympathetic Neural Regulation of Airway Tone

Electrical stimulation of the vagus nerve in animals causes bronchoconstriction, which is potentiated by inhibitors of cholinesterase and blocked by atropine (Nadel, 1980). Substantial topographical variation in responsiveness to cholinergic-mediated bronchoconstriction exists. Maximal airway contraction parallels vagal innervation, so that the central and intermediate-sized bronchi have the greatest narrowing after vagal stimulation compared to peripheral airways or lung strips (Nadel et al., 1971; Drazen and Schneider, 1977; Shioya et al., 1987). Even within the central airways, contraction increases through the first five generations of bronchi, both in vitro (Shioya et al, 1987b) and in vivo (Shioya et al., 1987; Leff et al., 1988b) in canines. Such studies suggest that there are substantial differences in cholinergic innervation in airways located <1 cm from the mainstem bronchi.

Modulation of cholinergic transmission to airway smooth muscle and mucous glands may be effected at the ganglia, along the short postganglionic fibers from the ganglia to end-organs, or at the neuroeffector junction. One or more of these mechanisms may account for increased cholinergic tone in asthma. In studies of ferret trachea, Skoogh (1986) has demonstrated that parasympathetic ganglia filter high-frequency electrical stimulation applied via an electrode to the vagus nerve, and this filtering can be modified by β-adrenergic agonists such as isoproterenol. This may occur via inhibition of acetylcholine release from preganglionic nerve terminals (Baker et al., 1983; Skoogh and Svedmyr, 1989) via circulating epinephrine or release of norepinephrine from those sympathetic fibers that synapse in airway ganglia. β-Adrenergic-mediated inhibition of cholinergic ganglionic transmission in ferret trachea is most effective at higher transmission frequencies and is both time- and concentration-dependent (Skoogh and Svedmyr, 1989). Along the postganglionic fiber and at the neuromuscular junction, adrenergic stimulation may inhibit cholinergic transmission by a prejunctional mechanism (Rhoden et al., 1988). It has been suggested that inflammatory mediators also may decrease sympathetic drive to the cholinergic ganglia and thus up-regulate cholinergic transmission (Barnes, 1986). Thus, several putative mechanisms exist by which adrenergic stimulation may regulate cholinergic transmission at any or several points along the efferent tract. However, these have yet to be demonstrated clearly in human airways.

Mediators derived from other cells within the airway also may alter postganglionic transmission. Mast cells can release histamine, thromboxane, and platelet-activating factor that up-regulate cholinergic responses (Benson and Graff, 1977; Loring et al., 1978; Munoz et al., 1986; Leff et al., 1987) (Fig. 1). Likewise, epithelial cells can release prostaglandin E_2, which could attenuate cholinergic tone (Walters et al., 1984). The airway epithelium may release other factors that inhibit cholinergic-stimulated airway tone in canine smooth muscle (Barnes et al., 1985; Gao and Vanhouette, 1988), though the physiological importance of this mechanism recently has been questioned (Strek et al., 1993).

In the past several years, there have been important advances in understanding muscarinic receptor pharmacology in the lung, which likely will lead to a better understanding of the modulation of cholinergic tone in airways. In parallel with cholinergic nerve fiber distribution in the central and peripheral airways, mapping

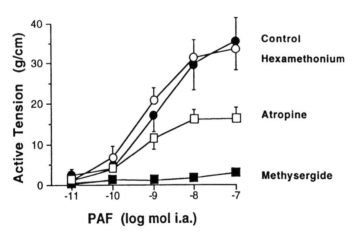

Figure 1 Parasympathetic postganglionic stimulation of airway contraction by platelet-activating factor (PAF) in a canine in situ model of airway reactivity. Direct intra-arterial infusion of platelet-activating factor to the trachea stimulated contraction of canine tracheal smooth muscle that was attenuated by atropine but not by hexamethonium. One of the mechanisms, then, of airway contraction elicited by PAF is via up-regulation of postganglionic cholinergic transmission, though another (perhaps more important) mechanism is via serotonin receptors. (From Leff et al., 1987, with permission of the publisher.)

of the muscarinic receptors on airway smooth muscle by autoradiography has demonstrated a decreasing distribution of muscarinic receptors from central airways to peripheral lung (Van Koppen et al., 1987; Mak and Barnes, 1990). Five distinct molecular forms of the muscarinic receptor have been cloned and expressed, and pharmacological characterization of the first three subtypes has been accomplished (Kubo et al., 1986; Bonner et al., 1987; Goyal, 1989). Selective agonists and antagonists available for the M_1, M_2, and M_3 subtype receptors are listed in Table 1 (Goyal, 1989). Each of these three receptors has been found in human airway smooth muscle (Bloom et al., 1987; Mak and Barnes, 1990; Mak et al., 1992). A fourth receptor type (M_4) has been demonstrated in rabbit but not in human lung tissues (Dörje et al., 1991; Mak et al., 1992). A proposed schema for location and function of these receptors in airway smooth muscle is shown in Figure 2. As with functional studies of airways, there are marked species differences in the distribution of muscarinic receptor subtypes.

Each muscarinic receptor subtype has a different physiological role in controlling airway tone, and each may be regulated differently and have distinctive signal

Table 1 Classification of Muscarinic Receptor Subtypes Found in Airways

Feature	Receptor subtype		
	M_1	M_2	M_3
Molecular form	m_1	m_2	m_3
Main tissues of expression	Neural	Neural, smooth muscle	Smooth muscle, exocrine glands, neural
Selective antagonist	Pirenzepine Telenzepine Dicyclomine Trihexyphenidyl	AFDX-116 Himbacine Gallamine Pancuronium Methoctramine Stercuronium	4-DAMP HHSD
Dissociation constants for antagonists (nM)			
Atropine	1	1	1
Pirenzepine	10	50	200
HHSD	40	200	2
AFDX-116	800	100	3000

Dissociation constants are the reciprocal of the affinity compound for a receptor. The lower the value of the constant, the higher the affinity or potency. Atropine has the highest potency and the lowest dissociation constant with regard to all subtypes of receptors.
4-DAMP, 4-diphenyl acetyoxy-*N*-methylpiperidine methiodide; HHSD, hexahydrosiladifenidol.
Source: Adapted from Goyal (1989), with permission of the publisher.

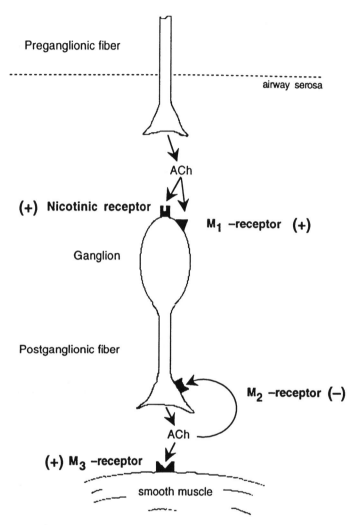

Figure 2 Schematic of muscarinic receptor subtypes in cholinergic ganglia and nerves fibers in airways, and how these receptors modulate airway tone. Stimulation of either the nicotinic or the M_1-receptor stimulates the ganglion and leads to postganglionic transmission. Stimulation of the M_3-receptor initiates smooth muscle contraction. Stimulation of the M_2-receptor by acetylcholine (ACh) released at the neuroeffector junction feeds back to inhibit further ACh release. Substances that block the M_2-receptor interrupt the feedback inhibition and may facilitate increased cholinergic tone. (+), excitatory response after receptor stimulation; (−), inhibitory response after receptor stimulation.

transduction mechanisms. M_3-receptors are localized to smooth muscle and submucosal glands in human airways (Mak and Barnes, 1990). Activation of M_3-receptors on airway smooth muscle causes phosphoinositide hydrolysis, calcium release, and smooth muscle contraction (Roffel et al., 1990). M_1-receptors are localized to parasympathetic ganglia and to sympathetic nerve terminals in airways (Barnes, 1993). M_1-receptors may have a facilitory role in the ganglia, enhancing the transmission after a nicotinic stimulus. M_1-receptor antagonists such as pirenzepine then would reduce vagal tone, thus ameliorating asthma symptoms. Blockade of M_1-receptors may enhance β-adrenergic relaxation of airways by blocking neural cholinergic transmission in canine tracheal smooth muscle (Mitchell et al., 1993). However, the effects of pirenzepine and telenzepine in human studies are contradictory: these antagonists may block reflex cholinergic bronchoconstriction elicited by sulfur dioxide (Lammers et al., 1989), but may also act on peripheral, as opposed to central, airways (Cazzola et al., 1987), and may not block reflex cholinergic tone at all in patients with chronic obstructive pulmonary disease (Ukena et al., 1992).

M_2-receptors are relatively few in number and are found within the post-ganglionic nerves. These prejunctional receptors are inhibitory and provide negative feedback to the postganglionic fiber (Fryer and Maclagan, 1984; Blaber et al., 1985). M_2-receptor agonists such as pilocarpine inhibit, and specific antagonists such as gallamine potentiate, vagally mediated bronchoconstriction (Minnette and Barnes, 1988). A dysfunctional M_2-receptor could potentiate vagal tone and thereby lead to increased bronchoconstriction, and this has developed into an attractive hypothesis by which to explain airways hyperreactivity in asthma. Inhalation of pilocarpine by nonasthmatic volunteers blocks reflex cholinergic bronchoconstriction elicited by sulfur dioxide but has no effect in asthmatic subjects (Minnette et al., 1989). Similar results are seen after histamine challenge (Ayala and Ahmed, 1991). M_2-receptors are damaged by parainfluenza virus infection in a guinea pig model (Fryer and Jacoby, 1991). Antigen challenge in a similar model also causes M_2-receptor dysfunction; this can be reversed by polyanionic substances such as heparin (Fryer and Wills-Karp, 1991; Fryer and Jacoby, 1992). Inflammatory mediators such as the major basic protein from eosinophils block the M_2-receptor by acting as an allosteric antagonist (Jacoby et al., 1993 (Fig. 3). Indomethacin blocks M_2-receptor function in guinea pigs, though exactly which prostaglandin is required for proper M_2-receptor activation is not yet known (Fryer and Okanlami, 1993). Finally, loss of M_2-receptor function could lead to worsened bronchoconstriction after challenge with β-adrenergic antagonists; as β-adrenoceptor stimulation normally suppresses cholinergic transmission (see below), M_2-function normally would be expected to block any increase in cholinergic transmission (Barnes, 1989; Ind et al., 1989). Taken together, these data suggest strongly that M_2-receptor dysfunction, as postulated to occur in asthma, could lead to substantial changes in airway tone mediated by the

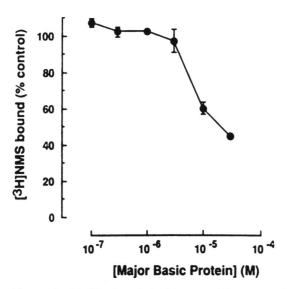

Figure 3 Modification of the M_2-muscarinic receptor by the inflammatory mediator eosinophil major basic protein in guinea pig cardiac muscle membranes. Major basic protein functioned as an allosteric inhibitor of the receptor, an action not seen with other eosinophil cationic proteins. These data suggest a mechanism by which mediators from inflammatory cells infiltrating airways can enhance vagally mediated bronchoconstriction. (From Jacoby et al., 1993, with permission of the publisher.)

cholinergic innervation. This hypothesis remains to be demonstrated in appropriate human studies.

Other neural systems in the airways may influence cholinergic regulation of airway caliber. Evidence has accumulated suggesting that afferent nerves containing tachykinin mediators such as substance P and neurokinin A may interact with postganglionic cholinergic neurons. These tachykinins are stored in the nerve endings of capsaicin-sensitive afferent C-fibers of several species, including humans (Lundberg et al., 1984). Substance P elicits airway smooth muscle contraction in several species that is due, at least in part, to muscarinic receptor stimulation, as it can be attenuated by atropine (Tanaka and Grunstein, 1984; Sekizawa et al., 1987; Joos et al., 1988; Myers and Undem, 1991). Tachykinins potentiate contraction elicited by vagal nerve stimulation or electrical field stimulation in vitro (Hall et al., 1989). Substance P can depolarize parasympathetic ganglion neurons in guinea pig bronchi (Undem et al., 1990), and stimulation of capsaicin-sensitive nerves in guinea pigs leads to contraction via both pre- and postganglionic cholinergic stimulation (Myers and Undem, 1991) (Fig. 4).

Other diseases of chronic airflow obstruction may be influenced by alterations

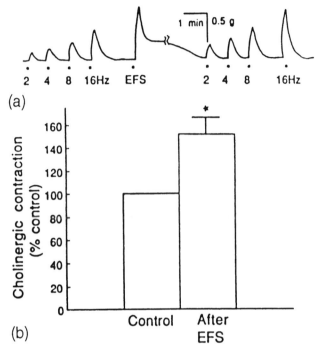

Figure 4 Stimulation of capsaicin-sensitive nerves in a guinea pig bronchial airway increases vagus nerve–evoked cholinergic bronchial contractions in an isolated, superfused model. (a) Representative recording of cholinergic-induced contraction evoked by vagal nerve stimulation. Direct electrical field stimulation (EFS) of the bronchus at settings that stimulated capsaicin-sensitive nerves selectively augmented the response to subsequent vagal nerve stimulation. (b) Summary of these data in 10 animals. Data are normalized to the control period before EFS. *$p < 0.05$ versus control. (From Myers and Undem, 1991, with permission of the publisher.)

in cholinergic tone, but there is little conclusive evidence for an intrinsic defect in the parasympathetic nervous system. In chronic obstructive pulmonary disease, cholinergic vagal tone may be the only reversible neural element, and anticholinergic drugs such as ipratropium have been recognized as the agents of choice in this disease (reviewed in Gross, 1988). This may be simply because dilation of the narrowed, diseased airways leads to a greater increase in airflow than similar dilation in the normal airway with a larger starting radius (Fig. 5). One recent report suggests that cholinergic tone is increased in proportion to the severity of airway disease, but as this was measured by the percent increment in FEV_1 after treatment with atropine, it may be subject to the same geometrical considerations

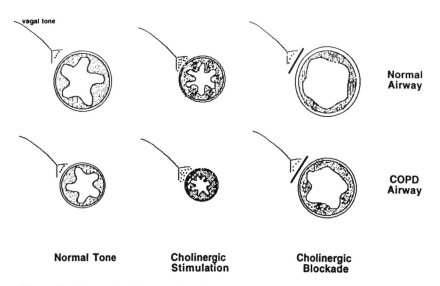

Figure 5 Schematic of how cholinergic tone leads to worsened bronchoconstriction in patients with chronic obstructive pulmonary disease (COPD). Patients with diseased, narrowed airways will have a greater relative decrease in diameter after cholinergic stimulation than normal subjects, since resistance is inversely proportional to the fourth power of the radius. Conversely, cholinergic blockade (e.g., with ipratropium) has a larger effect on narrowed airways in COPD than on normal airways. These changes would occur even if baseline vagal tone were the same in both groups of patients.

(Gross et al., 1989). No specific dysfunction of ganglionic, postganglionic, or muscarinic receptor subtypes has been demonstrated in chronic obstructive pulmonary disease. Approximately 50% of patients with cystic fibrosis have increased responsiveness to inhaled methacholine, but this may indicate only a nonspecific airways hyperreactivity (Larsen et al., 1979).

In summary: there is no direct, convincing evidence that cholinergic tone is increased in asthma. Several putative mechanisms, such as increased transmission through airway ganglia, increased postganglionic transmission, decreased adrenergic inhibition of cholinergic transmission, preferential blockade of the M_2 receptor, and/or an increased sensory afferent receptor stimulation of cholinergic fibers, may lead to increased cholinergic drive. The increased reactivity seen to cholinergic agonists in asthmatic subjects suggests an increase in cholinergic drive, but such hyperreactivity also is seen with a variety of other mediators. Therefore, increased cholinergic responsiveness of airway smooth muscle is unlikely to be a primary abnormality in asthma. Anticholinergic therapy with currently available drugs may be of some benefit in asthma, but the degree of benefit can be disputed (Kerstjens et al., 1992). It is possible that more selective therapy based on muscarinic receptor subtype stimulation/blockade may lead to better control of asthma. The beneficial effect of cholinolytic agents in chronic obstructive pulmonary disease likely is due to the geometry of the diseased airways and not to an underlying change in airway tone in these patients. Few studies are available in patients with cystic fibrosis, and no change in cholinergic drive can be demonstrated clearly in these patients.

C. Parasympathetic Neural Regulation of Other Airway Functions

The parasympathetic nervous system regulates glycoprotein and mucus secretion in airways. Secretory cells are innervated by a plexus of postganglionic fibers derived from both parasympathetic and sympathetic nerves (Murlas et al., 1980). The viscosity and elasticity characteristics of tracheobronchial mucus are determined primarily by mucus glycoproteins, which account for 2% of mucus by mass (95% being water) (Havez and Roussel, 1976). Normal control of mucus secretion is affected by both cholinergic and adrenergic innervation of secretory cells and submucosal gland cells (Kaliner et al., 1988). Stimulation of parasympathetic vagal nerve fibers stimulates both fluid and macromolecule secretion by an atropine-sensitive mechanism (Kokin, 1896; Florey et al., 1932; Gallagher et al., 1975). Mucous glycoprotein secretion is increased either by cholinergic neural stimulation (Phipps, 1981) or by exogenous muscarinic agonists such as acetylcholine (Shelhamer et al., 1980b). Cholinergic-stimulated secretion is greatly augmented in volume and maintains the original physical properties of spontaneously secreted mucus (Ueki et al., 1980; Leikauf et al., 1984). Cholinergic fibers innervating secretory cells can be stimulated by environmental factors such as cold, antigen, or ozone, with subsequent fluid and mucus glycoprotein secretion (Wanner et al., 1975; Ueki et al., 1980; Borson et al., 1980). Muscarinic receptors on human airway submucosal glands are of both M_1 and M_3 subtype in a ratio of 1:2 (Mak and Barnes, 1990), but whether selective stimulation of these different receptors leads to differences in secretion by these glands is unknown. Structural studies have demonstrated the presence of cholinergic nerve fibers near ciliated epithelial cells in large airways. However, cholinergic control of baseline ciliary function has not been demonstrated clearly (Wanner, 1988).

Mast cell release of preformed mediators such as histamine in airways is under cholinergic control. Substantial heterogeneity of mast cell function in different tissues has been recognized (Pearce, 1986). Stimulation of muscarinic receptors augments histamine release from isolated respiratory mast cells in vitro (Austen and Orange, 1975), but not from isolated human basophils in vitro (Sullivan et al., 1975). A more recent study examined bronchial airway mast cell secretion in a

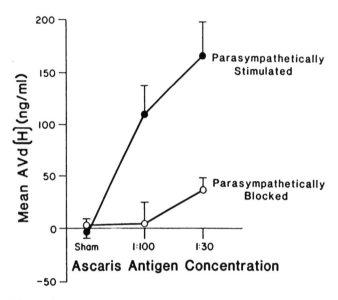

Figure 6 Cholinergic neural modulation of canine bronchial mast cell secretion in an in situ model. Histamine secretion by mast cells was measured as the arteriovenous difference (AVd) across a third-generation bronchus after direct intra-arterial antigen challenge. Parasympathetic stimulation by vagal nerve stimulation combined with β-adrenoceptor blockade with propranolol augmented substantially histamine secretion after antigen challenge with *Ascaris* antigen compared to controls (antigen challenged and ganglion blocked with hexamethonium). Parasympathetic stimulation followed by sham challenge did not stimulate histamine secretion. (From Leff et al., 1986, with permission of the publisher.)

canine in situ preparation after vagus nerve stimulation. Vagal stimulation alone did not change basal histamine release but augmented substantially histamine release stimulated by antigen (Leff et al., 1986) Fig. 6). Such histamine release was sufficient to elicit smooth muscle contraction (Garrity et al., 1983) and may represent another mechanism by which inappropriate parasympathetic neural stimulation in asthma and other diseases increases bronchial reactivity.

III. THE SYMPATHETIC NERVOUS SYSTEM

A. Sympathetic Innervation of the Lung

The sympathetic nervous innervation to the lungs originates from the first six thoracic preganglionic fibers that end in the lower cervical and first four thoracic ganglia. Postganglionic fibers enter the lung at the hila accompanying the vagus nerves. Histochemical studies have demonstrated catecholamine-containing fibers in a plexus surrounding bronchial blood vessels and submucosal glands. Catecholamine-positive nerve fibers are found from the lobar bronchi to the terminal bronchioli (Partanen et al., 1982). There is no significant innervation of human bronchial smooth muscle by sympathetic nerves as demonstrated by fluorescence histochemistry (Doidge and Satchell, 1982; Partanen et al., 1982; Sheppard et al., 1983) and electron microscopy (Latinen et al., 1985), though such innervation is seen in canines (Richardson, 1979) and guinea pigs (Smith and Satchell, 1985). Sympathetic nerve fibers are found in close association with airway ganglia and cholinergic nerves within the airway, and close apposition of adrenergic and cholinergic nerve varicosities, often enclosed within the same Schwann's cell sheath, can be seen (Partanen et al., 1982; Daniel et al., 1986). Norepinephrine is the major adrenergic neurotransmitter released by these fibers; release is mediated by feedback inhibition of its activity in the extravesicular space. A second endogenous catecholamine with major effects on airway function, epinephrine, is released by the adrenal medulla.

B. Sympathetic Neural Regulation of Airway Tone

Exogenous administration of epinephrine relaxes airway smooth muscle precontracted with acetylcholine (Davis et al., 1982) and bronchodilates human subjects previously bronchconstricted with methacholine (Sands et al., 1985). Inhalation of more specific β_2-receptor agonists in humans causes consistent, reproducible bronchodilation (reviewed in Tattersfield, 1987). Norepinephrine relaxes precontracted airway smooth muscle in canines through stimulation of the β-adrenoceptor; this relaxation is converted to contraction if the animals are treated with propranolol, so that the entire effect of norepinephrine is α-adrenoceptor stimulation (Leff and Munoz, 1981). Infusion of norepinephrine into human subjects at concentrations within the physiological range does not elicit bronchodilation (Berkin et al., 1985), suggesting that this catecholamine does not function as a hormone. Circulating norepinephrine, then, represents the spillage of neurotransmitter from sympathetic neural activity, whereas circulating epinephrine is a hormone secreted directly by sympathetic stimulation.

Functional studies of adrenergic innervation in the lung have demonstrated major species differences. Thus, while electrical field stimulation and pharmacological stimulation of smooth muscle demonstrate noradrenergic modulation of canine (Russell, 1980) and guinea pig (Doidge and Satchall, 1982) tracheal smooth muscle, similar field stimulation of human airway smooth muscle causes contraction followed by relaxation, both of which can be blocked by tetrodotoxin (demonstrating a potent nonadrenergic, noncholinergic inhibitory innervation) but is unaffected by propranolol (Davis et al., 1982; Palmer et al., 1986). Adrenergic nerves may influence bronchomotor tone indirectly: first, by modulation of cholinergic nerve fibers within the airway, and second, by secretion of epinephrine from the adrenal medulla.

It has been assumed previously that parasympathetic resting smooth muscle tone in airways is counterbalanced by sympathetic neural innervation and secretion of catecholamines (Innes and Nickerson, 1975). As discussed above, defective inhibition of sympathetic tone on cholinergic tone in the parasympathetic, paratracheal ganglia then would lead to "unbalanced" parasympathetic tone in airways, increased cholinergic transmission, and subsequent bronchoconstriction. Several lines of evidence have been suggested to support this hypothesis of a primary defect in sympathetic tone. Bronchoconstriction can be elicited in asthmatic subjects after β-adrenergic blockade with propranolol (Zaid and Beall, 1966; Richardson and Sterling, 1969), and some asthmatic subjects demonstrate extraordinary sensitivity to β-adrenergic blockade, including ophthalmological instillation of β-adrenergic blockers such as timolol (McMahon et al., 1979; Dunn et al., 1986) (Fig. 7). β-Adrenergic blockade under these circumstances might unmask existing contractile influences in the airway—either parasympathetic or

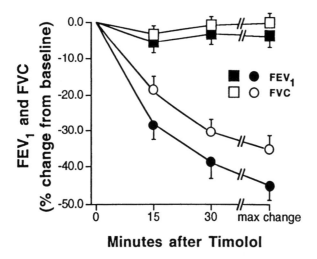

Figure 7 Effect of β-adrenergic blockade in asthmatic subjects with timolol ophthalmological solution. A cohort of 14 subjects with sensitivity to β-adrenergic antagonists bronchoconstricted substantially within 30 min of instillation of timolol; however, 10 other subjects with similar degrees of asthma had no demonstrable sensitivity to timolol. Normal subjects do not bronchoconstrict after dosing with ophthalmological, oral, or inhaled β-adrenergic antagonists (not shown). These data demonstrate the exquisite sensitivity of some asthmatic patients to β-adrenergic blockade. (From Dunn et al., 1986, with permission of the publisher.)

α-adrenergic (Boushey et al., 1980; Leff, 1988). Normal subjects do not broncho-constrict after challenge with propranolol (McNeill, 1966; Richardson and Sterling, 1969; Tattersfield et al., 1973). The mechanism of β-adrenoceptor-blockade-induced bronchoconstriction may be enhancing cholinergic tone, as the effect can be blocked by anticholinergic agonists in human subjects (Ind et al., 1989) and ganglionic blockers such as hexamethonium in guinea pigs (Herxheimer, 1967). As noted above, blockade of β-adrenoceptors on cholinergic nerves increases ACh release; coupled with a defect in the M_2-muscarinic receptor in asthmatic but not normal subjects (and thus a further unopposed increase in ACh release), substantially greater cholinergic tone and bronchoconstriction could result. Even a small amount of β-adrenoceptor antagonist could release the inhibition supplied by adrenergic fibers to cholinergic transmission, and this potentially could explain the exquisite sensitivity of some asthmatic subjects to β-antagonists. The sensitivity to β-adrenoceptor blockade differs from the nonspecific bronchial reactivity associated with asthma due to methacholine, histamine, isocapnic hyperventilation, and exercise challenge, suggesting a different mechanism (Ind et al., 1989). Thus, a potential mechanism by which sympathetic nerves regulate airway tone is by modulation of cholinergic transmission to smooth muscle. If the cholinergic regulatory neural pathway is defective in asthmatic patients, such deficiency then could be unmasked by β-adrenergic blockade (Fig. 8). The importance of this mechanism is controversial and remains to be defined completely (Leff, 1988).

A defect in β-adrenergic receptors on airway smooth muscle has also been suggested to explain airways hyperreactivity. However, considerable controversy exists regarding the properties of α- and β-adrenergic receptors in asthmatic subjects (Insel and Wasserman, 1990). Human bronchial preparations harvested from the lungs of either normal or asthmatic subjects at autopsy relax similarly to isoproterenol (Svedmyr et al., 1977). Both normal volunteers and subjects with stable, mild asthma bronchodilate similarly to inhaled albuterol (Tattersfield et al., 1983). However, at least two studies have suggested some degree of β-adreno-ceptor hypofunction in asthmatic subjects with severe disease (Goldie et al., 1986; Cerrina et al., 1986). Analysis of binding properties of receptors on airway smooth muscle has not been possible in living subjects. In vitro autoradiography demonstrates no substantial decrease in the number of β-adrenergic receptors in airways from asthmatic subjects, though there may be more receptors localized to smooth muscle in asthmatic versus nonasthmatic subjects (Spina et al., 1989). Inflammatory mediators such as platelet-activating factor can down-regulate β-adrenergic receptors in human lung in vitro (Agarwal and Townley, 1987), but the extent to which such mediators modulate adrenergic receptor density and function in asthma is largely unknown. Several studies have suggested the presence of autoantibodies to the β-adrenoceptor in humans; however, in most of these studies, the titer is low compared to other specific antibodies (frequently ≤1:100) and the antibodies can be demonstrated in only a small percentage of patients (generally

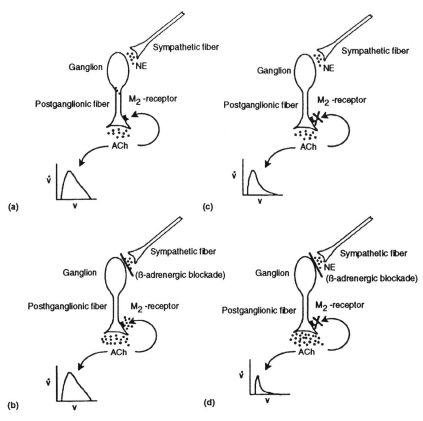

Figure 8 Schematic demonstrating how the combination of β-adrenergic blockade and a selective M_2-muscarinic receptor defect could lead to bronchoconstriction in asthmatic subjects. (a) *Normal*. In normal subjects, acetylcholine (ACh) release at the neuroeffector junction is regulated both by sympathetic nerve terminals that synapse on the cholinergic ganglion and release norepinephrine (NE), and by feedback inhibition of the M_2-muscarinic receptor on the cholinergic postganglionic fiber. These subjects have a normal flow-volume curve. (b) *Normal after β-blockade*. Administration of β-adrenoceptor antagonists releases sympathetic inhibition of the ganglion. However, some of the increased ACh release causes increased feedback inhibition, shutting down further release. A normal flow-volume curve results. (c) *Asthma*. In asthmatic subjects, a postulated defect in the M_2-muscarinic receptor would lead to increased ACh release and increased cholinergic tone. These subjects would demonstrate some airflow obstruction even at baseline. (d) *Asthma after β-blockade*. Release of sympathetic inhibition, combined with a defective M_2-receptor, leads to substantially increased bronchoconstriction and airflow obstruction after challenge with a β-adrenoceptor antagonist.

≤15%) (reviewed in Insel and Wasserman, 1990). It seems unlikely that the majority of asthmatic patients possess antibodies to the β-adrenoceptor.

Given the difficulty of studying β-adrenoceptor density and function in human airways tissues, investigators have turned to the study of the adrenergic system in other organ systems in humans (reviewed in Kaliner et al., 1982). Asthmatic subjects demonstrate diminished responses in pulse pressure to isoproterenol (Shelhamer et al., 1980) and pupillary response to phenylephrine (Henderson et al., 1979) compared to normal subjects. Several investigations have used peripheral blood leukocytes or lymphocytes as mirrors of β-adrenergic receptors in airways, but problems exist in separating differences between normal and asthmatic subjects and normal variation in receptor density among different cell subpopulations (Maisal et al., 1989; Insel and Wasserman, 1990). Furthermore, β-adrenergic agonist treatment may itself down-regulate β-adrenoceptors in lymphocytes (Maisal et al., 1989), and this effect may vary among lymphocyte subpopulations (Maisal et al., 1990). β-Adrenergic receptor density on leukocytes is decreased in asthmatic subjects, but this too may be dependent on previous β-agonist therapy (Sano et al., 1983). A recent study suggests that local lung changes in β-adrenoceptor density may be more important; in this study, Rudchenko and associates (1993) demonstrated that while there were no differences in β-adrenoceptor affinity in blood monocytes in asthmatics versus patients with chronic bronchitis, macrophages obtained at bronchoalveolar lavage from the same asthmatic subjects had substantially higher numbers of β-receptors compared to the alveolar macrophages of chronic bronchitics. These data taken together imply that the use of peripheral blood cells as a mirror for β-adrenergic receptor changes in airways may be problematic, and that better studies of cell receptor number and affinity on human airway smooth muscle and inflammatory cells are needed.

One further area of investigation in determining the potential β-adrenergic defect in asthma has been studies of exercise in normal and asthmatic subjects. A clear association between asthma and exercise has been recognized for centuries (Willis, 1679). Asthmatic, but not normal, subjects develop airflow obstruction immediately after cessation of exercise, which may persist for 15–30 min (this topic is discussed in another chapter by Dr. Solway). Several investigations in the past 15 years have suggested that asthmatic subjects have lower plasma epinephrine concentrations during and immediately after exercise compared to normal volunteers, who manifest a brisk increase in plasma epinephrine (Barnes et al., 1981; Warren et al., 1982, 1984). A more recent study, however, demonstrates that the increase in plasma epinephrine and norepinephrine is similar in normal and asthmatic subjects after exercise, even as FEV_1 and FVC fall substantially in the asthmatic subjects (Dosani et al., 1987) (Fig. 9). Repeated exercising of patients with exercise-induced asthma at relatively short intervals leads to a diminution of the severity of the bronchoconstriction, and it has been suggested that this

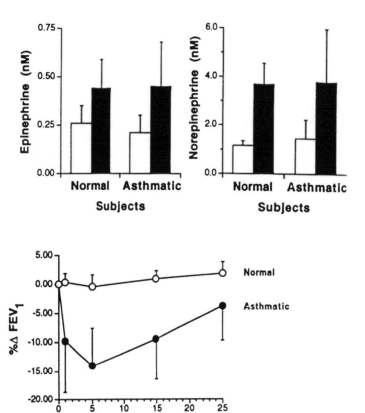

Figure 9 Catecholamine secretion and pulmonary function in six normal and seven asthmatic subjects after exercise. Subjects were exercised to achieve an O_2 consumption $\geqslant 50\%$ of predicted maximum. Both epinephrine and norepinephrine concentrations were equivalent in both groups at baseline and immediately after cessation of exercise. These data suggest that there is no demonstrable defect in catecholamine secretion in patients with asthma, even as FEV_1 and FVC decrease. (Figure drawn from data presented by Dosani et al., 1987, with permission of the publisher.)

diminution may be related to the elevated circulating plasma catecholamine concentrations elicited by the first exercise trial (Stearns et al., 1981). These circulating catecholamines would prevent bronchoconstriction in subsequent exercise trials. These data suggest that (1) sympathetic responsiveness is not blunted in asthmatic subjects during exercise, and (2) circulating catecholamines secreted by the sympathetic nervous system can protect the airways against other concomitant bronchoconstricting stimuli.

Data thus has accumulated that suggests that an underlying defect in β-adrenergic neural function or receptors may be present in asthma, and that such a defect may lead to a heightened response to other stimuli, whether exogenous or endogenous (i.e., cholinergic). The question then is whether the sympathetic nervous system opposes parasympathetic (or other neurally mediated) tone in airways, and whether it responds in a homeostatic manner to exogenous bronchoconstriction. Circulating plasma norepinephrine concentrations do not elicit bronchoconstriction in swine (White et al., 1987), and infusion of norepinephrine does not bronchodilate human subjects (Berkin et al., 1985). As previously noted, there is virtually no direct sympathetic innervation to human airway smooth muscle (Richardson, 1979). Therefore, only circulating epinephrine released by the adrenal medulla could oppose bronchoconstriction elicited by vagal tone. Plasma epinephrine concentrations are low in the normal state in both normal and asthmatic volunteers (<20 pg/ml, or <0.5 nmol/L) (Barnes et al., 1982; Sands et al., 1985). Plasma catecholamine concentrations are not correlated to the severity of bronchoconstriction or the plasma histamine concentrations in asthmatic subjects, suggesting that the sympathetic nervous system is not responding to changes in airway tone (Barnes et al., 1982). In a study of both normal and asthmatic subjects, Sands et al. (1985) demonstrated that bronchoconstriction elicited by inhaled methacholine did *not* lead to sympathetic stimulation and secretion of plasma catecholamines, even when airway specific conductance (SGaw) fell to 25% of baseline (Fig. 10). Infusion of epinephrine to achieve a plasma concentration equivalent to that found in mild hypotension promptly returned SGaw to baseline. In a study using swine as a model for sympathetic responses [swine, like human, lack sympathetic innervation to airway smooth muscle (Richardson, 1979)], bronchoconstriction elicited by either methacholine or prostaglandin $F_{2\alpha}$ did not elicit catecholamine secretion, even when lung resistance was >500% of baseline (White et al., 1987) (Fig. 11). Plasma catecholamine concentrations do not increase in asthmatic subjects even when bronchoconstriction is severe (Barnes et al., 1982). In both humans and swine, catecholamine secretion can be provoked by other physiological maneuvers such as hypotension, hypoxemia, or acidosis, and epinephrine secretion from these stimuli is sufficient to dilate airways (Sands et al., 1985; White et al., 1987; Brofman et al., 1990). Histamine release from mast cells causes epinephrine secretion from the adrenal medulla, and this secretion is sufficient to reverse bronchoconstriction elicited by mast cell degranulation (White et al., 1989). Taken together, these data suggest that the sympathetic nervous system is not coupled to airway proprioception, and that bronchoconstriction per se does not elicit a homeostatic sympathetic bronchodilator response in either normal or asthmatic human subjects. Mediators that stimulate the sympathetic nervous system may elicit catecholamine secretion, which then is sufficient to dilate airways. It is possible that alterations in β-adrenoceptor density or function in other constitutive cells in airways, or in inflammatory cells

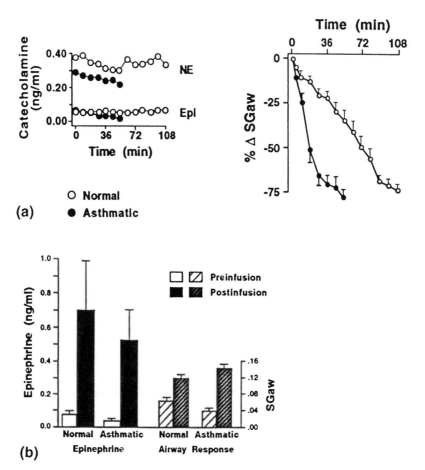

Figure 10 Catecholamine secretion and pulmonary function in normal and asthmatic subjects after induction of bronchoconstriction. Plasma catecholamine concentrations and specific conductance (SGaw) were measured as subjects inhaled increasing concentrations of methacholine. (a) As SGaw decreased to 25% of original baseline in both cohorts, there was no change in either plasma epinephrine (Epi) or plasma norepinephrine (NE) concentrations. These data suggest that in both normal and asthmatic subjects, the sympathetic nervous system does not defend airway caliber. (b) Infusion of epinephrine promptly reversed the induced bronchoconstriction in both normal and asthmatic subjects. These data demonstrate that human airways were capable of responding to exogenous catecholamines and would respond to endogenous β-adrenergic secretion. (From Sands et al., 1985, with permission of the publisher.)

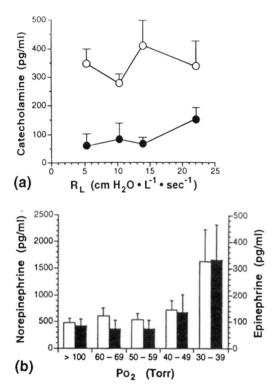

Figure 11 Catecholamine secretion in swine after induction of bronchoconstriction with aerosolized methacholine. Swine were used as a model of human sympathetic responses because swine, like humans, lack direct sympathetic innervation of airway smooth muscle, so that sympathetic regulation of airway caliber is via circulating epinephrine. (a) Plasma catecholamine concentrations do not change significantly as lung resistance is increased to >500% of baseline resistance by methacholine. (b) Plasma catecholamine concentrations increase dramatically as swine are made hypoxemic by substitution of nitrogen for oxygen. These data suggest that the sympathetic nervous system does not respond to acute changes in airway caliber, though it remains capable of responding to other physiological stimuli. (From White et al., 1987, with permission of the publisher.)

invading airways in asthma or other diseases, may lead to altered airway responsiveness and worsened inflammation. Release of inhibition of cholinergic transmission, by either exogenous blockade or endogenous alteration in sympathetic tone, may account for some aspects of airway hyperreactivity, but is unlikely to be the entire explanation. Such changes in β-adrenoceptor number and function will have to be reconciled with the demonstrated lack of homeostatic regulation of airway tone by the sympathetic nervous system.

It has been suggested that a dysfunction in α-adrenergic function may explain airways hyperreactivity in asthmatic subjects. Allergic asthmatic subjects with near-normal pulmonary function require less phenylephrine to produce mydriasis than healthy volunteers (Kaliner et al., 1982; Davis et al., 1985). Asthmatic subjects also have increased α_1-adrenoceptor activity in peripheral blood vessels (Henderson et al., 1979). α-Adrenergic receptors exist in the lung, but neural α-adrenergic stimulation is not likely to be of major importance in regulating airway caliber (Barnes, 1986). α-Adrenergic stimulation does not contract human airway smooth muscle, but can contract smooth muscle collected from human airways exposed to endotoxin (Simonsson et al., 1972). This suggests that α-adrenergic function could be induced in certain conditions. Prazocin, a potent antagonist for the α_1-adrenoceptor, given by aerosol to human subjects protects against exercise-induced bronchoconstriction, though it has no effect in asthmatic subjects (Barnes et al., 1981b). Since neither epinephrine nor norepinephrine causes bronchoconstriction (Berkin et al., 1985), however, it is difficult to see how α-adrenoceptors in airway smooth muscle could be activated endogenously. Overall, there is sparse α-adrenergic innervation to airway smooth muscle and little evidence that enhanced α-adrenergic receptor responses are present in asthmatic airways (Spina et al., 1989b). Therefore, demonstration of a specific mechanism by which α-adrenergic pathways modulate human asthma is lacking at this time, and specific therapy with α-adrenergic antagonists cannot be recommended.

Comparatively little work has been done to establish the role of adrenergic innervation in other airways diseases. Adequate control subjects with lung impairment similar to that of patients with chronic obstructive pulmonary disease do not exist, so determination of changes in adrenergic function is difficult. Patients with cystic fibrosis have abnormalities in β-adrenergic regulation of blood pressure, ion transport, and lymphocyte function, and abnormalities in α-adrenergic regulation of pupillary reflexes (reviewed in Davis, 1988). It is not clear whether similar abnormalities are present in the airways of patients with cystic fibrosis.

In summary: there is little sympathetic innervation directly to airway smooth muscle, and the influence of the sympathetic nervous system in airway caliber may be indirect, through modulation of cholinergic airway tone and by secretion of circulating epinephrine. The sympathetic nervous system is not coupled to airway proprioception and does not defend airway tone in a homeostatic manner. While some evidence exists for a defect in either α- or β-adrenergic function in asthma, the topic remains controversial, and new evidence is needed.

C. Sympathetic Neural Regulation of Other Airway Functions

As noted above, both the cholinergic and sympathetic nervous systems innervate airway mucous glands and ciliated epithelial cells. As with cholinergic innervation, it is not clear that sympathetic innervation regulates baseline, homeostatic mucociliary function. Direct application of sympathomimetic agents has a number of stimulatory effects both in human and in animal in vitro and cell culture models. It is now established clearly that β-adrenergic stimulation, particularly of β_2-receptors, increases ciliary beating in the central airways of several species, including humans, in a dose-dependent manner (Wanner, 1988). α-Adrenergic stimulation seems to have little effect on ciliary beat frequency, but α-adrenergic stimulation of tracheal mucus glands causes a high volume of fluid secretion with a low protein concentration (Ueki et al., 1980; Leikauf et al., 1984). The effect of β-adrenergic stimulation is the opposite: stimulation causes secretion that has a high macromolecule concentration, high viscosity, and less water (Ueki et al., 1980; Leikauf et al., 1984; Borson et al., 1984). Treatment of bovine tracheal glands with β-adrenergic agonists in vitro results in release of chondroitin-sulfate proteoglycans (Paul et al., 1988). Serous cells within the epithelium contain more α-adrenergic receptors and fewer β-adrenergic receptors than mucus cells, whereas muscarinic receptors are distributed equally in both cell types (Barnes and Basbaum, 1983; Basbaum et al., 1984). Unlike the postulated defect in sympathetic regulation of airway caliber in asthma, there is no specific evidence that either α- or β-adrenergic neural regulation of mucus secretion or cilia function is impaired in asthma or other obstructive airways diseases, though a global defect in β-adrenergic function in asthmatic subjects presumably would affect the composition and transport of airway secretions as well as airway smooth muscle.

Airway mast cell secretion also is modulated by sympathetic innervation. Exogenous β-adrenergic stimulation inhibits degranulation of mast cells in passively sensitized human lung strips (Orange et al., 1971), in canine peripheral lung after antigen challenge (Barnett et al., 1978), and in canine tracheal smooth muscle after non-immune-stimulated degranulation (Brown et al., 1982). Canine models of in situ determination of mast cell response in large airways to neurally mediated sympathetic stimulation, followed by antigen challenge, demonstrate that β-adrenergic stimulation suppresses mast cell release of histamine (Garrity et al., 1985) (Fig. 12). Neural β-adrenergic stimulation blocks both histamine release and resulting contraction of bronchial smooth muscle and suggests an additional indirect mechanism by which the sympathetic nervous system might modulate airway tone. A recent study demonstrates that in guinea pig airways, stimulation of mast cell release by antigen sufficient to cause contraction can be blocked by electrical field stimulation; such blockade can be reversed partially with propranolol, suggesting that the responsible pathway is sympathetic (Undem et al., 1989). Peripheral sympathetic innervation and mast cell density are highly correlated in rats (Levine et al., 1990). These data suggest strongly that mast cell density and secretion are closely regulated by β-adrenergic innervation. Exogenous β-adrenergic agonists then may modulate airways reactivity by direct effects on airway smooth muscle, by suppression of cholinergic tone, and by suppression of mast cell secretion (and perhaps even mast cell density).

α-Adrenergic agonists facilitate histamine release from human lung fragments

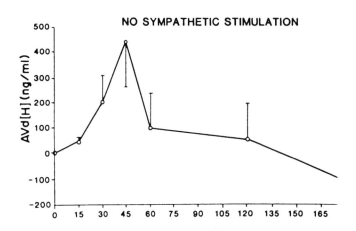

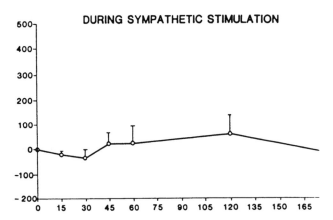

Figure 12   Sympathetic neural modulation of canine bronchial mast cell secretion in an in situ model. Histamine secretion by mast cells was measured as the arteriovenous difference (AVd) across a third-generation bronchus after direct intra-arterial antigen challenge. Stimulation with the autonomic nicotinic agonist dimethylphenylpiperazinium combined with muscarinic blockade elicited maximal sympathetic stimulation, which blocked completely histamine secretion after antigen challenge compared to controls (antigen challenged with no prior sympathetic stimulation). Compare this suppression of mast cell secretion to histamine secretion augmented by cholinergic stimulation in a similar model (Fig. 6). (From Garrity et al., 1985, with permission of the publisher.)

in vitro, suggesting the presence of α-adrenoceptors on mast cells (Kaliner et al., 1972). However, more recent studies in canines using cultured mastocytoma cells demonstrate a lack of α-adrenoceptors on these cells (Phillips et al., 1985). Selective stimulation of the α-adrenergic innervation to a bronchial airway has little effect on mast cell release in a canine model (White et al., 1989). These data suggest little α-adrenergic regulation of mast cell secretion in animal models. Little human data are available, and it is not known if human airway mast cells have α-adrenoceptors.

IV. SUMMARY

Autonomic neural regulation of airway function is complex, and even the classic cholinergic and adrenergic innervations are incompletely understood. Neither system functions in isolation, and better approaches to the study of their integrated function in airways—with each other, with the NANC systems, and with other constitutive and inflammatory cells within the airways—in humans are required before we can identify the exact mechanisms in the genesis of airways hyperreactivity in human asthma. Parasympathetic innervation is the major excitatory innervation for airway tone, and its regulation by other neural systems and by inflammatory mediators may occur at several sites. The recent hypothesis that a specific defect in inhibitory M_2-receptors may lead to airways hyperreactivity in asthmatic subjects requires testing in humans; if this is true, therapy with selective M_2-receptor agonists may hold promise for treatment of this disease. Sympathetic regulation of airways caliber occurs principally via suppression of cholinergic tone, and the demonstration of sensitivity to β-adrenergic antagonists in asthmatic subjects suggests that release of this inhibition is part of the mechanism of airways hyperreactivity. This must be counterbalanced by the fact that the sympathetic nervous system does not defend airways caliber after bronchoconstriction. Defects in β-adrenoceptor function and exaggerations in α-adrenergic responsiveness are likely to have little clinical significance in asthma. It is not clear whether the defects in cholinergic and adrenergic regulation are the same in all asthmatic subjects; given the diversity of this disease syndrome, it is easy to speculate that different subjects may have different, and perhaps subtle, defects in autonomic regulation. Further testing of these systems in human subjects combined with delineation of the NANC and neuropeptide systems in human airways may lead to better therapeutic approaches in asthma and chronic airflow obstruction.

ACKNOWLEDGMENTS

I thank Richard Mitchell, Ph.D., and Alan Leff, M.D., for their helpful advice and comments. This work is supported in part by Grants HL-02484 and HL-48696 from the National Heart, Lung, and Blood Institute.

REFERENCES

Agarwal, D.K., and Townley, R.G. (1987). Effect of platelet-activating factor in β-adreno-ceptors in human lung. *Biochem. Biophys. Res. Commun. 143*:1–6.

Austen, K.F., and Orange, R. (1975). Bronchial asthma: the possible role of chemical mediators of immediate hypersensitivity in the pathogenesis of subacute chronic disease. *Am. Rev. Respir. Dis. 112*:423–436.

Ayala, L.E., and Ahmed, T. (1989). Is there loss of a protective muscarinic receptor in asthma? *Chest 96*:1285–1291.

Baker, D., Basbaum, C.B., Herbert, D.A., and Mitchell, R.A. (1983). Transmission in airway ganglia of ferrets: inhibition by norepinephrine. *Neurosci. Lett. 41*:139–143.

Barnes, P.J. (1986). Neural control of human airways in health and disease. *Am. Rev. Respir. Dis. 134*:1289–1314.

Barnes, P.J. (1989). Muscarinic receptor subtypes: implications for lung disease. *Thorax 44*:161–167.

Barnes, P.J. (1993). Muscarinic receptor subtypes in airways. *Life Sci. 52*:521–527.

Barnes, P.J., and Basbaum, C.B. (1983). Mapping of adrenergic receptors in the trachea by autoradiography. *Exp. Lung Res. 5*:183–192.

Barnes, P.J., Brown, M.J., Silverman, M., and Dollery, C.T. (1981a). Circulating cate-cholamines in exercise and hyperventilation induced asthma. *Thorax 36*:435–440.

Barnes, P.J., Wilson, N.M., and Vickers, H. (1981b). Prazocin, an alpha1-adrenoceptor antagonist partially inhibits exercise-induced asthma. *J. Allergy Clin. Immunol. 68*: 411–419.

Barnes, P.J., Ind, P.W., and Brown, M.J. (1982). Plasma histamine and catecholamines in stable asthmatic subjects. *Clin. Sci. 62*:661–665.

Barnes, P.J., Cuss, F.M., and Palmer, J.B. (1985). The effect of airway epithelium on smooth muscle contraction in bovine trachea. *Br. J. Pharmacol. 86*:685–691.

Barnett, D.B., Chesrown, S.E., Zbinden, A.F., Nisam, M., Reed, B.R., Bourne, H.R., and Gold, W.M. (1978). Cyclic AMP and cyclic GMP in peripheral lung: regulation in vivo. *Am. Rev. Respir. Dis. 118*:723–733.

Basbaum, C.B., Griso, M.A., and Widdicombe, J.H. (1984). Muscarinic receptors: evidence for a nonuniform distribution in tracheal smooth muscle and exocrine glands. *J. Neurosci. 4*:508–520.

Benson, M.K., and Graff, P.D. (1977). Bronchial reactivity: interaction between vagal stimulation and inhaled histamine. *J. Appl. Physiol. 43*:643–647.

Berkin, K.G., Inglis, G.C., Ball, S.G., and Thomson, N.C. (1985). Airway responses to low concentrations of adrenaline and noradrenaline in normal subjects. *Q.J. Exp. Physiol. 70*:203–209.

Blaber, L.C., Fryer, A.D., and Maclagan, J. (1985). Neuronal muscarine receptors attenuate vagally-induced contraction of feline bronchial smooth muscle. *Br. J. Pharmacol. 86*: 723–728.

Bloom, J.W., Halonen, M., Lawrence, L.J., Rould, E., Seaver, N.A., and Yamamura, H.I. (1987). Characterization of high affinity [3H]pirenzepine and (−)-[3H] quinuclidinyl benzilate binding to muscarinic cholinergic receptors in rabbit peripheral lung. *J. Pharmacol. Exp. Ther. 240*:51–58.

Bonner, T.I., Buckley, N.J., Young, A.C., and Brann, M.R. (1987). Identification of a family of muscarinic acetylcholine receptor genes. *Science (Wash. DC) 237*:527–532.

Borson, D.B., Chinn, R.A., Davis, B., and Nadel, J.A. (1980). Adrenergic and cholinergic nerves mediate fluid secretion from tracheal glands of ferrets. *J. Appl. Physiol. 49*:1027–1031.

Boushey, H.A., Holtzman, M.J., Sheller, J.R., Nadel, J.A. (1980). Bronchial hyperreactivity. *Am. Rev. Respir. Dis. 121*:389–413.

Brofman, J.D., Leff, A.R., Munoz, N.M., Kirchhoff, C., and White, S.R. (1990). Sympathetic secretory response to hypercapnic acidosis in swine. *J. Appl. Physiol. 69*: 710–717.

Brown, J.K., Leff, A.R., Frey, M.J., Reed, B.R., Lazarus, S.C., Shields, R., and Gold, W.M. (1982). Characterization of tracheal mast cell reactions in vivo: inhibition by a beta-adrenergic agonist. *Am. Rev. Respir. Dis. 126*:842–848.

Cazzola, M., Rano, S., DeSantis, D., Principe, P.J., and Marmo, E. (1987). Respiratory responses to pirenzepine in healthy subjects. *Int. J. Clin. Pharmacol. Ther. Toxicol. 25*: 105–109.

Cerrina, J., Ladurie, M.L.R., Labat, C., Raffestin, B., Bayol, A., and Brink, C. (1986). Comparison of human bronchial muscle responses to histamine in vivo with histamine and isoproterenol agonists in vitro. *Am. Rev. Respir. Dis. 123*:156–160.

Daniel, E.E., Kannon, M., Davis, C., and Pusey-Daniel, V. (1986). Ultrastructural studies on the neuromuscular control of human tracheal and bronchial smooth muscle. *Respir. Physiol. 63*:109–128.

Davis, C., Kannan, M.S., Jones, T.R., and Daniel, E.E. (1982). Control of human airway smooth muscle: in vitro studies. *J. Appl. Physiol. 53*:1080–1087.

Davis, P.B. (1988). Autonomic function in patients with airway obstruction. In: *The Airways: Neural Control in Health and Disease*. (Kaliner, M.A., and Barnes, P.J., eds.). Marcel Dekker, New York, pp. 87–118.

Davis, P.B., Paget, G.L., and Turi, V. (1985). Alpha-adrenergic responses in asthma. *J. Lab. Clin. Med. 105*:164–169.

Doidge, J.M., and Satchall, D.G. (1982). Adrenergic and non-adrenergic inhibitory nerves in mammalian airways. *J. Autonom. Nerv. Sys. 5*:83–99.

Dörje, F., Levey, A.I., and Brann, M.R. (1991). Immunological detection of muscarinic receptor subtype proteins (m_1–m_5) in rabbit peripheral tissues. *Mol. Pharmacol. 40*: 459–462.

Dosani, R., Van Loon, G.R., and Burki, N.R. (1987). The relationship between exercise-induced asthma and plasma catecholamines. *Am. Rev. Respir. Dis. 136*:973–978.

Drazen, J.M., and Schneider, M.W. (1977). Comparative response of tracheal spiral and parenchymal strips to histamine and carbachol. *J. Clin. Invest. 61*:1441–1447.

Dunn, T.L., Gerber, M.J., Shen, A.S., Fernandez, E., Iseman, M.D., and Cherniack, R.M. (1986). The effect of topical ophthalmic instillation of timolol and betaxolol on lung function in asthmatic subjects. *Am. Rev. Respir. Dis. 133*:264–268.

Florey, H., Carleton, H.M., and Wells, A.Q. (1932). Mucus secretion in the trachea. *Br. J. Exp. Pathol. 13*:269–284.

Fryer, A.D., and Jacoby, D.B. (1991). Parainfluenza virus infection damages inhibitory M_2

muscarinic receptors on pulmonary parasympathetic nerves in the guinea pig. *Br. J. Pharmacol. 102*:267–271.

Fryer, A.D., and Jacoby, D.B. (1992). Function of pulmonary M_2 muscarinic receptors in antigen-challenged guinea pigs is restored by heparin and poly-L-glutamate. *J. Clin. Invest. 90*:2292–2298.

Fryer, A.D., and Maclagan, J. (1984). Muscarinic inhibitory receptors in pulmonary parasympathetic nerves in the guinea pig. *Br. J. Pharmacol. 83*:973–978.

Fryer, A.D., and Okanlami, O.A. (1993). Neuronal M_2 muscarinic receptor function in guinea-pig lungs is inhibited by indomethacin. *Am. Rev. Respir. Dis. 147*:559–564.

Fryer, A.D., and Wills-Karp, M. (1991). Dysfunction of M_2-muscarinic receptors in pulmonary parasympathetic nerves after antigen challenge. *J. Appl. Physiol. 71*:2255–2261.

Gallagher, J.T., Kent, P.W., and Passatore, M. (1975). The composition of tracheal mucus and the nervous control of its secretion in the cat. *Proc. R. Soc. Lond. (Biol.) 192*:49–76.

Gao, Y., and Vanhouette, P.M. (1988). Removal of the epithelium potentiates acetylcholine in depolarizing canine bronchial smooth muscle. *J. Appl. Physiol. 65*:2400–2405.

Garrity, E.R., Stimler, N.P., Munoz, N.M., Fried, R., and Leff, A.R. (1983). Response of bronchial smooth muscle to immune and non-immune degranulation in situ. *J. Appl. Physiol. 55*:1803–1810.

Garrity, E.R., Stimler, N.P., Munoz, N.M., Tallet, J., David, A.C., and Leff, A.R. (1985). Sympathetic modulation of biochemical and physiological response to immune degranulation in canine bronchial airways in vivo. *J. Clin. Invest. 75*:2038–2046.

Goldie, R.G., Spina, D., Henry, P.J., Lulich, K.M., and Paterson, J.W. (1986). In vitro responsiveness of human asthmatic bronchus to carbachol, histamine, β-adrenoceptor agonists and theophylline. *Br. J. Clin. Pharmacol. 22*:669–676.

Goyal, R.K. (1989). Muscarinic receptor subtypes. Physiology and clinical implications. *N. Engl. J. Med. 321*:1022–1029.

Gross, N.J. (1988). Drug therapy. Ipratropium bromide. *N. Engl. J. Med. 319*:486–494.

Gorss, N.M., Co, E., and Skorodin, M.S. (1989). Cholinergic bronchomotor tone in COPD. Estimates of its amount in comparison with that in normal subjects. *Chest 96*:984–987.

Hall, A.K., Barnes, P.J., Meldrum, L.A., and Maclagan, H.J. (1989). Facilitation by tachykinins of neurotransmission in guinea-pig pulmonary parasympathetic nerves. *Br. J. Pharmacol. 97*:274–280.

Havez, R., and Roussel, P. (1976). Bronchial mucus: physical and biochemical features. In: *Bronchial Asthma: Mechanisms and Therapeutics*. (Weiss, E.B., and Segal, M.S., eds.). Little, Brown, Boston, pp. 409–422.

Henderson, W.R., Shelhamer, J.H., Reingold, D.B., Smith, L.J., Evans, R., and Kaliner, M. (1979). Alpha-adrenergic hyper-responsiveness in asthma. Analysis of vascular and pupillary responses. *N. Engl. J. Med. 300*:642–647.

Herxheimer, H. (1967). The bronchoconstrictor action of propranolol aerosol in the guinea pig. *J. Physiol. (Lond.) 190*:41–42.

Ind, P.W., Nixon, C.M.S., Fuller, R.W., and Barnes, P.J. (1989). Anticholinergic blockade of beta-blocker-induced bronchoconstriction. *Am. Rev. Respir. Dis. 139*:1390–1394.

Innes, I.R., and Nickerson, M. (1975). Norepinephrine, epinephrine, and the sympathomimetic amines. In: *The Pharmacological Basis of Therapeutics*. (Goodman, L.S., and Gilman, A., eds.). 5th ed. Macmillan, New York, pp. 477–513.

Insel, P.A., and Wasserman, S.I. (1990). Asthma: a disorder of adrenergic receptors? *FASEB J. 4*:2732–2736.

Jacoby, D.B., Gleich, G.J., and Fryer, A.D. (1993). Human eosinophil major basic protein is an endogenous allosteric antagonist at the inhibitory muscarinic M2 receptor. *J. Clin. Invest. 91*:1314–1318.

Joos, G., Pauwells, R., and van der Straeten, M. (1988). The mechanism of tachykinin-induced bronchconstriction in the rat. *Am. Rev. Respir. Dis. 137*:1038–1044.

Kaliner, M. (1982). Autonomic nervous system abnormalities and allergy. *Ann. Intern. Med. 96*:349–357.

Kaliner, M., Orange, R.P., and Austen, K.F. (1972). Immunological release of histamine and slow reacting substance of anaphylaxis from human lung. IV. Enhancement by cholinergic and alpha-adrenergic stimulation. *J. Exp. Med. 136*:556–567.

Kaliner, M.A., Shelhamer, J.H., Borson, D.B., Patow, C.A., Marom, Z., and Nadel, J.A. (1988). Respiratory mucus. In: *The Airways: Neural Control in Health and Disease*. (Kaliner, M.A., and Barnes, P.J., eds.). Marcel Dekker, New York, pp. 575–593.

Kerstjens, H.A.M., Brand, P.L.P., Hughes, M.D., Robinson, N.J., Postma, D.S., Sluiter, H.J., Bleecker, E.R., Dekhuijzen, P.N.R., DeJong, P.M., Mengelers, H.J.J., Overbeek, S.E., Schoonbrood, D.F.M.E., and the Dutch Chronic Non-Specific Lung Disease Study Group (1992). A comparison of bronchodilator therapy with or without inhaled cortico-steroid therapy for obstructive airways disease. *N. Engl. J. Med. 327*:1413–1419.

Kokin, P. (1896). Ueber die secretorischen nerven der Kelkhopf und Luftrohrenschleim-drusen. *Pflugers Arch. 63*:622–630.

Kubo, T., Fukuda, K., Mikami, A., Maeda, A., Takahawa, H., Mishina, M., Haga, T., Haga, K., Ichiyama, A., Kanagawa, K., Kojima, A., Matsuo, H., Hirose, T., and Numa, S. (1986). Cloning, sequencing and expression of complementary DNA encoding the muscarinic acetylcholine receptor. *Nature (Lond.) 323*:411–418.

Laitinen, A., Partanen, M., Hervonen, A., and Laitinen, L.A. (1985). Electron micro-scopic study on the innervation of human lower respiratory tract. Evidence of adrenergic nerves. *Eur. J. Respir. Dis. 67*:209–215.

Lammers, J.-W., Minette, P., McCusker, M., and Barnes, P.J. (1989). The role of pirenz-epinesensitive (M_1) muscarinic receptors in vagally mediated bronchoconstriction in humans. *Am. Rev. Respir. Dis. 139*:446–449.

Larsen, G.L., Barron, R.J., Cotton, E.K., and Brooks, J.G. (1979). A comparative study of inhaled atropine sulfate and isoproterenol hydrochloride in cystic fibrosis. *Am. Rev. Respir. Dis. 119*:339–407.

Leikauf, G.D., Ueki, I.F., and Nadel, J.A., (1984). Autonomic regulation of viscoelasticity of cat tracheal gland secretions. *J. Appl. Physiol. 56*:426–430.

Leff, A.R. (1988). Endogenous regulation of bronchomotor tone. *Am. Rev. Respir. Dis. 137*: 1198–1216.

Leff, A.R., and Munoz, N.M. (1981). Interrelationship between alpha- and beta-adrenergic agonists and histamine in canine airways. *J. Allergy Clin. Immunol. 68*:300–309.

Leff, A.R., Stimler, N.P., Munoz, N.M., Shioya, T., Tallet, J., and Dame, C. (1986). Augmentation of respiratory mast cell secretion of histamine caused by vagus nerve stimulation during antigen challenge. *J. Immunol. 136*:1066–1073.

Leff, A.R., White, S.R., Munoz, N.M., Popovich, K.J., and Stimler-Gerard, N.P. (1987).

Parasympathetic mediation of platelet-activating factor induced contraction in canine trachealis in-vivo. *J. Appl. Physiol. 62*:599–605.

Leff, A.R., Munoz, N.M., and Shioya, T. (1989). Comparative distribution of cholinomimetic and parasympathetic contractile responses in the major diameter airways of dogs. *Pulm. Pharmacol. 1*:107–112.

Levine, J.D., Coderre, T.J., Covinsky, K., and Basbaum, A.I. (1990). Neural influences on synovial mast cell density in rat. *J. Neurosci. Res. 26*:301–307.

Loring, S.H., Drazen, J.M., Snapper, J.R., and Ingram, R.H., Jr. (1978). Vagal and aerosol histamine interactions on airway responses in dogs. *J. Appl. Physiol. 45*:40–44.

Lundberg, J.M., Hokfelt, T., Martling, C.R., Saria, A., and Cuello, C. (1984). Substance P–immunoreactive sensory nerves in the lower respiratory tract of various mammals including man. *Cell. Tissue Res. 235*:251–261.

Maisal, A.S., Fowler, P., Rearden, A., Motulsky, H.J., and Michel, M.C. (1989). A new method for isolation of human lymphocyte subsets reveals differential regulation of β-adrenergic receptors by terbutaline treatment. *Clin. Pharmacol. Ther. 46*:429–439.

Maisal, A.S., Knowlton, K.U., Fowler, P., Rearden, A., Ziegler, M.G., Motulsky, H.J., Insel, P.A., and Michel, M.C. (1990). Adrenergic control of circulating lymphocyte subpopulations: effects of congestive heart failure, dynamic exercise and terbutaline treatment. *J. Clin. Invest. 85*:462–467.

Mak, J.C.W., and Barnes, P.J. (1990). Autoradiographic visualization of muscarinic receptor subtypes in human and guinea pig lung. *Am. Rev. Respir. Dis. 141*:1559–1568.

Mak, J.C.W., Baraniuk, J.N., and Barnes, P.J. (1992). Localization of muscarinic receptor subtype mRNAs in human lung. *Am. J. Respir. Cell Mol. Biol. 7*:344–348.

McMahon, C.D., Shaffer, R.N., Hoskins, H.D., and Hetherington, J. (1979). Adverse effects experienced by patients taking timolol. *Am. J. Ophthalmol. 88*:736–738.

McNeill, R.S. (1966). Effect of propranolol on ventilatory function. *Am. J. Cardiol. 18*: 473–475.

Minnette, P.A., and Barnes, P.J. (1988). Prejunctional inhibitory muscarinic receptors on cholinergic nerves in human and guinea pig airways. *J. Appl. Physiol. 64*:2532–2537.

Minnette, P.A., Lammers, J.-W., Dixon, C.M.S., McCusker, M.C., and Barnes, P.J. (1989). A muscarinic agonist inhibits reflex bronchoconstriction in normal but not in asthmatic subjects. *J. Appl. Physiol. 67*:2461–2465.

Mitchell, R.W., Koenig, S.M., Popovich, K.J., Kelly, E., Tallet, J., and Leff, A.R. (1993). Pertussis toxin augments β-adrenergic relaxation of muscarinic contraction in canine trachealis. *Am. Rev. Respir. Dis. 147*:327–331.

Munoz, N.M., Shioya, T., Murphy, T.M., Primack, S., Dame, C., and Leff, A.R. (1986). Potentiation of vagal efferent contractile response of canine tracheal smooth muscle by thromboxane mimetic, U-46619. *J. Appl. Physiol. 61*:1173–1179.

Murlas, C., Nadel, J.A., and Basbaum, C.B. (1980). A morphometric analysis of the autonomic innervation of cat tracheal glands. *J. Auton. Nerv. Syst. 2*:23–27.

Myers, A.C., and Undem, B.J. (1991). Functional interactions between capsaicin-sensitive and cholinergic nerves in the guinea pig bronchus. *J. Pharmacol. Exp. Ther. 259*: 104–109.

Nadel, J.A. (1977). Autonomic control of airway smooth muscle and airway secretions. *Am. Rev. Respir. Dis. 115* (Pt. 2):117–126.

Nadel, J.A. (1980). Autonomic control of airway smooth muscle. In: *Physiology and Pharmacology of the Airways*. (Nadel, J.A. ed.). Marcel Dekker, New York, pp. 217–257.

Nadel, J.A., Cabezas, G.A., and Austin, J.H.M. (1971). In vivo roentgenographic examination of parasympathetic innervation of small airways: use of powdered tantulum and a fine focal spot x-ray. *Invest. Radiol. 6*:9–17.

Orange, R.P., Austen, W.G., and Austen, K.F. (1971). Immunological release of histamine and slow-reacting substance of anaphylaxis from human lung. I. Modulation by agents influencing cellular levels of cyclic $3',5'$ adenosine monophosphate. *J. Exp. Med. 134*:136s–138s.

Partanen, M., Laitinen, A., Hervonen, A., Toivanen, M., and Laitinen, L.A. (1982). Catecholamine and acetylcholinesterase containing nerves in human lower respiratory tract. *Histochemistry 76*:175–188.

Paul, A., Picard, J., Mergey, M., Veissiere, D., Finkbeiner, W.-E., and Basbaum, C.B. (1988). Glycoconjugates secreted by bovine tracheal serous cells in culture. *Arch. Biochem. Biophys. 260*:75–84.

Phillips, M.J., Barnes, P.J., and Gold, W.M. (1985). Characterization of purified dog mastocytoma cells: autonomic membrane receptors and pharmacologic modulation of histamine release. *Am. Rev. Respir. Dis. 132*:1019–1026.

Phipps, R.J. (1981). The airway mucociliary system. In: *International Review of Physiology. Respiratory Physiology III*. (Widdicombe, J.C., ed.). 23rd ed., University Park Press, Baltimore, pp. 213–260.

Phipps, R.J., Nadel, J.A., and Davis, B. (1980). Effect on alpha-adrenergic stimulation on mucus secretion and on ion transport in cat trachea in vitro. *Am. Rev. Respir. Dis. 121*: 359–365.

Poch, G., and Holzman, S. (1980). Quantitative estimation of overadditive and underadditive drug effects means of theoretical additive dose-response curves. *J. Pharmacol. Meth. 4*:179–188.

Rhoden, K.J., Meldrum, L.A., and Barnes, P.J. (1988). Inhibition of cholinergic neurotransmission in human airways by beta-2-adrenoceptors. *J. Appl. Physiol. 65*:700–705.

Richardson, J.B. (1979). Nerve supply to the lungs. *Am. Rev. Respir. Dis. 119*:785–802.

Richardson, J.B. (1988). Innervation of the lung. In: *The Airways: Neural Control in Health and Disease*. (Kaliner, M.A., and Barnes, P.J., eds.). Marcel Dekker, New York, pp. 23–33.

Richardson, J.B., and Béland, J. (1976). Nonadrenergic inhibitory nervous system in human airways. *J. Appl. Physiol. 41*:764–771.

Richardson, P.S., and Sterling, G.M. (1969). Effects of β-adrenergic receptor blockade on airway conductance and lung volume in normal and asthmatic subjects. *Br. Med. J. 3*: 143–145.

Roffel, A.F., Elzinga, C.R.S., and Zaagsma, J. (1990). Muscarinic M_3-receptors mediate contraction of human central and peripheral airway smooth muscle. *Pulm Pharmacol. 3*:47–52.

Rudchenko, S.A, Korichneva, I.L., Tupykin, V.G., Tkachuk, V.A., and Chuchanlin, A.G. (1993). Beta-adrenergic receptors on alveolar macrophages and lymphocytes: studies with cells from asthmatics. In: *Signal Transduction in Lung Cells*. (Brody, J.S., Center, D.M., and Tkachuk, V.A., eds.). Marcel Dekker, New York, pp. 555–576.

Russell, J.A. (1980). Noradrenergic inhibitory innervation of canine airways. *J. Appl. Physiol.* *48*:16–22.

Sands, M.F., Douglas, F.L., Green, J., Banner, A., Robertson, G.L., and Leff, A.R. (1985). Homeostatic regulation of bronchomotor tone by sympathetic activation during bronchoconstriction in normal and asthmatic humans. *Am. Rev. Respir. Dis.* *131*:995–998.

Sano, Y., Watt, G., and Townley, R.G. (1983). Decreased mononuclear cell β-adrenergic receptors in bronchial asthma: parallel studies of lymphocyte and granulocyte desensitization. *J. Allergy Clin. Immunol.* *72*:495–503.

Sekizawa, K., Tamaoki, J., Graf, P.D., Basbaum, C.B., and Nadel, J.A. (1987). Enkephalinase inhibitor potentiates mammalian tachykinin-induced contraction in ferret trachea. *J. Pharmacol. Exp. Ther.* *243*:1211–1217.

Shelhamer, J.H., Metcalfe, D.D., Smith, L.J., and Kaliner, M. (1980a). Abnormal adrenergic responsiveness in allergic subjects: analysis of isoproterenol-induced cardiovascular and plasma cyclic adenosine monophosphate responses. *J. Allergy Clin. Immunol.* *66*:52–61.

Shelhamer, J.H., Marom, Z., and Kaliner, M. (1980b). Immunologic and neuropharmacologic stimulation of mucus glycoprotein release from human airways in vitro. *J. Clin. Invest.* *66*:1400–1408.

Sheppard, M.N., Kurian, S.S., Henzen Logmans, S.C., Michetti, F., Cocchia, D., Cole, P., Rush, R.A., Marangos, P.J., Bloom, S.R., and Polak, J.M. (1983). Neuron-specific enolase and S-100. New markers for delineating the innervation of the respiratory tract in man and other animals. *Thorax 38*:333–340.

Shioya, T., Solway, J., Munoz, N.M., Mack, M., and Leff, A.R. (1987a). Distribution of airway contractile responses within the major diameter bronchi during exogenous bronchoconstriction. *Am. Rev. Respir. Dis.* *135*:1105–1111.

Shioya, T., Pollack, E.R., Munoz, N.M., and Leff, A.R. (1987b). Distribution of airway contractile responses in major resistance airways of the dog. *Am. J. Pathol.* *129*:102–117.

Simonsson, B.G., Svedmyr, N., Skoogh, B.E., Anderson, R., and Bergh, N.P. (1972). In vivo and in vitro studies on alpha-adrenoceptors in human airways; potentiation with bacterial endotoxin. *Scand. J. Respir. Dis.* *53*:227–236.

Skoogh, B-E. (1986). Parasympathetic ganglia in the airways. *Bull. Eur. Physiopathol. Respir. 22* (Suppl. 7):1289–1314.

Skoogh, B-E. (1988). Airway parasympathetic ganglia. In: *The Airways: Neural Control in Health and Disease.* (Kaliner, M.A, and Barnes, P.J., eds.). Marcel Dekker, New York, pp. 217–240.

Skoogh, B-E., and Svedmyr, N. (1989). Beta$_2$-adrenoceptor stimulation inhibits ganglionic transmission in ferret trachea. *Pulm. Pharmacol. 1*:167–172.

Smith, R.V., and Satchell, D.G. (1985). Extrinsic pathways of the adrenergic innervation of the guinea-pig trachealis muscle. *J. Auton. Nerv. Sys. 14*:61–73.

Spina, D., Rigby, P.J., Paterson, J.W., and Goldie, R.G. (1989a). Autoradiographic localization of β-adrenoceptors in asthmatic human lung. *Am. Rev. Respir. Dis. 140*:1410–1415.

Spina, D., Rigby, P.J., Paterson, J.W., and Goldie, R.G. (1989b). α_1-adrenergic function and autoradiographic distribution in human asthmatic lung. *Br. J. Pharmacol. 97*:701–708.

Stearns, D.R., McFadden, E.R., Jr., Breslin, F.J., and Ingram, R.H., Jr. (1981). Reanalysis of the refractory period in exertional asthma. *J. Appl. Physiol. 50*:503–508.

Strek, M.E., White, S.R., Munoz, N.M., Williams, F.S., Vita, A.J., Leff, A.R., Ndukwu, M.I., and Mitchell, R.W. (1993). Physiological significance of epithelium removal in guinea pig tracheal smooth muscle. *Am. Rev. Respir. Dis. 147*:1477–1482.

Sullivan, T.J., Parker, P., Eisen, S.A., and Parker, C.W. (1975). Modulation of cyclic AMP in purified rat mast cells. II. Studies on the relationship between intracellular cyclic AMP concentrations and histamine release. *J. Immunol. 114*:1480–1485.

Svdemyr, N.L., Larsson, S.A., and Thiringer, G.K. (1977). Development of "resistance" in beta-adrenergic receptors of asthmatic patients. *Chest 69*:479–483.

Tanaka, D.T., and Grunstein, M.M. (1984). Mechanisms of substance P–induced contraction of rabbit airway smooth muscle. *J. Appl. Physiol. 57*:1551–1557.

Tattersfield, A.E. (1987). Effect of beta-agonists and anticholinergic drugs on bronchial reactivity. *Am. Rev. Respir. Dis. 136*:S64–S67.

Tattersfield, A.E., Leaver, D.G., and Pride, N.B. (1973). Effects of β-adrenergic blockade and stimulation on normal human airways. *J. Appl. Physiol. 35*:613–619.

Tattersfield, A.E., Holgate, S.T., Harvey, J.E., Gribbin, H.R. (1983). Is asthma due to partial beta-blockade of airways? *Agents Actions 13* (Suppl.):265–271.

Ueki, I., German, V.F., and Nadel, J.A. (1980). Micropipette measurement of airway submucosal gland secretion. Autonomic effects. *Am. Rev. Respir. Dis. 121*:351–357.

Ukena, D., Wehinger, C., Engelstatter, R., Steinijans, V., and Sybrecht, G.W. (1992). The muscarinic M_1-receptor-selective antagonist, telenzepine, had no bronchodilatory effects in COPD patients. *Eur. Respir. J. 6*:378–382.

Undem, B.J., Adams, G.K., III, and Buckner, C.K. (1989). Influence of electrical field stimulation on antigen-induced contraction and mediator release in the guinea pig isolated superfused trachea and bronchus. *J. Pharmacol. Exp. Ther. 249*:23–30.

Undem, B.J., Myers, A.C., Barthlow, H., and Weinreich, D. (1990). Vagal innervation of the guinea pig bronchus. *J. Appl. Physiol. 69*:1336–1346.

Van Koppen, C.J., Blankenstein, W.M., Klaasen, B.M., Rodrigues de Miranda, F., Beld, A.J., and Van Ginneken, C.A.M. (1987). Autoradiographic visualization of muscarinic receptors in human bronchi. *J. Pharmacol. Exp. Ther. 244*:760–764.

Walters, E.H., O'Byrne, P.M., Fabbri, L.M., Graf, P.D., Holtzman, M.J., and Nadel, J.A. (1984). Control of neurotransmission by prostaglandins in canine trachealis smooth muscle. *J. Appl. Physiol. 57*:129–134.

Wanner, A. (1988). Autonomic control of mucociliary function. In: *The Airways: Neural Control in Health and Disease*. (Kaliner, M.A., and Barnes, P.J., eds.). Marcel Dekker, New York, pp. 551–573.

Wanner, A., Zarzecki, S., Hirsch, J., and Epstein, S. (1975). Tracheal mucus transport in experimental canine asthma. *J. Appl. Physiol. 39*:950–957.

Warren, J.B., Keynes, R.J., Brown, M.J., Jenner, D.A., and McNicol, M.W. (1982). Blunted sympathoadrenal response to exercise in asthmatic subjects. *Br. J. Dis. Chest 76*:147–150.

Warren, J.B., Jennings, S.J., and Clark, T.J.H. (1984). Effect of adrenergic and vagal blockade on the normal human airway response to exercise. *Clin. Sci. 66*:79–85.

White, S.R., Sands, M.F., Munoz, N.M., Murphy, T.M., Popovich, K.J., Shioya, T., Blake, J.S., Mack, M., and Leff, A.R. (1987). Homeostatic regulation of airway smooth muscle tone by catecholamine secretion in swine. *J. Appl. Physiol. 62*:972–977.

White, S.R., Blake, J.S., Murphy, T.M., Mack, M.M., Munoz, N.M., and Leff, A.R. (1989a). Effects of vasomotor and mediator-induced hypotension on bronchomotor tone in swine. *J. Appl. Physiol. 66*:1852–1859.

White, S.R., Stimler-Gerard, N.P., Munoz, N.M., Popovich, K.J., Murphy, T.M., Blake, J.S., Mack, M.M., and Leff, A.R. (1989b). Effect of beta-adrenergic blockade and sympathetic stimulation on canine bronchial mast cell response to immune degranulation in vivo. *Am. Rev. Respir. Dis. 139*:73–79.

Widdicombe, J.G. (1977). Studies on afferent innervation. *Am. Rev. Respir. Dis. 115* (Pt. 2): 99–105.

Widdicombe, J.G. (1981). Nervous receptors in the respiratory tract and lungs. In: *Lung Biology in Health and Disease. Part 1, Regulation of Breathing* (T. Hornbein, ed.). Marcel Dekker, New York, pp. 429–472.

Willis, T. (1679). *Pharamceutica Rationales.* Part 2. Dring, Harper, and Leigh, London.

Zaid, G., and Beall, G.N. (1966). Bronchial response to beta-adrenergic blockade. *N. Engl. J. Med. 275*:580–584.

DISCUSSION

Kaliner: You raise several contradictory points. You suggest that there is no firm evidence for a defect in β-adrenergic responses because the airways are not defended by epinephrine secretion during airway constriction. However, the observation that β-adrenergic blockade causes a dose-related increase in airflow obstruction in essentially all asthmatics suggests that the β-adrenergic system is critical to maintaining an open airway. We believe that the data support the concept that a defect in β-adrenergic function is not sufficient by itself to cause asthma, but certainly contributes to the propensity that a patient exhibits toward developing asthma.

Linnoila: The data are somewhat contradictory, though some of the conflict may be in the dose of propranolol employed. The lack of a sympathetic secretion to airflow obstruction may simply indicate that proprioception is not coupled to the adrenergic system.

Kaliner: The lack of epinephrine secretion may be a design defect in humans but does not indicate that beta responses are not important in asthma.

White: The circulating concentrations of epinephrine in resting humans, both normal and asthmatics, is quite low—5–25 ng/ml. These low concentrations do not elicit bronchodilation, so it is difficult to reconcile this observation with the actions of beta blockers in asthmatics. If such a concentration did modulate airway tone, then normal subjects should constrict to beta blockers, and they do not.

Kaliner: It is more likely that the β-adrenergic action takes place at the level of the cholinergic ganglion, where beta actions negatively modulate cholinergic tone. Thus, normal subjects would not bronchodilate to epinephrine or constrict to beta blockade, as their cholinergic tone and responses to cholinergic input are negligible. However, asthmatics respond to enhanced cholinergic tone after beta blockade with bronchoconstriction. The confirmation of this hypothesis is the inhibitory action of atropine on beta-blocker-induced asthma.

Belvisi: It has been shown by Allison Fiyer's group that cationic proteins increase functional cholinergic contractile responses, and this action has been assumed to be due to a malfunction of the muscarinic M_2-autoreceptor. Could another answer be that cationic proteins (which have been shown to release tachykinins) are stimulating sensory nerves in guinea pigs to release neuropeptides, which in turn enhance cholinergic neurotransmission?

White: The delineation of muscarinic receptor subtypes may now allow us to address the whole issue of cholinergic overstimulation. I agree that inflammatory modulation of the M_2-receptor may work as you suggest.

Kummer: You suggest that there is no direct "sympathetic" innervation of the airway smooth muscle in humans. Does this really include all sympathetic fibers or only those that are nonadrenergic?

White: I agree with your point. Dr. Richardson showed that there is no α- or β-adrenergic sympathetic innervation of the human airway smooth muscle. There certainly is NANC inhibitor innervation in human muscle.

Nadel: Why are cholinergic antagonists effective as bronchodilators in chronic bronchitis? Is it due to cholinergic tone or geometry of the airway wall?

Barnes: The reasons why anticholinergics are effective in chronic obstructive pulmonary disease may be due to blocking normal vagal tone, which has an exaggerated effect owing to geometrical narrowing of the airways.

3

Neuropeptides in the Lower Airways Investigated by Modern Microscopy

David R. Springall and Julia M. Polak
Royal Postgraduate Medical School, London, England

I. INTRODUCTION

The innervation of the respiratory system is one of the many control systems involved in modulating lung function. It is now fully recognized that neural control acts via both local nerves and the central nervous system to integrate the control of respiration, gas exchange, mucus secretion, blood flow, and the response to various exogenous stimuli.

The pulmonary system is innervated by autonomic and sensory nerves (1). Autonomic nerves are functionally of cholinergic, adrenergic, and nonadrenergic, noncholinergic (NANC) types. NANC nerves are of both excitatory and inhibitory types. Axon dendrites are found closely associated with vascular smooth muscle and airway smooth muscle. Nerve fibers are found frequently to penetrate the airway epithelium. The origin of these nerve fibers is either intrinsic (postganglionic parasympathetic nerves) from local ganglia found in the wall of airways or extrinsic from the sympathetic chain or from sensory neurons found in the dorsal root and vagal ganglia.

Numerous well-established histological methods have been used for the demonstration of the lung innervation. Enzyme histochemical methods (e.g., cholinesterase) may be employed for the demonstration of parasympathetic nerves and fluorescence histochemical methods for the demonstration of sympathetic nerves. The sensory nervous system is recognized in part by the morphological distribution of axon dendrites (e.g., penetrating airway epithelium or other sensory areas) and also by characteristic ultrastructural features.

Table 1 Major Lung Peptides

Peptide[a]	No. of amino acids	Main actions	Nerve type	Abundance[b] Human	Rat/gpig
CGRP	37	Vasodilatation, bronchoconstriction	Sensory	±	+++
Substance P	11	Vasodilatation, bronchoconstriction	Sensory	±	+++
Galanin	29	Muscle constriction	Parasympathetic	±	±
NPY	36	Vasoconstriction	Sympathetic	+++	++
PHI/PHM	27	Muscle relaxation, secretion	Parasympathetic	+++	++
VIP	28	Vasodilatation, muscle relaxation, secretion	Parasympathetic	+++	++

[a]CGRP, calcitonin gene-related peptide; NPY, neuropeptide Y; PHI/PHM, peptide histidine isoleucine or (in humans) peptide histidine methionine; VIP, vasoactive intestinal polypeptide.
[b]Abundance of the peptide-containing nerves in humans compared with rat and guinea pig (gpig) graded from ± (sparse) to +++ (dense).

The main neuropeptides found in the respiratory tract of humans and animals include neuropeptide tyrosine (NPY), vasoactive intestinal polypeptide (VIP), the neurokinins, and calcitonin gene-related peptide (CGRP) (2). However, there are marked interspecies variations in their relative frequency (see Table 1). In rodents and small mammals CGRP and substance P predominate (3), whereas in humans, NPY and VIP are the major neuropeptides (4).

II. NPY

NPY is a 36-amino-acid peptide originally discovered by Tatemoto and Mutt in extracts of amidated peptides from the gastrointestinal tract (5). It was later found to be present in many neurons of the central and peripheral nervous systems (6,7), closely or almost exclusively associated with the sympathetic branch of the autonomic nervous system (8). NPY originates from a larger precursor, the structure of which (see Fig. 1) has been disclosed by cloning of the NPY gene from an adrenal tumor that contained large quantities of NPY by Minth and co-workers (9). Three receptor subtypes for NPY and related peptides have been identified, designated Y1, Y2, and Y3 according to their affinity for NPY, PYY, and analogs or fragments (10). Although several studies have demonstrated NPY receptors in peripheral mammalian tissues, few have reported the localization of ligand-binding sites in tissue sections and none concerning the lung. Furthermore, there

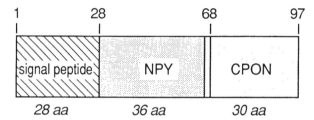

Figure 1 Diagrammatic representation of the precursor molecule of NPY. The peptide is flanked by another (called CPON) at its C-terminus.

is an apparent difficulty in demonstrating specific NPY-binding sites in human tissues. The recent cloning of human and bovine NPY receptor cDNA sequences (11,12) may circumvent the problems by allowing the use of in situ hybridization techniques to determine the localization of mRNA-encoding NPY receptors.

NPY-containing nerves in the respiratory tract are found closely associated with blood vessels and airway smooth muscle (Fig. 2). The demonstration of the presence of NPY-containing nerve fibers in the sympathetic nervous system of the

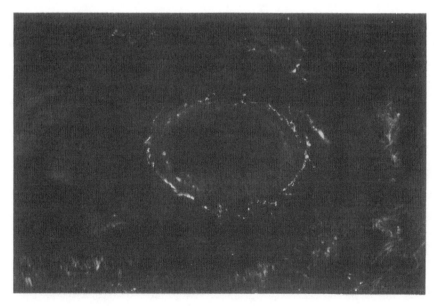

Figure 2 Immunoreactivity for NPY in human lung. Staining is evident in nerve fibers around a pulmonary vessel. Tissue fixed in Zamboni's fluid and stained by indirect immunofluorescence (magnification ×240).

lung (7,13) has been established by a variety of techniques, including the use of the sympathetic neurotoxin 6-hydroxydopamine (6-ODHA) in animals, by reserpine treatment, by sympathectomy, and by retrograde tracing methods that showed that the cell bodies of NPY-containing nerve fibers are situated in the sympathetic chain. In humans, NPY is a major neuropeptide, and extrinsic denervation of the lung, by transplantation, leads to a depletion of NPY-containing nerve fibers in that tissue, suggesting that their origin is indeed extrinsic and possibly from neurons in the sympathetic chain (4). Coexistence of NPY with catecholamines, using antibodies to catecholamine-converting enzymes, also suggests a close anatomical association of catecholamine-containing nerve fibers and NPY-containing nerve fibers. The main action of NPY is vasoconstriction (14), but it is also reported to be a bronchoconstrictor (15).

III. VIP

VIP is a 28-amino-acid peptide whose existence was postulated after Said's observations of a peptide vasodilator agent in the respiratory tract (16). However, the peptide was first isolated from gut extracts by the classical Karolinska Institutet

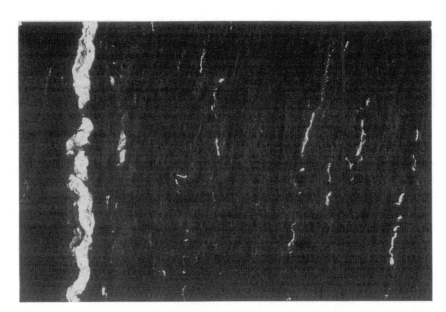

Figure 3   Immunoreactivity for VIP in human lung. Staining is present in nerve fibers and nerve bundles in bronchial smooth muscle. Tissue fixed in Zamboni's fluid and stained by indirect immunofluorescence (magnification ×240).

technique of Mutt (17). Again, it was found to be a widely distributed neuropeptide, not only in the central nervous system, but also in many peripheral organs, including the respiratory tract (18). VIP is a potent vasodilatory, mucosecretory, and muscle relaxant neuropeptide (19). Pre-pro VIP encodes another neuropeptide called peptide histidine isoleucine (PHI) of 27 amino acids and of more limited expression in humans and animals (20,21). VIP-containing nerves in the lung are associated with blood vessels, secretory structures, and airway smooth muscle (Fig. 3), and local ganglia containing VIP are found in the trachea and around the intrapulmonary airways (4,18). VIP-containing nerve fibers are also cholinesterase positive. The extrinsically denervated human lung shows a persistence of VIP-containing nerve fibers, hence supporting studies in animals suggesting the origin of VIP-containing nerve fibers from local parasympathetic ganglia (4).

Binding sites for VIP are reported to occur in human and guinea pig lung, localized to airway epithelium, smooth muscle of blood vessels and large, but not small, airways (22), and alveolar wall (23). This matches with the distribution of VIP-containing nerves. The human VIP receptor has been cloned (24) and the mRNA found to be highly expressed in lung tissue.

IV. NEUROKININS

The neurokinins are a family of peptides encoded in two related genes (25), one of which encodes substance P. These are more abundantly present in animals, and the abundance of substance P or other neurokinin-containing nerve fibers in humans is still contentious (4,26). Substance P and other neurokinins are weak bronchoconstrictors and potent vasodilators (27,28). This is in keeping with the finding that substance P–binding sites in human and guinea pig respiratory tract are predominantly on the microvessels (29) (Fig. 4). The origin of substance P–containing nerve fibers is similar to that of CGRP (see below) from sensory neurons located in the dorsal root ganglia and the nodose/jugular ganglion in the vagus nerve (30). This sensory nature can be further demonstrated in experimental animals by tracing methods, use of the sensory neurotoxin capsaicin, and denervation procedures. In humans, tachykinin-immunoreactive nerve fibers are found only sparsely in the respiratory tract, occasionally penetrating the airway epithelium (4). In rodents, the nerves are seen surrounding smooth muscle, blood vessels, and glands, and in airway epithelium (Fig. 5).

V. CGRP

CGRP is a peptide of 37 amino acids encoded in the same gene as calcitonin (31). The messenger RNA for calcitonin is expressed principally in thyroid tissue, whereas CGRP messenger RNA is found in the sensory nervous system, including

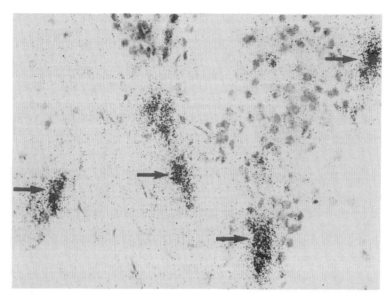

Figure 4 Photomicrograph of emulsion-dipped human lung sections, showing high silver grain density (arrows), representing ^{125}I–Bolton Hunter–substance P binding to sub-epithelial microvessels (magnification ×200).

Figure 5 Immunoreactivity for substance P in rat lung. Numerous fibers are evident, some penetrating the airway epithelium (arrow). Tissue fixed in Zamboni's fluid and stained by indirect immunofluorescence (magnification ×400).

that of the respiratory tract. CGRP-containing nerve fibers are found around blood vessels, penetrating the airway epithelium and airway smooth muscle (3) (Fig. 6). The fibers have a similar distribution to those containing substance P (3), and the two peptides are largely colocalized in the same nerves (32). Receptors for CGRP are found in the respiratory tract at sites that match the distribution of nerves; they are present on vascular smooth muscle in guinea pigs and humans (33), but on endothelium in the rat (34), and to a lesser extent on airway smooth muscle in both species. This is in keeping with the strong vasodilator but weak bronchoconstrictor actions of CGRP (35) and also suggests that the peptide acts on vessels by direct and, in some species, endothelium-dependent mechanisms. The origin of CGRP-containing nerves, like that of substance P, is sensory neurons of the vagal ganglia and the dorsal root ganglia (36). This sensory nature is demonstrated by chemical (capsaicin) and surgical (vagal ligation) denervations, which result in disappearance of pulmonary CGRP-immunoreactive tissue (3). Retrograde tracing studies show that tracheal CGRP-containing nerves originate principally from the right vagal ganglia, whereas those in the lung arise from the ipsilateral vagal ganglia and from dorsal root ganglia at spinal levels T_1–T_6 (36) (Figs. 7, 8).

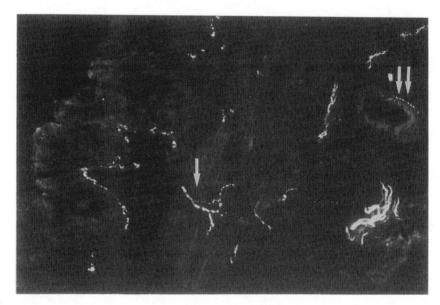

Figure 6 Immunoreactivity for CGRP in rat lung. Nerve fibers can be seen in bronchial smooth muscle (arrow) and around a blood vessel (double arrow). Tissue fixed in Zamboni's fluid and stained by indirect immunofluorescence (magnification ×400).

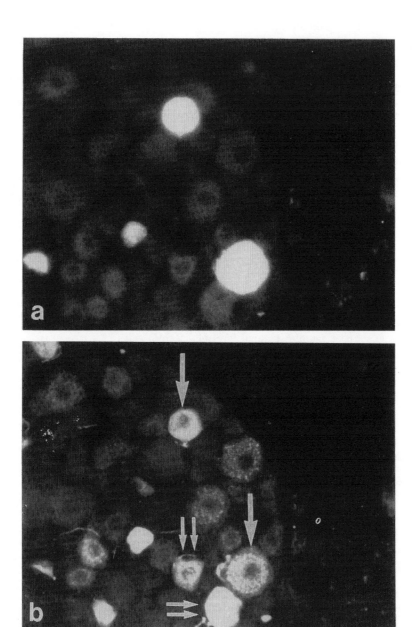

Figure 7 Retrograde tracing in section of the ipsilateral dorsal root ganglion (level T_2) from rat injected with True blue in the left lung and photographed with light excited at different wavelengths to show True blue fluorescence (a) and immunofluorescence staining for CGRP (b). A moderate number of cells are retrogradely labeled. Many CGRP immunoreactive cells are evident, some colocalized with True blue labeled cells (arrows), while others show CGRP immunoreactivity but no retrograde labeling (double arrows) (magnification ×400).

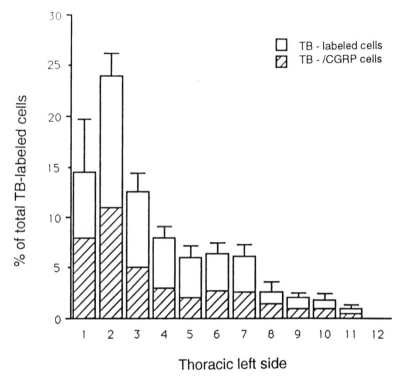

Figure 8 Results of retrograde tracing after injection of the tracer True blue into the left lung of a rat. The diagram shows the mean percentages of the total number of retrogradely labeled cells and the proportion of CGRP immunoreactive cells ($\square$) in all dorsal root ganglia that were present in individual ganglia on the thoracic left side at spinal levels T_1–T_{12}.

VI. GALANIN

Galanin is a 29-amino-acid peptide originally found in the gut (37) and later shown, like the other neuropeptides, to have a much wider distribution in many central and peripheral nerve fibers. Galanin is found in mammalian lung nerves with low frequency when compared with lower species such as chickens (38,39) and originates, like VIP, from local ganglion cells.

VII. NITRIC OXIDE IN THE LUNG INNERVATION

Nitric oxide (NO) is a short-lived molecule that has been suggested to play a major role in cell-to-cell communication (40). It is generated from L-arginine by NO synthase, an enzyme that occurs in different constitutive or inducible isoforms

(40,41). Constitutive enzymes are present in the central and peripheral nervous systems and in endothelium (40,42); the inducible enzyme has been identified in many cell types, including vascular smooth muscle, endothelial cells (40), and mouse macrophages (43). In the lung, NO causes bronchodilatation (44) and pulmonary vasodilatation (45). Furthermore, it can be detected in exhaled air (46). It may also act as a neurotransmitter of inhibitory NANC nerves including those in humans (47,48), and it modulates adrenergic neuronal vasoconstriction (49).

Because NO cannot be detected and thus localized in tissue sections, sites of synthesis are demonstrated by immunohistochemistry with antisera raised to NO synthase or by substrate binding using radiolabeled inhibitors. Immunohisto-chemistry allows the identification of specific isoforms but cannot distinguish whole active enzyme and fragments, whereas substrate binding cannot distinguish isoforms but only detects enzyme having an intact active site. In human, rat, pig, and guinea pig lung, antiserum to the neuronal isoform of NO synthase stains nerves in many tissues including blood vessels and airway smooth muscle (50–52) (Fig. 9). In guinea pig lung, the neuronal enzyme is reported to be present mainly in VIP-containing fibers and not in sensory or noradrenergic nerves (51).

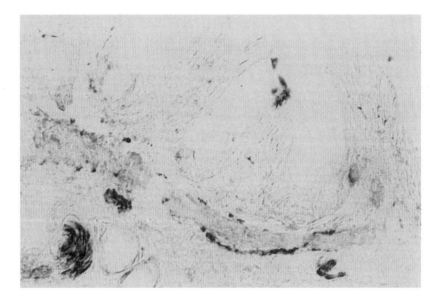

Figure 9   Immunoreactivity for neuronal NO synthase in pig lung. Staining is evident in nerve bundles and in nerve fibers present in bronchial smooth muscle. Tissue fixed in 1% paraformaldehyde and stained by avidin-biotin-complex immunoperoxidase (magnification ×400).

The precise functions of NO in the lung and the separate roles played by neuronal and endothelial sources are not fully known. The vasodilator actions of NO suggest that it may have a role in postnatal adaptation of the lung, which changes at birth from being vasoconstricted to vasodilated. In keeping with this possibility is the finding that there is a decrease in both neural and endothelial NO synthase immunoreactivity in pigs from 2 hr to 10 days postpartum (52,53).

VIII. ENDOPEPTIDASES

The actions of neuropeptides are modulated by many factors, such as their rate of release, the presence and levels of receptors on target cells, and breakdown of the peptides. Different enzymes are involved in this breakdown, depending on the peptide; considerable interest has been shown in the enzyme that is responsible for the catabolism and inactivation of many peptides, particularly the tachykinins, called neutral endopeptidase (enkephalinase, endopeptidase 24.11). Most morphological studies have used autoradiographic detection of radiolabeled antagonist binding or histochemical and immunostaining methods to localize the enzyme. In the respiratory tract, this enzyme is present on membranes of airway epithelial cells and in fibroblasts, blood vessels, glands, and alveolar septa (54–56). It is present even in fetal lung, where it may act to modulate and direct the action of peptide growth factors (57).

The enzyme activity is known to be increased by dexamethasone treatment (58) and decreased, possibly owing to internalization, by the inhalation of toxic agents such as hypochlorous acid (58) or cigarette smoke (59) and by respiratory infections. Decreased enzyme activity leads to prolongation of the vasodilator and bronchoconstrictor actions of tachykinins, which has been suggested as a mechanism of bronchial hyperreactivity and could result in local edema and inflammation (59). Such effects could be of great relevance to asthma, a condition characterized by reversible bronchial hyperreactivity.

IX. QUANTIFICATION AND CONFOCAL MICROSCOPY

Immunocytochemistry can be used to show the distribution of nerve fibers overall, when using antisera to anatomical markers of nerves generally, such as PGP 9.5, or to specific subsets of nerves using antisera to the various neuropeptides or to enzymes that synthesize classical neurotransmitters, such as tyrosine hydroxylase for catecholaminergic nerves and cholineacetyltransferase for cholinergic nerves. By using image analysis, the morphometric data from tissue sections may be quantified. Thus, the density of nerve fibers overall or immunoreactive for a given peptide can be determined in specific tissues (such as the smooth muscle, blood vessels, glands) in various regions of the respiratory tract. The density and pharmacological properties of peptide-binding sites (receptors) obtained using radiolabeled ligand binding and autoradiography in the same tissues can be quantified similarly (see below). Thereby, function and anatomy can be closely related.

The nerve density is usually assessed by measuring an area occupied by specific nerve terminals around a particular tissue structure (60,61). The area of nerve terminals may be related to a field area defined either by the area of the structure under investigation or, if this is large, by the area of the image in which the measurements are made. A measure of the density of innervation is thereby obtained as the ratio of neural area to field area and permits modifications in disease to be assessed.

A confocal microscope allows fully focused images of thin horizontal planes in a thick section of tissue to be obtained by using a small aperture at the detector to reject out-of-focus information from reflectance or fluorescence generated by single-point illumination of the specimen (62). The microscope is capable of scanning the specimen in separate parallel horizontal planes throughout its depth and storing the images of each plane individually (Figs. 10, 11). Several images can then be amalgamated to produce an integrated two-dimensional picture (Fig. 12). This is particularly advantageous for images of cells with tortuous pathways (e.g., neural axons). An ordinary microscope will only be able to capture focused images of portions of these structures. In addition, by the use of appropriate software, a series of images may be reconstructed as a three-dimensional view; this model can then be rotated in any plane to obtain useful information on the relationship of tissue structures, for example airways and nerves or nerves immunostained for two neuropeptides using separate fluorochromes. Such techniques are also of great value for three-dimensional (volume) quantification of nerve density.

Several methods are available for investigating receptor/binding sites. To obtain numerical assessment of binding sites, the classical method is to determine radiolabeled ligand binding to membranes isolated from tissue homogenates. For analysis of the mRNA coding for a particular receptor, classical molecular biology

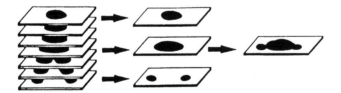

Figure 10   Diagrammatic representation of confocal microscopy. Fully focused images of thin parallel horizontal planes, throughout the depth of a thick section of tissue, can be stored and then several of these images can be amalgamated to produce a composite two-dimensional picture.

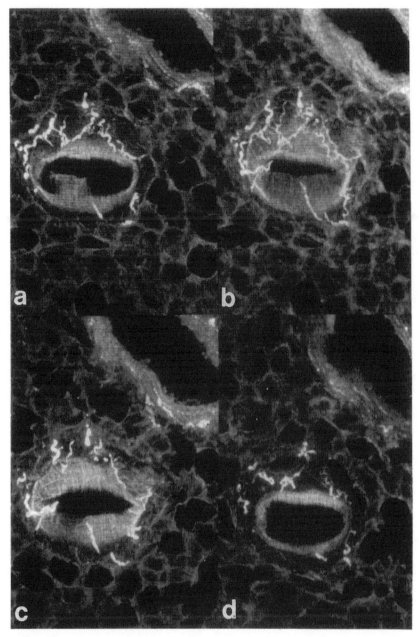

Figure 11 Immunofluorescence staining for PGP 9.5 around a pulmonary vessel. Confocal images showing four "slices" (a–d) taken at different levels.

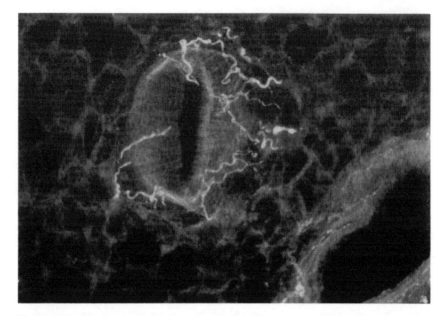

Figure 12 This image shows the summed projection obtained by adding parts a–d of Figure 11. In the conventional fluorescence microscope, blurring due to the detection of out-of-focus information would be highly significant. The confocal microscope discriminates against out-of-focus information to produce a sharp image. Section = 100 μm thick.

techniques including in situ hybridization are used. However, for precise localization of active binding sites within any tissue, the technique of choice is in vitro receptor autoradiography with image analysis (63). In vitro receptor autoradiography was first introduced in 1979 (64) and its use has been increasing steadily. The same principles apply to both in vitro autoradiography and isolated-membrane preparations; both use radioligands, but the former employs tissue sections, and the detection of bound radioactivity is carried out using autoradiographic methods (Fig. 13). The autoradiography may be performed using films (Fig. 14) to give results suitable for quantification, or by emulsion dipping of the slides, which may be used for microscopic analysis (see Fig. 4). Image analysis is used to obtain quantitative data by comparing the integrated optical density of specific areas of the autoradiograms with those of coexposed standards containing known quantities of radioactive material (Fig. 15).

X. CONCLUSIONS

Using these quantitation techniques, it is possible to evaluate the potential functional role and importance of neuropeptides in the airways. The amount of

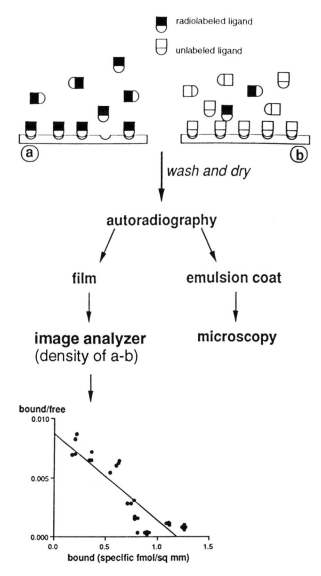

Figure 13 Diagram showing the principle of receptor localization by in vitro auto-radiography. Serial sections are radiolabeled with (^{125}I or ^{3}H) ligand without (a) or with (b) an excess of unlabeled "cold" ligand (to detect nonspecific binding by competitive displacement). Visualization of the radiolabel is performed either by apposing the slides to autoradiography film or by dipping the slides in liquid emulsion (this can also be performed following exposure to film). After exposure and development, optical densities of specific areas on the film can be quantified by image analysis in comparison to coexposed standards; hence specific binding may be assessed and kinetic data derived (e.g., by Scatchard plot, as illustrated). Emulsion-coated slides may be used for microscopic localization of binding sites (see Fig. 4), which is difficult to achieve with a film autoradiogram.

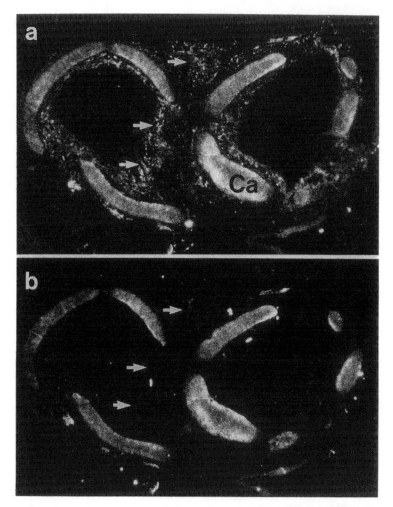

Figure 14   Reversed-tone photomicrograms showing (a) total and (b) nonspecific ^{125}I–Bolton Hunter–substance P binding in human lung, representing specific binding (arrows) to subepithelial and extrabronchial microvessels (magnification ×5). Ca, cartilage.

various neuropeptides and the localization of nerves containing them may be determined. This can then be correlated with the distribution of binding sites/receptors to show that peptides released from nerve terminals can act locally and whether they could act directly on smooth muscle or via a mediator released from another cell type. The increasing use of confocal microscopy will permit better visualization, and thus quantification, or nerves and allow measurements to be

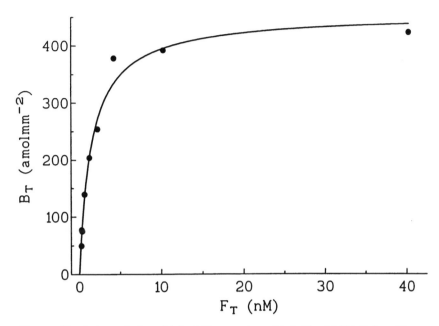

Figure 15 Guinea pig bronchi "cold" saturation with 0.25 mM ^{125}I–Bolton Hunter–substance P and 0–40 mM unlabeled substance P. B_T, total bound ligand; F_T, total free ligand.

made in three dimensions, thereby giving a more accurate picture of events that occur in vivo and how they are modified in disease.

REFERENCES

1. Richardson, J.B. (1979). Nerve supply to the lungs. *Am. Rev. Respir. Dis. 119*: 785–802.
2. Springall, D.R., Bloom, S.R., and Polak, J.M. (1991). Neural, endocrine and endothelial regulatory peptides of the respiratory tract. In: *The Lung Scientific Foundation*. (Barnes, P.J., Cherniack, N.S., and Weibel, E.R. eds.). Raven Press, New York, pp. 69–90.
3. Cadieux, A., Springall, D.R., Mulderry, P.K., Rodrigo, J., Ghatei, M.A., Terenghi, G., Bloom, S.R., and Polak, J.M. (1986). Occurrence, distribution and ontogeny of CGRP immunoreactivity in the rat lower respiratory tract: effect of capsaicin treatment and surgical denervations. *Neuroscience 19*:605–627.
4. Springall, D.R., Polak, J.M., Howard, L., Power, R.F., Krausz, T., Manickam, S., Banner, N.R., Khagani, A., Rose, M., and Yacoub, M.H. (1990). Persistence of intrinsic neurones and possible phenotypic changes after extrinsic denervation of

human respiratory tract by heart-lung transplantation. *Am. Rev. Respir. Dis. 141*: 1538–1546.

5. Tatemoto, K., Carlquist, M., and Mutt, V. (1982). Neuropeptide Y: a novel brain peptide with structural similarities to peptide YY and pancreatic polypeptide. *Nature 296*:659–660.

6. Allen, Y.S., Adrian, T.E., Allen, J.M., Tatemoto, K., Crow, T.J., Bloom, S.R., and Polak, J.M. (1983). Neuropeptide Y distribution in rat brain. *Science 221*:877–879.

7. Sheppard, M.N., Polak, J.M., Allen, J.M., and Bloom, S.R. (1984). Neuropeptide tyrosine (NPY): a newly discovered peptide is present in the mammalian respiratory tract. Thorax 39:326–330.

8. Lundberg, J.M., Terenius, L., Hökfelt, T., and Goldstein, M. (1983). High levels of neuropeptide Y in peripheral noradrenergic neurons in various mammals including man. *Neurosci. Lett. 42*:167–172.

9. Minth, C.D., Bloom, S.R., Polak, J.M., and Dixon, J.E. (1984). Cloning characterization and DNA sequence of a human cDNA encoding neuropeptide tyrosine. *Proc. Natl. Acad. Sci. USA 81*:4577–4581.

10. Michel, M.C. (1991). Receptors for neuropeptide Y: multiple subtypes and multiple second messengers. *Trends Pharmacol. Sci. 12*:389–394.

11. Herzog, H., Hort, Y.J., Ball, H.J., Hayes, G., Shine, J., and Selbie, L.A. (1992). Cloned human neuropeptide Y receptor couples to two different second messenger systems. *Proc. Natl. Acad. Sci. USA 89*:5794–5798.

12. Rimland, J., Xin, W., Sweetnam, P., Saijoh, K., Nestler, E.J., and Duman, R.S. (1991). Sequence and expression of a neuropeptide Y receptor cDNA. *Mol. Pharmacol. 40*:869–875.

13. Uddman, R., Sundler, F., and Emson, P. (1984). Occurrence and distribution of neuropeptide-Y-immunoreactive nerves in the respiratory tract and middle ear. *Cell Tissue Res. 237*:321–327.

14. Mutt, V., Fuxe, K. Hökfelt, T., and Lundberg, J.M., eds. (1989). *Neuropeptide Y.* Karolinska Institute Nobel Conference Series. Raven Press, New York, 1989.

15. Cadieux, A., Benchekroun, M.T., St-Pierre, S., and Fournier, A. (1989). Bronchoconstrictor action of neuropeptide Y (NPY) in isolated guinea pig airways. *Neuropeptides 13*:215–219.

16. Said, S.I. (1988). Vasoactive intestinal peptide in lung. In: *Vasoactive Intestinal Peptide and Related Peptides*. (Said, S.I., and Mutt, V., eds.). New York Academy of Sciences, New York, pp. 450–464.

17. Mutt, V. (1988). Vasoactive intestinal polypeptide and related peptides. Isolation and chemistry. In: *Vasoactive Intestinal Peptide and Related Peptides*. (Said, S.I., and Mutt, V. eds.). New York Academy of Sciences, New York, pp. 1–19.

18. Dey, R.D., Shannon, W.A., and Said, S.I. (1981). Localisation of VIP-immunoreactive nerves in airways and pulmonary vessels of dogs, cats and human subjects. *Cell Tissue Res. 220*:231–238.

19. Barnes, P.J. (1987). Vasoactive intestinal peptide and pulmonary function. In: *Current Topics in Pulmonary Pharmacology and Toxicology*. (Hollinger, M.A., ed.). Elsevier, New York, pp. 156–173.

20. Yamagami, T., Ohsawa, K., Nishizawa, M., Inoue, C., Gotoh, E., Yanaihara, N.,

Yamamoto, H., and Okamoto, H. (1988). Complete nucleotide sequence of human vasoactive intestinal peptide/PHM-27 gene and its inducible promotor. In: *Vasoactive Intestinal Peptide and Related Peptides*. New York Academy of Sciences, New York, pp. 87–102.

21. Christofides, N.D., Yiangou, Y., Piper, P.J., Ghatei, M.A., Sheppard, M.N., Tatemoto, K., Polak, J.M., and Bloom, S.R. (1984). Distribution of peptide histidine isoleucine in the mammalian respiratory tract and some aspects of its pharmacology. *Endocrinology 115*:1958–1963.

22. Carstairs, J.R., and Barnes, P.J. (1986). Visualization of vasoactive intestinal peptide receptors in human and guinea pig lung. *J. Pharmacol. Exp. Ther. 239*:249–255.

23. Leys, K., Morice, A.H., Madonna, O., and Sever, P.S. (1986). Autoradiographic localisation of VIP receptors in human lung. *FEBS Lett. 199*:198–202.

24. Sreedharan, S.P., Robichon, A., Peterson, K.E., and Goetzl, E.J. (1991). Cloning and expression of the human vasoactive intestinal peptide receptor. *Proc. Natl. Acad. Sci. USA 88*:4986–4990.

25. Nawa, H., Kotani, H., and Nakanishi, S. (1984). Tissue specific generation of two preprotachykinin mRNAs from one gene by alternative RNA splicing. *Nature 312*: 729–734.

26. Martling, C.R. (1987). Sensory nerves containing tachykinins and CGRP in the lower airways. Functional implications for bronchoconstriction, vasodilatation and protein extravasation. *Acta Physiol. Scand. 563* (Suppl.):1–57.

27. Lundberg, J.M., Saria, A., Theodorsson-Norheim, E., Brodin, E., Hua, X., Martling, C.R., Gamse, R., and Hökfelt, T. (1985). Multiple tachykinins in capsaicin-sensitive afferents; occurrence, release and biological effects with special reference to irritation of the airways. In: *Tachykinin Antagonists*. (Håkanson, R., and Sundler, F., eds.). Elsevier, Amsterdam, pp. 159–169.

28. Rogers, D.F., Belvisi, M.G., Aursudkij, B., Evans, T.W., and Barnes, P.J. (1988). Effects and interactions of sensory neuropeptides on airway microvascular leakage in guinea pigs. *Br. J. Pharmacol. 95*:1109–1116.

29. Walsh, D.A., Salmon, S., Wharton, J., and Polak, J.M. (1993). Regional substance P binding in guinea pig and human lungs. *Neuropeptides 24*(4):P15.

30. Terenghi, G., MacGregor, G.P., Bhuttarchaji, S., Wharton, J., Bloom, S.R., and Polak, J.M. (1983). Vagal origin of substance P–containing nerves in guinea pig lung. *Neurosci. Lett. 36*:229–236.

31. Rosenfeld, M.G., Mermod, J.J., Amara, S.G., Swanson, L.W., Sawchenko, P.E., Rivier, J., Vale, W.W., and Evans, R.M. (1983). Production of a novel neuropeptide encoded by the calcitonin gene via tissue specific RNA processing. *Nature 304*: 129–135.

32. Martling, C.R., Saria, A., Fischer, J.A., Hökfelt, T., and Lundberg, J.M. (1988). Calcitonin gene-related peptide and the lung: neuronal coexistence with substance P, release by capsaicin and vasodilatory effect. *Regul. Pept. 20*:125–139.

33. Mak, J.C., and Barnes, P.J. (1988). Autoradiographic localization of calcitonin gene-related peptide (CGRP) binding sites in human and guinea pig lung. *Peptides 9*: 957–963.

34. Springall, D.R., Mannan, M.M., Moradoghli-Haftvani, A., and Polak, J.M. (1993).

Calcitonin gene-related binding sites are up-regulated in neonatal rat lung after capsaicin treatment. *Am. Rev. Respir. Dis. 147*:A725.

35. Barnes, P.J. (1991). Sensory nerves, neuropeptides, and asthma. In: *Advances in the understanding and treatment of asthma*. (Piper, P.J., and Krell, R.D., eds.). *Ann. NY Acad. Sci. 629*:359–370.

36. Springall, D.R., Cadieux, A., Oliveira, H., Su, H., Royston, D., and Polak, J.M. (1987). Retrograde tracing shows that CGRP-immunoreactive nerves of rat trachea and lung originate from vagal and dorsal root ganglia. *J. Auton. Nerv. Syst. 20*: 155–166.

37. Tatemoto, K., Rökaeus, A., Jornvall, H., McDonald, T.J., and Mutt, V. (1983). Galanin-novel biologically active peptide from porcine intestine. *FEBS Lett. 164*:124.

38. Cheung, A., Polak, J.M., Bauer, F.E., Cadieux, A., Christofides, N.D., Springall, D.R., and Polak, J.M. (1985). Distribution of galanin immunoreactivity in the respiratory tract of pig, guinea pig, rat and dog. *Thorax 40*:889–896.

39. Luts, A., Uddman, R., and Sundler, F. (1989). Neuronal galanin is widely distributed in the chicken respiratory tract and coexists with multiple neuropeptides. *Cell Tissue Res. 256*:95–103.

40. Moncada, S., Palmer, R.M.J., and Higgs, E.A. (1991). Nitric oxide: physiology, pathophysiology, and pharmacology. *Pharm. Rev. 43*:109–142.

41. Förstermann, U., Schmidt, H., Pollock, J.S., Sheng, H., Mitchell, J.A., Warner, T.D., Nakane, M., and Murad, F. (1991). Isoforms of nitric oxide synthase. Characterization and purification from different cell types. *Biochem. Pharmacol. 42*:1849–1857.

42. Bredt, D., Hwang, P., and Snyder, S. (1990). Localization of nitric oxide synthase indicating a neural role for nitric oxide. *Nature 347*:768–770.

43. Lyons, C.R., Orloff, G.J., and Cunningham, J.M. (1992). Molecular cloning and functional expression of an inducible nitric oxide synthase from a murine macrophage cell line. *J. Biol. Chem. 267*:6370–6374.

44. Dupuy, P.M., Shore, S.A., Drazen, J.M., Frostell, C., Hill, W.A., and Zapol, W.M. (1992). Bronchodilator action of inhaled nitric oxide in guinea pigs. *J. Clin. Invest. 90*:421–428.

45. Fineman, J.R., Chang, R., and Soifer, S.J. (1991). L-Arginine, a precursor of EDRF in vitro, produces pulmonary vasodilation in lambs. *Am. J. Physiol. 261*:H1563–1569.

46. Gustafsson, L.E., Leone, A.M., Persson, M.G., Wiklund, N.P., and Moncada, S. (1991). Endogenous nitric oxide is present in the exhaled air of rabbits, guinea pigs and humans. *Biochem. Biophys. Res. Commun. 181*:852–857.

47. Kannan, M.S, and Johnson, D.E. (1992). Nitric oxide mediates the neural non-adrenergic, noncholinergic relaxation of pig tracheal smooth muscle. *Am. J. Physiol. 262*:L511–514.

48. Belvisi, M.G., Stretton, C.D., Yacoub, M., and Barnes, P.J. (1992). Nitric oxide is the endogenous neurotransmitter of bronchodilator nerves in humans. *Eur. J. Pharmacol. 210*:221–222.

49. Liu, S.F., Crawley, D.E., Evans, T.W., and Barnes, P.J. (1991). Endogenous nitric

oxide modulates adrenergic neural vasoconstriction in guinea-pig pulmonary artery. *Br. J. Pharmacol. 104*:565–569.

50. Springall, D.R., Riveros-Moreno, V., Buttery, L.K., Suburo, A., Bishop, A.E. Merrett, A., Moncada, S., and Polak, J.M. (1992). Immunological detection of nitric oxide synthase(s) in human tissues using heterologous antibodies suggesting different isoforms. *Histochemistry 98*:259–266.

51. Kummer, W., Fischer, A., Mundel, P., Mayer, B., Hoba, B. Philippin, B., and Preissler, U. (1992). Nitric oxide synthase in VIP-containing vasodilator nerve fibres in the guinea-pig. *Neuroreport 3*:653–655.

52. Springall, D.R., Buttery, L.D.K., Hislop, A., Riveros-Moreno, V., Moncada, S., Haworth, S.G., and Polak, J.M. (1993). Nitric oxide synthase immunoreactive nerves in pig lung decrease during development. *Am. Rev. Respir. Dis. 147*:A939.

53. Hislop, A.A., Buttery, L.K.D., Springall, D.R., Pollock, J., Polak, J.M., and Haworth, S.G. (1993). Postnatal changes in localisation of endothelial nitric oxide synthase in the porcine pulmonary vasculature. *Am. Rev. Respir. Dis. 147*:A224.

54. Johnson, A.R., Ashton, J., Schulz, W.W., and Erdos, E.G. (1985). Neutral metallo-endopeptidase in human lung tissue and cultured cells. *Am. Rev. Respir. Dis. 132*:564–568.

55. Sales, N., Dutriez, I., Maziere, B., Ottaviani, M., and Roques, B.P. (1991). Neutral endopeptidase 24.11 in rat peripheral tissues: comparative localization by "ex vivo" and "in vitro" autoradiography. *Regul. Pept. 33*:209–222.

56. Kummer, W., and Fischer, A. (1991). Tissue distribution of neutral endopeptidase 24.11 ("enkephalinase") activity in guinea pig trachea. *Neuropeptides 18*:181–186.

57. Sunday, M.E., Hua, J., Torday, J.S., Reyes, B., and Shipp, M.A. (1992). CD10/ neutral endopeptidase 24.11 in developing human fetal lung. Patterns of expression and modulation of peptide-mediated proliferation. *J. Clin. Invest. 90*:2517–2525.

58. Lang, Z., and Murlas, C.G. (1992). Neutral endopeptidase of a human airway epithelial cell line recovers after hypochlorous acid exposure: dexamethasone accelerates this by stimulating neutral endopeptidase mRNA synthesis. *Am. J. Respir. Cell Mol. Biol. 7*:300–306.

59. Nadel, J.A. (1991). Neutral endopeptidase modulates neurogenic inflammation. *Eur. Respir. J. 4*:745–754.

60. Cowen, T. (1984). Image analysis of FITC-immunofluorescence histochemistry in perivascular substance P-positive nerves. *Histochemistry 81*:609–610.

61. Terenghi, G., Bunker, C.B., Liu, Y.F., Springall, D.R., Cowen, T., Dowd, P.M., and Polak, J.M. (1991). Image analysis quantification of peptide-immunoreactive nerves in skin of patients with Raynaud's phenomenon and systemic sclerosis. *J. Pathol. 164*:245–252.

62. White, J.G., Amos, W.B., and Fordham, M. (1987). An evaluation of confocal versus conventional imaging of biological structures by fluorescence light microscopy. *J. Cell. Biol. 105*:41–48.

63. Palacios, J.M., and Dietl, M.M. (1989). Regulatory peptide receptors: visualisation by autoradiography. In: *Regulatory Peptides*. (Polak, J.M., ed.). Birkhauser Verlag, Basel, pp. 70–97.

64. Young, W.S., III, and Kuhar, M.J. (1979). A new method for receptor autoradiography: [³H]opioid receptors in rat brain. *Brain Res. 179*:255–270.

DISCUSSION

Widdicombe: You showed that CGRP was increased in the bronchial epithelium in patients with chronic cough, yet the bronchi are rather insensitive to cough stimuli compared to the larynx and trachea. Could the shearing forces of coughing stimulate increased bronchial CGRP nerve fiber expression?

Polak: Laryngeal biopsies are not available from chronic coughers. However, the increase in CGRP is consistent with capsaicin hypersensitivity in these patients.

Barnes: Preliminary studies with Peter Jeffries have demonstrated increased epithelial nerve profiles in chronic coughers.

Nadel: It would be interesting to know if CGRP causes cough. Substance P may induce cough in guinea pigs.

White: Endothelin and nitric oxide synthetase were induced in asthmatic subjects. Could they balance each others' effects?

Barnes: The inducible NOS produces much more NO than the constitutive enzyme. This may have proinflammatory or cytotoxic effects. The constitutive enzyme may generate NO that acts locally to cause vasodilation or bronchoconstriction.

Kaliner: We have been unsuccessful in trying to extract nitric oxide synthase from respiratory tissues. Perhaps there is very little constitutively expressed enzyme present, and we have not induced the inducible NOS.

McDonald: Where did you localize substance P binding sites?

Polak: Substance P bound to endothelium and vascular smooth muscle of the submucosa and lamina propria.

4

Neural Control of the Upper Respiratory Tract

James N. Baraniuk
Georgetown University Medical Center, Washington, D.C.

I. INTRODUCTION

Human nasal mucosa is a complex, dynamic structure designed to protect the airways from injury due to inhaled factors and infection (Fig. 1). Sensory, parasympathetic, and sympathetic nerves regulate epithelial, vascular, and glandular processes in the human nasal mucosa (Raphael et al., 1988b, 1989a, 1991a; Baraniuk, 1991; Baraniuk and Kaliner, 1991; Barnes et al., 1991a,b). Each population of nerves contains a unique combination of classical neurotransmitters (e.g., acetylcholine, norepinephrine) and neuropeptides (Table 1). When the nerves are depolarized, the same combination of neurotransmitters is released from all peripheral (mucosal) and central (CNS) neurosecretory sites. With such a wide variety of neurotransmitters, it is difficult to ascertain the functions of individual neurotransmitters released from individual nerve populations. Investigations using exogenous addition of neurotransmitters in vitro and in vivo, capsaicin to trigger nociceptive sensory nerves, and the autoradiographic determination of the receptor distributions for each neurotransmitter have been successful and very insightful. The recent introduction of specific neuropeptide antagonists will again revolutionize our understanding of neural effects in human disease. This approach has been foreshadowed by the widespread use of cholinergic and adrenergic blocking agents that led to the "discovery" of nonadrenergic, noncholinergic (NANC) actions of nerves. Pharmacologically defined NANC actions were once thought to define a new system of nerves; the situation is now known to be much

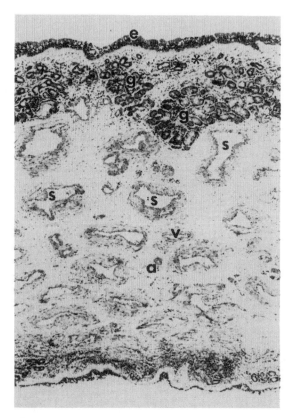

Figure 1   Human nasal mucosa is covered by a respiratory epithelium (e) of ciliated cells, goblet cells, basal cells, infiltrating leukocytes, and nociceptive nerves. Beneath the basement membrane is an area (∗) of fenestrated capillaries and postcapillary venules that are the site of plasma extravasation. This lamina propria is the site of inflammatory leukocytic infiltrations. Submucosal glands (g) containing serous and mucous cells are connected by a tubuloalveolar ductal system to the nasal cavity. Additional fenestrated capillaries and concentrations of IgA plasma cells are found within 50 μm of the glands. Deep in the mucosa is the vascular erectile tissue. Arteries (a) carry blood to deep arteriovenous anastomoses that regulate flow into venous sinusoids (s). The sinusoids are collapsed in this figure. Draining veins carry blood from the deep sinusoids and superficial vascular plexus. Nociceptive sensory, parasympathetic, and sympathetic neurons enter with the arterial vessels in neurovascular bundles.

Table 1 Neurotransmitters and Additional Peptides

Trigeminal sensory neurons	
Tachykinins	
Substance P	SP
Neurokinin A	NKA
Neuropeptide K	
Neurokinin B	NKB
Calcitonin gene-related peptide	CGRP
Gastrin-releasing peptide	GRP
Postganglionic parasympathetic neurons	
Acetylcholine	ACh
Vasoactive Intestinal Peptide	VIP
*P*eptide with *H*istidine at the N-terminal and *M*ethionine at the C-terminal	PHM
*P*eptide with *H*istidine at the N-terminal and *I*soleucine at the C-terminal	PHI
*P*eptide with *H*istidine at the N-terminal and *V*aline at the C-terminal	PHV
Helodermin	
Postganglionic sympathetic neurons	
Noradrenaline	
Neuropeptide tyrosine (Y)	NPY
Other neuropeptides that may occur in respiratory mucosa	
Galanin	GAL
Atrial natriuretic peptide	ANP
Neurotensin	NT
Enkephalins	ENK
Dynorphin	DYN
Cholecystokinin octapeptide	CCK-8
Pituitary adenylate cyclase–Activating Peptide	PACAP
	PACAP-27
Somatostatin	SOM
Calcitonin	CALC
Endothelin I, II, and III	ET_1, ET_2, ET_3
Proteolytic peptide products from plasma and mucosal proteins	
Bradykinin	BK
Kallidin (lysyl-bradykinin)	Lys-BK
Complement fragments	C3a, C5a

Table 2 Innervation of Human Nasal Mucosa

	Sensory	Parasympathetic	Sympathetic
Nerves	I, V	VII	Thoracic
Ganglia	Trigeminal	Sphenopalatine	Stellate
Transmitters	Substance P	Acetylcholine	Norepinephrine
	Neurokinin A	VIP	Neuropeptide Y
	CGRP, (GRP)	(PHM)	
Structures innervated	Arterial and venous vessels	Arterial and venous vessels	Arterial and venous vessels
	Glands	Glands	Anastomoses
	Epithelium		Glands
Predominant functions	Afferent limb of reflexes	Reflex gland secretion and vasodilation	Constriction of vessels
	Axon response		

more complex since the anatomically defined "sensory," "parasympathetic," and "sympathetic" neural systems contain heterogeneous populations of nerve fibers with often unique combinations of neurotransmitters and unique functions in tissue-specific locations. These combinations of neurotransmitters have been reasonably well identified in nasal mucosa since trigeminal, sphenopalatine, stellate, and other ganglia can be obtained for investigation (Table 2).

II. NOCICEPTIVE SENSORY NERVES

Nasal sensory nerves originate in the trigeminal ganglion and innervate the nose via ethmoidal and posterior nasal nerves. These highly branched neurons innervate vessels, glands, and the epithelium where they extend between basal cells (Fig. 2). Some endings have fine terminal extensions that reach up between epithelial cells to the region of the tight junctions. Nociceptive (pain message) nerves are nonmyelinated bare fibers (type C) that do not have specialized sensory organs. The turbinate, septal, and sinus mucosa does not appear to have other specialized sensory organs like the skin or tendons. The olfactory mucosa, which covers the superior portion of the nasal cavity, is innervated by cranial nerve I, which will not be discussed at this time. The nervus terminalis, or cranial nerve XIII, also innervates this region (Demski and Schwanzel-Fukuda, 1987).

Trigeminal nociceptive sensory nerves can be stimulated by inflammatory mediators such as histamine, bradykinin, serotonin, K^+, and H^+ that are released following mucosal injury or mast cell degranulation after allergen exposure (Baraniuk et al., 1990b; Baraniuk, 1991; Baraniuk and Kaliner, 1991; Martins et al., 1991; Raphael et al., 1991b). Prostaglandins and peptidoleukotrienes

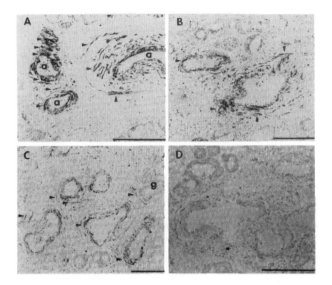

Figure 2 CGRP nerve fibers. CGRP-immunoreactive sensory nerve fibers are shown innervating vessels (A–C). A single fiber is seen near a glandular acinus in (C). This is the typical immunohistochemical distribution for nociceptive neurotransmitters. Adsorption of antiserum with excess CGRP ablated the immunoreactivity (D).

modulate sensory nerve function by altering the threshold to depolarization making it easier to stimulate these nerves. Other inhaled agents such as SO_2, O_3, formaldehyde, nicotine, cigarette smoke, and capsaicin can also stimulate these nerves. Depolarization of the peripheral nerve ending generates a wave of depolarization that extends throughout the entire length of the nerve axon and all the extensively branched rami. The wave of depolarization releases colocalized combinations of neuropeptides from neurosecretory swellings (varicosities) that are found near glands and vessels. The neuropeptides including calcitonin gene-related peptide (CGRP), gastrin-releasing peptide (GRP), the tachykinins substance P (SP) and neurokinin A (NKA), and possibly other as yet unidentified neuropeptides (Tables 1 and 2) Lundblad, 1984; Uddman and Sundler, 1986; Holzer, 1988; Baraniuk et al., 1990c,d; 1991a). The varicosities are strung like beads on a string along the extensively branched peripheral sensory nerve. The same combination of neuropeptides is packaged in vesicles in all the varicosities and released from all the central and peripheral varicosities of a single neuron (Dale principle) (Dale, 1935; Eccles, 1986). Sensory neuropeptides act on specific receptors on target cells to initiate and amplify inflammatory responses to mucosal injury. This "axon response" mechanism mediates cutaneous wheal-and-flare reactions (Lundblad, 1984; Hua, 1986) and in the respiratory tract initiates and

Table 3 Radioligand Binding Sites for Various Mediators in Human Nasal Mucosa

	Sensory				Parasympathetic				Sympathetic			Inflammatory		
	SP	NKA	CGRP	GRP	VIP	M$_1$	M$_2$	M$_3$	α	β	NPY	BK	Hist	ET1
Epithelium	+	0	0	++	+	+	0	+[a]	?	+	0	0	0	0
Serous cells	+	0	0	++	+	+	0	+	?	?	0	0	0	0[b]
Mucous cells	+	0	0	++	+	+	0	+	?	?	0	0	0	0[b]
Vessels resistance (artery, AVA)	+	+	++	0	0	+	0	+	?	?	++	++	+	+
Capacitance (sinusoids)	+	0	±	0	0	+	0	+	?	?	0	++	+	+
Permeable (postcap ven)	+	0	±	0	0	+	0	+	?	?	0	++	+	±

[a]M$_3$ autoradiography results confirmed by in situ hybridization for m3 receptor mRNA.
[b]Myoepithelial cell binding sites appear to be present.
?, not reported; 0, not present; ±, weak binding suggesting low density of binding sites; +, binding sites present; ++, dense binding suggesting high density of binding sites.

amplifies local mucosal vasodilation, vascular permeability, and glandular exocytosis and may facilitate vascular wall leukocyte adhesion (Holzer, 1988; McDonald 1988). SP likely plays a central role in this inflammatory process since ^{125}I-SP binding sites are widely distributed (Table 3) and SP stimulates many proinflammatory effects in various models in vivo and in vitro (Tables 4 and 5). CGRP may be an important long-acting vasodilator and contribute to filling of venous sinusoids (Baraniuk et al., 1990c).

Sensory nerve stimulation and central reception of nociceptive nerve impulses in the brainstem and higher centers leads to the appreciation of sensations of itch, burning, and congestion, and the initiation of important central reflexes such as sneeze, cough, and parasympathetic secretory reflexes (Fig. 3) (Raphael et al., 1988b, 1991a).

Neuropeptide actions are limited by the enzyme neutral endopeptidase (NEP) (Nadel, 1990, 1992; Borson, 1991). It is thought that all cells bearing peptide receptors also express NEP. Destruction of NEP activity may lead to prolonged,

Table 4 Putative Actions of Neurotransmitters

System	Transmitter	Receptor	Actions
Sensory	SP > NKA	NK_1	Arterial dilation, vascular permeability, mucus secretion, leukocyte infiltration
	NKA > SP	NK_2	Dilation, permeability, bronchoconstriction
	CGRP	CGRP	Arterial dilation
	GRP	GRP	Serous secretion, mucous secretion, trophic factor
Parasympathetic	Acetylcholine	M_1	Glandular secretion
		M_2	Inhibitory autoreceptor
		M_3	Glandular secretion, vasodilation, smooth muscle contraction
		M_4	?
		M_5	?
	VIP	VIP	Arterial dilation, serous secretion, mucous secretion
Sympathetic	Norepinephrine	α_1, α_2	Vasoconstriction
		β	Dilate (minor effect)
	NPY	Y_1	Postsynaptic receptors stimulate vasoconstriction
		Y_2	Presynaptic receptors inhibit neurotransmitter release

Table 5 Secretory Responses During Human Nasal Provocations In Vivo

Agent	Sensory nerve stimulation (itch, sneeze)	Parasymp. reflex gland secretion	Directly stimulate glands	Plasma leak	Reduce nasal airflow
Methacholine	0	0	+	0	+
Histamine	+	+	+	+	+
Capsaicin	+	+	0	0	+
Nicotine	+	+	?	0	+
Serotonin	+	+	?	?	+
Substance P					
Normal	0	0	?	?	0
Rhinitis	0	0	?	+	+
CGRP	0	0	0	0	0
Bombesin	0	0	+	0	0
Bradykinin					
Normal	0	0	0	+	+
Rhinitis	+	+	+	+	+
Allergen	+	+	+	+	+
Eating Spicy Food	+	+	0	0	+

+, significant stimulation within 10 min of administration; 0, no significant effect; ?, not measured.

unopposed inflammatory effects of neuropeptides and may contribute to respiratory hyperresponsiveness that can develop in some inflammatory rhinitic conditions.

III. PARASYMPATHETIC REFLEXES

Nociceptive sensory nerve stimulation leads to central recruitment of parasympathetic reflexes. This is possibly the most important function of nociceptive nerves in the human nose. Parasympathetic efferent connections are formed by preganglionic seventh nerve motor fibers from the superior salivatory nucleus that synapse in the sphenopalatine ganglion. Postganglionic fibers innervate glands and some vessels in the nasal mucosa. Parasympathetic reflexes release acetylcholine, vasoactive intestinal peptide (VIP) (Baraniuk et al., 1990e), and other VIP-related peptides (Barnes et al., 1991a,b). The cholinergic component of parasympathetic reflexes is a predominant stimulus for mucus secretion in allergic rhinitis and perhaps other respiratory diseases (Druce et al., 1985; Raphael et al., 1988a, 1989a, 1991a).

At least five muscarinic receptor genes (m1–m5) have been cloned (Barnes, 1989; Levine et al., 1991). However, since there are partially selective pharmacological ligands for only three receptors (M_1–M_3) (Mak and Barnes, 1990), it is not known with certainty which of these receptor subtypes mediates acetylcholine's

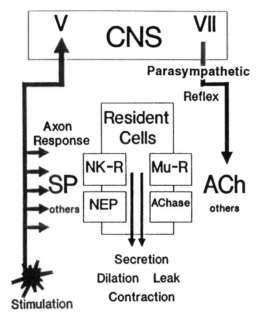

Figure 3 Stimulation of sensory nerve fibers releases SP and other sensory neuropeptides in the periphery that act on receptors on mucosal resident cells (vessels, glands, epithelium, connective tissue cells) to stimulate acute effects. CNS stimulation by trigeminal fibers recruits parasympathetic neural reflexes, which release ACh, VIP, and other peptides that act on their specific receptors. The actions of neuropeptides and ACh are limited in part by degradation by neutral endopeptidase (NEP) and acetylcholinesterase (AChase), respectively.

actions in vivo. In addition, the kinetic binding affinities of cloned receptors expressed in cell lines do not correlate with the apparent pharmacological subtypes of receptors found in vivo and in vitro (Dorje et al., 1991). Initial evidence (reviewed below) implicates the M_3 receptor as the preeminent subtype in human nasal mucosa (Mullol et al., 1992a; Okayama et al., 1993). Sensory nerve stimulation and parasympathetic reflexes contribute to the pathology of inflammatory conditions such as allergic rhinitis, vasomotor rhinitis, infectious rhinitis (common cold), chronic bronchitis, asthma, and infectious bronchitis.

IV. SYMPATHETIC NERVES

Sympathetic reflexes have been poorly studied, yet probably play a crucial role in the normal homeostasis of nasal mucosa and the responses to exercise and other stresses that require a patent nasal airway (Baraniuk et al., 1990a). Sympathetic nerve fibers originate in the sympathetic chain and stellate ganglion. Sympathetic nerves contain either noradrenaline or noradrenaline plus neuropeptide Y (NPY) (Potter, 1988; Lacroix et al., 1990). Both are potent vasoconstrictors. Agonists of α_1- and α_2-adrenergic receptors are popular "nasal decongestants" that effectively reduce mucosal thickness in vivo.

V. NEURAL REGULATION OF NASAL SECRETORY PROCESSES

The nose is an organ designed to clean and condition inspired air. The mucus that lines the nasal cavity is a mixture of products exocytosed from glands and a plasma exudate. Mucus contains an abundance of antimicrobial defense molecules and provides water to humidify inspired air. Nerves regulate the glandular and vascular processes that generate nasal mucus and govern nasal airflow (Raphael et al., 1988b, 1991a; Baraniuk, 1991; Baraniuk and Kaliner, 1991). Sensory nerve–parasympathetic reflex arcs release neurotransmitters that stimulate glandular exocytosis from the tubuloalveolar submucosal glands (Fig. 4) (Raphael et al., 1989a, 1991a). Mucous cells that contain high-molecular-weight mucins occupy central locations in the gland acini, while protein-secreting serous cells occupy more peripheral locations. Serous cells exocytose lysozyme, lactoferrin, secretory leukocyte protease inhibitor (SLPI), secretory component (SC), secretory IgA (dimers or trimers of IgA with SC), and other enzymes and nonspecific anti-microbial defense proteins. Contraction of pariacinar myoepithelial cells squeezes the exocytosed material through the ducts and into the nasal cavity. Mucous cells, serous cells, and vascular permeability contribute approximately equal amounts of protein to nasal secretions (Table 6). Mucins, lysozyme, sIgA, and albumin are the most abundant constituents of nasal mucus.

Epithelial cells regulate ion and water fluxes across the epithelial tight junctions (Al-Bazzaz, 1986; Wanner, 1986; Welsch, 1987). Epithelial goblet cells also secrete mucins. Ciliated cells sweep the mucinous gel phase of nasal "mucus" over the 5-μm-thick, sol-phase epithelial lining fluid to the back of the throat where the mucus and any adsorbed inhaled material are swallowed.

Plasma extravasation occurs from postcapillary venules and fenestrated capillaries. Fenestrated capillaries (Cauna and Hinderer, 1969; Cauna, 1970) are localized to the subepithelial and periglandular regions of human nasal mucosa. These fenestrated capillaries are similar to those of the kidney and choroid plexus, and so suggest a filtering function. They likely permit flow of plasma water and ions but not larger proteins or cells (Renkin, 1992). It is unclear if flow is unidirectional or bidirectional. Postcapillary venules are the sites of mediator-induced plasma and plasma protein flux and diapedesis (McDonald, 1988; Michel, 1992). When postcapillary venule endothelial cells are stimulated during neurogenic inflammation, they contract and leave bare the vascular basement membrane

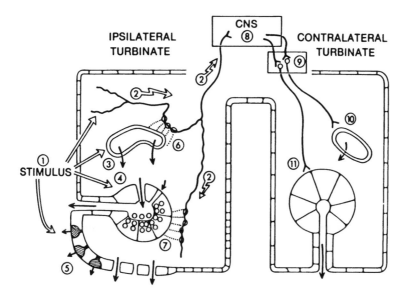

Figure 4 Schematic view of nasal mucosal injury and neurogenic inflammation with sensory nerve axon responses and parasympathetic nerve reflexes. Inhaled noxious stimuli (1) may directly act on sensory nerves (2), vessels (3), glands (4), or epithelial cells (5) to stimulate pain, vasodilation, and vascular permeability, or exocytosis. Stimulation of the sensory nerves causes a wave of depolarization to spread over the entire nerve (2) and for the release of prepackaged neurotransmitters from varicosities in vessel walls (6) and gland acini (7). The actions of sensory neuropeptides at these sites is terms the "axon response." Stimulation of central registries (8) recruits reflexes including sneeze and bilateral parasympathetic responses (9). The parasympathetic reflexes cause vasodilation (10) and glandular secretion (11).

(McDonald, 1988). This permits adherence of circulating leukocytes and fluid extravasation that delivers water, ions, albumin, IgG, and other plasma proteins to the interstitial space. When superficial nasal blood flow and intravascular pressures are increased, the transvascular flux of fluid is increased, and this in turn increases the interstitial hydrostatic pressure. When this pressure change rises above about 5 cm H_2O, the interstitial fluid is forced between epithelial cells and through tight junctions into the nasal cavity (Persson, 1991; Persson et al., 1991). This pressure-driven flux of interstitial fluid does not damage the epithelium. Blood flow to the superficial, permeable postcapillary vessels is regulated by arterioles. Neurotransmitters may regulate endothelial cell expression of adhesion molecules that permit and promote leukocyte adhesion, activation, and infiltration.

Deeper in the mucosa, blood flows into a separate plexus of venous sinusoids via arteriovenous anastomoses (Cauna and Hinderer, 1969; Cauna, 1970; Baraniuk

Table 6 Concentrations of Proteins in Human Nasal Secretions Collected 10 min After Saline Challenge Using the Method of Raphael et al. (contents of nasal lavage fluid with 4 ml of saline 10 min after saline challenge, $n = 148$)

	Mean (μg/ml)	(95% confidence limits)	% total protein
Vascular origin			
Albumin	12	(0–27)	15
IgG	3	(1–6)	4
Monomeric IgA	2	(0–4)	2
IgM	<2		<2
Serous cell origin			
Lysozyme	12	(9–14)	14
Secretory IgA	12	(4–27)	14
Lactoferrin	3	(1–8)	4
Mucous cell origin			
Glycoconjugate[a]	93	(55–130)	
Protein content[b]	23	(15–32)	28
Total protein	82	(38–126)	

[a]Concentration compared to a standard mucin.
[b]Concentration converted to equivalent μg human serum albumin protein.

et al., 1990a). Dilation of these vessels with coincident closure of draining muscular veins leads to massive sinusoidal filling, which increases the thickness of the nasal mucosa. This, in turn, determines the cross-sectional area of the air-filled nasal cavity and the nasal resistance to airflow. Nasal patency is produced by the collapse of the sinusoids. Two mechanisms may cause sinusoidal collapse. Contraction of the arteriovenous anastomoses by active sympathetic influences may halt blood flow into the sinusoids. This, combined with relaxation of the draining veins and passive tension of interstitial elastic fibers, may permit passive collapse of the sinusoidal vessels (Wright, 1910; Dawes and Pritchard, 1953). Alternatively, sympathetic influences may directly stimulate myoepithelial cells in the sinusoidal walls and induce vasoconstriction. This theory is supported by recent work in rat nasal sinusoids that indicates that topical administration of α-adrenergic agonists induced significant increases in intrasinusoidal blood pressures, relative stasis of blood flow, and reduced tissue blood volume (Kristiansen et al., 1993). The increased pressure could force blood out of the sinusoid and into draining vessels. α_1-Adrenergic effects predominate. It is unclear if this mechanism occurs only in rats, or if humans have a similar mechanism of sinusoidal collapse.

In most normal humans, these processes of filling and collapse of the sinusoids and regulation of nasal resistance are coordinated on the two sides of the nose so that while one nostril is patent, the other has a high nasal resistance to airflow (Cole, 1990). Over a 4–6-hr period, the patent nostril will become obstructed as the opposite nostril becomes patent. The net nasal airflow resistance from both nostrils does not change, so the process goes unnoticed by most people. This carefully orchestrated process is termed the nasal cycle and is under neural control. Disruption of neural pathways by Vidian neurectomy (Konno and Togawa, 1979) or sympathetic dysfunction (e.g., Horner's syndrome) disrupts the nasal cycle and can lead to intractible nasal congestion, obstruction to airflow, and rhinorrhea.

Nasal provocation has proven to be a successful method to demonstrate the effects of neuropeptides on nasal secretion in vivo (Table 5) (Baraniuk, 1991; Raphael et al., 1991a). Agents such as capsaicin, histamine, bradykinin, serotonin, and nicotine stimulate the nociceptive sensory nerves and lead to itch or pain, and central reflexes such as sneeze and parasympathetic motor reflexes that induce glandular secretion and obstruction to nasal airflow (Baraniuk, 1991).

VI. TACHYKININS

Stimulation of nociceptive sensory nerves leads to the release of the tachykinins SP and NKA, as well as the release of CGRP and probably GRP. SP and NKA are the products of the preprotachykinin A (or PPT-I) gene (Nakanishi, 1987; Holzer, 1988; Helke et al., 1990). The PPT-I gene contains seven exons. SP is coded by exon 3 and NKA by exon 6. PPT-I generates three distinct mRNAs: alpha-PPT-I lacks exon 6 and codes for SP alone; beta-PPT-I contains all the exons and codes for both NKA and SP; while gamma-PPT-I mRNA excludes exon 4 and codes for SP, NKA, and neuropeptide gamma. Nerve growth factor and noxious stimuli may upregulate PPT-I transcription, while glucocorticoids may down-regulate transcription (Lindsay et al., 1989; MacLean et al., 1989; Helke et al., 1990; Kageyama et al., 1991). In rats, the delta-PPT gene codes for a theoretical, novel tachykinin called NP-delta (Harmar et al., 1990). NKB is produced from the PPT-II gene but has not yet been described in peripheral mammalian tissue (Helke et al., 1990).

Three tachykinin receptors, NK_1, NK_2, and NK_3, have been cloned (Yokata et al., 1989; Gerard et al., 1990; Ohkubo and Nakanishi, 1991; Takahashi et al., 1992). SP is the preferred ligand for NK_1 receptors, while NKA is the preferred ligand for NK_2 receptors. NKB binds preferentially to NK_3 receptors. Pharmacological data suggest that there may be additional SP receptors and that there may be additional subtypes of NKA-preferring receptors (Petitet et al., 1992; Astolfi et al., 1993). Tachykinin receptors belong to the seven-transmembrane-region, rhodopsin-like receptor gene superfamily (Fig. 5). These proteins have a glycosylated extracellular NH_2-terminal, seven transmembrane regions, three extracellular loops, three intracellular loops, and an intracellular COOH tail. The extracellular domains determine the specificity of ligand binding (see below). The

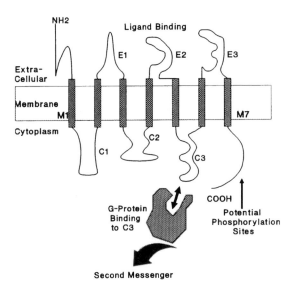

Figure 5 Prototypic G-protein-related receptor showing extracellular amino-terminal, three extracellular loops (E1, E2, and E3), seven transmembrane helical regions (M1–M7), three intracellular loops (C1, C2, and C3), and the intracellular carboxy-terminal. Binding of receptor-specific G proteins is determined by ligand binding with consequent conformational changes of C3 and phosphorylation of the C-terminal region.

third intracellular loop binds to specific G proteins that are stimulated when the ligand binds the receptor and transduce that signal into the cytoplasm (Boyd et al., 1991). The COOH tails contain phosphorylation sites that regulate the binding of G proteins, and so the activity of the receptor. Phosphorylation of this tail after receptor binding may be one mechanism of receptor "tachyphylaxis." Over 40 structurally and evolutionarily related G-protein-associated receptors have been cloned including α- and β-adrenergic, dopamine, serotonin, histamine, bradykinin, GRP, NPY, ANP, CCK, and IL-8 receptors. The nature of the G proteins associated with each receptor is under intense scrutiny, but is beyond the scope of this review (Birnbaumer and Brown, 1990).

The distribution of NK_1-receptors has been suggested by autoradiographic studies of ^{125}I-SP binding sites. ^{125}I-SP binds to epithelium, vessels, and glands in the upper respiratory tract (Table 3). This distribution is consistent with the known ability of SP to induce secretion, vasodilation, and vascular permeability (Tables 4 and 5) (Baraniuk et al., 1991a). SP nasal provocation apparently has minimal effects in normal subjects, but induces albumin secretion and obstructs nasal airflow in allergic rhinitis subjects (Devillier et al., 1988; Braunstein et al., 1991). These data suggest that in humans, only subjects with established inflammatory syndromes such as allergic rhinitis or asthma experience clinically evident bronchoconstrictor or secretory responses after provocation with substance P or NKA, and that expression of tachykinin receptors may be increased in vivo after inflammation has been initiated. This conjecture has yet to be demonstrated in vivo. SP induces serous and mucus glandular secretion from human nasal mucosal explants in vitro (Baraniuk et al., 1990a). In pig nasal mucosa, SP increases nasal blood flow and the thickness of the mucosa (Stjarne et al., 1991). SP-immunoreactive material can be collected in nasal lavage fluid (B. Mossiman, personal communication) and bronchoalveolar lavage fluid (Neiber et al., 1992) from allergic subjects after allergen challenge, suggesting that SP is released during this process. It is most likely that nociceptive nerves are the source of this SP, although other sources such as inflammatory cells have been suggested in the past (Payan et al., 1984).

Actions of NKA in vitro (Baraniuk et al., 1990a) and in vivo (Braunstein et al., 1991) suggest that NKA may also play a proinflammatory role in nasal mucosa. However, NKA likely has greater relevance as a bronchoconstrictor in human airways (Barnes et al., 1991a,b).

Capsaicin, the hot spicy essence of chili peppers (Holzer, 1991), specifically stimulates and depolarizes nociceptive sensory nerves and induces the sensation of burning pain. Capsaicin has proved invaluable for the investigation of these nerves (Lundblad, 1984; Hua, 1986). Capsaicin has been used to estimate the magnitude of the axon response in pig nasal mucosa (Stjarne et al., 1991). Capsaicin stimulates an increase in nasal blood flow and mucosal volume. This effect is ablated by ganglion-blocking drugs, indicating that the recruited parasympathetic reflexes mediate most of the vascular reaction induced by capsaicin. When higher doses of capsaicin are combined with ganglion blockade, then the axon response can be detected as a small increase in superficial blood flow (Stjarne et al., 1991). This indicates that in higher animals, the axon response mechanism exists, but that the major effect of sensory nerve stimulation is recruitment of parasympathetic nerves. The effects of axon responses on glandular secretion are suggested, but have not been definitively shown. Evaluation of the precise role of SP and other neuropeptides in the axon response and overall inflammatory responses of mucosae will require the development and use of SP-specific antagonists (Maggi et al., 1991; Regoli et al., 1991; Advenier et al., 1992; Watling, 1992).

These data suggest that SP released from nociceptive sensory nerves may induce vasodilation, vascular permeability, and glandular secretion in human nasal mucosa, but that tachykinin effects may be more prominent during inflammation. This hypothesis requires the demonstration of induction of tachykinin receptors by inflammation or proinflammatory biological response modifiers. The most important effects of tachykinins may be their actions on interneurons in the brainstem and spinal cord that lead to the recruitment of systemic and parasympathetic reflexes and the supratentorial appreciation of nasal discomfort. It is conceivable that by the publication date of this volume, many of the questions raised about the roles of sensory neuropeptides and nociceptive nerves will have preliminary answers as a result of studies involving specific tachykinin antagonist drugs in animal models and human disease.

The molecular actions of tachykinin antagonists (Table 7) (Watling, 1992; Presti and Gardner, 1993) have been elegantly demonstrated by kinetic binding studies using chimeric receptors (Gether et al., 1993; Fong et al., 1993). Nonpeptide antagonists like CP 96345 appear to bind to His[197] located at the junction of the second extracytoplasmic loop of the NK_1 receptor and the extracellular side of the fifth transmembranous region (Fig. 6). This amino acid does not appear to be necessary for SP binding to the receptor, or for the activation of associated G proteins. However, when CP 96345 binds to His[197], it fills the "well" formed between the seven transmembranous loops of the receptor and prohibits SP "docking" into that well. By blocking access, CP 96345 denies SP the opportunity to interact with its receptor and activate its associated G protein. The specificity of the SP binding appears to reside in the positively charged N-terminal region of the peptide and interactions with the second and third extracellular loops of the receptor. Cross-reactivity between tachykinin peptides is likely to be determined by the shared C-terminal regions, while specificity is determined by the vastly different N-terminal regions. These antagonist actions of CP 96345 are probably prototypic of other receptor antagonists (Beinborn et al., 1993). Modification of these compounds is very likely to lead to the development of receptor-subtype-specific antagonists, which will be of great use for defining pharmacological events, and which will likely have utility as very specific treatments for neurotransmitter-mediated conditions. The explosion in the number of nonpeptide receptor agonists and antagonists (Table 7) is in stark contrast to recent opinions that the complexity of the peptide-binding sites could demand unique peptide

Table 7 Tachykinin Agonist and Antagonist Drugs

	NK_1	NK_2	NK_3
Agonists	[Sar⁹]SP GR 73632	[βALA⁸]NKA(4-10)	Senktide
Antagonists	CP 96345	GR 100679	GR 138676
	CP 99994	L-659,877	R487
	FK224	MEN 10,208	
	FK888	MEN 10,207	
	L-668,169	MEN 10,282	
	R454	MEN 10,376	
	R455	R396	
	RP 67580	SR 48968	
	SR 140333		

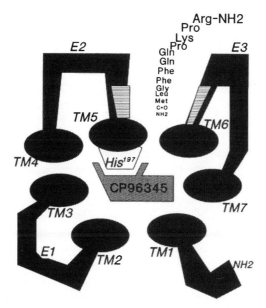

Figure 6 NK$_1$ receptor. Substance P is shown attempting to bind to the NK$_1$ receptor. This view of the extracellular surface of the cell shows the receptor's three extracellular loops and N-terminal chain above the surface of the plasma membrane (represented by the plane of the paper). The peptide, which has adapted a linear conformation for receptor binding, is attempting to insert its relatively hydrophobic C-terminal tail into the central "well" that is formed by the seven transmembranous regions (TM1–TM7). Receptor specificity is generated by specific interactions of the N-terminal and central peptide regions of substance P with amino acid side chains of the E2 and E3 regions (hatched areas). However, in this instance, substance P is not able to bind the receptor since the NK$_1$ receptor antagonist CP 96345 has bound to histidine[197], the amino acid at the junction of TM5 and E2, and blocked the entrance to the central "well." Although His[197] is the binding site for the receptor antagonist, it is not required for substance P binding or receptor activation.

derivatives that could be difficult and expensive to produce and be subject to proteolysis. The lipophilicity of the nonpeptide antagonist compounds suggests that they may readily pass through the blood-brain barrier and be active in the central nervous system. These drugs may unlock many of the mysteries of the human psyche, the actions of peripheral and central nerves, neurotransmitters, and other inflammatory mediators that bind to seven-transmembrane-type receptors.

VII. CALCITONIN GENE-RELATED PEPTIDE

CGRP is 37 amino acid residues long and is produced in two forms: CGRP-I (CGRP-α, A-CGRP) and CGRP-II (CGRP-β, B-CGRP), which differ by three amino acids. Both forms are vasodilators. CGRP is produced from the same genes as calcitonin (Steenbergh et al., 1986). CGRP is the predominant gene product in neural tissue and either or both forms may be produced. CGRP is also produced in tracheobronchial neuroendocrine cells (Springall et al., 1988). In the fetal human, CGRP is present in trigeminal ganglion cells either alone or colocalized with substance P. The percentage of ganglion cells staining for CGRP or substance P reaches adult levels in the immediate preterm period (Springall et al., 1988). The roles of these neurons during development of the upper airways are not known. Recently CGRP has been reported to be present in rat stomach lamina propria mononuclear cells (Jakab et al., 1993) and rat tracheal epithelial serous cells (Baluk et al., 1993). CGRP has not been reported in epithelial cells in human nasal mucosa. The function of nonneural CGRP is open to speculation.

In the adult human nasal mucosa, CGRP-containing nerve fibers densely innervate arterial vessels, but also innervate venous vessels and some gland acini (Fig. 2) (Uddman and Sundler, 1986; Baraniuk et al., 1990c). CGRP fibers extend to the epithelium. ^{125}I-CGRP binding sites are concentrated on arterioles (Table 7) (Baraniuk et al., 1990c). These distributions of CGRP nerve fibers and binding sites are consistent with the physiological role of CGRP as a long-acting and potent arterial vasodilator (Gamse and Saria, 1985; Brain and Williams, 1988).

CGRP receptor genes have not been characterized as yet, but two or possibly three receptor subtypes have been suggested by pharmacological studies using the CGRP$_1$-selective agonist CGRP(8-37), the CGRP$_2$-selective agonist [Cys(acet-amidomethyl)^{2,7}]-hCGRPα, and amylin, a CGRP-like peptide with about 50% sequence homology to CGRPα and CGRPβ (Chin et al., 1993). These studies may prove difficult since CGRP and CGRP-related peptides may interact with tachy-kinin and CCK receptors as well as CGRP receptor subtypes. Further pharmacological studies are required to evaluate these possibilities. Studies in nasal airways have not been undertaken as yet.

CGRP nasal provocation has been attempted (Geppetti et al., 1988; S. Guarnac-cia, personal communication) (Table 3). Topically applied CGRP was found to have no effects on protein secretion, although this peptide could have some difficulty gaining entry to the region of its receptors deep in the mucosa or could be degraded by proteases. CGRP could still play a role in vivo, since it can synergistically interact with other vasomotor agents to increase the blood flow (Brain and Williams, 1985, 1988; Gamse and Saria, 1985). CGRP agonists may have some utility as arterial dilators, while CGRP antagonists could act as decongestants of CGRP-induced arterial dilation.

CGRP-binding sites have not been found on glandular cells (Baraniuk et al., 1990c). In short-term explant cultures of human nasal mucosa, CGRP failed to stimulate mucus release (Baraniuk et al., 1990c). Although CGRP-immunoreactive nerve fibers are present in gland acini, the release of CGRP from nerves in glands probably does not lead to exocytosis since there are no CGRP-binding sites on those cells. In contrast, CGRP stimulated nasal glandular secretion from guinea pig nasal mucosa (Gawin et al., 1993) indicating that there are significant differences in the responses to peptides in different species, and that conclusions drawn from animal experiments may not be representative or extrapolated to human conditions.

VIII. GASTRIN-RELEASING PEPTIDE

GRP is a 27-amino-acid peptide that shares sequence homology with neuromedin C (GRP[18–27]), bombesin (14-amino-acid amphibian peptide), and neuromedin B. Bombesin-related peptides have been classified into three families: the bombesin family (includes GRP, neuromedin C, and other less well-characterized peptides), the ranatensin family (neuromedin B), and the litorin family (no known mammalian peptides). These peptides share a common C-terminal sequence (WAVGHLM-NH$_2$) that is the active part of the peptide (Willey et al., 1984; Sunday et al., 1988; Lundgren et al., 1990). Bombesin-immunoreactive peptides are expressed by normal human bronchial epithelial lung neuroendocrine cells (Tsutsumi, 1988) and can be detected in bronchoalveolar lavage fluids (Aguayo et al., 1989). GRP gene transcripts are processed differently in endocrine and neural tissues (Spindel et al., 1987a,b), suggesting that mRNA splicing may contribute to the diversity of bombesin-immunoreactive peptides. In addition, GRP and related peptides are NEP substrates; some of the "diversity" may represent proteolytic digestion in vivo.

Two distinct bombesin receptor subtypes have been cloned from the human small cell lung carcinoma cell line NCI-H345 (Corjay et al., 1991). GRP-receptors (Spindel et al., 1990; Battey et al., 1991) preferentially bind GRP, neuromedin C, and bombesin, while NMB-receptors (Wada et al., 1991) prefer neuromedin B and bombesin over GRP. The distributions of each receptor subtype in normal respiratory mucosa have not been determined. Since these receptors may play autocrine, trophic roles in small cell carcinoma in vivo (Sunday et al., 1988), it is possible that GRP has trophic properties in respiratory mucosa.

In human nasal mucosa, GRP nerve fibers are present in the walls of arterial and venous vessels and around glandular acini (Baraniuk et al., 1990d). Their distribution is very similar to that of CGRP and NKA (Uddman and Sundler, 1986; Raphael et al., 1991a) and suggests, but does not prove, that GRP is colocalized in the same trigeminal sensory neurons. ^{125}I-GRP binding sites are localized to the epithelium and submucosal glands of human nasal and tracheal mucosa (Table 3) (Baraniuk et al., 1990d). In vitro, GRP induces both serious and mucous cell exocytosis from human nasal mucosa (Baraniuk et al., 1990d; Mullol et al., 1992b). In vivo, bombesin stimulates mucous glycoconjugate and lysozyme secretion from human nasal mucosa, but does not stimulate albumin secretion (Baraniuk et al., 1992c), suggesting an effect on submucosal (and possibly goblet cell) secretion without increases in vascular permeability. The C-terminal region is essential for function since GRP, bombesin, and GRP[20–27] stimulate secretion

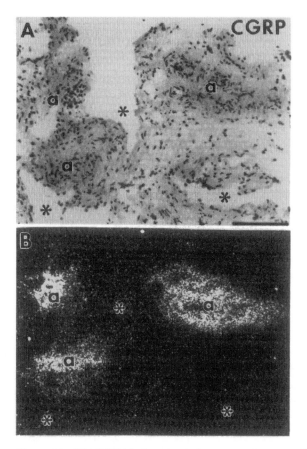

Figure 7   125I-CGRP binding sites in human nasal mucosa. 125I-CGRP bound to arterial vessels (a), which were seen in brightfield in (A) and darkfield in (B). Venous vessels did not bind 125I-CGRP. Excess cold CGRP ablated the binding (not shown).

in human nasal and feline tracheal mucosa in vitro and guinea pig nasal mucosa in vivo, while GRP[1–16], an N-terminal fragment, is inactive (Baraniuk et al., 1990d; Gawin et al., 1990; Lundgren et al., 1990). These varied effects suggest that GRP may be released from sensory nerves and may act as a serous and mucous cell secretagogue and potential growth factor in human nasal mucosa.

IX. NEUTRAL ENDOPEPTIDASE

The actions of neuropeptides are limited by their degradation by neutral endopeptidase E.C.3.4.24.11 (NEP) (Nadel, 1990; Borson, 1991), with contributions by other enzymes (angiotensin-converting enzyme, aminopeptidase M, carboxypeptidase N, and others) in some systems (Desmazes et al., 1992; Nadel, 1992). NEP is active on SP, NKA, CGRP, GRP, bombesin, enkephalins, endothelin, atrial naturetic peptide (ANP), bradykinin, and many other peptides. NEP is postulated to be present on all cells that possess peptide receptors (Nadel, 1990; Borson, 1991).

The inhibition of NEP enzyme activity with phosphoramidon or thiorphan augments many peptide-induced effects, including: epithelial cell ciliary beat frequency; goblet cell secretion; mucous secretion from ferret and feline tracheas; airway microvascular leak induced by tachykinins, bradykinin, depolarization of the vagal nerve, electrical field stimulation, and capsaicin; and neutrophil adhesion to the endothelial surface of vessels (Holzer, 1988; McDonald, 1988; Nadel, 1990; Piedemonte et al., 1990; Borson, 1991). These functional data indicate that NEP activity is present on epithelial, glandular, vascular, and smooth muscle cells. The distributions of NEP immunoreactive material and mRNA are consistent with these activities. In situ hybridization of NEP mRNA and immunohistochemistry of NEP-immunoreactive material have revealed NEP in epithelium, glands, and vessels of human nasal mucosa (Baraniuk et al., 1993), as well as tracheobronchial smooth muscle cells (Baraniuk et al., 1991b). As already shown, these cells are innervated by neuropeptide-containing nerves (Table 2) and possess neuropeptide-binding sites (Table 3). These data are consistent with the hypothesis that NEP is present in all cells that bear neuropeptide receptors and that NEP plays a role in regulating neuropeptide-mediated responses (Nadel, 1990; Borson, 1991).

NEP may have many diverse functions in vivo. In gut, the highest activity is in jejunal mucosa, with the NEP-immunoreactive material and mRNA present in epithelium. NEP may assist with peptide digestion (Bunnett et al., 1993). Occasionally, longitudinal and circular muscle were found to contain mRNA, but vessels in the lamina propria and ganglion cells were negative. Renal proximal tubule brush border contains abundant NEP. NEP may degrade urinary filtrate peptides so that their amino acids can be absorbed. NEP may also help regulate renal blood flow by actions on ANP, renin, bradykinin, or other vasoactive peptides. NEP may play a role in vascular remodeling in rats exposed to chronic hypoxia (Winter et al., 1991) and regulate other vascular events through actions on arterial and venous endothelial cells (Llorens-Cortes et al., 1992). NEP may play a regulatory role in the developing human fetal lung since it is expressed at the growing front of bronchial epithelium (Sunday et al., 1992).

There has been intense speculation regarding the role of NEP in neurogenic inflammation (Fig. 2). Inhibition of neutral endopeptidase potentiates inflammation and bronchial smooth muscle contraction induced by SP, capsaicin, and vagal nerve stimulation, suggesting that tachykinins and/or other NEP substrate peptides can cause neurogenic inflammation and that their effects are regulated by NEP (Umeno et al., 1989; Honda et al., 1991). Decreases in NEP activity may underlie the increased responses of respiratory mucosa to neurogenic inflammation and some agonists ("hyperresponsiveness") found during and after viral and *Mycoplasma* infections (Jacoby et al., 1988; Piedemonte et al., 1990; McDonald, 1992) and after exposure to cigarette smoke (Dusser et al., 1989), ozone (Yeadon, Wilkinson and Payan, 1990), and hypochlorous acid (Murlas, Murphy and Lang, 1990). The release of peptides into areas with reduced NEP activity could lead to enhanced peptide-induced epithelial cell function, glandular secretion, and vascular permeability. Cell surface NEP activity may also be modulated by rapid internalization and proteolytic degradation as occurs in human neutrophils that are activated by phorbol ester or diacylglycerol (Erdos et al., 1989). Although alterations of NEP activity or expression in human diseases have yet to be demonstrated in vivo, inhalation of thiorphan, an NEP inhibitor, permits inhaled NKA to exert a mild bronchoconstrictor effect in nonasthmatic subjects who otherwise do not respond to inhaled NKA (Cheung et al., 1992). This suggests that tonic peptide and NEP activities exist in the normal airway, and that the balance between the two influences can be altered to produce bronchoconstriction in an otherwise normal airway.

Glucocorticoid treatment, on the other hand, has been shown to increase NEP mRNA and enzyme activity in a cultured, SV-40 transformed human epithelial cell line (Borson and Gruenert, 1991) and Calu-1 cells (human lung epidermoid carcinoma cell line) (Lang and Murlas, 1992). This enhancement of NEP expression may explain the beneficial effects of glucocorticoids on neurogenic plasma extravasation in virus-infected rat trachea (Piedemonte et al., 1990; McDonald, 1992). Glucocorticoid treatment also alters the size of NEP mRNA transcripts (Borson and Gruenert, 1991), suggesting complex regulation at the level of mRNA transcription and/or pretranslational processing. The NEP gene of 24 exons codes for a 749-amino-acid, membrane-bound protein (Shipp et al., 1988; D'Adamio et al., 1989). Alternative mRNA splicing generates several different sized mRNAs including one with a deletion of exon 16 that yields a nonfunctional protein (Iijima et al., 1992) and a hypothetical product with a deletion of exons 5–18 (Llorens-Cortes et al., 1990). NEP expression by lung fibroblasts in culture can be increased by incubation with cytokines such as IL-1α, TNF-α, transforming growth factor, IL-6, and GM-CSF (Kondepudi and Johnson, 1993). Cyclic nucleotide and prostaglandin-dependent mechanisms may mediate these effects. These cytokines may play a role in the induction of NEP in perivascular cells of synovial vessels from rheumatoid arthritis and osteoarthritis patients (Mapp et al., 1992). Synovium from normal joints demonstrates no immunohistochemical staining, indicating that some aspect of the disease state can induce NEP immunoreactivity. Induction or suppression of NEP would be expected to alter the responses of cells to other peptides and could represent an efficient method of autoregulation (Stefano et al., 1992). These studies indicate that glucocorticoids and cytokines can modulate transcription or posttranscription processing (Mattox et al., 1992) of

NEP expression and neurogenic inflammation in some cells, models, and diseases. While these investigations of NEP and its complex regulation continue to direct speculations regarding the significance of neurogenic inflammation in human disease, there is still a need to measure actual NEP activities in vivo in human tissues to clearly define whether NEP or its absence plays any significant pathogenic role in human disease.

X. MUSCARINIC MECHANISMS

The parasympathetic nervous system regulates glandular and vasomotor processes in the nasal mucosa (Konno and Togawa, 1979; Baraniuk, 1991). Five muscarinic receptor genes (m1–m5) (Barnes, 1989; Levine et al., 1991) have been cloned. Like the tachykinin receptors, they have seven transmembrane regions and are associated with G proteins (Fig. 5). Relatively selective receptor ligands are known for only M_1 (pirenzepine), M_2 (gallamine, AF DX-116), and M_3 (4-DAMP) subtypes (Barnes, 1989). Because these ligands are selective only at nanomolar concentrations, conclusions drawn from their use must be interpreted with caution (Barnes, 1989; Levine et al., 1991). M_1- and M_3-receptors are associated with G proteins that stimulate phosphoinositol metabolism (Barnes, 1989). In human nasal mucosa, total muscarinic (^{3}H-QNB) and M_1 (^{3}H-pirenzepine) binding sites have been enumerated by Scatchard analysis (Okayama et al., 1993). Using ^{3}H-QNB, total muscarinic receptor binding was found to be 688.4 ± 49.6 fmol/mg protein (B_{max}). In competition experiments, 4-DAMP displaced $[^3H](-)$QNB with the lowest IC_{50}, followed by PZ and AF-DX 116. ^{3}H-PZ binding sites had B_{max} of 303.0 ± 27.3 fmol/mg protein and represent 45% of the total QNB binding sites. Autoradiography indicates that human nasal mucosa (Table 3) and human and guinea pig bronchial mucosa have M_1- and M_3-binding sites on glands, epithelium, endothelium, and smooth muscle (Mak and Barnes, 1990).

In situ hybridization indicates that human nasal mucosa has m3 mRNA in epithelial cells, submucosal glands, and endothelial cells of vessels (Fig. 8) (Baraniuk et al., 1992a). Human bronchi have m1 mRNA in unidentified alveolar cells, but not bronchi (Mak et al., 1992). m3 mRNA is found in bronchial epithelium, endothelium, glands, and smooth muscle. m2 mRNA was found in smooth muscle and ganglion cells, but m4 and m5 mRNA were not detectable.

In humans, stimulation of parasympathetic reflexes leads to glandular secretion that can be reduced or eliminated by the nonselective muscarinic antagonists atropine and ipratropium bromide (Konno and Togawa, 1979; Raphael et al., 1988a, 1989a; Baroody et al., 1992). After nasal allergen challenge, cholinergic reflexes are the predominant stimulus for mucus secretion (Raphael et al., 1991b). The subtype(s) responsible for glandular secretion have not been determined with certainty since the muscarinic antagonists are not totally selective. Initial data indicate that both pirenzepine and 4-DAMP can block methacholine-induced

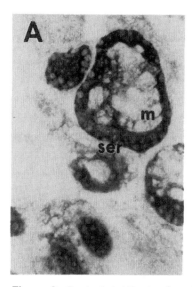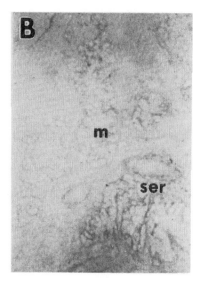

Figure 8 In situ hybridization for m3 muscarinic receptor subtype. (A) Biotin-labeled antisense m3 receptor oligonucleotide probes bound to the cytoplasm for serous cells in human nasal mucosal glands. Mucous cells also appeared to bind a small amount of the probe in cytoplasmic areas outside mucous granules. (Courtesy of *American Journal of Rhinology.*)

secretion, suggesting roles for both M_1- and M_3-receptor subtypes (Okayama et al., 1993). However, substantially higher concentrations of pirenzepine are required, leaving open the possibility that the pirenzepine effect is due to a nonselective action on M_3-receptors. The vagaries of these conclusions are the result of the semiselective nature of current muscarinic receptor subtype antagonists and are the best results that can be expected with this type of pharmacological investigation. This situation is typified by Dorje et al. (1991), who demonstrated that binding constants for several muscarinic antagonists differed by only two logs when tested on cell lines that express only individual cloned muscarinic receptors. In addition, each ligand had its own pattern of relative affinities, indicating that inconsistencies in the apparent receptor subtype mediating a reaction could appear in vitro or in vivo in pharmacological investigations. Detection of the muscarinic receptor mRNAs expressed in tissues by individual cell types would be a valuable indication of the receptor subtype(s) expressed and modulated in vivo.

Muscarinic receptor activation is also a strong stimulus for bronchial smooth muscle contraction, and M_3-receptors are likely responsible for this effect (Barnes,

1989). Muscarinic receptors contribute to parasympathetic vasodilation (Anggard, 1974; Lundberg et al., 1981), and since vasodilation and filling of venous sinusoids determines the thickness of the respiratory mucosa, these parasympathetic vasomotor processes contribute to the regulation of airflow (Cole, 1990).

M_2 binding sites are apparently not present in human nasal mucosa (Okayama et al., 1993). The actions of inhibitory M_2 autoreceptors are described elsewhere in this volume.

Muscarinic receptors may play important roles in the hypersecretion and airflow limitation of rhinitis, chronic bronchitis, and asthma. Among the abnormalities associated with these conditions is hyperresponsiveness to methacholine (Druce et al., 1985; Devillier et al., 1988; Stjarne et al., 1989). The mechanism of this hyperresponsiveness is not known. Significant changes in total muscarinic receptors numbers (B_{max} by Scatchard analysis) have not been found (van Megen et al., 1991a), but, because there are multiple receptor subtypes, it is possible that numerically small, but functionally significant, changes in the relative number of each subtype (e.g., M_3-receptors) on the cell surface could occur in disease states. Alternatively, changes in intracellular signaling mechanisms may be at fault.

The hypothesis that M_3-receptors mediate parasympathetic reflexes suggests that M_3 antagonists may act as selective inhibitors of glandular secretion and vasodilation in human nasal mucosa and bronchoconstriction in human bronchi, and that M_3-selective antagonists would be free of effects on M_2 and other subtype receptors. Such drugs may be of clinical benefit for the treatment of allergic and nonallergic rhinitis, asthma, and bronchitis.

XI. VASOACTIVE INTESTINAL PEPTIDE

VIP is colocalized with acetylcholine in parasympathetic neurons in the nose (Uddman and Sundler, 1986; Klaassen et al., 1988; Baraniuk et al., 1990e). It may also be present in some subsets of sensory nerves (Hartschuh et al., 1983). Posttranslational modification of prepro-VIP (Yiangou et al., 1986) leads to the production of PHM (*P*olypeptide with *H*istidine at the N-terminal and *M*ethionine at the C-terminal). PHM is the human form of PHI (*I*soleucine at the C-terminal), which was previously described in porcine neural tissues (Christiophides et al., 1984). Other VIP-like peptides have also been described, including PHV (*P*eptide with *H*istidine at the N-terminal and *V*aline at the C-terminal), helodermin, helospectin, pituitary adenylate cyclase–activating peptide (PACAP and PACAP-27), extendin-3, growth hormone releasing factor, and secretin. These peptides share sequence homology, may interact with VIP receptors, and have similar vasodilatory, bronchodilatory, and secretomotor effects. Separate VIP-preferring (Sreedharan et al., 1991; Ishihara et al., 1992) and secretin-preferring (Ishihara et al., 1991) receptors have been cloned, and it is likely that separate PACAP and extendin-3 receptors will also be discovered.

Standard autoradiographic receptor localization studies with ^{125}I-VIP have discovered VIP binding sites on epithelium, glands, and vessels (Table 3, Fig. 9) (Baraniuk et al., 1990e). VIP dilates both resistance and capacitance vessels in nasal mucosa (Malm et al., 1980). In vitro, VIP stimulates serous cell exocytosis of lactoferrin and mucin-immunoreactive material from human nasal mucosal explants (Mullol et al., 1992b). VIP and acetylcholine did not have additive effects in this system. VIP increases mucus secretion from cat trachea (Shimura et al., 1988) and epithelial cell secretion in dog and ferret trachea (Richardson and Webber, 1987). In contrast, VIP may inhibit mucus secretion from human bronchial explants (Coles et al., 1981; Marom and Goswami, 1991). Anti-VIP antibodies may play a pathological role in some cases of chronic bronchitis (Marom and Goswami, 1991). VIP has been touted as the NANC bronchodilator, that is, a bronchodilating neurotransmitter released after vagal nerve stimulation in the presence of adrenergic and cholinergic blockade. As will be discussed elsewhere

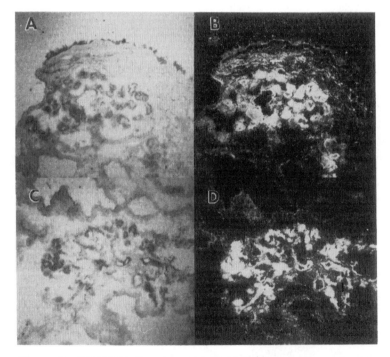

Figure 9 ^{125}I-VIP binding sites. (A and C) Brightfield images showing epithelium, glands, and vessels. (B) Darkfield image of A showing ^{125}I-VIP binding to glands, epithelium, and vessels. The binding to glands is more intense than to sinusoidal vessels or epithelium. The addition of excess VIP ablates the binding (not shown).

in this text, it is likely that nitric oxide (NO) released from parasympathetic and/or other nerve populations or stimulated end-organs that express the constitutive NO synthetase have more potent bronchodilating effects (Fisher et al., 1993). The role of NO in the nose is currently under investigation.

Increased immunohistochemical staining of VIP-immunoreactive material has been reported in vasomotor rhinitis (Kurian et al., 1983). Increased glandular secretion and vasodilation may contribute to chronic vascular congestion and hypersecretion. An absence of VIP-immunoreactive nerve fibers has been reported in cystic fibrosis (Heinz-Erian et al., 1985) that could contribute to a decrease in VIP-induced serous cell secretion and formation of thick, mucoid secretions.

In vivo, VIP may be more active as a vasodilator than as a secretagogue since atropine blocks essentially all gland secretion but only partially blocks neurogenically induced blood flow (Eccles and Wilson, 1974; Malm et al., 1980; Lundberg et al., 1981; Larsson et al., 1986; Stjarne et al., 1991). Nasal provocation studies are needed to determine the effects of VIP in normal human subjects and in subjects with defined pathological diagnoses. Effective VIP antagonists will also be required to determine the relative importance of acetylcholine and VIP in parasympathetically mediated secretion.

The rate of nerve depolarization may vary the amounts of acetylcholine, VIP, and PHM released from parasympathetic nerves (Hokfelt et al., 1987). At low rates, ACh is released, while at high rates, acetylcholine + VIP are released. VIP may augment the postsynaptic acetylcholine-induced secretory response in glands (e.g., cat salivary glands) (Lundberg et al., 1981), but may also have presynaptic inhibitory effects that act to limit neuropeptide release. This mechanism would conserve the amount of stored peptide since there are no reuptake mechanisms, and peptides can only be resupplied by axonal transport from the neural cell body. Pharmacological regulation of nasal parasympathetic nerve function by these autoreceptors has not yet been reported in humans. Evaluation of the full range of VIP-related effects will result from the use of peptide and nonpeptide VIP antagonists (Presti and Gardner, 1993).

XII. ADRENERGIC SYMPATHETIC NERVES

Sympathetic nerves induce vasoconstriction and reduce mucosal thickness (Wasicko et al., 1991). They mediate the increase in nasal patency that occurs during exercise (Richerson and Seebohm, 1968). Sympathetic tone generated by the rapidly acting, short-duration noradrenaline and slow-onset, long-lasting NPY likely contribute to the normal nasal cycle of periodic, coordinated unilateral mucosal swelling with nasal obstruction followed by mucosal shrinkage and mucus secretion. The regulation of sympathetic reflex effects and the role of sympathetic nerves in the nasal cycle and responses to inhaled agents have been less well appreciated compared to the proinflammatory neural circuits. Damage of this system (e.g., Horner's syndrome) leads to the absence of this regulatory mechanism and the onset of unabating congestion and edema. Sympathetic dysfunction may contribute to other nonallergic rhinopathies.

Sympathetic nerves exit the superior cervical ganglion as a plexus that runs with the carotid artery to the bony carotid canal and form the deep petrosal nerve. This merges with preganglionic parasympathetic fibers to form the Vidian nerve (Uddman and Sundler, 1986). Postganglionic sympathetic fibers contain either noradrenaline or noradrenaline plus neuropeptide Y (NPY) (Lacroix et al., 1990). Noradrenergic nerve fibers extensively innervate arterial vessels and arteriovenous anastomoses, but only rarely are found in glands (Uddman and Sundler, 1986; Klaassen et al., 1988).

Noradrenaline acts on vascular α_1- and α_2-adrenergic receptors (Lacroix, 1989), which again belong to the rhodopsin superfamily of receptors (Fraser et al., 1988). Multiple subtypes of each adrenergic receptor have been cloned: α_{1A}, α_{1B}, and α_{1C}; α_{2A}, α_{2B}, α_{2C}, and possibly α_{2D}; and β_1, β_2, and β_3 (Harrison et al., 1991; Bylund, 1992). In general, there is more similarity in the sequence of one receptor subtype between two species (e.g., rat vs. human α_{1B}) than between two receptors of a single species (e.g., α_{1A} vs. α_{1B}). α_{1A}-Receptors stimulate influx of Ca^{2+} from extracellular sources, while α_{1B}-receptors stimulate formation of inositol triphosphate and release of Ca^{2+} from intracellular stores. α_{1B}-Receptors may be more prone to desensitization by repeated administration of α-agonists. α_{1C}-Receptors have been demonstrated in bovine brain, while the status of the α_{1D}-receptor requires confirmation. α_2-Receptors stimulate K^+ influx and attenuate cAMP formation by inhibiting adenylate cyclase activity. The pharmacology of cloned gene products and tissues is complex, with poor correlations between cloned subtypes, receptor functions, and ligand-binding characteristics (Harrison et al., 1991; Bylund, 1992). These problems may be due to mixtures of receptor subtypes in vivo, membrane microenvironments, or local extracellular mediators.

The distribution of each subtype protein and mRNA in nasal mucosa, and other human tissues, is not known.

Vasoconstriction is one of the most important functions of sympathetic nerves (Richerson and Seebohn, 1968; Baraniuk, 1991). Norepinephrine and α-agonist drugs may act at several sites in human nasal mucosa. Stimulation of α-receptors on superficial plexus arterioles may reduce blood flow to the postcapillary venules that are the sites of vascular permeability and so reduce the volume of extravasated plasma, edema, and rhinorrhea. Deeper in the mucosa, α-adrenergic agents may act on the arteries and arteriovenous anastomoses (AVA) that regulate blood flow into venous sinusoids. Dilation of the AVAs with closure of draining muscular veins increases blood flow into the sinusoids and causes them to swell. This increases the thickness of the nasal mucosa, which in turn reduces the cross-sectional area of the airway and increases the resistance to nasal airflow (Fig. 10).

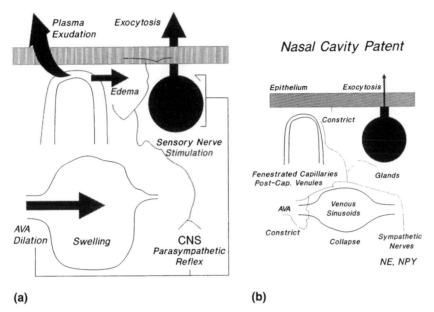

Figure 10 Nasal mucosa in inflammation and after sympathetic stimulation. Prototypic comparison of parasympathetic actions during nasal inflammation and "anti-inflammatory" actions of sympathetic nerves and agonists. (a) Prototypic pathophysiological processes are demonstrated in the nasal mucosa. Mediators have caused increased superficial blood flow with plasma extravasation, edema formation, and exudation across the epithelium. Glandular exocytosis has been stimulated. Arteriovenous anastomoses are dilated and blood has swollen the sinusoids and so increased the thickness of the nasal mucosa. Nociceptive nerves have been stimulated and released their neuropeptides, which may be caused by axon response effects. More potent and significant are the parasympathetic reflexes, which are recruited from the central nervous system and regulate the glandular secretion and contribute to the vasodilation. (b) Sympathetic stimulation, or the application of vasoconstricting α-adrenergic agonists or NPY, reduces blood flow to superficial vessels and arteriovenous anastomoses and thus reduces plasma exudation and permits sinusoidal collapse. This improves nasal patency and reduces nasal airflow resistance. Norepinephrine may be a weak glandular secretagogue. It is not known if sympathetic nerves have direct inhibitory effects on sensory or parasympathetic neural transmission in peripheral nerves in vivo.

Sympathetically mediated constriction of AVAs reduces blood flow into the sinusoids and permits passive collapse of these vessels, a decrease in mucosal thickness, and improvement in nasal patency. Direct stimulation and contraction of myoepithelial cells of the venous sinusoids may also contribute to the collapse of these vessels (Kristiansen et al., 1993). For these reasons, α-adrenergic agonists are useful nasal decongestants (Lockhart et al., 1992).

α_1-Adrenergic agonists may stimulate glandular secretion by direct actions on glands (Mullol et al., 1992a). In addition, they may bind to autoreceptors on sensory or parasympathetic nerves and modulate the actions of these nerves (Matran et al., 1989). Stimulation of sensory nerves leads to the appreciation of itch, congestion, and fullness in the nose and recruits central reflexes including sneeze and the parasympathetic reflexes that are responsible for most mucosal glandular secretion (Baraniuk, 1991). It is not known which α_1- or α_2-receptor subtypes are involved in any of these vital processes.

Prolonged use of topically applied α-adrenergic agonists leads to a refractory condition of unremitting vascular congestion ("rhinitis medicamentosa") (Naclerio, 1991). The mechanism of this paradoxical reaction is unresolved, but could involve induction or repression of individual α-receptor mRNAs and proteins, absent sympathetic tone with unopposed vasodilation and venous congestion, or increased sensitivity of nociceptive sensory nerves with increased sensations of nasal blockage without actual obstruction. Receptor densities have not been evaluated in rhinitis medicamentosa, but allergic rhinitis and control nasal mucosal tissues contain identical numbers of α_1- and α_2-receptors by Scatchard analysis, indicating that modulation of this system does not occur in allergic rhinitis (van Megen et al., 1991a). Without more selective ligands, it is not possible to fully characterize the receptor subtypes. Surprisingly, autoradiographic studies to identify the cells that express α_1- and α_2-adrenergic receptors in the respiratory tract, and alterations that may occur in rhinitis medicamentosa, have not been reported.

β-Adrenergic agonists are mild vasodilators (McLean et al., 1976) and have no effect on glandular or vascular secretory processes in human nasal mucosa in vivo or glandular secretion in vitro (Mullol et al., 1992a). ^{125}I-Cyanopindolol binding sites have only been reported on epithelial cells (van Megen et al., 1991a).

XIII. NEUROPEPTIDE Y

Neuropeptide tyrosine (neuropeptide Y, NPY) is present with noradrenaline in a population of sympathetic neurons (Uddman and Sundler, 1986; Pernow, 1988; Potter, 1988; Baraniuk et al., 1990a). This peptide has many of the same actions as noradrenaline. NPY-induced vasoconstriction is slower in onset but longer in duration than that of noradrenaline. The walls of arterioles and arteriovenous anastomoses are densely innervated by NPY nerve fibers (Uddman and Sundler, 1986; Baraniuk et al., 1990a). The fibers intercalate between smooth muscle cells. Individual fibers are also located in venous vessels. A few fibers are present in glandular acini.

In pigs (Lacroix et al., 1990) there is a population of NPY nerve fibers that innervate glands, originate in the sphenopalatine ganglion, and contain VIP. They may represent a separate population of parasympathetic fibers. Their function is not known. This makes it difficult to be dogmatic about the localization of specific neuropeptides in specific neural systems. Furthermore, it indicates that there may be great species-to-species and organ-to-organ variability in the distributions and functions of autonomic and sensory nerves and their neuropeptides. The functions of colocalized neuropeptides such as NPY + VIP are difficult to predict since complex synergistic and antagonistic interactions can occur when mixtures of peptides are released (Brain and Williams, 1985, 1988; Gamse and Saria, 1985; Geppetti et al., 1988; Khalil et al., 1988).

^{125}I-NPY binding sites have been localized by autoradiography on the smooth muscle of arterioles and the arteriolar portion of arteriovenous anastomoses, with less dense representation on the adjacent venous portions (Table 3, Fig. 11)

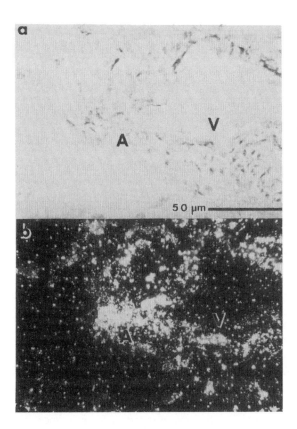

Figure 11 ^{125}I-NPY binding sites. (a) Phase contrast view of an arteriovenous anastomosis showing the arterial (A) and venous (V) portions. (b) Darkfield autoradiography shows silver grains indicative of ^{125}I-NPY binding over the arterial (A) region, with a lower density over the venous portion (V).

(Baraniuk et al., 1990a). There is a strong correlation between the locations of NPY nerve fibers, NPY binding sites on vessels, and the vasoconstrictor function of this peptide.

To test if NPY could be a vasoconstrictor in humans in vivo, nasal provocations have been performed (Baraniuk et al., 1992b). NPY (2.3 nmol applied topically to the nasal mucosa) reduced the resistance to inspiratory airflow by 57% ± 18% ($p < 0.001$) in 10 normal subjects and by 50% ± 17% ($p < 0.05$) in 12 subjects with perennial rhinitis, suggesting that NPY reduced the thickness of the nasal mucosa. NPY in doses of 0.1–10 nmol had no effect on exudation or secretion of proteins of vascular (albumin) or glandular origin (lysozyme, glycoconjugate), indicating that NPY had no effect on plasma extravasation or exudation or glandular exocytosis in the baseline state. Since NPY's vasoconstrictor properties could be latent and only unmasked in the presence of increased vascular permeability and albumin exudation, bradykinin (BK) nasal provocation was performed (Baraniuk et al., 1990b; Baraniuk et al., 1992b). BK, which is a known vasodilator that induces vascular permeability, significantly increased total protein (10- to 20-fold), albumin (10- to 30-fold), and glycoconjugate (two- to five-fold) concentrations in lavage fluid. When NPY (2.3 nmol) was administered with BK (500 nmol), the total protein secreted was reduced by 59% ± 15% ($p < 0.05$), and albumin was reduced by 63% ± 17% ($p < 0.02$). There was no significant effect on glandular secretion. Therefore, NPY potently blocked the vascular permeability induced by BK. Nasal administration of effective doses of NPY was safely tolerated without systemic hemodynamic consequences. Since exogenous administration of NPY to the human nasal mucosa reduced nasal airflow resistance and albumin exudation but without affecting submucosal gland secretion, NPY agonists may be useful for the treatment of mucosal diseases characterized by vasodilation, vascular permeability, and plasma exudation. Because NPY is a long-acting vasoconstrictor, NPY agonists may offer advantages over α-agonists for the treatment of chronic vascular congestion. Further study on the mechanisms of NPY's effects, its effects in vivo, and the development of appropriate peptides for therapy are required before NPY agonists replace α-adrenergic agonists.

NPY receptor subtypes have been suggested. Postsynaptic Y_1-receptors may be linked to phosphatidylinositol hydrolysis (Hakanson and Wahlestadt, 1987; Hinson et al., 1988) and may potentiate the postjunctional effects of norepinephrine (Potter, 1991). Prejunctional Y_2-receptors are coupled to adenylate cyclase inhibition (Hakanson and Wahlestadt, 1987). Prejunctional Y_2-receptors on sympathetic nerves may inhibit norepinephrine release and prevent depletion of these neurotransmitters (Potter, 1991). NPY inhibits cholinergic transmission in guinea pig trachea by a direct effect on NPY receptors on postganglionic cholinergic nerves or other unresolved presynaptic sites (Stretton and Barnes, 1988). An analogous effect on sensory nerves has been suggested for opiates (Rogers and Barnes, 1989). The roles of autoreceptors in nasal mucosal nerves have not been investigated. The diversity and distributions of NPY receptors and their different relative affinities for NPY fragments (Potter, 1991) suggest that it may be possible to selectively design NPY analog peptides that bind to peripheral vascular NPY receptors and perform as potent, long-acting vasoconstricting agents. However, NPY is a substrate for neutral endopeptidase in human nasal mucosa (Baraniuk et al., 1990a), so NPY peptide and nonpeptide analogs that are relatively resistant to proteolytic degradation may be required for therapeutic use.

XIV. SUMMARY

Sensory nerves detect the conditions of inspired air and respond to inhaled noxious agents and mucosal injury. Different forms of insults may generate a wide variety of mediators, including H^+, K^+, prostaglandins, leukotrienes, bradykinin, and histamine, that may stimulate or modulate nociceptive sensory nerves by acting on specific receptors, ion channels, or other macromolecules in nerves. These mediators may also act directly on vessels to induce vascular permeability and venous sinusoidal filling. Release of sensory neuropeptides such as SP, NKA, CGRP, GRP, and others causes the local axon response. The magnitude of the axon response may be subtle in large animals, as demonstrated by the modest increase in superficial blood flow found in the pig (Stjarne et al., 1991), and be less potent than the effects found in rodent airways (McDonald, 1988).

The most important effects of sensory nerve stimulation in the nose are the recruitment of parasympathetic and other central reflexes and the induction of the sensation of pain. These lead to sneezing and other avoidance behaviors that attempt to rapidly clear the upper airway of offending agents while protecting the lower airways. Parasympathetic reflexes cause rapid and copious secretion of submucosal gland mucus and vasodilation, which increases nasal airflow resistance (Table 7). Muscarinic M_3, and possibly M_1, receptors most likely mediate glandular secretion. The sensory nerve-parasympathetic reflex arc is of paramount importance in the normal nose and contributes significantly to the pathology of allergic, infectious, and other nonallergic rhinitis. Thus, nociceptive sensory nerves and the recruitment of parasympathetic reflexes are potent proinflammatory mechanisms in the nasal mucosa.

The sympathetic nervous system acts to reduce mucosal blood flow, sinusoidal filling, and mucosal thickness, and thus induces restoration of nasal air flow and nasal patency. Short-acting noradrenaline and long-acting NPY are potent vasoconstrictors. Loss of sympathetic influences may contribute to some chronic, nonallergic rhinopathies.

Neutral endopeptidase may also act in an "anti-inflammatory" fashion by degrading neuropeptides, thus limiting the durations of their actions in vivo. Destruction or down-regulation of NEP may contribute to the "hyperresponsiveness" of this mucosa that can occur after infection or exposure to cigarette smoke

Table 8   Major Nasal Mucosal Functions that Are Regulated by Neural Pathways

Airflow resistance	
Determined by mucosal thickness (sinusoidal filling)	
Vasodilation:	Sensory (SP, CGRP)
	Parasympathetic (acetylcholine, VIP)
Vasoconstriction:	Sympathetic (noradrenaline, NPY)
Vascular permeability	
Plasma extravasation:	Sensory (SP)
Vasoconstriction:	Sympathetic (noradrenaline, NPY)
Glandular secretion of antimicrobial defense factors, mucins, and other products	
Exocytosis and expulsion:	Sensory (SP, GRP)
	Parasympathetic (muscarinic M_3, possibly M_1, VIP)
	Sympathetic (α-adrenergic)

or other toxic gases. Restoration of NEP activity may be a beneficial effect of glucocorticoid use.

As has been shown, nasal nerves play key, coordinated roles in nasal homeostasis and in pathological situations (Table 8). Drugs based on neurotransmitters, such as the α-adrenergic agonists and anticholinergic agents, have been among the most successful medications marketed. Future investigations with neuropeptide analogs, such as SP antagonists, will reveal more information about the roles of these nerves in health and disease and may have clinical utility in rhinitis.

ACKNOWLEDGMENT

Dr. Baraniuk has been awarded the Edward Livingston Trudeau Scholar Award by the American Lung Association.

REFERENCES

Advenier, C., Naline, E., Toty, L., Bakdach, H., Emonds-Alt, X., Vilain, P., Breliere, J.C., and Le Fur, G. (1992). Effects on the isolated human bronchus of SR 48968, a potent and selective nonpeptide antagonist of the neurokinin A (NK2) receptors. *Am. Rev. Respir. Dis. 146*:1177–1181.

Aguayo, S.M., Kane, M.A., King, T.E., Jr., Schwartz, M.I., Grauer, L., and Miller, Y.E. (1989). Increased levels of bombesin-like peptides in the lower respiratory tract of asymptomatic cigarette smokers. *J. Clin. Invest. 84*:1105–1113.

Al-Bazzaz, F. (1986). Regulation of salt and water transport across airway mucosa. *Clin. Chest Med. 7*:259–272.

Atsolfi, M., Meini, S., Treggiari S., Maggi, C.A., and Manzini, S. (1993). Characteristics of the NK-2 receptor which mediates the motor response to tachykinins in humans isolated bronchus. *Neuropeptides 24*:198–199A.

Baluk, P., Nadel, J.A., and McDonald, D.M. (1993). Calcitonin gene related peptide in secretory granules of serous cells in the rat tracheal epithelium. *Am. J. Respir. Cell Mol. Biol. 8*:446–453.

Baraniuk, J.N. (1991). Neural control of human nasal secretion. *Pulm. Pharmacol. 4*:20–31.

Baraniuk, J.N., and Kaliner, M.A. (1991). Neuropeptides and nasal secretion. *Am. J. Physiol. 261 (Lung Cell Mol. Physiol. 5)*:L223–L235.

Baraniuk, J.N., Castellino, S., Goff, J., Lundgren, J.D., Mullol, J., Merida, M., et al. (1990a). Neuropeptide Y (NPY) in human nasal mucosa. *Am. J. Respir. Cell Mol. Biol. 3*:165–173.

Baraniuk, J.N., Lundgren, J.D., Goff, J., Gawin, A.Z., Mizuguchi, H., Peden, D., Merida, M., Shelhamer, J.H., and Kaliner, M.A. (1990b). Bradykinin receptor distribution in human nasal mucosa, and analysis of in vitro secretory responses in vitro and in vivo. *Am. Rev. Respir. Dis. 141*:706–714.

Baraniuk, J.N., Lundgren, J.D., Goff, J., Mullol, M., Castellino, S., Merida, M., Shelhamer, J.H., and Kaliner, M.A. (1990c). Calcitonin gene related peptide (CGRP) in human nasal mucosa. *Am. J. Physiol. 258*:L81–L88.

Baraniuk, J.N., Lundgren, J.D., Goff, J., Peden, D., Merida, M., Shelhamer, J., and Kaliner, M.A. (1990d). Gastrin releasing peptide (GRP) in human nasal mucosa. *J. Clin. Invest. 85*:998–1005.

Baraniuk, J.N., Okayama, M., Lundgren, J.D., Mullol, M., Merida, M., Shelhamer, J.H., and Kaliner, M.A. (1990e). Vasoactive intestinal peptide (VIP) in human nasal mucosa. *J. Clin. Invest. 86*:825–831.

Baraniuk, J.N., Lundgren, J.D., Okayama, M., Goff, J., Mullol, M., Merida, M., Shelhamer, J.H., and Kaliner, M.A. (1991a). Substance P and neurokinin A (NKA) in human nasal mucosa. *Am. J. Respir. Cell Mol. Biol. 4*:228–236.

Baraniuk, J.N., Mak, J., Letarte, M., Davies, R., Twort, C., and Barnes, P.J. (1991b). Neutral endopeptidase mRNA expression. *Am. Rev. Respir. Dis. 143*:A40.

Baraniuk, J.N., Kaliner, M.A., and Barnes, P.J. (1992a). Muscarinic m3 receptor mRNA in situ hybridization in human nasal mucosa. *Am. J. Rhinol. 6*:145–148.

Baraniuk, J.N., Silver, P.B., Kaliner, M.A., and Barnes, P.J. (1992b). Neuropeptide Y (NPY) is a vasoconstrictor in human nasal mucosa. *J. Appl. Physiol. 73*:1867–1872.

Baraniuk, J.N., Silver, P.B., Lundgren, J.D., Cole, P., Kaliner, M.A., and Barnes, P.J. (1992c). Bombesin stimulates mucous cell and serous cell secretion in human nasal provocation tests. *Am. J. Physiol. 262 (Lung Cell Mol. Physiol. 6)*:L48–L52.

Baraniuk, J.N., Ohkubo, K., Kwon, O.J., Mak, J., Rohde, J., Durham, S.R., and Barnes, P.J. (1993). Localization of neutral endopeptidase mRNA in human nasal mucosa. *J. Appl. Physiol. 74*:272–279.

Barnes, P.J. (1989). Muscarinic receptor subtypes: implications for lung disease. *Thorax 44*:161–167.

Barnes, P.J., Baraniuk, J.N., and Belvisi, M.G. (1991a). Neuropeptides in the respiratory tract. Part 1. *Am. Rev. Respir. Dis. 144*:1187–1198.

Barnes, P.J., Baraniuk, J.N., and Belvisi, M.G. (1991b). Neuropeptides in the respiratory tract. Part 2. *Am. Rev. Respir. Dis. 144*:1391–1399.

Baroody, F.M., Majchel, A.M., Roecker, M.M., Roszko, P.J., Zegrelli, E.C., Wood, C.C., and Naclerio, R.M. (1992). Ipratropium bromide (Atrovent nasal spray) reduces the nasal response to methacholine. *J. Allergy Clin. Immunol. 89*:1065–1075.

Battey, J.F., Way, J.M., Corjay, M.H., Shapira, H., Kusano, K., Harkins, R., Wu, J.M., Slattery, T., Mann, E., and Feldman, R.I. (1991). Molecular cloning of the bombesin/ gastrin releasing peptide receptor from Swiss 3T3 cells. *Proc. Natl. Acad. Sci. USA 88*: 395–399.

Beinborn, M., Lee, Y.M., McBride, E.W., Quinn, S.M., and Kopin, A.S. (1993). A single amino acid of the cholecystokinin-B/gastrin receptor determines specificity for nonpeptide antagonists. *Nature 362*:348–350.

Birnbaumer, L., and Brown, A.M. (1990). G proteins and the mechanisms of action of hormones, neurotransmitters, and autocrine and paracrine regulatory factors. *Am. Rev. Respir. Dis. 141*:S106–S114.

Borson, D.B. (1991). Roles of neutral endopeptidase in airways. *Am. J. Physiol. 260 (Lung Cell. Mol. Physiol. 4)*:L212–L225.

Borson, D.B., and Gruenert, D.C. (1991). Glucocorticoids induce neutral endopeptidase in transformed human trachea epithelial cells. *Am. J. Physiol. 260 (Lung Cell. Mol. Physiol. 4)*:L83–L89.

Boyd, N.D., MacDonald, S.G., Kage, R., Luber-Narod, J., and Leeman, S.E. (1991). Substance P receptor: biochemical characterization and interactions with G proteins. *Ann. NY Acad. Sci. 632*:79–93.

Brain, S.D., and Williams, T.J. (1985). Inflammatory oedema induced by synergism between calcitonin gene-related peptide (CGRP) and mediators of increased vascular permeability. *Br. J. Pharmacol. 86*:855–861.

Brain, S.D., and Williams, T.J. (1988). Substance P regulates the vasodilator activity of calcitonin gene related peptide. *Nature 335*:73–75.

Braunstein, G., Fajac, I., Lacronique, J., and Frossard, N. (1991). Clinical and inflammatory responses to exogenous tachykinins in allergic rhinitis. *Am. Rev. Respir. Dis. 144*: 630–636.

Bunnett, N.W., Wu, V., Sternini, C., Klinger, J., Shimomaya, E., Payan, D., Kobayashi, R., and Walsh, J.H. (1993). Distribution and abundance of neutral endopeptidase (EC 3.4.24.11) in alimentary canal of the rat. *Am. J. Physiol. 264 (Gastrintest. Liver Physiol. 27)*:G497–G508.

Bylund, D.B. (1992). Subtypes of α_1- and α_2-adrenergic receptors. *FASEB J. 7*:832–839.

Cauna, N. (1970). Electron microscopy of the nasal vascular bed and its nerve supply. *Ann. Otol. Rhinol. Laryngol. 79*:443–450.

Cauna, N., and Hinderer, K.H. (1969). Fine structure of blood vessels of the human nasal respiratory mucosa. *Ann. Otol. Rhinol. Laryngol. 78*:865–879.

Cheung, D., Bel, E.H., Den Hartigh, J., Dijkman, J.J., and Sterk, P.J. (1992). The effect of an inhaled neutral endopeptidase inhibitor, thiorphan, on airway responsiveness to neurokinin A in normal humans in vivo. *Am. Rev. Respir. Dis. 145*:1275–1280.

Chin, S.Y., Hall, J.M., Brain, S.D., and Morton, I.K.M. (1993). Do amylin and CGRP interact with similar receptors in the perfused rat isolated kidneys? *Neuropeptides 24*:207.

Christiophides, N.D., Polak, J.M., and Bloom, S.R. (1984). Studies on the distribution of PHI in mammals. *Peptides 5*:261.

Cole, P. (1990). Nasal airflow resistance. In: *Rhinitis and Asthma: Similarities and Differences*. (Mygind, N., Pipkorn, U., and Dahl, R., eds.). Munsgaard, Copenhagen, pp. 391–414.

Coles, S.J., Said, S.I., and Reid, L.M. (1981). Inhibition by vasoactive intestinal peptides of glycoconjugate and lysozyme secretion by human airways in vitro. *Am. Rev. Respir. Dis.* *124*:531–536.

Corjay, M.H., Dobrzanski, D.J., Way, J.M., Viallet, J. Shapira, H., Worland, P., Sausville, E.A., and Battey, J.F. (1991). Two distinct bombesin receptor subtypes are expressed and functional in human lung carcinoma cells. *J. Biol. Chem.* *266*:18771–18779.

D'Adamio, L., Shipp, M.A., Masteller, E.L., and Reinherz, E.L. (1989). Organization of the gene encoding common acute lymphoblastic leukemia antigen (neutral endopeptidase 24.11): multiple miniexons and separate 5' untranslated regions. *Proc. Natl. Acad. Sci. USA* *86*:7103–7107.

Dale, H.H. (1935). Pharmacology and nerve endings. *Proc. Roy. Soc. Med.* *68*:319–324.

Dawes, J.D.K., and Pritchard, M.M.J. (1953). Studies of the vascular arrangements of the nose. *J. Anat.* *87*:311–322.

Demski, L.S., and Schwanzel-Fukuda, M., eds. (1987). The terminal nerve (nervus terminalis). *Ann. NY Acad. Sci.* *519*:1– 213.

Desmazes, N., Lockhart, A., Lacroix, A., and Dusser, D.J. (1992). Carboxypeptidase M-like enzyme modulates the noncholinergic bronchoconstrictor response in guinea pig. *Am. J. Respir. Cell Mol. Biol.* *7*:477–484.

Devillier, P., Dessanges, J.F., Rakatosihanaka, J., Ghaem, A., Boushey, H.A., and Lockhart, H. (1988). Nasal response to substance P and methacholine in subjects with and without allergic rhinitis. *Eur. Respir. J.* *1*:356–361.

Dorje, F., Wess, J., Lambrecht, G., Tacke, R., Mutschler, E., and Brann, M.R. (1991). Antagonist binding profiles of five cloned human muscarinic receptor subtypes. *J. Pharmacol. Exp. Ther.* *256*:727–733.

Druce, H.M., Wright, R.H., Kossoff, D., and Kaliner, M.A. (1985). Cholinergic nasal hyperreactivity in atopic subjects. *J. Allergy Clin. Immunol.* *76*:445–452.

Dusser, D.J., Djoric, T.D., Borson, D.B., and Nadel, J.A. (1989). Cigarette smoke induces bronchoconstrictor hyperresponsiveness to substance P and inactivates airway neutral endopeptidase in the guinea pig. *J. Clin. Invest.* *84*:900–906.

Eccles, J.C. (1986). Chemical transmission and Dale's principle. *Prog. Brain Res.* *68*:3014–3020.

Eccles, R., and Wilson, H. (1974). The autonomic innervation of the nasal blood vessels of the cat. *J. Physiol.* *238*:549–560.

Erdos, E.G., Wagner, B., Harbury, C.B., Painter, R.G., Skidgell, R.A., and Fa, X.G. (1989). Down-regulation and inactivation of neutral endopeptidase 24.11 (enkephalinase) in human neutrophils. *J. Biol. Chem.* *263*:9456–9461.

Filorentin, D., Sassi, A., and Rocques, B.P. (1984). A highly sensitive fluorometric assay for "enkephalinase," a neutral metalloendopeptidase that releases tyrosine-glycine-glycine from enkephalins. *Anal. Biochem.* *141*:62–69.

Fisher, J.T., Anderson, J.W., and Waldron, M.A. (1993). Nonadrenergic noncholinergic neurotransmitter of feline trachealis: VIP or nitric oxide? *J. Appl. Physiol.* *74*:31–39.

Fong, T.M., Cascieri, M.A., Yu, H., Banasal, A., Swain, C., and Strader, C.D. (1993). Amino-aromatic interaction between histidine 197 of the neurokinin-1 receptor and CP 96345. *Nature* *362*:350–353.

Fraser, C.M., Potter, P., and Venter, J.C. (1988). Adrenergic agents. In: *Allergy: Principles and Practise Vol. 1.* (Middleton, E., Jr., Reed, C.E., Ellis, C., Adkinson, N.F., Jr., and Yuninger, J.W., eds.). C.V. Mosby, St. Louis, pp. 636–647.

Gamse, R., and Saria, A. (1985). Potentiation of tachykinin induced plasma protein extravasation by calcitonin gene related peptide. *Eur. J. Pharmacol. 114*:61–66.

Gawin, A., Baraniuk, J.N., and Kaliner, M.A. (1990). The effects of gastrin releasing peptide (GRP) and analogues on guinea pig nasal mucosa. *Am. Rev. Respir. Dis. 141*:A173.

Gawin, A., Baraniuk, J.N., and Kaliner, M. (1993). Effects of substance P and calcitonin gene related peptide (CGRP) on guinea pig nasal mucosal secretion in vivo. *Acta Otolaryngol. (Stockh.) 113*:533–539.

Geppetti, P., Fusco, B.M., Marabini, S., Maggi, C.A., Faniullacci, M., and Sicuteri, F. (1988). Secretion, pain and sneezing induced by the application of capsaicin to the nasal mucosa in man. *Br. J. Pharmacol. 93*:509–514.

Gerard, N.P., Eddy, R.L., Shows, T.B., and Gerard, C. (1990). The human neurokinin A (substance K) receptor. Molecular cloning of the gene, chromosome localization, and isolation of cDNA from tracheal and gastric tissues. *J. Biol. Chem. 265*:20455–20462.

Gether, U., Johanson, T.E., Snider, R.M., Lowe, J.A., Nakanishi, S., and Schwarz, T.W. (1993). Different binding epitopes on the NK1 receptor for substance P and a non-peptide antagonist. *Nature 362*:345–348.

Hakanson, R., and Wahlestadt, C. (1987). Neuropeptide Y acts via prejunctional (Y2) and postjunctional (Y1) receptors. *Neuroscience 22*:S679.

Harmar, A.J., Hyde, V., and Chapman, C. (1990). Identification and cDNA of delta-preprotachykinin, a fourth splicing variant of the rat substance P precursor. *FEBS Lett. 275*:22–24.

Harrison, J.K., Pearson, W.R., and Lynch, K.R. (1991). Molecular characterization of α_1- and α_2-adrenoceptors. *TIPS 12*:62–67.

Hartschuh, W., Weihle, E., and Reinecke, M. (1983). Peptidergic (neurotensin, VIP, substance P) nerve fibres in the skin. Immunohistochemical evidence of an involvement of neuropeptides in nociception, pruritus, and inflammation. *Br. J. Dermatol. 109* (Suppl. 25):14.

Heinz-Erian, P., Dey, R.D., and Said, S.I. (1985). Deficient vasoactive intestinal peptide innervation in sweat glands of cystic fibrosis patients. *Science 229*:1407–1409.

Helke, C.J., Drause, J.E., Mantyh, P.W., Couture, R., and Bannon, M.J. (1990). Diversity in mammalian tachykinin peptidergic neurons: multiple peptides, receptors, and regulatory mechanisms. *FASEB J. 4*:1606–1615.

Hinson, J., Rauh, C., and Coupet, J. (1988). Neuropeptide Y stimulates inositol phospho-lipid hydrolysis in rat brain miniprisms. *Brain Res. 446*:379–382.

Hokfelt, T., Fuxe, K., and Pernow, B. (1987). Coexistence of neuronal messengers: a new principle in chemical transmission. *Prog. Brain Res. 68*:1–37.

Holzer, P. (1988). Local effector functions of capsaicin-sensitive sensory nerve endings: involvement of tachykinins, calcitonin gene related peptide and other neuropeptides. *Neuroscience 24*:739–768.

Holzer, P. (1991). Capsaicin: cellular targets, mechanisms of action, and selectivity for thin sensory neurons. *Pharmacol. Rev. 43*:143–201.

Honda, I., Kohrogi, H., Yamaguchi, T., Ando, M., and Araki, S. (1991). Enkephalinase inhibitor potentiates substance P- and capsaicin-induced bronchial smooth muscle contractions in humans. *Am. Rev. Respir. Dis. 143*:1416–1418.

Hua, X.Y. (1986). Tachykinins and calcitonin gene related peptide in relation to peripheral functions of capsaicin-sensitive sensory nerves. *Acta Physiol. Scand. 127* (Suppl. 551): 1–45.

Iijima, H., Gerard, N.P. Squassoni, C., Ewig, J., Face, D., and Drazen, J.M. (1992). Exon 16 del: a novel form of human neutral endopeptidase (CALLA). *Am. J. Physiol. 262 (Lung Cell. Mol. Physiol. 6)*:L725–L729.

Ishihara, T., Nakamura, S., Kaziro, Y., Takahashi, T., Takahashi, K., and Nagata, S. (1991). Molecular cloning and expression of a cDNA encoding the secretin receptor. *EMBO J. 10*:1635–1641.

Ishihara, T., Shigemoto, R., Mori, K., Takahashi, K., and Nagata, S. (1992). Functional expression and tissue distribution of a novel receptor for vasoactive intestinal polypeptide. *Neuron 8*:811–819.

Jacoby, D.B., Tamaoki, J., Borson, D.B., and Nadel, J.A. (1988). Influenza infection increases airway smooth muscle responsiveness to substance P in ferrets by decreasing enkephalinase. *J. Appl. Physiol. 64*:2653–2658.

Jakab, G., Webster, H. deF., Salamon, I., and Mezey, E. (1993). Neural and nonneural origin of calcitonin gene related peptide (CGRP) in the gastric mucosa. *Neuropeptides 24*:117–122.

Kageyama, R., Sasai, Y., and Nakanishi, S. (1991). Molecular cloning of transcription factors that bind to the cAMP responsive region of the SP precursor gene: cDNA of a novel c/EBP-related factor. *J. Biol. Chem. 266*:15525–15531.

Khalil, Z., Andrews, P.V., and Helme, R.D. (1988). VIP modulates substance P induced plasma extravasation in vivo. *Eur. J. Pharmacol. 151*:281–287.

Klaassen, A.B.M., van Megen, Y.J.B., Kuipers, W., and van den Broek, P. (1988). Autonomic innervation of the nasal mucosa. *O.R.L. 50*:32–41.

Kondepudi, A., and Johnson, A. (1993). Cytokines increase neutral endopeptidase activity in lung fibroblasts. *Am. J. Respir. Cell Mol. Biol. 8*:43–49.

Konno, A., and Togawa, K. (1979). Role of the vidian nerve in nasal allergy. *Ann. Otol. Rhinol. Laryngol. 88*:258–266.

Kristiansen, A.B., Heyeraas, K.J., and Kirkebo, A. (1993). Increased pressure in venous sinusoids during decongestion of rat nasal mucosa induced by adrenergic agonists. *Acta. Physiol. Scand. 147*:151–161.

Kurian, S.S., Blank, M.A., and Sheppard, M.N. (1983). Vasoactive intestinal polypeptide (VIP) in vasomotor rhinitis. *Clin. Biochem. 11*:425–427.

Lacroix, J.S. (1989). Adrenergic and non-adrenergic mechanisms in sympathetic vascular control of the nasal mucosa. *Acta Physiol. Scand. 136*(Suppl. 581):1–49.

Lacroix, J.S., Anggard, A., Hokfelt, T., O'Hare, M.M., Fahrenkrug, J., and Lundberg, J.M. (1990). Neuropeptide Y: presence in sympathetic and parasympathetic innervation of the nasal mucosa. *Cell Tissue Res. 259*:119–128.

Lang, Z., and Murlas, C.G. (1992). Neutral endopeptidase of a human airway epithelial cell line recovers after hypochlorous acid exposure: dexamethasone accelerates this by stimulating neutral endopeptidase mRNA synthesis. *Am. J. Respir. Cell Mol. Biol. 7*: 300–306.

Larsson, O., Duner-Engstrom, M., Lundberg, J.M., Freholm, B.B., and Anggard, A. (1986). Effects of VIP, PHM and substance P on blood vessels and secretory elements of the human submandibular gland. *Regul. Pept. 13*:319–326.

Levine, R.R., Birdsall, N.J.M., North, R.A., Holman, M., Watanabe, A., and Iversen, L.L. (1991). Subtypes of muscarinic receptors III. *Trends Pharmacol. Sci. 9*(Suppl.):1–93.

Lindsay, R.M., Lockett, C., Sternberg, J., and Winter, J. (1989). Substance P and calcitonin gene related peptide levels are regulated by nerve growth factor. *Neuroscience 33*:53–65.

Llorens-Cortes, C., Giros, B., and Schwartz, J.C. (1990). A novel potential metallopeptidase derived from the enkephalinase gene by alternate splicing. *J. Neurochem. 55*: 2146–2148.

Llorens-Cortes, C., Huang, H., Vicart, P., Gase, J.M., Paulin, D., and Corvol, P. (1992). Identification and characterization of neutral endopeptidase in endothelial cells from venous or arterial origins. *J. Biol. Chem. 267*:140112–14018.

Lockhart, A., Dinh-Xuan, A.T., Regnard, J., Cabanes, L., and Matran, R. (1992). Effect of airway blood flow on airflow. *Am. Rev. Respir. Dis. 146*:S19–S23.

Lundberg, J.M., Anggard, A., and Fahrenkrug, J. (1981). Complementary role of vasoactive intestinal peptide (VIP) and acetylcholine for cat submandibular gland blood flow and secretion. *Acta Physiol. Scand. 113*:329–336.

Lundblad, L. (1984). Protective reflexes and vascular effects in the nasal mucosa elicited by activation of capsaicin-sensitive substance P-immunoreactive trigeminal neurons. *Acta Physiol. Scand. 529* (Suppl):1–42.

Lundgren, J., Ostrowski, N., Baraniuk, J.N., Shelhamer, J.H., and Kaliner, M. (1990). Gastrin releasing peptide stimulates glycoconjugate release from feline tracheal explants. *Am. J. Physiol. 258*:L68–L74.

MacLean, D.B., Bennett, B., Morris, M., and Wheeler, F.B. (1989). Differential regulation of calcitonin gene related peptide an substance P in cultured neonatal rat vagal sensory neurons. *Brain Res. 478*:349–355.

Maggi, C.A., Patacchini, R., Astolfi, M., Rovero, P., Guilliani, S., and Giachetti, A. (1991). NK2-receptor agonists and antagonists. *Ann. NY Acad. Sci. 632*:184–190.

Mak, J.C.W., and Barnes, P.J. (1990). Autoradiographic visualization of muscarinic receptor subtypes in human and guinea pig lung. *Am. Rev. Respir. Dis. 141*:1559–1568.

Mak, J.C.W., Baraniuk, J.N., and Barnes, P.J. (1992). Localization of muscarinic receptor subtype mRNAs in human lung. *Am. J. Respir. Cell Mol. Biol. 7*:344–348.

Malm, L., Sundler, F., and Uddman, R. (1980). Effects of vasoactive intestinal peptide (VIP) on resistance and capacitance vessels in nasal mucosa. *Acta Otolaryngol. (Stockh.) 90*:304–308.

Mapp, P.I., Walsh, D.A., Kidd, B.L., Cruwys, S.C., Polak, J.M., and Blake, D.R. (1992). Localization of the enzyme neutral endopeptidase to the human synovium. *J. Rheumatol. 19*:1838–1844.

Marom, Z., and Goswami, S.K. (1991). Respiratory mucus hypersecretion (bronchorrhea): a case discussion—possible mechanism(s) and treatment. *J. Allergy Clin. Immunol. 87*:1050–1055.

Martins, M.A., Shore, S.A., and Drazen, J.M. (1991). Release of tachykinins by histamine, methacholine, PAF, LTD4, and substance P from guinea pig lungs. *Am. J. Physiol. 261 (Lung Cell Mol. Physiol. 5)*:L449–L455.

Matran, R., Martling, C.R., and Lundberg, J.M. (1989). Inhibition of cholinergic and nonadrenergic noncholinergic bronchoconstriction in the guinea pig mediated by neuropeptide Y and α_2-adrenoceptors and opiate receptors. *Eur. J. Pharmacol. 163*: 15–23.

Mattox, W., Ryner, L., and Baker, B.S. (1992). Autoregulation and multifunctionality among trans-acting factors that regulate alternative pre-mRNA processing. *J. Biol. Chem. 267*:19023–19026.

McDonald, D.M. (1988). Neurogenic inflammation in the rat trachea. I. Changes in venules, leukocytes and epithelial cells. *J. Neurocytol. 17*:605–628.

McDonald, D.M. (1992). Infections intensify neurogenic plasma extravasation in the airway mucosa. *Am. Rev. Respir. Dis. 146*:S40–S44.

McLean, J., Mathews, K., Ciarkowski, A., Brayton, P.R., and Solomon, R.W. (1976). The effects of topical saline and isoproterenol on nasal airway resistance. *J. Allergy Clin. Immunol. 58*:563–574.

Michel, C.C. (1992). The transport of albumin: a critique of the vesicular system in transendothelial transport. *Am. Rev. Respir. Dis. 146*:S32–S36.

Mullol, J., Raphael, G.D., Lundgren, J.D., Baraniuk, J.N., Mérida, M., Shelhamer, J.H., and Kaliner, M.A. (1992a). Comparison of human nasal mucosal secretion in vivo and in vitro. *J. Allergy Clin. Immunol. 89*:584–592.

Mullol, J., Rieves, R.D., Lundgren, J.D., Baraniuk, J.N., Mérida, M., Hausfeld, J.H., Shelhamer, J.H., and Kaliner, M.A. (1992b). The effects of neuropeptides on mucous glycoprotein secretion from human nasal mucosa in vitro. *Neuropeptides 21*:231–238.

Murlas, C.G., Murphy, T.P., and Lang, Z. (1990). HOCl causes airway substance P hyperresponsiveness and neutral endopeptidase hypoactivity. *Am. J. Physiol. 258 (Lung Cell. Mol. Physiol. 2)*:L361–L368.

Naclerio, R.M. (1991). Allergic rhinitis. *N. Engl. J. Med. 325*:860–869.

Nadel, J.A. (1990). Decreased neutral endopeptidases: possible role in inflammatory diseases of airways. *Lung 123* (Suppl.):123–127.

Nadel, J.A. (1992). Membrane-bound peptidases: endocrine, paracrine, and autocrine effects. *Am. J. Respir. Cell Mol. Biol. 7*:469–470.

Nakanishi, S. (1987). Substance P precursor and kininogen: their structures, gene organizations, and regulation. *Physiol. Rev. 67*:1117–1142.

Nieber, K., Baumgartern, C.R., Rathsack, R., Furkert, J. Oehme, P., and Kunkel, G. (1992). Substance P and β-endorphan-like immunoreactivity in lavage fluids of subjects with and without allergic asthma. *J. Allergy Clin. Immunol. 90*:646–652.

Ohkubo, H., and Nakanishi, S. (1991). Molecular characterization of the three tachykinin receptors. *Ann. NY Acad. Sci. 632*:53–62.

Okayama, M., Baraniuk, J.N., Merida, M., and Kaliner, M.A. (1993). Autoradiographic localization of muscarinic receptor subtypes in human nasal mucosa. *Am. J. Respir. Cell Mol. Biol. 8*:176–185.

Payan, G.P., Levine, J.D., and Goetzl, E.J. (1984). Modulation of immunity and hypersensitivity by sensory neuropeptides. *J. Immunol. 132*:1601.

Pernow, J. (1988). Co-release and functional interactions of neuropeptide Y and noradrenaline in peripheral sympathetic vascular control. *Acta Physiol. Scand. 568* (Suppl.):1–55.

Persson, C.G.A. (1991). Mucosal exudation in respiratory defence: neural or nonneural control? *Int. Arch. Allergy Appl. Immunol. 94*:222–226.

Persson, C.G.A., Erjefalt, I., Alkner, U., Baumgarten, C., Greiff, L., Gustafsson, B., Luts, A., Pipkorn, U., Sundler, F., Svensson, C., and Wollmer, P. (1991). Plasma exudation as a first line respiratory mucosal defence. *Clin. Exp. Allergy 21*:17–24.

Petitet, F., Saffroy, M., Torrens, Y., Lavielle, S., Chassaing, G., Loeuillet, D., Glowinski, J., and Beaujouans, J.C. (1992). Possible existence of a new tachykinin receptor subtype in the guinea pig ileum. *Peptides 13*:383–398.

Piedemonte, G., McDonald, D.M., and Nadel, J.A. (1990). Glucocorticoids inhibit neurogenic plasma extravasation and prevent virus-potentiated extravasation in the rat trachea. *J. Clin. Invest. 86*:1409–1415.

Potter, E.K. (1988). Neuropeptide Y as an autonomic neurotransmitter. *Pharmacol. Ther. 37*:251–273.

Presti, M.E., and Gardner, J.D. (1993). Receptor antagonists for gastrointestinal peptides. *Am. J. Physiol. 264 (Gastrointest. Liver Physiol. 27)*:G399–G406.

Raphael, G.D., Druce, H.M., Baraniuk, J.N., and Kaliner, M.A. (1988a). Pathophysiology of rhinitis. I. Assessment of the sources of protein in methacholine-induced nasal secretions. *Am. Rev. Respir. Dis. 138*:413–420.

Raphael, G.D., Meredith, S.D., Baraniuk, J.N., and Kaliner, M.A. (1988b). Nasal reflexes. *Am. J. Rhinol. 2*:8–12.

Raphael, G.D., Hauptschein Raphael, M., and Kaliner, M.A. (1989a). Gustatory rhinitis: a syndrome of food-induced rhinorrhea. *J. Allergy Clin. Immunol. 83*:110–115.

Raphael, G.D., Meredith, S.D., Baraniuk, J.N., Druce, H.M., Banks, S.M., and Kaliner, M.A. (1989b). The pathophysiology of rhinitis. II. Assessment of the sources of protein in histamine-induced nasal secretions. *Am. Rev. Respir. Dis. 139*:791–800.

Raphael, G.R., Baraniuk, J.N., and Kaliner, M.A. (1991a). How and why the nose runs. *J. Allergy Clin. Immunol. 87*:457–467.

Raphael, G.D, Igarashi, Y., White, M.V., and Kaliner, M.A. (1991b). The pathophysiology of rhinitis. V. Sources of protein in allergen-induced nasal secretions. *J. Allergy Clin. Immunol. 88*:33–43.

Regoli, D., Nantel, F., Tousidgnant, C., Jukic, D., Rouissy, N., and Rhaleb, E. (1991). Neurokinin agonists and antagonists. *Ann. NY Acad. Sci. 632*:170–183.

Renkin, E.M. (1992). Cellular and intercellular transport pathways in exchange vessels. *Am. Rev. Respir. Dis. 146*:S28–S31.

Richardson, P.S., and Webber, S.E. (1987). The control of mucous secretion in the airways by peptidergic mechanisms. *Am. Rev. Respir. Dis. 136*:S72–S77.

Richerson, H.B., and Seebohm, P.B. (1968). Nasal airway response to exercise. *J. Allergy 41*:269–284.

Rogers, D.F., and Barnes, P.J. (1989). Opioid inhibitions of neurally mediated mucus secretion in human bronchi. *Lancet 1*:930–932.

Shimura, S., Sasaki, T., Ekeda, K., Sasaki, K., and Takishima, T. (1988). VIP augments cholinergic-induced glycoconjugate secretion in tracheal submucosal glands. *J. Appl. Physiol. 65*:2537–2544.

Shipp, M.A., Richardson, N.E., Sayre, P.H., Brown, N.R., Masteller, E.L., and Clayton, L.K. (1988). Molecular cloning of the common acute lymphoblastic leukemia antigen

(CALLA) identifies a type of type II integral membrane protein. *Proc. Natl. Acad. Sci. USA 85*:4819–4823.

Spindel, E.R., Sunday, M.E., and Hofler, H. (1987a). Transient elevation of mRNAs encoding gastrin releasing peptide (GRP), a putative pulmonary growth factor, in human fetal lung. *J. Clin. Invest. 80*:1172–1177.

Spindel, E.R., Zilberberg, M.D., and Chin, W.W. (1987b). Analysis of the gene and multiple messenger ribonucleic acids (mRNAs) encoding human gastrin releasing peptide: alternate RNA splicing occurs in neural and endocrine tissue. *Mol. Endocrinol. 1*:224–230.

Spindel, E., Giladi, R., Brehm, E., Goodman, P., and Segerson, T.P. (1990). Cloning and functional characterization of a complementary DNA encoding the murine fibroblast bombesin/gastrin releasing peptide receptor. *Mol. Endocrinol. 4*:1956–1962.

Springall, D.R., Collins, G., Barer, G., Suggett, A.J., Bee, D., and Polak, J.M. (1988). Increased intracellular levels of calcitonin gene related peptide-like immunoreactivity in pulmonary endocrine cells of hypoxic rats. *J. Pathol. 155*:259–267.

Sreedharan, S.P., Robichon, A., Peterson, K.E., and Goetzl, E.J. (1991). Cloning and expression of the human vasoactive intestinal peptide receptor. *Proc. Natl. Acad. Sci. USA 88*:4986–4990.

Steenbergh, P.H., Hoppener, J.W., Zandberg, Q., Visser, A., Lips, C.J., and Jansz, H.S. (1986). Structure and expression of the human calcitonin/CGRP genes. *FEBS Lett. 209*: 97–103.

Stefano, G.B., Paemen, L.R., and Hughes, T.K. (1992). Autoimmunoregulation: differential modulation of CD10/neutral endopeptidase 24.11 by tumor necrosis factor and neuropeptides. *J. Neuroimmunol. 41*:9–14.

Stjarne, P., Lundblad, L., Lundberg, J.M., and Anggard, A. (1989). Capsaicin and nicotine sensitive afferent neurones and nasal secretion in healthy human volunteers and in patients with vasomotor rhinitis. *Br. J. Pharmacol. 96*:693–701.

Stjarne, P., Lacroix, J.S., Anggard, A., and Lundberg, J.M. (1991). Compartment analysis of vascular effects of neuropeptides and capsaicin in the pig nasal mucosa. *Acta Physiol. Scand. 141*:335–342.

Stretton, C.D., and Barnes, P.J. (1990). Modulation of cholinergic neurotransmission in guinea pig trachea by neuropeptide Y. *Br. J. Pharmacol. 93*:672–678.

Sunday, M.E., Kaplan, L.M., Motoyama, E., Chin, W.W., and Spindel, E.R. (1988). Gastrin releasing peptide (mammalian bombesin) gene expression in health and disease. *Lab. Invest. 59*:5–24.

Sunday, M.E., Hua, J., Torday, J.S., Reyes, B., and Shipp, M.A. (1992). CD10/neutral endopeptidase 24.11 in developing human fetal tissue. *J. Clin. Invest. 90*:2517–2525.

Takahashi, K., Tanaka, A., Hara, M., and Nakanishi, S. (1992). The primary structure and gene organization of human substance P and neuromedin K receptors. *Eur. J. Biochem. 204*:1025–1033.

Tsutsumi, Y. (1988). Immunohistochemical localization of gastrin releasing peptide in normal and diseased human lung. *Ann. NY Acad. Sci. 547*:336–350.

Uddman, R., and Sundler, F. (1986). Innervation of the upper airways. *Clin. Chest. Med. 7*: 201–209.

Umeno, E., Nadel J.A., Huang, H.T., and McDonald, D.M. (1989). Inhibition of neutral endopeptidase potentiates neurogenic inflammation in the rat trachea. *J. Appl. Physiol.* 66:2647–2653.

van Megen, Y.J.B., Klaassen, A.B.M., Rodrigues de Miranda, J.F., van Ginneken, C.A.M., and Wentges, B.T.R. (1991a). Alterations of muscarinic acetylcholine receptors in the nasal mucosa of allergic patients in comparison with nonallergic individuals. *J. Allergy Clin. Immunol.* 87:521–529.

van Megen, Y.J.B., Klaassen, A.B.M., Rodrigues de Miranda, J.F., van Ginneken, C.A.M., and Wentges, B.T.R. (1991b). Alterations in adrenoceptors in the nasal mucosa of allergic patients in comparison with nonallergic individuals. *J. Allergy Clin. Immunol.* 87:530–540.

Wada, E., Way, J., Shapira, H., Kusano, K., Lebacqu-Verheyden, A.M., Coy, D.H., Jensen, R.T., and Battey, J. (1991). cDNA cloning, characterization and brain region-specific expression of a neuromedin-B-preferring bombesin receptor. *Neuron 6:* 421–430.

Wanner, A. (1986). Mucociliary clearance in the trachea. *Clin. Chest Med.* 7:247–258.

Wasicko, M.J., Leiter, J.C., Erlichman, J.S., Strobel, R.J., and Bartlett, D., Jr. (1991). Nasal and pharyngeal resistance after topical mucosal vasoconstriction in normal humans. *Am. Rev. Respir. Dis.* 144:1048–1052.

Watling, K.J. (1992). Nonpeptide antagonists herald new era in tachykinin research. *Trends Pharmacol. Sci.* 13:266–269.

Welsch, M.J. (1987). Electrolyte transport by airway epithelia. *Physiol. Rev.* 67:1143–1183.

Willey, J.C., Lechner, J.F., and Harris, C.C. (1984). Bombesin and the C-terminal tetradecapeptide of gastrin releasing peptide are growth factors for normal human bronchial epithelial cells. *Exp. Cell Res.* 153:245–248.

Winter, R.J.D., Zhoa, L., Krausz, T., and Hughes, J.M.B. (1991). Neutral endopeptidase 24.11 inhibition reduces pulmonary vascular remodeling in rats exposed to chronic hypoxia. *Am. Rev. Respir. Dis.* 144:1342–1346.

Wright, J. (1910). The contractile elements in the connective tissue in the elastic fibres in the nasal mucosa in health and disease. *J. Med. NY 91:*729–734.

Yeadon, M., Wilkinson, D., and Payan, A.N. (1990). Ozone induces bronchial hyperreactivity to inhaled substance P by functional inhibition of enkephalinase. *Br. J. Pharmacol.* 99:191.

Yiangou, Y., Requejo, F., Polak, J.M., and Bloom, S.R. (1986). Characterization of a novel prepro VIP derived peptide. *Biochem. Biophys. Res. Comm. 139:*1142.

Yokata, Y., Sasai, S., Tanaka, K., Fujiwara, T., Tsuchida, K., Shigemoto, R., and Nakanishi, S. (1989). Molecular characterization of a functional cDNA for rat substance P receptor. *J. Biol. Chem.* 264:17649–17652.

DISCUSSION

McDonald: Would you expect inhaled glucocorticosteroids to influence the synthesis of tachykinins in nerves of the airway mucosa, considering that synthesis of the peptides is limited to sensory ganglia of the vagus nerve and dorsal root ganglia?

Baraniuk: Inhaled glucocorticoids are unlikely to affect tachykinin gene transcription in sensory neuron cell bodies. Oral steroids may affect this process, however.

Widdicombe: We discussed earlier the effects of hyperosmolality on lower airway nerves and mucosa, but the osmolality changes in the nose may be greater than in the lower airways. Is anything known about the effects of osmolality changes in the nose?

Baraniuk: Togias has performed hypertonic saline nasal provocation. This causes pain (sensory nerve stimulation) and parasympathetic reflex secretions.

Kunkel: What antibody did you use for NEP detection in nasal tissue? We could not demonstrate NEP immunoreactivity in the epithelium or glands of the airways.

Baraniuk: Several antibodies are available, and each appears to give slightly different immunohistochemical staining patterns. This may relate to differences in glycosylation. Genentech's MEK antisera worked best, but I understand these are no longer available.

5

Peptide Biosynthesis and Secretion: Some Recent Developments and Unresolved Issues

Hugh P. J. Bennett
Royal Victoria Hospital and McGill University, Montreal, Quebec, Canada

I. INTRODUCTION

Peptide hormones and neuropeptides must undergo a complex and highly regulated process of biosynthesis prior to secretion (Bennett, 1985). They are initially synthesized as precursor molecules, which must be subject to proteolytic cleavage and enzymatic processing to give rise to the functional molecule. Furthermore, they must be enriched into a secretory granule fraction and there await the appropriate signal that initiates the process known as regulated secretion. Many fundamental mechanisms are found to be common to simple eukaryotic cells (e.g., yeast) and mammalian endocrine and neuroendocrine cells. Peptide hormones and neuropeptides and a multitude of other secreted proteins, such as collagen, integral membrane proteins, as well as viral proteins, initially share a common intracellular biosynthetic compartment, the endoplasmic reticulum (ER)/Golgi complex. Within this intracellular locus, all secreted proteins are subject to a cascade of sorting and processing events. While many of the processes are quite well understood (e.g., glycosylation), the manner in which proteins destined for regulated secretion via the secretory granule fraction are segregated from those destined for unregulated or constitutive secretion are poorly understood (Kelly, 1985; Burgess and Kelly, 1987; Brion et al., 1992). Recent articles have extensively reviewed protein biosynthesis in general and peptide hormone and neuropeptide biosynthesis in particular (Rothman and Orci, 1992; Mains et al., 1990). For the purposes of this review I will discuss some recent developments that have begun to

125

address outstanding issues. The focus of much of the review will be on pro-opiomelanocortin (POMC). This multihormone precursor is subject to many of the known posttranslational modifications (Mains and Eipper, 1990). It is subject to sorting to the regulated pathway of secretion and can act as a precursor to both classical hormones within the pituitary and neuropeptides within the hypothalamus (Eipper and Mains, 1980; Emeson and Eipper, 1986; Mains and Eipper, 1990). Furthermore, POMC is subject to some of the best-characterized examples of tissue specific processing. For most of these processes and events, the precursor structure is a major factor in determining the manner in which they are carried out.

II. POSTTRANSLATIONAL PROCESSING

With recent advances in the cell biology of protein biosynthesis, we are reaching a better understanding of the synthesis and secretion of proteins (Burgess and Kelly, 1987; Rothman and Orci, 1992). Like other proteins destined for secretion, peptide hormones and neuropeptides are synthesized as precursors containing amino-terminal extensions known as signal or presequences. These structures promote the cotranslational binding of ribosomes to the outer membrane of the ER and insertion of the growing peptide chain into the lumen of this subcellular compartment. Within the ER and subsequently the Golgi apparatus and secretory granule fraction, newly synthesized precursors are subject to a multitude of processing and sorting events, which ultimately yields the mature hormone or neuropeptide. Posttranslational events will be considered in the order in which they occur as precursors travel through the secretory pathway. Emphasis is placed on recent advances that have clarified certain mechanisms or raised further questions.

A. Disulfide Bridge Formation

The formation of correct disulfide bridging is a critical event in the biosynthesis of secretory and integral membrane proteins. Many peptide and protein hormones bear sulfur bridges that are critical for function. Correct bridging is ultimately determined by thermodynamic drive but is catalyzed intracellularly by the enzyme protein disulfide isomerase (PDI), which is a soluble enzyme found in the lumen of the rough ER. The cell as a whole is a reducing environment, and under the redox conditions prevailing in the cytoplasm, formation of the disulfide bridges is not favored. This apparent paradox was resolved in a recent article by Hwang et al. (1992), who reasoned that the secretory pathway must have a different redox potential from the rest of the cell. They examined the fate of a small synthetic peptide containing both the -Asn-X-Ser- signal for N-glycosylation and a lone cysteine. Upon diffusion into the ER, the peptide was trapped in this organelle as a consequence of glycosylation. The lone cysteine was found to be oxidized through conjugation with glutathione. In this study, it was shown that although the redox potential of the cytoplasm is about -230 mV, in the secretory pathway it is about -160 mV (Hwang et al., 1992), which is close to the optimum for PDI-catalyzed disulfide bridge formation of -110 mV (Hawkins et al., 1991). It was also concluded that the main redox buffer in the ER is most likely glutathione. The cytoplasmic ratio of reduced:oxidized glutathione is between 30:1 and 100:1 (i.e., an overwhelming reducing environment), but in the ER the ratio is about 1:1 to 3:1. Human POMC, the common precursor for ACTH and β-endorphin, contains five cysteines at positions 2, 8, 20, 24, and 88 (Cochet et al., 1982). POMCs of other species have only four cysteines with disulfide bridges linking residues 4 with 24 and 8 with 20 (Bennett et al., 1984). We have investigated what happens to the fifth cysteine at position 88 in human POMC as it traverses the ER/Golgi/secretory granule pathway. The presence of the lone cysteine has no effect on the bridging of the other cysteines; however, human joining peptide, the processed peptide containing the lone cysteine of human POMC, was found mainly conjugated to glutathione (Robinson et al., 1991). Oxidation clearly occurs in the secretory pathway, with oxidized glutathione acting as the redox agent, implying, as the results of Hwang et al. (1992) predict, that the secretory pathway must be considerably more oxidizing than the cytoplasm. Subsequently, the joining peptide/glutathione heterodimer has a tendency to undergo thiol exchange, giving rise to joining peptide homodimers, both in the secretory granule and upon manipulation in vitro (Robinson et al., 1991).

The role that the cysteines at positions 2, 8, 20, and 24 play in the intracellular fate of porcine POMC has been investigated recently (Roy et al., 1991). Site-directed and deletion mutagenesis technology was used to create mutant POMC in which one or both disulfide bridges were disrupted. Upon expression in a neuroendocrine cell line (neuro 2A), the mutant in which both bridges were disrupted yielded POMC that was poorly segregated to the secretory granule fraction and inefficiently processed, at least with respect to endorphin bio-synthesis. These findings imply that proper folding of the amino-terminal region of POMC is important for efficient processing and targeting.

B. Glycosylation

Glycosylation of proteins destined for secretion is a common posttranslational event. Since attachment of oligosaccharides occurs prior to segregation to the secretory granule, this modification is a prime candidate for participation in sorting and targeting events. It has been known since its discovery that POMC is a glycoprotein (Eipper and Mains, 1980). POMC is both O- and N-glycosylated at threonine 45 and asparagine 65, respectively. An additional N-glycosylation site is present within the carboxyl-terminal region of the ACTH domain of rat and mouse POMC. The biosynthetic origins of O- and N-glycans are quite distinct (Abeijon and Hirschberg, 1992). N-linked glycosylation is initiated by the transfer en bloc of a preformed oligosaccharide from a dolichal lipid carrier to selected asparagine residues. Linkage to asparagine is via N-acetylglucosamine. This process is a cotranslational event occurring within the lumen of the ER. The initial high mannose structure is subject to a series of trimming and capping events as the maturing glycoprotein traverses the ER/Golgi compartment. The resultant N-linked oligosaccharide can be of the high mannose, complex, or hybrid type depending on the nature of the capping sugars added. In contrast, O-linked oligosaccharides are assembled within the Golgi compartment by the sequential addition of sugars following the initial transfer of *N*-acetylgalactosamine to selected threonine or serine residues. No lipid carrier is involved in this process. The maturation of both types of glycan structure is catalyzed by a series of sugar transferases located at discrete loci within the Golgi compartment. Structural cues within the amino acid sequence of glycoproteins determine which amino acids become modified. N-glycans are attached to the amide nitrogen of asparagine residues found within the linear sequence -Asn-X-Thr/Ser-. The structures that promote O-glycosylation are less well understood. However, O'Connell et al. (1992) have identified the presence of multiple proline residues close to the site of O-glycosylation as an important determinant in this process. This implies that a disordered structure surrounding the glycosylation site is critical. O'Connell et al. (1991) also provided experimental evidence that a polar residue at position -1 relative to the glycosylation site prevented modification. The N- and O-glycosylation sites within POMC conform to these structural requirements.

Previous studies have shown that the N- and O-glycans within POMC are sulfated (Hoshina et al., 1982; Bourbonnais and Crine, 1985; Bennett, 1986). We have recently completed a detailed mass spectrometric analysis of both the N- and O-linked oligosaccharides of the 16K fragment of bovine POMC (Siciliano et al., 1993, 1994). Studies of the N-glycan at asparagine 65 indicated that sulfated *N*-acetylgalactosamine capped antennae were present. Similar structures have previously been identified in the pituitary glycoprotein hormones (Parsons and Pierce, 1980). However, the POMC oligosaccharide differs from that of the glycoprotein hormones because the sulfate is exclusively located on the 3-arm of the biantennary structure. The terminal *N*-acetylglucosamine is fully fucosylated. A substantial proportion of the N-glycan also bore fucose on the 6-arm linked to *N*-acetyl-glucosamine. Structural analysis of the O-glycan linked to threonine 45 indicated that sulfated *N*-acetylgalactosamine capped antennae were also present looking much like the structure found on the 3-arm of the N-linked structure. This raises the possibility that the same *N*-acetylgalactosamine transferase is involved in both N- and O-glycan biosynthesis. The sulfotransferase is a Golgi enzyme that is specific for oligosaccharide structures terminating in *N*-acetylgalactosamine. The enzyme will not sulfate oligosaccharide structures terminating in galactose (Skelton et al., 1991). The N-linked oligosaccharide chains of pituitary glycoprotein hormones lutropin and thyrotropin are modified in this way (Parsons and Pierce, 1980; Green and Baenziger, 1988). Thus addition of N-acetylgalactosamine is critical in determining whether a given glycan structure will be sulfated. It has been proposed that a primary amino acid sequence is recognized by the N-acetylgalactosamine transferase in question. It has been suggested that a tripeptide -Pro-X-Arg/Lys- positioned six to nine residues to the amino-terminus of the glycosylation site constitutes the structural cue for the addition of terminal N-acetylgalactosamine (Smith and Baenziger, 1992). The N-glycosylation site within the corticotropin segment of rat and mouse POMC is located close to such a sequence and has been shown to bear sulfated N-acetylgalactosamine (Skelton et al., 1992). The amino acid sequences adjacent to the N- and O-glycans found toward the amino-terminal of bovine POMC do not contain the putative sugar-transfer signal sequence. Whether these sulfated structures are generated in the same manner remains to be determined. A liver endothelial receptor has recently been discovered that is specific for sulfated N-acetylgalactosamine. Circulating sulfated glycoproteins have shortened half-lives resulting from uptake by the liver through a receptor-mediated endocytic pathway (Fiete et al., 1991). Paradoxically, the N-terminal fragment bearing sulfated N- and O-glycans has been shown to have a relatively prolonged circulating half-life (Lu et al., 1983). It can be concluded that these sulfated structures may not be recognized by the liver receptor.

Chromogranin A, a member of the granin family of proteins found copackaged with hormones and neuropeptides, has been shown to have sulfated O-glycan structures (Rosa et al., 1985). The novel sulfated O-glycan found at threonine 45 of POMC may be a prototypic structure possibly to be found in chromogranin A and other proteins and peptides destined for regulated secretion. Interestingly, variants of peptidylglycine α-amidating mono-oxygenase (PAM) resulting from alternative mRNA splicing have been shown to harbor both sulfated O-glycans and tyrosine (Yun and Eipper, 1993).

C. Phosphorylation and Sulfation

These posttranslational modifications like glycosylation are confined to the ER/Golgi compartment. These events are dictated by structural cues that are now quite well understood. POMC from various species contain such amino acid sequences. Phosphorylation is catalyzed by physiological casein kinase thought to be confined to the Golgi apparatus, which utilizes ATP as cosubstrate (Capasso et al., 1989). The enzyme recognizes the serine and threonine residues found within the sequence -Ser/Thr-X-Glu-. This posttranslational modification is typified by the phosphorylation of the serine residue 31 found in the carboxyl-terminal region of human corticotropin ACTH (Bennett et al., 1983). The extent of phosphorylation of ACTH varies from tissue to tissue. While ACTH is partially phosphorylated in the pituitary, modification of corticotropin-like intermediate lobe (CLIP), a biosynthetic derivative of ACTH, is almost complete within the rat hypothalamus (Emeson and Eipper, 1986). Predicted phosphorylation sites are found within many prohormone sequences. The function of this posttranslational event is not understood. The action of the physiological casein kinase found in the secretory pathway may be fortuitously reflected in the modification of POMC. Its major role may be in the modification and regulation of the activity of as yet uncharacterized proteins or proteins found in the secretory pathway.

Sulfation of selected tyrosine residues is a common posttranslational modification of many secretory proteins (Huttner, 1988). This modification is also prompted by protein structure. The sulfation of tyrosine residue 28 within the β-lipotropin (βLPH) segment of bovine POMC provides an example of the environment required for tyrosine sulfation (Bateman et al., 1990). An acidic residue, usually glutamic acid, must be present immediately to the amino-terminus of the tyrosine in question. Several other acidic residues must also be present close to the site of sulfation as well as an absence of modifications that might interfere with the action of the sulfotransferase enzyme. Tyrosylprotein sulfotransferase activity has been located in the Golgi fraction that catalyzes the transfer of sulfate from 3'-phosphoadenosine 5'-phosphosulfate (PAPS) to secretory proteins (Lee and Huttner, 1983). The pH optimum of this enzyme was found to be approximately 6 and enzymatic activity could be stimulated with Mg^{2+} and Mn^{2+}. Modification of tyrosine residues has been localized to the same compartment as terminal glycosylation events (i.e., galactosylation and sialylation). Thus tyrosine sulfation occurs in the trans-Golgi compartment and is one of the last posttranslational events to occur prior to secretory granule formation (Baeuerle and Huttner, 1987).

A role for tyrosine sulfation has been suggested in constitutive protein secretion. Tyrosine sulfation has been shown to facilitate the transit of a secretory protein (a yolk protein of *Drosophila melanogaster*) from the trans-Golgi to the cell surface (Friederich et al., 1988). Members of the granin family of proteins found in secretory granules of many endocrine and neuroendocrine residues have been shown to be sulfated on tyrosine (Rosa et al., 1985). Variants of PAM have also been found to be tyrosine-sulfated (Yun and Eipper, 1993). Tyrosine sulfation of cholecystokinin has been shown to be essential for the full expression of the biological activity of this gut hormone (Jensen et al., 1982).

D. Prohormone Cleavage

With the cloning, sequencing, and expression of several candidate prohormone cleavage enzymes or convertases, many aspects of peptide biosynthesis have been clarified (Barr, 1991; Lindberg, 1991; Steiner et al., 1992; Seidah and Chrétien, 1992; Smeekens, 1993). Two groups of researchers have independently identified two convertases that show specificity for pairs of basic residues consistent with their probable role in hormone maturation. Both Smeekens and Steiner (1990, 1991) and Seidah et al. (1990, 1991) employed similar polymerase chain reaction (PCR) strategies to facilitate the cloning and sequencing of two candidate convertases. Extensive homology was noted between the two convertases (PC-1, also known as PC-3, and PC-2) and other members of the subtilisin family of serine proteases. They are proteases with neutral pH optima and are dependent on calcium for full activity. The convertase family includes the KEX-2 gene product (kexin) proven to be involved in α-factor biosynthesis in yeast (Fuller et al., 1989) and furin, a Golgi-localized endoprotease expressed in many mammalian cells, which has been implicated in the processing of several constitutively secreted polypeptides and proteins (Van de Ven et al., 1990). Kexin and furin are membrane-bound proteases bearing membrane-spanning sequences. These convertase enzymes act mainly within the constitutive pathway of secretion. Localization of PC-1/3 and PC-2 by in situ hybridization to distinct regions of the brain and pituitary has strongly implicated these proteases as important participants in prohormone processing (Day et al., 1992; Schafer et al., 1993). The distribution of the two convertases within the mouse pituitary may provide the biochemical basis for tissue-specific processing of POMC. Using Northern analysis technique PC-1/3 but not PC-2, mRNA has been localized to the anterior pituitary while mRNA for both PC-1/3 and PC-2 has been localized to the intermediate lobe (Seidah et al., 1991b). Expression of the two convertases in a variety of cell lines together with POMC has revealed much about their substrate specificity (Benjannet et al., 1991; Thomas et al., 1991). Proinsulin processing can also be accounted for in terms of the actions of PC-1/3 and PC-2 (Smeekens et al., 1992; Bennett et al., 1992; Rhodes et al., 1992).

In many of these studies, convertase and substrate have been coexpressed within model cell lines using the vaccinia virus system. PC-1/3 has been shown to be active with respect to POMC processing when coexpressed in BSC-40 cells, a constitutively secreting cell line lacking secretory granules (Benjannet et al., 1991; Thomas et al., 1991). Cleavage of the precursor occurred at -Lys-Arg- and -Arg-Arg- sequences to yield products typical of anterior lobe processing (i.e., ACTH, β-lipotropin, and β-endorphin). When PC-2 was expressed with POMC in BSC-40, further processing of the ACTH domain to give $ACTH_{1-13}$ amide was evident. This particular cleavage event can be considered diagnostic for pituitary intermediate lobe processing. The cumulative evidence indicates that both PCs cleave predominantly to the carboxyl-terminal side of the pairs of basic residues (Barr, 1991). These amino acids are subsequently removed by the action of a carboxypeptidase (see Sec. II.E.1). A critical role for PC-1 in POMC processing was evident when the expression of this convertase enzyme was suppressed in the mouse anterior pituitary AtT-20 cell line using antisense RNA technology (Bloomquist et al., 1991). Suppression of PC-1 activity in this corticotroph cell line significantly inhibited POMC processing, resulting in the accumulation of uncleaved precursor. A critical series of experiments has recently been undertaken to determine whether PC-1/3 and PC-2 can account for all the cleavage events typical of intermediate lobe processing (Zhou et al., 1993). Mutant strains of AtT-20 cells were developed that stably expressed the PC-2 convertase. The mutant cells acquired the ability to undertake all the additional cleavages typical of the intermediate pituitary. This included the generation of Lys^1γ_3MSH, a product generated through cleavage of the 16K or amino-terminal fragment of POMC as well as carboxyl-terminally truncated forms of β-endorphin. These latter cleavage events are known to occur late in the biosynthetic processing of POMC. Thus the products of POMC processing and the kinetics of their generation that are typical of the intermediate lobe are duplicated in AtT-20 cells expressing PC-2 as well as the endogenous PC-1/3 convertase. These experiments and those from other laboratories suggest that the combined action of PC-1/3 and PC-2 will account for most of the cleavage events that give rise to biologically active peptide hormones and neuropeptides. It will be critical to determine which convertase is responsible for each cleavage event since there are clearly some cleavages that can be catalyzed by both convertases. Converting enzymes must be demonstrated within the appropriate cellular locus to qualify as a candidate in each processing event. The availability of sequence specific antisera will facilitate the intracellular localization of PC-1/3 and PC-2.

E. Secretory Granule Events

There has always been some ambiguity concerning where prohormone cleavage begins. In view of the fact that precursor derivatives usually become packaged together in an efficient stoichiometric fashion, it has been assumed that both cleavage and subsequent processing events occur almost exclusively within the secretory granule fraction (Gainer et al., 1985). However, there has been clear evidence that processing of the parathyroid hormone precursor is initiated in the Golgi apparatus (Habener, 1979). More recently, it has become clear that not only do cleavage events occur in the Golgi system, but also trimming of basic residues and amidation can occur in this organelle (Schnabel et al., 1989). However, for the purposes of this review, the processes traditionally associated with the secretory granule will be briefly discussed together. A comprehensive review of these processing activities can be found elsewhere (Mains et al., 1990).

1. Carboxypeptidase H (CPH)

Following the endoproteotytic action of the convertases, basic residues must be removed from the carboxyl-termini of the peptide products. A carboxypeptidase-B-like enzyme has been purified from several endocrine tissues and shown to be approximately 50 kDa with a pH optimum of between 5.5 and 6 (Fricker, 1988). It is a zinc metallopeptidase found in both soluble and membrane-associated forms. cDNAs encoding carboxypeptidase H have been cloned, and the predicted protein sequence reveals no classical transmembrane sequence (Fricker et al., 1986). Both Northern blot analysis of carboxypeptidase H mRNA and enzymatic assays for the enzyme itself show a wide distribution of this exopeptidase among endocrine and neuroendocrine cells (Fricker et al., 1986; Fricker, 1988).

2. Peptidylglycine-α-Amidating Mono-oxygenase (PAM)

Many peptide hormones and neuropeptides have amidated carboxyl-termini. It has been appreciated for some time that when peptides are generated through the concerted actions of convertase and carboxypeptidase H enzymes to reveal glycine-extended termini, they almost without exception become amidated (Bradbury and Smyth, 1987). Glycine-extended structures are known to give rise to α-amidated carboxyl-termini in a two-step process involving sequential actions of a peptidyl-glycine α-hydroxylating mono-oxygenase followed by a peptidyl-α-hydroxyglycine α-amidating lyase (Katopodis et al., 1990; Eipper et al., 1991). The mono-oxygenase activity is a copper-dependent enzyme with a pH optimum of about 6.5 (Eipper et al., 1992). Both enzymatic activities are encoded within a common precursor bearing a transmembrane domain at its carboxyl-terminus. Alternative mRNA splicing yields soluble forms of the hydroxylating mono-oxygenase and the lyase activities (Eipper et al., 1992).

III. AGGREGATION AND SORTING

Current evidence from gene transfer experiments tell us that the intracellular fate of precursors to propeptide hormones and neuropeptides is dictated by their structures (Thomas and Thorne, 1988). Many studies have provided clear evidence of prohormone structures promoting sorting to the regulated pathway of secretion. For instance, the proregion of the somatostatin precursor has been shown to be responsible for efficient segregation of this prohormone (Sevarino et al., 1989; Stoller and Shields, 1989). The expression of a chimeric fusion protein consisting of the proregion of the somatostatin precursor and α-globin was studied in the endocrine GH_4 cell line. Despite the fact that α-globin is normally constitutively secreted, the chimeric protein was efficiently segregated to the regulated pathway. Thus the proregion of the somatostatin precursor contains information that redirects -globin from the constitutive pathway to the secretory granule fraction (Stoller and Shields, 1989). The study of the biosynthetic fate of egg-laying hormone (ELH) has revealed that sorting to the regulated pathway follows initial cleavage of the precursor (Fisher and Scheller, 1988). Pro-ELH is processed in the trans-Golgi fraction of the bag cells of the mollusc *Aplysia* into amino- and carboxyl-terminal fragments. Only the carboxyl-terminal fragment bearing the ELH sequence is segregated to the dense core granule fraction for subsequent processing (Fisher et al., 1988). Thus structural features absent from the carboxyl-terminal fragment but present in the amino-terminal fragment are implicated as bringing about efficient packaging of ELH within the secretory granule fraction. Evidence for sorting following processing was also found when the biosynthetic fate of POMC in AtT-20 cells was investigated using an electron microscopic immunogold labeling technique (Schnabel et al., 1989). Both ACTH and the joining peptide (JP) fragment were found in the trans-Golgi fraction, indicating that processing has already begun in this organelle prior to sorting. ACTH and JP together with other products of POMC processing are efficiently cosegregated to the granule fraction. Thus the sorting and segregation signal, if it exists, must be present in both ACTH and JP. The nature of this redundant signal has yet to be determined.

The characteristics of acidic proteins, the granins, found to be copackaged within the secretory granule fraction of almost all endocrine and neuroendocrine cells may hold the key to the sorting process. The granins are a family of acidic secretory proteins of which chromogranin A is the best characterized and most frequently studied (Scammell, 1993). First found in the chromaffin cells of the adrenal medulla, chromogranin A was subsequently found to be identical to the parathyroid secretory protein. Like other members of the granin family (i.e., chromogranin B and secretogranin II), the chromogranin has an extremely high proportion of acidic amino acids resulting in a low pI of about 5. Under conditions prevailing within the trans-Golgi fraction (i.e., acidic pH and elevated calcium concentrations), chromogranin A will form aggregates. High calcium (1–10 mM) and low pH (5–6) causes chromogranin A to aggregate with high efficiency, whereas constitutively secreted proteins do not aggregate under these conditions (Gorr et al., 1989; Gerdes et al., 1989; Chanet and Huttner, 1991; Thompson et al., 1992). This process is reversible if the calcium concentration is lowered or if the pH is raised (Yoo and Albanesi, 1990; Yoo and Lewis, 1992). It has been suggested that this process mimics events taking place in the trans-Golgi fraction. Indeed, parathyroid hormone has been shown to coprecipitate with chromogranin A in vitro in a process that may simulate granule formation (Gorr et al., 1988). Whether such a universal mechanism explains how proteins of the regulated pathway are efficiently segregated from constitutively secreted proteins remains to be determined.

We have recently shown that an isomerization reaction occurs within the JP sequence of mouse and porcine POMC (Toney et al., 1993). At -aspartyl-glycine-sequences, a succinimide structure forms resulting from the nucleophilic attack of the amide nitrogen of the glycine residue on the β-carboxyl group of the aspartic acid residue. The resulting succinimide structure was found upon hydrolysis to partially give rise to the original -Asp-Gly- sequence and an isomerized peptide in which the aspartic acid is linked to the glycine through the β-carboxyl group. This posttranslational event may be initiated in the trans-Golgi fraction with a time course similar to that of granule formation. This phenomenon has also been shown to occur at -Asp-Gly- within human JP and the amino-terminal fragment of β-lipotropin isolated from rat and mouse pituitaries. We are currently investigating whether succinimide formation within JP is occurring as a consequence of aggregation to chromogranin A under conditions of lower pH and high calcium.

IV. SECRETION

The transport of proteins through the secretory pathway is a vectoral process. This suggests that each step along the way is mediated by specific targeting events (Bennett and Scheller, 1993). These processes are most likely mediated by proteins localized to transport vesicles. The ultimate step in this process is the fusion of secretory vesicles with the cell membrane followed by exocytosis. The critical role that calcium influx plays in this process has long been appreciated. While extracellular calcium concentrations are in the millimolar range, intracellular concentrations are maintained at nanomolar levels. This is achieved through sequestration of calcium bound to proteins within intracellular organelles including the ER. Upon stimulation of secretory cells, voltage-gated channels are opened and calcium flows into the cell across a very large concentration gradient. The manner in which elevated calcium concentrations mediate membrane fusion and exocytosis is poorly understood. A number of vesicular and cell membrane proteins implicated in calcium-mediated fusion have been characterized recently. One such protein is the synaptic vesicle protein synaptagmin (Perin et al., 1990). This protein has been implicated in the calcium-dependent interaction with membrane phospholipids. In this sense synaptagmin may act as a calcium sensor and form a critical link between stimulation of calcium influx and membrane fusion. α-Latrotoxin, a component of the venom of the black widow spider, will promote calcium-independent fusion of synaptic residues with presynaptic plasma membranes (Petrenko et al., 1991). α-Latrotoxin binds to a cell surface protein that is a member of the neurexin family of integral membrane proteins (Ushkaryov et al., 1992). It has been proposed that a neurexin brings about a change in conformation of synaptotagmin and other proteins that mimics that resulting from calcium influx. The formation of a fusion pore initiates the process of fusion of synaptic vesicles and the plasma membrane (White, 1992). The nature of this pore and the proteins involved have yet to be established.

We recently characterized a novel cystine-rich polypeptide purified from extracts of the skin of the amphibian *Xenopus laevis* (James et al., 1993). Using an isolated guinea pig villus enteroyte bioassay (MacLeod and Hamilton, 1990), we have demonstrated that this peptide has L-type calcium channel agonist activity (MacLeod, Nembessis, Bell, James, and Bennett, manuscript in preparation). The peptide promotes calcium influx, and this in turn activates separate chloride and potassium conductances leading to shrinkage of the villus cells, which is monitored electronically. Sequence analysis revealed that this cystine-rich polypeptide had the following primary structure: LKCVNLQANGIKMTQECAKEDTKCLT-

LRSLKKTLKFCASGRTCTTMKIMSLPEGQITCCGENMCNA. The structure revealed extensive homology with the neurotoxin/cytotoxin family of peptides found in the venom of poisonous snakes. Many biologically active peptides found in the skin of amphibians have structural homologs in higher vertebrates. This new class of calcium channel agonist may prove useful in characterizing calcium-dependent events including those occurring during peptide hormone secretion. We are currently investigating whether this new cystine-rich peptide represents a prototypic form of an as yet undiscovered mammalian peptide.

ACKNOWLEDGMENTS

The author thanks the Medical Research Council of Canada for financial assistance and Lenora Naimark for preparing the manuscript.

REFERENCES

Abeijon, C., and Hirschberg, C.B. (1992). Topography of glycosylation reactions in the endoplasmic reticulum. *Trends Biochem. Sci. 17*:32.

Baeuerle, P.A., and Huttner, W.B. (1987). Tyrosine sulfation is a trans-Golgi specific protein modification. *J. Cell Biol. 105*:2655.

Barr, P.J. (1991). Mammalian subtilisins: the long-sought dibasic processing endoproteases. *Cell 66*:1.

Bateman, A., Solomon, S., and Bennett, H.P.J. (1990). Post-translational modification of bovine pro-opiomelanocortin: tyrosine sulfation and pyroglutamate formation, a mass spectrometric study. *J. Biol. Chem. 265*:22130.

Benjannet, S., Rondeau, N., Day, R., Chrétien, M., and Seidah, N.G. (1991). PC1 and PC2 are pro-protein convertases capable of cleaving POMC at distinct pairs of basic residues. *Proc. Natl. Acad. Sci. USA 88*:3564.

Bennett, H.P.J. (1984). Isolation and characterization of the 1 to 49 amino-terminal sequence of pro-opiomelanocortin from bovine posterior pituitaries. *Biochem. Biophys. Res. Commun. 125*:229.

Bennett, H.P.J. (1985). Peptide hormone biosynthesis—recent developments. *Recent Results Cancer Res. 99*:34.

Bennett, H.P.J. (1986). Biosynthetic fate of the amino-terminal fragment of pro-opiomelanocortin within the intermediate lobe of the mouse pituitary. *Peptides 7*:615.

Bennett, M.K., and Scheller, R.H. (1993). The molecular machinery for secretion is conserved from yeast to neurons. *Proc. Natl. Acad. Sci. USA 90*:2559.

Bennett, D.L., Bailyes, E.M., Nielsen, E., Guest, P.C., Rutherford, N.G., Arden, S.D., and Hutton, J.C. (1992). Identification of the type 2 proinsulin processing endopeptidase as PC2, a member of the eukaryote subtilisin family. *J. Biol. Chem. 267*:15229.

Bennett, H.P.J., Brubaker, P.L., Seger, M.A., and Solomon, S. (1983). Human phosphoro-serine-31-corticotropin 1-39. Isolation and characterization. *J. Biol. Chem. 258*:8108.

Bloomquist, B.T., Eipper, B.A., and Mains, R.E. (1991). Prohormone-converting enzymes: regulation and evaluation of function using antisense RNA. *Mol. Endocrinol. 5*:2014.

Bourbonnais, Y., and Crine, P. (1985). Post-translational incorporation of (^{35}S) sulfate into oligosaccharide side chains of pro-opiomelanocortin in rat intermediate lobe cells. *J. Biol. Chem. 260*:5832.

Bradbury, A.F., and Smyth, D.G. (1987). Biosynthesis of the C-terminal amide in peptide hormones. *Biosci. Rep. 7*:907.

Brion, C., Miller, S.G., and Moore, H.-P.H. (1992). Regulated and constitutive secretion: differential effects of protein synthesis arrest on transport of glycosaminoglycan chains to the two secretory pathways. *J. Biol. Chem. 267*:1477.

Burgess, T.L., and Kelly, R.B. (1987). Constitutive and regulated secretion of proteins. *Annu. Rev. Cell Biol. 3*:243.

Capasso, J.M., Keenan, T.W., Abeijan, C., and Hirschberg, C.B. (1989). Mechanism of phosphorylation in the lumen of the Golgi apparatus. *J. Biol. Chem. 264*:5233.

Chanat, E., and Huttner, W.B. (1991). Milieu-induced, selective aggregation of regulated secretory proteins in the *trans*-Golgi network. *J. Cell Biol. 115*:1305.

Cochet, M., Chang, A.C.Y., and Cohen, S.N. (1982). Characterization of the structural gene and putative 5-regulatory sequences for human pro-opiomelanocortin. *Nature 297*:335.

Day, R., Schafer, M.K.-H., Watson, S.J., Chrétien, M., and Seidah, N.G. (1992). Distribution and regulation of the prohormone convertases PC1 and PC2 in the rat pituitary. *Mol. Endocrinol. 6*:485.

Eipper, B.A., and Mains, R.E. (1980). Structure and biosynthesis of proadrenocorticotropin/endorphin and related peptides. *Endocr. Rev. 1*:1.

Eipper, B.A., Perkins, S.N., Husten, E.J., Johnson, R.C., Keutmann, H.T., and Mains, R.E. (1991). Peptidyl-α-amidating lyase: purification, characterization, and expression. *J. Biol. Chem. 266*:7827.

Eipper, B.A., Stoffers, D.A., and Mains, R.E. (1992). The biosynthesis of neuropeptides: peptide α-amidation. *Annu. Rev. Neurosci. 15*:57.

Emeson, R.B., and Eipper, B.A. (1986). Characterization of pro-ACTH/endorphin-derived peptides in rat hypothalamus. *J. Neurosci. 6*:837.

Fiete, D., Srivastava, V., Hindsgaul, O., and Baenziger, J.U. (1991). A hepatic reticuloendothelial cell receptor specific for SO_4-4GalNAcβ1,2Manα that mediates rapid clearance of lutropin. *Cell 67*:1103.

Fisher, J.M., and Scheller, R.H. (1988). Prohormone processing and the secretory pathway. *J. Biol. Chem. 263*:16515.

Fisher, J.M., Sossin, W., Newcomb, R., and Scheller, R.H. (1988). Multiple neuropeptides derived from a common precursor are differentially packaged and transported. *Cell 54*:813.

Friederich, R., Fritz, H.-J., and Huttner, W.B. (1988). Inhibition of tyrosine sulfation in the trans-Golgi retards the transport of a constitutively secreted protein to the cell surface. *J. Cell Biol. 107*:1655.

Fuller, R.S., Brake, A., and Thorne, J. (1989). Yeast prohormone processing enzyme (Kex2 gene product) is a Ca^{2+}-dependent serine protease. *Proc. Natl. Acad. Sci. USA 86*:1434.

Gainer, H., Russell, J.T., and Loh, P. (1985). The enzymology and intracellular organization of peptide precursor processing: the secretory vesicle hypothesis. *Neuroendocrinology 40*:171.

Gerdes, H-H., Rosa, P., Phillips, E., Baeuerle, P.A., Frank, R., and Argos, P. (1989). The primary structure of human secretogranin II, a widespread tyrosine-sulfated secretory granule protein that exhibits low pH- and calcium-induced aggregation. *J. Biol. Chem.* *264*:12009.

Gorr, S.-U., Dean, W.L., Radley, T.S., and Cohn, D.V. (1988). Calcium-binding and aggregation properties of parathyroid secretory protein-1 (chromogranin A). *Bone Miner.* *4*:17.

Green, E.D., and Baenziger, J.U. (1988). Asparagine linked oligosaccharides on lutropin, fallitropin and thyrotropin. 1. Structural elucidation of the sulfated and sialated oligosaccharides on bovine, ovine and human pituitary glycoprotein hormones. *J. Biol. Chem.* *263*:25.

Habener, J.F., Amherdt, M., Ravazzola, M., and Orci, L. (1979). Parathyroid hormone biosynthesis. Correlation of conversion of biosynthetic precursors with intracellular migration as determined by electron microscope autoradiography. *J. Cell. Biol.* *80*:715.

Hawkins, H.C., Nardi, M., and Freedman, R.B. (1991). Redox properties and cross-linking of dithiol/disulphide active sites of mammalian protein disulphide isomerase. *Biochem. J.* *275*:341.

Hoshina, M., Hortin, G., and Boime, I. (1982). Rat pro-opiomelanocortin contains sulfate. *Science 217*:63.

Huttner, W.B. (1988). Tyrosine sulfation and the secretory pathway. *Annu. Rev. Physiol.* *50*:363.

Hwang, C., Sinskey, A.J., and Lodish, H.F. (1992). Oxidized redox state of glutathione in the endoplasmic reticulum. *Science 257*:1496.

James, S., Gibbs, B., Toney, K., and Bennett, H.P.J. Purification of antimicrobial peptides from an extract of the skin of *Xenopus laevis* using heparin-affinity HPLC: characterization by ion-spray mass spectrometry. *Anal. Biochem.* (submitted).

Jensen, R.T., Kemp, G.F., and Gardner, J.D. (1982). Interactions of COOH-terminal fragments of cholecystokinin with receptors on dispersed acini from guinea pig pancreas. *J. Biol. Chem.* *257*:5554.

Katopodis, A.G., Ping, D., and May, S.W. (1990). A novel enzyme from bovine neurointermediate pituitary catalyzes dealkylation of α-hydroxyglycine derivatives, thereby functioning sequentially with peptidylglycine α-amidating monooxygenase in peptide amidation. *Biochemistry 29*:6115.

Kelly, R.B. (1985). Pathways of protein secretion in eukaryotes. *Science 230*:25.

Lee, R.W.H., and Huttner, W.B. (1983). Tyrosine O-sulfated proteins of PC 12 pheochromocytoma cells and their sulfation by tyrosylprotein sulfotransferase. *J. Biol. Chem. 258*:11326.

Lindberg, I. (1991). The new eukaryotic precursor processing proteinases. *Mol. Endocrinol.* *5*:1361.

Lu, C.-L., Chan, J.S.D., de Lean, A., Chen, A., Seidah, N.G., and Chrétien, M. (1983). Metabolic clearance rate and half-time of disappearance of human N-terminal and adrenocorticotropin of pro-opiomelanocortin in the rat: a comparative study. *Life Sci.* *33*:2599.

MacLeod, R.J., and Hamilton, J.R. (1990). Regulatory volume increases in mammalian jejunal villus cells is due to bumetanide-sensitive NaKC12 cotransport. *Am. J. Physiol.* *258*:G665–674.

Mains, R.E., and Eipper, B.E. (1990). The tissue-specific processing of pro-ACTH/ endorphin: recent advances and unsolved problems. *Trends Endocrinol. Metab. 1*:388.

Mains, R.E., Nickerson, I.M., May, V., Stoffers, D.A., Perkins, S.N., Ouafik, L'H., Husten, E.J., and Eipper, B.A. (1990). Cellular and molecular aspects of peptide hormone biosynthesis. *Frontiers Neuroendocrinol. 11*:52.

Noel, G., Keutmann, H.T., and Mains, R.E. (1991). Investigation of the structural requirements for peptide precursor processing in AtT-20 cells using site-directed mutagenesis of proadrenocorticotropin/endorphin. *Mol. Endocrinol. 5*:404.

O'Connell, B.C., Hagen, F.K., and Talsa, L.A. (1992). The influence of flanking sequence on the O-glycosylation of threonine in vitro. *J. Biol. Chem. 267*:2506.

Parsons, T.F., and Pierce, J.G. (1990). Oligosaccharide moieties of glycoprotein hormones: bovine lutropin resists enzymatic deglycosylation because of O-sulfated *N*-acetylhexosamines. *Proc. Natl. Acad. Sci. USA 77*:7089.

Perin, M.S., Fried, V.A., Mignery, G.A., Jahn, R., and Sudhof, T.C. (1990). Phospholipid binding by a synaptic vesicle protein homologous to the regulatory region of protein kinase C. *Nature 345*:260.

Petrenko, A.G., Perin, M.S., Davletov, B.A., Ushkaryov, Y.A., Geppert, M., and Sudhof, T.C. (1991). Binding of synaptotagmin to the alpha-latrotoxin receptor implicated both in synaptic vesicle exocytosis. *Nature 353*:65.

Rhodes, C.J., Lincoln, B., and Shoelson, S.E. (1992). Preferential cleavage of des-31,32-proinsulin over intact proinsulin by the insulin secretory granule type II endopeptidase. *J. Biol. Chem. 267*:22719.

Robinson, P., Bateman, A., Mulay, S., Spencer, S.J., Jaffe, R.B., Solomon, S., and Bennett, H.P.J. (1991). Isolation and characterization of three forms of joining peptide from adult human pituitaries: lack of adrenal androgen-stimulating activity. *Endocrinology 129*:859.

Rosa, P., Hille, A., Lee, R.W.H., Zanini, A., DeCamilli, P., and Huttner, W.B. (1985). Secretogranins I and II: two tyrosine-sulfated secretory proteins common to a variety of cells secreting peptides by the regulated pathway. *J. Cell Biol. 101*:1977.

Rothman, J.E., and Orci, L. (1992). Molecular dissection of the secretory pathway. *Nature 355*:409.

Roy, P., Chevrier, D., Fournier, H., Racine, C., Zollinger, M., Crine, P., and Boileau, G. (1991). Investigation of a possible role of the amino-terminal pro-region of proopiomelanocortin in its processing and targeting to secretory granules. *Mol. Cell. Endocrinol. 82*:237.

Scammell, J.G. (1993). Granins. Markers of the regulated secretory pathway. *Trends Endocrinol. Metabol. 4*:14.

Schnabel, E., Mains, R.E., and Farquhar, M.G. (1989). Proteolytic processing of pro-ACTH/endorphin begins in the Golgi complex. *Mol. Endocrinol. 3*:1223.

Schafer, M.K.-H., Day, R., Cullinan, W.E., Chrétien, M., Seidah, N.G., and Watson, S.J. (1993). Gene expression of prohormone and proprotein convertases in the rat CNS: A comparative in situ hybridization analysis. *J. Neurosci. 13*:1258.

Seidah, N.G., and Chrétien, M. (1992). Proprotein and pro-hormone convertases of the subtilisin family: recent developments and future perspectives. *Trends Endocrinol. Metab. 3*:133.

Seidah, N.G., Gaspar, L., Mion, P., Marcinkiewicz, M., Mbikay, M., and Chrétien, M.

(1990). cDNA sequence of two distinct pituitary proteins homologous to Kex2 and furin gene products: tissue specific mRNAs encoding candidates for pro-hormone processing proteinases. *DNA Cell Biol. 9*:415.

Seidah, N.G., Marcinkiewicz, M., Benjannet, S., Gaspar, L., Beaubien, G., Mattei, M.G., Lazure, C., Mbikay, M., and Chrétien, M. (1991). Cloning and primary sequence of a mouse candidate prohormone convertase PC1 homologous to PC2, furin, and Kex2: distinct chromosomal localization and messenger RNA distribution in brain and pituitary compared to PC2. *Mol. Endocrinol. 5*:111.

Sevarino, K.A., Stork, P., Ventimiglia, R., Mandel, G., and Goodman, R.H. (1989). Amino-terminal sequences of prosomatostatin direct intracellular targeting but not processing specificity. *Cell 57*:11.

Siciliano, R.A., Morris, H.R., Bennett, H.P.J., and Dell, A. (1994). O-glycosylation mimics N-glycosylation in the 16K fragment of bovine pro-opiomelanocortin: the major O-glycan attached to Thr-45 carries SO$_4$-4GalNAcβ1-4GlcNAcβ1- which is the archetypal nonreducing epitope in the N-glycans of pituitary glycohormones. *J. Biol. Chem.* (in press).

Siciliano, R.A., Morris, H.R., McDowell, R.A., Azadi, P., Rogers, M.E., Bennett, H.P.J., and Dell, A. (1993). The Lewis x epitope is a major non-reducing structure in the sulphated N-glycans attached to Asn-65 of bovine pro-opiomelanocortin. *Glycobiology 3*:225.

Skelton, T.P., Hooper, L.V., Srivastava, V., Hindsgaul, O., and Baenziger, J.U. (1991). Characterization of a sulfotransferase responsible for the 4-O-sulfation of terminal β-*N*-acetyl-D-galactosamine on the asparagine-linked oligosaccharides of glycoprotein hormones. *J. Biol. Chem. 266*:17142.

Skelton, T.P., Kumar, S., Smith, P.L., Beranek, M.C., and Baenziger, J.U. (1992). Proopiomelanocortin synthesized by corticotrophs bears asparagine-linked oligosaccharides terminating with SO$_4$-4GalNAcβ1,4GlcNAcβ1,2Manα. *J. Biol. Chem. 265*:12998.

Smeekens, S.P. (1993). Processing of protein precursors by a novel family of subtilisin-related mammalian endoproteases. *Biotechnology 11*:182.

Smeekens, S.P., and Steiner, D.F. (1990). Identification of a human insulinoma cDNA encoding a novel mammalian protein structurally related to the yeast dibasic processing protease Kex2. *J. Biol. Chem. 265*:2997.

Smeekens, S.P., Avruch, A.S., LaMendola, J., Chan, S.J., and Steiner, D.F. (1991). Identification of a cDNA encoding a second putative prohormone convertase related to PC2 in AtT20 cells and islets of Langerham. *Proc. Natl. Acad. Sci. USA 88*:340.

Smeekens, S.P., Montag, A.G., Thomas, G., Albiges-Rizo, C., Carroll, R., Benig, M., Phillips, L.A., Martin, S., Ohagi, S., Gardner, P., Swift, H.H., and Steiner, D.F. (1992). Proinsulin processing by the subtilisin-related proprotein convertases furin, PC2, and PC3. *Proc. Natl. Acad. Sci. USA 89*:8822.

Smith, P.L., and Baenziger, J.U. (1992). Molecular basis of recognition by the glycoprotein hormone-specific *N*-acetylgalactosamine-transferase. *Proc. Natl. Acad. Sci. USA 89*:329.

Steiner, D.F., Smeekens, S.P., Ohagi, S., and Chan, S.J. (1992). The new enzymology of precursor processing endoproteases. *J. Biol. Chem. 267*:23435.

Stoller, T.J., and Shields, D. (1989). The propeptide of preprosomatostatin mediates intracellular transport and secretion of α-globin from mammalian cells. *J. Cell Biol.* *108*:1647.

Thomas, G., and Thorne, B.A. (1988). Gene transfer techniques to study neuropeptide processing. *Annu. Rev. Physiol. 50*:323.

Thomas, L., Leduc, R., Thorne, B.A., Smeekens, S.P., Steiner, D.F., and Thomas, G. (1991). Kex2-like endoproteases PC2 and PC3 accurately cleave a model prohormone in mammalian cells: evidence for a common core of neuroendocrine processing enzymes. *Proc. Natl. Acad. Sci. USA 88*:5297.

Thompson, M.E., Zimmer, W.E., Haynes, A.L., Valentine, D.L., Forss-Petter, S., and Scammell, J.G. (1992). Prolactin granulogenesis is associated with increased secretogranin expression and aggregation in the golgi apparatus of GH_4C_1 cells. *Endocrinology 131*:318.

Toney, K., Bateman, A., Gagnon, C., and Bennett, H.P.J. (1993). Aspartimide formation in the joining peptide sequence of porcine and mouse pro-opiomelanocortin. *J. Biol. Chem. 268*:1024.

Ushkaryov, Y.A., Petrenko, A.G., Geppert, M., and Sudhof, T.C. (1992). Neurexins: synaptic cell surface proteins related to the α-latrotoxin receptor and laminin. *Science 257*:50.

Van de Ven, W.J.M., Voorberg, J., and Fontijn, R. (1990). Furin is a subtilisin-like proprotein processing enzyme in higher eukaryotes. *Mol. Biol. Rep. 14*:265.

White, J.M. (1992). Membrane fusion. *Science 258*:917.

Yoo, S.H., and Albanesi, J.P. (1990). Ca^{2+}-induced conformational change and aggregation of chromogranin A. *J. Biol. Chem. 265*:14414.

Yoo, S.H., and Lewis, M.S. (1992). Effects of pH and Ca^{2+} on monomer-dimer and monomer-tetramer equilibria of chromogranin A. *J. Biol. Chem. 267*:11236.

Yun, H-Y., and Eipper, B.A. (1993). Sulfation of tyrosine or oligosaccharide governed by alternative splicing of peptidylglycine α-amidating mono-oxygenase (PAM) gene. *FASEB Meeting Abs. 742*, San Diego, May 30–June 3, 1993.

Zhou, A., Bloomquist, B.T., and Mains, R.E. (1993). The prohormone convertases PC1 and PC2 mediate distinct endoproteolytic cleavages in a strict temporal order during proopiomelanocortin biosynthetic processing. *J. Biol. Chem. 268*:1763.

DISCUSSION

Erdös: What is the importance of glycosylation outside the cell?

Bennett: A number of liver receptor-mediated endocytotic pathways have been characterized. These pathways can be critical in determining the circulating half-life of glycoproteins. A liver hepatocyte sulfo-galactosamine receptor has recently been characterized that has been implicated in the clearance of luteinizing hormone and has been shown to bear sulfated N-glycans.

Polak: Since new proteases and amidating enzymes are being discovered and characterized, it may soon be possible to visualize intracellular events of cleavage and packaging of neuropeptides using, for instance, double-fold electron immuno-cytochemistry and specific corresponding antibody.

Bennett: To be implicated in a given biosynthetic event, processing enzymes must be shown to be localized to the appropriate intracellular organelle. This information has been lacking in some instances in the past. This situation will no doubt be rectified with the availability of sequence-specific antibodies for use in immunolocalization studies.

Elde: Your new peptide with agonist activity at calcium channels is very interesting. Most toxins characterized to date have *antagonist* activity at ion channels. Do you know of toxins in addition to your new *Xenopus* skin peptide that are *agonist* at ion channels?

Bennett: Russell's viper venom has been shown to promote the release of insulin from isolated rat islets of Langerhans in vitro (Jones and Mann, *J. Endocrinol. 136*:27, 1993). This has been shown to be a calcium-dependent process. The active component in the crude venom has not been identified, however.

6

Vasoactive Intestinal Polypeptide in the Respiratory Tract

Sami I. Said

State University of New York at Stony Brook, Stony Brook, and Northport VA Medical Center, Northport, New York

I. DISCOVERY IN THE LUNG

Though first isolated from small intestine, vasoactive intestinal polypeptide (VIP) had earlier been discovered in the lung as a vasodilator peptide (Said, 1967; Said and Mutt, 1969). Now recognized as a neuropeptide with wide distribution, VIP is a likely neurotransmitter or neuromodulator. In the lung, VIP influences all aspects of airway function. Through its deficiency, VIP may be related to the pathogenesis of bronchial asthma, and because of its anti-inflammatory and tracheobronchial relaxant properties, it offers new promise in the management of several lung disorders.

II. LOCALIZATION AND DISTRIBUTION

VIP-containing neurons are localized in the tracheobronchial smooth muscle layer, around submucosal mucous and serous glands, in the lamina propria, and in the walls of pulmonary and bronchial arteries (Dey et al., 1981). Neuronal cell bodies of airway microganglia are strongly immunoreactive to VIP and provide an intrinsic source of innervation of pulmonary structures with the peptide (Dey et al., 1981). VIP-containing nerves are undiminished in denervated human lungs removed for transplantation (Springall et al., 1990). VIP-like immunoreactivity is also present in mast cells of rat lungs and is released from rat peritoneal cells by mast-cell degranulators (Cutz et al., 1978). Several VIP-like peptides have been chemically characterized in rat basophilic leukemia cells, which resemble mast cells (Goetzl et al., 1988). VIP is also present in eosinophils (Aliakbari et al., 1987) and may be present in sensory nerves in the lung, as suggested by its release, together with substance P, by capsaicin (Pakbaz et al., 1993).

A. Coexistence with Other Neurotransmitters and Neuropeptides

Numerous reports document the colocalization of VIP with other neuropeptides or neurotransmitters (Furness et al., 1992; Table 1). These neuropeptides include peptide histidine methionine (in human tissues) or peptide histidine isoleucine (in other mammalian species), opioid peptides, galanin, substance P, calcitonin gene-related peptide, and neuropeptide Y. Among the nonpeptide neurotransmitters that may coexist with VIP in the same neurons are acetylcholine, norepinephrine, and nitric oxide, the latter recognized by the presence of nitric oxide synthase, the enzyme that catalyzes its formation from L-arginine (Dey et al., 1993; Fig. 1). These colocalizations facilitate multiple physiological interactions, including: (a) the innervation of exocrine glands with acetylcholine, and VIP ensures greater secretion and blood flow responses than if the two transmitters were present alone (Lundberg et al., 1980); (b) VIP modulates cholinergic transmission in respiratory airways (Ellis and Farmer, 1989; Martin et al., 1990); (c) the copresence of VIP and substance P in cholinergic nerves to the airways (Dey et al., 1988) provides a balance between the bronchoconstrictor, proinflammatory substance P and the bronchial-relaxant, anti-inflammatory VIP; and (d) VIP and nitric oxide act cooperatively to bring about smooth muscle relaxation in blood vessels, respiratory airways, and other tissues (Said et al., 1992; Grider et al., 1992).

Table 1 Examples of Coexistence Between VIP and Other Neuropeptides and Neurotransmitters: Sites of Coexistence

Peptide/transmitter	Sites of coexistence
PHM/PHI	Cosynthesized and co-released with VIP in human (PHM) and other mammalian species (PHI)
Acetylcholine	Most VIP-containing neurons are also cholinergic
Neuropeptide Y	Sympathetic ganglia
Opioid peptides	Lung, intestine, adrenal chromaffin cells
CGRP, substance P	Lung, sweat glands
Nitric oxide	Airway ganglia, enteric neurons

CGRP, calcitonin gene-related peptide; PHI, peptide histidine isoleucineamide; PHM, peptide histidine methionine.

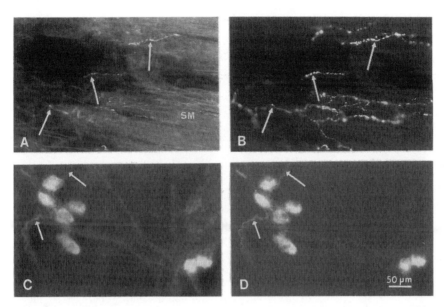

Figure 1 Colocalization of immunoreactive VIP (B,D) and nitric oxide synthase (NOS; A,C) in ferret trachealis muscle and plexus. (A,B) Nerve fibers containing NOS (A) and VIP (B) are in close association with tracheal smooth muscle (SM). Arrows indicate examples of identical nerve varicosities containing both immunoreactivities. (C,D) Neuronal cell bodies in a whole mount of the superficial muscular plexus. Several neurons contain both NOS and VIP. Nerve fibers emanate from some nerve cell bodies (arrows). (Courtesy Dr. Richard D. Dey, West Virginia University Medical Center, Morgantown, WV.)

III. RECEPTORS AND SECOND MESSENGERS

Specific receptors for VIP have been identified in membrane preparations of normal mammalian and human lungs (Paul and Said, 1987; Patthi et al., 1988) and in human lung tumor cells, including small cell carcinoma cell lines (Luis and Said, 1990). The VIP receptor is coupled to the adenylyl cyclase–stimulating protein G_s.

The ability of VIP to increase tissue cyclic AMP content (Lazarus et al., 1986), and use of autoradiography (Leroux et al., 1984), have permitted the localization of the VIP receptors in the lung, in submucosal serous and mucous glands, tracheal and bronchial epithelial cells, bronchial smooth muscle, and alveolar cells. Upon binding to its receptors, the VIP-receptor complex is internalized by endocytosis (Muller et al., 1985). In addition to membrane receptors, nuclear receptors for VIP have been demonstrated in HT-29 human colonic adenocarcinoma cell lines.

VIP receptors in the lung have been characterized by a number of techniques, including photoaffinity labeling, cross-linking of [125]VIP to its receptor, and detergent solubilization (Paul and Said, 1987; Patthi et al., 1988). Two polypeptides (Mr 53 and 18 kd) that bind VIP were recently purified, respectively, from porcine liver (Couvineau et al., 1990) and guinea pig lung (Paul and Said, 1987). More recently, the nucleotide sequence and amino acid sequence of the VIP-receptor cDNA was reported (Ishihara et al., 1992). This receptor, isolated from a rat cDNA library, mediated VIP-induced cAMP accumulation and was expressed in brain and other tissues. The cloned receptor, with 459 amino acid residues and a calculated Mr of 52 kd, contains seven transmembrane segments and shows sequence similarity to the secretin, calcitonin, and parathyroid hormone receptors (Ishihara et al., 1992). An earlier report of the cloning and expression of the cDNA encoding the VIP receptor (Sreedharan et al., 1991) proved incorrect.

A. Cyclic AMP

Strong evidence points to adenylyl cyclase stimulation and cyclic AMP production as the dominant mode of action of VIP in most instances. This certainly applies to the relaxation of vascular and nonvascular smooth muscle (Shreeve et al., 1992), the stimulation of pancreatic exocrine (Robberecht et al., 1976) and intestinal secretion (Dupont et al., 1980), various endocrine actions (Gourdi et al., 1979; Birnbaum et al., 1980; Chik et al., 1988), and the modulation of T-cell functions (Ottaway, 1987). Studies on the gene expression of VIP (Gozes and Brenneman, 1989) suggest a two-way relationship between VIP and cyclic AMP: VIP biosynthesis is promoted by higher cyclic AMP levels, and thus the peptide is potentially capable of stimulating its own generation.

B. Calcium and Phosphoinositides

With few exceptions, the evidence for other second messengers is less compelling. In the superior cervical ganglion (Audigier et al., 1988) and adrenal chromaffin cells (Malhotra et al., 1988), relatively high (10^{-6} M) concentrations of VIP increased the breakdown of phosphoinositides to inositol phosphates, enhancing the intracellular mobilization of Ca^{2+}. The effect on adrenal medulla was linked to the induction of catecholamine secretion (Malhotra et al., 1988). In concentrations as low as 10^{-10} M, VIP increased intracellular $[Ca^{2+}]$ in at least some rat cortical astrocytes (Brenneman, D. E., personal communication) and, at higher concentrations, in cultured rat hippocampal neurons (Tatsuno et al., 1992). In the latter preparation, pituitary adenylyl cyclase-activating peptide (PACAP), a VIP-related peptide, exerted the same action, at greater potency (Tatsuno et al., 1992). The VIP-induced rise in intracellular $[Ca^{2+}]$ in astrocytes was correlated with increased secretory activity of these cells, including the production of trophic factors (Gressens et al., 1993). Similar, though more subtle, changes in intracellular $[Ca^{2+}]$ due to VIP have been described in prolactin-producing rat anterior pituitary cells (Sand et al., 1989).

Despite these observations, the significance of intracellular $[Ca^{2+}]$ and inositol phosphates in mediating the physiological or pharmacological actions of VIP in different cell types remains uncertain. Further, as with other agents, the actions of VIP on target cells are probably mediated by more than one signal transduction pathway. Possible interactions between these pathways have been described (Meisheri and Rüegg, 1985; Pfitzer et al., 1985; Szewczak et al., 1990; Saito et al., 1992; Calderano et al., 1993), but are still incompletely understood.

C. Role of Calmodulin

Isolated guinea pig lung membranes contain a VIP-binding protein (Mr 18 kd) that is similar in electrophoretic and chromatographic behavior, binding of VIP, and amino acid sequence to porcine brain calmodulin (Stallwood et al., 1992). It has been proposed that this protein may be identical to calmodulin, and that calmodulin may be the VIP receptor in the lung (Stallwood et al., 1992). The identity of this low-affinity binding protein with calmodulin has been confirmed (Andersson et al., 1993), but it is premature to conclude that calmodulin is the VIP receptor in the lung (Said, 1993a).

IV. PHARMACOLOGICAL ACTIONS

VIP relaxes airway smooth muscle both in vitro and in vivo: it relaxes isolated tracheal or bronchial segments from guinea pigs, rabbits, dogs, and humans and prevents or attenuates the constrictor effect of histamine, prostaglandin $F_{2\alpha}$, kallikrein, leukotriene D_4, neurokinins A and B, and endothelin. This relaxation is relatively long-lasting and is independent of adrenergic or cholinergic receptors or of cyclooxygenase activity. Inhaled VIP protects against the bronchoconstriction induced by histamine or prostaglandin $F_{2\alpha}$ in dogs, and by histamine in guinea pigs, and infused VIP reverses serotonin-induced bronchoconstriction in cats.

Other effects of VIP include: relaxation of pulmonary vascular smooth muscle, with greater potency than prostacyclin (Saga and Said, 1984); stimulation of water and chloride ion secretion and, in some preparations, of macromolecular secretion; inhibition of mast cell degranulation (Said, 1988, 1991a,b,c); and inhibition of airway smooth muscle cell proliferation (Maruno and Said, 1993a).

Several VIP-related peptides, including PHI, PHM, helodermin, and PACAP, also bind to VIP receptors, resulting in varying degrees of smooth muscle relaxation. Their potency is generally lower than that of VIP but, for helodermin, the duration of action is considerably longer (Foda and Said, 1989).

V. DEGRADATION

VIP is degraded by proteases that are present at or near the airway mucosa, including mast cell tryptase and chymase (Caughey et al., 1988), and by neutral endopeptidase ("enkephalinase," Goetzl et al., 1989; Trotz et al., 1990) and possibly other enzymes.

VI. VIP AS A REGULATOR OF LUNG FUNCTION

A. Smooth Muscle Relaxation

Among the principal likely functions of VIP is the relaxation of smooth muscle tone. In particular, VIP has been considered a probable transmitter of the dominant nonadrenergic, noncholinergic component of neurogenic relaxation of airway, vascular, and other smooth muscle (Said, 1988, 1991b). Within the past two years, it has become increasingly apparent that nitric oxide also is a major transmitter of this relaxation. The two transmitters, acting together, via different second messengers, bring about the relaxation (Li and Rand, 1991; Said, 1992). The relative contributions of VIP and nitric oxide, and the interactions between them in different tissues, are under active investigation (Boeckxstaens et al., 1992; Grider, 1993; He and Goyal, 1993).

B. Airway Smooth Muscle Cell Proliferation

VIP stimulates the proliferation of some types of normal, nonneoplastic cells, including keratinocytes (Haegerstrand et al., 1989) and neuronal cells (Pincus et al., 1990). On the other hand, it inhibits the growth and proliferation of other cell types, including vascular smooth muscle (Hultgårdh-Nilsson, 1988) and bronchial smooth muscle (Maruno and Said, 1993a), as well as small cell lung cancer cells (Maruno and Said, 1993b). The inhibition of airway smooth muscle cell proliferation is of special interest to bronchial asthma, where proliferation of smooth muscle is an important factor in increased airway resistance (Hossain, 1973).

C. Airway Secretion

VIP stimulates water and ion transport in canine tracheal epithelium (Nathanson et al., 1983) and stimulates the secretion of sulfated macromolecules by tracheal explants from ferrets (Peatfield et al., 1983) but, paradoxically, inhibits macromolecular secretion in human tracheal explants (Coles et al., 1981).

D. Pulmonary Vasodilation

VIP dilates the vessels supplying the nose, upper airways (Malm et al., 1980), and trachea and bronchi (Widdicombe et al., 1987), as well as the pulmonary vessels

(Hamasaki et al., 1983; Nandiwada et al., 1985). As a pulmonary vasodilator, VIP is 50 times as potent as prostacyclin (Saga and Said, 1984). The pulmonary vasodilator action of VIP is independent of endothelium (Sata et al., 1986).

E. Trophic Action

VIP promotes the survival of neuronal cells and their precursors in culture (Brenneman and Eiden, 1986; Pincus et al., 1990) and protects neurons from the toxic effect of the envelope protein of the human immunodeficiency virus (Brenneman et al., 1988). In a follow-up study, the same group of investigators reported that VIP, acting via specific receptors, has growth-promoting functions in whole cultured mouse embryos (Gressens et al., 1993). This trophic action of VIP may be comparable to its protective effect against lung injury described below.

F. Anti-inflammatory Action

Recent data suggest that VIP has potent anti-inflammatory activity in the lung (Said, 1993b). This notion was suggested by the ability of the peptide to (a) modulate the function of inflammatory cells, including lymphocytes, alveolar macrophages, and mast cells, and (b) antagonize major humoral mediators of inflammation, including histamine, prostaglandin $F_{2\alpha}$, leukotrienes C_4 and D_4, platelet-activating factor (PAF), neurokinins A and B, and endothelin. Direct evidence that VIP can prevent or attenuate acute edematous lung injury has been obtained in several experimental models of adult respiratory distress syndrome (ARDS): VIP dose-dependently protects perfused, ventilated rat or guinea pig lungs against injury caused by (a) intratracheal instillation of HCl, (b) infusion of PAF or of the oxidant herbicide paraquat, (c) addition to the perfusate of xanthine and xanthine oxidase (Fig. 2a and b), which produce injury through the generation of toxic oxygen species, and (d) prolonged perfusion of the lungs, a form of injury that also results from free oxygen radicals. The protective effect of VIP is selective, since of its related peptides only helodermin gave equal protection, while secretin and glucagon were ineffective.

In addition to the ability of exogenously administered VIP to reduce or prevent lung injury, we noted that untreated injured lungs released large amounts of the peptide, suggesting that it may be an endogenous modulator of inflammatory lung injury (Berisha et al., 1990).

The mechanisms of lung protection by VIP are not yet fully known. As well as attenuating injury, the administration of VIP also decreased or abolished the release of cyclooxygenase products triggered by the oxidant injury (Berisha et al., 1990). Later work revealed that VIP inhibits the activity of phospholipase A_2 (Trotz et al., 1990), the key enzyme in the biosynthesis of the proinflammatory arachidonate products (both cyclooxygenase- and lipoxygenase-catalyzed), and PAF. Further, VIP has been demonstrated to have direct antioxidant activity, as it scavenges the toxic singlet oxygen radical (Misra and Misra, 1990).

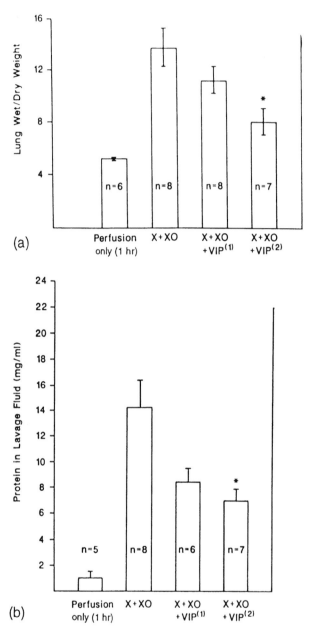

Figure 2 (a) VIP dose-dependently reduced wet-to-dry weight ratios in rat lungs injured with xanthine + xanthine oxidase (X + XO). VIP[1], 1 μg·kg^{-1}·min^{-1}; VIP[2], 10 μg·kg^{-1}·min^{-1}. *Significantly lower than in X + XO group ($p < 0.001$) and no different from perfusion-only group. (From Berisha et al., 1990; reprinted with permission from *American Journal of Physiology*.) (b) VIP dose-dependently reduced protein leakage into alveoli in rat lungs injured with X + XO. *Significantly lower than in X + XO group ($p < 0.03$) but higher than in perfusion-only group ($p < 0.01$). (From Berisha et al., 1990; reprinted with permission from *American Journal of Physiology*.)

Finally, VIP attenuates injury to cultured bronchial epithelial cells caused by activated eosinophils (Sakakibara et al., 1991) and is present in relatively high concentrations in eosinophils (Aliakbari et al., 1987). These findings are particularly relevant to bronchial asthma, in view of the importance of bronchial epithelial injury (Laitinen et al., 1985) and of activated eosinophils in the pathogenesis of the disease (Venga and Håkansson, 1991).

VII. IMPORTANCE IN LUNG DISEASE

The selective lack of VIP innervation in certain tissues has been linked to the pathogenesis of several disease processes. In *achalasia of the esophagus*, the loss of normal peristalsis of the esophageal body and the failure of the lower sphincter to relax with swallowing are correlated with the virtual absence in esophageal smooth muscle of VIP-containing nerves (Gridelli et al., 1982). In *Hirschsprung's disease*, or congenital megacolon, the normally rich VIP innervation of the small intestine is strikingly reduced in the affected, aganglionic segments of the bowel, possibly contributing to their persistent contraction (Tsuto et al., 1989). In *diabetic impotence*, the abnormally sparse VIP innervation of penile cavernous tissue may be responsible for the failure of effective vasodilation required for penile erection (Gu et al., 1984). In this, as in the preceding conditions, a simultaneous deficiency of nitric oxide synthase, and thus of nitric oxide, may well be an important pathogenetic factor. In *cystic fibrosis*, the marked deficiency of VIP nerves around sweat-gland acini and ducts may be related to the defect in chloride ion transport and to other aspects of the exocrinopathy of this disorder (Heinz-Erian et al., 1985). In the *acquired immune deficiency syndrome*, the killing of neurons by glycoprotein gp120, the envelope protein of the human immunodeficiency virus, may be related in some way to interference with the neurotrophic effects of VIP in the central nervous system (Brenneman et al., 1988).

A deficiency of VIP may be a cause of bronchial hyperreactivity and possibly other changes in *asthmatic airways*. This hypothesis rests on the findings that (1) VIP is a likely neurotransmitter, with nitric oxide, of the dominant neurogenic component of airway relaxation in humans, the nonadrenergic relaxant system (Matsuzaki et al., 1980), and (2) sections of airways from subjects with asthma show a selective absence of nerves immunoreactive to VIP, in contrast to airways of control subjects without asthma, which have a rich distribution of VIP nerves (Ollerenshaw et al., 1989). The lack of VIP-containing nerves in asthmatic airways does not *establish* whether the VIP deficiency is a cause of the disease. To prove that the VIP deficiency is a primary defect, the finding should be shown to antedate the onset of the disease, and direct evidence of impaired gene expression of VIP should be demonstrated in VIP-containing neurons within the lung. If the lack of VIP in asthmatic airways is significantly related to the pathogenesis of the disease, then supplemental VIP may prove to be rational replacement therapy, much like insulin for type I diabetics. The possible significance in asthma of circulating high-affinity VIP-binding antibodies, some of which are capable of catalyzing the hydrolysis of the peptide (Paul et al., 1989), is unknown.

VIII. THERAPEUTIC POTENTIAL: VIP AS A BRONCHODILATOR AND ANTI-INFLAMMATORY AGENT

As a candidate bronchodilator, VIP exhibits several desirable features that may also be shared by related peptides or analogs:

1. It is naturally present as a neuropeptide in the airways and other tissue sites.
2. It binds to specific, high-affinity receptors in the lung and relaxes both large and small airways, including lung parenchymal strips, from human subjects (Saga and Said, 1984).
3. It is a potent relaxant of airway smooth muscle and is effective in preventing or reversing bronchoconstriction due to a wide variety of pharmacological agents (see above).
4. It produces tracheobronchial relaxation that is relatively long-lasting and independent of adrenergic receptors.
5. By promoting water and ion transport in airway epithelium, VIP should make bronchial secretions easier to mobilize and eliminate.
6. Given by inhalation as aerosol, VIP is poorly absorbed into the general circulation and therefore elicits no systemic side effects, such as tachycardia or arterial hypotension.
7. VIP inhibits antigen-induced release of histamine from guinea pig lung and, in view of its presence normally in mast cells, may act as a natural modulator of mast cell function.
8. VIP protects the lung against acute increase in airway pressure due to anaphylaxis (Takamatsu et al., 1991) or to the release of proinflammatory peptides by capsaicin (Pakbaz et al., 1993).
9. VIP has distinct and potent *anti-inflammatory actions*, exerted on both cellular and chemical mediators of inflammation (Said, 1991c). Because asthma is a disease of chronic inflammation of the airways and not merely a state of airway constriction (Kirby et al., 1987; Azzawi et al., 1990), these anti-inflammatory properties enhance the potential usefulness of VIP as an antiasthma agent.
10. The inhibitory action of VIP on airway smooth muscle cell proliferation is an additional beneficial activity in the long-term management of asthma.

A. Clinical Trials in Asthma

In view of the favorable experimental data on VIP as a relaxant of airway smooth muscle in vitro and as a bronchodilator in animals in vivo, and in light of the other features just cited, it was logical to investigate the effectiveness of this peptide in asthmatic patients. This was done even before the anti-inflammatory properties of VIP became known. Given by intravenous infusion (1, 3, or 6 pmol/kg·min^{-1} for 15 min), VIP had no effect on specific airway conductance in six normal subjects (Palmer et al., 1986). In an earlier study, however, given in the same manner, VIP caused significant bronchodilation (improvement in forced expiratory air flow) in seven atopic asthmatics and attenuated histamine-induced bronchoconstriction in the same subjects (Morice et al., 1983). In both studies, a moderate tachycardia and decrease in arterial blood pressure were observed at the higher dose rates.

Results with inhaled VIP have been similarly inconclusive. In six atopic subjects with mild asthma, aerosolized VIP (total dose 100 μg, i.e., approximately 1.4 μg/kg) reduced bronchial reactivity to histamine but did not significantly increase specific airway conductance on its own (Barnes and Dixon, 1984). Nor did VIP pretreatment prevent exercise-induced bronchoconstriction in six adult asthmatics (Bundgaard et al., 1983). Of 15 moderately severe asthmatics who received 5 μg/kg of sterile VIP by inhalation, 10 showed significant improvement in spirometric data or specific airway conductance, but the observed bronchodilation was less pronounced than that resulting from inhalation of the β-agonist albuterol (Mojarad et al., 1985).

The most likely explanation why inhaled VIP was less effective in asthmatic airway obstruction than had been expected is that the peptide is rapidly degraded by airway peptidases.

B. VIP Analogs with Greater Therapeutic Potential in Asthma

The above conclusion suggested two possible approaches toward enhancing the therapeutic effectiveness of VIP in asthma: (a) to combine the peptide with one or more selective peptidase inhibitors, or (b) to use a related peptide that has similar bronchial-relaxant activity but is more resistant to inactivation by airway proteases. The former approach was validated by the demonstration that the addition of phosphoramidon, a neutral endopeptidase inhibitor, potentiates and prolongs VIP-induced tracheal relaxation in vitro (Liu et al., 1987). The second, simpler, approach requires the availability of naturally occurring peptides or synthetic analogs of VIP that are more resistant to digestion. One such peptide is helodermin, originally isolated from the lizard Gila monster *Heloderma*, but also present in mammalian tissues. Helodermin exhibits close sequence similarity to VIP, but has a carboxy-terminal extension that may account for its greater resistance to enzymatic degradation (Liu et al., 1991) and thus for its longer-lasting tracheal relaxation (Foda and Said, 1989). Synthetic analogs have been reported that are more potent than VIP as airway smooth muscle relaxants, are longer-acting, or both (Bolin et al., 1992). The special advantages these VIP analogs may offer in the management of asthma will be put to the test in clinical trials.

IX. CONCLUSIONS

It is now 20 years since VIP was discovered and isolated. Its role in lung physiology and disease is still being defined. But it is clear that the work on VIP has stimulated, and continues to stimulate, directly or indirectly, the exciting new field of neuropeptides in the respiratory tract.

ACKNOWLEDGMENTS

I thank Mary Messina-Stalhut for help with the preparation of the manuscript. This work was supported by NIH Research Grant HL-30450 and by the Department of Veterans Affairs. S.I.S. is a Medical Investigator of the Department of Veterans Affairs.

REFERENCES

Aliakbari, J., Sreedharan, S.P., Turck, C.W., and Goetzl, E.J. (1987). Selective localization of vasoactive intestinal peptide and substance P in human eosinophils. *Biochem. Biophys. Res. Commun. 148*:1440–1445.

Andersson, M., Carlquist, M., Maletti, M., and Marie, J.C. (1993). Simultaneous solubilization of high-affinity receptors for VIP and glucagon and of a low-affinity binding protein for VIP, shown to be identical to calmodulin. *FEBS Lett. 318*:35.

Audigier, S., Barberis, C., and Jard, S. (1988). Vasoactive intestinal polypeptide increases inositol phosphate breakdown in the rat superior cervical ganglion. *Ann. NY Acad. Sci. 527*:579.

Azzawi, M., Bradley, B. Jeffery, P.K., Frew, A.J., Wardlaw, A.J., Knowles, G., Assoufi, B., Collins, J.V., Durham, S., and Kay, A.B. (1990). Identification of activated T lymphocytes and eosinophils in bronchial biopsies in stable atopic asthma. *Am. Rev. Respir. Dis. 142*:1407.

Barnes, P.J., and Dixon, C.M.S. (1984). The effect of inhaled vasoactive intestinal peptide on bronchial reactivity to histamine in humans. *Am. Rev. Respir. Dis. 130*:162.

Berisha, H., Foda, H., Sakakibara, H., Trotz, M., Pakbaz, H., and Said, S.I. (1990). Vasoactive intestinal peptide prevents lung injury due to xanthine/xanthine oxidase. *Am. J. Physiol. 259 (Lung Cell. Mol. Physiol. 3)*:L151.

Birnbaum, R.S., Alfonzo, M., and Kowal, J. (1980). Vasoactive intestinal peptide- and adrenocorticotropin-stimulated adenyl cyclase in cultured adrenal tumor cells: evidence for a specific vasoactive intestinal peptide receptor. *Endocrinology 106*:1270.

Boeckxstaens, G.E., Pelckmans, P.A., De Man, J.G., Bult, H., Herman, A.G., and Van Maercke, Y.M. (1992). Evidence for a differential release of nitric oxide and vasoactive intestinal polypeptide by nonadrenergic noncholinergic nerves in the rat gastric fundus. *Arch. Int. Pharmacodyn. Ther. 318*:107.

Bolin, D.R., Cottrell, J., Michalewsky, J., Garippa, R., O'Neill, N., Simko, B., and O'Donnell, M. (1992). Degradation of vasoactive intestinal peptide in bronchial alveolar lavage fluid. *Biomed. Res. 13*:25.

Brenneman, D.E., and Eiden, L.E. (1986). Vasoactive intestinal peptide and electrical activity influence neuronal survival. *Proc. Natl. Acad. Sci. USA 83*:1159.

Brenneman, D.E., Westbrook, G.L., Fitzgerald, S.P., Ennist, D.L., Elkins, K.L., Ruff, M.R., and Pert, C.B. (1988). Neuronal cell killing by the envelope protein of HIV and its prevention by vasoactive intestinal peptide. *Nature 335*:639.

Bundgaard, A., Enehjelm, S.D., and Aggestrup, S. (1983). Pretreatment of exercise-induced asthma with inhaled vasoactive intestinal peptide (VIP). *Eur. J. Respir. Dis. 64* (Suppl. 128):427.

Calderano, V., Chiosi, E., Greco, R., Spina, A.M., Giovane, A., Quagliuolo, L., Servillo, L., Balestrieri, C., and Illiano, G. (1993). Role of calcium in chloride secretion mediated by cAMP pathway activation in rabbit distal colon mucosa. *Am. J. Physiol. 264 (Gastrointest. Liver Physiol. 27)*:G252.

Caughey, G.H., Leidig, F., Viro, N.F., and Nadel, J.A. (1988). Substance P and vasoactive intestinal peptide degradation by mast cell tryptase and chymase. *J. Pharmacol. Exp. Ther. 244*:133.

Chik, C.L., Ho, A.K., and Klein, D.C. (1988). α_1-Adrenergic potentiation of vasoactive intestinal peptide stimulation of rat pinealocyte adenosine $3',5'$-monophosphate and guanosine $3',5'$-monophosphate: evidence for a role for calcium and protein kinase-C. *Endocrinology 122*:702.

Coles, S.J., Said, S.I., and Reid, L.M. (1981). Inhibition by vasoactive intestinal peptide of glycoconjugate and lysozyme secretion by human airways in vitro. *Am. Rev. Respir. Dis. 124*:531.

Couvineau, A., Voisin, T., Guijarro, L., and Laburthe, M. (1990). Purification of vasoactive intestinal peptide receptor from porcine liver by a newly designed one-step affinity chromatography. *J. Biol. Chem. 265*:13386.

Cutz, E., Chan, W., Track, N.S., Goth, A., and Said, S.I. (1978). Release of vasoactive intestinal polypeptide in mast cells by histamine liberators. *Nature 275*:661.

Dey, R.D., Shannon, W.A., and Said, S.I. (1981). Localization of VIP-immunoreactive nerves in airways and pulmonary vessels of dogs, cats and human subjects. *Cell Tissue Res. 220*:231.

Dey, R.D., Hoffpauir, J., and Said, S.I. (1988). Co-localization of vasoactive intestinal peptide- and substance P-containing nerves in cat bronchi. *Neuroscience 24*:275.

Dey, R.D., Mayer, B., and Said, S.I. (1993). Colocalization of vasoactive intestinal peptide and nitric oxide synthase in neurons of ferret trachea. *Neuroscience 54*:839–843.

Dupont, C., Laburthe, M., Broyart, J.P., Bataille, D., and Rosselin, G. (1980). Cyclic AMP production in isolated colonic epithelial crypts: a highly sensitive model for the evaluation of vasoactive intestinal peptide action in human intestine. *Eur. J. Clin. Invest. 10*:67.

Ellis, J.L., and Farmer, S.G. (1989). Modulation of cholinergic neurotransmission by vasoactive intestinal peptide and peptide histamine isoleucine in guinea-pig tracheal smooth muscle. *Pulmon. Pharmacol. 2*:107.

Foda, H.D., and Said, S.I. (1989). Helodermin, a C-terminally extended VIP-like peptide, evokes long-lasting tracheal relaxation. *Biomed. Res. 10*:107.

Furness, J.B., Bornstein, J.C., Murphy, R., and Pompolo, S. (1992). Roles of peptides in transmission in the enteric nervous system. *Trends Neurosci. 15*:66.

Goetzl, E.J., Sreedharan, S.P., and Turck, S.W. (1988). Structurally distinctive vasoactive intestinal peptides from rat basophilic leukemia cells. *J. Biol. Chem. 263*:9083.

Goetzl, E.J., Sreedharan, S.P., Turck, C.W., Bridenbaugh, R., and Malfroy, B. (1989). Preferential cleavage of amino- and carboxyl-terminal oligopeptides from vasoactive intestinal polypeptide by human recombinant enkephalinase (neutral endopeptidase, EC 3.4.24.11). *Biochem. Biophys. Res. Commun. 158*:850.

Gourdi, D., Bataille, D., Vauclin, N., Grouselle, D., Rosselin, G., and Tixier-Vidal, A. (1979). Vasoactive intestinal peptide (VIP) stimulates prolactin (PRL) release and cAMP production in a rat pituitary cell line (GH3/B6). Additive effects of VIP and TRH on PRL release. *FEBS Lett. 104*:165.

Gozes, I., and Brenneman, D.E. (1989). VIP: molecular biology and neurobiological function. *Mol. Neurobiol. 3*:201.

Gressens, P., Hill, J.M., Gozes, I., Fridkin, M., and Brenneman, D.E. (1993). Growth factor function of vasoactive intestinal peptide in whole cultured mouse embryos. *Nature 362*:155.

Gridelli, B., Buffa, R., Salvini, P., De Rai, F., Uggeri, Fiocca, R., and Said, S.I. (1982). Lack of VIP-immunoreactive nerves in oesophageal achalasia. *Ital. J. Gastroenterol. 14*:211.

Grider, J.R. (1993). Interplay of VIP and nitric oxide in regulation of the descending relaxation phase of peristalsis. *Am. J. Physiol. 264 (Gastrointest. Liver Physiol. 27)*:G334.

Grider, J.R., Murthy, K.S., Jin, J.-G., and Makhlouf, G.M. (1992). Stimulation of nitric oxide from muscle cells by VIP: prejunctional enhancement of VIP release. *Am. J. Physiol. 262 (Gastrointest. Liver Physiol. 25)*:G774.

Gu, J., Polak, J.M., Lazarides, M., Pryor, J.P., Blank, M.A., Polak, J.M., Morgan, R., Marangos, P.J., and Bloom, S.R. (1984). Decrease of vasoactive intestinal polypeptide (VIP) in the penises from impotent men. *Lancet 2*:315.

Haegerstrand, A., Jonzon, B., Dalsgaard, C.J., and Nilsson, J. (1989). Vasoactive intestinal polypeptide stimulates cell proliferation and adenylate cyclase activity of cultured human keratinocytes. *Proc. Natl. Acad. Sci. USA 86*:5993.

Hamasaki, Y., Mojarad, M., and Said, S.I. (1983). Relaxant action of VIP on cat pulmonary artery: comparison with acetylcholine, isoproterenol, and prostaglandin E$_1$. *J. Appl. Physiol. 54*:1607.

He, X.D., and Goyal, R.K. (1993). Nitric oxide involvement in the peptide VIP-associated inhibitory junction potential in the guinea-pig ileum. *J. Physiol. 461*:485.

Heinz-Erian, P., Dey, R.D., Flux, M., and Said, S.I. (1985). Deficient vasoactive intestinal peptide innervation in sweat glands of cystic fibrosis patients. *Science 229*:1407.

Hossain, S. (1973). Quantitative measurement of bronchial muscle in asthma. *J. Pathol. 110*:319.

Hultgårdh-Nilsson, A., Nilsson, J., Jonzon, B., and Dalsgaard, C.-J. (1988). Growth-inhibitory properties of vasoactive intestinal polypeptide. *Regul. Peptides 22*:267.

Ishihara, T., Shigemoto, R., Mori, K., Takahashi, K., and Nagata, S. (1992). Functional expression and tissue distribution of a novel receptor for vasoactive intestinal polypeptide. *Neuron 8*:811.

Kirby, J.G., Hargreave, F.E., Gleich, G.J., O'Byrne, P.M. (1987). Bronchoalveolar cell profiles of asthmatic and nonasthmatic patients. *Am. Rev. Respir. Dis. 136*:379.

Laitinen, L.A., Heino, M., Laitinen, A., Kava, T., and Haahtela, T. (1985). Damage of the airway epithelium and bronchial reactivity in patients with asthma. *Am. Rev. Respir. Dis. 131*:599.

Lazarus, S.C., Basbaum, C.B., Barnes, P.J., and Gold, W.M. (1986). Mapping of VIP receptors by use of an immunocytochemical probe for the intracellular mediator cyclic AMP. *Am. J. Physiol. 251*:C115.

Leroux, P., Vaudry, H., Fournier, A., St.-Pierre, S., and Pelletier, G. (1984). Characterization and localization of vasoactive intestinal peptide receptors in the rat lung. *Endocrinology 114*:1506.

Li, C.G., and Rand, M.J. (1991). Evidence that part of the NANC relaxant response of guinea-pig trachea to electrical field stimulation is mediated by nitric oxide. *Br. J. Pharmacol. 102*:91.

Liu, L.-W., Sata, T., Kubota, E., Paul, S., and Said, S.I. (1987). Airway relaxant effect of vasoactive intestinal peptide (VIP): selective potentiation by phosphoramidon, an enkephalinase inhibitor. *Am. Rev. Respir. Dis. 135* (Suppl.):A86.

Liu, L.-W., Trotz, M., Erdös, E.G., and Said, S.I. (1991). Vasoactive intestinal peptide (VIP) and helodermin degradation by airway enzymes. *Am. Rev. Respir. Dis. 143*:A618.

Luis, J., and Said, S.I. (1990). Characterization of VIP- and helodermin-preferring receptors on human small cell carcinoma cell lines. *Peptides 11*:1239.

Lundberg, J.M., Änggård, A., Emson, P., Fahrenkrug, J., Hökfelt, T., and Mutt, V. (1980). Vasoactive intestinal polypeptide in cholinergic neurons of exocrine glands: functional significance of coexisting transmitters of vasodilation and secretion. *Proc. Natl. Acad. Sci. USA 77*:1651.

Malhotra, R.K., Wakade, T.D., and Wakade, A.R. (1988). Vasoactive intestinal polypeptide and muscarine mobilize intracellular Ca^{2+} through breakdown of phosphoinositides to induce catecholamine secretion. *J. Biol. Chem. 263*:2123.

Malm, L., Sundler, F., and Uddman, R. (1980). Effects of vasoactive intestinal polypeptide (VIP) on resistance and capacitance vessels in the nasal mucosa. *Acta Otolaryngol. (Stockholm) 90*:304.

Martin, J.G., Wang, A., Zacour, M., and Biggs, D.F. (1990). The effects of vasoactive intestinal polypeptide on cholinergic neurotransmission in an isolated innervated guinea pig tracheal preparation. *Respir. Physiol. 79*:111.

Maruno, K., and Said, S.I. (1993a). Inhibition of human airway smooth muscle cell proliferation by vasoactive intestinal peptide (VIP). *Am. Rev. Respir. Dis. 147*:A253.

Maruno, K., and Said, S.I. (1993b). Small cell lung carcinoma: inhibition of proliferation by vasoactive intestinal peptide and helodermin and enhancement of inhibition by anti-bombesin antibody. *Life Sci. (Pharmacol. Lett.) 52*:PL267.

Matsuzaki, Y., Hamasaki, Y., and Said, S.I. (1980). Vasoactive intestinal peptide: a possible transmitter of nonadrenergic relaxation of guinea pig airways. *Science 210*:1252.

Meisheri, K.D., and Rüegg, J.C. (1983). Dependence of cyclic-AMP induced relaxation on Ca^{2+} and calmodulin in skinned smooth muscle of guinea pig *Taenia coli*. *Pflügers Arch. 399*:315.

Misra, B.R., and Misra, H.P. (1990). Vasoactive intestinal peptide, a singlet oxygen quencher. *J. Biol. Chem. 265*:15371.

Mojarad, M., Grode, T.L., Cox, C.P., Kimmel, G., and Said, S.I. (1985). Differential responses of human asthmatics to inhaled vasoactive intestinal peptide (VIP). *Am. Rev. Respir. Dis. 131*:A281.

Morice, A., Uunwin, R.J., and Sever, P.S. (1983). Vasoactive intestinal peptide causes bronchodilatation and protects against histamine-induced bronchoconstriction in asthmatic subjects. *Lancet 2*:1225.

Muller, J-M., El Battari, A., Ah-Kye, E., Luis, J., Ducret, F., Pichon, J., and Marvaldi, J. (1985). Internalization of the vasoactive intestinal peptide (VIP) in human adenocarcinoma cell line (HT29). *Eur. J. Biochem. 152*:107.

Nandiwada, P.A., Kadowitz, P.J., Said, S.I., Mojarad, M., and Hyman, A.L. (1985). Pulmonary vasodilator responses to vasoactive intestinal peptide in the cat. *J. Appl. Physiol. 58*:1723.

Nathanson, I., Widdicombe, J.H., and Barnes, P.J. (1983). Effect of vasoactive intestinal peptide on ion transport across dog tracheal epithelium. *J. Appl. Physiol. 55*:1844.

Ollerenshaw, S., Jarvis, D., Woolcock, A., Sullivan, C., and Scheibner, T. (1989). Absence of immunoreactive vasoactive intestinal polypeptide in tissue from the lungs of patients with asthma. *N. Engl. J. Med. 320*:1244.

Ottaway, C.A. (1987). Selective effects of vasoactive intestinal peptide on the mitogenic response of murine T cells. *Immunology 62*:291.

Pakbaz, H., Berisha, H., Absood, A., Foda, H.D., and Said, S.I. (1993). VIP in sensory nerves of the lung: capsaicin-induced release of immunoreactive vasoactive intestinal peptide (VIP) from guinea pig lungs. *Am. Rev. Respir. Dis. 147*:A477.

Palmer, J.B.D., Cuss, F.M.C., Warren, J.B., Blank, M., Bloom, S.R., and Barnes, P.J. (1986). Effect of infused vasoactive intestinal peptide on airway function in normal subjects. *Thorax 41*:663.

Patthi, S., Somerson, S., and Veliçelebi, G. (1988). Solubilization of rat lung vasoactive intestinal peptide receptors in the active state. Characterization of the binding properties and comparison with membrane-bound receptors. *J. Biol. Chem. 263*:19363.

Paul, S., and Said, S.I. (1987). Characterization of receptors for vasoactive intestinal peptide solubilized from the lung. *J. Biol. Chem. 262*:158.

Paul, S., Volle, D.J., Beach, C.M., Johnson, D.R., Powell, M.J., and Massey, R.J. (1989). Catalytic hydrolysis of vasoactive intestinal peptide by human autoantibody. *Science 244*:1158.

Peatfield, A.C., Barnes, P.J., Bratcher, C., Nadel, J.A., and Davis, B. (1983). Vasoactive intestinal peptide stimulates tracheal submucosal gland secretion in ferret. *Am. Rev. Respir. Dis. 128*:89.

Pfitzer, G. Rüegg, J.C., Zimmer, M., and Hofmann, F. (1985). Relaxation of skinned coronary arteries depends on the relative concentrations of Ca^{2+}, calmodulin and active cAMP-dependent protein kinase. *Pflügers Arch. 405*:70.

Pincus, D.W., Dicicco-Bloom, E.M., and Black, I.B. (1990). Vasoactive intestinal peptide regulates mitosis, differentiation and survival of cultured sympathetic neuroblasts. *Nature 343*:564.

Robberecht, P., Conlon, T.P., and Gardner, J.D. (1976). Interaction of porcine vasoactive intestinal peptide with dispersed pancreatic acinar cells from the guinea pig: structural requirements for effects of VIP and secretin on cellular cyclic AMP. *J. Biol. Chem.* *251*:4635.

Saga, T., and Said, S.I. (1984). Vasoactive intestinal peptide relaxes isolated strips of human bronchus, pulmonary artery, and lung parenchyma. *Trans. Assoc. Am. Physicians 97*:304.

Said, S.I. (1967). Vasoactive substances in the lung. In: *Proceedings of the Tenth Aspen Emphysema Conference*, June 7–10. Aspen, CO. U.S. Public Health Serv. Publication, 1787, p. 223.

Said, S.I. (1988). Vasoactive intestinal peptide in the lung. *Ann. NY Acad. Sci. 527*:450.

Said, S.I. (1991a). Vasoactive intestinal polypeptide: Biological role in health and disease. *Trends Endocrinol. Metab. 2*:107.

Said, S.I. (1991b). Vasoactive intestinal polypeptide (VIP) in asthma. *Ann. NY Acad. Sci. 629*:305.

Said, S.I. (1991c). VIP as a modulator of lung inflammation and airway constriction. *Am. Rev. Respir. Dis. 143*:S22.

Said, S.I. (1992). Nitric oxide and vasoactive intestinal peptide as co-transmitters of smooth muscle relaxation. *News Physiol. Sci. 7*:181.

Said, S.I. (1993a). Vasoactive intestinal peptide: its interactions with calmodulin and catalytic antibodies. *Neurochem. Int. 23*:215–219.

Said, S.I. (1993b). Vasoactive intestinal peptide (VIP) and related peptides as antiasthma and anti-inflammatory agents. *Biomed. Res. 13* (Suppl. 2):257.

Said, S.I., and Mutt, V. (1969). Long acting vasodilator peptide from lung tissue. *Nature 224*:699.

Saito, K., Yamatani, K., Manaka, H., Takahashi, K., Tominaga, M., and Sasaki, H. (1992). Role of Ca^{2+} on vasoactive intestinal peptide-induced glucose and adenosine 3′,5′-monophosphate production in the isolated perfused liver. *Endocrinology 130*:2267.

Sakakibara, H., Takamatsu, J., and Said, S.I. (1991). Eosinophil-mediated injury of cultured bronchial epithelial cells: attenuation by vasoactive intestinal peptide (VIP). *Am. Rev. Respir. Dis. 143*:A44.

Sand, O., Chen, B., Li, Q., Karlsen, H.E., Bjøro, T., and Haug, E. (1989). Vasoactive intestinal peptide (VIP) may reduce the removal rate of cytosolic Ca^{2+} after transient elevations in clonal rat lactotrophs. *Acta Physiol. Scand. 137*:113.

Sata, T., Misra, H.P., Kubota, E., and Said, S.I. (1986). Vasoactive intestinal polypeptide relaxes pulmonary artery by an endothelium-independent mechanism. *Peptides 7*:225.

Shreeve, S.M., DeLuca, A.W., Diehl, N.L., and Kermode, J.C. (1992). Molecular properties of the vasoactive intestinal peptide receptor in aorta and other tissues. *Peptides 13*:919.

Springall, D.R., Polak, J.M., Howard, L., Power, R.F., Krausz, T., Manickam, S., Banner, N.R., Khagani, A., Rose, M., and Yacoub, M.H. (1990). Persistence of intrinsic neurons and possible phenotypic changes after extrinsic denervation of human respiratory tract by heart-lung transplantation. *Am. Rev. Respir. Dis. 141*:1538.

Sreedharan, S.P., Robichon, A., Peterson, K.E., and Goetzl., E.J. (1991). Cloning and expression of the human vasoactive intestinal peptide receptor. *Proc. Natl. Acad. Sci. USA 88*:4986.

Stallwood, D., Brugger, C.H., Baggenstoss, B.A., Stemmer, P.M., Shiraga, H., Landers, D.F., and Paul, S. (1992). Identity of a membrane-bound vasoactive intestinal peptide-binding protein with calmodulin. *J. Biol. Chem. 267*:19617.

Szewczak, S.M., Behar, J., and Billett, G. (1990). VIP-induced alterations in cAMP and inositol phosphates in the lower esophageal sphincter. *Am. J. Physiol. 259 (Gastrointest. Liver Physiol. 22)*:G239.

Takamatsu, J., Shima, K., Sakakibara, H., Trotz, M., and Said, S.I. (1991). Anaphylaxis in guinea pig lungs: attenuation of physiologic and biochemical consequences by vaso-active intestinal peptide. *Am. Rev. Respir. Dis. 143*:A44.

Tatsuno, I., Yada, T., Vigh, S., Hidaka, H., and Arimura, A. (1992). Pituitary adenylate cyclase activating polypeptide and vasoactive intestinal peptide cytosolic free calcium concentration in cultured rat hippocampal neurons. *Endocrinology 131*:73.

Trotz, M.E., Luis, J., and Said, S.I. (1990). Vasoactive intestinal peptide (VIP) inhibits phospholipase A_2: a mechanism of anti-inflammatory activity. *FASEB J. 4*:A1124.

Tsuto, T., Obata-Tsuto, H.L., Iwai, N., Takahashi, T., and Ibata, Y. (1989). Fine structure of neurons synthesizing vasoactive intestinal peptide in the human colon from patients with Hirschsprung's disease. *Histochemistry 93*:1.

Venge, P., and Håkansson, L. (1991). The eosinophil and asthma. In: *Asthma: Its Pathology and Treatment.* (Kaliner, M.A., Barnes, P.J., and Persson, C.G.A., eds.). Marcel Dekker, New York.

Widdicombe, J., Laitinen, A., and Laitinen, L.A. (1987). Effects of inflammatory and other mediators on airway vascular beds. *Am. Rev. Respir. Dis. 135* (Suppl.):S67.

DISCUSSION

McDonald: Are VIP receptors localized to endothelial cells of postcapillary venules?

Said: This has not been clearly demonstrated, although it is of importance in explaining the antiinflammatory effects of VIP.

Erdös: Why is helodermin a more effective bronchodilator than VIP in vivo?

Said: Helodermin has 35 amino acid residues, compared to 28 for VIP. Heloder-min is more resistant than VIP to both NEP and bovine tracheal homogenate peptidases. This difference in sensitivity to peptidases could explain helodermin's increased potency.

Erdös: Does VIP antagonize the mitogenic effects of angiotensin on vascular smooth muscle?

Said: We have not done this experiment, but have found that VIP counteracts the mitogenic actions of histamine on airway smooth muscle cells. VIP also inhibits proliferation of small cell carcinoma cells in vitro.

7

Neuropeptide Y in the Airways

Jan M. Lundberg and Agnes Modin
Karolinska Institute, Stockholm, Sweden

I. INTRODUCTION

A. Presence of NPY in Nasal Mucosa

The nasal mucosa receives a very dense sympathetic innervation, mainly of vascular structures (Dahlström and Fuxe, 1965; Änggård and Densert, 1974). The sympathetic nerve supply is of importance both for nutritive blood flow and tone of the large venous sinusoids and thereby for nasal airway resistance. Noradrenaline (NA) has for long been considered a sympathetic vascular neurotransmitter acting via both α_1- and α_2-adrenoceptors as a contractile agent of smooth muscle and via β-adrenoceptors as a relaxant (Andersson and Bende, 1984). Avian pancreatic polypeptide (APP)-like immunoreactivity (Li) was also observed in adrenergic nerves, mainly around arteries (Lundberg et al., 1980). Subsequently, it became clear that the APP-like substance in perivascular nerves was most likely identical to the structurally related peptide neuropeptide Y (NPY) (Lundberg et al., 1982), a 36-amino-acid peptide originally isolated from porcine brain (Tatemoto, 1982). In certain species, like the rat (Leblanc et al., 1987) and pig (Lacroix et al., 1990), a population of the parasympathetic neurons in the sphenopalatine ganglion innervating exocrine elements of the nasal mucosa also contains NPY Li (see Lundberg et al., 1988b). In the pig, about 20% of the nasal mucosal NPY Li has a nonsympathetic origin as revealed by denervation experiments (Lacroix et al., 1990) (Table 1). In humans, most of the NPY nerves are perivascular and mainly of sympathetic origin (Lundberg et al., 1988; Lacroix et al., 1990). NPY fibers in the pig and human are found both around arteries and

Table 1 Pig Nasal Mucosal Content of NPY and NA

	NPY (pmol/g)	NA (nmol/g)
Control	11.4 ± 1.2	14.0 ± 1.6
Reserpine	6.3 ± 0.5*	0.09 ± 0.01**
Reserpine + preganglionic transection	14.5 ± 1.5	0.08 ± 0.02**
2 weeks after sympathectomy	3.1 ± 0.1**	0.10 ± 0.03**

Control animals are compared with pigs pretreated 24 hr before with reserpine (1 mg/ kg i.v.) with or without simultaneous transection of the cervical sympathetic trunk and with pigs whose superior cervical sympathetic ganglion was removed 2 weeks before sacrifice. Data are given as means ± SEM. ($n = 8$). *$p < 0.05$; **$p < 0.01$.

venous sinusoids and in particularly high density around presumable arterio-venous shunt vessels (see also Baraniuk et al., 1990; Lacroix et al., 1990).

B. Presence of NPY in Lower Airways

In the trachea and bronchi NPY Li fibers are present both around blood vessels, glandular tissue, and within the smooth muscle layers (Lundberg et al., 1983; Uddman et al., 1984; Lundberg et al., 1988b). Although the exact nature of these fibers has not been worked out in detail, the perivascular NPY fibers in, e.g., tracheobronchial circulation are probably of mainly sympathetic origin. The presence of NPY in cell bodies of local peribronchial ganglia also indicates a parasympathetic nature possibly innervating, e.g., glands. NPY Li is thus in the pig colocalized with vasoactive intestinal polypeptide (VIP) in tracheal periglandular nerve fibers as well as in fibers innervating the tracheobronchial smooth muscle layer (Martling et al., 1990). Furthermore, there is an abundant perivascular innervation with NPY IR fibers in the pulmonary circulation.

II. NPY RELEASE IN AIRWAYS

Sympathetic stimulation using a high frequency is associated with profound vascular effects (Fig. 1) and increased overflow of NPY Li into the nasal venous effluent, suggesting release (Lundblad et al., 1987; Lacroix et al., 1989). The NPY release seems to be inhibited via NA acting on prejunctional α_2-receptors as revealed by increased NPY overflow after α-adrenoceptor blockade or NA depletion after reserpine treatment. Furthermore, increasing the local extraneuronal concentrations of NA by the reuptake blocker desipramine reduced the NPY Li overflow. Reserpine pretreatment per se seems to deplete not only NA from local tissue stores, but also NPY, probably because of increased release in excess of resupply (Lundblad et al., 1987; Lacroix et al., 1988) (Table 1). Therefore,

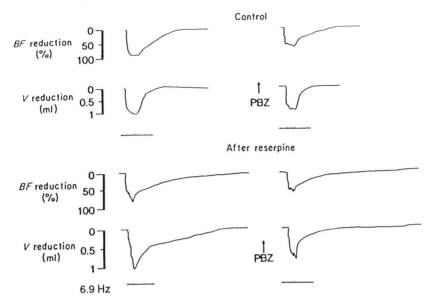

Figure 1 Recordings of pig nasal arterial blood flow (BF) and nasal mucosal volume (V, representing mainly capacitance vessel function) on sympathetic nerve stimulation with irregular bursts (recorded from the natural firing of human vasoconstrictor nerves) at an average frequency of 6.9 Hz for 30 sec in a control animal (top) and 24 hr after reserpine pretreatment (1 mg/kg i.v.) combined with preganglionic transection of the cervical sympathetic trunk (lower panel). The effects before and after local i.a. pretreatment with the α-adrenoceptor antagonist phenoxybenzamine (PBZ) are also illustrated. (Modified from Lacroix et al., 1988.)

preganglionic transection of the cervical sympathetic trunk should be performed simultaneously with reserpine administration to create a situation where NA release is absent while NPY overflow is enhanced (see Lundblad et al., 1987). Since large nasal and long-lasting vasoconstrictor effects to sympathetic stimulation remain after reserpine (Fig. 1), there is suggestive evidence for a role of a nonadrenergic transmitter like NPY both in the vascular control of the nasal mucosa (Lundblad et al., 1987; Lacroix et al., 1988) and trachea (Matran et al., 1993).

III. NPY RECEPTORS AND EFFECTS IN AIRWAYS

A. Vasoconstriction

NPY is a potent vasoconstrictor agent in the nasal (Lundblad et al., 1987; Lacroix et al., 1988) (Fig. 2) and tracheobronchial mucosa (Salonen et al., 1988; Matran,

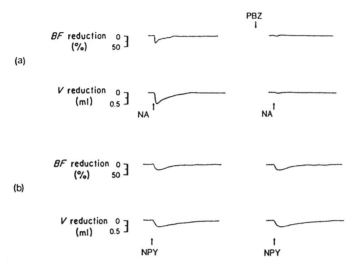

Figure 2 Typical recordings of reductions in pig nasal arterial blood flow (BF, %) and nasal mucosa volume (V, ml) caused by i.a. bolus injections of (a) NA (5×10^{-9} moles) and (b) NPY (6×10^{-9} moles). After local i.a. pretreatment with phenoxybenzamine (PBZ) the NA response was abolished while the NPY-evoked vasoconstriction was unaffected. (Modified from Lacroix et al., 1988.)

1991; Matran et al., 1993) (Table 2). In the cat, NPY seems to have preferential effects on blood flow compared to nasal mucosal volume (Lundblad et al., 1987). In contrast, in the pig (Lacroix et al., 1988) (Fig. 2) and humans (Baraniuk et al., 1992), NPY also influenced capacitance vessels in agreement with the innervation pattern (Lacroix et al., 1990) and presence of ^{125}I-NPY binding sites on venous vessels as well (Baraniuk et al., 1990). In contrast to the NA response, the nasal vascular NPY effect is long-lasting and occurs independently of pretreatment with adrenoceptor antagonists (Lacroix et al., 1988) (Fig. 2). Two weeks after surgical

Table 2 Summary of Major Effects of NPY in Airways

Arterial vasoconstriction
Capacitance vessel constriction
Inhibition of bradykinin-evoked plasma exudation
Inhibition of cholinergic bronchoconstriction
Inhibition of sensory NK2 receptor-mediated bronchoconstriction

Data are compiled from studies in experimental animals and humans. For references see text.

sympathetic denervation, there is a supersensitivity development for the vascular NPY effects in the nasal mucosa (as well as for α_2-agonists) (Lacroix and Lundberg, 1989), suggesting that the NPY receptors normally are down-regulated by release of endogenous peptide (Fig. 3).

Two major subtypes of NPY receptors (Y_1 and Y_2) have been postulated by pharmacological criteria (Wahlestedt et al., 1986, 1987). Thus, the C-terminal fragment of NPY, NPY(13–36), evokes only prejunctional inhibitory effects on transmitter release from sympathetic nerves, a classical effect of NPY (see Lundberg et al., 1982), while lacking vasoconstrictive effects in vitro. The NPY receptors were named Y_1 for the postjunctional and Y_2 for the prejunctional type. However, this subdivision was complicated by the observations that NPY(13–36) in fact caused vasoconstriction in vivo in pig spleen (Lundberg et al., 1988a) and nasal mucosa (Lacroix et al., 1988). Recently, it was suggested that the NPY analog Leu[31], Pro[34] NPY was selective for Y_1-receptors (Fuhlendorff et al., 1990). When the vascular effects of the Y_2- and Y_1-agonists were compared in the pig nasal vascular bed, it was demonstrated that the Y_1 analog had a similar response as NPY(1–36) while the Y_2-agonist NPY(13–36) was considerably less active, suggesting that Y_1-receptor mechanisms dominated in the nasal vascular bed, at least regarding control of blood flow (Modin et al., 1991) (Fig. 4).

One NPY receptor (Y_1), belonging to the G-protein-coupled receptor super-family, has recently been cloned and structurally identified (Herzog et al., 1992; Larhammar et al., 1992). This receptor can evidently couple to two different second-messenger systems; a pertussis toxin–sensitive G protein that mediates

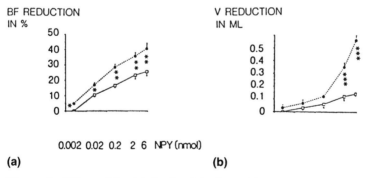

(a) **(b)**

Figure 3 Effects of local bolus i.a. injections of increasing doses of neuropeptide Y (NPY) on (a) the nasal arterial blood flow (BF, reduction in %) and (b) the volume (V, ml) of the nasal cavity. Control conditions (open circles) are compared with data obtained 2 weeks after surgical sympathectomy ($n = 6$, broken line, filled symbols). Data are given as means ± SEM. *$p < 0.05$; **$p < 0.01$; ***$p < 0.001$. (Modified from Lacroix and Lundberg, 1989.)

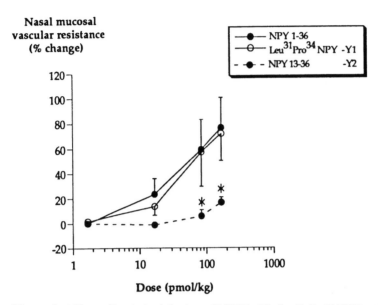

Figure 4   Effects of i.a. bolus injections of NPY(1–36), Leu[31], Pro[34]-NPY and NPY(13–36) on nasal mucosal vascular resistance expressed as percent change from basal values. Data = means ± SEM, $n = 5$, *$p < 0.05$. (Modified from Modin et al., 1991.)

inhibition of cyclic AMP accumulation and, alternatively, an elevation in intracellular Ca^{2+} occurs. In the absence of cloned Y_2-receptor, the final establishment of vascular Y-receptor subtypes awaits development of specific receptor antagonists.

B.   Prejunctional Regulation of Transmitter Release

NPY counteracts contractions of electrically stimulated tracheal rings (Stretton and Barnes, 1988), the vagally stimulated guinea pig trachea (Grundemar et al., 1988, 1990), and on-field stimulation-evoked bronchial contractions (Matran et al., 1989). Furthermore, NPY exerts similar effects in vivo (Stretton et al., 1990). These effects are most likely due to activation of presynaptic Y_2-receptors inhibiting the release of both acetylcholine and tachykinins (see Matran et al., 1989; Satoh et al., 1992) (Fig. 5).

It has been postulated that the NPY-evoked inhibition of transmitter release from sensory nerves in the airways, like the α_2-adrenoceptor effect, involves high conductance Ca^{2+}-activated K^+-channels (Stretton et al., 1992). However, this view has been challenged, at least for the α_2-receptor modulation (Lou and Lundberg, 1993).

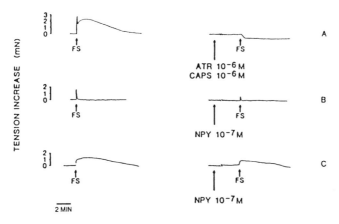

Figure 5 Typical examples of responses elicited by electrical field stimulation (FS, 5 Hz for 10 sec) of the isolated guinea pig hilus bronchi. (Left panel) (A) Control effect, (B) reduction of the cholinergic, and (C) nonadrenergic, noncholinergic (NANC) responses in the presence of NPY 10^{-7} M. (Right panel) (A) Bronchial response after capsaicin tachyphylaxis (CAPS) and atropine (ATR), (B) cholinergic contraction after capsaicin tachyphylaxis, (C) NANC contraction after atropine. (Modified from Matran et al., 1989.)

The prejunctional effect of NPY may be mediated by Y_2-receptors since the fragment NPY(13–36) is a relatively potent inhibitor of the electrical field stimulation-evoked contraction in the guinea pig trachea (Grundemar et al., 1990). In addition to a prejunctional inhibitory effect on transmitter release, NPY exerts a weak and sometimes inconsistent contraction of tracheobronchial smooth muscle in the guinea pig (Matran et al., 1989), an effect that at least in the trachea seems to be mediated via arachidonic acid metabolites (Cadieux et al., 1989).

IV. NPY IN HUMANS—CLINICAL IMPLICATIONS

The airway mucosa in humans receives a dense innervation by NPY-containing perivascular nerves (Lundberg et al., 1988; Baraniuk et al., 1990; Lacroix et al., 1990). Since local administration of exogenous NPY to the human nasal mucosa reduced nasal airflow resistance as well as albumin exudation evoked by bradykinin (Baraniuk et al., 1992), this peptide, perhaps in the form of more metabolically stable analogs, has a potential interest for the treatment of nasal mucosal disease characterized by congestion and plasma exudation. Since bradykinin activates capsaicin-sensitive sensory nerves, and NPY, at least in experimental animals, inhibits peptide release from these nerves, another potential action of NPY agonists is present at the prejunctional level in addition to the vasoconstrictor effect. The relative effects of NPY at the prejunctional level on sympathetic,

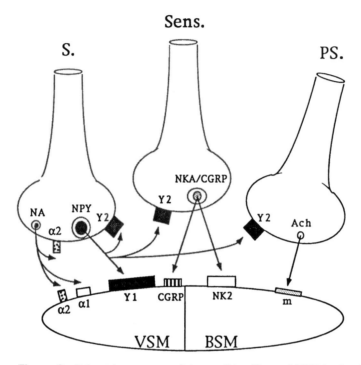

Figure 6 Schematic summary of the possible effects of NPY in the airways. NPY is coreleased with NA from sympathetic nerves (S.), causing vasoconstriction via Y_1 receptors and prejunctional regulation of transmitter release via Y_2 receptors from sympathetic, parasympathetic (PS.), and sensory nerves (Sens.). In addition, NPY is present in parasympathetic nerves in some species. Vascular smooth muscle (VSM) in the airways also contain α_1- and α_2-adrenoceptors and calcitonin gene-related peptide (CGRP). Bronchial smooth muscles (BSM) have neurokinin 2 (NK2) and muscarinic (m) receptors mediating contractions from released neurokinin A (NKA) and acetylcholine (ACh), respectively.

parasympathetic, and sensory nerves have to be established, however (Fig. 6). Furthermore, since the sensitivity of NPY-receptors seems to be regulated by endogenous mechanisms (Lacroix and Lundberg, 1989), possibly long-term changes regarding vascular Y-receptors upon repeated applications of NPY agonists should be considered.

ACKNOWLEDGMENTS

This chapter summarizes data obtained with support from the Swedish Medical Research Council (Grant 14X-6554). For excellent secretarial help we are grateful to Mrs. Ylva Jerhamre.

REFERENCES

Andersson, K.-E., and Bende, M. (1984). The role of adrenoceptors in the control of human nasal mucosal blood flow. *Ann. Otol. Rhinol. Laryngol. 93*:179.

Änggård, A., and Densert, O. (1974). Adrenergic innervation of the nasal mucosa in cat. A historical and physiological study. *Acta Oto-Laryngol. 78*:232.

Baraniuk, J.N.S., Castellino, J., Goff, J., Lundgren, J.D., Mullol, J., Merida, M., Shelhamer, J.H., and Kaliner, M.A. (1990). Neuropeptide Y (NPY) in human nasal mucosa. *Am. J. Respir. Cell Mol. Biol. 3*:165.

Baraniuk, J.N., Silver, P.B., Kaliner, M.A., and Barnes, P.J. (1992). Neuropeptide Y is a vasoconstrictor in human nasal mucosa. *J. Appl. Physiol. 73*(5):1867.

Cadieux, A., Benchekroun, M.T., St-Pierre, S., and Fournier, A. (1989). Bronchoconstrictor action of neuropeptide Y (NPY) in isolated guinea-pig airways. *Neuropeptides 13*:215.

Dahlström, A., and Fuxe, K. (1965). The adrenergic innervation of the nasal mucosa in certain mammals. *Acta Oto-Laryngol. 59*:65.

Fuhlendorff, J., Gether, U., Aakerlund, L., Langeland-Johansen, N. Thogersen, H., Melberg, S.G., Olsen, U.B., Thastrup, O., and Schwartz, T.W. (1990). [Leu^{31},Pro^{34}] Neuropeptide Y: a specific Y$_1$ receptor agonist. *Proc. Natl. Acad. Sci. USA 87*:182.

Grundemar, L., Widmark, E., Waldeck, B., and Håkansson, R. (1988). Neuropeptide Y: prejunctional inhibition of vagally induced contractions in the guinea-pig trachea. *Regul. Pept. 23*:309.

Grundemar, L., Grundström, N., Johansson, I.G.M., Andersson, R.G.G., and Håkansson, R. (1990). Suppression by neuropeptide Y of capsaicin-sensitive sensory nerve-mediated contraction in guinea-pig airways. *Br. J. Pharmacol. 99*:473.

Herzog, H., Hort, Y.J., Ball, H.J., Hayes, G., Shine, J., and Selbie, L.A. (1992). Cloned human neuropeptide Y receptor couples to two different second messenger systems. *Proc. Natl. Acad. Sci. USA 89*:5794.

Lacroix, J.S., and Lundberg, J.M. (1989). Adrenergic and neuropeptide Y supersensitivity in denervated nasal mucosa vasculature of the pig. *Eur. J. Pharmacol. 169*:125.

Lacroix, J.S., Stjärne, P., Änggård, A., and Lundberg, J.M. (1988). Sympathetic vascular control of the pig nasal mucosa. II. Reserpine-resistant, non-adrenergic nervous responses in relation to neuropeptide Y and ATP. *Acta Physiol. Scand. 133*:183.

Lacroix, J.S., Stjärne, P., Änggård, A., and Lundberg, J.M. (1989). Sympathetic vascular control of the pig nasal mucosa. III. Co-release of noradrenaline and neuropeptide Y. *Acta Physiol. Scand. 135*:17.

Lacroix, J.S., Änggård, A., Hökfelt, T., O'Hare, T., Fahrenkrug, J., and Lundberg, J.M. (1990). Neuropeptide Y: presence in sympathetic and parasympathetic innervation of the nasal mucosa. *Cell Tissue Res. 259*:119.

Larhammar, D., Blomqvist, A.G., Yee, F., Jazin, E., Yoo, H., and Wahlestedt, C. (1992). Cloning and functional expression of a human neuropeptide Y/peptide YY receptor of the Y1 type. *J. Biol Chem. 267*(16):10935.

Leblanc, G.G., Trimmer, B.A., and Landis, S.C. (1987). Neuropeptide Y-like immunoreactivity in rat cranial parasympathetic neurons: coexistence with vasoactive intestinal peptide and choline acetyltransferase. *Proc. Natl. Acad. Sci. USA 84*:3511.

Lou, Y.-P., and Lundberg, J.M. (1993). Different effects of the K$^+$ channel blockers 4-aminopyridine and charybdotoxin on sensory nerves in guinea-pig lung. *Pharmacol. Toxicol. 72*:139.

Lundberg, J.M., Hökfelt, T., Änggård, A., Kimmel, J., Goldstein, M., and Markey, K. (1980). Coexistence of an avian pancreatic polypeptide (APP) immunoreactive substance and catecholamines in some peripheral and central neurons. *Acta Physiol. Scand.* *110*:107.

Lundberg, J.M., Terenius, L., Hökfelt, T., Martling, C.-R., Tatemoto, K., Mutt, V., Polak, J., Blom, S.R., and Goldstein, M. (1982). Neuropeptide Y (NPY)-like immunoreactivity in peripheral noradrenergic neurons and effects of NPY on sympathetic function. *Acta Physiol. Scand.* *116*:477.

Lundberg, J.M., Terenius, L., Hökfelt, T., and Goldstein, M. (1983). High levels of neuropeptide Y in peripheral noradrenergic neurons in various mammals including man. *Neurosci Lett.* *42*:167.

Lundberg, J.M., Hemsén, A., Larsson, O., Rudehill, A., Saria, A., and Fredholm, B.B. (1988a). Neuropeptide Y receptor in pig spleen: binding characteristics, reduction of cyclic AMP formation and calcium antagonist inhibition of vasoconstriction. *Eur. J. Pharmacol.* *145*:21.

Lundberg, J.M., Martling, C.-R., and Hökfelt, T. (1988b). Airways, oral cavity and salivary glands: classical transmitters and peptides in sensory and autonomic motor neurons. In: *Handbook of Chemical Neuranatomy.* (Björklund, A., Hökfelt, T., and Owman, C., eds.). Elsevier, B.V., Amsterdam, p. 391.

Lundblad, L., Änggård, A., Saria, A., and Lundberg, J.M. (1987). Neuropeptide Y and non-adrenergic sympathetic vascular control of the cat nasal mucosa. *J. Auton. Nerv. Syst.* *20*:189.

Martling, C.-R., Matran, R., Alving, K., Hökfelt, T., and Lundberg, J.M. (1990). Innervation of lower airways and neuropeptide effects on bronchial and vascular tone in the pig. *Cell Tissue Res.* *260*:223.

Matran, R. (1991). Neural control of lower airway vasculature. *Acta Physiol. Scand.* *142* (Suppl. 601):1.

Matran, R., Martling, C.-R., and Lundberg, J.M. (1989). Inhibition of cholinergic and non-adrenergic, non-cholinergic bronchoconstriction in the guinea pig mediated by neuropeptide Y and α_2-adrenoceptors and opiate receptors. *Eur. J. Pharmacol.* *163*:15.

Matran, R., Franco-Cereceda, A., Alving, K., and Lundberg, J.M. Sympathetic vascular control of the laryngeo-tracheal, bronchial and pulmonary circulation in the pig: evidence for non-adrenergic mechanisms involving neuropeptide Y. *Acta Physiol. Scand.* (submitted).

Modin, A., Pernow, J., and Lundberg, J.M. (1991). Evidence for two neuropeptide Y receptors mediating vasoconstriction. *Eur. J. Pharmacol.* *203*:165.

Salonen, R.O., Webber, S.E., and Widdicombe, J.G. (1988). Effects of neuropeptides and capsaicin on the canine tracheal vasculature in vivo. *Br. J. Pharmacol.* *95*:1262.

Satoh, H., Lou, Y.-P., Lee, L.-Y., and Lundberg, J.M. (1992). Inhibitory effects of capsazepine and the NK2 antagonist SR48968 on bronchoconstriction evoked by sensory nerve stimulation in guinea-pigs. *Acta Physiol. Scand.* *146*:535.

Stretton, C.D., and Barnes, P.J. (1988). Modulation of cholinergic neurotransmission in guinea pig trachea by neuropeptide Y. *Br. J. Pharmacol.* *93*:672.

Stretton, C.D., Belvisi, M.G., and Barnes, P.J. (1990). Neuropeptide Y modulates non-adrenergic, non-cholinergic neural bronchoconstriction in vivo and in vitro. *Neuropeptides* *17*:163.

Stretton, C.D., Miura, M., Belvisi, G., and Barnes, P.J. (1992). Calcium-activated potassium channels mediate prejunctional inhibition of peripheral sensory nerves. *Proc. Natl. Acad. Sci. USA 89*:1325.

Tatemoto, K. (1982). Neuropeptide Y: complete amino acid sequence of the brain peptide. *Proc. Natl. Acad. Sci. USA 79*:5485.

Uddman, R., Sundler, F., and Emson, P. (1984). Occurrence and distribution of neuropeptide Y-immunoreactive nerves in the respiratory tract and middle ear. *Cell Tissue Res. 237*:321.

Wahlestedt, C., Yanaihara, N., and Håkansson, R. (1986). Evidence for different pre- and post-junctional receptors for neuropeptide Y and related peptide. *Regul. Pept. 13*:307.

Wahlestedt, C., Edvinsson, L., Ekblad, E., and Håkansson, R. (1987). Effects of neuropeptide Y at sympathetic neuroeffector junction: existence of Y_1 and Y_2 receptors. In: *Neuronal Messenger in Vascular Function*, Fernström Symp. No. 10. (Nobin, A., and Owman, C.H., eds.). Elsevier, Amsterdam, p. 231.

DISCUSSION

Polak: There are many NPY-immunoreactive nerve fibers in the human lung, both around blood vessels and in airway smooth muscle.

Lundberg: Yes, but they are probably not of sympathetic origin. In the smooth muscle, NPY may regulate transmitter release.

Belvisi: Stretton and Barnes (unpublished data) have shown that NPY does not inhibit cholinergic neurotransmission in human airways, in contrast to data in guinea pigs.

Barnes: Are there any NPY antagonists that are useful?

Lundberg: We have not found any NPY-blocking effect of the proposed NPY antagonist PP56 in the pig vasculature.

8

Sensory Nerves and Tachykinins

Stefano Manzini, Cristina Goso, and Arpad Szallasi
Menarini Ricerche Sud, Rome, Italy

I. INTRODUCTION

Asthma is a chronic inflammatory disease characterized by transient episodes of airflow obstruction and involvement of inflammatory cells such as mast cells and eosinophils (Asthma Consensus, 1992). In the last decade, a series of cell types and mediators have been proposed to be major determinants of the pathophysiology of this disease, although it is simplicistic to think of asthma as a derangement in the function of a single mediator. It is interesting to note that an important consistent finding in asthmatic subjects is airway hyperreactivity, i.e., an exaggerated pulmonary response to aerosol challenge with provocative agents as autacoids, irritants, or antigen. This notation could indicate the involvement, among other factors, of sensory nerves monitoring the airway environment in such a phenomenon. This hypothesis, first proposed by Barnes in 1986, is particularly attractive since a subset of sensory nerves innervating the airways (the capsaicin-sensitive ones) have the peculiarity to exert a dual sensory-efferent function. Appropriate stimulation of these nerves produces both sensory information to the central nervous system (eventually leading to protective reflex responses, such as cough) and a local release of their neurotransmitter contents (Paintal, 1973; Coleridge and Coleridge, 1984; Buck and Burks, 1986; Maggi and Meli, 1988; Fuller, 1990; Holzer, 1991). Released sensory neuropeptides are potent inducers of a series of motor and inflammatory effects collectively called "neurogenic inflammation" that have an impressive similarity to several features of asthma pathology.

The aim of this chapter is to give an update on sensory nerves in the mammalian airways (with special reference to the capsaicin-sensitive subset) and on the biological actions of the tachykinins, which are the major mediators released from these nerves. Finally, a brief outline of the new therapeutic perspectives opened by these studies will be presented.

II. THE SENSORY (AFFERENT) INNERVATION OF THE AIRWAYS

The lungs and the airways have a dual afferent innervation (cf. Coleridge and Coleridge, 1984). Some afferent fibers travel in sympathetic nerve branches; therefore, they are referred to as sympathetic afferents (Kostreva et al., 1975), whereas others join the vagal nerve (vagal afferents) (Paintal, 1973). Although sympathetic afferents arising from the lung and airways are responsive to chemical irritants (disturbances in breathing) (cf. Coleridge and Coleridge, 1984), they, unlike those that innervate the heart, do not appear to transmit pain perception (Morton et al., 1951). By contrast, a distinct subpopulation of vagal afferents respond to chemical irritants with the peculiar dual sensory-efferent response previously described in the Introduction. These vagal afferents share the trait of being susceptible to the triple (excitant, sensory blocking, and neurotoxic) actions of capsaicin (cf. Virus and Gebhart, 1979; Buck and Burks, 1986; Holzer, 1988); therefore, they are generally termed capsaicin-sensitive vagal afferents.

Recent findings suggest that the capsaicin-sensitive innervation of the airways is more complex than previously thought (Myers and Undem, 1991; Canning and Undem, 1993). At least in the guinea pig, there is experimental evidence for the existence of a capsaicin-sensitive relaxant innervation, which seems to be anatomically distinct from the contractile innervation (Canning and Undem, 1993): relaxant fibers are suggested to have somata in either the brain or the jugular ganglia (Canning and Undem, 1993), whereas the perykarya of the majority of the contractile fibers are likely to reside in the nodose ganglion (Widdicombe, 1964). Moreover, the relaxant, but not the contractile, innervation appears to be somehow associated with the esophagus (in vitro, relaxation could only be achieved when trachea preparations with intact esophagus are used) (Canning and Undem, 1993). This capsaicin-sensitive relaxant innervation of the guinea pig trachealis, however, appears not to use tachykinins as mediators. As yet, the mediator of this pathway is unknown; nitric oxide, however, may be an attractive candidate.

To make the picture even more complicated, recently, a close bidirectional link between capsaicin-sensitive and cholinergic fibers has also been described (Myers and Undem, 1991). Electric field stimulation (EFS) of the guinea pig bronchus results in a biphasic constriction (Grundstrom et al., 1981): the rapid first phase is abolished by hexamethonium or atropine (i.e., it is cholinergic in nature), whereas the second prolonged phase can be prevented by capsaicin pretreatment (i.e., it is linked to the activation of capsaicin-sensitive afferents). Interestingly, it has been found that atropine but not hexamethonium, an agent selective for nicotinic receptors, can reduce (by approximately 30%) the magnitude of the capsaicin-sensitive contractions (Myers and Undem, 1991). It has also been observed that after stimulation of capsaicin-sensitive afferents the cholinergic contractions increased (Myers and Undem, 1991). Among other neuropeptides, tachykinins can prejunctionally facilitate the release of neurotransmitters from cholinergic nerves (Barnes, 1992). Bronchospasm induced in anaesthetized guinea pigs by a synthetic selective agonist of tachykinin NK-2 receptors is reduced by atropine and enhanced by acetylcholinesterase inhibitors (Ballati et al., 1992a), suggesting that this cholinergic facilitation could be operative in vivo, as well. Since capsaicin-sensitive airway afferents are thought to possess nicotinic receptors (Kizawa and Takayanagi, 1985), acetylcholine may also have a positive feedback action.

III. THE CAPSAICIN-SENSITIVE PRIMARY AFFERENT

Capsaicin has been in use for decades to study the physiological function of these pathways and their contribution to disease states of the airways (cf. Buck and Burks, 1986; Szallasi and Blumberg, 1993a). A consistent picture is, however, hampered by the striking species differences, which make the extrapolation from animal experimentation to human beings extremely complicated (Glinsukon et al., 1980). At least five factors are thought to contribute to these species differences:

1. Species-related differences in the anatomical organization of capsaicin-sensitive vagal fibers that innervate the airways (cf. Paintal, 1973; Coleridge and Coleridge, 1984)
2. Species-related neurochemical differences (i.e., differences in neurotransmitters stored in and released from these nerves) (cf. Holzer, 1988)
3. Species-related differences in the expression of receptors for these neurotransmitters, as exemplified by the tachykinin receptors (cf. Maggi et al., 1993a)
4. Species-related differences in the expression of vanilloid (capsaicin) receptors (cf. Szallasi and Blumberg, 1993b) and/or pharmacodynamic differences (i.e., contrasting results depending on the route of administration and the chemical nature of the agonist) (Szolcsanyi et al., 1990, 1991)
5. Species-related differences in other neural pathways that can interact with capsaicin-sensitive sensory vagal fibers (Myers and Undem, 1991)

Before dealing with these issues, we briefly summarize our current knowledge on how capsaicin and related compounds act at the subcellular level.

Capsaicin interacts at a specific membrane recognition site, referred to as the "vanilloid receptor" (Szallasi and Blumberg, 1990a), shared with the ultrapotent agonist resiniferatoxin (Szallasi and Blumberg, 1989, 1990b) and the competitive antagonist capsazepine (Bevan et al., 1992). Upon binding to the vanilloid receptor, capsaicin opens a cation conductance, which is permeable to both divalent and monovalent cations (Bevan et al., 1987; Marsh et al., 1987; Bevan and Szolcsanyi, 1990). The functional capsaicin antagonist ruthenium red, which does not interfere with capsaicin binding (Szallasi and Blumberg, 1990a), is supposed to block the channel function (cf. Amann and Maggi, 1991). As yet, it is not known whether the vanilloid receptor is operated by an endogenous ligand. The resulting influx of cations leads to impulse generation (afferent function) and to a release of neuromediators (efferent function) (cf. Holzer, 1991). Excitation by capsaicin is followed by a refractory state, which can be either reversible (traditionally termed "desensitization") (Jancso, 1968) or irreversible.

Desensitization is presumably a very complex process (cf. Holzer, 1991; Dray, 1992), which, among others, includes a block of intra-axonal transport (Miller et al., 1982; Taylor et al., 1984, 1985) that deprives somata of nerve growth factor (NGF) (in the absence of NGF, cultured DRG neurons lose their ability to respond to capsaicin) (Winter et al., 1988), a calcium-dependent activation of calcineurin (Yeats et al., 1992), a neuronal phosphatase, which, in turn, may inactivate either the binding site or the connected conductance, and a generalized impairment of cellular functions, which is likely to follow calcium accumulation (Bevan et al., 1987). Acute and chronic desensitization are likely to involve distinct mechanisms: chronic, but not acute, desensitization was shown to be associated with a loss of vanilloid receptors (Szallasi and Blumberg, 1992). The irreversible loss of responsiveness most likely reflects gross neurotoxicity, which is thought to be due to a combined effect of rising calcium and sodium levels and the osmotic damage that follows intracellular NaCl formation (cf. Bevan and Szolcsanyi, 1990). In addition to its "specific" neural actions, capsaicin also exerts a variety of "aspecific" nonneuronal effects (cf. Holzer, 1991). It is noteworthy that a nonneural capsaicin action, i.e., the block of potassium conductance in ventricular myocytes, which has been reevaluated by using the ultrapotent capsaicin analog resiniferatoxin (RTX), is insensitive to RTX, suggesting that these actions are, in fact, either "aspecific" or are mediated by receptors that possess structure-activity relations distinct from that of the neural vanilloid receptor (Castle, 1992). We anticipate that further use of RTX or other selective agonists will make a better distinction between specific and nonspecific capsaicin-like actions.

Capsaicin-sensitive sensory nerves characteristically use tachykinins, such as substance P and neurokinin A, as neurotransmitters (cf. Buck and Burks, 1986; Holzer, 1988). In addition to the tachykinins, they also store and release a large variety of other neuropeptides, such as calcitonin gene-related peptide (CGRP), vasoactive intestinal peptide (VIP), and galanin (cf. Holzer, 1988; Maggi and Meli, 1988). The biological and pathobiological functions of these neuropeptides have recently been covered in detail (Joos, 1989; Barnes, 1990; Holzer, 1991; Maggi et al., 1993a). It is important, however, to note that neuropeptides colocalized in capsaicin-sensitive nerves represent impressive examples for biological cooperativity: CGRP, for example, not only potentiates release of substance P in the dorsal horn, but also inhibits the endopeptidase that cleaves substance P (cf. Hokfelt, 1991).

Although surprisingly little is known about the physiological role of capsaicin-sensitive neural pathways in the airways, increasing evidence suggests that they are activated by a great variety of endogenous and exogenous agents and thus may be involved in the pathobiology of various disease states (cf. Joos, 1989; Barnes, 1990; Maggi, 1991). Since the initial report of Lundberg and Saria (1983) that capsaicin pretreatment protects against irritancy by cigarette smoke in rat airways, numerous exogenous agents have been reported to activate capsaicin-sensitive vagal afferents, including toluene diisocyanate (Mapp et al., 1990), an important cause of occupational asthma. Among endogenous agents, protons have recently gained special attention as candidates for being endogenous operators of the vanilloid receptor (cf. Dray, 1992). Protons have been shown to open a cation conductance with properties similar to the vanilloid-operated cation channel (Bevan and Yeats, 1991). Nevertheless, whether vanilloid- and proton-activated conductances are identical or only similar is still hotly debated. Bradykinin (Kaufman et al., 1980; Qian et al., 1993) and lipoxin A_4 (Meini et al., 1992) represent other important endogenous agents that can activate capsaicin-sensitive airway afferents. It should be emphasized, however, that neither low pH (Szallasi and Blumberg, 1993a) nor bradykinin (Szallasi, unpublished observation) nor lipoxin A_4 (Meini et al., 1992) appears to interact at vanilloid receptors.

Since the guinea pig is a frequently used species to model human allergic/hyperreactive airway diseases, it is hardly surprising that capsaicin-sensitive nerves are best characterized in guinea pig airways. The existence of substance P–immunoreactive nerve fibers in the guinea pig airways was established by Nilsson in 1977; the actual capsaicin sensitivity of these fibers was shown in the early eighties by two groups working independently, Janos Szolcsanyi's group in Hungary (Szolcsanyi and Bartho, 1982; Szolcsanyi, 1983) and Jan M. Lundberg's team at the Karolinska Institute, Sweden (Lundberg and Saria, 1983; Dahlsgaard and Lundberg, 1983; Lundberg et al., 1983a,b). It was also found that the capsaicin-sensitive innervation of the guinea pig trachea and the stem bronchi is asymmetrical (Lundberg et al., 1983b): these areas receive vagal afferents mainly from the right vagus nerve. The somata of capsaicin-sensitive vagal afferents innervating the trachea and the stem bronchi are thought to reside in the nodose ganglion (Widdicombe, 1964). By contrast, the lungs receive capsaicin-sensitive afferents symmetrically both from the vagus nerve and from a nonvagal source (Lundberg et al., 1983b). The latter fibers, which account for approximately 40% of the substance P–immunoreactive fibers in the lungs, are thought to originate from thoracic dorsal root ganglia (Lundberg et al., 1983b).

Less is known about the capsaicin-sensitive afferent innervation of the human airways. Capsaicin appears to induce a similar pattern of responses in human and guinea pig airways; the threshold dose is, however, higher in the human (Lundberg et al., 1983c; Joos, 1989). Whether this difference reflects the existence of a "less sensitive" vanilloid receptor subtype in human airways or is predominantly due to pharmacokinetical differences remains to be established. It should also be kept in mind that dose-response curves for capsaicin in human airways usually have been determined using surgical specimens obtained from older smokers, and it is known that long-term smoking can reduce airway substance P levels (cf. Barnes, 1990). Capsaicin-induced upper airway reflexes are better studied in human beings (cf. Fuller, 1990). Topical capsaicin administration to the nasal mucosa provokes sneezing (Geppetti et al., 1988) and an increase in mucus secretion that can be blocked by both antimuscarinic agents and local anesthetics (Stjarne et al., 1989). There is, however, little change in nasal resistance (Geppetti et al., 1988). Inhalation of capsaicin from a nebulizer results in cough and a reflex increase in airway resistance (Collier and Fuller, 1984; Fuller et al., 1985; Choudry et al., 1993). It should be noted that prostaglandin E_2 (PGE_2), as in animal experimentation, enhanced the sensitivity of the human cough reflex to capsaicin (Choudry et al., 1989). The mechanism by which PGE_2 sensitizes capsaicin-sensitive nerves is unclear. It appears to be unrelated to the vanilloid receptor since PGE_2 had no apparent effect on RTX binding (Szallasi, unpublished observation). Capsaicin when injected i.v. produced a "raw, burning" sensation in the chest and, at least in some subjects, paroxysmal coughing (Winning et al., 1986).

IV.   THE VANILLOID (CAPSAICIN) RECEPTOR IN AIRWAYS: TISSUE- AND SPECIES-RELATED DIFFERENCES

Although capsaicin and its ultrapotent analog RTX differ dramatically in structure (Fig. 1), they share a homovanillyl moiety, which appears to play a pivotal role in capsaicin-like activity. Chemical modification of the homovanillic acid in capsaicin leads to loss of potency (Szolcsanyi and Jancso-Gabor, 1975, 1976). Likewise, replacement of the homovanillic acid moiety of RTX with a hydroxy-phenyl group, as found in tinyatoxin, results in an approximately 10-fold drop in potency (Szallasi et al., 1991), whereas the complete removal of the homovanillyl moiety (resiniferonol 9,13,14-orthophenylacetate) leads to a complete loss of capsaicin-like activity (Szallasi et al., 1989). Therefore, capsaicin-like molecules appear to be best termed vanilloids, and the receptor at which they interact may be referred to as the vanilloid receptor (Szallasi and Blumberg, 1990a).

A combination of high lipophilicity and relatively poor potency (in vitro capsaicin displays an affinity for most of the capsaicin-like responses in the micromolar-submicromolar range) prevented the use of radiolabeled capsaicin to demonstrate the existence of specific capsaicin binding sites. On the other hand, by using [^{3}H]RTX we have been able to demonstrate specific, saturable binding to membranes obtained from dorsal root ganglia as well as spinal cord of a variety of species (Szallasi and Blumberg, 1990a, 1991a, 1993a). This specific binding

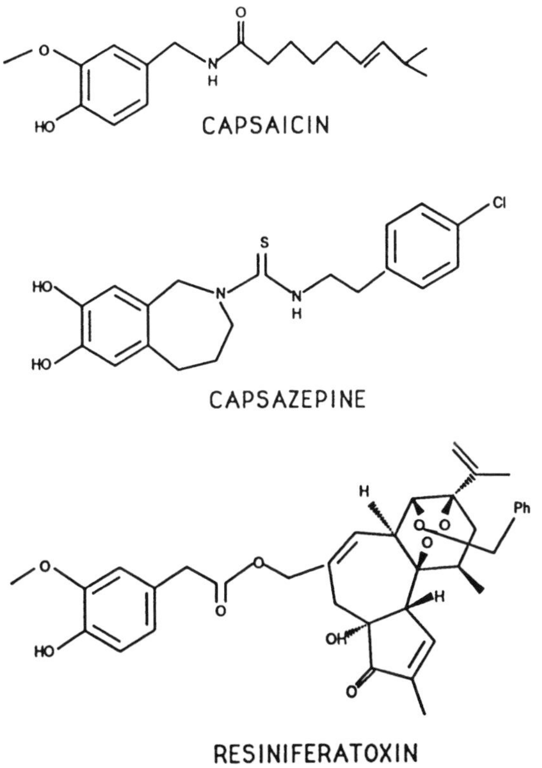

CAPSAICIN

CAPSAZEPINE

RESINIFERATOXIN

Figure 1 Structure of capsaicin, its ultrapotent agonist, resiniferatoxin, and their competitive antagonist, capsazepine.

showed appropriate tissue distribution (no specific binding in tissues, such as cerebellum, unresponsive to capsaicin), appropriate species (no specific binding in species, such as chicken, regarded insensitive to capsaicin actions) as well as pharmacological specificity to represent the vanilloid receptor (Szallasi and Blumberg, 1990a,b). Moreover, surgical or chemical ablation of capsaicin-sensitive nerves abolished specific RTX binding (Szallasi et al., 1990, 1993a,b).

Specific [³H]RTX binding is fully inhibited by capsaicin or biologically active capsaicin and RTX analogs, but not by biologically inactive or marginally active structural analogs (Szallasi and Blumberg, 1990a; Szallasi et al., 1991). Although both RTX and capsaicin show striking differences in affinity to produce different in vivo and in vitro responses, the underlying mechanism of which is unclear, in general, RTX is orders of magnitude more potent than capsaicin (Szallasi and Blumberg, 1989, 1990b; Maggi et al., 1990; Winter et al., 1990). In accord, cold

RTX inhibits specific binding of [³H]RTX with a 1000- to 10,000-fold higher potency than does capsaicin (Szallasi and Blumberg, 1990a; Szallasi et al., 1993a). The recently developed capsaicin antagonist capsazepine (James et al., 1992; Szallasi et al., 1993c), but not the functional antagonist ruthenium red (Szallasi and Blumberg, 1990a), competes for specific RTX binding sites.

Radiation inactivation analysis of the porcine vanilloid receptor yielded an apparent molecular weight (radiation inactivation size) of 270 kDa (Szallasi and Blumberg, 1991b). This size corresponds to the molecular target size of receptor complexes, such as, for example, the benzodiazepine/GABA receptor complex (Chang and Barnard, 1982).

In rat sensory ganglia as well as in spinal cord, but not in peripheral tissues, specific [³H]RTX binding displays apparent positive cooperativity of binding (Szallasi and Blumberg, 1993a; Szallasi et al., 1993b,d). Moreover, limited experimental evidence suggests the existence of a redox modulatory site of binding (Szallasi et al., 1993d), as exemplified by the NMDA receptor complex (Aizenman et al., 1989). Since positive cooperativity of binding is usually indicative of the existence of more than one binding site on the same receptor (Fels et al., 1982), it is tempting to speculate that central vanilloid receptors comprise a receptor complex with at least two interacting binding sites, a functionally linked redox modulatory site and a separate recognition site at which ruthenium red exerts its inhibitory action (this latter site is presumably the channel itself).

While central vanilloid receptors are relatively well characterized, the biochemical means to study peripheral vanilloid receptors have only recently become available. In the traditional RTX binding assay, the high nonspecific binding completely masks specific binding by peripheral tissues (Szallasi and Blumberg, 1990a). Nonspecific RTX binding can, however, be diminished by adding alpha₁-acid glycoprotein (AGP), also known as orosomucoid, to the assay mixture after the binding assay has been terminated (Szallasi et al., 1992). While specific binding of [³H]RTX to the vanilloid receptor shows a very profound temperature dependence (on ice, both association and dissociation are unmeasurably slow) (Szallasi and Blumberg, 1993a), AGP readily binds RTX even at 0°C (Szallasi et al., 1992). This difference in temperature dependence allows us to use AGP to reduce nonspecific RTX binding without compromising specific binding. In this modified assay, nonspecific binding that without adding AGP at the K_d value represents approximately 50% of the total when using rat DRG or spinal cord membranes (Szallasi and Blumberg, 1990a) comprises less than 10% of the total binding after AGP has been added (Szallasi et al., 1992). This modified RTX binding assay is now able to detect peripheral vanilloid receptors in a variety of peripheral tissues including the airways (Szallasi et al., 1993c). Interestingly, specific [³H]RTX binding to both rat urinary bladder and airway membranes differs from RTX binding to rat DRG and spinal cord membranes (Fig. 2): (1) in peripheral tissues, binding is clearly noncooperative (Szallasi et al., 1993a,b), and

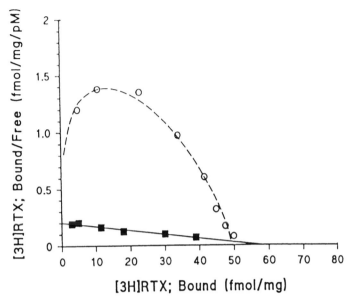

Figure 2 Scatchard plot of specific [³H]resiniferatoxin (RTX) binding to particulate preparations obtained from spinal cord (○), and airways (■) of the rat. Points represent mean values from a typical experiment. For spinal cord membranes, theoretical curves were calculated from the values for K_d, B_{max}, and Hill number determined by computer fit to the allosteric Hill equation. For airway membranes, the line was fitted using the LIGAND program. Note the convex Scatchard plots characteristic of positive cooperativity of ligand binding in spinal cord membranes as opposed to the straight Scatchard plot in airways indicating noncooperative binding. In rat spinal cord, the affinity of the vanilloid receptor to RTX is approximately 10-fold higher than in the airways (K_d values are 20 pM and 290 pM, respectively).

(2) the binding affinity of RTX to peripheral tissues is somewhat lower; K_d values are 18–19 pM for DRG (Szallasi and Blumberg, 1993a; Szallasi et al., 1993a), 16 pM for spinal cord (Szallasi et al., 1993a), 30–90 pM for urinary bladder (Szallasi et al., 1993a,b), and 250 pM for airways (Szallasi et al., submitted). If, in fact, positive cooperativity of binding in rat DRG and spinal cord reflects the existence of two or more binding subunits, the noncooperative binding in peripheral tissues implies a single binding site. Radiation inactivation analysis of peripheral tissues may confirm the validity of this assumption.

Vanilloid receptors in rat urinary bladder (Szallasi et al., 1993a) and airway membranes (Szallasi et al., submitted) bind capsaicin with similar potencies. By contrast, capsazepine binds to vanilloid receptors in airways with a 40-fold higher affinity (K_i values are 5 μM for urinary bladder, and 0.12 μM for airway membranes) (Szallasi et al., submitted). We surmise that this difference may be indicative of heterogeneity of peripheral vanilloid receptors. As to the urinary bladder, the 5 μM binding affinity of capsazepine is in excellent agreement with its biological potency to block capsaicin-induced responses (Maggi et al., 1993b). Unfortunately, the biological potency of capsazepine to inhibit capsaicin-induced responses in rat airways has not yet been determined. In guinea pig airways, RTX displayed an affinity of approximately 1 nM (Szallasi et al., 1993c). In accord with the previous functional and neurochemical findings (Manzini et al., 1989), we found an approximately 60% higher binding in the bronchi than in the trachea (Szallasi et al., 1993c). Capsaicin and capsazepine inhibited specific binding of [^{3}H]RTX to guinea pig airway membranes with K_i values of 500 nM and 100 nM, respectively (Szallasi et al., 1993c). These binding affinities are in fair agreement with the biological potencies of these compounds (Belvisi et al., 1992).

In sharp contrast to the rat or guinea pig, hamsters are almost insensitive to capsaicin actions (Glinsukon et al., 1980). Therefore, our findings indicating the existence of vanilloid receptors in hamster DRG were unexpected (Szallasi and Blumberg, 1993a). Interestingly, unlike in DRG, specific RTX-binding sites are virtually absent in hamster airways (Szallasi, unpublished observation). If the assumption that hamsters synthesize vanilloid receptors in DRG but for some obscure reason do not transport them to the peripheral nerve terminals turns out to be valid, this mechanism may provide a mechanistic understanding of the capsaicin insensitivity of the hamster.

In conclusion, our rather preliminary binding data confirmed the existence of vanilloid receptors in the airways of the rat and the guinea pig, species sensitive to airway irritation by capsaicin, while in the hamster, a species almost resistant to capsaicin, specific binding remained under the detection limit of the current binding assay. In addition, binding data have addressed some important questions, such as heterogeneity between central and peripheral vanilloid receptors on the one hand and among peripheral vanilloid receptors on the other hand, which remain to be answered.

V. BIOLOGICAL EFFECTS OF TACHYKININS IN THE RESPIRATORY TRACT

Although activation of the endings of capsaicin-sensitive nerves produces the corelease of several neurotransmitters, the vast majority of the motor and inflammatory effects in the airways can be explained by the actions of tachykinins, and this will be the subject of this section. Of course, CGRP also can elicit impressive biological effects mainly through vasodilation of the pulmonary circulation and reduction of metabolic degradation of tachykinins. In addition, some direct effects of CGRP on gland secretion, leukocyte adhesion, and inflammatory cell (T lymphocyte and neutrophil) function have been proposed. However, the biological effects of CGRP in the airways has recently been reviewed (Maggi, 1993).

A. Motor Effects

Tachykinins are powerful and effective endogenous bronchoconstrictor substances (Advenier et al., 1987; Martling et al., 1987; Maggi, 1990; Ballati et al., 1992a). This has also been confirmed in human bronchus, with neurokinin A being more potent than substance P (Advenier et al., 1987; Naline et al., 1989), suggesting a major involvement of NK-2 receptors. Both inhaled and infused neurokinin A, but not substance P, generate bronchoconstriction in human volunteers (Fuller et al., 1987; Evans et al., 1988; Crimi et al., 1988), and this response is greater in asthmatics than in healthy subjects (Joos et al., 1987). Inhaled neurokinin A–induced bronchoconstriction in humans can be further enhanced by the inhalation of a neutral endopeptidase inhibitor such as thiorphan (Cheung et al., 1992). It is important to understand that in the airways, NK-1 receptors can mediate both bronchoconstriction (Devillier et al., 1988; Maggi et al., 1991) and bronchodilatation (Manzini, 1992), depending on their localization (smooth muscle vs. epithelium), while activation of NK-2 receptors generally leads to bronchoconstriction (Advenier et al., 1987; Devillier et al., 1988). Consequently, the same neuropeptide, substance P, can exert a contractile or, alternatively, relaxant action depending on which tachykinin receptor is expressed predominantly in a given species. In the guinea pig, bronchoconstriction by vagal afferents is supposed to be mediated by an interaction of substance P and neurokinin A at both NK-1 and NK-2 receptors (Devillier et al., 1988). By contrast, in the mouse activation of the same pathway leads to bronchodilatation presumably by an interaction of substance P at NK-1 receptors located on epithelial cells (Manzini, 1992).

B. Effects on Vascular Permeability

Tachykinins and capsaicin increase vascular permeability in rat and guinea pig airways, likely through the opening of endothelial gaps at postcapillary venules (McDonald et al., 1988; Rogers et al., 1988). By using receptor-selective synthetic tachykinins agonists, the involvement of NK-1 receptor has been suggested (Abelli et al., 1991). This hypothesis has recently been confirmed by the demonstration that a selective NK-1 nonpeptide antagonist (CP 96,345) blocked neurogenic, substance P, or cigarette smoke–induced increase in plasma exudation and edema in guinea pig airways (Eglezos et al., 1991; Delay-Goyet and Lundberg, 1991; Lei et al., 1992).

C. Secretory Effects

Substance P is a potent stimulant of myoepithelial cells of submucosal gland (Coles et al., 1984), and in this way it increases mucus secretion (Barnes et al., 1991). By using selective agonist and antagonists, the involvement of NK-1 receptors in tachykinin-induced airways mucus secretion has recently been defined (Gentry, 1991; Meini et al., 1992b; Geppetti et al., 1992). Substance P– and capsaicin-sensitive vagal nerve stimulation excite the secretory function of goblet cells, as well (Kuo et al., 1990). The potent secretagogue properties of substance P have been confirmed in human bronchi and nasal mucosa (Rogers et al., 1989; Mullol et al., 1992).

D. Effects on Inflammatory Cells

Tachykinins can modulate the function of several inflammatory cells. Some of these actions (e.g., mast cell and eosinophil degranulation) are unrelated to stimulation of specific tachykinin receptors and probably due to direct stimulation of G proteins (Repke and Bienert, 1987, 1988; Kroegel et al., 1990). However, specific receptorial tachykinin effects on inflammatory cells have also been described. For example, alveolar macrophages are activated by extremely low concentrations of tachykinins (especially selective agonists of the NK-2 receptor) (Brunelleschi et al., 1990), and this response seems to be amplified in cells taken from antigen-sensitized and -boosted guinea pigs (Brunelleschi et al., 1992). Of relevance is the recent anatomical evidence suggesting a close proximity of sensory nerves and alveolar macrophages (Weihe, 1992). Evidence in favor of and against a receptorial stimulating effect of tachykinins on human neutrophil function is present in the literature (Wozniak et al., 1989; Wiedermann et al., 1989; Iwamoto et al., 1990). A priming effect rather than a direct stimulating action could also be involved (Perianin et al., 1989; Brunelleschi et al., 1991). Neutrophil chemotactic effects of tachykinins have been shown, and this action is due to the C-terminal portion of tachykinin sequence (Marasco et al., 1981). McDonald (1988) has demonstrated a massive recruitment of neutrophils in rat trachea during neurogenic inflammation. The expression of an endothelial leukocyte adhesion molecule (ELAM-1) by microvascular endothelium following substance P exposure (Matis et al., 1990) could be relevant in such migration. In addition, an important role of tachykinins as modulators of proliferative and physiological responses of T lymphocytes has been proposed (Payan et al., 1983, 1984; Casini et al., 1989; McGillis et al., 1990).

VI. CHANGES IN SENSORY NERVE FUNCTION IN AIRWAY ALLERGIC MODELS

If tachykinins are relevant for the pathogenesis of asthma, their depletion should exert a protective effect in experimental models mimicking asthmatic response to antigen challenge. Various studies in capsaicin-pretreated guinea pigs have been carried out, and evidence in favor of (Saria et al., 1983; Manzini et al., 1987; Ladenius and Biggs, 1989; Lundberg et al., 1991) and against (Ahlstedt et al., 1986; Lai, 1991; Ingenito et al., 1991) this hypothesis has been published. At least in the guinea pig, protection was generally observed when capsaicin treatment was carried out after the immunological sensitization and before antigen challenge. Recent findings indicate that blockade of tachykinin metabolism by neutral endopeptidase inhibitors produces a significant enhancement of the bronchomotor response to antigen challenge (Kohrogi et al., 1991) and that capsaicin pretreatment antagonizes the increase in bronchial hyperreactivity elicited by repeated antigen aerosol challenges in guinea pigs (Matsuse et al., 1991). In guinea pig isolated trachea, Ellis and Undem (1992) provided functional evidence that antigen challenge, by releasing histamine, can enhance noncholinergic contractions due to the release of tachykinins from capsaicin-sensitive nerves. Further, in guinea pigs sensitized and repeatedly challenged with antigen, an increase of bronchial reactivity to neurokinin A (Ballati et al., 1992b), substance P (Boichot et al., 1992), and nonadrenergic, noncholinergic vagal stimulation (Ballati et al., 1992b) has recently been reported.

Capsaicin pretreatment has been shown to abolish the bronchial hyperresponsiveness (but not the hypotension) induced by i.v. administration of platelet-activating factor (PAF) in anesthetized guinea pigs (Perretti and Manzini, 1992). Interestingly, even in the rabbit (a species relatively resistant to the effects of capsaicin), the capsaicin pretreatment both in the adult (Spina et al., 1991) and in the newborn (Riccio et al., 1993) can prevent the development of bronchial hyperreactivity to PAF or antigen. These protective effects of capsaicin pretreatment were uncoupled to a clear depletion of tachykinins in bronchial tissues. Further studies are necessary to determine whether these findings could be due to: (a) a depletion of a small pool of sensory peptides releasable by capsaicin and undetected by current methodology, (b) a peculiar long-lasting sensory neuron blocking action of capsaicin, and (c) some effects of capsaicin unrelated to sensory nerves.

In humans, very little is known about the interaction between sensory nerves and allergic processes; however, Iwamoto et al., (1990) demonstrated that in allergic asthmatic patients substance P, but not neurokinin A, produced greater erythemas and wheals than in normal subjects. Recently, a substantial increase in the number and length of substance P–immunoreactive nerves has been found in bronchial specimens obtained from asthmatic subjects as compared to controls or to patients with nonasthmatic chronic airflow obstruction (Ollerenshaw et al., 1991), even though contradictory results have been reported by Howarth et al. (1991). An overexpression of the gene encoding NK-1 binding sites has been shown in bronchial samples obtained from asthmatic subjects (Peters et al., 1992).

Further studies are needed to explain the vicious circle between allergic processes and activation of sensory nerves in human lung.

VII. THERAPEUTIC PERSPECTIVES

The great increase in our knowledge of the potential role of sensory nerves and tachykinins in the physiology and pathology of airways has boosted the pharmaceutical research to identify and synthesize specific antagonists of tachykinin receptors (see Table 1) as possible new therapeutic agents for respiratory diseases. Recently, potent, selective, and bioavailable tachykinin receptor antagonists have been developed and are now undergoing phase I and II clinical trials. These compounds can be regarded as suitable tools to investigate in humans the physiological and pathophysiological role of tachykinins, and it is reasonable to forecast that within 2–3 years their potential as new therapeutic agents in respiratory disorders will be exploited. Interestingly, a recent clinical report indicates the effectiveness of FK224 in protecting asthmatic patients from bradykinin aerosol-induced bronchoconstriction and cough (Ichinose et al., 1992).

In addition, the novel findings suggesting a heterogeneity of the vanilloid

Table 1 Tachykinin Receptor Antagonists

A. Unselective peptide tachykinin receptor antagonists

[D-Pro^2, D-Trp^{7,9}] substance P

Spantide I: [D-Arg^1, D-Trp^{7,9}, Leu^{11}] substance P

B. Selective peptide tachykinin receptor antagonists

NK-1 selective

L 668,169: cyclo (Gln- D-Trp- (NMe) Phe - (R)Gly [ANC-2] -Leu- Met)$_2$

Spantide II: [D-NicLys^1, Pal^3,D-Cl$_2$Phe^5,Asn^6-,DTrp^{7,9},Nle^{11}] substance P

GR 82,334: [D-Pro^9 [spiro-γ-lactam] Leu^{10},Trp^{11}]physalaemin (1-11)

FR 113,680: Ac-Thr-D-Trp(CHO)-Phe-NMeBzl

FK888: (3-(N-Me)indolil)-CO-Hyp-2 Nal-NMeBzl

NK-2 selective

L 659,877: cyclo (Gln-Trp-Phe-Gly-Leu-Met)

MEN 10376: [Tyr^5, D-Trp^{6,8,9},Lys^{10}] Neurokinin A (4-10)

R396: Ac-Leu-Asp-Gln-Trp-Phe-GlyNH$_2$

MDL 29,913: cyclo [Leu-Ψ (CH$_2$NCH$_3$)-Leu-Gln-Trp-Phe-Gly]

MEN 10627: c(Met-Asp-Trp-Phe-Dap-Leu)

C. Selective nonpeptide tachykinin receptor antagonists

NK-1 selective

CP 96,345: [(2S, 3S)-*cis*-2-(diphenylmethyl)-*N*-[(2-methoxyphenyl)-methyl]-1-azabicyclo[2,2,2] octan-3-amine]

RP67,580: {(3aR, 7aR) 7,7-diphenyl-2- [imino-2-(2 methoxyphenyl) ethyl] perhydroisoindol-4-one}

NK-2 selective

SR 48,968: (S)-*N*-methyl-*N*-[4- (4-acetylamino-4-phenylpiperidino)-2-(3,4-dichlorophenyl) butyl] benzamide

D. Equipotent antagonists of NK-1 and NK-2 receptors

FK 224: {*N*-[*N*^{2}-[*N*-[*N*-[*N*-[2,3-didehydro-*N*-methyl-*N*-[*N*-[3-(2-pentylphenyl)-propionyl]-L-threonil]tyrosyl-L-leucynyl]-D-phenylalanyl]-L-*allo*-threonyl]-L-asparaginyl]-L-serine-v-lactone}

---

receptor and some peculiarities of its expression in the airways could represent a starting point for the identification of new selective molecular entities that can inhibit at the prejunctional level the excitability of this subset of sensory nerves, thus preventing the occurrence of neurogenic inflammation.

REFERENCES

Abelli, L., Maggi, C.A., Rovero, P., Del Bianco, E., Regoli, D., Drapeau, G., and Giachetti, A. (1991). Effect of synthetic tachykinin analogues on airway microvascular leakage in rats and guinea-pigs: evidence for the involvement of NK-1 receptors. *J. Auton. Pharmacol. 11*:267.

Advenier, C., Naline, E., Drapeau, G., and Regoli, D. (1987). Relative potencies of neurokinins in guinea pig trachea and human bronchus. *Eur. J. Pharmacol. 139*:133.

Ahlstedt, S.K., Alving, B., Hesselmar, B., and Olafsson, E. (1986). Enhancement of the bronchial reactivity in immunized rats by neonatal treatment with capsaicin. *Int. Arch. Allergy Appl. Immunol. 80*:262.

Aizenman, E., Lipton, S.A., and Loring, R.H. (1989). Selective modulation of NMDA responses by reduction and oxidation. *Neuron 2*:1257.

Amann, R., and Maggi, C.A. (1991). Ruthenium red as a capsaicin antagonist. *Life Sci. 49*:849.

Asthma Consensus. (1992). *Eur. Respir. J. 5*:601.

Ballati, L., Evangelista, S., Maggi, C.A., and Manzini, S. (1992a). Effects of selective tachykinin receptor antagonists on capsaicin- and tachykinin-induced bronchospasm in anaesthetized guinea-pigs. *Eur. J. Pharmacol. 214*:215.

Ballati, L., Evangelista, E., and Manzini, S. (1992b). Repeated antigen challenge induced airway hyperresponsiveness to neurokinin and vagal non-adrenergic, non-cholinergic (NANC) stimulation in guinea pigs. *Life Sci. (Pharmacol. Lett.) 51*:PL119.

Barnes, P.J. (1986). Asthma as an axon reflex. *Lancet 1*:242.

Barnes, P.J. (1990). Neuropeptides as modulators of airway function. In: *Mediators in Airway Hyperreactivity.* (Nijkamp, F.P., Engels, F., Henricks, P.A.J., and Oosterhout, A.J.M., eds.). Birkhauser Verlag, Basel, p. 175.

Barnes, P.J. (1992). Modulation of neurotransmission in airways. *Physiol. Rev. 72*(3):699.

Barnes, P.J., Baraniuk, J.M., and Belvisi, M.G. (1991). Neuropeptides in the respiratory tract. Part I. *Am. Rev. Respir. Dis. 144*:1187.

Belvisi, M.G., Miura, M., Stretton, D., and Barnes, P.J. (1992). Capsazepine as a selective antagonist of capsaicin-induced activation of C-fibres in guinea-pig bronchi. *Eur. J. Pharmacol. 215*:341.

Bevan, S., and Szolcsanyi, J. (1990). Sensory neuron-specific actions of capsaicin: mechanisms and applications. *Trends Pharmacol. Sci. 11*:330.

Bevan, S., and Yeats, J. (1991). Protons activate a cation conductance in a subpopulation of rat dorsal root ganglion neurons. *J. Physiol. (Lond.) 433*:145.

Bevan, S., James, I.F., Rang, H.P., Winter, J., and Wood, J.N. (1987). The mechanism of action of capsaicin—a sensory neurotoxin. In: *Neurotoxins and Their Pharmacological Applications.* (Jenner, P., ed.). Raven Press, New York, p. 261.

Bevan, S., Hothi, S., Hughes, G., James, I.F., Rang, H.P., Shah, K., Walpole, C.S.J., and

Yeats, J.C. (1992). Capsazepine: a competitive antagonist of the sensory neurone excitant capsaicin. *Br. J. Pharmacol. 107*:544.

Boichot, E., Lagente, V., LeGail, G., Carré, C., Mencia-Huerta, J.M., Braquet, P., Lockhart, A., and Frossard, N. (1992). Bronchial responses to substance P after antigen challenge in the guinea-pig in vivo and in vitro studies. *Mediat. Inflamm. 1*:207.

Brunelleschi, S., Vanni, L., Ledda, F., Giotti, A., Maggi, C.A., and Fantozzi, R. (1990). Tachykinins activate guinea-pig alveolar macrophages: involvement of NK-2 and NK-1 receptors. *Br. J. Pharmacol. 100*:417.

Brunelleschi, S., Tarli, S., Giotta, A., and Fantozzi, R. (1991). Priming effects of mammalian tachykinins on human neutrophils. *Life Sci. (Pharmacol. Lett.) 48*:PL1.

Brunelleschi, S., Parenti, A., Ceni, E., Giotti, A., and Fantozzi, R. (1992). Enhanced responsiveness of ovalbumin-sensitized guinea-pig alveolar macrophages to tachy-kinins. *Br. J. Pharmacol. 107*:964.

Buck, S.H., and Burks, T.F. (1986). The neuropharmacology of capsaicin: review of some recent observations. *Pharmacol. Rev. 38*:179.

Canning, B.J., and Undem, B.J. (1993). Relaxant innervation of the guinea-pig trachealis: demonstration of capsaicin-sensitive and -insensitive vagal pathways. *J. Physiol. (Lond.) 460*:719.

Casini, A., Geppetti, P., Maggi, C.A., and Surrenti, C. (1989). Effects of calcitonin gene-related peptide (CGRP) neurokinin A and neurokinin A (4-10) on the mitogenic response of human peripheral blood mononuclear cells. *Naunyn Schmiedeberg's Arch. Pharmacol. 339*:354.

Castle, N.A. (1992). Differential inhibition of potassium currents in rat ventricular myocytes by capsaicin. *Cardiovasc. Res. 26*:1137.

Chang, L-R., and Barnard, E.A. (1982). The benzodiazepine/GABA receptor complex: molecular size in brain synaptic membranes and in solution. *J. Neurochem. 39*:1507.

Cheung, D., Bel, E.H., Hartigh, J.D., Dijkman, J.H., and Sterk, P.J. (1992). The effect of an inhaled neutral endopeptidase inhibitor, thiorphan, on airway responses to neurokinin A in normal humans in vivo. *Am. Rev. Respir. Dis. 145*:1275.

Choudry, N.B., Fuller, R.W., and Pride, N.B. (1989). Sensitivity of the human cough reflex: effect of inflammatory mediators prostaglandin E_2, bradykinin, and histamine. *Am. Rev. Respir. Dis. 140*:137.

Choudry, N.B., Studham, J., Harland, D., and Fuller, R.W. (1993). Modulation of capsaicin induced airway reflexes in humans: effect of monoamine oxidase inhibition. *Br. J. Clin. Pharmacol. 35*:184.

Coleridge, J.C.G., and Coleridge, H.M. (1984). Afferent vagal C fibre innervation of the lungs and airways and its functional significance. *Rev. Physiol. Biochem. Pharmacol. 99*:1.

Coles, S.J., Neil, K.H., and Reid, L.M. (1984). Potent stimulation of glycoprotein secretion in canine trachea by substance P. *J. Appl. Physiol. 57*:1323.

Collier, J.G., and Fuller, R.W. (1984). Capsaicin inhalation in man and the effects of sodium cromoglycate. *Br. J. Pharmacol. 81*:113.

Crimi, N., Palermo, F., Oliveri, R., et al. (1988). Effect of nedocromil on bronchospasm induced by inhalation of substance P in asthmatic airways. *Clin. Allergy 18*:375.

Delay-Goyet, P., and Lundberg, J.M. (1991). Cigarette smoke-induced airway oedema is blocked by the NK1 antagonist, CP-96,345. *Eur. J. Pharmacol. 203*:157.

Dray, A. (1992). Neuropharmacological mechanisms of capsaicin and related substances. *Biochem. Pharmacol. 44*:611.

Devillier, P., Advenier, C., Drapeau, G., Marsac, J., and Regoli, D. (1988). Comparison of the effects of epithelium removal and of enkephalinase inhibitor on the neurokinin-induced contractions of guinea-pig isolated trachea. *Br. J. Pharmacol. 94*:675.

Eglezos, A., Giuliani, S., Viti, G., and Maggi, C.A. (1991). Direct evidence that capsaicin-induced plasma protein extravasation is mediated through tachykinin NK-1 receptors. *Eur. J. Pharmacol. 209*:277.

Ellis, J.L., and Undem, B.J. (1992). Antigen-induced enhancement of noncholinergic contractile responses to vagus nerve and electrical field stimulation in guinea pig isolated trachea. *J. Pharmacol. Exp. Ther. 262*:646.

Evans, T.W., Dixon, C.M.S., Clarke, B., Conradson, T.B., and Barnes, P.J. (1988). Comparison of neurokinin A and substance P on cardiovascular and airway function in man. *Br. J. Pharmacol. 25*:273.

Fels, G., Wolff, E.K., and Maelicke, A. (1982). Equilibrium binding of acetylcholine to the membrane-bound acetylcholine receptor. *Eur. J. Biochem. 127*:31.

Fuller, R.W. (1990). The human pharmacology of capsaicin. *Arch. Int. Pharmacodyn. 303*:147.

Fuller, R.W., Dixon, C.M.S., and Barnes, P.J. (1985). Bronchoconstrictor response to inhaled capsaicin in humans. *J. Appl. Physiol. 58*:1080.

Fuller, R.W., Maxwell, D.L., Dixon, C.M.S., and Barnes, P.J. (1987). The effects of substance P on cardiovascular and respiratory function in human subjects. *J. Appl. Physiol. 62*:1473.

Gentry, S.E. (1991). Tachykinin receptors mediating airway macromolecular secretion. *Life Sci. 48*:1609.

Geppetti, P., Fusco, B.M., Marabini, S., Maggi, C.A., Fanciullacci, M., and Sicuteri, F. (1988). Secretion, pain, and sneezing induced by the application of capsaicin to the nasal mucosa in man. *Br. J. Pharmacol. 93*:509.

Geppetti, P., Bertrand, C., Bacci, E., Snider, R.M., Maggi, C.A., and Nadel, J.A. (1992). The NK-1 receptor antagonist CP 96,345 blocks substance P-evoked mucus secretion from the ferret trachea in vitro. *Neuropeptides 22*:25.

Glinsukon, T., Stitmunnaithum, V., Toskulkao, C., Buranawuti, T., and Tangkrisanavinont (1980). Acute toxicity of capsaicin in several animal species. *Toxicon 18*:215.

Grundstrom, Anderson, R.G.G., and Wikberg, J.E.S. (1981). Pharmacological characterization of the autonomous innervation of the guinea-pig tracheobronchial smooth muscle. *Acta Pharmacol. Toxicol. 49*:150.

Hokfelt, T. (1991). Neuropeptides in perspective: the last ten years. *Neuron 7*:867.

Holzer, P. (1988). Local effector functions of capsaicin-sensitive sensory nerve endings: involvement of tachykinins, calcitonin gene-related peptide and other neuropeptides. *Neuroscience 24*:739.

Holzer, P. (1991). Capsaicin: cellular targets, mechanisms of action, and selectivity for thin sensory neuron. *Pharmacol. Rev. 43*:144.

Howarth, P.H., Djukanovic, R., Wilson, J., Springall, D., Polak, J., and Holgate, S.T. (1991). The influence of asthma on the presence and distribution of neuropeptide containing nerves within the endobronchial mucosa. *Am. Rev. Respir. Dis. 143*:A622.

Ichinose, M., Nakajima, N., Takahashi, T., Yamamauchi, H., Inque, H., and Takishima, T. (1992). Protection against bradykinin-induced bronchoconstriction in asthmatic patients by neurokinin receptor antagonist. *Lancet 340*:1248.

Ingenito, E.P., Pliss, L.B., and Ingram, R.H. (1991). Effects of capsaicin on mechanical, cellular and mediator responses to antigen in sensitized guinea pigs. *Am. Rev. Respir. Dis. 143*:572.

Iwamoto, I., Yamazaki, H., Nakagawa, N., Kimura, A., Tomioka, H., and Yoshida, S. (1990a). Differential effects of two C-terminal peptides of substance P on human neutrophils. *Neuropeptides 16*:103.

Iwamoto, I., Kimura, A., Tanaka, M., Tomioka, H., and Yoshida, S. (1990b). Skin reactivity to substance P, not to neurokinin A, is increased in allergic asthmatics. *Int. Arch. Allergy Appl. Immunol. 93*:120.

James, I.F., Hothi, S.K., Slack, I.J., Bevan, S., Donoghue, J., Walpole, C.S.J., and Winter, J. (1992). Binding of [^{3}H]resiniferatoxin to capsaicin-sensitive sensory neurons: regulation by nerve growth factor. *Soc. Neurosci. Abstr. 18*:60.11.

Jancso, N. (1968). Desensitization with capsaicin and related acylamides as a tool for studying the function of pain receptors. In: *Pharmacology of Pain.* (Lin, K., Armstrong, D., and Pardo, E.G., eds.). Pergamon Press, Oxford, p. 33.

Joos, G.F. (1989). The role of sensory neuropeptides in the pathogenesis of bronchial asthma. *Clin. Exp. Allergy 19*(S1):9.

Joos, G., Pauwels, R., and Van der straeten, M. (1987). The effect of inhaled substance P and neurokinin A on the airways of normal and asthmatic subjects. *Thorax 42*:779.

Kaufman, M.P., Coleridge, H.M., Coleridge, J.C.G., and Baker, D.G. (1980). Bradykinin stimulates afferent vagal C-fibres in intrapulmonary airways of dogs. *J. Appl. Physiol. 48*:511.

Kizawa, Y., and Takayanagi, I. (1985). Possible involvement of substance P immunoreactive nerves in the mediation of nicotine-induced contractile responses in isolated guinea pig bronchus. *Eur. J. Pharmacol. 113*:319.

Kohrogi, H., Kamaguchi, T., Kawano, O., Honda, I., Ando, M., and Araki, S. (1991). Inhibition of neutral endopeptidase potentiates bronchial contraction induced by immune response in guinea pigs in vitro. *Am. Rev. Respir. Dis. 144*:636.

Kostreva, D.R., Zuperku, E.J., Hess, G.L., Coon, R.L., and Kampine, J.P. (1975). Pulmonary afferent activity recorded from sympathetic nerves. *J. Appl. Physiol. 39*:37.

Kroegel, C., Giembycz, M.A., and Barnes, P.J. (1990). Characterization of eosinophil cell activation by peptides. Differential effects of substance P, melittin and fMet-Leu-Phe. *J. Immunol. 145*:2581.

Kuo, H.P., Rohde, J.A.L., Barnes, P.J., and Rogers, D.F. (1990). Cigarette smoke induced goblet cell secretion: neural involvement in guinea pig trachea. *Eur. Respir. J. 3*:189.

Ladenius, A.R.C., and Biggs, D.F. (1989). Capsaicin prevents the induction of airway hyperresponsiveness in a guinea pig model of asthma. *Am. Rev. Respir. Dis. 139*:A232.

Lai, Y.L. (1991). Endogenous tachykinins in antigen-induced acute bronchial responses of guinea pigs. *Exp. Lung Res. 17*:1047.

Lei, Y-H., Barnes, P.J., and Rogers, D.F. (1992). Inhibition of neurogenic plasma exudation in guinea-pig airways by CP-96,345, a new non peptide NK1 receptor antagonist. *Br. J. Pharmacol.* *105*:261.

Lundberg, J.M., and Saria, A. (1983). Capsaicin-induced desensitization of airway mucosa to cigarette smoke, mechanical and chemical irritants. *Nature 302*:251.

Lundberg, J.M., Saria, A., Brodin, E., Rosell, S., and Folkers, K. (1983a). A substance P antagonist inhibits vagally induced increase in vascular permeability and bronchial smooth muscle contraction in the guinea pig. *Proc. Natl. Acad. Sci. USA* *80*:1120.

Lundberg, J.M., Brodin, E., and Saria, A. (1983b). Effects and distribution of vagal capsaicin-sensitive substance P neurons with special reference to the trachea and lungs. *Acta Physiol. Scand. 119*:143.

Lundberg, J.M., Martling, C-R., and Saria, A. (1983c). Substance P and capsaicin-induced contraction of human bronchi. *Acta Physiol. Scand. 119*:149.

Lundberg, J.M., Alving, K., Karlsson, J.A., Martan, R., and Nilsson, G. (1991). Sensory neuropeptide involvement in animal models of airway irritation and of allergen-evoked asthma. *Am. Rev. Respir. Dis. 143*:1429.

Maggi, C.A. (1990). Tachykinin receptors in the airways and lung: what should we block? *Pharmacol. Res. 22*:527.

Maggi, C.A. (1991). The pharmacology of the efferent function of sensory nerves. *J. Auton. Pharmacol. 11*:173.

Maggi, C.A., and Meli, (1988). The sensory-efferent function of capsaicin-sensitive sensory neurons. *Gen. Pharmacol. 19*:1.

Maggi, C.A., Patacchini, R., Tramontana, M., Amann, R., Giuliani, S., and Santicioli, P. (1990). Similarities and differences in the action of resiniferatoxin and capsaicin on central and peripheral endings of primary sensory neurons. *Neuroscience 37*:531.

Maggi, C.A., Patacchini, R., Quartara, L., Rovero, P., and Santicioli, P. (1991). Tachykinin receptors in the guinea-pig isolated bronchi. *Eur. J. Pharmacol. 197*:167.

Maggi, C.A., Patacchini, R., Rovero, P., and Giachetti, A. (1993a). Tachykinin receptors and tachykinin receptor antagonists. *J. Auton. Pharmacol. 13*:23.

Maggi, C.A., Bevan, S. Walpole, C.S.J., Rang, H.R., and Giuliani, S. (1993b). A comparison of capsazepine and ruthenium red as capsaicin antagonists in the rat isolated urinary bladder and vas deferens. *Br. J. Pharmacol. 108*:801.

Manzini, S. (1992). Bronchodilatation by tachykinins and capsaicin in the mouse main bronchus. *Br. J. Pharmacol. 105*:968.

Manzini, S., Maggi, C.A., Geppetti, P., and Bacciarelli, C. (1987). Capsaicin desensitization protects from antigen-induced bronchospasm in conscious guinea-pigs. *Eur. J. Pharmacol. 138*:307.

Manzini, S., Conti, S., Maggi, C.A., Abelli, L., Somma, V., Del Bianco, E., and Geppetti, P. (1989). Regional differences in the motor and inflammatory responses to capsaicin in guinea pig airways. *Am. Rev. Respir. Dis. 140*:936.

Mapp, C.E., Chitano, P., Fabbri, L.M., Patacchini, R., Santicioli, P., Geppetti, P., and Maggi, C.A. (1990). Evidence that toluene diidocyanate activates the efferent function of capsaicin-sensitive primary afferents. *Eur. J. Pharmacol. 180*:113.

Marasco, W.A., Showell, H.J., and Becker, E.L. (1981). Substance P binds to the formyl peptide chemotaxis receptor on the rabbit neutrophil. *Biochem. Biophys. Res. Commun.* *99*:1065.

Marsh, S.H., Stansfeld, C.E., Brown, D.A., Davey, R., and McCarthy, D. (1987). The mechanism of action of capsaicin on sensory C-type neurons and their axons in vitro. *Neuroscience 23*:275.

Martling, C.R., Theodorsson-Norheim, E., and Lundberg, J.M. (1987). Occurrence and effects of multiple tachykinins: substance P, neurokinin A and neuropeptide K in human lower airways. *Life Sci. 40*:1633.

Matis, W.L., Lavker, R.M., and Murphy, G.F. (1990). Substance P induces the expression of an endothelial-leukocyte adhesion molecule by microvascular endothelium. *J. Invest. Dermatol. 94*:492.

Matsuse, T., Thomson, R.J., Chen, X-R., Salari, H., and Schellenberg, R.R. (1991). Capsaicin inhibits airway hyperresponsiveness but not lipoxygenase activity or eosinophilia after repeated aerosolized antigen in guinea pigs. *Am. Rev. Respir. Dis. 144*:368.

McDonald, D.M. (1988). Neurogenic inflammation in the rat trachea. I. Changes in venules, leucocytes and epithelial cells. *J. Neurocytol. 17*:583.

McDonald D.M., Mitchell, R.A., Gabella, G., and Haskell, A. (1988). Neurogenic inflammation in the rat trachea. II. Identity and distribution of nerves mediating the increase in vascular permeability. *J. Neurocytol. 17*:605.

McGillis, J.P., Mitsuhashi, M., and Payan, D.G. (1990). Immunomodulation by tachykinin neuropeptides. *Ann. NY Acad. Sci. 594*:85.

Meini, S., Evangelista, S., Geppetti, P., Szallasi, A., Blumberg, P.M., and Manzini, S. (1992a). Pharmacological and neurochemical evidence for the activation of capsaicin-sensitive nerves by lipoxin A4 in guinea pig bronchus. *Am. Rev. Respir. Dis. 146*:930.

Meini, S., Manzini, S., Barnes, P.J., and Rogers, D.F. (1992b). Effect of selective tachykinin agonists on mucus secretion and smooth muscle contraction in ferret airways. *Am. Rev. Respir. Dis. 145*:A370.

Miller, M.S., Buck, S.H., Sipes, I.G., Yamamura, H.I., and Burks, T.F. (1982). Regulation of substance P by nerve growth factor: disruption by capsaicin. *Brain Res. 250*:193.

Morton, D.R., Klassen, K.P., and Curtis, G.M. (1951). The clinical physiology of the human bronchi. II. The effect of vagus section upon pain of tracheobronchial origin. *Surgery 30*:800.

Mullol, J., Rieves, R.D., Baraniuk, J.N., Lundgren, D., Merida, M., Hausfeld, J.H., Shelhamer, J.H., and Kaliner, M.A. (1992). The effects of neuropeptides on mucous glycoprotein secretion from human nasal mucosa in vitro. *Neuropeptides 21*:231.

Myers, A.C., and Undem, B.J. (1991). Functional interactions between capsaicin-sensitive and cholinergic nerves in the guinea pig bronchus. *J. Pharmacol. Exp. Ther. 259*:104.

Naline, E., Devillier, P., Drapeau, G., Toty, L., Bakdach, H., Regoli, D., and Advenier, C. (1989). Characterization of neurokinin effects and receptor selectivity in human isolated bronchi. *Am. Rev. Respir. Dis. 140*:679.

Nilsson, G., Dahlberg, K., Brodin, E., Sundler, F., and Strandberg, K. (1977). Distribution and constrictor effect of substance P in guinea pig tracheobronchial tissues. In: *Substance P.* (von Euler, U.S., and Pernow, B., eds.). Raven Press, New York, p. 57.

Ollerenshaw, S.L., Jarvis, D., Sullivan, C.E., and Woolcock, A.J. (1991). Substance P immunoreactive nerves in airways from asthmatics and nonasthmatics. *Eur. Respir. J. 4*:673.

Paintal, A.S. (1973). Vagal sensory receptors and their reflex effects. *Physiol. Rev. 53*:159.

Payan, D.G., Brewster, D.R., and Goetzl, E.J. (1983). Specific stimulation of human T-lymphocytes by substance P. *J. Immunol. 131*:1613.

Payan, D.G., Brewster, D.R., Missirian-Bastian, A., and Goetzl, E.J. (1984). Substance P recognition by a subset of human T-lymphocytes. *J. Clin. Invest. 74*:1532.

Perianin, A., Snyderman, R., and Malfroy, B. (1989). Substance P primes human neutrophil activation: a mechanism for neurological regulation of inflammation. *Biochem. Biophys. Res. Commun. 2*:520.

Perretti, F., and Manzini, S. (1993). Activation of capsaicin-sensitive sensory fibers modulates PAF-induced bronchial hyperresponsiveness in anaesthetized guinea pigs. *Am. Rev. Respir. Dis. 148*:927.

Peters, M.J., Adcock, I.M., Gelder, C.M., Shirasaki, H., Belvisi, M.G., Yacoub, M., and Barnes, P.J. (1992). NK1 Receptor gene expression is increased in asthmatic lung and reduced by corticosteroids. *Am. Rev. Respir. Dis. 145*:A835.

Qian, Y., Emonds-Alt, X., and Advenier, C. (1993). Effects of capsaicin, CP-96,345 and SR 48968 on the bradykinin-induced airways microvascular leakage in guinea-pigs. *Pulm. Pharmacol. 6*:63.

Repke, H., and Bienert, M. (1987). Mast cell activation—a receptor-independent mode of substance P action? *FEBS Lett. 221*:236.

Repke, H., and Bienert, M. (1988). Structural requirements for mast cell triggering by substance P-like peptides. *Agents Actions 23*:207.

Riccio, M.M., Manzini, S., and Page, C.P. (1993). The effect of neonatal capsaicin on the development of bronchial hyperresponsiveness in allergic rabbits. *Eur. J. Pharmacol. 232*:89.

Rogers, D.F., Belvisi, M.G., Aursudkij, B., Evans, T.W., and Barnes, P.J. (1988). Effects and interactions of sensory neuropeptides on airway microvascular leakage in guinea-pigs. *Br. J. Pharmacol. 95*:1109.

Rogers, D.F., Aursudkij, B., and Barnes, P.J. (1989). Effects of tachykinins on mucus secretion in human bronchi in vitro. *Eur. J. Pharmacol. 174*:283.

Saria, A., Lundberg, J.M., Skofitsch, G., and Lembeck, F. (1983). Vascular protein leakage in various tissues induced by substance P, capsaicin, bradykinin, serotonin, histamine and by antigen challenge. *Naunyn-Schmiedeberg's Arch. Pharmacol. 324*:212.

Spina, D., Mckenniff, M.U., Coyle, A.J., Seeds, E.A.M., Perretti, F., Tramontana, M., Manzini, S., and Page, C.P. (1991). Effect of capsaicin on PAF-induced bronchial hyperresponsiveness and pulmonary cell accumulation in the rabbit. *Br. J. Pharmacol. 103*:1268.

Stjarne, P., Lundblad, L., Lundberg, J.M., and Anggard, A. (1989). Capsaicin- and nicotine-sensitive afferent neurones and nasal secretion in healthy human volunteers and in patients with vasomotor rhinitis. *Br. J. Pharmacol. 96*:693.

Szallasi, A., and Blumberg, P.M. (1989). Resiniferatoxin, a phorbol-related diterpene, acts as an ultrapotent analog of capsaicin, the irritant constituent in red pepper. *Neuroscience 30*:515.

Szallasi, A., and Blumberg, P.M. (1990a). Minireview. Resiniferatoxin and its analogs provide novel insights into the pharmacology of the vanilloid (capsaicin) receptor. *Life Sci.* *47*:1399.

Szallasi, A., and Blumberg, P.M. (1990b). Specific binding of resiniferatoxin, an ultra-potent capsaicin analog, by dorsal root ganglion membranes. *Brain Res.* *524*:106.

Szallasi, A., and Blumberg, P.M. (1991a). Characterization of vanilloid receptors in the dorsal horn of pig spinal cord. *Brain Res.* *547*:335.

Szallasi, A., and Blumberg, P.M. (1991b). Molecular target size of the vanilloid (capsaicin) receptor in pig dorsal root ganglia. *Life Sci.* *48*:1863.

Szallasi, A., and Blumberg, P.M. (1992). Vanilloid receptor loss in rat sensory ganglia associated with long term desensitization to resiniferatoxin. *Neurosci. Lett.* *136*:51.

Szallasi, A., and Blumberg, P.M. (1993a). Mechanisms and therapeutic potential of vanilloids (capsaicin-like molecules). *Adv. Pharmacol.* (in press).

Szallasi, A., and Blumberg, P.M. (1993b) [^{3}H]resiniferatoxin binding by the vanilloid receptor: species-related differences, effects of temperature and sulfhydryl reagents. *Naunyn-Schmiedeberg's Arch. Pharmacol.* *347*:84.

Szallasi, A., Sharkey, N.A., and Blumberg, P.M. (1989). Structure-activity analysis of resiniferatoxin analogs. *Phytother. Res.* *3*:253.

Szallasi, A., Szallasi, Z., and Blumberg, P.M. (1990). Permanent effects of neonatally administered resiniferatoxin in the rat. *Brain Res.* *537*:182.

Szallasi, A., Szolcsanyi, J., Szallasi, Z., and Blumberg, P.M. (1991). Inhibition of [^{3}H]resiniferatoxin binding to rat dorsal root ganglion membranes as a novel approach in evaluating compounds with capsaicin-like activity. *Naunyn-Schmiedeberg's Arch. Pharmacol.* *344*:551.

Szallasi, A., Lewin, N.E., and Blumberg, P.M. (1992). Identification of alpha$_1$-acid glycoprotein (orosomucoid) as a major vanilloid-binding protein in serum. *J. Pharmacol. Exp. Ther.* *262*:883.

Szallasi, A., Conte, B., Goso, C., Blumberg, P.M., and Manzini, S. (1993a). Characterization of a peripheral vanilloid (capsaicin) receptor in the rat urinary bladder. *Life Sci./ Pharmacol. Lett.* *52*:PL221.

Szallasi, A., Conte, B., Goso, C., Blumberg, P.M., and Manzini, S. (1993b). Vanilloid receptors in the urinary bladder: regional distribution, localization on sensory nerves, and species-related differences. *Naunyn-Schmiedeberg's Arch. Pharmacol.* *347*:624.

Szallasi, A., Goso, C., Blumberg, P.M., and Manzini, S. (1993c). Characterization by specific [^{3}H]resiniferatoxin binding of vanilloid (capsaicin) receptors in guinea pig airways: regional differences, and affinity for capsazepine. *Neuropeptides* *24*:C4.

Szallasi, A., Lewin, N.E., and Blumberg, P.M. (1993d). Vanilloid (capsaicin) receptor in the rat: positive cooperativity of resiniferatoxin binding and its modulation by reduction and oxidation. *J. Pharmacol. Exp. Ther.* (in press).

Szolcsanyi, J. (1983). Tetrodotoxin-resistent non-cholinergic contraction evoked by capsaicinoids and piperine on the guinea-pig trachea. *Neurosci. Lett.* *42*:83.

Szolcsanyi, J., and Bartho, L. (1982). Capsaicin-sensitive non-cholinergic excitatory innervation of the guinea pig tracheobronchial smooth muscle. *Neurosci. Lett.* *34*:247.

Szolcsanyi, J., and Jancso-Gabor, A. (1975). Sensory effects of capsaicin congeners. I. Relationship between chemical structure and pain-producing potency. *Drug Res.* *25*:1877.

Szolcsanyi, J., and Jancso-Gabor, A. (1976). Sensory effects of capsaicin congeners. II. Importance of chemical structure and pungency in desensitizing activity of capsaicin-type compounds. *Drug Res.* 26:33.

Szolcsanyi, J., Szallasi, A., Szallasi, Z., Joo, F., and Blumberg, P.M. (1990). Resinifera-toxin, an ultrapotent selective modulator of capsaicin-sensitive primary afferent neurons. *J. Pharmacol. Exp. Ther.* 255:923.

Szolcsanyi, J., Bartho, L., and Petho, G. (1991). Capsaicin-sensitive bronchopulmonary receptors with dual sensory-efferent function: mode of action of capsaicin antagonists. *Acta Physiol. Hung.* 77:293.

Taylor, D.C.M., Pierau, F-K., and Szolcsanyi, J. (1984). Long lasting inhibition of horseradish peroxidase (HPR) transport in sensory nerves induced by capsaicin treatment of the receptive field. *Brain Res.* 298:45.

Taylor, D.C.M., Pierau, F-K., and Szolcsanyi, J. (1985). Capsaicin-induced inhibition of axoplasmic transport is prevented by nerve growth factor. *Cell Tissue Res.* 240:569.

Virus, R.M., and Gebhart, G.F. (1979). Minireview. Pharmacologic actions of capsaicin: apparent involvement of substance P and serotonin. *Life Sci.* 25:1273.

Weihe, E. (1992). Significance of neuropeptides at the neuroimmune connection in inflammatory pain. *Neuropeptides* 22:69.

Widdicombe, J.G. (1964). Respiratory reflexes. In: *Handbook of Physiology*, Vol. 1. (Fenn, W.O., and Rahn, H., eds.). American Physiology Society, Washington, D.C., p. 585.

Wiedermann, C.J., Wiedermann, F.J., Apperi, A., Kiselbach, G., Konwalinka, G., and Braunsteiner, H. (1989). In vitro human polymorphonuclear keukocyte chemokinesis and human monocyte chemotaxis are different activities of aminoterminal and carboxyterminal substance P. *Naunyn-Schmiedeberg's Arch. Pharmacol.* 340:185.

Winning, A.J., Hamilton, R.D., Shea, S.A., and Guz, A. (1986). Respiratory and cardiovascular effects of central and peripheral intravenous injections of capsaicin in man: evidence for pulmonary chemosensitivity. *Clin. Sci.* 71:519.

Winter, J., Forbes, C.A, Sternberg, J., and Lindsay, R.M. (1988). Nerve growth factor (NGF) regulates adult rat cultured dorsal root ganglion neuron responses to the excitotoxin capsaicin. *Neuron* 1:973.

Winter, J., Dray, A., Wood, J.N., Yeats, J.C., and Bevan, S. (1990). Cellular mechanism of action of resiniferatoxin: a potent sensory neuron excitotoxin. *Brain Res.* 520:131.

Wozniak, A., Mclennan, G., Betts, W.H., Murphy, A., and Scicchitano, R. (1989). Activation of human neutrophils by substance P: effect on FMLP-stimulated oxidative and arachidonic acid metabolism and on antibody-dependent cell-mediated cytotoxicity. *Immunology* 68:359.

Yeats, J.C., Boddeke, H.W.G.M., and Docherty, R.J. (1992). Capsaicin desensitization in rat dorsal root ganglion neurones is due to activation of calcineurin. *Br. J. Pharmacol.* 107 (Proc. Suppl.):238P.

DISCUSSION

Widdicombe: Apart from the C-fiber sensory nerves, where do you find capsaicin receptors? Can capsaicin cause cough? Have the capsaicin-sensitive cough fibers been identified?

Manzini: In the rat, RTX binding sites disappear after neonatal capsaicin pretreatment. Therefore, capsaicin-sensitive sites may be restricted to nerves. We have also obtained data on neonatally capsaicin-treated rats showing that the RTX nerves in the urinary bladder were totally ablated. However, it is impossible to rule out that some of these capsaicin-sensitive sites might be on nonneural sites.

Kowalski: Are there data on the possible presence of capsaicin-sensitive sites on cholinergic nerve fibers?

Manzini: No capsaicin receptors have been described on cholinergic nerves thus far. There is data that stimulation of capsaicin-sensitive nerves can cause the increased release of acetylcholine, however.

Kowalski: Our data in rats failed to show any effect of capsaicin desensitization on antigen-driven vascular permeability, suggesting that capsaicin-sensitive nerves were not involved in the vascular permeability seen after antigen challenge.

Manzini: As you know, evidence both for and against involvement of capsaicin-sensitive sensory nerves in allergic responses has been published, depending on animal species and/or method of treatment.

Nadel: You have shown effects of neurogenic inflammation in the postcapillary venules in rats. Balck could find no direct innervation of postcapillary venules. What is the source of the tachykinins for this effect? Do they derive from varicosities located in the epithelium or are they released upstream in the vascular system and then act by a cascade action at this site?

Manzini: I agree that capsaicin-sensitive nerves cannot be directly demonstrated to innervate the postcapillary venule. The vascular permeability could be in response to stimulation of other structures (epithelium, endothelium). However, NK-1 receptor antagonists prevent this reaction, suggesting that tachykinins mediate the response.

Polak: I was delighted to hear Dr. Manzini speak of species variations. In support of his statement, we found little evidence of neurokinins in sensory fibers in the human lung. Hence, investigations of animal models should be interpreted cautiously in regard to extrapolation to humans.

9

Pulmonary Endocrine Cells In Vivo and In Vitro

R. Ilona Linnoila

Biomarkers and Prevention Research Branch, National Cancer Institute, National Institutes of Health, Rockville, Maryland

I. INTRODUCTION

Pulmonary endocrine cells (PECs) are part of the diffuse endocrine system (DES) distributed throughout the body (Nylen and Becker, 1990). Other components of DES include cell populations in the hypothalamus and adenohypophysis, pineal gland, paraganglia, cell populations in the thymus, pancreatic islets, thyroid C cells, parathyroid glands, endocrine cells of the breast, gastrointestinal and genitourinary tracts, melanocytes, and Merkel cells of the skin. PECs were first described by Feyrter (1946, 1954) and Fröhlich (1949), who referred to them as clear cells (*helle Zelle*) because of their lucid cytoplasm in hematoxylin-eosin-stained sections. Since then many names have been given to them (Table 1). Some names are eponymes, such as Feyrter cell and Kultschitzky cell; others refer to the possible functions of PECs or their staining characteristics since special techniques are required to identify PECs in pulmonary epithelium.

The purpose of this chapter is to briefly review recent advances in the PEC research and to present the interested reader with tools that might be helpful in future studies on the possible role of these cells in the lung.

II. MARKERS FOR PULMONARY ENDOCRINE CELLS

The major function of PECs is the formation, packaging, and secretion of specific peptide and amine products. PECs share characteristics of neural and endocrine

Table 1 Various Names for the Pulmonary Endocrine Cell

AFG cell (*a*rgyrophilic, *f*luorescent, *g*ranulated)
APUD cell (*a*mine *p*recursor *u*ptake and/or *d*ecarboxylation)
Argentaffin cell
Argyrophilic cell
Chromafffin cell or chromaffin-type cell
Clear cell
Dense-core granulated cell
Endocrine cell or endocrine-like cell[a]
Enterochromaffin-like cell
Feyrter cell
Granulated cell
Kultschitzky (-like) cell
Neuroendocrine cell[a]
Neuroepithelial endocrine (NEE) cell
Neurosecretory cell
P cell
Paraneuron
Small granule or small granulated cell
Small granule endocrine cell
Small granule neuroendocrine cell
Terms applied to innervated clusters
  Corpuscle
  Islet receptor
  Neuroepithelial body

[a]Currently most commonly used names.

cells, collectively referred to as neuroendocrine (NE) features. The markers for NE properties can be divided into *general NE markers* and *specific products*. The expression of specific products is highly variable, is limited to subgroups of NE cells and their tumors, and can be species dependent (Polak et al., 1993), such as calcitonin (CT) in thyroid C cells and in a subset of human PECs and insulin or glucagon in pancreatic islet cells. In addition to the specific products, PECs share characteristics common to the entire DES. These general NE markers include: dense-core granules (the storage site of the amine and peptide products), high levels of the key amine-handling enzyme L-dopa decarboxylase, and certain surface markers, other enzymes, and proteins (Table 2).

The advantage of using general NE markers is that it allows investigators to study the entire population of PECs at a time. It also facilitates interspecies comparisons.

Table 2 Markers of Pulmonary Endocrine Cells

Marker	Comments
A. General markers	
Dense-core granules[a]	Storage site of specific products, seen on electron microscopy
Chromogranin A[a]	Matrix protein of granules
Synaptophysin	Structural protein of cytoplasmic clear vesicles of neurons and neuroendocrine cells
PGP 9.5[a]	Cytoplasmic protein of unknown function in neurons and neuronendocrine cells
Cluster I membrane antigens of small cell lung carcinoma	Antigens related to neural cell adhesion molecules (N-CAM) such as Leu-7, NKH-1, and MOC-1, shared by neurons and neuronendocrine cells
L-Dopa decarboxylase Tyrosine hydroxylase	Key amine-producing enzymes, seen also by formaldehyde-induced fluorescence
Neuron-specific enolase[a]	Glycolytic isoenzyme found in neurons and neuroendocrine cells
Brain isoenzyme of creatine kinase	High levels detected in brain and endocrine tissues
B. Specific products[a]	
Peptides and amines	More than 10 specific peptide and amine products identified

[a]Currently most commonly used markers for pulmonary endocrine cells.

III. PULMONARY ENDOCRINE CELLS AND FETAL LUNG DEVELOPMENT

PECs can occur as solitary cells or innervated organoid clusters called neuro-epithelial bodies (NEBs; Lauweryns et al., 1972) throughout the conducting airway epithelium, but are rarely seen in the alveolar region of the lung (Figs. 1, 2). They are most numerous during the fetal and neonatal period. Putative precursors of PECs are demonstrable in the proximal airways of human fetal lungs at 8 weeks of gestation, and definite cells can be seen about 2 weeks later (Fig. 3). Thus PECs are the first specialized cells to appear in the developing lung epithelium, preceding other secretory cells, ciliated cells, basal cells, and alveolar cells. According to the current, widely accepted view, PECs along with all the other cells in the pulmonary epithelium originate from a common stem cell. This stem cell gives rise to differentiation-committed cells that then develop to fully differentiated pulmonary epithelial cells, which include PECs, other secretory cells, ciliated and basal cells (Gazdar et al., 1981; Baylin, 1990).

Human fetal PECs contain abundant gastrin-releasing peptide (GRP, a mam-

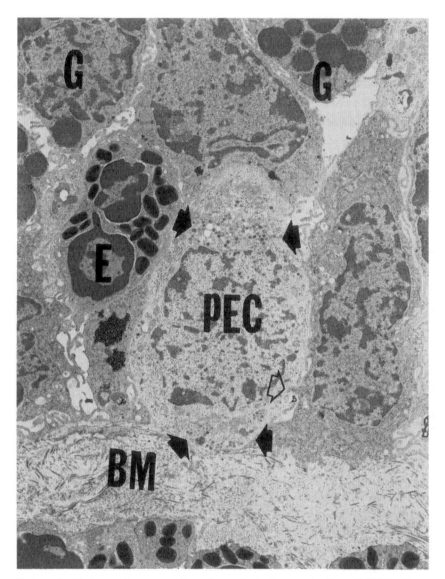

Figure 1 Electron micrograph of a guinea pig tracheal endocrine cell (PEC). Large quantities of characteristic dense-core vesicles (solid arrows) can be seen in the apical as well as basal cytoplasmic compartments. Overall, the PEC has lighter-appearing cytoplasm and is located on the basement membrane (BM). Note a specialized cell-to-cell attachment (open arrow), typical of epithelial cells. Guinea pig trachea frequently contains numerous eosinophils (E) and goblet cells (G). (×9000.)

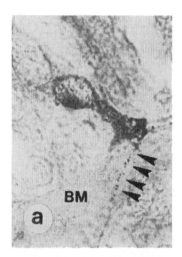

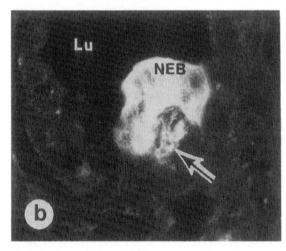

Figure 2   Photomicrographs of a solitary pulmonary endocrine cell (PEC) and a neuro-epithelial body (NEB). (a) Solitary guinea pig tracheal PEC with innervation (arrowheads). (Argyrophilic silver impregnation, ×1160.) BM = basement membrane. (b) Hamster NEB bulging to bronchial lumen (Lu) is innervated by a bundle of fluorescent nerve fibers (arrow). (Formaldehyde-induced fluorescence for catecholamines, ×360.)

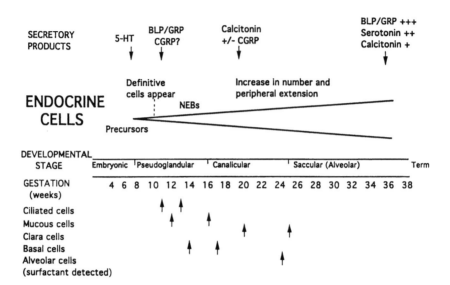

Figure 3 Development and differentiation of human pulmonary endocrine cells (PECs) in the fetal lung. Appearance of neuroendocrine products and PECs is indicated against time in weeks from ovulation to term and stages of lung development. Precursor cells for PECs appear at about 8 weeks of gestation, followed by definite cells at 10 weeks. Serotonin (5-HT) is the first secretory product by PECs, followed by bombesin-like peptides (BLP)/ gastrin-releasing peptide (GRP). Calcitonin is demonstrable at 20 weeks, and there are conflicting reports when calcitonin gene-related peptide (CGRP) will appear. Ciliated, mucous, Clara, basal, and alveolar cells appear later than PECs, as indicated by arrows at the bottom. The first arrow on the line for any given cell type indicates the appearance of a precursor, while the second one marks the mature form. [Adapted with modifications from Cutz (1987) and Gosney (1992).]

malian bombesin-like peptide, BLP; Wharton et al., 1978) (Fig. 4). Addition of bombesin has been shown to result in increased growth and maturation of lungs in mice in utero, and in human and murine organ cultures. Moreover, blocking by antibombesin MAb 2A11 (Cuttitta et al., 1985a) inhibited over 50% of lung automaturation in serum-free organ cultures (Sunday et al., 1990) suggesting that

Figure 4 Photomicrographs of pulmonary endocrine cells (PECs) from human fetal lung. (a) Neuroepithelial body (NEB, arrow) and solitary cell (arrowhead) with positive staining by an antibody against form I of human pro-GRP molecule (Cuttitta et al., 1988). (Immunoperoxidase stain, ×450.) (b) The same NEB (arrow) in a serial section demonstrating immunoreactivity detected by an antibody against the N-terminal portion of human GRP molecule. (Immunoperoxidase stain, ×450.) Lu = bronchial lumen.

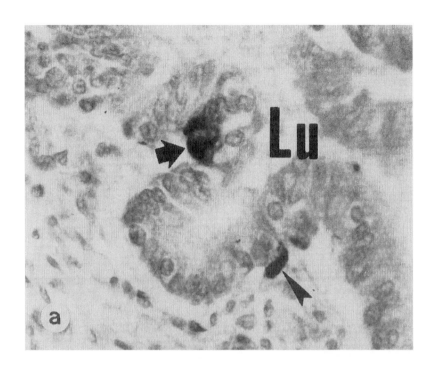

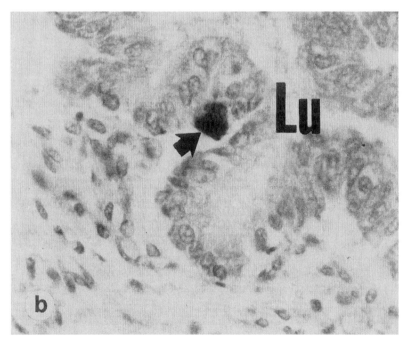

endogenous BLPs such as GRP might play an important role in the normal maturation process. In a recent study, Sunday et al. (1993) showed that two ligand-receptor systems, BLPs and the epidermal growth factor (EGF), may induce both growth and maturation in a reciprocal fashion during later stages of lung development.

In fetal hamster lung, it has been shown that nonendocrine cells abutting NEBs divided more often than those at a distance. In established airways, NEBs thus emerged as foci of growth in the epithelium (Hoyt et al., 1993b). Moreover, calcitonin gene-related peptide (CGRP), the prevalent peptide in NEBs of prenatal

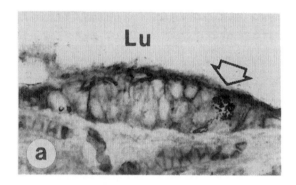

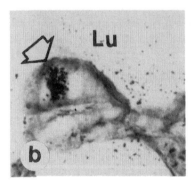

Figure 5   Photomicrographs of thymidine labeling in neuroepithelial bodies (NEBs). (a) Paraffin section of the lung was obtained from an adult hamster that received ^{3}H-thymidine 60 min before sacrifice. The section was stained for argyrophilia according to the method of Grimelius and processed for autoradiography. Note a heavily labeled cell (arrow) immediately adjacent to the NEB, while the cells composing the NEB remained unlabeled. ($\times$710.) (b) Corresponding section from a hamster that received ^{3}H-thymidine 24 hr before sacrifice. Now one of the NEB cells is heavily labeled (arrow). ($\times$920.) Lu = bronchial lumen.

hamsters, rats, and mice, was recently shown to raise the BrdU labeling index of the tracheobronchial lining in 14-day fetal rat lung explants as well as to stimulate growth of guinea pig pulmonary epithelial cells (Hoyt et al., 1993a; White et al., 1993). This not only supports the idea that NEBs exert a local mitogenic effect, but suggests a possible mechanism for their action.

Under normal conditions, adult PECs rarely divide, while following exposure to the systemic carcinogen diethylnitrosamine, known to cause PEC hyperplasia, these cells label readily with tritiated thymidine (Linnoila, 1982) (Fig. 5).

IV. DISEASED LUNGS

In PEC populations of unremarkable postnatal human lungs, minimal variation occurs from childhood to old age (Gosney, 1993). However, increase in PEC numbers has been described in association with multiple diseases or pathological conditions (Table 3, Fig. 6) that appear heterogeneous, but in many cases involve inflammation (Gould et al., 1983; Gosney, 1992). Phenotypical PEC changes are also common. While in adult human lung the relation of BLP-containing PECs to CT-containing PECs is approximately 2:1, the normal predominance of BLP-containing PECs can be lost in chronic bronchitis and emphysema so that cells containing CT outnumber them by a ratio of almost 2:1 (Gosney et al., 1989). Areas of pneumonia were shown to be associated with the highest numbers of CT-containing cells, which were increased up to 17 times. In addition, colocalization of BLP, CT, and CGRP has rarely been found in normal fetuses, but is frequently present in infants with acute or chronic lung diseases (Stahlman and Gray, 1993). Other peptides, such as ACTH, β-endorphin, α-subunit of human chorionic gonadotropin, somatostatin, vasoactive intestinal polypeptide, and growth hormone, may also appear only in proliferating PECs associated with various abnormalities (Table 4).

Table 3 Conditions Associated with Increased Numbers of Pulmonary Endocrine Cells

Asthma	Cystic fibrosis
Pneumonia	Eosinophilic granuloma
Chronic bronchitis, emphysema	Plexogenic pulmonary arteriopathy
Cigarette smoking	Neonatal conditions:
Pulmonary fibrosis	Bronchopulmonary dysplasia
Bronchiectasis	Wilson-Mikity syndrome
Neoplasms (surrounding lung)	Asphyxia with brainstem injury
Mechanical ventilation with O_2	Sudden infant death syndrome

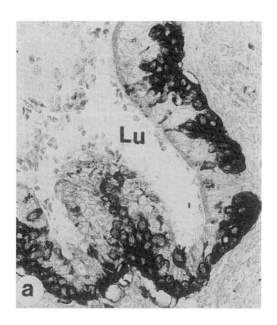

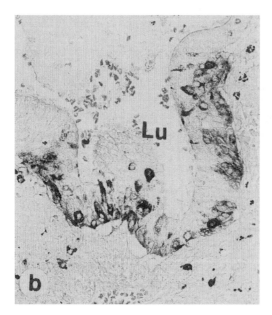

Figure 6 Photomicrographs of pulmonary endocrine cell (PEC) hyperplasia in human lung. (a) Linear and nodular hyperplasia of chromogranin A–positive PECs in a nonneoplastic bronchial epithelium adjacent to a papillary adenocarcinoma. (Immunoperoxidase stain, ×360.) (b) Serial section revealing many of the same cells positive for Leu-7, another general marker for neuroendocrine differentiation (Table 2). (Immunoperoxidase stain, ×290.) Lu = bronchial lumen.

Table 4 Specific Products of Human Pulmonary Endocrine Cells

	Healthy lungs	Abnormal lungs
"Eutopic"		
Serotonin	+ +	+ +
Bombesin-like peptides	+ + + +	+ + +
Calcitonin	+	+ + +
Calcitonin gene-related peptide	+	+
Cholecystokinin	+	N.D.
"Ectopic"		
Adrenocorticotrophin	—	+ +
β-Endorphin	—	+
Somatostatin	—	+
Vasoactive intestinal polypeptide	—	+
Growth hormone	—	+
α-Subunit of human chorionic gonadotrophin	—	+

N.D. = not done

V. PHYSIOLOGY AND IN VITRO MODELS

The physiological role of PECs in normal lung is not fully understood, but multiple functions have been proposed: (1) promotion and regulation of the growth of developing airways by stimulating the proliferation of local endoderm and airway branching, (2) regulation of the development of alveolar capillaries, (3) release of substances in response to increasing fetal hypoxia near term in order to maintain vasoconstriction in the pulmonary circuit, and (4) neonatal respiratory adaptation. Solitary pulmonary endocrine cells may not always be innervated, but experimental data suggest that the innervation of NEBs plays a crucial role in their action. There is morphological evidence that both efferent and afferent nerves are connected to NEBs, hence conceiving a modulation of the responses of the PECs by central nervous mechanisms (Fig. 7).

The physiological studies have been hampered by the fact that PECs make up only a small fraction of pulmonary cells, are diffusely distributed throughout the bronchopulmonary epithelium, and are heterogeneous. Successful isolation and culture of PECs is essential for the investigation of cellular and membrane properties of these cells. Several in vitro methods have been described, including organ cultures, microexplants, and enriched cell populations.

Carabba et al. (1985) and Sorokin et al. (1993) defined a fetal rat lung organ culture system that gave rise to appreciable numbers of PECs retaining normal relationships with surrounding cells, with the exception of their sensory innervation (Fig. 8). The method used lungs from fetal rats at 14–16 days of gestation, which were cultured in the presence of 40% fetal bovine serum, 1.5% agar, and 6.3% glucose for 1–8 days. Over this period, morphogenetic patterns for PECs were similar in vivo and in vitro. In this system, the early appearance of precursor

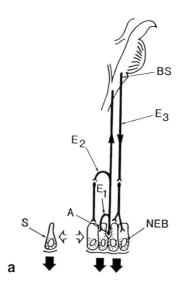

a

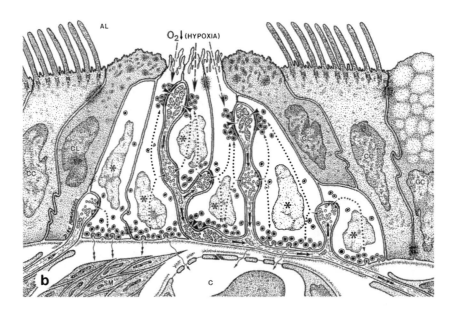

b

Figure 7 Diagrammatic representations of the receptor and effector functions of pulmonary endocrine cells and a typical neuroepithelial body (NEB). (a) All solitary cells (S) are probably not innervated. They, as well as the cells in NEBs, may react to changes in their local environment (open arrow) to release their secretory products in a paracrine fashion (solid arrow). NEBs have a complex innervation (see also b), but are also open to stimulation by changes in their local environment (open arrow). For instance, reduction in the local level

PECs in the largest bronchi and their subsequent progression into more distal airway branches by centrifugally directed differentiation could be monitored. It is of importance that PECs in this in vitro model were highly concentrated in the airway epithelium owing to a rapid buildup of cells during the gestational period, which allowed their distinct development while the lungs were cultured. In addition, development of nerves and ganglia could be followed. Ebina et al. (1993) demonstrated that the release of CGRP from the PECs in this system could be evoked by calcium influx across the plasma membrane, when calcium and the calcium ionophore A23187 were added to the culture medium.

A modification of the organ culture approach is to use lung epithelial fragments and microexplants. Sonstegard et al. (1979) used small lung explants (1–2 mm²) from fetal rabbits at 20 days or near term (term = 31 days) of gestation, which were plated on Gelfoam sponges and maintained for up to 22 days. The morphology and ultrastructure of the NEBs in vitro was comparable with noncultured controls, and experiments on secretion suggested that the process is calcium dependent, as in other secretory cells. However, isolation and separation of intact NEBs both from fetal rabbits and from humans have encountered several limitations, including the sparsity and uneven distribution of NEBs (Sonstegard et al., 1982; Dey et al., 1986). Moreover, the procedures aimed at the preservation of NEBs appeared quite complex.

Another, very promising, approach involves the use of single-cell preparations

---

of oxygen causes degranulation at points of contact with afferent nerve endings (A). This afferent activity feeds back via efferent endings in close continuity with them (E₁) and via local (probably intrapulmonary) reflex arcs (E₂). Together with a vagal efferent input (E₃) descending from the brainstem (BS), these probably moderate the function of the NEB and determine the nature of the afferent signal that eventually reaches vagal sensory nuclei. Degranulation at afferent nerve endings in NEBs is accompanied by basal exocytosis (solid arrows) in the same paracrine fashion as in solitary cells. [From Gosney (1992) with permission of the publisher.] (b) Hypoxic air stimulates (broken line arrows) the NEB cells (asterisks) to discharge the contents of their DCVs at efferent synaptic sites of sensory nerve endings (large arrowheads). This could result in depolarization and the development of a generator potential that spreads over the nerve fiber (arrows). Supposing that afferent and efferent nerve terminals occur along the same nerve process, as was indeed demonstrated in serial sections of newborn rabbit NEBs, the efferent endings would develop synaptic activity (small arrowheads). This may in turn influence the physiological state of the receptor cell (dotted line arrows), thus modulating the transduction of stimuli or regulating local paracrine secretion on nearby blood vessels (wavy line arrows) or smooth muscle bundles (double-headed wavy arrows). Some nerve terminals appear to be exclusively efferent (double arrowhead). CL, Clara-like cells; CC, ciliated cells; GC, goblet cell; C, capillary; SM, smooth muscle bundle; AL, airway lumen. [From Adriansen and Scheuermann (1992) with permission of the publisher.]

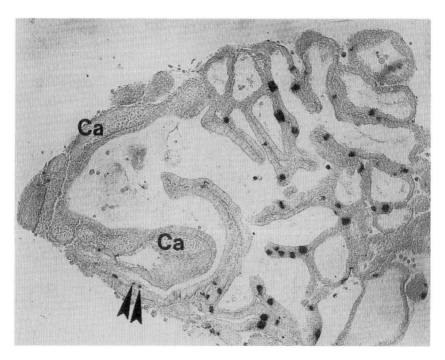

Figure 8 Photomicrograph of a representative section through a fetal rat lung organ culture. The section is through a 15-day fetal lung, kept in culture for 8 days, showing pulmonary endocrine cells with calcitonin gene-related peptide–like immunoreactivity in 40 foci composed of multicellular clusters. Most of the population occurs in intermediate-sized conducting airways. A few solitary cells are present in the largest airways (arrowheads). (Immunoperoxidase stain, ×85.) Ca = cartilage. [From Sorokin et al. (1993) with permission of the publisher.]

enriched for PECs. This permits the study of a specific cell type in a more controlled environment and avoids the problems of cell targeting associated with organ/explant-type cultures. Enrichment of PECs in the culture can be achieved in various ways. Cutz et al. (1985) used fetal rabbit lung because it contains a relatively large number of PECs. First they combined enzymatic procedures with either density gradient centrifugation for single PECs or differential adherence to plastic to obtain intact NEB clusters (Speirs and Cutz, 1993). Later they developed an immunomagnetic cell separation procedure to isolate PECs (Speirs et al., 1992). A two- to threefold increase in PEC recovery over the gradient isolation was obtained by using magnetic beads coupled with MOC-1 antibody, directed against a membrane antigen that is shared by SCLC and PECs (de Leij et al., 1985; Souhami et al., 1987) and related to the neural cell adhesion moleculer, N-CAM (Carbone et al., 1991). The predominant cells in the culture were derived from

NEB cells as defined by nuclear morphology; the solitary PECs were rare (Cutz et al., 1993). Exposure of NEB cultures to hypoxia resulted in decreased intracellular content of serotonin (5-HT) accompanied by increased exocytosis of dense core vesicles (DCV). Exposure of cultures to Ca^{2+} ionophore also resulted in reduction of 5-HT content, and increased exocytosis, confirming that the Ca^{2+} signaling pathway is involved in stimulus-secretion coupling. The demonstration of an O_2-sensitive K^+ current and an O_2-binding protein in NEB cells further suggested that they are transducers of the hypoxia stimulus and hence could function as airway chemoreceptors (Youngson et al., 1993).

The experience of in vitro propagation of lung tumors with NE features has suggested that defined culture media may play a major role in successful long-term cultures (Simms et al., 1980; Carney et al., 1981, 1985; Gazdar et al., 1990). Linnoila et al. (1993) used the chemically defined media originally developed for the selective growth of SCLC to establish long-term, PEC-enriched primary cultures of newborn (1 day old) hamster PECs. Up to 80% of the cultured cells contained characteristic DCV. These cells were maintained in culture up to 12 months (Fig. 9). The cultured hamster PECs contained immunoreactive CT (iCT) and released it through cholinergic-nicotinic stimulus (Fig. 10). The iCT had a chromatographic elution profile identical to that of intact hamster lung, which is known to contain high levels of iCT (Linnoila et al., 1984; Nylen et al., 1987).

In the hamster PEC cultures, a 3-week exposure to nicotine or bombesin caused an increase in total number of cells and total incorporation of tritiated thymidine (Nylen et al., 1993). In rabbit PECs there was an increase in the total protein of the cultures in response to bombesin (Speirs et al., 1993). However, there was no increase in ^{3}H-thymidine labeling of rabbit PECs in the bombesin-treated cultures, although there was an increase in labeling of the nonneuroendocrine population. The authors concluded that bombesin might have an *indirect* effect by perhaps causing a recruitment of either undifferentiated stem cells or immature PECs not yet expressing NE characteristics. Moreover, preliminary data with ^{125}I-GRP autoradiography on whole cells suggested that bombesin receptors were located mostly on fibroblasts and less on epithelial cells (Fig. 11). The authors hypothesized that in rabbit, GRP could cause the release of an unidentified factor, produced by fibroblasts. This may act on undifferentiated stem cells causing them to gradually acquire NE features. This hypothesis would also be in accordance with the preliminary findings on the localization of the mRNA for the recently cloned GRP receptor in nonepithelial compartment in human fetal lung tissue (Wang and Cutz, 1993).

VI. PULMONARY ENDOCRINE CELLS AND LUNG CANCER

Lung cancer continues to be a serious health problem. In the United States there will be approximately 170,000 new lung cancer cases in 1993. Lung cancer has a very poor prognosis with an overall 5-year survival of about 13% and is currently

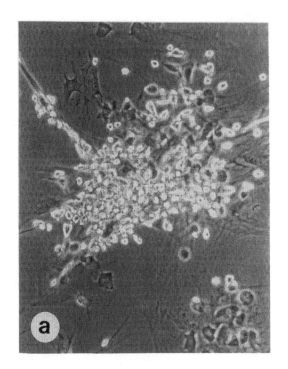

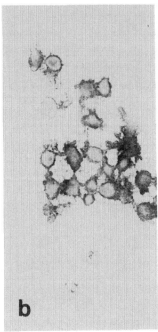

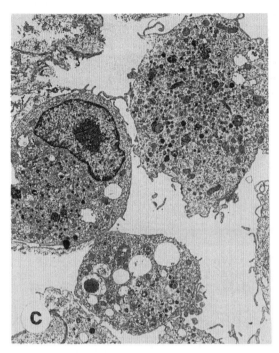

Figure 9 Growth pattern and characteristic features of a long-term selective culture of enriched hamster pulmonary endocrine cells. (a) Phase-contrast micrograph of a colony composed of adherent and floating, highly refractile small cells at 3 months of culture in defined (HITES) media with 5% serum. (×120.) (b) Cytospin preparation of the cultured PECs at 6 weeks stained for immunoreactive calcitonin reveals positivity in the majority of the cells. (Immunoperoxidase stain, ×330.) (c) Electron micrograph of the cultured cells demonstrates the presence of characteristic dense-core (endocrine) granules in 80% of the cells. (×3200.)

the leading cause of cancer deaths in both men and women in the United States. Approximately 30–35% of all lung cancers have NE characteristics, which is in sharp contrast to the very low number of PECs found in the nonneoplastic human lung (Table 5). Owing to current poor treatment results, future efforts need to be directed to prevention, new treatment strategies, and new methods for early detection.

Pulmonary NE tumors include the following distinct histological types: small cell lung carcinoma (SCLC); carcinoid, non-small-cell lung cancers with NE features (Matthews and Linnoila, 1988); and a small percentage of other NE tumors that are difficult to classify into the existing categories. All these tumors express a characteristic spectrum of the same *general* and *specific NE markers* as PECs (Linnoila et al., 1988; Gazdar et al., 1988).

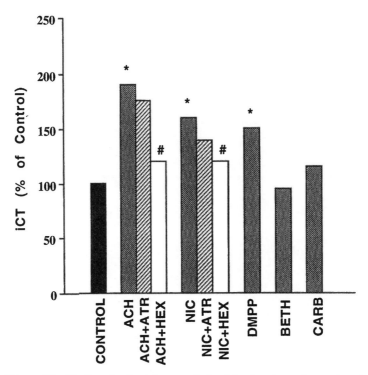

Figure 10 Cholinergic-nicotinic control of calcitonin release from cultured pulmonary endocrine cells. The amount of immunoreactive calcitonin following 1 hr of incubation is compared to control (black bar, 100%). The dose range of the agonists tested was 10^{-4}–10^{-6} M (gray bars). The antagonist dose was equimolar to the agonist (hatched and white bars). Maximal stimulations shown include acetylcholine (ACH) 10^{-6} M, nicotine (NIC) 10^{-5} M, dimethylphenyl piperazinium (DMPP) 10^{-5} M. ATR = atropine; HEX = hexamethonium. *$p < 0.05$ versus control; #$p < 0.05$ versus respective agonist. [Data from Nylen et al. (1993).]

Figure 11 Photomicrograph of gastrin-releasing peptide (GRP) binding into fibroblasts in a short-term pulmonary endocrine cell (PEC) culture. (a) Autoradiograph of a 3-day unfixed culture of fetal rabbit PECs, incubated with ^{125}I-GRP. Note the heavy concentration of the label over what appeared to be fibroblasts (open arrow). (×100.) (b) Parallel culture of (a) immunostained for cytokeratin to show islands of epithelial cells (solid arrow) surrounded by unstained fibroblasts. (×220.) [From Speirs et al. (1993) with permission of the publisher.]

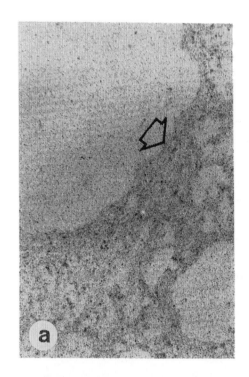

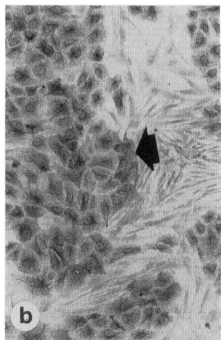

215

Table 5 Differentiation in Pulmonary Carcinogenesis

Event	Neuroendocrine	Peripheral airway	Squamous
Normal lung	+	+ + +	—
Hyperplasia	+ +	+ + + +	—
Metaplasia	?	+	+ + +
Dysplasia	+	+ +	+ + +
CA in situ	?	?	+ + +
Invasive CA	+ + + (30%)	+ + + (35%)	+ + + (30%)

The establishment of large collections of well-characterized lung cancer cell lines with NE differentiation, primarily derived from human SCLCs (Carney et al., 1985), has facilitated the study of NE-related functions that may also be relevant to PECs (Gazdar et al., 1988; Moody and Cuttitta, 1993; Sorenson et al., 1984). One such concept is the autocrine growth of SCLC stimulated by bombesin/GRP (Cuttitta et al., 1985a; Lebacq-Verheyden et al., 1990). According to this concept, selected SCLCs secrete GRP, which binds to specific GRP receptors in these tumors causing stimulation of growth. A monoclonal antibody binding to the C-terminal heptapeptide of either bombesin or GRP can be used as a bombesin antagonist to disrupt the growth both in vitro and in vivo. In the SCLC system, clonogenic growth in soft agar is more sensitive to stimulation by bombesin than mass culture growth. It is conceivable that such a loop exists also in PECs. While the currently available in vitro models for PECs utilize rodent systems where other specific products such as 5-HT, CGRP, and CT appear to dominate, there is recent evidence that CGRP can stimulate growth of tracheobronchial epithelial cells (Hoyt et al., 1993a; White et al., 1993). Further studies are needed to examine the potential role of autocrine loops in nonneoplastic PECs, since it is equally conceivable that the establishment of an autocrine growth mechanism may be a premalignant change, and not a feature of normal PECs.

Unlike in squamous cell carcinomas, where squamous cell metaplasia, dysplasia, and carcinoma in situ are thought to precede the formation of invasive squamous cell carcinoma, morphological precursor lesions for pulmonary NE tumors are poorly understood (Table 5). In addition to NE tumors, precursor lesions for other tumors, such as adenocarcinomas, with peripheral airway cell differentiation are not well characterized either.

The similarity in appearance, i.e., the presence of cytoplasmic DCV (endocrine granules), led to the suggestion that NE tumors, such as carcinoids and SCLC, may actually originate from PECs, which share the same characteristics (Bensch et al., 1965; Gmelich et al., 1967). While a spectrum ranging from hyperplasias and dysplasias of PECs, tumorlets, and minute pulmonary carcinoids to NE carcinomas of varying degree of differentiation has been described, it is not known whether this truly represents a meaningful premalignant *sequence* or another example of the "field cancerization" phenomenon (Slaughter et al., 1954; Strong et al., 1984). It is important to remember that similarities in phenotype, antigen expression, or differentiation do not necessarily mean common histogenesis (Gould et al., 1983). According to the concept of field cancerization, much or all of the epithelium has undergone one or more genetic (and morphological) alterations as a result of repeated exposure to the carcinogens, such as tobacco, and is at increased risk for developing multiple, *independent* morphological alterations and malignancies.

It is currently widely believed that all the bronchopulmonary epithelial cells, including PECs, as mentioned earlier, and their tumors are derived from a common stem cell origin. The stem cell undergoes progressive differentiation, through morphologically undifferentiated intermediate stages that are committed to certain pathways of differentiation (differentiation-committed cells) to fully differentiated cells (Gazdar et al., 1981). The neoplastic process frequently disrupts normal differentiation processes, and as a result, signs toward multidirectional differentiation in tumors or even in a single tumor cell are not uncommon (McDowell et al., 1981).

To further elucidate what might regulate pulmonary NE differentiation, Mabry et al. (1991) have developed an interesting in vitro model. The combination of two genetic abnormalities that can occur in lung cancer (amplification of c- or N-*myc* genes and mutations in *ras* gene) resulted in transition of typical SCLC cells to large cell undifferentiated carcinoma. In one SCLC this transition was also consistent with a fall in cellular neuropeptide (GRP) production. The authors then postulated that there exists an early proliferative cell that has *transient* endocrine features. This cell would give rise to a pluripotent undifferentiated cell, which, in turn, can differentiate into each of the mature bronchial epithelial cell types, including mature PECs.

It has been a distinct drawback that rodents and other potential experimental animals tend not to get SCLC, which is the major form of NE lung cancer in humans. However, selected chemical carcinogens do induce marked hyperplasias of PECs in various species although the resulting tumors may not have NE features (Fig. 12; Linnoila et al., 1981, 1982, 1984). Such animal models provide another tool to study PECs, which are rare in normal adult lung, and the mechanisms of NE differentiation (Sunday and Willet, 1992). Endocrine neoplasms perhaps most closely resembling those occurring in humans have been induced in hamsters by administration of nitrosamines in the presence of hyperoxia by Schüller et al. (1988) while nitrosamines under eupoxic conditions caused PEC hyperplasia and non-NE tumors. It is conceivable that two separate mechanisms are at work: (1) a reactive, essentially nonspecific proliferation of PECs in response to pulmonary injury, and (2) a separate process of carcinogenesis involving a different cell type,

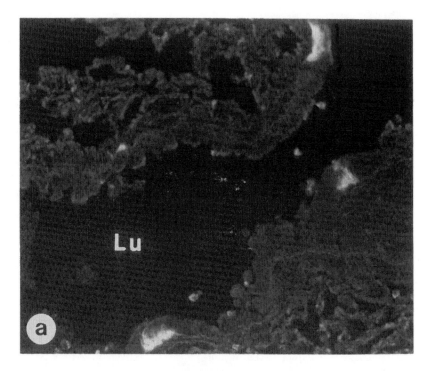

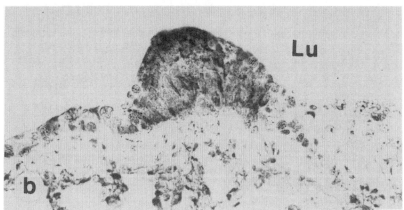

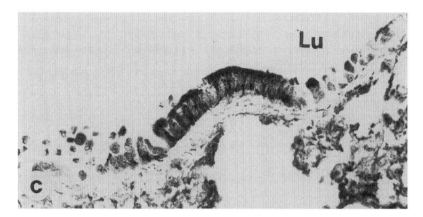

Figure 12 Photomicrographs of experimental pulmonary endocrine cell (PEC) hyperplasia. Syrian golden hamsters were exposed to the systemic carcinogen diethylnitrosamine (DEN) by subcutaneous injections. (a) Three neuroepithelial bodies (NEBs) and a solitary PEC in the same bronchus at 12 weeks of exposure. (Formaldehyde-induced fluorescence for catecholamines, ×230.) (b) Prominent, hyperplastic NEB protruding to the lumen at 16 weeks of exposure. (Argyrophilic silver impregnation according to Grimelius, ×290.) (c) Linear hyperplasia of PECs in the same lung. (Argyrophilic silver impregnation according to Grimelius, ×290.) The DEN-induced PEC hyperplasia consisted of both increased numbers of cells in NEBs and linear hyperplasia, which is reminiscent of the PEC hyperplasia in humans associated with neoplasia (see Fig. 6). Also, the number of foci (NEBs and clusters) was increased (Linnoila et al., 1981). Lu = bronchial lumen.

probably a precursor that has the potential for variable patterns of differentiation, depending on the environment in which carcinogenesis occurs.

VII. SUMMARY AND CONCLUSIONS

PECs are a small, highly heterogeneous, very versatile cell population, which share features of epithelial as well as neural cells. The major function of PECs is the synthesis and storage of various bioactive amines and neuropeptides. Proposed physiological roles include chemoreception and trophic actions. While few in number, PECs are strategically located throughout the bronchopulmonary tree where they can govern most of the lung through axon reflexes and paracrine mechanisms. They can quickly respond to pathological conditions both by increasing in number and by varying their contents, thus leading to potential amplification/ modulation of their actions.

Recently developed general NE markers and enrichment methods of PECs will greatly facilitate solving the remaining mysteries of their functions. Potentially important future explorations should include research on interactions of PECs also with other (nonneuropeptide) growth factors, mesenchymal factors, and immuno-regulation. PECs can be viewed as pluripotent players at the critical interphase of the respiratory and nonrespiratory lung.

ACKNOWLEDGMENT

Great appreciation is extended to Ms. Ruth Miehm-Rowan for her skillful typing of the manuscript and for assistance in preparing the references.

REFERENCES

Adriaensen, D., and Scheuermann, D.W. (1993). Neuroendocrine cells and nerves of the lung. *Anat. Rec. 236*:70.

Baylin, S.B. (1990). "APUD" cells: fact and fiction. *Trends Endocrinol. Metab. 1*:198.

Bensch, K.G., Gordon, G.B., and Miller, L.R. (1965). Studies on the bronchial counter-part of the Kultschitzky (argentaffin) cell and innervation of bronchial glands. *J. Ultra. Res. 12*:668.

Carabba, V.H., Sorokin, S.P., and Hoyt, R.F., Jr. (1985). Development of neuroepithelial bodies in intact and cultured lungs of fetal rats. *Am. J. Anat. 173*:1.

Carbone, D.P., Koros, A.M.C., Linnoila, R.I., Jewett, P., and Gazdar, A.F. (1991). Neural cell adhesion molecule expression and messenger RNA splicing patterns in lung cancer cell lines are correlated with neuroendocrine phenotype and growth morphology. *Cancer Res. 51*:6142.

Carney, D.N., Gazdar, A.F., Bepler, G., Guccion, J.G., Marangos, P.J., Moody, T.W., Zweig, M.H., and Minna, J.D. (1985). Establishment and identification of small cell lung cancer cell lines having classic and variant features. *Cancer Res. 45*:2913.

Cuttitta, F., Carney, D.N., Mulshine, J., Moody, T.W., Fedorko, J., Fischler, A., and Minna, J.D. (1985a). Autocrine growth factors in human small cell lung cancer. *Cancer Surv. 4*:707.

Cuttitta, F., Carney, D.N., Mulshine, J., Moody, T.W., Fedorko, J., Fischler, A., and Minna, J.D. (1985b). Bombesin-like peptides can function as autocrine growth factors in human small cell lung cancer. *Nature 316*:823.

Cuttitta, F., Fedorko, J., Gu, J., Lebacq-Verheyden, A.M., Linnoila, R.I., and Battey, J.F. (1988). Gastrin-releasing peptide gene-associated peptides are expressed in normal human fetal lung and small-cell lung cancer: a novel peptide family found in man. *J. Clin. Endocrinol. Metab. 67*:576.

Cutz, E. (1987). Cytomorphology and differentiation of airway epithelium in developing human lung. In: *Lung Carcinomas.* (McDowell, E.M., ed.). Churchill Livingstone, Edinburgh, p. 1.

Cutz, E., Yeger, H., Wong, V., Bienkowski, E., and Chan, W. (1985). In vitro characteris-tics of pulmonary neuroendocrine cells isolated from rabbit fetal lung. I. Effects of culture media and nerve growth factor. *Lab. Invest. 53*:672.

Cutz, E., Speirs, V., Yeger, H., Newman, C., Wang, D., and Perrin, D.G. (1993). Cell biology of pulmonary neuroepithelial bodies-validation of an in vitro model. 1. Effects of hypoxia and Ca^{2+} ionophore on serotonin content and exocytosis of dense core vesicles. *Anat. Rec. 236*:41.

Dey, R.D., Snyder, J.M., Speciale, S.G., and Price, J. (1986). Characterization of human pulmonary endocrine cells maintained in vitro. *Exp. Lung Res. 10*:369.

Ebina, M., Hoyt, R.F., Jr., Sorokin, S.P., and McNelly, N.A. (1993). Calcium and ionophore A23187 lower calcitonin gene-related peptide-like immunoreactivity in endocrine cells of organ cultured fetal rat lungs. *Anat. Rec. 236*:226.

Feyrter, F. (1946). Über die These von den peripheren endokrinen Drusen. *Wien Z. Innere Med. Grenzgeb. 10*:9.

Feyrter, F. (1954). Über die Argyrophilie des HelleZellen-Systems im Bronchialbaum des Menschen. *Z. Mikrosk. Anat. Forsch. 761*:73.

Fröhlich, F. (1949). Die HelleZelle der Bronchialschleimhaut und ihre Beziehungen zum Problem der Chemoreceptoren. *Frankfurter Z. Pathol. 60*:517.

Gazdar, A.F., Carney, D.N., Guccion, J.G., and Baylin, S.B. (1981). Small cell carcinoma of the lung: cellular origin and relationship to other pulmonary tumors. In: *Small Cell Lung Cancer.* (Greco, F.A., Oldham, R.K., and Bunn, P.A., Jr., eds.). Grune & Stratton, New York, p. 145.

Gazdar, A.F., Helman, L.J., Israel, M.A., Russell, E.K., Linnoila, R.I., Mulshine, J.L., Schuller, H.M., and Park, J.G. (1988). Expression of neuroendocrine cell markers L-dopa decarboxylase, chromogranin A, and dense core granules in human tumors of endocrine and nonendocrine origin. *Cancer Res. 48*:4078.

Gmelich, J.T., Bensch, K.G., and Liebow, A.A. (1967). Cells of Kultchizky type in bronchioles and their relation to the origin of peripheral carcinoid tumor. *Lab. Invest. 17*:88.

Gosney, J.R. (1992). *Pulmonary Endocrine Pathology: Endocrine Cells and Endocrine Tumours of the Lung.* Butterworth-Heinemann, Oxford, pp. 1–184.

Gosney, J.R. (1993). Neuroendocrine cell populations in postnatal human lungs: minimal variation from childhood to old age. *Anat. Rec. 236*:177.

Gosney, J.R., Sissons, M.C.J., Allibone, R.O., and Blakely, A.F. (1989). Pulmonary endocrine cells in chronic bronchitis and emphysema. *J. Pathol. 157*:127.

Gould, V.E., Linnoila, R.I., Memoli, V.A., and Warren, W.H. (1983). Neuroendocrine components of the bronchopulmonary tract: Hyperplasias, dysplasias, and neoplasms. *Lab. Invest. 49*:519.

Hoyt, R.F., Jr., McNelly, N.A., and Sorokin, S.P. (1993a). Calcitonin gene-related peptide (CGRP) as regional mitogen for tracheobronchial epithelium of organ cultured fetal rat lungs. *Am. Rev. Respir. Dis. 147*:A498.

Hoyt, R.F., Jr., Sorokin, S.P., McDowell, E.M., and McNelly, N.A. (1993b). Neuroepithelial bodies and growth of the airway epithelium in developing hamster lung. *Anat. Rec. 236*:15.

Lauweryns, J.M., Cokelaere, M., and Theunynek, P. (1972). Neuro-epithelial bodies in the respiratory mucosa of various mammals. *Zell. 135*:569.

Lebacq-Verheyden, A.-M., Trepel, J., Sausville, E.A., and Battey, J.F. (1990). Bombesin and gastrin-releasing peptide: neuropeptides, secretogogues, and growth factors. In: *Peptide Growth Factors and Their Receptors II.* (Sporn, M.B., and Roberts, A.B., eds.). Springer-Verlag, Berlin, p. 71.

Leij, L.D., Poppema, S., Nulend, J.K., Haar, A., Schwander, E., Ebbens, F., Postmus, P.E., and The, T.H. (1985). Neuroendocrine differentiation antigen on human lung carcinoma and Kulchitski cells. *Cancer Res. 45*:2192.

Linnoila, R.I. (1982). Effects of diethylnitrosamine on lung neuroendocrine cells. *Exp. Lung Res. 3*:225.

Linnoila, R.I., Nettesheim, P., and DiAugustine, R.P. (1981). Lung endocrine-like cells in hamsters treated with diethylnitrosamine: alterations in vivo and in cell culture. *Proc. Natl. Acad. Sci. USA 78*:5170.

Linnoila, R.I., Becker, K.L., Silva, O.L., Snider, R.H., and Moore, C.F. (1984). Calcitonin as a marker for diethylnitrosamine-induced pulmonary endocrine cell hyperplasia in hamsters. *Lab. Invest. 51*:39.

Linnoila, R.I., Mulshine, J.L., Steinberg, S.M., Funa, K., Matthews, M.J., Cotelingam, J.D., and Gazdar, A.F. (1988). Neuroendocrine differentiation in endocrine and non-endocrine lung carcinomas. *Am. J. Clin. Pathol. 90*:641.

Linnoila, R.I., Gazdar, A.F., Funa, K., and Becker, K.L. (1993). Long-term selective culture of hamster pulmonary endocrine cells. *Anat. Rec. 236*:231.

Mabry, M., Nelkin, B.D., Falco, J.P., Barr, L.F., and Baylin, S.B. (1991). Transitions between lung cancer phenotypes—implications for tumor progression. *Cancer Cells 3*:53.

Matthews, M.J., and Linnoila, R.I. (1988). Pathology of lung cancer—an update. In: *Lung Cancer: A Comprehensive Treatise*. (Little, A.G., and Weichselbaum, R.R., eds.). Grune & Stratton, Chicago, p. 5.

McDowell, E.M., and Trump, B.F. (1981). Pulmonary small cell carcinoma showing tripartite differentiation in individual cells. *Hum. Pathol. 12*:286.

Moody, T.W., and Cuttitta, F. (1993). Growth factor and peptide receptors in small cell lung cancer. *Life Sci. 52*:1161.

Nylen, E.S., and Becker, K.L. (1990). The diffuse endocrine systems. In: *Principles and Practice of Endocrinology and Metabolism*. (Becker, K.L., ed.). J.B. Lippincott, Philadelphia, p. 1276.

Nylen, E.S., Linnoila, R.I., Snider, R.H., Tabassian, A.R., and Becker, K.L. (1987). Comparative studies of hamster calcitonin from pulmonary endocrine cells in vitro. *Peptides 8*:977.

Nylen, E.S., Becker, K.L., Snider, R.H., Tabassian, A.R., Cassidy, M.M., and Linnoila, R.I. (1993). Cholinergic-nicotinic control of growth and secretion of cultured pulmonary neuroendocrine cells. *Anat. Rec. 236*:129.

Polak, J.M. (1993). Lung endocrine cell markers, peptides, and amines. *Anat. Rec. 236*:169.

Schüller, H.M., Becker, K.L., and Witschi, H.P. (1988). An animal model for neuroendocrine lung cancer. *Carcinogenesis 9*:293.

Slaughter, D.P., Southwick, H.W., and Smejkal, W. (1954). "Field cancerization" in oral stratified squamous epithelium: clinical implications of multicentric origin. *Cancer 6*:963.

Sonstegard, K., Wong, V., and Cutz, E. (1979). Neuro-epithelial bodies in organ cultures of fetal rabbit lungs. *Cell Tissue Res. 199*:159.

Sonstegard, K.S., Mailman, R.B., Cheek, J., Tomlin, T., and DiAugustine, R.P. (1982). Morphological and cytochemical characterization of neuroepithelial bodies in fetal rabbit lung. I. Studies of isolated neuroepithelial bodies. *Exp. Lung Res. 3*:349.

Sorenson, G.D., Pettengill, O.S., Cate, C.C., Ghatei, M.A., and Bloom, S.R. (1984). Modulation of bombesin and calcitonin secretion in cultures of small cell carcinoma of the lung. In: *The Endocrine Lung in Health and Disease*. (Becker, K.L., and Gazdar, A.F., eds.). W.B. Saunders, Philadelphia, p. 596.

Sorokin, S.P., Ebina, M., and Hoyt, R.F., Jr. (1993). Development of PGP 9.5- and calcitonin gene-related peptide-like immunoreactivity in organ cultured fetal rat lungs. *Anat. Rec. 236*:213.

Souhami, R.L., Beverly, P.C., and Bobrow, L.G. (1987). Antigens of small-cell lung cancer: first international workshop. *Lancet 2*:325.

Speirs, V., and Cutz, E. (1993). An overview of culture and isolation methods suitable for in vitro studies on pulmonary neuroendocrine cells. *Anat. Rec. 236*:35.

Speirs, V., Wang, Y., Yeger, H., and Cutz, E. (1992). Isolation and culture of neuroendocrine cells from fetal rabbit lung using immunomagnetic techniques. *Am. J. Respir. Cell Mol. Biol. 6*:63.

Speirs, V., Bienkowski, E., Wong, V., and Cutz, E. (1993). Paracrine effects of bombesin/gastrin-releasing peptide and other growth factors on pulmonary neuroendocrine cells in vitro. *Anat. Rec. 236*:53.

Stahlman, M.T., and Gray, M.E. (1993). Colocalization of peptide hormones in neuroendocrine cells of human fetal and newborn lungs: an electron microscopic study. *Anat. Rec. 236*:206.

Strong, M.S., Incze, J., and Vaughan, C.W. (1984). Field cancerization in the aerodigestive tract: its etiology, manifestation, and significance. *J. Otolaryngol. 13*:1.

Sunday, M.E., and Willett, C.G. (1992). Induction and spontaneous regression of intense pulmonary neuroendocrine cell differentiation in a model of preneoplastic lung injury. *Cancer Res.* (Suppl. 52):2677s.

Sunday, M.E., Hua, J., Dai, H.B., Nusrat, A., and Torday, J.S. (1990). Bombesin increases fetal lung growth and maturation in utero and in organ culture. *Am. J. Respir. Cell Mol. Biol. 3*:199.

Sunday, M.E., Hua, J., Reyes, B., Masui, H., and Torday, J.S. (1993). Anti-bombesin monoclonal antibodies modulate fetal mouse lung growth and maturation in utero and in organ cultures. *Anat. Rec. 236*:25.

Wang, D., and Cutz, E. (1993). Expression of bombesin (BN)/gastrin releasing peptide (GRP) receptor gene in developing lung. *Lab. Invest. 68*:128A.

Wharton, J., Polak, J.M., Bloom, S.R., Ghatei, M.A., Solcia, E., Brown, M.R., and Pearse, A.G.E. (1978). Bombesin-like immunoreactivity in the lung. *Nature 273*:769.

White, S.R., Hershenson, M.B., Sigrist, K.S., Zimmerman, A., and Solway, J. (1993). Proliferation of guinea pig tracheal epithelial cells induced by calcitonin gene-related peptide. *Am. J. Respir. Cell Mol. Biol. 8*:592.

Youngson, C., Nurse, C., Yeger, H., and Cutz, E. (1993). Oxygen sensing in airway chemoreceptors. *Nature 365*:153.

DISCUSSION

Baraniuk: Are neuroendocrine cells innervated?

Linnoila: In normal lung, pulmonary neuroendocrine cells occur as solitary cells and in innervated clusters called neuroepithelial bodies (NEBs). NEBs are innervated by efferent and afferent fibers in a complex manner. There is also evidence that afferent and efferent nerve terminals can occur along the same nerve process, as demonstrated in serial sections of newborn rabbit NEBs. Solitary pulmonary endocrine cells are probably not always innervated.

McDonald: What is the functional significance of the nicotinic stimulation of secretion by neuroendocrine cells, considering that there is no evidence that the cells are innervated by cholinergic nerves?

Linnoila: The innervation problem is complex. Basically, there is evidence that afferent nerve endings feed back via efferent endings in close proximity and via local (probably intrapulmonary) reflex arcs. Together with a vagal efferent input descending from the brainstem, these modulate the function of NEBs and determine the nature of the afferent signal that eventually reaches vagal sensory nuclei. Degranulation of afferent nerve endings in NEBs is accompanied by basal exocytosis of NEB cells in a paracrine fashion. Much of the evidence of the nature of fibers is based on ultrastructural studies. Nicotinic stimulation may play a role in smoking-induced carcinogenesis.

Widdicombe: Barnes and his colleagues have shown that the airway epithelium has muscarinic receptors, but these do not seem to have a functional motor innervation. Therefore, the nicotinic receptors in the NEBs may also lack a functional cholinergic innervation.

10

Molecular Characterization of Autonomic and Neuropeptide Receptors

Claire M. Fraser and Norman H. Lee
The Institute for Genomic Research, Gaithersburg, Maryland

I. INTRODUCTION

For many years, it was believed that the control of airway function was dependent on the balance between the cholinergic (parasympathetic) and adrenergic (sympathetic) nervous systems. The cholinergic system is considered excitatory because it plays a role in maintaining airway tone and in mediating acute bronchospastic responses (Casale, 1993). The effect of acetylcholine to produce narrowing of the airways is blocked by atropine, indicating that this effect is mediated by muscarinic acetylcholine receptors (Colebatch and Halmagyi, 1963; Olsen et al., 1965). In contrast, the adrenergic system in the lung is considered inhibitory because stimulation of β-adrenergic receptors produces relaxation of bronchial smooth muscle. The beta-blockade theory of the pathogenesis of asthma from Szentivanyi (1968) proposed that asthma was related to an imbalance in the autonomic control of airway diameter due to a decrease in β-adrenergic sensitivity in bronchial smooth muscle, mucus glands, and mucosal blood vessels.

Recent advances in the study of neurotransmitter receptors have revealed that the neural control mechanisms of the airways are more complex than originally appreciated. For example, molecular cloning experiments have identified five subtypes of muscarinic receptors, termed m1–m5 (Kerlavage et al., 1987; Bonner et al., 1987). Three of these subtypes have been shown to be expressed in human lung where they subserve distinct physiological functions (Barnes, 1989a,b; Mak and Barnes, 1990). Human airway smooth muscle contains the m3-receptor subtype that is responsible for muscle contraction. The submucosal glands contain both m1 and m3 subtypes that are involved in mucus secretion. The m2-receptor subtype is present on cholinergic nerves and presumed to function as an autoreceptor, inhibiting acetylcholine release.

Human lung contains both β_1- and β_2-adrenergic receptor subtypes in a ratio of approximately 1:3 (Casale and Hart, 1987; Carstairs et al., 1985; Goldie et al., 1985). Most of the critical lung functions are mediated by β_2-adrenergic receptors, including relaxation of airway smooth muscle (Zaagsma et al., 1983), inhibition of antigen-induced mast cell mediator release (Orange et al., 1971a,b; Butchers et al., 1980), release of surfactant (Brown and Longmore, 1981), and inhibition of cholinergic neurotransmission (Vermiere and Vanhoutte, 1991; Rhoden et al., 1988). α-Adrenergic receptors have also been identified in the lung (Barnes et al., 1980; Casale and Hart, 1987). However, the exact role of α-adrenergic receptors in lung physiology is not entirely clear. Stimulation of α-adrenergic receptors has been reported to produce weak bronchoconstriction (Kneussl and Richardson, 1978; Goldie et al., 1985), mucus secretion (Culp et al., 1990; Lundgren and Shelhamer, 1990), and inhibition of cholinergic transmission (Starke, 1981; Grundstrom et al., 1981, 1984).

In addition to the cholinergic and adrenergic pathways, nonadrenergic, noncholinergic (NANC) pathways have more recently been demonstrated to be important in normal lung function (Polak and Bloom, 1982). The neurotransmitters for this system are peptides that mediate both excitatory (substance P, neurokinin A, neurokinin B, calcitonin gene-related peptide) and inhibitory (vasoactive intestinal polypeptide, peptide histidine methionine) responses. Many of these peptides were originally described in the gastrointestinal tract but have since been found to be localized throughout the body. As with the muscarinic and adrenergic receptors, subtypes of peptide receptors have recently been identified from gene-cloning studies. The presence of such a large number of receptors in the airways has therapeutic implications for the treatment of asthma and other respiratory diseases.

The purpose of this chapter is to review some of the molecular properties of autonomic and neuropeptide receptors that are important in normal lung function. The application of molecular biology to the study of receptors has had a profound impact on the understanding of receptor structure and function and receptor mechanisms.

II. PRIMARY STRUCTURE OF AUTONOMIC AND NEUROPEPTIDE RECEPTORS

Molecular biology has proven invaluable as a tool for the elucidation of the gene structure of neurotransmitter receptors and, thus, has provided the opportunity to analyze receptor subtypes, receptor structure, ligand binding, and receptor-effector coupling at the molecular level. Two early examples of how this approach advanced the understanding of receptor mechanisms are the β-adrenergic (Dixon et al., 1986; Chung et al., 1987) and muscarinic acetylcholine receptors (Kubo et al., 1986a,b). When the deduced amino acid sequences of these two pharmacologically unrelated receptors were compared, it was apparent that they were similar in structure to each other and also to rhodopsins (Ovchinnikov et al., 1983; Pappin et al., 1984), the visual pigments in the retina that function as light receptors. Although pharmacologically distinct, these receptors share a common mechanism of signal transduction involving guanine nucleotide regulatory proteins (G proteins), suggesting the existence of a family of membrane receptors (Table 1).

During the past 7 years, more than 200 G-protein-coupled receptors have been

Table 1 Membrane Receptors that Interact with Guanine Nucleotide Regulatory Proteins

Peptide hormone receptors	Neurotransmitter receptors
Adrenocorticotropin (ACTH)	Adenosine
Angiotensin	α-Adrenergic
Antidiuretic hormone	β-Adrenergic
Bombesin	Dopamine
Bradykinin	$GABA_B$
Calcitonin	Histamine
Cholecystokinin (CCK)	Muscarinic acetylcholine
Choriogonadotropin	Octopamine
Corticotropin-releasing hormone (CRF)	Serotonin
Endothelin	Tyramine
Follicle-stimulating hormone (FSH)	
Gastrin	Sensory systems
Glucagon	Olfaction
Gonadotropin-releasing hormone (GnRH)	Taste
Growth-hormone-releasing hormone (GRF)	Vision (rhodopsins)
Kinins (bradykinin, substance P, substance K)	
Luteinizing hormone (LH)	Other agents
Melanocyte-stimulating hormone (MSH)	C5a anaphylatoxin
Neurotensin	Cannabinoids
Opiates	IgE
Oxytocin	*Mas* oncogene
Parathyroid hormone	f-Met-Leu-Phe
Secretin	Platelet-activating factor
Somatostatin	Prostanoids (PGEs, leukotrienes)
Thyrotropin (TSH)	Thrombin
Thyrotropin-releasing hormone (TRH)	
Vasoactive intestinal polypeptide (VIP)	
Vasopressin	

Source: Adapted from Savarese and Fraser, 1992.

	TM1	TM2	TM3	TM4
hβ1AR	AGMGLLMALIVLLIVAGNVLVIVA	FIMSLASADLVMGLLVVPFGATIV	SFFCELWTSVDVLCVTASIETLCVIALDRY	ARGLVCTVWAISALVSFLPILM
hβ2AR	VGMGIVMSLIVLAIVFGNVLVITA	FITSLACADLVMGLAVVPFGAAHI	NFWCEFWTSIDVLCVTASIETLCVIAVDRY	ARVIILMVWIVSGLTSF.LPIQ
ham α1bAR	ISVGLVLGAFILFAIVGNILVILS	FIVNLAIADLLLSFTVLPFSAATLE	RIFCDIWAAVDVLCCTASILSLCAISIDRY	AILALLSVWLSTVISIG.PLL
hα2aAR	LTLVCLAGLLMLLTVFGNVLVIIA	FLVSLASADIIVATLVIPFSLANE	KTWCEIYLALDVLFCTSSIVHLCAISLDRY	KAI.IITCWVISAVISFPPLIS
hD1DR	ILTACFLSLLILSTLLGNTLVCAA	FVISLAVSDLLVAVLVMPWKAVAE	GSFCNIWVAFDIMCSTASILNLCVISVDRY	AFILISVAWTLSVLISFIPVQL
h5HT1aR	VITSLLLGTLIFCAVLGNACVVAA	LIGSLAVTDLMVSVLVLPMAALYQ	QVTCDLFIALDVLCCTSSILHLCAIALDRY	PRALISLTWLIGFLISIP.PML
hm1mAChR	AFIGITTGLLSLATVTGNLLVLIS	FLLSLACADLIIGTFSMNLYTTYL	TLACDLWLALDYVASNASVMNLLIISFDRY	AALMIGLAWLVSFVLWAPAILF
hm2mAChR	VFIVLVAGSLSLVTIIGNILVMVS	FLFSLACADLIIGVFSMNLYTLYT	PVVCDLWLALDYVVSNASVMNLLIISFDRY	AGMMIAAAWLSFILWAPAILF
mDOR	IAITALYSAVCAVGLLGNVLVMFG	YIFNLALADALATSTLPFQSAKYL	ELLCKAVLSIDYNMFTSIFTLTMMSVDRY	AKLINICIWLASGVGVPIMVM
hFMLPR	IITYLVFAVTFVLGVLGNGLVIWV	SYLNLAVADFCFTSTLPFFMVRKA	WFLCKFLTFIVDINLFGSVFLIALIALDRC	AKKVIIGPWVMALLLTLPVIIR
hSPR	VLWAAAYTVIVTSVVGNVVVMWI	FLVNLAFAEASMAAFNTVVNFTYA	LFYCKFHNFPIAAVFASIYSMTAVAFDRY	TKVVICVIWLALLLAFPQGYY
hSKR	ALWAPAYLALVLVAVTGNAIVIWI	FIVNLALADLCMAAFNAAFNFVYA	RAFCYFQNLPITAMFVSIYSMTAIAADRY	TKAVIAGIWLVALALASPQCFY
mTRHR	VVTLILVVICGLGVGNIMVVLV	YLVSLAVADMVLVAAGLPNITDS	YVGLCITYLQYLGINASCSITAFTTERY	AKKIIIFVWAFTSIYCMLEFFL
hLH/CGR	DFLRVLIWLINILAIMGNMTVLFL	LMCNLSFADFCMGLYLLLIASVDS	GSGGSTAGFFTVFASELSVTLTVTITLERW	AILIMLGGWLFSSLIAMLPLVG
hFSHR	NILRVLIWFISILAITGNIIVLVI	LMCNLAFADLCIGIYLLLIASVDI	GAGCDAAGFFTVFASELSVTLTAITLERW	AASVMVMGWIFAFAAALFPIFG
hTSHR	KFLRIVVWFVSLLALLGNVFVLLI	LMCNLAFADFCMGMYLLLIASVDL	GPGCNTAGFFTVFASELSVTLTVTITLERW	ACAIMVGGWVCCFLLALLPLVG
rmGlutR1	IIAIAFSCLGILVTLFVTLIFVLY	YIILAGIFLGYVC.PFTLIAKPTT	YLQRLLVGLSSAMCYSALVTKTNRIARILA	QVIIASILISVQLTLVVTLIIM
rmGlutR2	VGPVTIACLGALATLFVLGVFVRH	YILLGGVFLCY.CMTFVFIAKPST	TLRRLGLGTAFSVCYSALLTKTNRIARIFG	QVAICLALISGQLLIVAAWLVV

```
              TM5                         TM6                      TM7

hβ1AR     AYAIASSVVSFYVPLCIMAFVYL   TLGIIMGVFTLCWLPFFLANVV   DRLFVFFNWLGYANSAFNPIIYC
hβ2AR     AYAIASSIVSFYVPLVIMVFVYS   TLGIIMGTFTLCWLPFFIVNIV   KEVYILLNWIGYVNSGFNPLIYC
Ham α1bAR FYALFSSLGSFYIPLAVLVMYC    TLGIVVGMFILCWLPFFIALPL   DAVFKVVFWLGYFNSCLNPIIYP
hα2aAR    WYVISSCIGSFEAPCLIMILVYV   TLGIVVGMFILCWLPFFIALPL   DAVKFVVFWLGYFNSCLNPIIYP
hD1DR     TYAISSSVISFYIPVAIMIVTYT   TLSVIMGVFVCCWLPFFILNCI   SNTFDVFVWGFWANSSLNPIIYA
h5HT1aR   GYTIYSTFGAFYIPLLLMLVLYG   TLGIIMGTFILCWLPFFIVALV   TLLGAIINWLGYSNSLLNPVIYA
hm1mAChR  IITFGTAMAAFYLPVTVMCTLYW   TLSAILLAFILTWTPYNIMVLV   ETLWELGYWLCYVNSTINPMCYA
hm2mAChR  AVTFGTAIAAFYLPVIIMTVLYW   TILAILLAFIITWAPYNVMVLI   NTVWTIGYWLCYINSTINPACYA

mDOR      VTKICVFLFAFVVPILIITCYG    MVLVVVGAFVVCWAPIHIFVIV   VAALHLCIALGYANSSLNPVLYA
hFMLPR    VRGIIRFIIGFSAPMSIVAVSYG   VLSFVAAAFFLCWSPYQVVALI   GIAVDVTSALAFFNSCLNPMLYV

hSPR      VYHICVTVLIYELPLLVIGYAYT   MMIVVVCTFAICWLPFHIFFLL   QQVYLAIMWLAMSSTMYNPIIYC
hSKR      LYHLVVIALIYELPLAVMFVAYS   TMVLVVLTFAICWLPYHLYFIL   QQVVLALFWKAMSSTMYNPIIYC
mTRHR     PIYLMDFGVFYVMPMLATVLYG    MLAVVVILFALLWMPYRTLVVV   NWFLLFCRICIYLNSAINPVIYN

hLH/CGR   YILTLILNVVAFFICACYIKI     KMAILLFTDFTCMAPISFFAIS   TNSKVLLVLFYPINSCANPFLYA
hFSHR     YVMSLLVLNVLAFVVICGCYIHI   RMAMLIFTDFLCMAPISFFAIS   SKAKILLVLFHPINSCANPFLYA
hTSHR     YIVFLVTLNIVAFVIVCCCHVKI   RMAVLIFTDFICMAPISFYALS   SNSKILLVLFYPLNSCANPFLYA

rmGlutR1  LGVVAPVGYNGLLIMSCTYYAFK   AFTMYTTCIIWLAFVPIYFGSN   CFAVSLSVTVALGCMFTPKMYII
rmGlutR2  ASMLGSLAYNVLLIALCTLYAFK   GFTMYTTCIIWLAFLPIFYVTS   CVSVSLSGSVVLGCLFAPKLHII
```

Figure 1 Alignments of the seven transmembrane domains (TM1–TM7) and adjacent residues in representative G-protein-coupled receptors. The amino acids are represented by the single-letter code. Residues in boldface type represent highly conserved amino acids. Underlined residues represent conservative substitutions. Shaded residues represent amino acids conserved within a subfamily of receptors. The receptors sequences illustrated include: hβ1AR, human β₁-adrenergic receptor (Frielle et al., 1987); ham α1bAR, hamster α₁b-adrenergic receptor (Cotecchia et al., 1988); hα2AR, human α₂A-adrenergic receptor (Fraser et al., 1989); h5HT1aR, human 5HT₁ₐ receptor (Kobilka et al., 1987); mTRHR, mouse thyrotropin-releasing hormone (Straub et al., 1990); rmGlutR1, rat metabotropic glutamate receptor 1 (Tanabe et al., 1992); rmGlutR2, rat metabotropic glutamate receptor 2 (Tanabe et al., 1992). References for other sequences can be found in the legend to Figure 6. (From Lee and Kerlavage, 1993.)

cloned and sequenced (see Savarese and Fraser, 1992, for a review). Within this family are a number of receptors relevant to lung physiology, including adrenergic, muscarinic, and tachykinin receptors. G-protein-coupled receptors, which collectively represent the largest family of membrane receptors, bind such diverse ligands as small amine neurotransmitters, small peptides, glycoprotein hormones, and agents that mediate sensory responses such as light and odorants. Furthermore, they are important therapeutic targets in a large number of diseases. The rapid progress in G-protein-coupled receptor cloning has been based in large part on the conservation of primary structure among G-protein receptors, particularly within families, allowing for isolation of new cDNA and genomic clones by cross-hybridization. One important finding from molecular cloning studies is the existence of receptor subtypes that were not suspected or only weakly supported by earlier pharmacological data. This suggests that it may be possible to exploit structural information on receptors in the design of more selective therapeutic agents.

G-protein-coupled receptors are single polypeptides that range in size from ~400 to 1000 amino acids. The unifying feature of all G-protein-coupled receptors is the presence of seven domains of 20–28 hydrophobic amino acids that are presumed to represent transmembrane α-helical segments (TMS) that span the lipid bilayer (Figs. 1, 2). In all the G-protein receptors that have been cloned to date, it is the membrane-spanning domains that exhibit the greatest degree of amino acid sequence identity, ranging from 20–30% between unrelated receptors up to 90% between receptor subtypes (Kerlavage et al., 1987; Savarese and Fraser, 1992). The transmembrane helices are connected by more divergent, hydrophilic regions that are thought to be exposed extracellularly and intracellularly (Fig. 2). Because the amino-terminus (N-terminus) of a majority of G-protein-coupled receptors contains one or more sites for N-linked glycosylation, it is postulated to be located extracellularly, thus placing the carboxyl-terminus (C-terminus) in the intracellular environment. Immunological mapping supports this proposed membrane topology for the β_2-adrenergic receptor (Wang et al., 1989).

The current model for the secondary and tertiary structure of G-protein-coupled receptors is based on that of bacteriorhodopsin. Electron microscopy and high-resolution electron diffraction analysis of bacteriorhodopsin reveal seven closely packed helical segments that span the membrane and are arranged in a bundle perpendicular to the plane of the membrane (Henderson and Unwin, 1975; Henderson et al., 1986) (Fig. 2).

A detailed comparison of the amino acid sequences of G-protein-coupled receptors reveals that many, but not all, receptors in this family share a number of conserved amino acids or domains (Figs. 1, 3–5). It has been proposed that the conservation of certain critical amino acid sequences among a large number of diverse G-protein-coupled receptors is the result of evolutionary pressure to maintain a structure that is essential for G-protein-mediated signal transduction

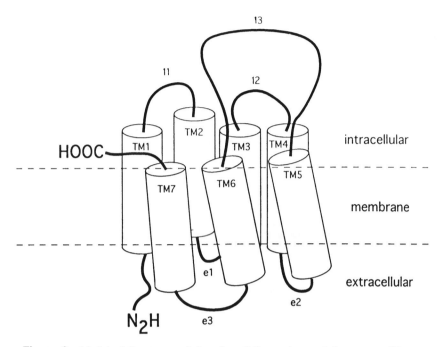

Figure 2 Model of the structural domains of G-protein-coupled receptors. The transmembrane domains are depicted as cylinders perpendicular to the plane of the plasma membrane. Transmembrane domains 1–7 (TM1–TM7) are proposed to traverse the membrane in an α-helical fashion and are connected by alternating extracellular (e1–e3) and intracellular (i1–i3) loops. The amino (NH_2) and carboxyl- (COOH) terminal regions of G-protein-coupled receptors are situated on the extracellular and intracellular sides of the plasma membrane. (From Lee and Kerlavage, 1993.)

(Atwood et al., 1991), whereas the conservation of structure observed within subfamilies may be required for specific receptor-ligand interactions.

Illustrated in Figures 3, 4, and 5 are the amino acids that are conserved among the three β-adrenergic receptor subtypes (β_1–β_3), five muscarinic acetylcholine receptor subtypes (m1–m5), and three tachykinin receptor subtypes (substance P, substance K, and neuromedin K), respectively, that have been cloned and sequenced to date. Several amino acids are found in all receptors from these subfamilies, including a cluster of glycine-asparagine-X-X-valine in transmembrane helix I, an aspartate in transmembrane helix II, a cluster of aspartate-arginine-tyrosine at the carboxyl terminal side of transmembrane helix III, tryptophans in transmembrane helix IV, VI, and VII, prolines in transmembrane helix V, VI, and VII, two cysteines in extracellular domains 1 and 2 that have been

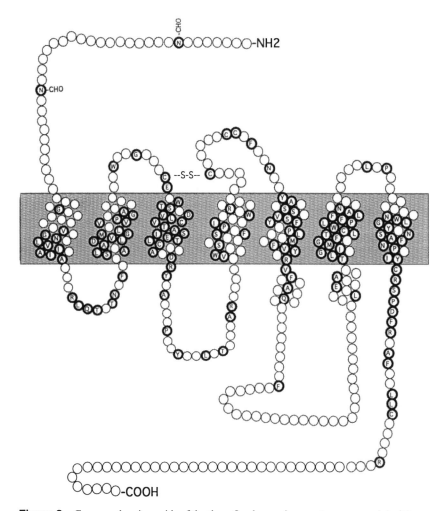

Figure 3 Conserved amino acids of the three β-adrenergic receptors on a model of the rat β₃-adrenergic receptor. Dark circles represent amino acids that are identical among the three β-receptor subtypes.

shown to be linked via a disulfide bond in several members of this family (Dixon et al., 1987a; Karnik et al., 1988; Kurtenbach et al., 1990), and a conserved cysteine in the carboxyl-terminal domain that has been shown to be palmitoylated in adrenergic receptors (O'Dowd et al., 1989).

Several unique sequence characteristics are also notable in the three receptor families. For example, both the adrenergic and muscarinic receptors contain a conserved tryptophan and an aspartate four residues apart in transmembrane helix

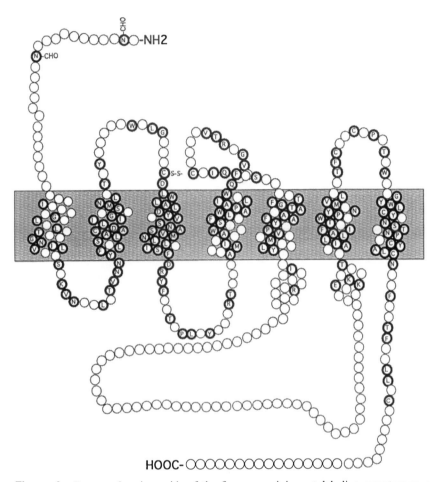

Figure 4 Conserved amino acids of the five muscarinic acetylcholine receptors on a model of the rat m1 muscarinic receptor. Dark circles represent amino acids that are identical among the five muscarinic receptor subtypes.

III that are absent in the tachykinin receptors (Figs. 3 and 4). As discussed below, these residues play a role in the binding of amine neurotransmitters to the adrenergic and muscarinic receptors. In the tachykinin receptors, a histidine residue, characteristic of this subfamily of receptors, is found in transmembrane helices V and VI (Fig. 5). Significant sequence conservation within each family of receptors is also found in all the intracellular domains that have been implicated in receptor–G protein interactions (see Savarese and Fraser, 1992, for a review).

Based on primary sequence analysis, members of the G-protein-receptor superfamily can be classified into distinct subfamilies (Figs. 1 and 6). These

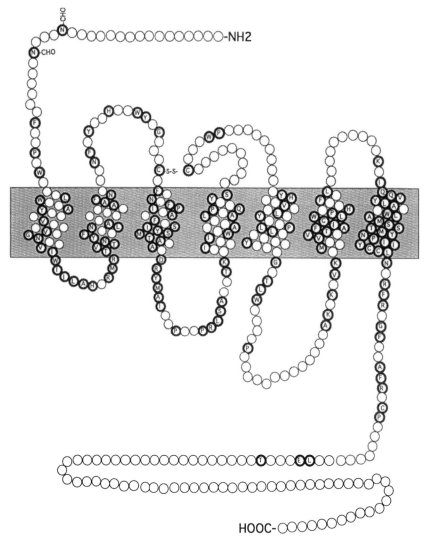

Figure 5   Conserved amino acids of the three tachykinin receptors on a model of the rat substance P receptor. Dark circles represent amino acids that are identical among the three tachykinin subtypes.

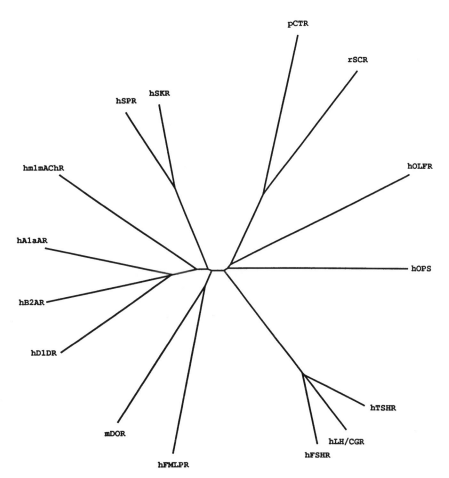

Figure 6 Relative homology of G-protein-coupled receptors. Sequences were aligned using CLUSTAL (Higgins and Sharp, 1989), and refinements to the alignment were made manually. The dendogram was created using the DeSoete Tree Fit (DeSoete, 1983) and Tree Tool (Mile Maciukenas, University of Illinois, unpublished). Only the aligned transmembrane regions depicted in Figure 1 were used in the distance calculations. The lengths of the lines are proportional to the percent difference between any two given sequences. All programs were run using the Genetic Data Environment (Steve Smith, Harvard University, unpublished). The receptors considered are as follows: hm1mAChR, human m1 muscarinic receptor (Peralta et al., 1988); hA1aAR, human α_{1a}-adrenergic receptor (Bruno et al., 1991); hB2AR, human β_2-adrenergic receptor (Chung et al., 1987); hD1DR, human D1 dopamine receptor (Zhou et al., 1990); mDOR, mouse δ-opiate receptor (Kieffer et al., 1992); hFMLPR, human *N*-formyl peptide receptor (Boulay et al., 1990); hSKR, human substance K receptor (Gerald et al., 1990); hSPR, human substance P receptor (Takeda et al., 1991); hFSHR, human follicle-stimulating hormone receptor (Minegish et al., 1991); hLH/CHR, human lutropin/choriogonadotropin receptor (Minegish et al., 1990); hTSHR, human thyrotropin receptor (Nagayama et al., 1989); hOLFR, human olfactory receptor (Buck and Axel, 1991); hOPS, human rhodopsin (Nathans and Hogness, 1984); pCTR, porcine calcitonin receptor (Lin et al., 1991); hSCR, human secretin receptor (Ishihara et al., 1991). (From Lee and Kerlavage, 1993.)

subclasses include receptors that bind biogenic amines (e.g., norepinephrine, dopamine, acetylcholine), glycoprotein hormones (e.g., thyrotropin, follicle-stimulating hormone, lutropin/choriogonadotropin), tachykinins (substance P, substance K/neurokinin A, neuromedin K/neurokinin B), sensory molecules (e.g., light, odorants), and other peptide hormones (e.g., calcitonin, secretin). The receptors within each of these subclasses display the greatest similarity to each other as compared with other receptors in the gene superfamily. In many instances, a receptor within a subfamily can be further divided into subtypes, each encoded by a separate gene. The finding of multiple subclasses within the larger G-protein-coupled receptor gene family suggests that one or a few early members of this protein family may have undergone a number of gene mutations, duplications, or rearrangements to create a diverse array of receptors. Many G-protein-coupled receptors have been identified in lower species such as *Drosophila* (Arakawa et al., 1990; Monnier et al., 1992; Saudou et al., 1992; Li et al., 1992), and a recent report has suggested that similar receptors may be expressed by the herpes virus (Nicholas et al., 1992). In support of the hypothesis of gene duplication is the finding that there are chromosome regions that are rich in receptor genes (e.g., chromosome 10_{q24-26}, which contains the genes for the α_{2A}-adrenergic and β_1-adrenergic receptors) (Yang-Feng et al., 1990).

III. MOLECULAR DETERMINANTS OF AUTONOMIC AND NEUROPEPTIDE RECEPTOR FUNCTION

During the past 5 years, considerable progress has been made in understanding the structure-function relationships of G-protein-coupled receptors through the construction and characterization of mutant receptor genes (Savarese and Fraser, 1992; Lee and Fraser, 1993). Inferences about normal receptor structure and function have been made based on the phenotypes of the mutant receptor proteins. Mutagenesis techniques have been utilized extensively to define (1) structural determinants involved in receptor-ligand interactions, (2) structural determinants involved in receptor–G protein interactions, (3) domains responsible for homologous and heterologous receptor desensitization and down-regulation, and (4) residues that are involved in posttranscriptional modifications such as glycosylation and palmitoylation. While complete elucidation of the mechanisms of receptor activation will be a formidable task, mutagenesis studies have already proven of tremendous utility in this regard.

A. β-Adrenergic Receptors

To date, a majority of the work on the structure-function relationships of G-protein-coupled receptors has been carried out using the β-adrenergic receptor as a model. Large regions of the intracellular and extracellular hydrophilic domains of the β_2-adrenergic receptor can be deleted without altering agonist and antagonist binding (see Fig. 7 for summary). Deletions of the N-terminus and the extracellular loops between TMS II and III (residues 99–102) and TMS VI and VII (residues 301–303) do not affect ligand binding (Dixon et al., 1987b). Likewise, truncation of the C-terminus (Dixon et al., 1987b; Kobilka et al., 1987) and deletions of the third cytoplasmic loop that connects TMS V and VI (residues 229–236 or 239–272) (Dixon et al., 1987b) do not alter binding properties. These data suggest that the determinants of ligand binding in β-adrenergic receptors may reside in one or more TMS. Deletions in the first cytoplasmic loop (residues 63–66) and the second cytoplasmic loop (residues 130–139, 136–144, or 140–150) produce receptors that are undetectable by immunoblotting, suggesting that these receptors are not correctly processed or inserted into the membrane (Dixon et al., 1987a).

The catecholamines, endogenous agonists for the β-adrenergic receptors, consist of a catechol ring and a protonated amine connected by a β-hydroxyethyl side chain. Based on studies utilizing synthetic adrenergic ligands, the amine group and substitutions on the β-hydroxyethyl side chain have been shown to be important for both agonist and antagonist binding, and the catechol ring has been demonstrated to be essential for agonist activity (Mukherjee et al., 1976). It has been suggested that the ligand-binding pocket of the β-adrenergic receptor may contain acidic amino acid residues that serve as counterions for the amine group of agonists and antagonists and polar amino acids that form hydrogen bonds with the catechol hydroxyl groups (Strader et al., 1988).

To begin to identify the amino acids involved in ligand binding to the β-adrenergic receptor, several strategies have been utilized, including the creation of chimeric receptors, substitution and deletion mutants, and site-directed mutagenesis. Chimeric receptors have proven useful for the identification of structural domains that regulate agonist and antagonist specificity, as well as G-protein coupling. Chimeras constructed from α_2-adrenergic and β_2-adrenergic receptors reveal that the seventh transmembrane domain is a major determinant of antagonist binding (Kobilka et al., 1988). Replacement of the N-terminus and several of the transmembrane domains (TMS I and II, TMS I–IV, or TMS I–V) of the β_2-adrenergic receptor with corresponding regions of the α_2-adrenergic receptor produces a hybrid receptor displaying neither β_2-adrenergic receptor nor α_2-adrenergic receptor agonist binding profiles (Kobilka et al., 1988).

An aspartate residue at position 113 (Asp[113]) in TMS III of the β_2-adrenergic receptor is conserved among several receptor subtypes that bind biogenic amines, including β-adrenergic (Frielle et al., 1987; Chung et al., 1987; Muzzin et al., 1991), α-adrenergic (Fraser et al., 1989a; Lomasney et al., 1990), dopaminergic (Grandy et al., 1989; Zhou et al., 1990), and muscarinic cholinergic receptors (Kubo et al., 1986; Bonner et al., 1987; Peralta et al., 1987; Fraser et al., 1989b), suggesting an interaction between the amine group of the ligand and the carboxylate side chain of Asp[113]. Substitution of Asp[113] in the β_2-adrenergic receptor with asparagine (Asn[113]) or glutamate (Glu[113]) significantly reduces receptor affinity for

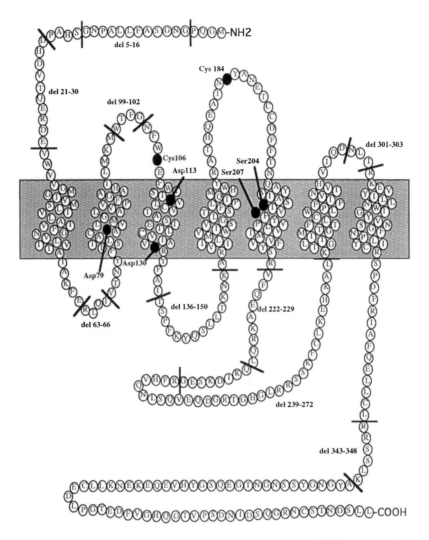

Figure 7 Mutational analysis of the β_2-adrenergic receptor. Highlighted on this schematic diagram of the β_2-adrenergic receptor are several key amino acids and domains that have been studied using mutagenesis techniques. The amino acid sequence of the receptor is given by the single-letter code. The area in the middle of the figure represents the plasma membrane; the areas above and below the membrane represent extracellular and intracellular space, respectively. (From Savarese and Fraser, 1992.)

antagonists (Strader et al., 1987). Furthermore, the Asn^{113} mutant receptor displays a 10^5-fold decrease in agonist potency for stimulation of adenylyl cyclase (Strader et al., 1988). Substitution of Asp^{113} with glutamate, which contains a carboxylate side chain, has a less marked effect on receptor activation, resulting in only a 10^3-fold decrease in agonist potency (Strader et al., 1988). These data suggest that the carboxylate side chain of Asp^{113} serves as a counterion for the amine group of β-adrenergic agonists and antagonists. Similar data have also been obtained from mutagenesis studies with other receptors that bind amine neurotransmitters, including α_2-adrenergic receptors (Wang et al., 1991), muscarinic receptors (Fraser et al., 1989b), and histamine receptors (Gantz et al., 1992), suggesting a common functional role for this conserved aspartate.

Although the aspartate at position 113 in the β_2-adrenergic receptor has been shown to play a role in binding the positively charged amino groups of β-adrenergic agonists and antagonists, a negatively charged amino acid at position 113 in the β_2-adrenergic receptor is not essential for agonist activation of the receptor. Strader et al. (1991) substituted a serine residue (Ser^{113}) for Asp^{113} in the β-receptor, replacing the carboxylate side chain of aspartate with the hydroxyl group of serine (Strader et al., 1991). A series of modified catecholamines were generated by substituting the amine-containing alkyl group with functional groups that could potentially interact with the hydroxyl group of serine (Ser^{113}) (Strader et al., 1991). Catechol derivatives capable of forming hydrogen bonds, such as catechol esters and ketones, were effective in mutant receptor activation but did not activate the wild-type β-adrenergic receptor (Strader et al., 1991). Hence, the negatively charged residue at this position (113) in the native receptor appears to relate primarily to the chemical nature of the endogenous ligands for the β_2-receptor and not to an absolute requirement for agonist activation (Strader et al., 1991).

Two additional aspartate residues have been shown to influence β-adrenergic agonist binding. Asp^{79}, located in the second transmembrane segment of the β_2-adrenergic receptor, and Asp^{130}, located in the third transmembrane segment of the β_2-adrenergic receptor, are also highly conserved among members of this gene family (Fig. 1). Replacement of Asp^{79} of the hamster β_2-adrenergic receptor with alanine produces a mutant receptor with a decreased affinity for agonists, as well as a decrease in adenylyl cyclase stimulation (Strader et al., 1987; Strader et al., 1988). Antagonist binding is not affected by this mutation (Strader et al., 1987; Strader et al., 1988). Similarly, substitution of Asp^{79} of the human β_2-adrenergic receptor with asparagine (Asn^{79}) results in significantly reduced agonist affinities and normal antagonist binding (Chung et al., 1988). This human mutant receptor (Asn^{79}) does not display guanine nucleotide-sensitive high-affinity binding of agonists, and more important, agonist binding produces no increase in intracellular cAMP levels (Chung et al., 1988). An aspartate corresponding to Asp^{79} in the β_2-adrenergic receptor is conserved in a large number of G-protein-coupled receptors (Fig. 1), and this residue has been demonstrated to be essential

for agonist-induced signal transduction with muscarinic, α_2-adrenergic, dopamine, and LH receptors (Fraser et al., 1989b; Wang et al., 1991; Neve et al., 1991;
Ji and Ji, 1991). It has been hypothesized that this highly conserved aspartate may
be involved in an agonist-induced conformational change that is essential for
receptor–G protein interactions (Fraser et al., 1988).

Substitution of Asp[130] of the human β_2-adrenergic receptor with asparagine
(Asn[130]) results in a receptor with normal antagonist binding but a significantly
higher affinity for agonists than the wild-type receptor (Fraser et al., 1988). While
this mutant receptor displays guanine nucleotide-sensitive agonist binding, it is
unable to mediate increases in cAMP (Fraser et al., 1988), suggesting that the
functional coupling of the Asn[130] β-receptor to G_s is altered. Furthermore, this
study demonstrates that guanine nucleotide effects on agonist affinity can be
dissociated from those on activation of G_s and adenylyl cyclase. From deletion
mutagenesis studies of the β_2-adrenergic receptor, Hausdorff et al. (1990) also
concluded that the molecular determinants of the β_2-adrenergic receptor involved
in the formation of the ternary complex are not identical to those that transmit the
agonist-induced stimulatory signal to G_s.

Structure-activity studies have demonstrated that β-adrenergic agonists require
the presence of a catechol ring containing hydroxyl groups at the *meta-* and *para-*
positions for full activity (Mukherjee et al., 1976). Two serine residues (Ser[204] and
Ser[207]) in transmembrane domain V of the β_2-adrenergic receptor have been
identified as potential hydrogen-bonding sites for the hydroxyl groups of the
catechol ring (Strader et al., 1989). Substitution of either serine residue with
alanine markedly reduces agonist binding affinity, without affecting antagonist
binding (Strader et al., 1989). Thus, it has been postulated that two hydrogen
bonds are involved in agonist binding to the β_2-adrenergic receptor: Ser[204] forms a
hydrogen bond with the *meta-*hydroxyl group of the catechol ring and Ser[207]
hydrogen bonds with the *para-*hydroxyl group of the catechol ring (Fig. 7)
(Strader et al., 1989).

This hypothesis is supported by the finding that agonists lacking either the
meta- or *para-*hydroxyl group display agonist-binding properties similar to those
of the mutant receptors lacking the serines at the corresponding loci (Strader et al.,
1989). These serine residues are conserved in all G-protein-coupled receptors that
bind catechol ligands (adrenergic and dopaminergic receptors), but are not found
in receptor subtypes whose ligands lack a catechol ring (muscarinic cholinergic
receptors and peptide hormone receptors).

B. Muscarinic Acetylcholine Receptors

Compared with the β-adrenergic receptor, less is known about the ligand-binding
domain of muscarinic acetylcholine receptors. However, the conserved aspartate
residue in TMS III appears to play a similar role in ligand binding as in the
β-adrenergic receptors. Site-directed mutagenesis of the rat m1 muscarinic recep-

tor demonstrated that substitution of Asp[105] with asparagine produces a receptor that fails to bind muscarinic ligands or activate carbachol-induced membrane phosphoinositide (PI) hydrolysis (Fraser et al., 1989b). These data are consistent with a role for this residue as a binding site for the cationic amines in muscarinic agonists and antagonists, similar to the role that Asp[113] plays in the β_2-adrenergic receptor (see above).

This hypothesis was confirmed in work from Hulme's laboratory (Curtis et al., 1989; Kurtenbach et al., 1990) using [3H]-propylbenzilylcholine mustard (PBCM) as an affinity label to identify regions of the muscarinic receptor responsible for binding muscarinic antagonists. The aziridine portion of PBCM corresponds to the positively charged onium group of muscarinic ligands that undergoes attack by nucleophilic amino acids and should theoretically label residue that acts as a counterion for the onium moiety. Purification and peptide sequence analysis of labeled rat brain muscarinic receptors indicated that [3H]-PBCM labeled Asp[105] in TMS III of the receptor, consistent with the results of the mutagenesis experiments.

The roles of conserved proline and tryptophan residues in TMS IV–VII of the m3 muscarinic receptors were recently studied using mutagenesis techniques (Wess et al., 1993). These residues are highly conserved among members of the family of G-protein-coupled receptors (Fig. 1). The effect of mutation of three of the proline residues at positions 242 in TMS V, 505 in TMS VI, and 540 in TMS VII to alanine was to reduce the levels of receptor expression in transfected cells by 35- to 100-fold compared with the wild-type receptor. While all three mutant receptors bound muscarinic ligands with affinities equal to or greater than the wild-type receptor, the Ala[540] mutant m3 receptor was severely impaired in its ability to stimulate carbachol-mediated PI hydrolysis. These data indicate that high affinity agonist binding is not sufficient for receptor activation, similar to data obtained in mutagenesis studies with the β_2-adrenergic receptor (Fraser et al., 1988).

In contrast to the effects of the proline-to-alanine mutations in the m3 muscarinic receptor, the Trp–Phe[192] (TMS IV) and Trp–Phe[503] (TMS VI) mutant receptors displayed a reduction in ligand affinities, suggesting that these residues may be involved in ligand binding (Wess et al., 1993). Consistent with this hypothesis are the results of molecular modeling studies that predict that these two tryptophan residues, along with other aromatic amino acids in the TMS, may be involved in muscarinic receptor-ligand interactions (Hibert et al., 1991). Because of the conservation of these amino acid residues, it seems reasonable to speculate that they may play similar roles in other G-protein-coupled receptors.

C. Tachykinin Receptors

Use of chimeric and point-mutated tachykinin receptors has begun to shed light on the domains involved in the binding of peptide agonists and nonpeptide antagonists to this class of receptors. As illustrated in Figures 3, 4, and 5, the primary and secondary structure of the tachykinin receptors is not unlike that of the adrenergic and muscarinic receptors; however, critical differences in these structures must exist in order to confer specificity for the binding of peptide versus small amine agonists.

The three tachykinins [substance P, substance K (neurokinin A), and neuromedin K (neurokinin B)] all share a common carboxyl-terminal sequence, Phe-X-Gly-Leu-Met-NH_2, and a similar range of biological activities. The receptors that bind the tachykinins differ in their affinities for the peptides as follows:

Neurokinin 1 receptor: substance P > neurokinin A > neurokinin B

Neurokinin 2 receptor: neurokinin A > neurokinin B > substance P

Neurokinin 3 receptor: neurokinin B > neurokinin A > substance P

It has been proposed that all three tachykinin receptors may recognize the common carboxyl domain of the peptides, whereas the divergent amino-termini may determine receptor subtype selectivity (Schwyzer, 1987; Buck et al., 1988).

Using chimeric NK-1/NK-2 receptors, Yokota et al. (1992) demonstrated that the specificity for substance P was determined primarily by the region of the receptors extending from TMS II to the second extracellular loop, together with a small contribution from the amino-terminal extracellular domain (Fig. 8). Fong et al. (1992a) further demonstrated with NK-1/NK-2 and NK-1/NK-3 chimeric receptors and point mutations that multiple extracellular domains in the receptors interact with peptide agonists; however, the three tachykinins do not interact with the same functional groups on each receptor. These conclusions are supported by the work of Gether et al. (1993), which indicates that several tachykinin receptor domains contribute to the binding specificity of the tachykinin agonists but in varying degrees for each peptide. Five residues conserved among the tachykinin receptors at positions 23, 24, 25 (amino terminal domain), 96, and 108 (first extracellular loop) have been postulated to interact with the common determinants on the three peptide agonists (Watling and Krause, 1993).

A number of nonpeptide tachykinin receptor antagonists, such as CP96345, CP99994, FK888, FK224, and RP67580, specific for the NK-1 receptor, and SR48968, GR100679, and GR103537, specific for the NK-2 receptor, have recently been discovered (Watling and Krause, 1993). The NK-1 antagonists represent potentially important new therapeutic agents in the treatment of inflammation, pain, and asthma. These compounds display a marked difference in affinity for the tachykinin receptor subtypes and between the same receptor subtype in different species (Watling and Krause, 1993).

Using site-directed mutagenesis, Fong and co-workers (1992b) identified two residues in the NK-1 receptor, Val[116] in TMS III and Ile[290] in TMS VI, that are responsible for the observed differences between rat and human NK-1 receptors

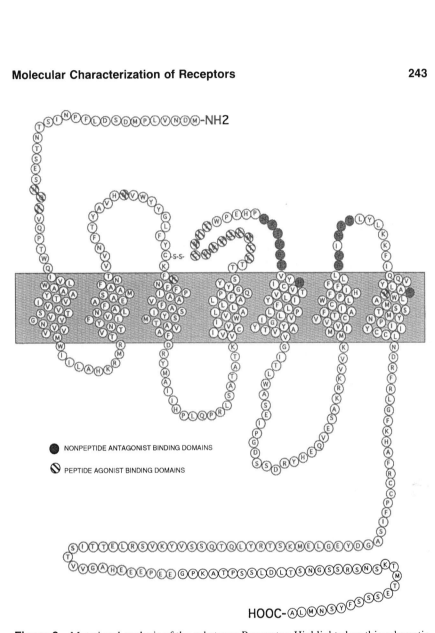

Figure 8 Mutational analysis of the substance P receptor. Highlighted on this schematic diagram of the substance P receptor are several key amino acids and domains that have been implicated in the binding of peptide agonists and nonpeptide antagonists. The amino acid sequence of the receptor is given by the single-letter code. The area in the middle of the figure represents the plasma membrane; the areas above and below the membrane represent extracellular and intracellular space, respectively.

in the binding affinities for CP96345 and RP67580. These amino acids presumably do not directly interact with the antagonist compounds but are likely involved in helical packing of the receptor proteins. Results from Sachais et al. (1993), Gether et al. (1993), and Fong et al. (1993) have suggested that residues in or near the second extracellular loop of the receptor are also involved in determining the affinity for nonpeptide antagonists. These findings indicate that the site of interaction of nonpeptide antagonists for the tachykinin receptors differs from that for the peptide agonists (Fig. 8). Furthermore, the interaction of the nonpeptide antagonists with the tachykinin receptors appears to be fundamentally different from the interaction of the antagonists for the amine neurotransmitter receptors.

IV. CONCLUSIONS

The cloning and expression of a large number of G-protein-coupled receptors to date have allowed the identification and examination of common versus unique structural motifs in this protein family. Although it appears now that there may be few, if any, domains other than the seven transmembrane helices that are shared by all members of this family, the conservation of secondary structure and the common mechanism of signal transduction establish an evolutionary relationship among these receptors (Savarese and Fraser, 1992). It has been estimated that the number of receptors in this family may be as large as 1000, which means that G-protein-coupled receptors may represent as much as 1–2% of all human genes.

One of the major goals of understanding G-protein-coupled receptor biology as a whole is to identify those protein sequences that distinguish one subclass of receptor from another in terms of ligand binding and receptor–G protein interactions and those DNA sequences that determine the regulation of receptor gene expression. As briefly reviewed in this chapter, mutagenesis techniques have provided, and will most likely continue to provide, important information in this regard. Understanding receptor diversity at a molecular level is a major goal of receptor research, as it would provide the framework for the design of new subtype-selective pharmacological agents.

REFERENCES

Arakawa, S., Gocayne, J.D., McCombie, W.R., et al. (1990). Cloning localization and permanent expression of a *Drosophila* octopamine receptor. *Neuron 4*:343.

Atwood, T.K., Eliopoulos, E.E., and Findlay, J.B.C. (1991). Multiple sequence alignment of protein families showing low sequence homology: a methodological approach using database pattern-matching discriminators for G protein-linked receptors. *Gene 98*:153.

Bonner, T.I., Young, A.C., Brann, M.R., et al. (1987). Identification of a family of muscarinic acetylcholine receptor genes. *Science 237*:527.

Barnes, P.J. (1989a). Muscarinic autoreceptors in airways: their possible role in airway disease. *Chest 96*:1220.

Barnes, P.J. (1989b). Muscarinic receptor subtypes: implications for lung disease. *Thorax 44*:161.

Barnes, P.J., Karliner, J.S., and Dollery, C.T. (1980). Human lung adrenoceptors studied by radioligand binding. *Clin. Sci. 58*:457.

Boulay, F., Tardif, M., Brouchon, L., et al. (1990). The human *N*-formyl peptide receptor. Characterization of two cDNA isolates and evidence for a new subfamily of G-protein-coupled receptors. *Biochemistry 29*:11123.

Brown, L.A.S., and Longmore, W.J. (1981). Adrenergic and cholinergic regulation of lung surfactant secretion in the isolated perfused rat lung and in the alveolar type II cell in culture. *J. Biol. Chem. 256*:66.

Bruno, J.F., Whittaker, J., Song, J., et al. (1991). Molecular cloning and sequencing of a cDNA encoding a human alpha1A adrenergic receptor. *Biochem. Biophys. Res. Commun. 179*:1485.

Buck, L., and Axel, R. (1991). A novel multigene family may encode odorant receptors: a molecular basis for odor recognition. *Cell 65*:175.

Buck, S.H., Pruss, R.M., Krstenansky, J.L., et al. (1988). A tachykinin receptor joins an elite club. *Trends Pharmacol. Sci. 9*:3.

Butchers, P.R., Skidmore, I.F., Vardey, C.J., et al. (1980). Characterization of the receptor mediating antianaphylactic effects of beta-adrenoceptor agonists in human lung tissue in vitro. *Br. J. Pharmacol. 71*:663.

Carstairs, J.R., Nimmo, A.J., and Barnes, P.J. (1985). Autoradiographic visualization of beta-adrenoceptor subtypes in human lung. *Am. Rev. Respir. Dis. 132*:541.

Casale, T.B. (1993). Neurogenic control of inflammation and airway function. In: *Allergy, Principles and Practice*, Vol. 4. (Middleton, E., Jr., Reed, C.E., Ellis, E.F., Adkinson, N.F., Jr., Yunginger, J.W., and Busse, W.W., eds.). Mosby, St. Louis, pp. 650–671.

Casale, T.B., and Hart, J.E. (1987). $(-)[^{125}I]$-pindolol binding to human peripheral lung beta-receptors. *Biochem. Pharmacol. 36*:2557.

Chung, F-Z., Lentes, K-U., Gocayne, J., et al. (1987). Cloning and sequence analysis of the human brain β-adrenergic receptor. Evolutionary relationship to rodent and avian β-receptors and porcine muscarinic receptors. *FEBS Lett. 211*:200.

Chung, F-Z., Wang, C-D., Potter, P.C., et al. (1988). Site-directed mutagenesis and continuous expression of the human β-adrenergic receptor. *J. Biol. Chem. 263*:4052.

Colebatch, H.J.H., and Halmagyi, D.F.J. (1963). Effect of vagotomy and vagal stimulation on lung mechanics and circulation. *J. Appl. Physiol. 18*:881.

Cottechia, S., Schwinn, D.A., Randall, R.R., et al. (1988). Molecular cloning and expression of the cDNA for the hamster alpha1-adrenergic receptor. *Proc. Natl. Acad. Sci. USA 85*:7159.

Culp, D.J., McBride, R.K., Graham, L.A., et al. (1990). α-Adrenergic regulation of secretion by tracheal glands. *Am. J. Physiol. 259*:L198.

Curtis, C.A.M., Wheatley, M., Bansal, S., et al. (1989). Propyl benzilylcholine mustard labels an acidic residue in transmembrane helix 3 of the muscarinic receptor. *J. Biol. Chem. 264*:489.

DeSoete, G. (1983). *Psychometrika 48*:621.

Dixon, R.A.F., Kobilka, B.K., Strader, D.J., et al. (1986). Cloning of the gene and cDNA for mammalian β-adrenergic receptor and homology with rhodopsin. *Nature 321*:75.

Dixon, R.A.F., Sigal, I.S., Candelore, M.R., et al. (1987a). Structural features required for ligand binding to the β-adrenergic receptor. *EMBO J. 6*:3269.

Dixon, R.A.F., Sigal, I.S., Rands, E., et al. (1987b). Ligand binding to the β-adrenergic receptor involves its rhodopsin-like core. *Nature 326*:73.

Fong, T.M., Huang, R-R.C., and Strader, C.D. (1992a). Localization of agonist and antagonist binding domains of the human neurokinin-1 receptor. *J. Biol. Chem. 267*:25664.

Fong, T.M., Yu, H., and Strader, C.D. (1992b). Molecular basis for the species selectivity of the neurokinin-1 receptor antagonists CP-96,345 and RP67580. *J. Biol. Chem. 267*:25668.

Fong, T.M., Cascieri, M.A., Yu, H., Bansal, A., Swain, C., and Strader, C.D. (1993). Amino-aromatic interaction between histidine 197 of the neurokinin-1 receptor and CP96345. *Nature 362*:350.

Fraser, C.M., Chung, F-Z., Wang, C-D., et al. (1988). Site-directed mutagenesis of human β-adrenergic receptors: substitution of aspartic acid-130 by asparagine produces a receptor with high-affinity agonist binding that is uncoupled from adenylate cyclase. *Proc. Natl. Acad. Sci. USA 85*:5478.

Fraser, C.M., Arakawa, S., McCombie, W.R., et al. (1989a). Cloning, sequence analysis, and permanent expression of a human α$_2$-adrenergic receptor in Chinese hamster ovary cells. *J. Biol. Chem. 264*:11754.

Fraser, C.M., Wang, C-D., Robinson, D.A., et al. (1989b). Site-directed mutagenesis of m1 muscarinic receptors: conserved aspartic acids play important roles in receptor function. *Mol. Pharmacol. 36*:840.

Frielle, T., Collins, S., Daniel, K.W., et al. (1987). Cloning of the cDNA for the human β$_1$-adrenergic receptor. *Proc. Natl. Acad. Sci. USA 84*:7920.

Gantz, I., Del Valle, J., Wang, L.D., Tashiro, T., et al. (1992). Molecular basis for the interaction of histamine with the histamine receptor. *J. Biol. Chem. 267*:20840.

Gerard, N.P., Eddy, R.L., Shows, T.B., et al. (1991). The human neurokinin A (substance K) receptor. Molecular cloning of the gene, chromosomal localization, and isolation of the cDNA from tracheal and gastric tissues. *J. Biol. Chem. 265*:20455.

Gether, U., Johansen, T.E., Snider, R.M., et al. (1993). Different epitopes on the NK1 receptor for substance P and a non-peptide antagonist. *Nature 362*:345.

Goldie, R.G., Lulich, K.M., and Paterson, J.W. (1985). Bronchial α-adrenoceptor function in asthma. *Trends Pharmacol. Sci. 6*:469.

Grandy, D.K., Marchionni, M.A., Makam, H., et al. (1989). Cloning of the cDNA and gene for a human D2 dopamine receptor. *Proc. Natl. Acad. Sci. USA 86*:9762.

Grundstrom, N., Andersson, R.G.G., and Wikberg, J.E.S. (1981). Prejunctional α$_2$-adrenoceptors inhibit contraction of tracheal smooth muscle by inhibiting cholinergic neurotransmission. *Life Sci. 28*:2981.

Grundstrom, N., Andersson, R.G.G., and Wikberg, J.E.S. (1984). Inhibition of the excitatory non-adrenergic, non-cholinergic neurotransmission in the guinea pig tracheobronchial tree mediated by α$_2$-adrenoceptors. *Acta Pharmacol. Toxicol. 54*:8.

Hausdorff, W.P., Hnatowich, M., O'Dowd, B.F., et al. (1990). A mutation of the β$_2$-adrenergic receptor impairs agonist activation of adenylyl cyclase without affecting high affinity agonist binding. *J. Biol. Chem. 265*:1388.

Henderson, R., and Unwin, P.N.T. (1975). Three-dimensional model of purple membrane obtained by electron microscopy. *Nature 257*:28.

Henderson, R., Baldwin, J.M., Downing, K.H., et al. (1986). Structure of purple membrane from *Halobacterium halobium*: recording, measurement and evaluation of electron micrographs at 3.5 Å resolution. *Ultramicroscopy 19*:147.

Hibert, M.F., Trumpp-Kallmeyer, S., Bruinvels, A., et al. (1991). Three-dimensional models of neurotransmitter G-binding protein-coupled receptors. *Mol. Pharmacol. 40*:8.

Higgins, D.G., and Sharp, P.M. (1988). Fast and sensitive multiple sequence alignments using a microcomputer. *Gene 73*:237.

Ishihara, T., Nakamura, S., Kaziro, Y., et al. (1991). Molecular cloning and expression of a cDNA encoding the secretin receptor. *EMBO J. 10*:1635.

Ji, I., and Ji, T.H. (1991). Asp383 in the second transmembrane domain of the lutropin receptor is important for high affinity hormone binding and cAMP production. *J. Biol. Chem. 266*:14953.

Karnik, S.S., Sakmar, T.P., Chen, H-B., et al. (1988). Cysteine residues 110 and 187 are essential for the formation of correct structure in bovine rhodopsin. *Proc. Natl. Acad. Sci. USA 85*:8459.

Kerlavage, A.R., Fraser, C.M., and Venter, J.C. (1987). Muscarinic cholinergic receptor structure: molecular biological support for subtypes. *Trends Pharmacol. Sci. 8*:426.

Kieffer, B.L., Befort, K., Gaveriaux-Ruff, C., et al. (1992). The delta-opioid receptor: isolation of a cDNA by expression cloning and pharmacological characterization. *Proc. Natl. Acad. Sci. USA 89*:12048.

Kneussl, M.P., and Richardson, J.B. (1978). Alpha-adrenergic receptors in human and canine tracheal and bronchial smooth muscle. *J. Appl. Physiol. 45*:307.

Kobilka, B.K., MacGregor, C., Daniel, K., et al. (1987). Functional activity and regulation of human β_2-adrenergic receptors expressed in *Xenopus* oocytes. *J. Biol. Chem. 262*:15796.

Kobilka, B.K., Kobilka, T.S., Daniel, K., et al. (1988). Chimeric α_2-, β_2-adrenergic receptors: delineation of domains involved in effector coupling and ligand binding specificity. *Science 240*:1310.

Kubo, T., Fukuda, K., Mikami, A., et al. (1986a). Cloning, sequencing, and expression of complementary cDNA encoding the muscarinic acetylcholine receptor. *Nature 323*:411.

Kubo, T., Maeda, A., Sugimoto, K., et al. (1986b). Primary structure of porcine cardiac muscarinic acetylcholine receptor deduced from the cDNA sequence. *FEBS Lett. 209*:367.

Kurtenbach, E., Curtis, C.A.M., Pedder, E.K., et al. (1990). Muscarinic acetylcholine receptors. Peptide sequencing identifies residues involved in antagonist binding and disulfide bond formation. *J. Biol Chem. 265*:13702.

Lee, N.H., and Fraser, C.M. (1993). Cross-talk between m1 muscarinic acetylcholine and β_2-adrenergic receptors. cAMP and the third intracellular loop of the m1 muscarinic receptor confer heterologous regulation. *J. Biol. Chem. 268*:7949.

Lee, N.H., and Kerlavage, A.R. (1993) Molecular biology of G protein-coupled receptors. *Drug News Perspect. 6*:488.

Li, X.J., Wu, Y.N., North, R.A., et al. (1992). Cloning, functional expression and developmental regulation of a neuropeptide Y receptor from *Drosophila* melanogaster. *J. Biol. Chem. 267*:9.

Lin, H.Y., Harris, T.L., Flannery, M.S., et al. (1991). Expression cloning of an adenylate cyclase-coupled calcitonin receptor. *Science 254*:1022.

Lomasney, J.W., Lorenz, W., Allen, L.F., et al. (1990). Expansion of the α_2-adrenergic receptor family: cloning and characterization of a human α_2-adrenergic receptor subtype, the gene for which is located on chromosome 2. *Proc. Natl. Acad. Sci. USA 87*:5094.

Lundgren, J.D., and Shelhamer, J.H. (1990). Pathogenesis of airway mucus hypersecretion. *J. Allergy Clin. Immunol. 85*:399.

Mak, J.C.W., and Barnes, P.J. (1990). Autoradiographic visualization of muscarinic receptor subtypes in human and guinea pig lung. *Am. Rev. Respir. Dis. 141*:1559.

Minegish, T., Nakamura, K., Takakura, Y., et al. (1990). Cloning and sequencing of the human LH/hCG receptor cDNA. *Biochem. Biophys. Res. Commun. 172*:1049.

Minegish, T., Nakamura, K., Takakura, Y., et al. (1991). Cloning and sequencing of the human FSH receptor cDNA. *Biochem. Biophys. Res. Commun. 175*:1125.

Monnier, D., Colas, J.F., Rosay, P., et al. (1992). NKD, a developmentally regulated tachykinin receptor in *Drosophila. J. Biol. Chem. 267*:1298.

Mukherjee, M.G., Caron, M.G., Mullikin, D., et al. (1976). Structure-activity relationships of adenylate cyclase-coupled beta adrenergic receptors: determination by direct binding studies. *Mol. Pharmacol. 12*:16.

Muzzin, P., Revelli, J-P., Kuhne, F., et al. (1991). An adipose tissue-specific β-adrenergic receptor. Molecular cloning and down-regulation in obesity. *J. Biol. Chem. 266*:24053.

Nagayama, Y., Kaufman, K.D., Seto, P., et al. (1989). Molecular cloning, sequence and functional expression of the cDNA for the human thyrotropin receptor. *Biochem. Biophys. Res. Commun. 165*:1184.

Nathans, J., and Hogness, D.S. (1984). Isolation and nucleotide sequence of the gene encoding human rhodopsin. *Proc. Natl. Acad. Sci. USA 81*:4851.

Neve, K.A., Cox, B.A., Henningsen, R.A., et al. (1991). Pivotal role for aspartate-80 in the regulation of dopamine D2 receptor affinity for drugs and the inhibition of adenylyl cyclase. *Mol. Pharmacol. 39*:733.

Nicholas, J., Cameron, K.R., and Honess, R.W. (1992). Herpes virus saimiri encodes homologues of G protein-coupled receptors and cyclins. *Nature 355*:362.

O'Dowd, B.F., Hnatowich, M., Caron, M.G., et al. (1989). Palmitoylation of the human β_2-adrenergic receptor. *J. Biol. Chem. 264*:7564.

Olsen, C.R., Colebatch, H.J.H., Mebel, P.E., et al. (1965). Motor control of pulmonary airways studied by nerve stimulation. *J. Appl. Physiol. 20*:202.

Orange, R.P., Austen, W.G., Austen, K.F., et al. (1971a). Immunological release of histamine and slow-reacting substance of anaphylaxis from human lung. I. Modulation by agents influencing cellular levels of cyclic 3'-5'-adenosine monophosphate. *J. Exp. Med. 134* (Suppl.):136.

Orange, R.P., Kaliner, M.A., LaRaia, P.J., et al. (1971b). Immunological release of histamine and slow-reacting substance of anaphylaxis from human lung. II. Influence of cellular levels of cyclic AMP. *Fed. Proc. 30*:1725.

Ovchinnikov, Y.A., Abdulaev, N.G., Feigina, M.Y., et al. (1983). Visual rhodopsin. Complete amino acid sequence and topography in a membrane. *Biorg. Khim. 9*:1331.

Pappin, D.J.C., Eliopoulos, E.E., Brett, M., et al. (1984). A structural model for ovine rhodopsin. *Int. J. Biol. Macromol. 6*:73.

Peralta, E.G., Ashkenazi, A., Winslow, J.W., et al. (1987). Distinct primary structures, ligand-binding properties and tissue-specific expression of four human muscarinic acetylcholine receptors. *EMBO J. 6*:3923.

Polak, J.M., and Bloom, S.R. (1982). Regulatory peptides in the respiratory tract of man and other animals. *Exp. Lung Res. 3*:313.

Rhoden, K.J., Meldrum, L.A., and Barnes, P.J. (1988). Inhibition of cholinergic neurotransmission in human airways by β_2-receptors. *J. Appl. Physiol. 65*:700.

Sachais, B.S., Snider, R.M., Lowe, J.A., et al. (1993). Molecular basis for the species selectivity of the substance P antagonist CP-96345. *J. Biol. Chem. 268*:2319.

Saudou, F., Boschert, U., Amlaiky, N., et al. (1992). A family of *Drosophila* serotonin receptors with distinct intracellular signalling properties and expression patterns. *EMBO J. 11*:7.

Savarese, T.M., and Fraser, C.M. (1992). In vitro mutagenesis and the search for structure-function relationships among G protein-coupled receptors. *Biochem. J. 283*:1.

Schwyzer, R. (1987). Membrane-assisted molecular mechanism of neurokinin receptor subtype selection. *EMBO J. 6*:2255.

Starke, K. (1981). Presynaptic receptors. *Annu. Rev. Pharmacol. Toxicol. 21*:7.

Strader, C.D., Sigal, I.S., Register, R.B., et al. (1987). Identification of residues required for ligand binding to the β-adrenergic receptor. *Proc. Natl. Acad. Sci. USA 84*:4384.

Strader, C.D., Sigal, I.S., Candelore, M.R., et al. (1988). Conserved aspartic acid residues 79 and 113 of the β-adrenergic receptor have different roles in receptor function. *J. Biol. Chem. 263*:10267.

Strader, C.D., Candelore, M.R., Hill, W.S., et al. (1989). Identification of two serine residues involved in agonist activation of the β-adrenergic receptor. *J. Biol. Chem. 264*:13572.

Strader, C.D., Gaffney, T., Sugg, E.E., et al. (1991). Allele-specific activation of genetically engineered receptors. *J. Biol. Chem. 266*:5.

Straub, R.E., Frech, G.C., Joho, R.H., et al. (1990). Expression cloning of a cDNA encoding the mouse pituitary thyrotropin-releasing hormone receptor. *Proc. Natl. Acad. Sci. USA 87*:9514.

Szentivanyi, A. (1968). The beta-adrenergic theory of atopic abnormality in bronchial asthma. *J. Allergy 42*:203.

Takeda, Y., Takeda, J., Sachais, B.S., et al. (1991). Molecular cloning, structural characterization and functional expression of the human substance P receptor. *Biochem. Biophys. Res. Commun. 179*:1232.

Tanabe, Y., Masu, M., Ishii, T., et al. (1992). A family of metabotropic glutamate receptors. *Neuron 8*:169.

Vermiere, P.A., and Vanhoutte, P.M. (1991). Inhibitory effects of catecholamines in isolated canine bronchial smooth muscle. *J. Appl. Physiol. 46*:787.

Wang, C-D., Buck, M.A., and Fraser, C.M. (1991). Alpha$_2$-adrenergic receptors: identification of amino acids involved in ligand binding and receptor activation by agonists. *Mol. Pharmacol. 40*:168.

Wang, H-Y., Lipfert, L., Malbon, C.C., et al. (1989). Site-directed antipeptide antibodies define the topography of the β_2-adrenergic receptor. *J. Biol. Chem.* *264*:14424.

Watling, K.J., and Krause, J.E. (1993). The rising sun shines on substance P and related peptides. *Trends Pharmacol. Sci.* *14*:81.

Wess, J., Nanavati, S., Vogel, Z., et al. (1993). Functional role of proline and tryptophan residues highly conserved among G protein-coupled receptors studied by mutational analysis of the m3 muscarinic receptor. *EMBO J.* *12*:331.

Yang-Feng, T.L., Xue, F.Y., Zhong, W.W., et al. (1990). Chromosomal organization of adrenergic receptor genes. *Proc. Natl. Acad. Sci. USA* *87*:1516.

Yokota, Y., Akazawa, C., Ohkubo, H., et al. (1992). Delineation of structural domains involved in the subtype specificity of tachykinin receptors through formation of substance P/substance K receptors. *EMBO J.* *11*:3585.

Zaagsma, J., van der Heijden, P.M.C.M., van der Schaar, M.W.G., et al. (1983). Comparison of functional β-adrenoceptor heterogeneity in central and peripheral airway smooth muscle of guinea pig and man. *J. Receptor Res.* *3*:8901.

Zhou, Q-Y., Grandy, D.K., Thambi, L., et al. (1990). Cloning and expression of human and rat D_1 dopamine receptors. *Science* *347*:76.

DISCUSSION

Skidgel: Can the serine-threonine-rich region that mediates sequestration/endocytosis be modified by phosphorylation or glycosylation?

Lee: Phosphorylation of G-protein receptors does not appear to be the primary mechanism for receptor internalization since Lefkowitz et al. have shown that inhibitors of PKA, PKC, and β-adrenergic receptor kinase do not impair this process. Other proteins may be responsible for internalization.

Skidgel: Muscarinic receptors can be down-regulated by stimulation of β-adrenergic receptors in cells transfected with both, except when mutant muscarinic receptors lacking the Ser/Thr-rich region are used. Could the Ser/Thr region be a localizing signal for receptor clustering?

Lee: I am not aware of G-protein receptor clustering prior to internalization.

McDonald: How can multiple receptors be internalized by stimulation of only one receptor?

Lee: Regional localization of receptors is one explanation, but there are no data to support this at the present time.

11

Autonomic Receptors in the Upper and Lower Airways

Judith Choi Wo Mak and Peter J. Barnes
National Heart and Lung Institute, London, England

I. INTRODUCTION

Autonomic innervation of the airways is much more complex than previously described (Richardson, 1979; Nadel and Barnes, 1984; Barnes, 1986). Not only a parasympathetic cholinergic system and sympathetic adrenergic system, but also a nonadrenergic, noncholinergic (NANC) system is present, which innervates airway smooth muscle and other structures in the lung and has inhibitory and excitatory components (Richardson, 1981; Barnes, 1984). Autonomic nerves influence airway tone by activating specific receptors on the target cells in the airways. In the cholinergic pathway, acetylcholine released from postganglionic nerve endings stimulates muscarinic cholinergic receptors. Adrenergic mechanisms include sympathetic nerves that release noradrenaline and circulating adrenaline secreted from the adrenal medulla; these catecholamines activate α- and β-adrenoceptors. The neurotransmitters of the NANC nervous system are not certain, but the most likely candidate for nonadrenergic inhibitory nerves is vasoactive intestinal peptide (VIP), whereas that of noncholinergic excitatory nerves is probably substance P (SP) or a related peptide. These neuropeptides interact with specific receptors on target cells. The different components of the autonomic nervous system interact with each other, by affecting release of neurotransmitter (via prejunctional receptors), at ganglia in the airways, and by interaction at postjunctional receptors (Barnes, 1989a).

II. LOCALIZATION AND FUNCTION OF MUSCARINIC RECEPTORS

Mammalian airways receive a rich cholinergic innervation (Richardson, 1979). Both histochemical and electron microscopic studies have demonstrated the presence of cholinergic nerve fibers in association with airway smooth muscle, submucosal glands, and airway ganglia, but no such fibers have been found in association with airway epithelium and bronchial or pulmonary blood vessels. The density of innervation has been shown to decrease in the smaller airways, and cholinergic nerves have not been found in association with the alveolar walls (Partanen et al., 1982; Laitinen et al., 1985; Laitinen and Laitinen, 1987).

Cholinergic nerves are the predominant neural bronchoconstrictor pathway in animal and human airways (Barnes, 1990). Cholinergic nerves arise in the brainstem and pass down the vagus nerve to form synapses in parasympathetic ganglia within the airway walls. From these ganglia, short postganglionic fibers travel to airway smooth muscle and submucosal glands. Stimulation of the vagus nerve releases acetylcholine (ACh), which activates muscarinic cholinergic receptors on target cells, such as airway smooth muscle and submucosal glands, to activate multiple second-messenger pathways including the release of IP_3 to elevate Ca^{2+} levels, activation of K^+ channels, and inhibition of adenylyl cyclase that results in bronchoconstriction and mucus secretion (Barnes, 1986). Muscarinic receptors regulate mucus secretion from both submucosal glands and airway epithelial goblet cells (Tokuyama et al., 1990). These effects are blocked by muscarinic antagonists such as atropine or ipratropium bromide. Autoradiographic mapping studies have shown that muscarinic receptors are predominantly localized to airway smooth muscle, to parasympathetic ganglia and nerve bundles, and to submucosal glands, with little or no labeling of airway epithelial cells or blood vessels in animal and human lung (Barnes et al., 1983a; van Koppen et al., 1988a; Mak and Barnes, 1990). In ferret, lung muscarinic receptors are localized predominantly to smooth muscle of large airways with few receptors in peripheral airways (Barnes et al., 1983b), whereas in guinea pig, rabbit, and human lung, muscarinic receptors appear to be distributed with equal density in large and small airways (Mak and Barnes, 1990; Mak et al., 1993). In rabbit and human lung, in contrast to ferret and guinea pig lung, labeling is also seen over alveolar walls.

Results obtained from both functional and radioligand binding studies have provided evidence for the existence of at least four pharmacological muscarinic receptor subtypes (M_1–M_4) (Hulme et al., 1990; Lazareno et al., 1990). However, five muscarinic receptor subtypes (m1–m5) have been cloned, sequenced, and expressed in mammalian cell lines (Kubo et al., 1986; Peralta et al., 1987; Bonner et al., 1987, 1988). The antagonist binding properties of m1–m4 and their patterns of expression in various tissues correspond closely to those of the pharmacologically defined M_1–M_4 receptors (Buckley et al., 1989; Dorje et al., 1991a; Bolden et al., 1992) (Table 1).

A. M₁-Receptors

In binding studies of lung membranes, M_1-receptors (pirenzepine-sensitive) appear to make up a significant proportion of the muscarinic receptor population in rabbit and human lung (Bloom et al., 1987a, 1988a; Casale and Ecklund, 1988; Gies et al., 1989; Mak and Barnes, 1989). Similar results are obtained whether the nonselective radioligand [³H]quinuclidinyl benzilate (QNB) is used with pirenzepine as a competing antagonist (indirect labeling) or with the selective radioligand [³H]pirenzepine (direct labeling) (Bloom et al., 1987a, 1988a; Mak and Barnes, 1989). Autoradiographic mapping of muscarinic receptor subtypes in human lung has shown that M_1-receptors are present over alveolar walls (Mak and Barnes, 1990). Other species, such as ferret and guinea pig, do not appear to have these parenchymal muscarinic receptors (Barnes et al., 1983a; Mak and Barnes, 1990), but their functional significance is far from clear as there is no evidence for cholinergic innervation of lung periphery. Recently the use of specific m1 oligonucleotide, cRNA, or cDNA probes has provided confirmation that these muscarinic receptors are of the m1-receptor subclass in human lung. Northern blot analysis of human lung parenchyma shows a prominent band corresponding to m1 mRNA, and in situ hybridization shows that m1 mRNA is localized to alveolar walls (Mak et al., 1992). Whether these alveolar M_1-receptors are functional remains to be determined.

M_1-receptors predominate in neuronal tissues, and there is evidence that M_1-receptors are localized to parasympathetic ganglia (Fig. 1) and to sympathetic

Table 1 Localization and Function of Muscarinic Receptor Subtypes in Lung

Receptor subtype	Localization	Function
M_1	Parasympathetic ganglia	Facilitation of neurotransmission
	Submucosal glands	?Increased mucus secretion
	Alveolar walls	?
M_2	Postganglionic cholinergic nerves	Inhibit acetylcholine release
	Airway smooth muscle	Antagonism of bronchodilatation
	Sympathetic nerves	Inhibit norepinephrine release
M_3	Airway smooth muscle	Contraction
	Submucosal glands	Increased mucus secretion
	?Goblet cells	Increased secretion
	Epithelial cells	?Increased ciliary beating
	Endothelial cells	Vasodilatation via release of NO
M_4	Alveolar walls) rabbit only	?
	Airway smooth muscle)	?
	Postganglionic cholinergic nerves?	Inhibit acetylcholine release

NO = nitric oxide.

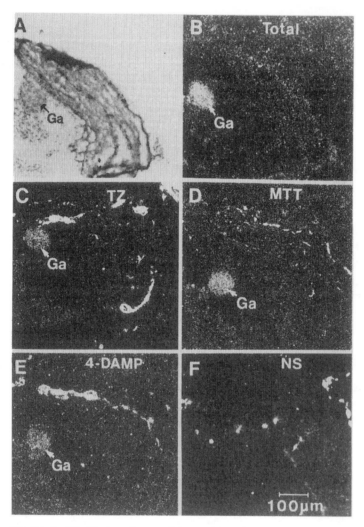

Figure 1 Distribution of muscarinic receptor subtypes in pig parasympathetic ganglion (Ga) of the airway. (A) Brightfield photomicrograph. (B–F) Darkfield photomicrographs of adjacent sections showing the distribution of autoradiographic grains after incubation with [^{3}H]QNB alone (B: total receptors) or in the presence of the M_1-selective antagonist telezepine (C: M_2- + M_3-receptors), the M_2-selective antagonist methoctramine (D: M_1- + M_3-receptors), the M_3-selective antagonist 4-DAMP (E: M_1- + M_2-receptors), or atropine (F: nonspecific labeling).

nerve terminals in the airways (Barnes, 1992). Pirenzepine has been shown to be effective in blocking bronchoconstriction due to vagus nerve stimulation in both dog and rabbit (Beck et al., 1987; Bloom et al., 1987b). Current evidence also suggests that M_1-receptors may be present in human airway cholinergic nerves. The effects of inhaled pirenzepine and the nonselective antagonist ipratropium bromide on cholinergic reflex bronchoconstriction, triggered by inhalation of sulfur dioxide (SO_2), were compared in atopic subjects. A dose of inhaled pirenzepine was found that failed to inhibit significantly the bronchoconstriction due to inhaled methacholine, whereas ipratropium bromide blocked this response. The same dose of pirenzepine, however, was as effective as ipratropium bromide in blocking SO_2-induced bronchoconstriction. Since pirenzepine could not be acting directly on airway smooth muscle receptors, it may be acting on some part of the cholinergic reflex pathway accessible to inhaled pirenzepine, which is likely to be parasympathetic ganglia in the airways (Lammers et al., 1989). In support of this possibility, a low concentration of pirenzepine has been shown to inhibit ganglionic transmission in rabbit bronchi in vitro (Bloom et al., 1988b). Furthermore, human airway parasympathetic ganglia have a high density of muscarinic receptors in autoradiographic studies (van Koppen et al., 1988a; Mak and Barnes, 1990).

Receptor autoradiographic mapping indicates that there are also M_1-receptors localized to submucosal glands in human larger airways (Mak and Barnes, 1990), although this has not yet been confirmed using molecular m1 probes. Functional studies of mucus secretion in human airways suggest that there are no functional M_1-receptors, since pirenzepine at low and selective concentrations has no inhibitory effect on secretion of mucus glycoproteins (Johnson et al., 1994).

B. M_2-Receptors

Radioligand binding studies in lung parenchymal membranes have revealed that the population of M_2-receptors appears to possess considerable species differences (Mak and Barnes, 1989; Fryer and El-Fakahany, 1990; Haddad et al., 1991). In contrast, airway smooth muscle may express a large proportion of M_2-receptors (van Koppen et al., 1985; Roffel et al., 1988). Furthermore, m2-receptor protein has recently been identified using immunoprecipitation assay in rabbit peripheral lung (Dorje et al., 1991b). m2 mRNA has also been detected in porcine tracheal smooth muscle and cultured human airway smooth muscle cells using Northern blot analysis (Maeda et al., 1988; Mak et al., 1992). This is consistent with the distribution of m2 mRNA to human airway smooth muscle by in situ hybridization (Mak et al., 1992).

There is evidence for prejunctional muscarinic receptors on postganglionic cholinergic nerves that inhibit the release of ACh and therefore function as feedback inhibitory receptors or autoreceptors in guinea pig (Fryer and Maclagan,

1984; Minette and Barnes, 1988), cat (Blaber et al., 1985), and dog airways (Ito and Yoshitomi, 1988). These prejunctional receptors have the characteristic of M_2-receptors and are selectively blocked by methoctramine (Watson et al., 1992). The presence of these M_2-receptors has recently been confirmed by direct measurement of ACh release in guinea pig trachea (Kilbinger et al., 1991). In human airways, activation of prejunctional M_2-receptors has an inhibitory effect on cholinergic nerve-induced contraction of airway smooth muscle in vitro (Minette and Barnes, 1988). In normal human subjects, inhalation of pilocarpine, a selective agonist of M_2-receptors, has an inhibitory effect on cholinergic reflex bronchoconstriction induced by SO_2, suggesting that these inhibitory receptors are present in vivo (Minette et al., 1989). Autoradiographic studies have demonstrated muscarinic receptors on human airway cholinergic nerves (van Koppen et al., 1988a; Mak and Barnes, 1990). In binding studies of human lung membranes, no significant population of M_2-receptors has been demonstrated (Bloom et al., 1988a; Gies et al., 1989; Mak and Barnes, 1989). Since cholinergic nerves make up only a trivial proportion of lung membranes, their contribution is not likely to be detectable.

Although the bronchoconstrictor responses to cholinergic agonists appear to involve the activation of M_3-receptors leading to phosphoinositide hydrolysis, binding studies have indicated a high proportion of M_2-receptors in airway smooth muscle (van Koppen et al., 1985; Roffel et al., 1988). Receptor mapping studies reveal the presence of M_2-receptors in peripheral airway smooth muscle in guinea pig (Mak and Barnes, 1990). m2 mRNA expression has also been detected in cultured human airway smooth muscle cells by Northern blot analysis (Mak et al., 1992). The physiological role of these airway smooth muscle M_2-receptors is uncertain. Recently it has been demonstrated that these M_2-receptors, by inhibition of adenylyl cyclase (Yang et al., 1991), may have a functional role in counteracting the bronchodilator response to β-agonists (due to activation of adenylyl cyclase) shown both in vitro (Meurs et al., 1993) and in vivo (Fernandes et al., 1992).

Sympathetic nerves interact with airway cholinergic nerves in some species, such as guinea pig. M_2-receptors on sympathetic nerve terminals may inhibit the release of norepinephrine from these nerves (Racke et al., 1992).

C. M_3-Receptors

Binding studies in guinea pig and human lung membranes indicate the presence of M_3-receptors (Mak and Barnes, 1989). Autoradiographic studies have demonstrated the distribution of M_3-receptors in airway smooth muscle of large and small human airways, which is consistent with the distribution of m3 mRNA-containing cells in these structures by in situ hybridization and the high level of m3 mRNA expression in cultured human airway smooth muscle cells by Northern blot analysis (Mak et al., 1992). In the airways, smooth muscle M_3-receptor activation results in rapid phosphoinositide hydrolysis (Grandordy et al., 1986; Chilvers et al., 1990; Roffel et al., 1990) and the formation of inositol (1,4,5)-triphosphate (Chilvers et al., 1989), which release calcium ions from intracellular stores.

M_3-receptors are also localized to submucosal glands in human airways (van Koppen et al., 1988a; Mak and Barnes, 1990), and there is high expression of m3 mRNA in these structures (Mak et al., 1992). M_3-selective antagonists potently inhibit mucus glycoprotein secretion from human airways in vitro, suggesting that M_3-receptors predominate (Johnson et al., 1992). M_3-receptors are only weakly expressed on airway epithelial cells (Mak and Barnes, 1990), in contrast to the high expression of m3 mRNA in these cells by in situ hybridization and Northern blot analysis, indicating that there may be a very rapid turnover of these receptors. A similar high expression of m3 mRNA in epithelial cells is found in human nasal mucosa (Baraniuk et al., 1992). M_3-receptors are also localized to endothelial cells of the bronchial circulation, which has been confirmed by in situ hybridization studies (Mak and Barnes, 1992), and presumably mediate the vasodilator response to cholinergic stimulation of the proximal airways (Matran et al., 1989). The vasodilator response to ACh in pulmonary vessels is mediated via M_3-receptors on endothelial cells (McCormack et al., 1988).

D. M_4-Receptors

In rabbit lung, there is evidence from binding studies for the existence of M_4-receptors, and this has been confirmed by the presence of m4 mRNA on Northern blot analysis (Lazareno et al., 1990) and a predominance of m4-receptor protein using immunoprecipitation assays (Dorje et al., 1991b). In situ hybridization has shown that m4 mRNA is localized to alveolar walls and airway smooth muscle, which is consistent with the visualization of M_4-receptors in these structures (Mak et al., 1993) (Fig. 2). There is also preliminary evidence that the muscarinic autoreceptors on postganglionic cholinergic nerves in guinea pig trachea may be M_4-receptors, rather than M_2-receptors (Kilbinger et al., 1993).

III. MUSCARINIC RECEPTOR FUNCTION IN ASTHMA

There is no convincing evidence of an increase in responsiveness to cholinergic agonists in asthmatic airways in vitro (Roberts et al., 1984; van Koppen et al., 1988b; Whicker et al., 1988). In sensitized guinea pigs exposed to inhaled ovalbumin, no significant changes in muscarinic receptors have been observed (Mita et al., 1983). There is also no change in muscarinic receptor density or affinity in bronchial smooth muscle from patients with chronic obstructive pulmonary disease compared with normal controls (van Koppen et al., 1989). In guinea pigs that become hyperresponsive to ACh after exposure to intravenous or inhaled

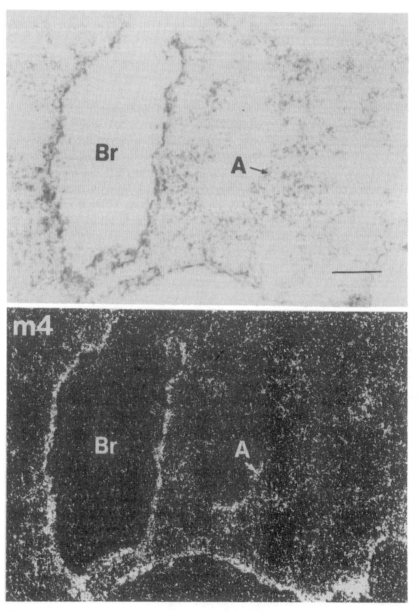

Figure 2 Distribution of m4 mRNA expression in rabbit lung. In situ hybridization of section using a [^{35}S]-labeled m4 cRNA probe. (Top) Brightfield photomicrograph. (Bottom) Positive hybridization signal to bronchiolar (Br) smooth muscle and alveoli (A).

platelet-activating factor, there is no parallel increase in contractile response of airways in vitro to ACh, no increase in muscarinic receptor density or affinity, and no change in coupling or biochemical consequences of receptor activation as measured by muscarinic stimulation of phosphoinositide hydrolysis (Robertson et al., 1988). All these studies tend to suggest that any increase in cholinergic responsiveness in vivo cannot be accounted for by any change in muscarinic receptor density or affinity or in their coupling.

In asthmatic patients, pilocarpine has no inhibitory action on SO_2-induced bronchoconstriction, indicating that there might be some dysfunction of muscarinic autoreceptors, which would result in exaggerated cholinergic reflex bronchoconstriction (Minette et al., 1989). Similarly, the inhaled nonselective muscarinic agonist methacholine has an inhibitory effect on histamine-induced bronchoconstriction in normal subjects but no such inhibitory effect in asthmatic subjects (Ayala and Ahmed, 1989). A functional defect in muscarinic autoreceptors may also explain why β-blockers produce such marked bronchoconstriction in asthmatic patients, since any increase in cholinergic tone due to blockade of inhibitory β-receptors on cholinergic nerves would normally be switched off by M_2-receptors in the nerves, and a lack of such receptors may lead to increased acetylcholine release, resulting in exaggerated bronchoconstriction (Barnes, 1989b). The protective effect of oxitropium bromide against propranolol-induced bronchoconstriction in asthmatic patients supports the idea (Ind et al., 1989).

The mechanism by which M_2-autoreceptors on cholinergic nerves may become dysfunctional is not yet certain. It is possible that chronic inflammation in airways may lead to down-regulation of M_2-receptors, which may have an important functional effect if the density of prejunctional muscarinic receptors is relatively low in comparison with airway smooth muscle muscarinic receptors. Experimental studies have recently demonstrated that influenza virus, neuraminidase, and major basic protein from eosinophils may inactivate M_2- rather than M_3-receptors (Fryer and Jacoby, 1991; Haddad and Gies, 1992; Jacoby et al., 1992). This may account for an increase in cholinergic reflex bronchoconstriction during an exacerbation of asthma, either due to a virus infection or due to allergen exposure.

IV. LOCALIZATION AND FUNCTION OF ADRENOCEPTORS

Adrenergic innervation to mammalian airways shows a considerable species variation (Richardson, 1979). In humans, a relatively sparse adrenergic innervation has been demonstrated by both fluorescence and electron microscopy. Nerve fibers have been found in close association with submucosal glands, bronchial arteries, parasympathetic ganglia, and, to a lesser extent, airway smooth muscle (Partanen et al., 1982; Pack and Richardson, 1984; Laitinen et al., 1985).

A. α-Adrenoceptors

There are relatively few α-receptors in lung (Barnes et al., 1979). α-Adrenoceptors that mediate bronchoconstriction have been demonstrated in airways of several species, including humans (Simonsson et al., 1972; Kneussl and Richardson, 1978), although it may only be possible to demonstrate their presence under certain conditions. There is now considerable doubt about the role of α-receptors in the regulation of airway tone in humans, however, since it has proved difficult to demonstrate their presence functionally or by autoradiography (Spina et al., 1989a), and α-blocking drugs do not appear to be effective as bronchodilators. Human peripheral lung strips contract with α-agonists, suggesting that small airways might have α-receptors, although the contractile response is more likely due to α-receptors on other contractile elements, such as pulmonary vessels (Black et al., 1981). Autoradiographic mapping of pulmonary α-receptors has demonstrated a very low density of α-receptors in smooth muscle of large airways, but a high density in small airways (Barnes et al., 1983b,c). α-Receptors are also localized to submucosal glands (Barnes and Basbaum, 1983).

The classical α-receptors that mediate contractile responses are α_1-receptors, which are selectively blocked by prazosin, whereas prejunctional α-receptors, mediating negative feedback of norepinephrine release, are α_2-receptors and selectively blocked by yohimbine (Hoffman and Lefkowitz, 1980). In canine trachea, the contractile response to both sympathetic nerve stimulation and exogenous α-agonists is mediated entirely by α_2-receptors postjunctionally, with very few α_1-receptors (Barnes et al., 1983d).

B. β-Adrenoceptors

Binding studies indicate a high density of β-receptors in lung of many species, including humans (Engel, 1981). Autoradiographic studies have revealed that β-receptors are localized to many different cell types within the lung, including smooth muscle of all airways from trachea to terminal bronchioles with increasing density of receptors, submucosal glands, airway epithelium with density exceeding the airway smooth muscle, and alveolar walls (Barnes et al., 1983; Carstairs et al., 1984). Activation of these receptors mediates a number of different functions in the lung, such as airway smooth muscle relaxation, inhibition of secretion of inflammatory mediators from mast cells and other inflammatory cells, fluid and ion transport across the airway epithelium, permeability and pulmonary blood vessels, mucus secretion, neurotransmission of sympathetic and parasympathetic nervous system, and surfactant secretion (Nijkamp et al., 1992) (Table 2).

1. Airway Smooth Muscle

Relaxation of human airway smooth muscle in vitro by β-agonists has been demonstrated in trachea, bronchi, bronchioles, and peripheral lung strips

Table 2 Localization and Function of β-Adrenoceptors in Lung

Localization	Function	β-receptor subtype
Airway smooth muscle	Relaxation	β_2
Epithelial cells	Increased ion and fluid transport	β_2
	Increased ciliary beating	β_2
Submucosal glands	Increased mucus secretion	β_1 (β_2)
Endothelial cells	Decreased microvascular leakage	β_2
Blood vessels	Relaxation	β_1 (β_2)
Alveolar type II cells	Increased surfactant secretion	β_2
Alveolar macrophages	Increased cAMP	β_2
Cholinergic ganglia and nerves	Decreased neurotransmission	β_2

(Zaagsma et al., 1983; Goldie et al., 1982, 1984). Binding studies in human lung indicate the predominance of β_2-receptors, the proportion of β_1- and β_2-receptors being approximately 30% and 70%, respectively (Engel, 1981). Autoradiographic studies have shown that the β-receptors of human airway smooth muscle are entirely β_2-receptors and that their density increases from trachea to terminal bronchioles (Carstairs et al., 1985). Recent studies have confirmed the expression of β_2-receptor mRNA in human airway smooth muscle by in situ hybridization (Hamid et al., 1991). The level of β_2-receptor mRNA in airway smooth muscle is high relative to the low receptor density, indicating a rapid turnover of β_2-receptors and the relative resistance of airway smooth muscle to tachyphylaxis. Functional studies also show that the relaxation of both central and peripheral human airways is mediated completely by β_2-receptors (Zaagsma et al., 1983; Goldie et al., 1984). In some species both β_1- and β_2-receptors have been demonstrated functionally in airway smooth muscle; the presence of β_1-receptors is related to the presence of sympathetic innervation of airway smooth muscle (Barnes et al., 1983e). Peripheral lung is devoid of sympathetic innervation, and relaxation to β-agonists is mediated by β_2-receptors (Lulich et al., 1976). The presence of β_3-receptors in airways has not yet been unequivocally demonstrated, although preliminary data suggest that the β_3-receptors may be involved in inhibition of NANC contractions in guinea pig bronchi (Itabashi et al., 1991).

2. Airway Epithelium

Even though no adrenergic innervation is present in the epithelium of mammalian species (Richardson, 1979), both α- and β-receptors can be demonstrated. Autoradiographic studies in animals (Barnes et al., 1982) and humans (Carstairs et al., 1984) revealed a high density of β-receptors in the airway epithelium from trachea to terminal bronchioles. In human airway epithelium, the β_2-receptors account for all β-receptors (Carstairs et al., 1985), which is consistent with the in situ hybridization studies showing the presence of β_2-receptor mRNA (Hamid et al., 1991). By means of an immunocytochemical method to examine the cellular localization of cAMP, it has been shown that β-receptors are present on ciliated epithelial cells but not on goblet cells (Lazarus et al., 1984). Human tracheal epithelial cells in suspension increase cAMP production dose dependently in the order of potency isoprenaline > epinephrine > norepinephrine, indicating a β_2-receptor-mediated effect (Davis et al., 1990).

3. Submucosal Glands

It has been shown by autoradiographic studies that human submucosal glands, which receive a sparse sympathetic nerve supply, have a small population of approximately 10% β_{-1}-receptors, while the remaining population are β_2-receptors, which is confirmed by in situ hybridization with the distribution of β_2-receptor mRNA in these structures (Hamid et al., 1991). With the use of autoradiography of the submucosal gland cells, β-receptors have been localized predominantly in the mucus cells (Barnes et al., 1982; Barnes and Basbaum, 1983).

The β-agonists increase airway mucus secretion in several animal species and in humans (Nadel et al., 1979; Peatfield and Richardson, 1982; Phipps et al., 1982). The effects found in human bronchi have been variable (Shelhamer et al., 1980; Phipps et al., 1982). Lazarus et al. (1984) have shown increases in immunoreactive cAMP in both cat serous and mucus cells of the submucosal glands after β-receptor stimulation with terbutaline and isoprenaline, indicating that both cells types possess functional β-receptors. Functional studies indicate that the β-receptors that mediate mucus secretion are mainly β_1-receptors (Peatfield and Richardson, 1982; Phipps et al., 1982). The β_2-receptors may be involved as well (Phipps et al., 1982).

4. Alveoli

The alveoli contain different cell types, including type I epithelial cells and capillary endothelial cells. Type II epithelial cells, which are less numerous, are involved in the synthesis and secretion of pulmonary surfactant. Autoradiographic studies have shown that approximately 90% of the β_2-receptors are localized to the alveolar walls (Carstairs et al., 1985). Binding studies indicate that alveolar type II cells possess both β_1- and β_2-receptors in the ratio of 1:3 (Das et al., 1987; Jones et al., 1987). The β-agonists stimulate the surfactant secretion in vivo (Brown and Longmore, 1981) and in vitro (Dobbs and Mason, 1979) from isolated type II cells. It is likely that the release of surfactant is mediated by β_2- rather than β_1-receptors (Chander and Fisher, 1990).

The β-receptors on alveolar type II cells account for no more than 5% of total lung β-receptors, which indicates that a major portion of the β-receptors is present on other cell types in the alveolus, like endothelial cells and type I pneumocytes.

Moreover, the finding that guinea pig alveolar macrophages possess a large number of β_2-receptors (Henricks et al., 1988) may have contributed to the high number of β_2-receptors on alveolar cells. The endothelial layer of pulmonary blood vessels of the guinea pig contains a high number of β_2-receptors (Engels et al., 1989). Because β-agonists reduce histamine-induced microvascular leakage and plasma extravasation in the lung and airways of guinea pig, the β-receptors in the alveolar walls might be functionally related to microvascular permeability (Persson et al., 1982). This effect is mediated by β_2-receptors, since inflammatory stimuli-induced airway plasma exudation in guinea pig is dose dependently inhibited by terbutaline (Erjefalt and Persson, 1991).

Pulmonary blood vessels have a rich sympathetic innervation (Richardson, 1979). The β-agonists induce relaxation of human pulmonary vessels, an effect that is probably mediated by β_1-receptors (Boe and Simonsson, 1980). In contrast, autoradiographic studies indicate that β-receptors on human pulmonary vessels are the β_2-receptors (Carstairs et al., 1985), in agreement with the distribution of β_2-receptor mRNA by in situ hybridization (Hamid et al., 1991).

5. Cholinergic Neurotransmission

Evidence has been found for a β-receptor-mediated inhibition of cholinergic transmission in human (Rhoden et al., 1988) and in dog airways (Danser et al., 1987; Ito, 1988). The prejunctional β-receptors that inhibit cholinergic transmission in human trachea (Bai et al., 1989) and bronchi (Rhoden et al., 1988), as well as in canine trachea, have been identified as β_2-receptors. In contrast, Danser et al. (1987) concluded that the prejunctional receptors in the canine airway were β_1-receptors.

V. β-ADRENOCEPTOR FUNCTION IN ASTHMA

Changes in β_2-receptor function have been reported to occur in lung diseases, such as asthma (Davis, 1988). Human isolated bronchial smooth muscle contains a homogeneous population of β_2-receptors (Goldie et al., 1984; Zaagsma et al., 1983). Whether β-receptor function in airway smooth muscle is abnormal in asthma is still uncertain because it is difficult to assess β-adrenergic responsiveness in asthmatic airways in vivo owing to baseline bronchodilatation. In guinea pigs sensitized to allergen and exposed to aerolized allergen, a reduction in β-receptor density has been reported (Barnes et al., 1980; Gatto et al., 1987), and in the Basenji greyhound model of airway hyperresponsiveness, there is evidence for reduced β-receptor function (Tobias et al., 1990). In vitro studies of airways from asthmatic patients who had died during asthma attacks or who had undergone lung surgery indicate that there is a decrease in the relaxant response to β-agonists (Goldie et al., 1986; Cerrina et al., 1986; Bai, 1990, 1991). However, this has not been found in all studies (Svedmyr et al., 1976; de Jongste et al., 1987; Whicker et al., 1988). Autoradiographic mapping has demonstrated a normal or even increased labeling of β-receptors in airway smooth muscle in fatal asthma (Spina et al., 1989b; Sharma and Jeffery, 1990; Bai et al., 1992). Indeed, there is an inverse relationship between reduced relaxation response to isoprenaline and the density and affinity of β-receptors (Bai et al., 1992). This suggests that airway smooth muscle β-receptors may become uncoupled in fatal asthma. The increase in β-receptor density may be the result of increased β_2-receptor gene transcription in response to a loss of negative feedback control of transcription.

Inflammatory mediators may impair the function of β-receptors. In guinea pigs, phospholipase A_2 causes a reduction and uncoupling of pulmonary β-receptors (Taki et al., 1986), suggesting that lipid mediators may influence β-receptor function. Platelet-activating factor (PAF) has been reported to reduce β-receptor binding in human lung membranes (Agrawal and Townley, 1987), although PAF does not reduce β-receptor function in animal airways after intravenous administration (Barnes et al., 1987). Leukotrienes B_4 and C_4 also reduce β-receptor binding, but in addition reduce isoprenaline-induced activation of adenylyl cyclase (Raaijmakers et al., 1989). Similarly, 15-lipoxygenase products may also impair pulmonary β-receptor function (Folkerts et al., 1984). Recently, it has been shown that cytokines may also influence the expression of β-receptors (van Oosterhout et al., 1992). However, the mechanism of β-receptor uncoupling in fatal asthma is not certain. Activation of inflammatory mediator receptors on airway smooth muscle cells leads to phosphoinositide hydrolysis (Hall and Chilvers, 1989) and the generation of diacylglycerol, which activates protein kinase C (PKC). PKC is capable of phosphorylating both β-receptors and the α-subunit of stimulatory guanine nucleotide regulatory proteins (Gs) that couple the β-receptor to adenylyl cyclase, thus resulting in either down-regulation or uncoupling of the receptor (Bouvier et al., 1987).

No significant differences in bronchodilator responses to inhaled β-agonists were found in mild asthmatics compared with normal subjects (Tattersfield et al., 1983). Barnes and Price (1983) have also investigated airway β-receptor function in asthma by comparing dose-response curves to inhaled salbutamol between severe asthmatics and normal subjects. It was found that larger doses of β-agonists were required to produce maximal bronchodilation in asthmatic compared with normal subjects. However, these findings could be explained by functional antagonism due to greater initial bronchoconstriction rather than a defect in β-receptor function per se. Interpretation of measurements of the effect of bronchodilator agonists in vivo is often complicated by variables, including the level of resting airway tone and the pathology of the airway obstruction.

VI. CONCLUSIONS

In this chapter an overview is presented of the localization and function of muscarinic receptors and adrenoceptors in the lungs of animal and humans, as well as their roles in asthma. These autonomic receptors can be found on a variety of cells within the airways. Many cell functions are mediated by muscarinic receptors and adrenoceptors, including smooth muscle contraction and relaxation, stimulation of mucus secretion, and so on. Importantly, β-receptors mediate inhibition of cholinergic transmission in the airways.

Recent advances in molecular biology of receptors, including cloning and sequencing of receptors and determination of mRNAs coding for the receptors, have provided a better understanding of muscarinic receptors and adrenoceptors. Several subtypes of muscarinic receptors, as well as the mRNAs encoding these subtypes, have now been identified and have been found in the airways of several species, including humans. It is now clear that there are marked differences between species in the distribution of muscarinic receptor subtypes in the airways. This may make clinical interpretation on the basis of animal studies somewhat difficult and emphasizes the need for more studies on human airways. However, these novel molecular techniques have not yet been applied to a great extent to adrenoceptors except β_2-receptors in the airways.

Increased sophistication of molecular tools will make it easier to elucidate the differential regulation of each receptor subtype in the airways in health and disease. The use of human primary cultured airway epithelium and smooth muscle and the development of potent and more selective agonists and antagonists and molecular probes for various receptor subtypes may lead to further understanding of their functions in the airways.

REFERENCES

Agrawal, D.K., and Townley, R.D. (1987). Effects of platelet activating factor on beta-adrenoceptors in human lung. *Biochem. Biophys. Res. Commun. 193*:1.

Ayala, L.E., and Ahmed, T. (1989). Is there a loss of a protective muscarinic receptor mechanism? *Chest 96*:1285.

Bai, T.R. (1990). Abnormalities in airway smooth muscle in fatal asthma. *Am. Rev. Respir. Dis. 141*:552.

Bai, T.R. (1991). Abnormalities in airway smooth muscle in fatal asthma. A comparison between trachea and bronchus. *Am. Rev. Respir. Dis. 143*:441.

Bai, T.R., Lam, R., and Frasad, F.Y.F. (1989). Effects of adrenergic agonists and adenosine on cholinergic neurotransmission in human tracheal smooth muscle. *Pulm. Pharmacol. 1*:193.

Bai, T.R., Mak, J.C.W., and Barnes, P.J. (1992). A comparison of beta-adrenergic receptors and in vitro relaxant responses to isoproterenol in asthmatic airway smooth muscle. *Am. J. Respir. Cell Mol. Biol. 6*:647.

Baraniuk, J.N., Kaliner, M.A., and Barnes, P.J. (1992). Muscarinic m3 receptor mRNA in situ hybridization in human nasal mucosa. *Am. J. Rhinol. 6*:145.

Barnes, P.J. (1984). The third nervous system in the lung: physiology and clinical perspectives. *Thorax 39*:561.

Barnes, P.J. (1986). Neural control of human airways in health and disease. *Am. Rev. Respir. Dis. 134*:1289.

Barnes, P.J. (1989a). Airway receptors. *Postgrad. Med. J. 65*:532.

Barnes, P.J. (1989b). Muscarinic autoreceptors in airways: their possible role in airway disease. *Chest 96*:1220.

Barnes, P.J. (1990). Muscarinic receptors in airways: recent developments. *J. Appl. Physiol. 68*:1777.

Barnes, P.J. (1992). Modulation of neurotransmission in airways. *Physiol. Rev. 72*:699.

Barnes, P.J. (1993). Muscarinic receptor subtypes in airways. *Life Sci. 52*:521.

Barnes, P.J., and Basbaum, C.B. (1983). Mapping of adrenergic receptors in the trachea by autoradiography. *Exp. Lung Res. 5*:183.

Barnes, J.C., and Pride, N.B. (1983). Dose-response curves to inhaled β-adrenoceptor agonists in normal and asthmatic subjects. *Br. J. Clin. Pharmacol. 15*:677.

Barnes, P.J., Nadel, J.A., Roberts, J.M., and Basbaum, C.B. (1983a). Muscarinic receptors in lung and trachea: autoradiographic localization using [^{3}H]quinuclidinyl benzilate. *Eur. J. Pharmacol. 86*:103.

Barnes, P.J., Basbaum, C.B., and Nadel, J.A. (1983b). Autoradiographic localization of autonomic receptors in airway smooth muscle: marked differences between large and small airways. *Am. Rev. Respir. Dis. 127*:758.

Barnes, P.J., Basbaum, C.B., Nadel, J.A., and Roberts, J.M. (1983c). Pulmonary α-adrenoceptors: autoradiographic localization using [^{3}H]prazosin. *Eur. J. Pharmacol. 88*:57.

Barnes, P.J., Skoogh, B.-E., Nadel, J.A., and Roberts, J.M. (1983d). Postsynaptic α_2-adrenoceptors predominate over α_1-adrenoceptors in canine tracheal smooth muscle and mediate neuronal and humoral α-adrenergic contraction. *Mol. Pharmacol. 23*:570.

Barnes, P.J., Nadel, J.A., Skoogh, B.-E., and Roberts, J.M. (1983e). Characterization of β-adrenoceptor subtypes in canine airway smooth muscle by radioligand binding and physiologic responses. *J. Pharmacol. Exp. Ther. 225*:456.

Barnes, P.J., Grandordy, B.M., Page, C.P., Rhoden, K.J., and Robertson, D.N. (1987). The effect of platelet activating factor on pulmonary beta-adrenoceptors. *Br. J. Pharmacol. 90*:709.

Beck, K.C., Vettermann, J., Flavahan, N.A., and Rehder, K. (1987). Muscarinic M_1 receptors mediate the increase in pulmonary resistance during vagus nerve stimulation in dogs. *Am. Rev. Respir. Dis. 137*:1135.

Blaber, L.C., Fryer, A.D., and Maclagan, J. (1985). Neuronal muscarinic receptors attenuate vagally-induced contraction of feline bronchial smooth muscle. *Br. J. Pharmacol. 86*:723.

Black, J., Turner, A., and Shaw, J. (1981). Alpha-adrenoceptors in human peripheral lung. *Eur. J. Pharmacol. 72*:83.

Bloom, J.W., Halonen, M., Lawrence, L.J., Rould, E., Seaver, N.A., and Yamamura, H.I. (1987a). Characterization of high affinity [^{3}H]pirenzepine and [^{3}H](−)-quinuclidinyl benzilate binding to muscarinic cholinergic receptors in rabbit peripheral lung. *J. Pharmacol. Exp. Ther. 240*:51.

Bloom, J.W., Yamamura, H.I., Baumgartner, C., and Halonen, M. (1987b). A muscarinic receptor with high affinity for pirenzepine mediates vagally induced bronchoconstriction. *Eur. J. Pharmacol. 133*:21.

Bloom, J.W., Halonen, M., and Yamamura, H.I. (1988a). Characterization of muscarinic cholinergic receptor subtypes in human peripheral lung. *J. Pharmacol. Exp. Ther. 244*:625.

Bloom, J.W., Baumgartner-Folkerts, C., Palmer, J.D., Yamamura, H.I., and Halonen, M. (1988b). A muscarinic receptor subtype modulates vagally stimulated bronchial contraction. *J. Appl. Physiol. 85*:2144.

Boe, J., and Simonsson, B.G. (1980). Adrenergic receptors and sympathetic agents in isolated human pulmonary arteries. *Eur. J. Respir. Dis. 61*:195.

Bolden, C., Cusack, B., and Richelson, E. (1992). Antagonism by antimuscarinic and neuroleptic compounds at the five cloned human muscarinic cholinergic receptors expressed in Chinese hamster ovary cells. *J. Pharmacol. Exp. Ther. 260*:576.

Bonner, T.I., Buckley, N.J., Young, A.C., and Brann, M.R. (1987). Identification of a family of muscarinic acetylcholine receptor genes. *Science 237*:527.

Bonner, T.I., Young, A.C., Brann, M.R., and Buckley, N.J. (1988). Cloning and expression of the human and rat M_5-muscarinic acetylcholine receptor genes. *Neuron 1*:403.

Bouvier, M., Leeb-Lundberg, L.M., Benovic, J.L., Caron, M.G., and Lefkowitz, R.J. (1987). Regulation of adrenergic receptor function by phosphorylation. II. Effects of agonist occupancy on phosphorylation of α_1- and β-adrenergic receptors by protein kinase C and cyclic AMP-dependent protein kinase. *J. Biol. Chem. 262*:3106.

Brown, L.A.S., and Longmore, W.J. (1981). Adrenergic and cholinergic regulation of lung surfactant secretion in the isolated perfused rat lung and in the alveolar type II cell in culture. *J. Biol. Chem. 256*:66.

Buckley, N.J., Bonner, T.I., and Brann, M.R. (1988). Localization of a family of muscarinic receptor mRNAs in rat brain. *J. Neurosci. 8*:4646.

Buckley, N.J., Bonner, T.I., Buckley, C.M., and Brann, M.R. (1989). Antagonist binding properties of five cloned muscarinic receptors expressed in CHO-K1 cells. *Mol. Pharmacol. 35*:469.

Carstairs, J.R., Nimmo, A.J., and Barnes, P.J. (1984). Autoradiographic localization of β-adrenoceptors in human lung. *Eur. J. Pharmacol. 105*:189.

Carstairs, J.R., Nimmo, A.J., and Barnes, P.J. (1985). Autoradiographic visualization of β-adrenoceptor subtypes in human lung. *Am. Rev. Respir. Dis. 132*:541.

Casale, T.B., and Ecklund, P. (1988). Characterization of muscarinic receptor subtypes in human peripheral lung. *J. Appl. Physiol. 65*:594.

Cerrina, J., Ladurie, M.L., Labat, C., Raffestin, B., Bayol, A., and Brink, C. (1986). Comparison of human bronchial muscle response to histamine in vivo with histamine and isoproterenol agonists in vitro. *Am. Rev. Respir. Dis. 134*:57.

Chander, A., and Fisher, A.B. (1990). Regulation of lung surfactant secretion. *Am. J. Physiol. (Lung Cell. Mol. Physiol.) 258*:L241.

Chilvers, E.R., Challiss, R.A., Barnes, P.J., and Nahorski, S.R. (1989). Mass changes of inositol (1,4,5) triphosphate in trachealis muscle following agonist stimulation. *Eur. J. Pharmacol. 164*:587.

Chilvers, E.R., Batty, I.H., Barnes, P.J., and Nahorski, S.R. (1990). Formation of inositol polyphosphates in airway smooth muscle after muscarinic receptor stimulation. *J. Pharmacol. Exp. Ther. 252*:786.

Danser, E.E., van Den Ende, A.H.J., Lorenz, R.R., Flavahan, N.A., and Vanhoutte, P.M.

(1987). Prejunctional β_1-adrenoceptors inhibit cholinergic transmission in canine bronchi. *J. Appl. Physiol. 62*:785.

Das, S.K., Sikpi, M.O., and Skolnick, P. (1987). Heterogeneity of β-adrenoceptors in guinea pig alveolar type II cells. *Biochem. Biophys. Res. Commun. 142*:898.

Davis, P.B. (1988). Autonomic function in patients with airway obstruction. In: *The Airways, Neural Control in Health and Disease.* (Kaliner, M.A., and Barnes, P.J., eds.). Marcel Dekker, New York, p. 87.

Davis, D.B., Silski, C.L., Kercomar, P., and Infeld, M. (1990). Beta-adrenergic receptors on human tracheal epithelial cells in primary culture. *Am. J. Physiol. (Cell Physiol.) 258*:C71.

de Jongste, J.C., Mons, H., Bonta, I.L., and Kerrebijn, K.F. (1987). Human asthmatic airway responses in vitro—a case report. *Eur. J. Respir. Dis. 70*:23.

Dobbs, L.G., and Mason, R.J. (1979). Pulmonary alveolar type II cells isolated from rats: release of phosphatidylcholine in response to β-adrenergic stimulation. *J. Clin. Invest. 63*:378.

Dorje, F., Wess, J., Lambrecht, G., Tacke, R., Mutschler, E., and Brann, M.R. (1991a). Antagonist binding profiles of five cloned human muscarinic receptor subtypes. *J. Pharmacol. Exp. Ther. 256*:727.

Dorje, F., Levey, A.I., and Brann, M.R. (1991b). Immunological detection of muscarinic receptor subtype proteins (m1–m5) in rabbit peripheral tissues. *Mol. Pharmacol. 40*:459.

Engel, G. (1981). Subclasses of beta-adrenoceptors—a quantitative estimation of beta$_1$- and beta$_2$-adrenoceptors in guinea pig and human lung. *Postgrad. Med. J. 57*:77.

Engels, F., Carstairs, J.R., Barnes, P.J., and Nijkamp, F.P. (1989). Autoradiographic localization of changes in pulmonary β-adrenoceptors in an animal model of atopy. *Eur. J. Pharmacol. 164*:139.

Erjefalt, I., and Persson, C.G.A. (1991). Pharmacologic control of plasma exudation into tracheobronchial airways. *Am. Rev. Respir. Dis. 143*:1008.

Fernandes, L.B., Fryer, A.D., and Hirschman, C.A. (1992). M_2 muscarinic receptors inhibit isoproterenol-induced relaxation of canine airway smooth muscle. *J. Pharmacol. Exp. Ther. 262*:119.

Folkerts, G., Nijkamp, F.P., and van Oosterholt, A.J.M. (1984). Induction in guinea pigs of airway hyperreactivity and decreased lung β-adrenoceptor number by 15-hydroxy-arachidonic acid. *Br. J. Pharmacol. 80*:597.

Fryer, A.D., and El-Fakahany, E.E. (1990). Identification of three muscarinic receptor subtypes in rat lung using binding studies with selective antagonists. *Life Sci. 47*:611.

Fryer, A.D., and Jacoby, D.B. (1991). Parainfluenza virus infection damages inhibitory M_2 muscarinic receptors on pulmonary parasympathetic nerves in the guinea-pig. *Br. J. Pharmacol. 102*:267.

Fryer, A.D., and Maclagan, J. (1984). Muscarinic inhibitory receptors in pulmonary parasympathetic nerves in the guinea pig. *Br. J. Pharmacol. 83*:973.

Gatto, C., Green, T.P., Johnson, M.G., Marchessault, R.P., Seybold, V., and Johnson, D.E. (1987). Localization of quantitative changes in pulmonary beta-receptors in ovalbumin-sensitized guinea pigs. *Am. Rev. Respir. Dis. 136*:150.

Gies, J.-P., Bertrand, C., Vanderheyden, P., Waeldele, F., Dumont, P., Pauli, G., and

Landry, Y. (1989). Characterization of muscarinic receptors in human, guinea pig and rat lung. *J. Pharmacol. Exp. Ther. 250*:309.

Goldie, R.G., Paterson, J.W., and Wale, J.L. (1982). A comparative study of β-adrenoceptors in human and porcine lung parenchyma strip. *Br. J. Pharmacol. 76*:523.

Goldie, R.G., Paterson, J.W., Spina, D., and Wale, J.L. (1984). Classification of β-adrenoceptors in human isolated bronchus. *Br. J. Pharmacol. 81*:611.

Goldie, R.G., Spina, D., Henry, P.J., Lulich, K.M., and Paterson, J.W. (1986). In vitro responsiveness of human asthmatic bronchus to carbachol, histamine, β-adrenoceptor agonists and theophylline. *Br. J. Clin. Pharmacol. 22*:669.

Grandordy, B.M., Cuss, F.M., Sampson, A.S., Palmer, J.B., and Barnes, P.J. (1986). Phosphatidylinositol response to cholinergic agonists in airway smooth muscle: relationship to contraction and muscarinic receptor occupancy. *J. Pharmacol. Exp. Ther. 238*:273.

Haddad, E.-B., and Gies, J.P. (1992). Neuraminidase reduces the super-high-affinity [^{3}H]oxotremorine-M binding sites in guinea pig. *Eur. J. Pharmacol. 211*:273.

Haddad, E.-B., Landry, Y., and Gies, J.-P. (1991). Muscarinic receptor subtypes in guinea pig airways. *Am. J. Physiol. (Lung Cell. Mol. Physiol.) 261*:L327.

Hall, I., and Chilvers, E.R. (1989). Inositol phosphates and airway smooth muscle. *Pulm. Pharmacol. 2*:113.

Hamid, Q.A., Mak, J.C.W., Sheppard, M.N., Corrin, B., Venter, J.C., and Barnes, P.J. (1991). Localization of beta$_2$-adrenoceptor messenger RNA in human and rat lung using in situ hybridization: correlation with receptor autoradiography. *Eur. J. Pharmacol. (Mol. Pharmacol. Sect.) 206*:133.

Henricks, P.A.J., van Esch, B., van Oosterhout, A.J.M., and Nijkamp, F.P. (1988). Specific and non-specific effects of β-adrenoceptor agonists on guinea pig alveolar macrophage function. *Eur. J. Pharmacol. 152*:321.

Hoffman, B.B., and Lefkowitz, R.J. (1980). Alpha-adrenergic receptor subtypes. *N. Engl. J. Med. 302*:1390.

Hulme, E.C., Birdsall, N.J.M., and Buckley, N.J. (1990). Muscarinic receptor subtypes. *Annu. Rev. Pharmacol. Toxicol. 30*:633.

Ind, P.W., Dixon, C.M.S., Fuller, R.W., and Barnes, P.J. (1989). Anticholinergic blockade of beta-blocker induced bronchoconstriction. *Am. Rev. Respir. Dis. 139*:1390.

Itabashi, S., Aikawa, T., Sekizawa, K., Sasakim, H., and Takishima, T. (1991). Beta-3 agonist selectively inhibits non-adrenergic non-cholinergic contraction in guinea pig bronchi (abstract). *Am. Rev. Respir. Dis. 143*:A750.

Ito, Y. (1988). Pre- and postjunctional actions of procaterol, a β$_2$-adrenoceptor stimulant, on dog tracheal tissue. *Br. J. Pharmacol. 95*:268.

Ito, Y., and Yoshitomi, T. (1988). Autoregulation of acetylcholine release from vagus nerve terminals through activation of muscarinic receptors in the dog trachea. *Br. J. Pharmacol. 93*:636.

Jacoby, D.B., Gleich, G.J., and Fryer, A.D. (1992). Human eosinophil major basic protein is an endogenous, allosteric antagonist at the inhibitory muscarinic M$_2$ receptor (abstract). *Am. Rev. Respir. Dis. 145*:A436.

Johnson, C.W., Rieves, R.O., Logun, C., and Shelhamer, J.H. (1994). Cholinergic stimulation of mucin type glycoprotein release from human airways in vitro is inhibited by an antagonist of the M_3 muscarinic receptor subtype. *Am. J. Respir. Cell Mol. Biol.* (in press).

Jones, L.G., Gray, M.E., Wood, A.J.J., and Lequire, V.S. (1987). Beta-adrenergic receptor properties of a pulmonary alveolar type II cell preparation form the adult rat. *Lung 165*:201.

Kilbinger, H., Schoreider, R., Siefken, H., Wolf, D., and D'Agostino, G. (1991). Characterization of prejunctional muscarinic autoreceptors in the guinea-pig trachea. *Br. J. Pharmacol. 103*:1757.

Kilbinger, H., von Barbeleben, R.S., and Siefken, H. (1993). Is the presynaptic muscarinic autoreceptor in the guinea-pig trachea an M_2 receptor? (abstract). *Life Sci. 52*:577.

Kneussl, M.P., and Richardson, J.B. (1978). Alpha-adrenergic receptors in human and canine tracheal and bronchial smooth muscle. *J. Appl. Physiol. 45*:307.

Kubo, T., Fukuda, K., Mikami, A., Maeda, A., Takahashi, H., Mishina, M., Haga, T., Haga, K., Ichiyama, A., Kangawa, K., Kojima, H., Matsuo, H., Hirose, T., and Numa, S. (1986). Cloning, sequencing, and expression of complementary DNA encoding the muscarinic acetylcholine receptor. *Nature 323*:411.

Laitinen, A., Partanen, M., Hervonen, A., and Laitinen, L.A. (1985). Electron microscopic study on the innervation of the human lower respiratory tract: evidence of adrenergic nerves. *Eur. J. Respir. Dis. 67*:209.

Laitinen, L.A., and Laitinen, A. (1987). Innervation of airway smooth muscle. *Am. Rev. Respir. Dis. 136*:S38.

Lammers, J.-W.J., Minette, P., MuCusker, M., and Barnes, P.J. (1989). The role of pirenzepine-sensitive (M_1) muscarinic receptors in vagally mediated bronchoconstriction in humans. *Am. Rev. Respir. Dis. 139*:446.

Lazareno, S., Buckley, N.J., and Roberts, F. (1990). Characterization of muscarinic M_4 binding sites in rabbit lung, chicken heart, and NG108-15 cells. *Mol. Pharmacol. 38*:805.

Lazarus, S.C., Basbaum, C.B., and Gold, W.M. (1984). Localization of cAMP in dog and cat trachea: effects of beta-adrenergic agonists. *Am. J. Physiol. (Cell Physiol.) 247*:C327.

Lulich, V.M., Mitchell, H.W., and Sparrow, M.P. (1976). The cat lung strip as an in vitro preparation of peripheral airways: a comparison of β-adrenoceptor agonists, autacoids and anaphylactic challenge of the lung strip and trachea. *Br. J. Pharmacol. 58*:71.

McCormack, D.G., Mak, J.C., Minette, P., and Barnes, P.J. (1988). Muscarinic receptor subtypes mediating vasodilation in the pulmonary artery. *Eur. J. Pharmacol. 158*:293.

Maeda, A., Kubo, T., Mishina, M., and Numa, S. (1988). Tissue distribution of mRNAs encoding muscarinic acetylcholine receptor subtypes. *FEBS Lett. 239*:339.

Mak, J.C.W., and Barnes, P.J. (1989). Muscarinic receptor subtypes in human and guinea pig lung. *Eur. J. Pharmacol. 164*:223.

Mak, J.C.W., and Barnes, P.J. (1990). Autoradiographic visualization of muscarinic receptor subtypes in human and guinea pig lung. *Am. Rev. Respir. Dis. 141*:1559.

Mak, J.C.W., Baraniuk, J.N., and Barnes, P.J. (1992). Localization of muscarinic receptor subtype messenger RNAs in human lung. *Am. J. Respir. Cell Mol. Biol. 7*:344.

Mak, J.C.W, Haddad, E.-B., Buckley, N.J., and Barnes, P.J. (1993). Visualization of muscarinic m4 mRNA and M_4 receptor subtype in rabbit lung. *Life Sci. 53*:1501.

Matran, R., Alving, K., Martling, C.-R., Lacroix, J.S., and Lundberg, J.M. (1989). Vagally mediated vasodilatation by motor and sensory nerves in the tracheal and bronchial circulation of the pig. *Acta Physiol. Scand. 135*:29.

Meurs, H., Roffel, A.F., Elzinga, C.R.S., de Boer, R.E.P., and Zaagsma, J. (1993). Muscarinic receptor-mediated inhibition of adenylyl cyclase and its role in the functional antagonism of isoprenaline in airway smooth muscle (abstract). *Br. J. Pharmacol. 108*:208P.

Minette, P.A., and Barnes, P.J. (1988). Prejunctional inhibitory muscarinic receptors on cholinergic nerves in human and guinea pig airways. *J. Appl. Physiol. 64*:2532.

Minette, P.A.H., Lammers, J., Dixon, C.M.S., McCuster, M.T., and Barnes, P.J. (1989). A muscarinic agonist inhibits reflex bronchoconstriction in normal but not in asthmatic subjects. *J. Appl. Physiol. 67*:2461.

Mita, H., Yui, Y., and Shida, T. (1983). Changes in $alpha_1$ and beta-adrenergic and cholinergic muscarinic receptors in guinea pig lung sensitized with ovalbumin. *Int. Arch. Allergy Appl. Immunol. 70*:225.

Nadel, J.A., and Barnes, P.J. (1984). Autonomic regulation of the airways. *Annu. Rev. Med. 35*:451.

Nadel, J.A., Davis, B., and Phipps, R.J. (1979). Control of mucus secretion and ion transport in airways. *Annu. Rev. Physiol. 41*:369.

Nijkamp, F.P., Engels, F., Henricks, P.A.J., and van Oosterhout, A.J.M. (1992). Mechanisms of β-adrenergic receptor regulation in lungs and its implications for physiological responses. *Physiol. Rev. 72*:323.

Pack, R.J., and Richardson, P.S. (1984). The aminergic innervation of the human bronchus: a light and electron microscopic study. *J. Anat. 138*:493.

Partanen, M., Laitinen, A., Hervonen, A., Toivanen, M., and Laitinen, L.A. (1982). Catecholamine- and acetylcholinesterase-containing nerves in human lower respiratory tract. *Histochemistry 76*:175.

Peatfield, A.C., and Richardson, P.S. (1982). The control of mucus secretion into the lumen of the cat by α- and β-adrenoceptors and their involvement during sympathetic nerve stimulation. *Eur. J. Pharmacol. 81*:617.

Peralta, E.G., Ashkenazi, A., Winslow, J.W., Smith, D.H., Ramachandran, J., and Capon, D.J. (1987). Distinct primary structure, ligand-binding properties and tissue-specific expression of four human muscarinic acetylcholine receptors. *EMBO J. 6*:3923.

Persson, C.G.A, Ejefalt, I., Grega, G.J., and Svensjo, E. (1982). The role of β-receptor agonists in the inhibition of pulmonary edema. *Ann. NY Acad. Sci. 384*:544.

Phipps, R.J., Williams, I.P., Richardson, P.S., Pell, J., Pack, R.J., and Wright, N. (1982). Sympathetic drugs stimulate the output of secretory glycoprotein from human bronchi in vitro. *Clin. Sci. 63*:23.

Raaijmakers, J.A.M., Beneker, C., van Geegen, E.C.G., Meisters, T.M.N., and Poven, P. (1989). Inflammatory mediators and β-receptor function. *Agents Actions 26*:45.

Racke, K., Hey, C., and Wessler, I. (1992). Endogenous noradrenaline release from guinea-pig isolated trachea is inhibited by activation of M_2 receptors. *Br. J. Pharmacol. 107*:3.

Rhoden, K.J., Meldrum, L.A., and Barnes, P.J. (1988). Inhibition of cholinergic neurotransmission in human airways by beta$_2$-adrenoceptors. *J. Appl. Physiol. 65*:700.

Richardson, J.B. (1979). Nerve supply to the lungs. *Am. Rev. Respir. Dis. 119*:785.

Richardson, J.B. (1981). Noradrenergic inhibitory innervation of the lung. *Lung 159*:315.

Roberts, J.A., Raeburn, D., Rodger, I.W., and Thomson, N.C. (1984). Comparison of in vivo airway responsiveness and in vitro smooth muscle sensitivity to methacholine. *Thorax 39*:837.

Robertson, D.N., Rhoden, K.J., Grandordy, B., Page, C.P., and Barnes, P.J. (1988). The effect of platelet activating factor on histamine and muscarinic receptor function in guinea-pig airways. *Am. Rev. Respir. Dis. 137*:1317.

Roffel, A.F., Elzinga, C.R.S., van Amsterdam, R.G.M., de Zeeuw, R.A., and Zaagsma, J. (1988). Muscarinic M_2 receptors in bovine tracheal smooth muscle: discrepancies between binding and function. *Eur. J. Pharmacol. 153*:73.

Roffel, A.F., Meurs, H., Elzinga, C.R.S., and Zaagsma, J. (1990). Characterization of the muscarinic receptor subtype involved in phosphoinositide metabolism in bovine tracheal smooth muscle. *Br. J. Pharmacol. 99*:293.

Sharma, R.K., and Jeffery, P.K. (1990). Airway β-adrenoceptor number in cystic fibrosis and asthma. *Clin. Sci. 78*:409.

Shelhamer, J., Marom, Z., and Kaliner, M. (1980). Immunologic and neuropharmacologic stimulation of mucous glycoprotein release from human airways in vitro. *J. Clin. Invest. 66*:1400.

Simonsson, B.G., Svedmyr, N., Skoogh, B.E., Anderson, R., and Bergh, N.P. (1972). In vivo and in vitro studies on alpha-adrenoceptors in human airways. Potentiation with bacterial endotoxin. *Scand. J. Respir. Dis. 53*:227.

Spina, D., Rigby, P.J., Paterson, J.W., and Goldie, R.G. (1989a). Alpha$_1$-adrenoceptor function and autoradiographic distribution in human asthmatic lung. *Br. J. Pharmacol. 97*:701.

Spina, D., Rigby, P.J., Paterson, J.W., and Goldie, R.G. (1989b). Autoradiographic localization of beta-adrenoceptors in asthmatic human lung. *Am. Rev. Respir. Dis. 140*:1410.

Svedmyr, N.L.V., Larsson, S.A., and Thiringer, G.K. (1976). Development of "resistance" in beta-adrenergic receptors of asthmatic patients. *Chest 69*:479.

Taki, F., Takagi, K., Satake, T., Sugiyama, S., and Ozawa, T. (1986). The role of phospholipase in reduced beta-adrenergic responsiveness in experimental asthma. *Am. Rev. Respir. Dis. 133*:362.

Tattersfield, A.E., Holgate, S.T., Harvey, J.E., and Gribbin, H.R. (1983). Is asthma due to partial beta-blockade of airways? In: *Pharmacology of Asthma, Agents Actions Supplement*, Vol. 13. (Morley, J., and Rainsford, K.D., eds.). Birkhauser Verlag, Stuttgart, p. 265.

Tobias, J.D., Sauder, R.A., and Hirshman, C.A. (1990). Reduced sensitivity to β-adrenergic agonists in basenji greyhound dogs. *J. Appl. Physiol. 69*:1212.

Tokuyama, K., Kuo, H.-P., Rohde, J.A.L., Barnes, P.J., and Rogers, D.F. (1990). Neural control of goblet cell secretion in guinea pig airways. *Am. J. Physiol. (Lung Cell. Mol. Physiol.) 259*:L108.

van Koppen, C.J., Rodrigues de Miranda, J.F., Beld, A.J., Hermanussen, M.W., Lammers, J.-W.J., and van Ginneken, C.A.M. (1985). Characterization of the muscarinic receptor in human tracheal smooth muscle. *Naunyn-Schmied. Arch. Pharmacol. 331*:247.

van Koppen, C.J., Blankesteijn, W.M., Klaasen, A.B.M., Rodrigues de Miranda, J.F., Beld, A.J., and van Ginneken, C.A.M. (1988a). Autoradiographic visualization of muscarinic receptors in human bronchi. *J. Pharmacol. Exp. Ther. 244*:760.

van Koppen, C.J., Rodrigues de Miranda, J.F., Beld, A.J., van Ginneken, C.A.M., Lammers, J.-W.J., and van Herwaarden, C.L.A. (1988b). Muscarinic receptor sensitivity in airway smooth muscle of patients with obstructive airway disease. *Arch. Int. Pharmacodyn. Ther. 295*:238.

van Koppen, C.J., Lammers, J.-W.J., Rodrigues de Miranda, J.F., Beld, A.J., van Herwaarden, C.L.A., and van Ginneken, C.A.M. (1989). Muscarinic receptor binding in central airway musculature in chronic airflow limitation. *Pulm. Pharmacol. 2*:131.

van Oosterhout, A.J.M., Stam, W.B., Vanderscheuren, R.G.J.R.A., and Nijkamp, F.P. (1992). The effects of cytokines on β-adrenoceptor function of human peripheral blood mononuclear cells and guinea pig trachea. *J. Allergy Clin. Immunol. 90*:304.

Watson, N., Barnes, P.J., and Maclagan, J. (1992). Actions of methoctramine, a muscarinic M_2 receptor antagonist, on muscarinic and nicotinic cholinoceptors in guinea-pig airways in vivo and in vitro. *Br. J. Pharmacol. 105*:107.

Whicker, S.D., Armour, C.L., and Black, J.L. (1988). Responsiveness of bronchial smooth muscle from asthmatic patients to relaxant and contractile agonists. *Pulm. Pharmacol. 1*:25.

Yang, C.M., Chow, S.-P., and Sung, T.-C. (1991). Muscarinic receptor subtypes coupled to generation of different second messengers in isolated tracheal smooth muscle cells. *Br. J. Pharmacol. 104*:613.

Zaagsma, J., van der Heijden, P.J.C.M., van der Schaar, M.W.G., and Bank, C.M.C. (1983). Comparison of functional β-adrenoceptor heterogeneity in central and peripheral airway smooth muscle of guinea pig and man. *J. Recept. Res. 3*:89.

DISCUSSION

Manzini: What are the functions of M_1-receptors in alveoli?

Mak: This is undetermined.

Kaliner: In human nasal mucosa, M_1- and M_3-receptors are present on glands and vessels, with 60% being M_3 subtype by Scatchard analysis (Mizoguchi et al., *J. Allergy Clin. Immunol.*, 1992).

Mak: In human bronchus, only m3 mRNA was found in glands by in situ hybridization.

Kunkel: Are drugs available to alter β-receptor number or sensitivity?

Mak: Glucocorticoids increase β_2-receptor mRNA and its receptor in human lung in vitro, rat lung in vivo, and cultured rat type II cells.

Erdos: Is there a steric relationship between muscarinic receptor and acetyl-cholinerase expression?

Mak: This has not been examined.

Baraniuk: What is the distribution of tachykinin-binding sites in bronchi and lungs?

Mak: In guinea pig lung, ^{125}I-SP and ^{3}H-FK-888 (NK-1-specific antagonist) bind to sites on smooth muscle of airways of all sizes, vascular smooth muscle, and submucosal glands.

12

Tachykinin Receptor Antagonists

Stefania Meini and Carlo A. Maggi
A. Menarini Pharmaceuticals, Florence, Italy

I. TACHYKININS AND TACHYKININ RECEPTORS

Tachykinins (TKs) are a family of peptides that share the common C-terminal sequence Phe-Xaa-Gly-Leu-Met-NH$_2$, where Xaa is an aromatic (Phe, Tyr) or a branched aliphatic (Val, Ile) amino acid (Table 1). TKs have been isolated from both mammalian and nonmammalian (e.g., eledoisin and kassinin) species (Erspamer, 1981). Although the discovery of substance P (SP) occurred more than 60 years ago (von Euler and Gaddum, 1931), the other mammalian tachykinins, neurokinin A (NKA, previously named substance K, neuromedin L, and neurokinin α) and neurokinin B (NKB, previously named neuromedin K or neurokinin β), were isolated just at the beginning of last decade (see Maggio, 1988, for review), leading to a new perspective of the physiology and pharmacology of mammalian TKs. Two N-terminally extended forms of NKA, termed neuropeptide K and neuropeptide γ (Takeda and Krause, 1989; van Giersbergen et al., 1992) have also been described, although their precise functional role has yet to be determined (Table 1). The common C-terminal sequence of TK is essential for their interaction with NK-1, NK-2, and NK-3 receptors, while some effects of SP (e.g., mast cell degranulation), not shared by other TKs, are mediated by its N-terminal region.

Throughout the peripheral (neuronal stimulation, smooth muscle contraction, vasodilation, increase in vascular permeability, secretion, etc.) and the central nervous system, the biological actions of TKs are mediated by the activation of at least three different receptors, termed NK-1, NK-2, and NK-3, for which SP, NKA, and NKB possess higher affinity, respectively (the nomenclature adopted is that recommended at the Montreal meeting on Substance P and Neurokinins, see

Table 1 Amnio Acid Sequence of Natural Tachykinins and Neurokinins

Tachykinin or neurokinin	Amino acid sequence
Substance P	H-Arg-Pro-Lys-Pro-Gln-Gln-Phe-Phe-<u>Gly-Leu-Met-NH$_2$</u>
Neurokinin A	H-His-Lys-Thr-Asp-Ser-Phe-Val-<u>Gly-Leu-Met-NH$_2$</u>
Neurokinin B	H-Asp-Met-His-Asp-Phe-Val-<u>Gly-Leu-Met-NH$_2$</u>
Kassinin	H-Asp-Val-Pro-Lys-Ser-Asp-Gln-Phe-Val-<u>Gly-Leu-Met-NH$_2$</u>
Eledoisin	Pyr-Pro-Ser-Lys-Asp-Ala-Phe-Ile-<u>Gly-Leu-Met-NH$_2$</u>
Physalaemin	Pyr-Ala-Asp-Pro-Asn-Lys-Phe-Tyr-<u>Gly-Leu-Met-NH$_2$</u>
Neuropeptide K	H-Asp-Ala-Asp-Ser-Ser-Ile-Glu-Lys-Gln-Val-Ala-Leu-Leu-Lys-Ala-Leu-Tyr-Gly-His-Gly-Gln-Ile-Ser-His-Lys-Arg-His-Lys-Thr-Asp-Ser-Phe-Val-<u>Gly-Leu-Met-NH$_2$</u>
Neuropeptide γ	H-Asp-Ala-Gly-His-Gly-Gln-Ile-Ser-His-Lys-Arg-His-Lys-Thr-Asp-Ser-Phe-Val-<u>Gly-Leu-Met-NH$_2$</u>

The C-terminal sequence common to the peptides of the tachykinin family is underlined.

Henry, 1987) (see Maggi et al., 1993, for review). Although each one of the natural TKs possesses higher affinity for its own or "preferred" receptor, some degree of "crosstalk" occurs at certain concentrations (e.g., SP acting on NK-2 receptors). Bioassay preparations that can be pharmacologically considered "monoreceptorial" (tissues bearing only one of the described different TK receptor types, Table 2) have been important for the characterization of different TK receptors.

Table 2 Pharmacology of NK-1, NK-2, and NK-3 Tachykinin Receptors

	NK-1	NK-2	NK-3
Agonists' order of potency	SP > NKA = NKB	NKA > NKB ≫ SP	NKB > NKA ≫ S
Selective agonists	SP methylester [Sar^9] SP sulfone [Pro^9] SP sulfone [Pro^9] SP Septide GR 73,632	[Nle^{10}] NKA (4-10) [βAla^8] NKA (4-10) [Lys^5, MeLeu^9, Nle^{10}] NKA (4-10) GR64,349	Senktide [MePhe^7] NKB
Monoreceptorial bioassays for tachykinin receptors	Dog carotid artery Guinea pig vas deferens Rabbit jugular vein Rabbit vena cava Mouse bronchus Guinea pig urethra	Rabbit pulmonary artery Rat vas deferens Hamster trachea Human bronchus Human colon Human urinary bladder	Rat portal vein

The three TK receptors have been isolated and cloned from various species: they all belong to the superfamily of rhodopsin-like G-protein-coupled receptors with seven hydrophobic transmembrane-spanning segments, which form the ligand recognition site (see Gerard et al., 1993, for review); phosphoinositide generation is the main second-messenger system linked to their activation (see Guard and Watson, 1991, for review).

Recognition of the biological importance of TKs in regulating bodily functions has stimulated great interest in the possibility of modulating or blocking the action of these peptides through the development of specific receptor antagonists. In the

Table 3 The Three Generations of Tachykinin Antagonists

First generation: insertion of D-Trp on the backbone of SP	
[D-Pro^2,D-Trp^{7,9}]SP	Leander et al., 1981
Spantide I	Folkers et al., 1984
Second generation: peptide antagonists	
NK-1 selective	
L668,169	McKnight et al.,1988
Spantide II	Folkers et al., 1990
GR71,251	Hagan et al., 1990
GR82,334	Hagan et al., 1991
FR113680	Morimoto et al., 1992
FK888	Fujii et al., 1992
NK-2 selective	
L659,877	McKnight et al., 1988
MEN10,207	Rovero et al., 1990
R396	Dion et al., 1990
MDL28,564	Harbeson et al., 1990
MDL29,913	van Giersbergen et al., 1991
MEN10,376	Maggi et al., 1991
MEN10,573	Quartara et al., 1992
MEN10,612	Quartara et al., 1992
NK-3 selective	
GR138676	Stables et al., 1993
Third generation: nonpeptide antagonists	
NK-1 selective	
CP96,345	Snider et al., 1991
RP67,580	Garret et al., 1991
WIN51708	Appell et al., 1992
WIN62,577	Appell et al., 1992
CP99,994	McLean et al., 1992
SR140,333	Emonds-Alt et al., 1993
NK-2 selective	
SR48,968	Emonds-Alt et al., 1992

past few years, a number of receptor antagonists endowed with high potency and selectivity for only one of the three TK receptors have been developed. Receptor-selective antagonists have been extraordinarily useful in: (1) providing conclusive evidence for a physiological role of TKs as neurotransmitters in both the central and peripheral nervous system, (2) dissecting multiple components in the message carried by different TKs through different receptors, and (3) establishing a hierarchy between different TKs as mediators of a given physiological response.

A. First Generation of Tachykinin Receptor Antagonists

The first generation of TK receptor antagonists was developed in the early 1980s by the insertion of multiple D-amino acids in the sequence of SP ([D-Pro^2,D-Trp^{7,9}]SP) (Leander et al., 1981; Rosell and Folkers, 1982). These antagonists, the proto-type of which is Spantide I ([D-Arg^1,DTrp^{7,9},Leu^{11}]SP) (Folkers et al., 1984), have been extensively used to establish the transmitter role of TKs in peripheral preparations (Leander et al., 1981; Bjorkroth, 1983) and in the spinal cord (Otsuka and Yanagisawa, 1988).

However, Spantide I and its congeners are endowed with several drawbacks, including low potency, partial agonist activity, limited ability to discriminate between NK-1 and NK-2 receptors, neurotoxicity after intrathecal administration, local anesthetic activity, induction of mast cell degranulation, and ability to antagonize peptides unrelated to the TK family (Table 4).

B. Second Generation of Tachykinin Receptor Antagonists

The second generation of TK receptor antagonists (Table 3, amino acid sequence in Table 5) lists those ligands of *peptide nature* which, as compared to the first generation of antagonists, have an improved potency and selectivity for one of the three different TK receptors (Maggi et al., 1993). Besides this profile, they also show a specificity in blocking action of TKs as compared with unrelated peptides (e.g., bombesin) and lack of the above-described nonspecific effects.

The first examples of these compounds have been synthetized by the Merck Sharp & Dohme group that achieved improvements in affinity and selectivity by constraining the conformation of natural TK analogs through cyclization. L668,169 (McKnight et al., 1988), a cyclic dimer derived from a SP analog, displays a selective affinity for the NK-1 receptor, while L659,877 (McKnight et al., 1988), a cyclic hexapeptide, is a selective NK-2-receptor antagonist.

Regoli and co-workers developed a series of linear hexapeptide analogs of L659,877, the most interesting of which is R396 (Dion et al., 1990), which also possesses a high selectivity for NK-2 receptor.

The Glaxo group incorporated a spiro-γ-lactam structure into the sequence of SP and of the nonmammalian TK physalaemin, to obtain GR71,251 (Hagan et al., 1990) and GR82,344 (Hagan et al., 1991), respectively. Both compounds show a 1000-fold higher selectivity for the NK-1 than for NK-2 or NK-3 receptors, but

Table 4 First Generation of TK Antagonists and Drawbacks

Examples	
[D-Pro[2], D-Trp[7,9]] substance P	Leander et al., 1981
[D-Arg[1], D-Trp[7,9], Leu[11]] substance P (Spantide I)	Folkers et al., 1984
Drawbacks	
Low potency	Regoli, 1985
Neurotoxicity	Hokfelt et al., 1981; Rodriguez et al., 1983
Local anesthetic activity	Post et al., 1985
Mast cell degranulation	Hakanson et al., 1982
Partial agonism	Folkers et al., 1984; Bailey and Jordan 1984
Blockade of bombesin receptor	Jensen et al., 1984; 1988
Block of endothelin receptor	Fabregat and Rozengurt 1990; Gu et al., 1991
Blockade of growth-hormone-releasing peptide receptor	Bitar et al., 1991
Poor discrimination between NK-1 and NK-2 receptors	Watson, 1983; Buck and Shatzer, 1988

while GR71,251 degranulates mast cells and is extensively degraded by rat liver homogenates, the physalaemin analog, GR82,334, is improved in these respects.

The Menarini program was oriented to the development of selective NK-2 receptor antagonists. The introduction of D-Trp at positions 6, 8, and 9 into the sequence of the heptapeptide NKA(4–10) led to the development of MEN10,207 (Rovero et al., 1990), which is a potent and highly selective NK-2 receptor antagonist; this compound has been helpful in discriminating between the putative NK-2 subtypes (see below). MEN10,207 also has partial agonist activity. The substitution of Arg with Lys at the C-terminal end led to the development of MEN10,376 (Maggi et al., 1991), which has 100-fold less agonist activity than MEN10,207 while maintaining the same affinity for the NK-2 receptor. The two compounds display a significant higher affinity for the NK-2 receptor present in the rabbit pulmonary artery (NK-2A) as compared with their affinities for the NK-2 receptor present in the hamster trachea (NK-2B) (Maggi et al., 1990, 1991); the above-mentioned NK-2 antagonists L659,877 and R396 display an opposite pattern of affinity for the two bioassay systems (Tables 11 and 12). In this respect, it is interesting to mention a study correlating the lipophilicity of the C-terminal residue and subtype selectivity: in the presence of a C-terminal hydrophilic residue, the affinity is maximal for the NK-2A receptor, while lipophilic, bulky, and aromatic residues increase the affinity for the NK-2B receptor (Rovero et al., 1991). Merrell Dow introduced MDL28,564 (Harbeson et al., 1990), a pseudopep-

Table 5   Amino Acid Sequence of Peptide TK Receptor Antagonists
of Second Generation

NK-1 selective

    L 668,169: cyclo(Gln,D-Trp,(NMe)Phe(R)Gly[ANC-2]Leu,Met)$_2$

    GR 71,251: [D-Pro^9[spiro-γ-lactam]Leu^{10}, Trp^{11}] SP

    GR 82,334: [D-Pro^9[spiro-γ]lactam]Leu^{10}, Trp^{11}]physalaemin

    FR113,680: Ac-Thr-D-Trp(CHO)-Phe-NMeBzl

    FK 888: (2-(N-Me)indolil)-CO-Hyp-Nal-NMeBzl

NK-2 selective

    L 659,877: cyclo (Gln-Trp-Phe-Gly-Leu-Met)

    MEN 10,207: [Tyr^5, D-Trp$^{6.8.9}$, Arg^{10}] NKA (4-10)

    MEN 10,376: [Tyr^5, D-Trp$^{6.8.9}$, Lys^{10}] NKA (4-10)

    R 396: Ac-Leu-Asp-Gln-Trp-Phe-GlyNH$_2$

    MDL 29,913: cyclo[Leu-Ψ(CH$_2$NCH$_3$)-Leu-Gln-Trp-Phe-Gly]

    MEN 10,573: cyclo[Leu-Ψ(CH$_2$NH)-Asp(OBzl)-Gln-Trp-Phe-βAla]

    MEN 10,612: cyclo[Leu-Ψ(CH$_2$NH)-Cha-Gln-Trp-Phe-βAla]

---

tide derivative of NKA(4–10), bearing a reduced bond between positions 9 and 10.
This compound behaves as a partial agonist at the NK-2A receptor and antagonist
at the NK-2B receptor (see Sect. II.B).

A further development was the cyclic pseudopeptide MDL29,913 (Van Giersbergen et al., 1991), formally derived from L659,877, which is a highly selective
NK-2B receptor antagonist.

Two other cyclic pseudopeptides have recently been developed from Menarini,
MEN10,573 and MEN10,612 (Quartara et al., 1992), which, compared to
MDL29,913, have increased potency toward the NK-2B receptor (pK$_B$ 8.66, 9.06,
and 8.6 for MEN10,573, MEN10,612, and MDL29,913, respectively, in the
hamster trachea) and lower affinity for the NK-2A receptor (pK$_B$ 7.31, 7.41, and
7.60 for MEN10,573, MEN10,612, and MDL29,913, respectively, in the rabbit
pulmonary artery) (Maggi et al., 1993), thus resulting in increased overall
selectivity for the NK-2B receptor.

The Fujisawa group has introduced some short peptides having selective and
potent antagonist activity for the NK-1 receptor. The first compound is a tripeptide,
termed FR113,680 (Morimoto et al., 1992), formally derived from the octapeptide
([D-Pro^4,D-Trp^{7,9,10},Phe^{11}]SP(4–11) (Mizrahi et al., 1984). A further development
is the dipeptide FK888 (Fujii et al., 1992), which has high affinity for the NK-1
receptor, poor affinity for the NK-2 receptor (IC$_{50}$ in the guinea pig ileum being
0.64 nM compared to 0.11 μM in the rat vas deferens as NK-2 bioassay system),
and no activity at the NK-3 receptor (Fujii et al., 1992).

C. Third Generation of Tachykinin Receptor Antagonists

The third generation of TK receptor antagonists includes receptor-selective ligands of nonpeptide nature.

The first compound of this class, developed by Pfizer, is CP96,345, or [(2S,3S)-cis-2-(diphenylmethyl)-N-[(2-methoxyphenyl]-1-azabicyclo[2,2,2]octan-3-amine] (Fig. 1, Table 3), which displays a high potency (nM concentrations) and selectivity in blocking NK-1 receptors (Snider et al., 1991). CP96,345 possesses two chiral centers and four isomers, the most active being the 2S,3S form in cis configuration. Most studies have used a racemate ($\pm$) of the cis forms of CP96,345. In all the tests assayed, CP96,345 has shown an antagonist activity of competitive nature for NK-1 receptors but also, at μM concentrations, a non-specific smooth muscle depressant activity toward other stimulants (Lecci et al., 1991). Since an interaction of ($\pm$)CP96,345 with L-type voltage-sensitive calcium channels has been reported (Schmidt et al., 1992), this action might be responsible for its nonspecific depressant effect (unrelated to NK-1 receptor blockade).

A major advance stimulated by the introduction of CP96,345 has been the recognition of NK-1 receptor heterogeneity (see Sect. II.A). Thus, CP96,345 is approximately 100-fold more potent at NK-1 receptors in bovine, guinea pig, and human species as compared to that expressed in rat and mouse preparations (Sect. II.A) (Table 7).

Subsequently Pfizer developed CP99,994 (McLean et al., 1992) ([($\pm$)-(2S,3S)-3-methoxybenzylamino)2-phenylpiperidine]), which possesses a high affinity for the human IM9 cell line NK-1 receptor by inhibiting [^{125}I]BHSP with an IC_{50} of 0.17 nM, while possessing a 10,000-fold lower affinity for the NK-2 and NK-3 receptors. Contrary to CP96,345, CP99,994 lacks any appreciable affinity for verapamil-sensitive calcium channels. CP99,994 displays a selective NK-1 receptor blocking activity also in in vivo experiments to antagonize the capsaicin-induced plasma protein extravasation (EC_{50}: 0.01 mg/kg s.c.) in guinea pig airways (McLean et al., 1992). Nevertheless, an antinociceptive activity unrelated to NK-1 receptor blockade has been described for CP99,994 since its inactive enantiomer is equally active in inducing analgesia (3–30 mg/kg s.c.) (Boyce et al., 1993).

The Sterling-Winthrop group has described WIN51,708 and WIN62,577 (Appell et al., 1992), which are imidazo[4,5-b]quinoxaline cyanines derivatives: they possess NK-1 antagonist properties, although with lower potency and selectivity than CP96,345. These two compounds have been found to bind preferentially to the rat, as compared to guinea pig NK-1 receptor (Appell et al., 1992).

Another important nonpeptide NK-1 receptor antagonist was developed by Rhône-Poulenc, and termed RP67,580 or (3aR,7aR)-7,7-diphenyl-2-[imino-2-(2-methoxyphenyl)ethyl]perhydroisoindol-4-one) (Fig. 1) (Garret et al., 1991). This compound confirmed the species heterogeneity of the NK-1 receptor, by display-

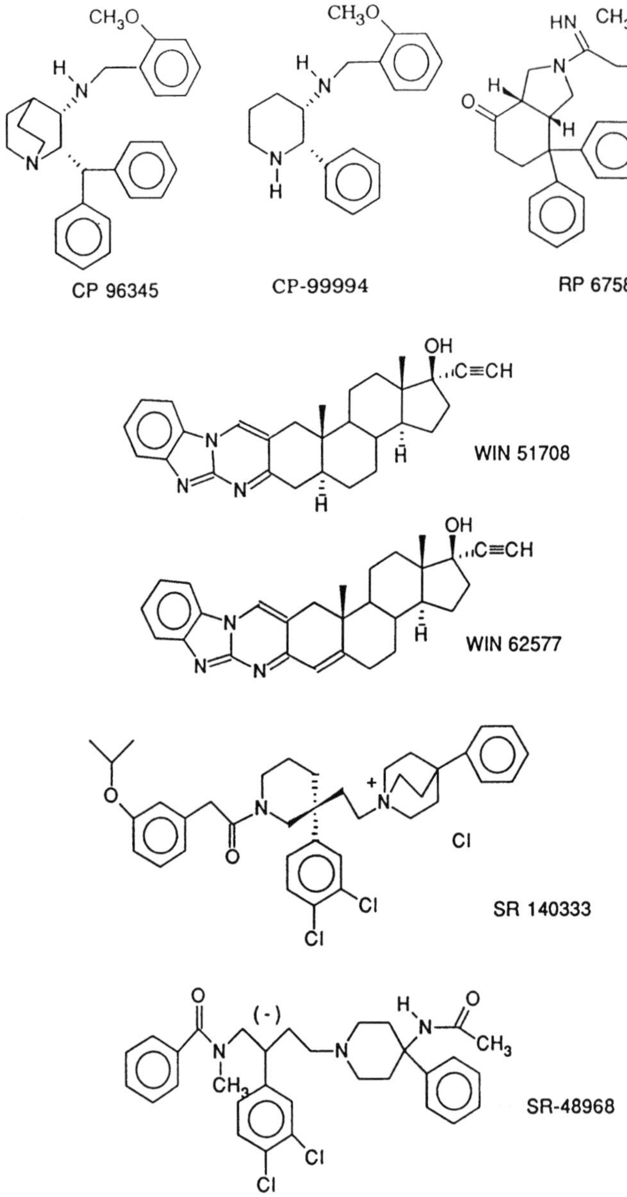

Figure 1 Chemical structures of nonpeptide TK receptor antagonists.

ing a complementary pharmacological profile compared to that of CP96,345. In other words, RP67,580 has a preferential affinity for the "rat/mouse" type of NK-1 receptor.

The Sanofi group has developed the first example of nonpeptide NK-2 receptor antagonist, SR48,968 or (S)-*N*-methyl-*N*[4-(4-acetylamino-4-phenylpiperidino)-2-(3,4-dichlorophenyl)butyl]benzamide (Emonds-Alt et al., 1992). This compound is a highly potent and selective NK-2 receptor antagonist (Ki 0.51 nM in rat duodenum membranes) and has negligible affinity for the NK-1 or NK-3 receptors.

More recently, the NK-1 antagonist SR140,333, or (S)-1-[2-(3,4-dichlorophenyl)-1-(3-isopropoxyphenylacetyl)piperidin-3-yl]ethyl]-4-phenyl-1-azoniabicyclo-[2.2.2]octane (Emonds-Alt et al., 1993; Jung et al., 1993; Oury-Donat et al., 1993), has been presented by the Sanofi group; this compound inhibits [^{125}I]SP binding to NK-1 receptor with an affinity of 0.02 and 0.01 nM to rat cortex and human IM9 cell line, respectively (Jung et al., 1993). In functional assays it is quite inactive (30,000-fold lower) at NK-2 and NK-3 receptors compared to its ability to inhibit the [Sar^9]SP sulfone-induced, NK-1-receptor-mediated, endothelium-dependent relaxation of the rabbit pulmonary artery (pA_2 10.5) (Emonds-Alt et al., 1993). In vivo studies have shown SR140,333 to inhibit [Sar^9]SP sulfone-induced plasma extravasation, salivation in the rat, hypotension in the dog, and septide-induced scratching behavior in the mouse (Emonds-Alt et al., 1993; Jung et al., 1993).

II. HETEROGENEITY OF TACHYKININ RECEPTORS

A. NK-1 Receptor

The pharmacology of the NK-1 receptor developed following the introduction of several receptor-selective agonists and antagonists. Among the compounds described and characterized as selective NK-1 receptor agonists (Table 2) are SP methylester (SPOMe) (Watson et al., 1983), septide ([Glp^6,Pro^9]SP(6–11)) (Wormser et al., 1986), [Sar^9]SP sulfone, [Pro^9]SP sulfone (Dion et al., 1987), and GR73,632 (Hagan et al., 1989). The availability of these ligands is important because the evidence for NK-1 receptor heterogeneity arises from: (1) different potencies in blocking the same agonist by different antagonists, (2) differences in affinities of agonists, or (3) different potencies in blocking various agonists by the same antagonist.

1. Species-Related Heterogeneity Detected by Nonpeptide Antagonists
The introduction of CP96,345 (Fig. 1, Table 3) (Snider et al., 1991) was the starting point for the recognition of species-related heterogeneity of the NK-1 receptor.

As shown in Table 6, Gitter et al. (1991) and Beresford et al. (1991) demonstrated in binding experiments that CP96,345 has a preferential affinity (100-fold

Table 6   Species-Related Differences in the Pharmacology of the
NK-1 Receptor Detected by Nonpeptide Antagonists

	Beresford et al., 1991 pIC_{50} of CP96,345 in $[^3H]SP$ binding
Rabbit brain	8.62
Guinea pig brain	8.50
Human brain	8.46
Bovine brain	8.86
Hamster brain	8.40
Gerbil brain	8.51
Rat brain	6.77
Mouse brain	6.92

	Gitter et al., 1991 Inhibition of CP96,345 in $[^{125}I]SP$ binding (nM)
Human U373	0.40
Human IM9	0.35
Guinea pig brain	0.32
Guinea pig lung	0.34
Rabbit brain	0.54
Mouse brain	32
Rat brain	35
Chicken brain	156

	Fardin et al., 1992 Inhibition of $[^3H]SP$ binding (nM)	
	RP67,580	CP96,345
Rat brain	7	117
Mouse brain	11.4	144
Human astrocytoma cell line (U373)	39.8	0.8

higher) for rabbit, guinea pig, bovine, hamster, gerbil, and human than for mouse, rat, or chicken NK-1 receptors. On the contrary, natural TK agonists like SP, NKA, NKB, and physalaemin do not reveal any difference in displacing SP binding from NK-1 receptors of different species. The high affinity of CP96,345 in blocking NK-1 receptors in different species can also be observed in peripheral preparations from dog, rabbit, and guinea pig as compared to rat (Table 7).

The introduction of RP67,580 (Garret et al., 1991) has extended the concept of these species-related differences: RP67,580 is (Fardin et al., 1992; Table 6) 10-fold

Table 7 Affinity of the Nonpeptide NK-1 Receptor Antagonist CP 96,345 Determined in Various Preparations for Tachykinin NK-1 Receptor Agonists

Preparation	Agonist	pA$_2$	Ref.
Dog			
Carotid artery	SP	8.70	Snider et al., 1991
	SP	9.50	Jukic et al., 1991
Rabbit			
Aorta	SP	8.15	Rubino et al., 1992
	SPOMe	8.81	Beresford et al., 1992
Iris sphincter	SPOMe	7.39	Beresford et al., 1992
Jugular vein	Sar^{9}SP	8.33	Patacchini et al., 1992
	SP	9.20	Jukic et al., 1991
Guinea pig			
Ileum	Sar^{9}SP	8.11	Lecci et al., 1991
	SP	9.50	Jukic et al., 1991
	SPOMe	8.89	Beresford et al., 1991
Circular muscle	Sar^{9}SP	8.14	Maggi et al., unpublished
	Septide	9.49	Maggi et al., unpublished
Urinary bladder	SPOMe	8.29	Beresford et al., 1992
	Sar^{9}SP	7.90	Longmore et al., 1992
Trachea	SPOMe	8.17	Beresford et al., 1992
Urethra	Sar^{9}SP	7.75	Maggi and Patacchini, 1992
Vas deferens	Sar^{9}SP	7.79	Patacchini et al., 1992
Colon	Sar^{9}SP	7.82	Giuliani et al., 1992
Rat			
Urinary bladder	SPOMe	6.30	Beresford et al., 1992
	Sar^{9}SP	6.16	Patacchini et al., 1992
	Septide	6.80	Meini et al., 1993
Spinal cord	SPOMe	7.13	Beresford et al., 1991

SP = substance P; SPOMe = substance P methylester; Sar^{9}SP = [Sar^9]substance P sulfone.

less potent than CP96,345 in binding to NK-1 receptor expressed in human astrocytoma cells (U373). The greater potency of RP67,580, compared to CP96,345, in blocking NK-1-mediated responses in rats can also be demonstrated in vivo. In urethane-anesthetized rats, comparable degrees of antagonism of the hypotensive response or plasma protein extravasation response to intravenously injected SP or NK-1-selective agonists can be demonstrated with lower doses of RP67,580 than CP96,345 (Garret et al., 1991; Lembeck et al., 1992; Santicioli et al., 1993; Maggi et al., 1993).

2. Central and Peripheral NK-1 Receptors in Rat and Guinea Pig Species

A few studies have described the possible existence of differences between central and peripheral NK-1 receptors in both rat and guinea pig species. Lew et al. (1990) (Table 8) showed that SP and [Sar⁹]SP sulfone are equipotent in displacing [^{125}I]Bolton Hunter [Sar⁹]SP sulfone from rat salivary glands or brain tissue. On the contrary, the NK-1 receptor selective agonist SPOMe and the other TKs, kassinin and NKB, were clearly more potent in the salivary glands than in the brain. To rule out that these differences could be due to a different metabolic breakdown of peptide agonists, Clerc et al. (1992) have used septide, which is resistant to enzymatic degradation. These authors showed that septide is about 10-fold more potent in displacing the binding of [^{3}H]SP from rat submaxillary glands (Ki 161 nM) than from rat brain (Ki 1157 nM) and about threefold more potent in guinea pig ileum (Ki 41 nM) than guinea pig brain (Ki 142 nM). A similar affinity of RP67,580 for NK-1 receptors in rat brain (Ki 3.3 nM) and in submaxillary glands (Ki 2.9 nM) (Fardin et al., 1992) was also reported.

3. Additional Intraspecies Differences Detected by Using Agonists

Beresford et al. (1992) have reported that CP96,345 is more potent in antagonizing endothelium-dependent relaxation produced by SPOMe in rabbit thoracic aorta than contractions produced in the rabbit iris sphincter muscle, and that the same profile occurs with the peptide antagonist GR82,334 (Table 9).

Recently, Petitet et al. (1992) have shown evidence for the possible existence of a new TK receptor in the guinea pig ileum. They found that various NK-1 receptor agonists, like SP, SPOMe, septide, and [Pro⁹]SP, were equipotent in functional studies (Table 9), but when tested in a binding assay, septide and

Table 8 Differences Between Central and Peripheral NK-1 Receptor in Rat and Guinea Pig Species

Ligand	K_D (nM) for [^{125}I]BH [Sar⁹]SP binding	
	Rat salivary glands	Rat brain
SP	0.25	0.38
[Sar⁹]SP sulfone	0.56	0.79
SPOMe	7.41	613
Kassinin	6.66	335
NKB	11.5	217

Source: Lew et al., 1990.

Table 9 Discrepancies in the Pharmacology of the NK-1 Receptor in the Guinea Pig Ileum Detected by Agonists (Binding (vs.) Functional Responses) and Antagonists (GR71,251 and GR82,334)

	Petitet et al., 1992 Functional studies EC_{50} [nM]	Binding studies IC_{50} [nM]
SP	2.5	1.1
Pro^{9}SP	2.5	1.2
SPOMe	3.5	250
Septide	2.0	250
Agonist	GR71,251 pK_B	
Septide	7.9	
Pro^{9}SP	6.5	
	Ireland et al., 1992 pK_B (vs SPOMe)	pKi (vs [^{3}H]SP)
GR82,334	7.64	5.58

Intraspecies Differences in NK-1 Receptors in Different Rabbit Smooth Muscle Preparations Detected by CP96,345 and GR82,334

	Beresford et al., 1992 pK_B	
	GR82,334	CP96,345
	(vs. SPOMe)	
Rabbit thoracic aorta	7.34	8.81
Rabbit isolated iris sphincter muscle	6.11	7.39

SPOMe were about 250-fold weaker than other agonists, SP and [Pro^9]SP. Furthermore, the NK-1 receptor antagonist GR71,251 was more potent in blocking septide than [Pro^9]SP sulfone in the guinea pig ileum (values reported in Table 9). Beresford et al. (1992) showed that the NK-1 receptor antagonist GR82,334 was 100-fold more potent in antagonizing the functional responses evoked by SPOMe than the binding of [^{3}H]SP in the guinea pig ileum (Table 9).

We have studied the possible existence of a septide-sensitive NK-1 receptor in rat urinary tract (Meini et al., 1993). All the NK-1 antagonists tested, with the rank order of potency RP67,580 > GR82,334 > CP96,345 showed a greater affinity in blocking contractile responses evoked by septide than [Sar^9]SP sulfone (Table 10). None of the synthetic NK-1 receptor selective agonists was affected by L659,877, thus indicating that NK-2 receptors are not involved.

B. NK-2 Receptor

The availability of "monoreceptorial" bioassays, selective agonists (Table 2), and antagonists (Table 3) has represented a crucial point for the recognition of heterogeneity of the NK-2 receptor.

The synthetic selective agonists utilized to characterize the NK-2 receptor are (Table 2): [Nle^{10}]NKA(4–10) (Dion et al., 1987), [βAla^8]NKA(4–10) (Rovero et al., 1989), GR63,349, and [Lys^3,Ser^5,[R-γ-lactam]Leu^9]NKA(3–10) (Hagan et al., 1990).

1. First Evidence for Heterogeneity on NK-2 Tachykinin Receptors

The first evidence of the possible heterogeneity of NK-2-receptor was provided in the guinea pig isolated trachea. McKnight et al. (1987) showed that L659,877, which has a good affinity for NK-2 receptor in the rat isolated vas deferens (pA$_2$ 8), was instead a poor antagonist in the guinea pig trachea (McKnight et al., 1988). They proposed the possible existence of a novel TK receptor type named NK-4. These findings can be reinterpreted now as the first evidence for NK-2-receptor heterogeneity.

2. Heterogeneity Detected by Using Peptide NK-2-Receptor Antagonists

The finding that different tissues recognize competitive antagonists with different affinities suggested the possibility of a pharmacological heterogeneity of the NK-2 receptor (Maggi et al., 1990).

Table 10 Antagonism by (±)CP 96,345, GR 82,334, and RP 67,580 of the Contractile Responses Produced by Septide or [Sar^9]SP Sulfone in the Rat Urinary Bladder

Antagonist		[Sar^9]SP sulfone	Septide
(±)CP 96,345	pK$_B$	<6	6.80 (6.63–6.97)
	slope	—	−1.05 (0.48–1.62)
GR 82,334	pK$_B$	5.93 (5.82–6.04)	7.01 (6.89–7.13)
	slope	−0.83 (0.52–1.14)	−0.79 (0.47–1.12)
RP 67,580	pk$_B$	6.99 (6.77–7.20)	7.57 (7.38–7.76)
	slope	−1.24 (0.69–1.79)	−0.86 (0.51–1.22)

95% confidence limits of pK$_B$ values and of slopes of Schild regression are shown in parentheses. Data were calculated from at least 12 experiments.

Testing of NKA(4–10) analogs bearing substituted D-Trp amino acid (see above) (Rovero et al., 1990) in different receptorial bioassays revealed strikingly different affinities of MEN10,207 and MEN10,376; these peptide antagonists possess a higher affinity for NK-2 receptors that are responsible for the contraction of the endothelium-deprived rabbit pulmonary artery (RPA) than for those expressed in the hamster trachea (HT) (Maggi et al., 1990). On the contrary, the linear peptide antagonist R396 displays an opposite pharmacological profile, having much higher affinity for the NK-2 receptor expressed in the hamster trachea (pA$_2$ 7.63) than that expressed in the rabbit pulmonary artery (pA$_2$ 5.42) (Table 11).

The affinity of other compounds with NK-2-receptor antagonist activity has been measured: the nonpeptide SR48,968 has higher affinity for the RPA than for the HT NK-2 receptor, while the cyclic peptides and pseudopeptides L659,877, MDL29,913, MEN10,573, and MEN10,612 are more potent at the HT than RPA NK-2 receptor (Table 11). Examinations of the rank order of potency of agonists revealed no evidence for receptor heterogeneity in these preparations (Maggi et al., 1990).

Further evidence of a heterogeneity of the NK-2 receptor has been presented by Ireland et al. (1991), who showed that L659,877 and R396 are more potent in rat colon muscolaris mucosae and less potent in guinea pig trachea and rabbit trachea and aorta, concluding that the NK-2 receptor in the latter preparations resembles

Table 11 Affinities of NK-2-Receptor-Selective Antagonists for Tachykinin NK-1, NK-2, and NK-3 Receptors

Antagonist	NK-1 receptor, GPI	NK-2 receptor RPA	NK-2 receptor HT	NK-3 receptor, RPV
MEN 10,207	5.52 (5.3–5.8)	7.89 (7.7–8.1)	5.94 (5.7–6.2)	4.90 (4.7–5.1)
MEN 10,376	5.66 (5.4–5.9)	8.08 (7.8–8.3)	5.64 (5.5–5.8)	Inactive[a]
L 659,877	5.60 (5.2–5.9)	6.72 (6.5–7.0)	7.92 (7.8–8.0)	5.40 (5.1–5.7)
R 396	Inactive[a]	5.42 (5.2–5.6)	7.63 (7.3–7.9)	Inactive[a]
MDL 29,913	5.37 (5.2–5.6)	7.77 (7.6–7.9)	8.65 (8.4–8.8)	Inactive[a]
SR 48,968	Inactive[b]	9.60 (9.4–9.8)	8.50 (8.3–8.6)	Inactive[b]
MEN 10,573	6.37 (5.9–6.8)	7.26 (7.1–7.4)	8.66 (8.5–8.8)	Inactive[b]
MEN 10,612	6.09 (5.7–6.4)	7.37 (7.2–7.5)	9.06 (8.9–9.2)	Inactive[b]

[a]Inactive at 10 μM.
[b]Inactive at 3 μM.
GPI = guinea pig ileum in the presence of atropine, substance P methylester as an agonist; RPA = endothelium-denuded rabbit pulmonary artery, neurokinin A as an agonist; HT = hamster trachea, neurokinin A as an agonist; RPV = rat portal vein, Arg-neurokinin B as an agonist. Affinities are presented as pA$_2$ values; 95% confidence limits are in parentheses.

that described in the rabbit pulmonary artery but is distinct from that found in hamster trachea and rat colon.

3. Heterogeneity Detected by the Differing Pharmacological Behavior (Agonist/Antagonist) of the Pseudopeptide MDL28,564

As pointed out in Sect. I.B, further evidence for NK-2 receptor heterogeneity has been provided by the use of the pseudopeptide analog MDL28,564. Buck et al. (1990) showed that MDL28,564 behaves as a full agonist in certain preparations with NK-2 receptors (e.g., guinea pig trachea) and as a competitive antagonist in others (e.g., rat vas deferens) and proposed the existence of NK-2-receptor subtypes. This pattern of activity has been found in the rabbit pulmonary artery, where MDL28,564 behaves as a full agonist (pD$_2$ 6.86), and in the hamster trachea, where it behaves as a competitive antagonist (pA$_2$ 6.21) (Fig. 2) (Patacchini et al., 1991).

4. Pharmacological Criteria to Distinguish NK-2 Receptors: NK-2a and N-2b

On the basis of the available evidence (Maggi et al., 1990; Buck et al., 1990; Patacchini et al., 1991), two forms of NK-2 receptors can be distinguished: NK-2A receptors recognize the antagonist rank order of potency MEN10,376 (or MEN10,207) > L659,877 > R396 and are also identified by the agonist character of MDL28,564; NK-2B receptors recognize the antagonist rank order of potency L659,877 > R396 > MEN10,376 (or MEN10,207) and are barely stimulated by MDL28,564, which, however, displays antagonist activity.

On the basis of these two criteria, the NK-2 receptor present in different tissues from different species have been classified. From Table 12, it is evident that preparations from rabbit, guinea pig, bovine, and human species fit into the NK-2A classification whereas those belonging to hamster and rat species adapt to the NK-2B. In terms of absolute potency, SR48,968 possesses the highest affinity for the NK-2A receptor (pA$_2$ 9.60) while MEN10,612 appears to be most potent at the NK-2B receptor (pA$_2$ 9.06) (see Table 11).

5. Further Evidence for Intraspecies Heterogeneity

The above-mentioned pharmacological heterogeneity of the NK-2 receptors may be entirely species-related: however, examples of intraspecies heterogeneity have been described as well, suggesting the existence of true receptor subtypes.

Although various smooth muscle preparations from guinea pig have been shown to express NK-2A receptors, guinea pig alveolar macrophages, which are activated through the NK-2 receptor (Brunelleschi et al., 1990a,b), are sensitive to the rank order of antagonist potency R396 (pK$_B$ 10.13) > L659,877 (pK$_B$ 9.27) > MEN10,376 (pK$_B$ 8.8) and to antagonism by MDL28,564, thus indicating they are activated through an NK-2B-like receptor (Brunelleschi et al., 1992).

Furthermore, the coexistence of NK-2A and NK-2B receptors has been shown

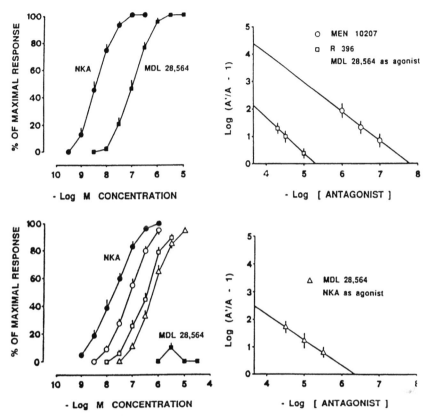

Figure 2 (Top left) Concentration-response curve to neurokinin A (circles) or MDL28,564 (squares) in the rabbit pulmonary artery (RPA). Each value is mean ± SEM of 8–12 experiments. (Top right) Schild's plots showing the antagonism by MEN10,207 (circles) and R396 (squares) of MDL28,564-induced contractions of the RPA. Each value is mean ± SEM of four experiments. (Bottom left) Concentration-response curve to neurokinin A in the hamster trachea (HT) in controls (filled circles, $n = 12$) or in the presence of various concentrations of MDL28,564 (open circles, 3 μM; open squares, 10 μM; open triangles 30 μM, $n = 4$ each). Filled squares indicate the effect of MDL28,564 per se. Each value is mean ± SEM of four to six experiments. (Bottom right) Schild's plot showing the antagonism by MD28,564 of the contractile response to neurokinin A in the HT. Each value is mean ± SEM of four experiments. (From Patacchini et al., 1992.)

Table 12 Pharmacological Criteria for Distinguishing NK-2A and NK-2B
Receptors

NK-2A receptor	NK-2B receptor
1. MEN 10,376 > L659,877 > R396	1. L659,877 > R 396 > MEN 10,376
2. MDL 28,564 agonist	2. MDL 28,564 antagonist
Rabbit pulmonary artery	Hamster trachea
Rabbit vas deferens	Hamster stomach
Rabbit bronchus	Hamster colon
Rabbit trachea	Hamster urinary bladder
Rabbit aorta	Rat colon
Rabbit urinary bladder	Rat urinary bladder
Guinea pig bronchi	Rat vas deferens
Guinea pig trachea	Rat ileum and colon
Guinea pig renal pelvis	Rat stomach
Guinea pig gallbladder	
Human ileum (circular muscle)	
Human colon (circular muscle)	
Human bronchus	
Bovine urinary bladder	
Bovine stomach	

by Nimmo et al. (1992) in the rat urinary bladder by using binding and autoradiographic techniques. They described two populations of binding sites in this tissue,
a minor one expressed at epithelial level (16%) for which MEN10,207 possesses a
high affinity (0.89 nM), and a second one (84%) located on the smooth muscle for
which MEN10,207 shows a lower affinity (0.28 μM).

C. NK-3 Receptor

Contrary to NK-1 and NK-2 receptors, the pharmacology of the NK-3 receptor has
not been well described because of the unavailability of potent and selective
antagonists, although highly selective synthetic agonists such as senktide (Wormser et al., 1986) and [MePhe7]NKB (Regoli et al., 1987) have been described
(Table 2).

The rat isolated portal vein is the only monoreceptorial bioassay for the NK-3
receptor in the periphery described thus far (Mastrangelo et al., 1987), although
contractile responses partly involving NK-3-receptor activation have been described in other organs (guinea pig ileum, rat isolated esophagus, and uterus).

Drapeau et al. (1990) found [Trp7,βAla8]NKA(4–10) to have an antagonist
activity in the rat portal vein (pA$_2$ 7.46). This molecule also has agonist activity at
NK-2 receptors, however.

Stables et al. (1993) presented a novel, conformationally constrained, analog of NKB named GR138,676 (Asp-Met-His-Asp-Phe-D-Pro[R-γ-lactam]-D-Leu-NHCH(CH$_2$OCH$_2$Ph)$_2$), which is a NK-3 receptor antagonist. In the rat portal vein, GR138,676 competitively antagonizes senktide-induced contractions with a pK$_B$ of 8.24 and senktide-evoked release of preloaded arachidonic acid from Chinese hamster ovary cells transfected with a human NK-3 receptor gene (pK$_B$ 8.27). However, in addition to having a high affinity for the NK-3 receptor, GR138,676 also possesses a similar antagonist activity at the NK-1 receptor, which limits its usefulness as a research tool to establish the functional roles of NK-3 receptors (Stable et al., 1993).

Recently, evidence of heterogeneity in terms of species differences has been presented in the pharmacology of the NK-3 receptor. Petitet et al. (1993) have observed species differences in NK-3 cortical binding sites by different affinities of the selective NK-2 antagonist SR48,968 and selective NK-3 agonists. In fact, SR48,968 is able to recognize NK-3 receptors in the guinea pig but not in the rat brain (IC$_{50}$ 320 and 28000 nM vs. [^{3}H]senktide, respectively) while [Pro^7]NKB has a 50-fold higher affinity for rat than guinea pig NK-3 cortical binding sites (IC$_{50}$ 14 and 700 nM vs. [^{125}I]BH eledoisin, respectively).

REFERENCES

Appell, K.C., Fragale, B.J., Loscig, J., Singh, S., and Tomczuk, B.E. (1992). Antagonists that demonstrate species differences in neurokinin 1 receptors. *Mol. Pharmacol. 41*: 772–778.

Bailey, S.J., and Jordan, C.C. (1984). A study of [D-Pro^2,D-Phe^7,D-Trp^9]substance P and [D-Trp^{7,9}]substance P as tachykinin partial agonists in the rat colon. *Br. J. Pharmacol. 82*:441–451.

Beresford, I.J.M., Ireland, S.J., Stables, J.M., Stubbs, C.M., Ball, D., Hagan, R.M., and Birch, P.J. (1992). Investigation into differences in tachykinin NK-1 receptors between and within species using a peptide and a non-peptide NK-1 receptor antagonist. *Br. J. Pharmacol. 106*:92P.

Bitar, K.G., Bowers, C.Y., and Coy, D.H. (1991). Effect of SP/bombesin antagonists on the release of growth hormone by GHRP and GHRH. *Biochem. Biophys. Res. Commun. 180*:156–161.

Bjorkroth, U. (1983). Inhibition of smooth muscle contractions induced by capsaicin and electrical transmural stimulation by a substance P antagonist. *Acta Physiol. Scand.* (Suppl. 515):11–16.

Boyce, S., Rupniak, N.M., Williams, A.R., Iversen, S.D., and Hill, R.G. (1993). Non-specific antinociceptive effects of CP-99,994. *Neuropeptides 24*:212.

Brunelleschi, S. (1992). Tachykinin actions on alveolar macrophages. *Neuropeptides 22*: 11–12.

Brunelleschi, S., Vanni, L., Ledda, F., Giotti, A., Maggi, C.A., and Fantozzi, R. (1990). Tachykinins activate guinea-pig alveolar macrophages: involvement of NK-2 and NK-1 receptors. *Br. J. Pharmacol. 100*:417–420.

Brunelleschi, S., Ceni, E., Fantozzi, R., and Maggi, C.A. (1992). Evidence for tachykinin NK-2B-like receptors in guinea-pig alveolar macrophages. *Life Sci. Pharmacol. Lett. 51*:PL177–181.

Buck, S.H., and Shatzer, S.A. (1988). Agonist and antagonist binding to tachykinin peptide NK-2 receptors. *Life Sci. 42*:2701–2708.

Buck, S.H., Harbeson, S.L., Hassmann, C.F., III, Shatzer, S.A., Rouissi, N., Nantel, F., and van Giersbergen, P.L.M. (1990). [Leu^9(CH$_2$NH$_2$)Leu^{10}]-neurokinin A(4–10) (MDL28,564) distinguishes tissue tachykinin peptide NK-2 receptors. *Life Sci. Pharmacol. Lett. 47*:PL37–PL41.

Clerc, F.F., Foucault, F., Ridoux-Silly, L., and Fardin, V. (1992). Affinity and stability of some substance P agonists in rat and guinea-pig: a comparative study in central and peripheral tissues. *Neuropeptides 22*:14.

Dion, S., D'Orleans-Juste, P., Drapeau, G., Rhaleb, N.E., Rouissi, N., Tousignant, C., and Regoli, D. (1987). Characterization of neurokinin receptors in various isolated organs by the use of selective agonists. *Life Sci. 41*:2269–2278.

Dion, S., Rouissi, N., Nantel, F., Jukic, D., Rhaleb, N.E., Tousignant, C., Telemaque, S., Drapeau, G., Regoli, D., Naline, E., Advenier, C., Rovero, P., and Maggi, C.A. (1990). Structure- activity study of neurokinins. Antagonists for the NK-2 receptor. *Pharmacology 41*:184–194.

Drapeau, G., Rouissi, N., Nantel, F., Rhaleb, N.E., Tousignant, C., and Regoli, D. (1990). Antagonists for the neurokinin NK-3 receptor evaluated in selective receptor systems. *Regul. Peptides 31*:125–135.

Emonds-Alt, X., Vilain, P., Goulaouic, P., Proietto, V., Van Broeck, D., Advenier, C., Naline, E., Neliat, G., Le Fur, G., and Breliere, J.C. (1992). A potent and selective nonpeptide antagonist of the neurokinin A (NK-2) receptor. *Life Sci. Pharmacol. Lett. 50*:PL101–106.

Emonds-Alt, X., Doutremepuich, J.D., Jung, M., Proietto, E., Santucci, V., Van Broeck, D., and Vilain, P. (1993). SR140333, a non-peptide antagonist of substance-P (NK-1) receptor. *Neuropeptides 24*:231.

Erspamer, V. (1981). The tachykinin peptide family. *Trends Neurosci. 4*:269–297.

Fabregat, I., and Rozengurt, E. (1990). [D-Arg^1,D-Phe^5,D-Trp^{7,9}Leu^{11}]substance P, a neuropeptide antagonist, blocks binding, calcium mobilizing and mitogenic effects of endothelin and vasoactive intestinal contractor in mouse 3T3 cells. *J. Cell Physiol. 145*: 88–94.

Fardin, V., and Garret, C. (1991). Species differences between [^{3}H] substance P binding in rat and guinea-pig shown by the use of peptide agonists and antagonists. *Eur. J. Pharmacol. 201*:231–234.

Fardin, V., Flamand, O., Foucault, F., Jolly, A., Bock, M.D., and Garret, C. (1992). RP67,580, a nonpeptide substance P antagonist: comparison with other antagonists and affinities in different animal species. *Neuropeptides 22*:22.

Folkers, K., Hakanson, R., Horig, H., Xu, J.C., and Leander, S. (1984). Biological evaluation of SP antagonists. *Br. J. Pharmacol. 83*:449–456.

Fujii, T., Murai, M., Morimoto, H., Hagiwara, D., Miyake, H., and Matsuo, A. (1992). Effect of novel SP antagonist, FK 888 on airway constriction and airway edema in guinea-pigs. *Neuropeptides 22*:24.

Garret, C., Carruette, A., Fardin, V., Moussaoui, S., Peyronel, J.F., Blanchard, J.C., and Laduron, P.M. (1991). Pharmacological properties of a potent and selective nonpeptide substance P antagonist. *Proc. Natl. Acad. Sci. USA 88*:10208–10211.

Gerard, N.P., Bao, L., Xiao-Ping, H., and Gerard, C. (1993). Molecular aspects of the tachykinin receptors. *Regul. Peptides 43*:21–35.

Gitter, B.D., Waters, D.C., Bruns, R.F., Mason, N.R., Nixon, J.A, and Howbert, J.J. (1991). Species differences in affinities of nonpeptide antagonists for substance P receptors. *Eur. J. Pharmacol. 197*:237–238.

Giuliani, S., Lecci, A., and Maggi, C.A. (1993). Tachykinin receptor antagonists and non-cholinergic activation of motility in the guinea-pig proximal colon. *Neuropeptides 22*:26.

Gu, X.H., Casley, D.J., and Nayler, W.G. (1991). The inhibitory effect of [D-Arg^1,D-Phe^6,D- Trp^{7,9},Leu^{11}]substance P on endothelin-1 binding sites in rat cardiac membranes. *Biochem. Biophys. Res. Commun. 179*:130–133.

Guard, S., and Watson, S.P. (1991). Tachykinin receptor types: classification and membrane signalling mechanisms. *Neurochem. Int. 18*:149–165.

Hagan, R.M., Ireland, S.J., Jordan, C.C., Bailey, F., Stephens-Smith, M., and Ward, P. (1989). Novel, potent and selective agonists at NK-1 and NK-2 receptors. *Br. J. Pharmacol. 98*:71P.

Hagan, R.M., Ireland, S.J., Jordan, C.C., Beresford, I.J.M., Stephens-Smith, M., Ewan, G., and Ward, P. (1990). GR71,251 a novel, potent and highly selective antagonist at neurokinin NK-1 receptor. *Br. J. Pharmacol. 99*:62P.

Hagan, R.M., Ireland, S.J., Bailey, F., McBride, C., Jordan, C.C., and Ward, P. (1991). A spirolactam conformationally-constrained analogue of physalaemin which is a peptidase resistant, selective NK-1 receptor agonist. *Br. J. Pharmacol. 102*:168P.

Hakanson, R., Horig, J., and Leander, S. (1982). The mechanism of action of a substance P antagonist [D-Pro^2,D-Trp^{7,9}]substance P. *Br. J. Pharmacol. 77*:697–700.

Harbeson, S.L., Buck, S.H., Hassmann, C.F., III, and Shatzer, S.A. (1990). Synthesis and biological activity of [Y(CH$_2$NH)] analogs of neurokinin A(4–10). In: *Peptides, Chemistry, Structure and Biology*. (Rivier, J., and Marshall, G.R., eds.). ESCOM, Leiden, pp. 180–181.

Henry, J.L. (1987). Discussion of nomenclature for tachykinins and tachykinin receptors. In: *Substance P and Neurokinins*. (Henry, J. L., Couture, R., Cuello, A.C., Pelletier, G., Quirion, R., and Regoli, D., eds.). Springer-Verlag, New York, p. xvii.

Hokfelt, T., Vincent, S., Hellsten, L., Rosell, S., Folkers, K., Marley, K., Goldstein, M., and Cuello, C. (1981). Immunohistochemical evidence for a 'neurotoxic' action of [D-Pro^2,D-Trp^{7,9}]substance P an analogue with substance P antagonistic activity. *Acta Physiol. Scand. 113*:571–573.

Ireland, S.J., Bailey, F., Cook, A., Hagan, R.M., Jordan, C.C., and Stephens-Smith, M. (1991). Receptors mediating tachykinin-induced contractile responses in guinea-pig trachea. *Br. J. Pharmacol. 103*:1463–1469.

Jensen, R.T., Jones, S.W., Folkers, K., and Gardner, J.D. (1984). A synthetic peptide that is a bombesin receptor antagonist. *Nature 309*:61–63.

Jensen, R.T., Heinz-Erian, P., Mantey, S., Jones, S.W., and Gardner, J.D. (1988). Characterization of the ability of various substance P antagonists to inhibit action of bombesin. *Am. J. Physiol. 254*:G883–G890.

Jukic, D., Rouissi, N., Laprise, R., Boussougou, M., and Regoli, D. (1991). Neurokinin receptors antagonists: old and new. *Life Sci. 49*:1463–1469.

Jung, M., Poncelet, M., Emonds-Alt, X., Le Fur, G., and Soubrie, P. (1993). Neuropharmacological profile of SR140,333, and non-peptide antagonist of substance-P (NK1) receptor. *Neuropeptides 22*:233.

Leander, S., Hakanson, R., Rosell, S., Folkers, K., Sundler, F., and Tornqvist, K. (1981). A specific substance P antagonist blocks smooth muscle contractions induced by noncholinergic nonadrenergic nerve stimulation. *Nature 294*:467–469.

Lecci, A., Giuliani, S., Patacchini, R., Viti, G., and Maggi, C.A. (1991). Role of NK-1 tachykinin receptors in thermonociception: effect of (±)-CP96,345, a non peptide substance P antagonist on the hot plate test in mice. *Neurosci. Lett. 129*:299–302.

Lembeck, F., Donnerer, J., Tsuchiya, M., and Nagahisa, A. (1992). The non peptide tachykinin antagonist, CP-96,345 is a potent inhibitor of neurogenic inflammation. *Br. J. Pharmacol. 105*:527–530.

Lew, R., Geraghty, D.P., Drapeau, G., Regoli, D., and Burcher, E. (1990). Binding characteristics of $[^{125}I]$Bolton-Hunter $[Sar^9,Met(O_2)^{11}]$ substance P, a new selective radioligand for the NK-1 receptor. *Eur. J. Pharmacol. 184*:97–108.

Longmore, J., Neal, R., Razzaque, Z., and Hill, R.G. (1992). Characterization of neurokinin receptors in guinea-pig urinary bladder smooth muscle: use of selective neurokinin antagonists. *Neuropeptides 22*:41–42.

Maggi, C.A., and Patacchini, R. (1992). Tachykinin NK-1 receptor in the guinea-pig isolated proximal urethra: characterization by receptor selective agonists and antagonists. *Br. J. Pharmacol. 106*:888–892.

Maggi, C.A., Patacchini, R., Giuliani, S., Rovero, P., Dion, S., Regoli, D., Giachetti, A., and Meli, A. (1990). Competitive antagonists discriminate between NK-2 tachykinin receptor subtypes. *Br. J. Pharmacol. 100*:588–592.

Maggi, C.A., Giuliani, S., Ballati, L., Lecci, A., Manzini, S., Patacchini, R., Renzetti, A.R., Rovero, P., Quartara, L., and Giachetti, A. (1991). In vivo evidence for tachykininergic transmission using a new NK-2 receptor selective antagonist, MEN 10376. *J. Pharmacol. Exp. Ther. 257*:1172–1178.

Maggi, C.A., Patacchini, R., Rovero, P., and Giachetti, A. (1993a). Tachykinin receptors and receptor antagonists. *J. Autonom. Pharmacol. 13*:23–93.

Maggi, C.A., Patacchini, R., Giuliani, S., and Giachetti, A. (1993b). In vivo and in vitro pharmacology of SR48,968, a non-peptide tachykinin NK_2 receptor antagonist. *Eur. J. Pharmacol. 234*:83–90.

Maggio, J.E. (1988). Tachykinins. *Annu. Rev. Neurosci. 11*:13–21.

Mastrangelo, D., Mathison, R., Huggel, H.J., Dion, S., Rovero, P., and Regoli, D. (1987). The rat portal vein: a preparation sensitive to neurokinins, particularly neurokinin B. *Eur. J. Pharmacol. 134*:321–326.

McLean, S., Ganong, A., Seymour, P.A., Snider, R.M., Desai, M.C., Rosen, T., Bryce, D.K., Longo, K.P., Reynolds, L.S., Robinson, G., Schmidt, A.W., Siok, C., and Heym, J. (1992). Pharmacology of CP-99,994; a nonpeptide antagonist of the tachykinin NK1 receptor. *Regul. Peptides* (Suppl. 1):S120.

McKnight, A.T., Maguire, J.J., Williams, B.J., Foster, A.C., Tridgett, R., and Iversen, L.L. (1988). Pharmacological specificity of synthetic peptides as antagonists at tachykinin receptors. *Regul. Peptides 22*:127.

Meini, S., Patacchini, R., and Maggi, C.A. (1993). Tachykinin NK1 receptor subtypes in the rat urinary bladder. *Br. J. Pharmacol. 108*:177P.

Mizrahi, J., Escher, E., D'Orleans-Juste, P., and Regoli, D. (1984). Undeca- and octapeptide antagonists for substance P, a study of the guinea-pig trachea. *Eur. J. Pharmacol. 99*: 193–202.

Morimoto, H., Murai, M., Maeda, Y., Hagiwara, D., Miyake, H., Matsuo, M., and Fujii, T. (1992). FR 113,680: a novel tripeptide substance P antagonist with NK-1 receptor selectivity. *Br. J. Pharmacol. 106*:123–126.

Nimmo, A., Carstairs, J.R., Maggi, C.A., and Morrison, J.F.B. (1992). Evidence for the coexistence of multiple NK-2 tachykinin receptor subtypes in rat bladder. *Neuropeptides 22*:48.

Otsuka, M., and Yanagisawa, M. (1988). Effect of a tachykinin antagonist on a nociceptive reflex in the isolated spinal cord-tail preparation of the newborn rat. *J. Physiol (Lond.) 395*:255–270.

Oury-Donat, F., Lefevre, I.A., Emonds-Alt, X., Le Fur, G., and Soubrie, P. (1993). SR140333, a novel and potent antagonist of the NK1 receptor: characterisation on the U373MG cell line. *Neuropeptides 24*:233.

Patacchini, R., Astolfi, M., Quartara, L., Rovero, P., Giachetti, A., and Maggi, C.A. (1991). Further evidence for the existence of NK-2 tachykinin receptor subtypes. *Br. J. Pharmacol. 104*:91–96.

Patacchini, R., Santicioli, P., Astolfi, M., Rovero, P., Giachetti, A., and Maggi, C.A. (1992). Activity of peptide and non-peptide antagonists at peripheral NK-1 tachykinin receptors. *Eur. J. Pharmacol. 215*:93–98.

Petitet, F., Saffroy, M., Torrens, Y., Lavielle, S., Chassaing, G., Loeuillet, D., Glowinski, J., and Beaujoan, J.C. (1992). Possible existence of a new tachykinin receptor subtype in the guinea-pig ileum. *Peptides 13*:383–388.

Petitet, F., Beaujouan, J.-C., Saffroy, M., Torrens, Y., and Glowinski, J. (1993). The nonpeptide NK-2 antagonist SR48968 is also a NK-3 antagonist in the guinea pig but not in the rat. *Biochem. Biophys. Res. Commun. 191*:180–187.

Post, C., Butterworth, J.F., Strichartz, G.R., Karlsson, J.A., and Persson, C.G.A. (1985). Tachykinin antagonists have potent local anaesthetic actions. *Eur. J. Pharmacol. 117*: 347–354.

Quartara, L., Patacchini, R., Giachetti, A., and Maggi, C.A. (1992). Novel cyclic pseudopeptides with high affinity for tachykinin NK-2 receptor. *Br. J. Pharmacol. 107*:473P.

Regoli, D., Mizrahi, J., D'Orleans-Juste, P., Dion, S., Drapeau, G., and Escher, E. (1985). Substance P antagonists showing some selectivity for different receptor types. *Eur. J. Pharmacol. 109*:121–125.

Regoli, D., Drapeau, G., Dion, S., and D'Orleans-Juste, P. (1987). Pharmacological receptors for substance P and neurokinins. *Life Sci. 40*:109–117.

Rodriguez, R.E., Salt, T.E., Cahusa, P.M.B., and Hill, R.G. (1993). The behavioural effects of intrathecally administered [D-Pro^2,D-Trp^{7,9}]substance P, an analogue with presumed antagonist actions in the rat. *Neuropharmacology 22*:173–176.

Rosell, S., and Folkers, K. (1982). Substance P antagonists: a new type of pharmacological tool. *Trends Pharmacol. Sci. 3*:211–212.

Rovero, P., Pestellini, V., Rhaleb, N.E., Dion, S., Rouissi, N., Tousignant, C., Telemaque,

S., Drapeau, G., and Regoli, D. (1989). Structure-activity studies of neurokinin A. *Neuropeptides 13*:263–270.

Rovero, P., Pestellini, V., Maggi, C.A., Patacchini, R., Regoli, D., and Giachetti, A. (1990). A highly selective NK-2 tachykinin receptor antagonist containing D-tryptophan. *Eur. J. Pharmacol. 175*:113–115.

Rovero, P., Quartara, L., Astolfi, M., Patacchini, R., Giachetti, A., and Maggi, C.A. (1991). Structure-activity study of the C-terminals residue of MEN10,207 tachykinin antagonist. *Peptides 13*:207–208.

Rubino, A., Thomann, H., Henlin, J.M., Schilling, W., and Criscione, L. (1992). Endothelium-dependent relaxant effect of neurokinins on rabbit aorta is mediated by the NK-1 receptor. *Eur. J. Pharmacol. 212*:237–240.

Santicioli, P., Giuliani, S., and Maggi, C.A. (1993). Failure of L-nitroarginine, a nitric oxide synthase inhibitor, to affect hypotension and plasma protein extravasation produced by tachykinin NK-1 receptor activation in rats. *J. Auton. Pharmacol. 13*:183–190.

Schmidt, A.W., McLean, S., and Heym, J. (1992). The substance P receptor antagonist CP96,345 interacts with Ca^{2+} channels. *Eur. J. Pharmacol. 219*:491–492.

Snider, R.M., Constantine, J.W., Lowe, J.A., III, Longo, K.P., Lebel, W.S., Woody, H.A., Drozda, S.E., Desai, M.C., Vinick, F.J., Spencer, R.W., and Hess, H.J. (1991). A potent nonpeptide antagonist of the substance P (NK-1) receptor. *Science 251*:435–437.

Stables, J.M., Arkinstall, S., Beresford, I.J.M., Seale, P., Ward, P., and Hagan, R.M. (1993). A novel peptidic tachykinin antagonist which is potent at NK_3 receptors. *Neuropeptides 24*:232.

Takeda, Y., and Krause, J.E. (1989). γ-Preprotachykinin-(72-79)-peptide amide potentiates substance P-induced salivation. *Eur. J. Pharmacol. 161*:267–271.

van Giersbergen, P.L.M., Shatzer, S.A., Henderson, A.K., Lai, J., Nakanishi, S., Yamamura, H.I., and Buck, S.H. (1991). Characterization of a tachykinin peptide NK-2 receptor transfected into murine fibroblast B82 cell. *Proc. Natl. Acad. Sci. USA 88*: 1661–1665.

van Giersbergen, P.L.M., Shatzer, S.A., Burcher, E., and Buck, S.H. (1992). Comparison of the effects of neuropeptide K and neuropeptide γ with neurokinin A at NK_2 receptors in the hamster urinary bladder. *Naunyn-Schmiedeberg's Arch. Pharmacol. 345*:51–56.

von Euler, U.S., and Gaddum, J.H. (1931). An unidentified depressor substance in certain tissue extracts. *J. Physiol. 72*:74–87.

Watson, S.P. (1983). Pharmacological characterization of a substance P antagonist [D-Arg^1,D-Pro^2,D-Trp$^{7.9}$,Leu^{11}]substance P. *Br. J. Pharmacol. 80*:205–209.

Wormser, U., Laufer, R., Hart, Y., Chorev, M., Gilon, C., and Selinger, Z. (1986). Highly selective agonists for substance P receptor subtypes. *EMBO J. 5*:2805–2808.

DISCUSSION

McDonald: Species differences in the NK-1 receptor are known to exist. Is there evidence of subtypes of NK-1 receptors within a given species?

Meini: The proposed existence of a septide-sensitive receptor seems to be independent from species-dependent heterogeneity, since evidence has been presented in guinea pig and in rat species. No differences have been noted in functional studies either.

Lee: Variations in ligand binding may be due to minor variations in amino acid sequences between species. For example, the affinity of binding to the serotonin receptor varies between species owing to single amino acid variations.

Manzini: At the present time, is it possible to affirm the presence of NK-2A and NK-2B receptor subtypes?

Meini: Subclassification of the NK-2 receptor into two subtypes seems to be species dependent. Nevertheless, some evidence demonstrating their coexistence in the same species have been presented: i.e., the guinea pig species expressing NK-2A receptor, has been shown to express NK-2B receptor on alveolar macrophages, while in the rat urinary bladder, both receptors seem to be present, at epithelial or at the smooth muscle level.

Kaliner: Can you please briefly summarize the clinical trials involving neuropeptide antagonists? Has asthma been studied?

Meini: The only asthma study that has been published involved the inhalation of FK-224, a cyclopeptide antagonist at both NK-1 and NK-2 receptors isolated from *Streptomyces violaceusniger*, followed by challenge with BK. FK-224 inhibited both the bronchospasm and cough usually elicited by BK.

Solway: In preliminary studies, we have found that CP-99994 causes a rightward shift of the dose-response to a substance P agonist in guinea pigs in vivo. This effect does not exhibit a dose-response above 300 μg/kg given intravenously. This could be interpreted to indicate a second NK-1 receptor, which is poorly antagonized by CP-99994.

13

Pulmonary Peptidases: General Principles of Peptide Metabolism and Molecular Biology of Angiotensin-Converting Enzyme, Neutral Endopeptidase 24.11, and Carboxypeptidase M

Randal A. Skidgel
University of Illinois College of Medicine, Chicago, Illinois

I. GENERAL PRINCIPLES OF PEPTIDE METABOLISM

Peptides are important mediators and modulators of many physiological and pathophysiological processes. Essentially all peptides are synthesized as large precursor propeptides that are enzymatically processed to smaller biologically active forms, generally ranging in length from 3 to 30 amino acids. The biological activities of these peptides are mediated via interaction with specific membrane receptors that stimulate a variety of second-messenger responses. Because they are highly potent, with significant activity at concentrations of 10^{-9} M and below, their actions are tightly controlled, primarily by peptidases—enzymes that hydrolyze one or more peptide bonds in the molecule. This limits the actions of these potent agents and thereby protects the organism from the deleterious and possibly lethal consequences of high levels of peptide hormones. However, peptide hydrolysis does not always lead to inactivation—in some cases alteration of peptide specificity (e.g., a metabolite binding to a different receptor) or even activation can result (e.g., conversion of angiotensin I to II).

A. Specificity of Peptidases

Although there has been a tendency to name peptidases after a specific substrate [e.g., angiotensin-converting enzyme (ACE), enkephalinase, etc.], these enzymes are not specific for a particular peptide, but rather recognize amino acids around the potential cleavage site. As a consequence, most enzymes cleave a variety of peptide substrates. For example, not only does ACE convert angiotensin I to II, it also cleaves enkephalins and other opioid peptides, substance P, fMet-Leu-Phe, bradykinin, neurotensin, LH-RH, and others (Skidgel and Erdös, 1993). The neutral endopeptidase 24.11 (NEP), because of its broad specificity (cleaving at the amino side of hydrophobic amino acids), hydrolyzes over 30 different peptides in vitro, including tachykinins, endothelins, ANF, enkephalins, angiotensin I and II, and bradykinin (Erdös and Skidgel, 1989). Conversely, a peptide can also be degraded by a variety of different peptidases, at least in vitro. For example, substance P is a substrate for peptidases such as prolylendopeptidase, dipeptidyl peptidase IV, endopeptidase 24.15, NEP, ACE, deamidase, calpain, chymase, and cathepsin G (Fig. 1).

B. Classification of Peptidases

Peptidases can generally be divided into two groups: endopeptidases and exopeptidases. Endopeptidases usually cleave an internal peptide bond whereas exopeptidases remove one or two amino acids at a time from the N- or C-terminus of a peptide. Prolylendopeptidase, endopeptidase 24.15, cathepsin G, chymase, and NEP would be considered endopeptidases while dipeptidyl peptidase IV and deamidase are exopeptidases. However, the metabolism of substance P illustrates some exceptions to these rules. For example, exopeptidases such as carboxypep-

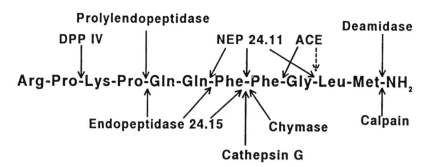

Figure 1 Hydrolysis of substance P by various peptidases. The primary sites of cleavage for each peptidase are shown by the arrows. NEP, neutral endopeptidase; ACE, angiotensin-converting enzyme; DPP IV, dipeptidyl peptidase IV.

tidases (deamidase) or peptidyl dipeptidases (ACE) usually require the presence of a free C-terminal amino acid in their peptide substrates. Nevertheless, both deamidase and ACE cleave substance P, which contains a blocked (amidated) C-terminal methionine (Fig. 1). In fact, deamidase was named for its ability to deamidate substance P and other tachykinins although it does cleave other peptides (angiotensin I, bradykinin, endothelin) by carboxypeptidase action (Jackman et al., 1990, 1992). ACE carries out an unusual cleavage, releasing the C-terminal tripeptide in contrast to most substrates (e.g., angiotensin I, bradykinin, enkephalins) where it removes the free C-terminal dipeptide (Skidgel and Erdös, 1993). In addition, although calpain was initially thought to be an endoprotease that cleaves only large protein substrates, it was subsequently found to hydrolyze a variety of small peptides and deamidate several blocked peptides including substance P (Hatanaka et al., 1985). Thus, although general rules can be established regarding the specificity of peptidases and their preferred cleavage sites, the actual pattern of hydrolysis must be determined experimentally with each potential peptide substrate.

C. Peptide Metabolism In Vivo

Although the specificity of peptidases and their potential peptide substrates can be determined in vitro, the ability of a particular enzyme to carry out this function in vivo will depend on its tissue and subcellular (e.g., cytosolic vs. membrane-bound) localization and its accessibility to a given peptide at that site. An illustration of this principle is given in Figure 2, which shows the localization of pulmonary enzymes that can inactivate bradykinin, a peptide known to cause bronchoconstriction. If bradykinin is generated in or injected into the bloodstream, plasma peptidases such as carboxypeptidase N (or kininase I) and endothelial membrane-bound enzymes such as ACE may inactivate a substantial portion of it before it reaches the alveoli and bronchioles. However, if bradykinin is administered by aerosol or generated in the airways, ACE or carboxypeptidase N would not have access to it. At these sites, it may be inactivated by membrane-bound carboxypeptidase M and/or NEP. Thus, inhibition of one enzyme may not significantly alter the response to the peptide, depending on the route of administration. This was the case in studies of the bronchoconstrictor response to bradykinin administered i.v. to guinea pigs (Chodimella et al., 1991). Although carboxypeptidase M and N readily inactivate bradykinin in vitro, administration of the carboxypeptidase inhibitor DL-2-mercaptomethyl-3-guanidinoethylthiopropanoic acid (MGTA) did not enhance the responsiveness of the guinea pigs to bradykinin, presumably because of the continued inactivation by ACE and NEP. However, when both ACE and NEP were inhibited with their respective specific inhibitors captopril and phosphoramidon (which increased responsiveness to bradykinin), MGTA caused a further enhancement of bronchoconstriction by the peptide (Chodimella et al., 1991).

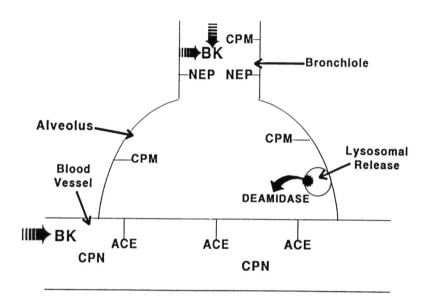

Figure 2 Schematic diagram showing the pulmonary localization of peptidases that can cleave peptides such as bradykinin. Bradykinin (BK) in the bloodstream can be inactivated by plasma carboxypeptidase N (CPN) or membrane-bound endothelial angiotensin-converting enzyme (ACE). Bradykinin in the airways or alveoli is cleaved by neutral endopeptidase 24.11 (NEP) or carboxypeptidase M (CPM). In inflammatory conditions or other disease states, lysosomal enzymes (e.g., deamidase) released from lung cells may also contribute to peptide degradation.

Another factor that must be taken into account when considering in vivo peptide metabolism is the effect of pathological processes on the levels of peptidases at a particular site of action. For example, inflammatory conditions may result in the release of lysosomal enzymes (e.g., deamidase) that could rapidly degrade peptides such as bradykinin or substance P (Fig. 2). Conversely, the levels of peptidases normally present could decrease under certain circumstances. For example, it has been shown that viral infection and lung injury caused by toluene diisocyanate or cigarette smoke decrease the levels of NEP in the lung (Nadel, 1992). This leads to a decreased degradation and enhanced responsiveness to its peptide substrates such as substance P (Nadel, 1992).

II. STRUCTURE AND MOLECULAR BIOLOGY OF PEPTIDASES

The recent increase in the use of molecular biological techniques has resulted in the cloning and sequencing of many different peptidases. The brief review that follows will focus on three enzymes that are highly expressed in the lung—ACE, NEP, and carboxypeptidase M.

A. Angiotensin-Converting Enzyme

Biochemical studies on ACE purified from various sources indicated that the enzyme has one active site based on the finding of one zinc atom and one inhibitor binding site per molecule (Skidgel and Erdös, 1993). It was thus surprising when Soubrier and colleagues (1988) reported that the full cDNA sequence of human endothelial (or somatic) ACE coded for a large protein of 1277 amino acids whose sequence contained two potential active sites located within two homologous domains (Fig. 3). Subsequent cloning of the shorter, testicular form of human ACE showed it contained only the C-terminal domain active site (Lattion et al., 1989) and was derived from the same gene as somatic ACE by use of a unique promoter in intron 12 that is normally spliced out (Hubert et al., 1991). These data indicate that somatic ACE arose via gene duplication and that testicular ACE represents the ancestral form of the enzyme.

The functional significance of the two active-site domains was not immediately apparent, and the possibility remained that only one active site was operative as the activities of the somatic and testicular forms of ACE were similar. However, recent investigations have revealed that ACE indeed has two zinc and two inhibitor binding sites per molecule (Ehlers and Riordan, 1991; Wei et al., 1992). Conclusive proof of the functionality of both domains was obtained by use of recombinant DNA technology. Wei et al. (1991a) transfected CHO cells with either truncated ACE cDNAs containing only one active site or the full-length cDNA containing point mutations that inactivated one of the sites. They found that both domains are independently functional and have similar K_m values with most substrates. However, the turnover number (k_{cat}) is about three- to 10-fold higher for the C-terminal domain than for the N-terminal domain (Wei et al., 1991a). The N-terminal domain is also less dependent on the presence of chloride ion than the C-terminal domain and is also largely responsible for the ability of ACE to carry out the unusual cleavage of the N-terminal tripeptide of LH-RH (Jaspard et al., 1993). Differences between the two domains indicate that each has unique structural characteristics that could play a role in determining their functions in vivo.

The overall sequence of ACE is not similar to that of any other known protein, despite enzymatic characteristics that are similar to carboxypeptidase A (whose structure was a model for the development of ACE inhibitors) and the presence of a consensus zinc-binding motif (HEXXH) at the two active sites (HEMGH; Fig. 3) that is found in many metallopeptidases, including NEP (Erdös and Skidgel, 1989; Soubrier et al., 1988). Within this region, the two histidine residues are involved in zinc binding and the glutamic acid plays a crucial role in catalysis, acting as a

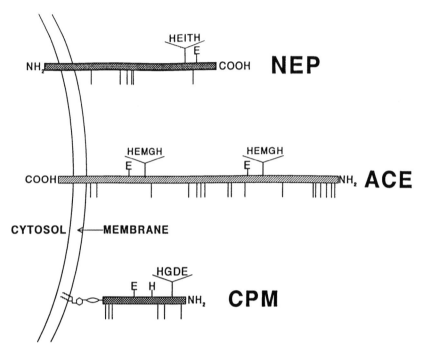

Figure 3 Diagrammatic representation of the primary structure and membrane anchoring of neutral endopeptidase 24.11 (NEP), angiotensin-converting enzyme (ACE), and carboxypeptidase M (CPM). All three enzymes are membrane-bound ectopeptidases (the majority of the protein and the active sites are extracellular) and are anchored by different mechanisms. NEP is bound through an N-terminal uncleaved hydrophobic signal peptide. ACE contains a hydrophobic transmembrane-spanning region near its C-terminus. CPM is covalently linked at its C-terminus to a phosphatidylinositol-glycan anchor. The locations of some of the catalytically important residues are shown in single-letter code (H = histidine, E = glutamic acid, I = isoleucine, T = threonine, M = methionine, G = glycine, D = aspartic acid). Sites for potential asparagine-linked glycosylation are denoted by vertical lines below the structures. For further details, see text.

general base in triggering the nucleophilic attack of water on the carbonyl group of the scissile peptide bond (Erdös and Skidgel, 1989; Soubrier et al., 1988). Another glutamic acid, about 30 residues C-terminal to the active site region, is probably the third zinc-binding residue (Fig. 3).

The deduced protein sequence of ACE contains 17 potential Asn-linked glycosylation sites (Fig. 3), consistent with the glycoprotein nature of the molecule determined in numerous studies of the protein itself (Skidgel and Erdös, 1993).

ACE is a membrane-bound enzyme and must be solubilized for purification with either detergent or by proteolysis with trypsin. Studies on the protein indicated that the enzyme is anchored to the cell membrane by a small anchor peptide located at the C-terminus (Skidgel and Erdös, 1993). The deduced protein sequence is consistent with this notion, containing only one hydrophobic region in the mature protein near the C-terminus. This was proven conclusively in recent studies in which mutant recombinant ACE lacking the C-terminal hydrophobic region was not membrane-bound, but rather secreted in soluble form from the transfected cells (Wei et al., 1991b).

B. Neutral Endopeptidase 24.11

The cloning and sequencing of NEP cDNA from rat, rabbit, and human sources revealed that the enzyme has been highly conserved in mammals (Malfroy et al., 1987; Devault et al., 1987; Malfroy et al., 1988). Of the 742 amino acids coded for by the human NEP cDNA, there are only six nonconservative changes in the sequence when compared with the rat enzyme (Malfroy et al., 1988). The rat and human enzymes have six potential Asn-linked glycosylation sites (Fig. 3) whereas the rabbit enzyme has five. It has been known for many years that the peptide bond specificity and inhibitor sensitivity of NEP are quite similar to those of the bacterial enzyme thermolysin. It was therefore surprising that the sequence of NEP has essentially no identity with thermolysin except for a few amino acids around putative active site residues, including the HEXXH sequence found in other metallopeptidases, as mentioned above (Fig. 3) (Erdös and Skidgel, 1989). By the use of site-directed mutagenesis, many of the important active-site residues in NEP have been identified. These include the zinc-binding residues His[583], His[587], and Glu[646] (Fig. 3), the catalytic Glu[584], and important ligand-binding amino acids Arg[747], Arg[102], Val[573], and His[711] (Erdös and Skidgel, 1989; Devault et al., 1988a,b; Bateman et al., 1989; Vijayaraghavan et al., 1990; Le Moual et al., 1991; Beaumont et al., 1991).

NEP is a transmembrane protein and is anchored via a hydrophobic region near the N-terminus that represents an uncleaved signal peptide (Fig. 3). This was proven by expressing recombinant NEP in which the hydrophobic N-terminal region was replaced by a cleavable signal peptide derived from the signal sequence of pro-opiomelanocortin (Lemay et al., 1989). This mutant NEP was secreted from the transfected cells in soluble form, indicating that the hydrophobic N-terminal portion was indeed responsible for membrane binding (Lemay et al., 1989).

Although NEP, as its name implies, is an endopeptidase, it possesses some characteristics of a peptidyl dipeptidase with certain substrates (such as enkephalins or bradykinin), preferentially cleaving bonds near the C-terminus over those located in the interior of the molecule. This is due to the presence of positively charged Arg[102] in the enzyme that binds to the negatively charged C-terminus in some substrates (Bateman et al., 1989; Beaumont et al., 1991). Beaumont et al. (1992) mutated the Arg^{102} to Glu^{102} in NEP to see whether charge reversal at this site would alter the substrate specificity. They found that Glu^{102} NEP hydrolyzed a C-terminally negatively charged substrate 16-fold slower than the native enzyme whereas cleavage of a positively charged substrate was increased by 29-fold, confirming the role of this active site substituent.

C. Carboxypeptidase M

Carboxypeptidase M was initially discovered in human placental microvilli and other tissues as a membrane-bound enzyme that cleaves C-terminal Arg or Lys from peptides (e.g., bradykinin, enkephalin hexapeptides, anaphylatoxins) at neutral pH (Skidgel et al., 1984; Skidgel, 1988). Recent studies using specific enzyme assays and Northern blot analysis revealed that the lung contains very high levels of the enzyme (Nagae et al., 1993). Immunohistochemical investigations showed it is localized on the surface of alveolar type I cells and in pulmonary macrophages (Nagae et al., 1993).

Purification and characterization of the enzyme demonstrated that it has unique properties, as well as certain characteristics that are similar to those of other "B-type" carboxypeptidases, i.e., carboxypeptidase N (kininase I), carboxypeptidase H (enkephalin convertase), and pancreatic carboxypeptidase B (Skidgel, 1988; Skidgel et al., 1989). The sequence of the enzyme, determined by cloning and sequencing its cDNA from a human placental library, also has similarities to those of other metallocarboxypeptidases, most notably the active subunit of carboxypeptidase N (41% identity) and carboxypeptidase H (43%) (Tan et al., 1989). Although it has much lower similarity to pancreatic carboxypeptidase A or B (15%), most of the active-site residues identified in carboxypeptidases A and B by X-ray crystallography have also been conserved in carboxypeptidase M (Fig. 3) (Tan et al., 1989). Although carboxypeptidase M is a zinc metalloenzyme, as are ACE and NEP, the sequence of the zinc-binding residues is different than that found in these and other metalloendopeptidases (Fig. 3). Thus, instead of the HEXXH motif found in ACE and NEP (consisting of two zinc-binding histidines and the catalytic glutamic acid), carboxypeptidase M has a HGDE sequence where the His^{66} and Glu^{69} are putative zinc-binding residues. The third zinc-binding ligand (His^{173} and the catalytic glutamic acid (Glu^{264}) are quite a distance removed in the primary sequence (Tan et al., 1989). Other putative active site residues in carboxypeptidase M include Arg^{137}, Tyr^{242}, and Gln^{249}, which are likely involved in ligand binding, and Lys^{118}, which may be important in the catalytic step (Tan et al., 1989). Confirmation of these putative functions will require the use of site-directed mutagenesis and expression of recombinant enzyme to assess the consequences of altering these potentially critical residues.

It was known from the initial discovery of the enzyme that carboxypeptidase M is tightly associated with membranes and could only be solubilized with detergent or trypsin (Skidgel et al., 1984, 1989, 1991). However, the deduced protein sequence did not contain a normal membrane-spanning region of 20 or more hydrophobic amino acids usually found in transmembrane proteins. Instead, the enzyme contained a mildly hydrophobic C-terminal sequence of 15 residues, which was a potential signal for attachment to a phosphatidylinositol-glycan (PI-G) membrane anchor (Tan et al., 1989; Skidgel et al., 1991). In this mode of attachment, the hydrophobic C-terminus temporarily anchors the newly synthesized protein in the Golgi membrane and the hydrophobic tail is cleaved off followed by attachment of the new C-terminal amino acid to the ethanolamine residue of a preformed PI-G anchor. In this case, the phosphatidylinositol portion of the anchor is incorporated into the outer leaflet of the membrane bilayer and the protein itself is not directly associated with the membrane (Fig. 3). Biochemical studies carried out on the enzyme in placental microvilli and in Madin Darby canine kidney (MDCK) cells proved that it is indeed attached to the membrane via a PI-G anchor (Skidgel et al., 1991; Deddish et al., 1990). Evidence included the ability of bacterial phosphatidylinositol-specific phospholipase C to release carboxypeptidase M from the membrane in a form that was freely soluble and the labeling of the enzyme in MDCK cells by [^{3}H]-ethanolamine (Skidgel et al., 1991; Deddish et al., 1990).

Carboxypeptidase M is a widely distributed ectoenzyme, present in many organs and tissues where it can participate in a variety of functions including controlling peptide hormone activity at the cell surface, extracellular protein and peptide degradation, and prohormone processing. Our recent findings of significant amounts of carboxypeptidase M in the brain and peripheral nerves (including the vagus) (Nagae et al., 1992) and especially high levels in lung (Nagae et al., 1993) open new avenues for investigation of the role of this enzyme. A recent study by Desmazes et al. (1992) showed that the specific carboxypeptidase inhibitor MGTA enhanced the noncholinergic bronchoconstrictor response to capsaicin and vagal nerve stimulation in guinea pigs. The exact mechanism of this response is not clear but could involve increased levels of one or more peptides that are substrates of carboxypeptidase M. Further studies in this area are clearly warranted.

ACKNOWLEDGMENT

The studies on carboxypeptidase M described here were supported in part by NIH Grant DK41431.

REFERENCES

Bateman, R.C., Jr., Jackson, D., Slaughter, C.A., Unnithan, S., Chai, Y.G., Moomaw, C., and Hersh, L.B. (1989). Identification of the active-site arginine in rat neutral endopeptidase 24.11 (enkephalinase) as arginine 102 and analysis of a glutamine 102 mutant. *J. Biol. Chem. 264*:6151–6157.

Beaumont, A., Le Moual, H., Boileau, G., Crine, P., and Roques, B.P. (1991). Evidence that both arginine 102 and arginine 747 are involved in substrate binding to neutral endopeptidase (EC 3.4.24.11). *J. Biol. Chem. 266*:214–220.

Beaumont, A., Barbe, B., Le Moual, H., Boileau, G., Crine, P., Fournie-Zaluski, M.-C., and Roques, B.P. (1992). Charge polarity reversal inverses the specificity of neutral endopeptidase-24.11. *J. Biol. Chem. 267*:2138–2141.

Chodimella, V., Skidgel, R.A., Krowiak, E.J., and Murlas, C.G. (1991). Lung peptidases, including carboxypeptidase, modulate airway reactivity to intravenous bradykinin. *Am. Rev. Respir. Dis. 144*:869–874.

Deddish, P.A., Skidgel, R.A., Kriho, V.B., Becker, R.P., and Erdös, E.G. (1990). Carboxypeptidase M in cultured Madin-Darby canine kidney cells: evidence that carboxypeptidase M has a phosphatidylinositol glycan anchor. *J. Biol. Chem. 265*: 15083–15089.

Desmazes, N., Lockhart, A., Lacroix, H., and Dusser, D.J. (1992). Carboxypeptidase M–like enzyme modulates the noncholinergic bronchoconstrictor response in guinea pig. *Am. J. Respir. Cell Mol. Biol. 7*:477–484.

Devault, A., Lazure, C., Nault, C., Le Moual, H., Seidah, N.G., Chretien, M., Kahn, P., Powell, J., Mallet, J., Beaumont, A., Roques, B.P., Crine, P., and Boileau, G. (1987). Amino acid sequence of rabbit kidney neutral endopeptidase 24.11 (enkephalinase) deduced from a complementary DNA. *EMBO J. 6*:1317–1322.

Devault, A., Nault, C., Zollinger, M., Fournie-Zaluski, M.-C., Roques, B.P., Crine, P., and Boileau, G. (1988a). Expression of neutral endopeptidase (enkephalinase) in heterologous COS-1 cells. *J. Biol. Chem. 263*:4033–4040.

Devault, A., Sales, V., Nault, C., Beaumont, A., Roques, B., Crine, P., and Boileau, G. (1988b). Exploration of the catalytic site of endopeptidase 24.11 by site-directed mutagenesis. Histidine residues 583 and 587 are essential for catalysis. *FEBS Lett. 231*: 54–58.

Ehlers, M.R.W., and Riordan, J.F. (1991). Angiotensin-converting enzyme: zinc- and inhibitor-binding stoichiometries of the somatic and testis isozymes. *Biochemistry 30*: 7118–7126.

Erdös, E.G., and Skidgel, R.A. (1989). Neutral endopeptidase 24.11 (enkephalinase) and related regulators of peptide hormones. *FASEB J. 3*:145–151.

Hatanaka, M., Sasaki, T., Kikuchi, T., and Murachi, T. (1985). Amidase-like activity of calpain I and calpain II on substance P and its related peptides. *Arch. Biochem. Biophys. 242*:557–562.

Hubert, C., Houot, A.-M., Corvol, P., and Soubrier, F. (1991). Structure of the angiotensin I-converting enzyme gene. Two alternate promoters correspond to evolutionary steps of a duplicated gene. *J. Biol. Chem. 266*:15377–15383.

Jackman, H.L., Tan, F., Tamei, H., Beurling-Harbury, C., Li, X.-Y., Skidgel, R.A., and Erdös, E.G. (1990). A peptidase in human platelets that deamidates tachykinins. Probable identity with the lysosomal "protective protein." *J. Biol. Chem. 265*:11265–11272.

Jackman, H.L., Morris, P.W., Deddish, P.A., Skidgel, R.A., and Erdös, E.G. (1992). Inactivation of endothelin I by deamidase (lysosomal protective protein). *J. Biol. Chem. 267*:2872–2875.

Jaspard, E., Wei, L., and Alhenc-Gelas, F. (1993). Differences in the properties and enzymatic specificities of the two active sites of angiotensin I-converting enzyme (kininase II). Studies with bradykinin and other natural peptides. *J. Biol. Chem. 268*: 9496–9503.

Lattion, A.-L., Soubrier, F., Allegrini, J., Hubert, C., Corvol, P., and Alhenc-Gelas, F. (1989). The testicular transcript of the angiotensin I-converting enzyme encodes for the ancestral, non-duplicated form of the enzyme. *FEBS Lett. 252*:99–104.

Lemay, G., Waksman, G., Roques, B.P., Crine, P., and Boileau, G. (1989). Fusion of a cleavable signal peptide to the ectodomain of neutral endopeptidase (EC 3.4.24.11) results in the secretion of an active enzyme in COS-1 cells. *J. Biol. Chem. 264*:15620–15623.

Le Moual, H., Devault, A., Roques, B.P., Crine, P., and Boileau, G. (1991). Identification of glutamic acid 646 as a zinc-coordinating residue in endopeptidase-24.11. *J. Biol. Chem. 266*:15670–15674.

Malfroy, B., Schofield, P.R., Kuang, W-J., Seeburg, P.H., Mason, A.J., and Henzel, W.J. (1987). Molecular cloning and amino acid sequence of rat enkephalinase. *Biochem. Biophys. Res. Commun. 144*:59–66.

Malfroy, B., Kuang, W-J., Seeburg, P.H., Mason, A.J., and Schofield, P.R. (1988). Molecular cloning and amino acid sequence of human enkephalinase (neutral endopeptidase). *FEBS Lett. 229*:206–210.

Nadel, J.A. (1992). Regulation of neurogenic inflammation by neutral endopeptidase. *Am. Rev. Respir. Dis. 145*:S48–S52.

Nagae, A., Deddish, P.A., Becker, R.P., Anderson, C.H., Abe, M., Tan, F., Skidgel, R.A., and Erdös, E.G. (1992). Carboxypeptidase M in brain and peripheral nerves. *J. Neurochem. 59*:2201–2212.

Nagae, A., Abe, M., Becker, R.P., Deddish, P.A., Skidgel, R.A., and Erdös, E.G. (1993). High concentration of carboxypeptidase M in lungs: presence of the enzyme in alveolar type I cells. *Am. J. Respir. Cell Mol. Biol. 9*:221–229.

Skidgel, R.A. (1988). Basic carboxypeptidases: regulators of peptide hormone activity. *Trends Pharmacol. Sci. 9*:299–304.

Skidgel, R.A., and Erdös, E.G. (1993). Biochemistry of angiotensin converting enzyme. In: *The Renin Angiotensin System. Vol. 1.* (Robertson, J.I.S., Nicholls, M.G., eds.), Gower Medical Publ., London, pp. 10.1–10.10.

Skidgel, R.A., Johnson, A.R., and Erdös, E.G. (1984). Hydrolysis of opioid hexapeptides by carboxypeptidase N. Presence of carboxypeptidase in cell membranes. *Biochem. Pharmacol. 33*:3471–3478.

Skidgel, R.A., Davis, R.M., and Tan, F. (1989). Human carboxypeptidase M. Purification and characterization of a membrane-bound carboxypeptidase that cleaves peptide hormones. *J. Biol. Chem. 264*:2236–2241.

Skidgel, R.A., Tan, F., Deddish, P.A., and Li, X.-Y. (1991). Structure, function and membrane anchoring of carboxypeptidase M. *Biomed. Biochim. Acta 50*:815–820.

Soubrier, F., Alhenc-Gelas, F., Hubert, C., Allegrini, J., John, M., Tregear, G., and Corvol, P. (1988). Two putative active centers in human angiotensin I-converting enzyme revealed by molecular cloning. *Proc. Natl. Acad. Sci. USA 85*:9386–9390.

Tan, F., Chan, S.J., Steiner, D.F., Schilling, J.W., and Skidgel, R.A. (1989). Molecular cloning and sequencing of the cDNA for human membrane-bound carboxypeptidase M. *J. Biol. Chem. 264*:13165–13170.

Vijayaraghavan, J., Kim, Y.-A., Jackson, D., Orlowski, M., and Hersh, L.B. (1990). Use of site-directed mutagenesis to identify valine-573 in the S′1 binding site of rat neutral endopeptidase 24.11 (enkephalinase). *Biochemistry 29*:8052–8056.

Wei, L., Alhenc-Gelas, F., Corvol, P., and Clauser, E. (1991a). The two homologous domains of human angiotensin I-converting enzyme are both catalytically active. *J. Biol. Chem. 266*:9002–9008.

Wei, L., Alhenc-Gelas, F., Soubrier, F., Michaud, A., Corvol, P., and Clauser, E. (1991b). Expression and characterization of recombinant human angiotensin I-converting enzyme. *J. Biol. Chem. 266*:5540–5546.

Wei, L., Clauser, E., Alhenc-Gelas, F., and Corvol, P. (1992). The two homologous domains of human angiotensin I-converting enzyme interact differently with competitive inhibitors. *J. Biol. Chem. 267*:13398–13405.

DISCUSSION

McDonald: Can peptidases mimic peptide receptors in autoradiography studies?

Skidgel: Although peptides bind less avidly to peptidases than to receptors, these interactions can be an important cause of false-positives in binding studies. Protease inhibitors that bind to proteases, but not receptors, can reduce this effect, but competitive inhibitors are not available for some proteases.

Moody: Are peptides more rapidly metabolized by exopeptidases or endopeptidases?

Skidgel: There is no general pattern. For angiotensin I–converting enzyme, the K_m for bradykinin is <1 μM, while it is about 1 mM for enkephalins. The K_m for both peptides with NEP is about 70–120 μM.

Nadel: How important is the stoichiometric ratio of NEP to peptide receptor(s) in determining the effects of peptides?

Skidgel: I believe there will usually be more NEP than receptor on cell surfaces, but electron microscopic immunohistochemical studies will be needed to define this interesting issue.

Elde: Given the great differences in relative affinities for receptors and peptidases, and the low concentrations of peptides in vivo, will peptidases even have a chance to degrade peptides?

Skidgel: There is a big difference between the peptide concentrations in vivo (likely in the subnanomolar range), and peptidase K_m values in vitro (μM). However, peptide release and metabolism are likely restricted to local sites where peptide concentrations are higher. In addition, protease inhibitors have profound effects in vivo, indicating the relevance of proteases.

14

Peptidases in the Respiratory Tract: The Localization and Actions of Carboxypeptidase M

Ervin G. Erdös
University of Illinois College of Medicine, Chicago, Illinois

I. INTRODUCTION

Enzymes that metabolize neuropeptides have long been of interest, ever since their peptide substrates (for example, bradykinin or angiotensin I) became available in synthetic form (Erdös, 1979). The actions of neuropeptides in the lung and their presence in normal and malignant tissues have been studied by many investigators, and some, but definitely not all, the pulmonary cell-types that contain major peptidases have been identified. For example, vascular endothelial cells have angiotensin I converting enzyme (ACE) or kininase II (Erdös, 1979; Johnson and Erdös, 1977; Johnson et al., 1982); bronchial epithelial cells and fibroblasts contain neutral endopeptidase 24.11 (enkephalinase; NEP; Johnson et al., 1985; Nadel, 1992a,b; Martins et al., 1990). Besides these stationary cells, other cell types that migrate into the lungs, such as macrophages, leukocytes, and platelets (Connelly et al., 1985; Holian and Scheule, 1990; Jackman et al., 1990), are rich in peptidases. NEP and cathepsin G in neutrophils (Skidgel et al., 1991), deamidase/cathepsin A in platelets (Jackman et al., 1990), and numerous other enzymes released from macrophages (Holian and Scheule, 1990) can contribute to peptide metabolism in the lung. Many aspects of the functions of these and other enzymes (Desmazes et al., 1992; Barnes et al., 1990; Lazarus et al., 1987; Martins et al., 1990; Said, 1984) are discussed in other chapters.

313

II. CARBOXYPEPTIDASE M

Recently, some of our efforts have been channeled into deciphering the importance of the presence of carboxypeptidase M (CPM) in lungs. This enzyme was discovered in our laboratory (Johnson et al., 1984; Skidgel et al., 1984) and purified and cloned by Skidgel and colleagues (Skidgel et al., 1989; Tan et al., 1989). It is membrane-bound and widely distributed in the body on the plasma membrane of several cell types. At neutral pH, CPM cleaves basic C-terminal amino acids (Deddish et al., 1989; Skidgel, 1988) from, for example, bradykinin, anaphylatoxins, and Arg^6/Lys^6-enkephalins (Skidgel, 1988). Hypothetically, the arginine released from selected substrates could serve as a source for NO. Although plasma carboxypeptidase N and pancreatic carboxypeptidase B can hydrolyze the same peptide bonds, the homology in the primary amino acid sequence is low. There is only 41% homology with the active subunit of carboxypeptidase N and 15% homology with carboxypeptidase B (Tan et al., 1989). CPM is anchored to the plasma membrane by a phosphatidylinositol glycan tail attached to the C-terminal end. This is the mode of membrane attachment for CPM in many tissues, in cultured distal tubular kidney cells (Deddish et al., 1990), and in peripheral nerves (Nagae et al., 1992).

CPM mRNA has been detected by Northern blot in human lung tissue (Nagae et al., 1993). In homogenized, fractionated lung tissues after centrifugation, the bulk of enzymatic action is sedimented in the pellet (P_3) at $100,000 \times g$. Using dansyl-Ala-Arg hydrolysis as a measure of CPM activity, dog lung P_3 contained 484 nmol/hr/mg protein, human lung P_3 198 ± 30 nmol/hr/mg protein, and rat lung P_3 153 nmol/hr/mg protein.

As we have shown with ACE (Igić et al., 1973), CPM was also released from isolated, perfused lung tissue by agents that induce edema. These include the detergent deoxycholate, the peptide melittin, and the histamine-liberating compound 48/80 (Table 1). Besides CPM, we measured the release of NEP and ACE

Table 1   Release of Carboxypeptidase M (CPM) from the Perfused Rat Lung

	CPM activity (nmol/hr/ml)	
Perfused with PBS and	Perfusion fluid	Edema developed
Deoxycholate	0.9	50
Melittin	1.5	37
48/80	8.1	48
Ringer only[a]	0	—

[a]Phosphate-buffered Ringer's solution.
Source: Modified from Dragovíc et al., 1993.

into both the perfusate and the alveolar space in the developing edema (Dragović et al., 1993). The activity of NEP was determined with Glut-Ala-Ala-Phe-methoxynaphthylamide and that of ACE with ^{3}H-hippuryl-Gly-Gly substrate. The ratio of activity calculated per milliliter was much higher for NEP and CPM in the edema than in the perfusate, whereas ACE activity was higher in the perfusate when deoxycholate was used. Relatively more CPM than NEP was found in the perfusate. These comparisons are certainly in agreement with main cellular localization of the enzymes with ACE in endothelial cells (Johnson and Erdös, 1977; Johnson et al., 1982) and with CPM and NEP in the bronchial epithelium (Desmazes et al., 1992; Nadel, 1992a,b). Bronchoalveolar lavage (BAL) fluid of the lung also revealed the presence of NEP and CPM. Both enzymes appeared in the particulate membrane fraction after centrifugation of BAL fluid (P_3) and in the final supernatant (S_3) as well. NEP activity was about equally divided between the two fractions, S_3 and P_3, while CPM was more concentrated in the S_3 than in P_3 (85 vs. 25 nmol/hr/mg protein). Collected alveolar macrophages also had an appreciable amount of CPM and NEP content, but only in the pellet (Dragović et al., 1993).

III. RELEASE OF ENZYMES FROM LUNG CELLS

The release of the three enzymes from different lung cell types is of interest. All three enzymes are membrane-bound but each by a different anchoring mechanism. NEP is anchored at its N-terminal end by its signal peptide (Turner, 1987), ACE has a C-terminal hydrophobic anchor peptide (Skidgel and Erdös, 1993), and CPM is attached to the membrane by its phosphatidylinositol glycan tail (Deddish et al., 1990). Although the three enzymes may be released by detergent or protease action, CPM can be solubilized with phospholipase C as well; this involves splitting the tail portion at diacylglycerol (Deddish et al., 1990; Nagae et al., 1992).

CPM may also appear in nerves in the myelin layer (Nagae et al., 1992), in the plasma membrane of cultured vascular endothelial cells (Nagae et al., 1993), and in bronchial epithelial cells (Desmazes et al., 1992; Chodimella et al., 1991; Nadel, 1992a,b). The activity in the lung was due to authentic CPM, as shown by immunoprecipitation of the enzyme with polyclonal antiserum elicited against purified human placental CPM (Skidgel et al., 1989; Nagae et al., 1992). The antisera precipitated 98% of the CPM activity from human lung solubilized P_3 fraction, and 81% and 88% from baboon and dog lung. Since the antiserum to human enzyme had a great degree of cross-reactivity with other species, there probably exists a structural similarity of CPM in the three species. Even guinea pig lung cross-reacted to some degree, since 41% of the CPM activity from homogenized guinea pig lung disappeared when antiserum to human CPM was added.

We attempted to localize CPM in the alveolar space, using immunohistochemistry with cryosections of the human lung and fluorescent light microscopy. Staining was detected on the luminal surface of type I alveolar epithelial cells only, but not in other cells such as type II cells although macrophages contain CPM. Type I pneumocytes, with the overlying layer of surfactant, basement membrane, and capillary endothelial cells, form a gas-permeable membrane between inhaled air and blood. As no other marker enzyme is known for type I cells (Funkhouser et al., 1991; Schneeberger, 1991), CPM may be the first one found. This type of cell can also be involved in pulmonary diseases. For example, the opportunistic infection of the lung in AIDS patients is characterized by the attachment of *Pneumocystis carinii* to type I cells. Whether CPM would participate on the surface of cells in a receptor function for the parasite is not known.

CPM has other roles in the lung. It is an inactivator of bradykinin and anaphylatoxins because most of the actions of these peptides are abolished by the enzymatic release of their C-terminal arginine (Skidgel, 1988; Huey et al., 1983; Erdös, 1979). This can be important in shock or inflammation where des-Arg anaphylatoxins lose their main chemotactic activities. Finally, in those patients who are treated with an ACE inhibitor and develop a cough as a side effect (Morice et al., 1987), pulmonary CPM may have added significance. If bradykinin is really the major cause of the cough after the inhibition of its main inactivator ACE or kininase II, then CPM as a kininase I–type enzyme could ameliorate the developing side effects.

IV. SUMMARY

The goal of this brief review is to point out the possible roles pulmonary carboxypeptidase M can play in vasopeptide metabolism, accented by its recent localization in some lung cells. The presence of this enzyme in a variety of cell-types indicates that similar to other peptidases (e.g., NEP or ACE), it may have several functions that can depend on its location and its access to peptides with a scissible basic C-terminal amino acid.

REFERENCES

Barnes, P.J., Belvisi, M.G., and Rogers, D.F. (1990). Modulation of neurogenic inflammation: novel approaches to inflammatory disease. *TIPS 11*:185–189.

Chodimella, V., Skidgel, R.A., Krowiak, E.J., and Murlas, C.G. (1991). Lung peptidases, including carboxypeptidase, modulate airway reactivity to intravenous bradykinin. *Am. Rev. Respir. Dis. 144*:869–874.

Connelly, J.C., Skidgel, R.A., Schulz, W.W., Johnson, A.R., and Erdös, E.G. (1985). Neutral endopeptidase 24.11 in human neutrophils: cleavage of chemotactic peptide. *Proc. Natl. Acad. Sci. USA 82*:8737–8741.

Deddish, P.A., Skidgel, R.A., and Erdös, E.G. (1989). Enhanced cobalt activation and inhibitor binding of carboxypeptidase M at low pH: similarity to carboxypeptide H (enkephalin convertase). *Biochem. J. 261*:289–291.

Deddish, P.A., Skidgel, R.A., Kriho, V.B., Becker, R.P., and Erdös, E.G. (1990). Carboxypeptidase M in cultured Madin-Darby canine kidney cells: evidence that carboxypeptidase M has a phosphatidylinositol glycan anchor. *J. Biol. Chem. 265*: 15083–15089.

Desmazes, N., Lockhart, A., Lacroix, H., and Dusser, D.J. (1992). Carboxypeptidase M–like enzyme modulates the noncholinergic bronchoconstrictor response in guinea pig. *Am. J. Respir. Cell Mol. Biol. 7*:477–484.

Dragović, T., Igić, R., Erdös, E.G., and Rabito, S.F. (1993). Metabolism of bradykinin by peptidases in the lung. *Am. Rev. Respir. Dis. 147*:1491–1496.

Erdös, E.G. (1979). Kininases. In: *Bradykinin, Kallidin and Kallikrein*. Handbook of Experimental Pharmacology. Vol. XXV Suppl. (Erdös, E.G., ed.). Springer-Verlag, Heidelberg, pp. 427–487.

Funkhouser, J.D., Tangada, S.D., and Peterson, R.D.A. (1991). Ectopeptidases of alveolar epithelium: candidates for roles in alveolar regulatory mechanisms. *Am. J. Physiol. 260*: L381–L385.

Holian, A., and Scheule, R.K. (1990). Alveolar macrophage biology. *Hosp. Pract. 12*(15): 49–58.

Huey, R., Bloor, C.M., Kawahara, M.S., and Hugli, T.E. (1983). Potentiation of the anaphylatoxins in vivo using an inhibitor of serum carboxypeptidase N (SCPN). I. Lethality and pathologic effects on pulmonary tissue. *Am. J. Pathol. 112*:48–60.

Igić, R., Nakajima, T., Yeh, H.S.J., Sorrells, K., and Erdös, E.G. (1973). Kininases. In: Symposium on Kinin Peptides. Fifth International Congress on Pharmacology, 1972, San Francisco, CA. *Pharmacology and the Future of Man. Vol. 5.* (Acheson, G.H., ed.). Karger, Basel, pp. 307–319.

Jackman, H.L., Tan, F., Tamei, H., Beurling-Harbury, C., Li, X.-Y., Skidgel, R.A., and Erdös, E.G. (1990). A peptidase in human platelets that deamidates tachykinins: probable identity with the lysosomal "protective protein." *J. Biol. Chem. 265*:11265–11272.

Johnson, A.R., and Erdös, E.G. (1977). Metabolism of vasoactive peptides by human endothelial cells in culture: angiotensin I converting enzyme (kininase II) and angiotensinase. *J. Clin. Invest. 59*:684–695.

Johnson, A.R., John, M., and Erdös, E.G. (1982). Metabolism of vasoactive peptides by membrane-enriched fractions from human lung tissue, pulmonary arteries, and endothelial cells. *Ann. NY Acad. Sci. 384*:72–89.

Johnson, A.R., Skidgel, R.A., Gafford, J.T., and Erdös, E.G. (1984). Enzymes in placental microvilli: angiotensin I converting enzyme, angiotensinase A, carboxypeptidase, and neutral endopeptidase ("enkephalinase"). *Peptides 5*:789–796.

Johnson, A.R., Ashton, J., Schulz, W., and Erdös, E.G. (1985). Neutral metalloendopeptidase in human lung tissue and cultured cells. *Am. Rev. Respir. Dis. 132*:564–568.

Lazarus, S.C., Borson, D.B., Gold, W.M., and Nadel, J.A. (1987). Inflammatory mediators, tachykinins and enkephalinase in airways. *Int. Arch. Allergy Appl. Immunol. 82*: 372–376.

Martins, M.A., Shore, S.A., Gerard, N.P., Gerard, C., and Drazen, J.M. (1990). Peptidase modulation of the pulmonary effects of tachykinins in tracheal superfused guinea pig lungs. *J. Clin. Invest. 85*:170–176.

Morice, A.H., Brown, M.J., Lowry, R., and Higenbottam, T. (1987). Angiotensin-converting enzyme and the cough reflex. *Lancet* 2:1116–1118.

Nadel, J.A. (1992a). Membrane-bound peptidases: endocrine, paracrine, and autocrine effects. *Am. J. Respir. Cell Mol. Biol.* 7:469–470.

Nadel, J.A. (1992b). Regulation of neurogenic inflammation by neutral endopeptidase. *Am. Rev. Respir. Dis.* 145:S48–S52.

Nagae, A., Deddish, P.A., Becker, R.P., Anderson, C.H., Abe, M., Tan, F., Skidgel, R.A., and Erdös, E.G. (1992). Carboxypeptidase M in brain and peripheral nerves. *J. Neurochem.* 59:2201–2212.

Nagae, A., Abe, M., Becker, R.P., Deddish, P.A., Skidgel, R.A., and Erdös, E.G. (1993). High concentration of carboxypeptidase M in lungs: presence of the enzyme in alveolar type I cells. *Am. J. Respir. Cell Mol. Biol.* 9:221–229.

Said, S.I., (1984). Peptide hormones and neurotransmitters of the lung. In: *The Endocrine Lung in Health and Disease*. (Becker, K.L., and Gazdar, A., eds.). W.B. Saunders, Philadelphia, pp. 267–275.

Schneeberger, E.E. (1991). Alveolar type I cells. In: *The Lung: Scientific Foundations, Vol. 1*. (Crystal, R.G., West, J.B., Barnes, P.J., Chesniack, N.S., and Weibel, E.R., eds.). Raven Press, New York, pp. 229–234.

Skidgel, R.A. (1988). Basic carboxypeptidases: regulators of peptide hormone activity. *Trends Pharmacol. Sci.* 9:299–304.

Skidgel, R.A., and Erdös, E.G. (1993). Biochemistry of angiotensin converting enzyme. In: *The Renin Angiotensin System. Vol. 1*. (Robertson, J.I.S., and Nicholls, M.G., eds.). Gower Medical Publ., London, pp. 10.1–10.10.

Skidgel, R.A., Johnson, A.R., and Erdös, E.G. (1984). Hydrolysis of opioid hexapeptides by carboxypeptidase N. Presence of carboxypeptidase in cell membranes. *Biochem. Pharmacol.* 33:3471–3478.

Skidgel, R.A, Davis, R.M., and Tan, F. (1989). Human carboxypeptidase M. Purification and characterization of a membrane-bound carboxypeptidase that cleaves peptide hormones. *J. Biol. Chem.* 264:2236–2241.

Skidgel, R.A., Jackman, H.L., and Erdös, E.G. (1991). Metabolism of substance P and bradykinin by human neutrophils. *Biochem. Pharmacol.* 41:1335–1344.

Tan, F., Chan, S.J., Steiner, D.F., Schilling, J.W., and Skidgel, R.A. (1989). Molecular cloning and sequencing of the cDNA for human membrane-bound carboxypeptidase M. *J. Biol. Chem.* 264:13165–13170.

Turner, A.J. (1987). Endopeptidase-24.11 and neuropeptide metabolism. In: *Neuropeptides and Their Peptidases*. (Turner, A.J., ed.). Chichester, Horwood, pp. 183–201.

DISCUSSION

Goetzl: We have found NEP antigen and activity in supernatants of lung fluids from ARDS and unilateral lung transplant patients, with 2/3 being membrane-associated and 1/3 resulting from proteolysis of NEP at the membrane. Have you seen changes in NEP expression in clinical disease?

Erdös: NEP activity is increased 50–60-fold in serum of patients with bacterial pneumonia (Johnson et al., *Am. Rev. Respir. Dis. 132*:1262–1267, 1985), but we do not know if this is from neutrophils or epithelial cells.

Manzini: Is the expression of NEP and peptide receptors regulated in the same fashion?

Erdös: This has not been investigated. However, some peptidases are also receptors. For example, aminopeptidase is a corornavirus receptor.

Nadel: Are surfactant apoproteins cleaved by carboxypeptidase M?

Erdös: I am not sure if these apoproteins have the proper C-terminal amino acids to be appropriate substrates.

15

The Concept of Neurogenic Inflammation in the Respiratory Tract

Donald M. McDonald

University of California at San Francisco, San Francisco, California

I. DEFINITION OF NEUROGENIC INFLAMMATION

The term "neurogenic inflammation" describes the increase in vascular permeability that is produced by substances released from sensory nerves. The existence of this phenomenon became apparent when experiments revealed that certain irritants, which stimulate sensory nerves, could cause an increase in vascular permeability typical of acute inflammation. Neurogenic inflammation was first observed in the skin and conjunctiva of rats and guinea pigs, and it is now known to occur in the respiratory tract of these species.

Neurogenic inflammation is only one of many consequences of sensory nerve stimulation. Furthermore, sensory-nerve-mediated changes are not the same in all organs, and species differences clearly exist, with the changes in humans likely to be different from those in rats and guinea pigs. Nonetheless, the possibility that neurogenic inflammation could play a role in human airway disease has prompted considerable interest in its mechanism, regulation, and pathophysiological consequences. Also, because of evidence that inflammatory diseases of the airways generally are accompanied by an increase in vascular permeability, neurogenic inflammation in the airways of rats and guinea pigs is a useful model for studying changes in vascular permeability associated with acute inflammation.

In this chapter I review the experiments that led to the concept of neurogenic inflammation and discuss neurogenic inflammation in the context of related phenomena such as antidromic vasodilatation and the axon reflex. Next I summarize the changes that are typical of neurogenic inflammation in the respiratory tract of rats and guinea pigs and address the issue of species differences. I also weigh the evidence for neurogenic inflammation in the human respiratory tract. Finally, I address the question of whether the term "neurogenic inflammation" should be used to describe all sensory-nerve-mediated changes in tissues or should apply only to the increase in vascular permeability resulting from sensory nerve stimulation.

II. THE CONCEPT OF NEUROGENIC INFLAMMATION

A. Jancsó's Discovery of Neurogenic Inflammation

The concept of neurogenic inflammation as we now know it stems from observations made by Jancsó and his colleagues in experiments involving capsaicin, the irritant in red peppers. These experiments, which were reported in three papers published in the 1960s (1–3), characterized the phenomenon of neurogenic inflammation and provided the conceptual framework for all subsequent studies of this phenomenon.

Jancsó and his colleagues used the term "neurogenic inflammation" to describe the increase in vascular permeability, plasma extravasation, and tissue swelling that occurred when capsaicin was applied to the conjunctiva, skin, or tongue of rats or guinea pigs (1–3). Irritating substances such as mustard oil, formaldehyde, and hypertonic saline had the same effect. Jancsó deduced that sensory nerves were involved in this phenomenon because irritants that caused plasma extravasation also caused intense discomfort, but when animals were "desensitized" by topical application or systemic injection of large doses of capsaicin, the irritants caused neither plasma extravasation nor discomfort (1). Jancsó demonstrated that vascular permeability was increased by using two intravascular tracers. Evans blue dye revealed the sites of plasma extravasation, and colloidal silver labeled the leaky blood vessels (1).

Further evidence of the involvement of sensory nerves came with the observation that antidromic electrical stimulation of sensory nerves could result in plasma extravasation and that this extravasation was abolished by surgical denervation. As with the effects of irritants, the plasma extravasation produced by electrical stimulation did not occur after desensitization with capsaicin. By comparison, the effect of substances such as dextran, which caused plasma extravasation but did not stimulate sensory nerves, was unaffected by desensitization (1–3).

Jancsó concluded that capsaicin and many other irritants increase vascular permeability through an effect on sensory nerves. He also reasoned that the inflammation produced by some irritants is "mediated entirely by the nervous system," which clearly was a departure from traditional views of inflammation (1).

These investigators used the term "neurogenic inflammation" because of the involvement of sensory nerves and because the plasma extravasation and edema that characterized the phenomenon are classic signs of inflammation. They recognized that this was the first experimental evidence that vascular permeability could be increased by sensory nerve stimulation and speculated that the nerves release a permeability-increasing "neurohumor" (2).

Jancsó and his colleagues emphasized that the increase in vascular permeability that they observed differed from the well-known phenomenon of "antidromic vasodilatation" (see below), which produced a change in blood flow but not an increase in vascular permeability (1,2).

B. Antidromic Vasodilatation and Axon Reflexes

Sensory nerves were known to have the potential of influencing blood vessels long before Jancsó discovered neurogenic inflammation, but the effect of the nerves was assumed to be limited to vasodilatation. In 1901 Bayliss reported that electrical stimulation of sensory nerves could increase blood flow in the limb of a dog, a phenomenon that was designated "antidromic vasodilatation" because the nerves were stimulated in a direction opposite to that of their normal conduction (4). Antidromic vasodilatation now is a well-documented effect of sensory nerve stimulation, but it has become tightly interwoven with another phenomenon, the "axon reflex."

The concept of the axon reflex was introduced in 1900 by Langley to describe a mechanism by which antidromic stimulation of preganglionic *sympathetic nerves* could evoke activity in efferent fibers from sympathetic ganglia located proximal to the site of stimulation (5). Specifically, Langley observed that electrical stimulation of the distal portion of the sympathetic trunk resulted in piloerection that was mediated by postganglionic sympathetic nerves from ganglia supplied by more proximal portions of the sympathetic trunk. Because the phenomenon did not involve the central nervous system as did true reflexes, Langley coined the terms "pseudo-reflex" and "axon-reflex" actions to describe the phenomenon.

The idea of the axon reflex was applied to *sensory nerves* by Bruce, who in 1910 proposed that sensory nerves could mediate vasodilatation through this mechanism. In this case, the term "axon reflex" described a phenomenon by which a stimulus to one branch of a sensory nerve fiber could trigger activity in another branch that innervated blood vessels and evoked vasodilatation (6). Thus, the axon reflex resembles antidromic vasodilatation, in that both terms describe mechanisms by which sensory nerve activation can result in vasodilatation. The axon reflex concept has the added feature of explaining how vasodilatation can spread outside the site directly affected by a stimulus.

Bruce adapted Langley's axon reflex concept to explain his finding that the redness and swelling produced in the rabbit eye by mustard oil was dependent on the presence of sensory nerves (6,7). In evidence of the involvement of sensory nerves, Bruce found that the redness and swelling could be blocked by a topical anesthetic or prevented by surgical destruction of the sensory (trigeminal) innervation. Bruce recognized that reflexes in the usual sense were not involved because mustard oil still produced the changes after the sensory nerves had been disconnected from the central nervous system, provided the irritant was applied before the nerves degenerated.

Some years later Lewis used the concept of the axon reflex to explain the control of blood flow in human skin at sites of injury (8). From studies of normal, anesthetized, and denervated regions of skin, Lewis learned that the sensory innervation had to be intact for vasodilatation (flare) to develop around an injury, but it was not essential for swelling (wheal) to form. Subsequent studies confirmed these observations and showed that the development of flare around an injury is mediated specifically by nociceptive C-fibers (9–11).

More recently, the concept of the axon reflex has been modified to address the question of how the flare around an injury can be twice as large as the receptive field of individual sensory axons (12). Evidently there is a mechanism by which activity can spread from axon to axon, not just from one collateral to another.

In addition to controlling the distribution of blood flow in the skin, axon reflexes were predicted many years ago to be involved in the regulation of blood flow of the airway mucosa, based on the branching pattern of the sensory nerves (13). The eye, skin, and respiratory mucosa may be special cases, however, because axon reflexes do not occur in all organs (9).

Despite the many studies of antidromic vasodilatation and axon reflexes, the possibility of a sensory-nerve-mediated increase in vascular permeability did not receive serious attention until Jancsó's experiments with capsaicin. In fact, from Lewis' finding that a wheal could form in denervated skin, it seemed that the increase in vascular permeability associated with injury of human skin was *not* mediated by sensory nerves (8). Furthermore, even though Bruce found that mustard oil caused swelling of the conjunctiva and eyelid, he apparently interpreted these changes as consequences of vasodilatation and did not specifically investigate the contribution of increased vascular permeability. Bruce's interpretation was consistent with the view at the time that vasodilatation was the key initial step in the inflammatory response (7), and "whenever injury to the tissues leads to vascular dilatation there is an increased effusion of fluid from the blood" (14).

III. NEUROGENIC INFLAMMATION IN THE RESPIRATORY TRACT

A. Studies of the Airways of Rats and Guinea Pigs

1. Stimuli that Evoke Neurogenic Inflammation

The phenomenon of neurogenic inflammation in the respiratory tract was first described by Lundberg and Saria, who discovered that capsaicin and other irritants can cause the extravasation of Evans blue dye in the airway mucosa of rats and guinea pigs (15,16). Just as Jancsó had found in the skin and eye, Lundberg and Saria learned that prior desensitization with capsaicin prevented irritant-induced plasma extravasation in the respiratory mucosa.

Lundberg and his associates also discovered that neurogenic inflammation can be produced by electrical stimulation of the vagus nerve or by exposure to cigarette smoke, ether, or formaldehyde (17–19). As in the case of capsaicin, these stimuli caused plasma extravasation in normal animals but not in desensitized animals. Furthermore, the investigators learned that the increase in vascular permeability produced by histamine and bradykinin is partly dependent on sensory nerves (16).

Subsequent studies by others have shown that neurogenic inflammation can be produced in the airways of rats and guinea pigs by an aerosol of hypertonic saline (20) or sodium metabisulfite (21) or by dry gas isocapnic hyperpnea (22,23).

2. Nerves that Mediate Neurogenic Inflammation

Sensory axons are though to evoke neurogenic inflammation by releasing tachykinins (substance P, neurokinin A) and perhaps other peptides. In evidence of this mechanism, unmyelinated nerve fibers that have tachykinin immunoreactivity and are capsaicin-sensitive are abundant in the epithelium and around arterioles in the respiratory mucosa of rats and guinea pigs (19,24–28). Furthermore, tachykinins are released from airway tissues by stimuli that evoke neurogenic inflammation (29–32).

Experiments using electrical stimulation, selective denervation, vagal ligation, and retrograde tracers have shown that most of the sensory nerve fibers involved in neurogenic inflammation have their cell bodies in vagal sensory ganglia and reach the respiratory tract via the vagus nerve (16,28,33–37). Sensory axons that arise from neurons in dorsal root ganglia also make a contribution (36,38,39).

Tachykinins released from sensory nerves increase vascular permeability by acting on NK-1 (substance P) receptors. This role of NK-1 receptors has been demonstrated in the airway mucosa of rats and guinea pigs through the use of selective agonists (19,40–42) and antagonists (43–47). NK-2 receptors may also be involved in distal bronchi of the guinea pig lung (48).

How do tachykinins increase vascular permeability? A straightforward mechanism would involve a direct action of tachykinins on NK-1 receptors on the endothelial cells of postcapillary venules. However, to test this hypothesis, one issue that still must be clarified is the distribution of NK-1 receptors, because it is uncertain whether NK-1 receptors are located on the endothelial cells of vessels that become leaky in response to tachykinins. This issue will eventually be settled by determining whether NK-1 receptors are present on postcapillary venules, by using immunohistochemistry with antibodies directed against the receptor protein (49). However, thus far, substance P binding sites have usually been localized in the airway mucosa by using iodinated Bolton-Hunter-coupled substance P (50–

52). From this approach, it appears that binding sites are located in the epithelium and on some small vessels in the lamina propria (52). It is uncertain whether the labeled vessels are arterioles or venules, but a study of rat skin showed Bolton-Hunter-coupled substance P binding sites on dermal arterioles and postcapillary venules (53). The Bolton-Hunter-coupled substance P also labels small blood vessels of the human fallopian tube (54) and smooth muscle in peripheral bronchi of the rabbit lung (55).

Alternatively, Kummer and his colleagues used an anti-idiotypic antibody (a rabbit polyclonal antibody raised against a rabbit C-terminal-specific anti–substance P antibody) to localize substance P binding sites in the guinea pig trachea by immunohistochemistry (56,57). Smooth muscle cells stained diffusely and epithelial cells stained along the basolateral membranes and at the apex, but blood vessels were not stained. Human airways had a similar distribution of staining (58). Because the staining was inhibited by preincubation with either substance P or neurokinin A, the antibody was assumed to bind both to NK-1 receptors and to NK-2 receptors.

Calcitonin gene-related peptide (CGRP) may participate in neurogenic inflammation through an effect on blood vessels. Most sensory nerve fibers in the airway mucosa that have substance P immunoreactivity also have CGRP immunoreactivity (35,59–62). CGRP receptors are present on some airway blood vessels (63–65), and CGRP causes long-lasting vasodilatation in the airways of some species (59,66–68). Although CGRP does not increase vascular permeability by itself, it can potentiate the edema-producing effect of substance P in the rat trachea (69) and in rat skin (70,71). However, CGRP has an *anti*-edema action in some systems (72), and its role in neurogenic inflammation in the respiratory mucosa is still unproved.

3. Sites of Extravasation in Neurogenic Inflammation

In rats and guinea pigs, neurogenic inflammation can occur at all levels of the respiratory tract, from the nose to fourth-order bronchi (15,16,73–75). Neurogenic plasma extravasation results from an increase in the permeability of postcapillary venules (diameter, 7–40 μm) and collecting venules (40–80 μm) located in the superficial portion of the airway mucosa just beneath the epithelium (74). In the rat trachea, these vessels constitute about half of the mucosal blood vessels by length and have two-thirds of the vascular surface area (74,76). The same vessels become leaky in response to inflammatory mediators such as histamine, bradykinin, and serotonin.

The endothelium of postcapillary venules and collecting venules becomes leaky in neurogenic inflammation because of the focal separations (endothelial gaps) that form at endothelial cell junctions (74). Such endothelial gaps are typically found in venules that have become leaky in response to mediators of acute inflammation (77). The endothelial gaps that form in venules of the rat trachea have an average diameter of 1.4 μm (range, 0.6–3 μm) (76). Gaps with this average diameter could be formed by a retraction of each of the apposing endothelial cells a distance of less than 2% of the cell length (76).

Within a minute of the onset of neurogenic inflammation, some 18 gaps per endothelial cell form in postcapillary venules and seven gaps per endothelial cell form in collecting venules (76). These values represent approximately 3% of the surface area of postcapillary venules and 0.7% of the surface area of collecting venules.

After the first minute, the number of gaps diminishes with a half-life of 3.2 min (76). The permeability increase has an even shorter duration, as evidenced by the 1.3-min half-life of the extravasation of Monastral blue, a particulate tracer (76). The cessation of the extravasation before the gaps close could result from a decrease in the driving force for the convective movement of Monastral blue out of the venules. Alternatively, sieve-like components of the endothelial glycocalyx could accumulate within the endothelial gaps and restrict the extravasation of Monastral blue while the gaps are still open (78,79).

4. Movement of Extravasated Plasma into the Airway Lumen

Fluid that leaks from postcapillary venules and collecting venules has the potential for producing mucosal edema. Whether sufficient fluid accumulates to cause tissue swelling is determined by the rates of extravasation and clearance. Potential routes for clearance of the extravasated fluid include the microvasculature, lymphatics, and airway lumen.

The process of plasma exudation into the airway lumen has received considerable attention recently and has been described as the first line of defense for neutralizing or washing away mucosal irritants (80–84).

The rapid movement of extravasated fluid from venules into the airway lumen apparently involves the unidirectional paracellular transit of unfiltered plasma across the epithelium (82,85–87). The hydrostatic pressure of the extravasated plasma is thought to create the route from the lamina propria into the airway lumen by opening epithelial cell tight junctions (83,84). How this occurs is unclear. One possibility is that the epithelial cells play an active part in the process, analogous to the retraction of endothelial cells in the formation of intercellular gaps. In any case, the process is rapidly reversible and is unaccompanied by appreciable absorption of solutes from the airway lumen (81,82).

5. Neutrophil Adhesion in Neurogenic Inflammation

The increase in vascular permeability that is characteristic of neurogenic inflammation is accompanied by the adhesion of neutrophils and eosinophils to the endothelium of the abnormally permeable venules (74,88–90). The number of adherent neutrophils is about 10 times the number of adherent eosinophils. It is unknown whether this adhesion results from changes in the endothelium or from the activation and chemotaxis of leukocytes by tachykinins (91–94).

The onset of the adhesion of leukocytes coincides with the increase in vascular permeability. However, leukocyte adhesion appears to be separate from the permeability increase, because the leukocyte attachment sites are distinct from endothelial gaps (76). Furthermore, the distributions of gaps and attached leukocytes in postcapillary venules overlap but are not identical: the gaps are most numerous in the smallest postcapillary venules, whereas leukocyte adhesion is maximal in the largest postcapillary venules (76).

In the case of neurogenic inflammation induced by capsaicin, most of the adherent neutrophils gradually reenter the circulation with a half-life of about 4 hr (89). However, if the activity of the tachykinin-degrading enzyme neutral endopeptidase is inhibited, a larger proportion of the neutrophils migrate into the tissue (89). Thus, the fate of the adherent neutrophils in neurogenic inflammation appears to be determined in part by the strength of a tachykinin-dependent chemotactic stimulus.

6.   Potentiation of Neurogenic Inflammation by Airway Infections

Respiratory tract infections can increase the sensitivity of animals to stimuli that evoke neurogenic inflammation (95). Naturally acquired infections due to *Mycoplasma pulmonis*, parainfluenza type I (Sendai) virus, and coronavirus (sialodacryoadenitis virus/rat coronavirus), which are among the most common respiratory pathogens in rats, result in an abnormally large neurogenic inflammatory response (96,97). Thus, when pathogen-free rats acquire these infections, they become abnormally sensitive to capsaicin, substance P, and vagal stimulation. As evidence of this abnormal sensitivity, a standardized stimulus produces an unusually large amount of plasma extravasation. Both an increase in the responsiveness of the blood vessels to tachykinins and an increase in the number of responsive blood vessels contribute to the augmented plasma extravasation (98).

Viral infections can potentiate neurogenic inflammation (99,100), but *M. pulmonis* infections have more severe, long-lasting consequences (98). Indeed, rats with airway infections caused by *M. pulmonis* develop a lifelong increased susceptibility to neurogenic inflammation and become so sensitive to capsaicin that they can die from a dose that is readily tolerated by pathogen-free rats (98).

M. pulmonis infections are exacerbated by viral infections and by inhalation of ammonia (101–103). When the severity of the infections increases, there is a corresponding increase in the sensitivity of the airway mucosa to capsaicin, as evidenced by the amount of plasma extravasation (98).

B.   Studies of Animals Other than Rats and Guinea Pigs

As mentioned above, neurogenic inflammation in the respiratory tract was first observed in rats and guinea pigs, and most of what is known about it has come from studies of these species (16,19,34,40). Neurogenic inflammation also occurs in the airways of the mouse, although the response is modest (J.J. Bowden,

P. Baluk, D.M. McDonald, unpublished observations). To the limited extent that it has been studied, neurogenic inflammation in other organs of the mouse resembles that in rats and guinea pigs (104–109).

However, many species do not develop neurogenic inflammation like that found in rats and guinea pigs. The first indication of species differences in neurogenic inflammation was the finding by Jancsó and his colleagues that capsaicin caused neither discomfort nor inflammation in the eyes or skin of frogs, chickens, and pigeons (2). Others have confirmed this observation (106,110,111).

The situation is more complicated in mammals. Capsaicin clearly stimulates unmyelinated axons and evokes physiological changes in many mammalian species (112–118). Furthermore, neurogenic inflammation can occur in some tissues of the rabbit (eye, skin) and cat (synovium of the knee joint) (119–122). Nonetheless, some species are much less sensitive to capsaicin than are guinea pigs, rats, and mice. For example, a capsaicin dose of 1.1 mg/kg is lethal to 50% of guinea pigs, and the corresponding doses for mice and rats are 6.5 mg/kg and 9.5 mg/kg, yet the lethal dose for rabbits is >50 mg/kg, and for hamsters it is >120 mg/kg (123). Similarly, capsaicin produces little or no neurogenic inflammation in hamsters, either in the respiratory tract (D.M. McDonald, unpublished observations) or in the urinary bladder (107).

To my knowledge, neurogenic inflammation, in the sense defined by Jancsó, has not been demonstrated in the respiratory tract of experimental animals other than guinea pigs, mice, or rats. Although the issue has not been examined systematically, it is known, for example, that a local injection of capsaicin does not cause plasma extravasation in the tracheal mucosa of anesthetized dogs (McDonald, Graf, and Nadel, unpublished observations), and substance P does not cause plasma extravasation in the airways of pigs (124).

It is not known why hamsters, rabbits, dogs, and pigs do not develop neurogenic plasma extravasation. There may be differences in the nerves, blood vessels, or both, and the lack of responsiveness in healthy animals could change in the presence of inflammatory airway diseases, as in the case of rats with respiratory tract infections.

C. Studies of Neurogenic Inflammation in the Human Respiratory Tract

The discovery of neurogenic inflammation in the respiratory tract of rats and guinea pigs prompted considerable speculation about the possible contribution of this phenomenon to the pathogenesis of airway diseases in humans. Numerous reviews have addressed this issue in the context of the mechanism, diversity, and consequences of sensory-nerve-mediated changes in the airways (125–134). Other reviews have dealt with the issue in the context of the actions of neuropeptides on the respiratory tract (135–144). Similarly, the question of whether neurogenic inflammation is involved in the development of edema in the human respiratory mucosa has been addressed in reviews of the mechanism and effects of increased vascular permeability in the airway mucosa (145–151).

Many of the approaches that have been used to characterize neurogenic inflammation in the airways of rats and guinea pigs obviously cannot be used in humans. Because of the difficulty of performing experiments in humans that would give definitive results, most of the speculation regarding the possible role of neurogenic inflammation in disease has necessarily been based on observations made on laboratory animals, primarily rats and guinea pigs.

Nonetheless, several strategies have been used to search for evidence of neurogenic inflammation in human airways. For example, the effect on the airway mucosa of tachykinins, capsaicin, hypertonic saline, or sodium metabisulfite has been tested; the release of tachykinins into airway fluids has been measured; the action of endogenous tachykinins has been blocked by specific antagonists; the effect of endogenous or exogenous tachykinins has been potentiated by inhibiting the activity of neutral endopeptidase; and the number of tachykinin-immuno-reactive nerves in nasal or bronchial biopsies from normal subjects has been compared with corresponding data from subjects with asthma.

1. Effects of Capsaicin and Tachykinins

Studies of the effects of capsaicin in humans have not detected an increase in vascular permeability in the nasal or tracheobronchial mucosa (152,153). Capsaicin can evoke cough, sneezing, mucus secretion, nasal congestion, and bronchoconstriction (154–164). However, the effect of capsaicin on airway resistance is reported to be small, transient, reversed by a muscarinic cholinergic antagonist, not exaggerated in asthmatics, and unlikely to result from plasma extravasation (163,165,166).

Substance P has little effect on the airway mucosa of normal subjects (167,168), but it may affect diseased airways differently. For example, it causes an abnormally large amount of nasal obstruction and increases the amount of albumin in the nasal lavage fluid from some patients with allergic rhinitis (168,169). Similarly, inhaled neurokinin A decreases airway conductance in subjects with asthma, whereas it has little effect on the airways of normal subjects (131,170,171). These observations will be easier to interpret when it is known where exogenous tachykinins act and whether the number, distribution, and function of tachykinin receptors in airway tissues change in the presence of diseases such as asthma and allergic rhinitis.

2. Amount of Substance P in Airway Fluids

The amount of substance P released from sensory nerves has been estimated from measurements of fluids obtained by nasal or bronchial lavage. The concentration of substance P is reported to be abnormally high in bronchial lavage fluid from some patients with atopic asthma and to increase further after antigen challenge

(172–174). Similarly, the concentration of substance P in sputum evoked by inhalation of hypertonic saline was found to be higher in patients with asthma or chronic bronchitis than in normal subjects (175).

The results of studies of nasal lavage fluid obtained from patients with allergic rhinitis are more variable. Some reports indicate that the substance P concentration is higher in allergic rhinitis (173,176), whereas others suggest that it is normal (177,178).

It is unknown whether the substance P in nasal or bronchial fluids comes exclusively from sensory nerves. Neuroendocrine cells of the airway epithelium are another potential source (179).

3. Effects of Tachykinin Antagonists

The discovery of nontoxic, specific antagonists of tachykinin receptors has created a powerful new strategy for investigating sensory-nerve-mediated changes in human airways. Although little information regarding the effects of these drugs in humans has been published thus far, one study reports that the neurokinin receptor antagonist FK-224 can block bradykinin-induced bronchoconstriction in asthmatic patients (180), suggesting the involvement of a tachykinin-dependent mechanism in this response. Although FK-224 is reported to be equally potent against NK-1 and NK-2 receptors, more specific antagonists are now being used in human trials.

4. Effect of Neutral Endopeptidase Inhibitors

Another strategy being used to determine the action of tachykinins on airway function in humans is to inhibit neutral endopeptidase (endopeptidase 24.11, enkephalinase). Inhibition of this peptidase would be expected to reduce the degradation of tachykinins and thereby potentiate their effects. Studies of rats and guinea pigs have shown that the inhibitors thiorphan and phosphoramidon can augment neurogenic inflammation (88,90,181–186). It is uncertain, however, whether this augmentation is mediated exclusively by tachykinins, because neutral endopeptidase degrades other peptides as well (187–190).

In normal humans, inhaled thiorphan can potentiate the bronchoconstrictor action of inhaled neurokinin A (171) and inhaled sodium metabisulfite (191). Similarly, phosphoramidon potentiates the constriction of human bronchi exposed in vitro to capsaicin or substance P (192). However, acetorphan, taken orally, does not change the bronchoconstrictor response of asthmatics to inhaled metabisulfite (193). I am not aware of studies in humans that have examined the effect of peptidase inhibitors on plasma extravasation evoked by sensory nerve stimuli.

5. Number of Substance P–Immunoreactive Nerves

The presence of an abnormally large number of substance P–containing nerve fibers in the airway mucosa of patients with asthma would implicate these nerves in the disease and would even suggest the involvement of neurogenic inflammation. Therefore, a report of such an increase (194) provoked considerable interest.

However, a subsequent study reexamined the issue and found no nerves with substance P or CGRP immunoreactivity in bronchial biopsies from eight atopic asthmatic subjects and eight nonatopic controls (195). Other types of nerve fibers clearly were present in the airway epithelium of both groups, as demonstrated by staining with the nonspecific neuronal marker antibody PGP 9.5 (195).

Similarly, other reports indicate that substance P–immunoreactive nerve fibers in human airways constitute no more than 1% of the intraepithelial nerve fibers that have PGP 9.5 immunoreactivity (196) and, in some cases, may even be absent (197). By comparison, in guinea pig airways, 60% of the nerve fibers have substance P immunoreactivity (196).

6. Unanswered Questions About Neurogenic Inflammation in Human Airways

A key issue concerning neurogenic inflammation in humans is whether neural factors can influence the permeability of airway blood vessels. Although most of the experimental approaches that have convincingly demonstrated neurogenic inflammation in the respiratory tract of rats and guinea pigs are not applicable to humans, it is possible to obtain indirect evidence of increased vascular permeability in the human nasal and bronchial mucosa under certain conditions (169,198). As these methods are perfected, it should become easier to determine whether sensory nerve stimuli can increase vascular permeability in human airways.

However, even when the vascular permeability issue is resolved, there will be questions about the identity of mediators released from the sensory nerves in the airways of humans. The observation that substance P–immunoreactive nerves constitute only a tiny proportion of the sensory nerve fibers in human airways raises the possibility that mediators other than substance P are involved. The identification of such hypothetical mediators would be a big step toward determining the effects on sensory nerves on airway blood vessels.

Regardless of what is learned from studies of normal humans, there will be questions about whether the sensory nerves or the targets of their mediators become abnormally responsive in the presence of inflammatory airway diseases.

Finally, there is the question of whether sensory nerves could play an important role in the pathogenesis of airway disease in humans even if neurogenic plasma extravasation does not occur, because of other changes that sensory nerves could mediate in the respiratory tract.

IV. OTHER SENSORY-NERVE-MEDIATED CHANGES IN THE RESPIRATORY TRACT

Neurogenic inflammation is only one of many consequences of stimulating sensory nerve fibers in the respiratory mucosa. Stimuli that activate sensory nerves can increase blood flow (68,116,199–202), cause smooth muscle contraction (131,203–206), facilitate cholinergic neurotransmission (207,208), stimulate secretion from epithelial secretory cells and submucosal glands (74,209–212), increase ciliary beat frequency (213–215), stimulate ion transport (216–219), and alter immunological functions (201,220–224).

The effects of capsaicin in rabbits and sheep illustrate the potential complexity of how sensory nerves can affect airway function. An acute dose of capsaicin does not produce neurogenic inflammation in the rabbit respiratory mucosa, in the sense defined by Jancsó and used by Lundberg and Saria, and capsaicin pretreatment does not reduce the amount of substance P immunoreactivity. However, capsaicin pretreatment does block the bronchial hyperresponsiveness produced by sensitization to antigen or by exposure to platelet-activating factor (225,226). Similarly, in sheep, capsaicin desensitization can prevent the development of bronchial hyperresponsiveness after antigen challenge (227).

V. IS THERE MORE TO NEUROGENIC INFLAMMATION THAN INCREASED VASCULAR PERMEABILITY?

Jancsó's definition of neurogenic inflammation as a sensory-nerve-mediated increase in vascular permeability seems straightforward. According to this definition, neurogenic inflammation occurs in the dura, conjunctiva, eyelid, middle ear, oral mucosa, dental pulp, salivary gland ducts, esophagus, biliary system, anal mucosa, ureter, urinary bladder, skin, joints, nose, larynx, trachea, and bronchi of rats and guinea pigs (228).

However, as new methods have been used to study the "efferent" effects of sensory nerves and as more tissues have been studied, changes other than increased vascular permeability have been identified, and some of these have been considered to be components of neurogenic inflammation. Thus, in some cases neurogenic inflammation in the respiratory tract has been regarded as a collection of changes, including plasma extravasation, neutrophil adhesion, vasodilatation, submucosal gland secretion, ion transport, bronchial smooth muscle contraction, increased cholinergic transmission, and cough (134,186,229).

Similarly, neurogenic inflammation in the eye is considered to be a combination of plasma extravasation, miosis, vasodilatation in the anterior uvea, breakdown of the blood-aqueous barrier, and increased intraocular pressure (119,230). In other organs, changes such as smooth muscle contraction, plasma extravasation, vasodilatation, mast cell degranulation, facilitation of transmitter release from nerve terminals, and recruitment of inflammatory cells have been included under the term "neurogenic inflammation" (231).

Appropriate terms obviously are needed for these sensory-nerve-mediated changes, but as Jancsó pointed out, all these changes cannot meaningfully be considered neurogenic inflammation. To avoid the ambiguity and confusion associated with multiple definitions for neurogenic inflammation, it seems desirable to restrict the term to increased vascular permeability resulting from sensory nerve stimulation.

VI. CONCLUSIONS

The concept of neurogenic inflammation has evolved from studies of the increase in vascular permeability produced by mediators released from sensory nerves. This phenomenon is well documented in the respiratory tract of rats and guinea pigs, as well as in several other organs of these species. Sensory nerve fibers that mediate neurogenic inflammation are unmyelinated, contain tachykinins and CGRP, and are stimulated by capsaicin and other irritants. Characteristically, these nerves are desensitized by repeated administration of capsaicin.

The increase in vascular permeability that occurs in neurogenic inflammation results mainly from the action of tachykinins on NK-1 receptors and selectively involves postcapillary venules and collecting venules of the airway mucosa. Intercellular gaps about a micrometer in diameter form in the endothelium of the affected vessels within seconds of the onset of a sensory nerve stimulus. The endothelial gaps, which can occupy as much as 3% of the luminal surface area of the venules, are transient and disappear spontaneously, with a half-life of about 3 min. During the period of increased permeability, plasma constituents and intravascular tracers as large as the gaps are extravasated into the respiratory mucosa. Some of the extravasated fluid in the mucosa crosses the epithelium and enters the airway lumen and may neutralize or wash away inhaled irritants.

Neurogenic inflammation is such a conspicuous phenomenon in the respiratory mucosa of rats and guinea pigs, one would expect it to be readily demonstrated in the airways of other species. This has not been the case. Despite the large body of information on the mechanism of neurogenic inflammation in rats and guinea pigs, little is known about why it does not occur in some species. Among the possible explanations are differences in the sensitivity of sensory nerves to irritants, differences in the mediators released from sensory nerves, and differences in the responsiveness of postcapillary venules and collecting venules to sensory nerve mediators.

Although neurogenic inflammation has not been demonstrated in the human respiratory mucosa, there is evidence that other sensory-nerve-mediated changes do occur and that these may be exaggerated in allergic rhinitis and asthma. From animal experiments, one would predict that inflammatory diseases of the airways would amplify the effects of sensory nerve stimulation.

Regardless of the outcome of the studies of sensory-nerve-mediated phenomena in humans, neurogenic inflammation in rats and guinea pigs continues to serve as a useful experimental model for examining the changes in endothelial cells that occur in acute inflammation, learning the consequences of plasma extravasation in the airway mucosa, and testing the antiedema action of drugs that may be useful for treating inflammatory diseases of the nose and bronchi.

As interest in neurogenic inflammation has expanded and as experimental approaches have diversified, the seemingly straightforward results of Jancsó's experiments have been overlaid by layers of complexity. The result has been a tendency for the concept of neurogenic inflammation to change. On the one hand, some investigators use the terms "axon reflex" and "neurogenic inflammation" interchangeably or regard several or even all of the sensory nerve-mediated changes in the respiratory tract as components of neurogenic inflammation. Such interpretations portray neurogenic inflammation as quite a different phenomenon from the one originally described. On the other hand, it seems advantageous for clarity, consistency, and historical accuracy to consider each sensory-nerve-mediated change separately and to recognize neurogenic inflammation as the distinct entity defined by Jancsó. Thus we will continue to benefit from what this neurally mediated increase in vascular permeability can teach us about sensory nerves, blood vessels, and inflammatory mediators.

ACKNOWLEDGMENTS

This work was supported in part by the National Heart, Lung, and Blood Institute Program Project Grant HL-24136 from the U.S. Public Health Service.

REFERENCES

1. Jancsó, N. (1960). Role of the nerve terminals in the mechanism of inflammatory reactions. *Bull. Millard Fillmore Hosp. (Buffalo, NY)* 7:53–77.
2. Jancsó, N., Jancsó-Gábor, A., and Szolcsányi, J. (1967). Direct evidence for neurogenic inflammation and its prevention by denervation and by pretreatment with capsaicin. *Br. J. Pharmacol. Chemother.* 31:138–151.
3. Jancsó, N., Jancsó-Gábor, A., and Szolcsányi, J. (1968). The role of sensory nerve endings in neurogenic inflammation induced in human skin and in the eye and paw of the rat. *Br. J. Pharmacol. Chemother.* 33:32–41.
4. Bayliss, W.M. (1901). On the origin from the spinal cord of the vaso-dilator fibres of the hindlimb, and on the nature of these fibres. *J. Physiol. (Lond.)* 26:173–209.
5. Langley, J.N. (1900). On axon-reflexes in the pre-ganglionic fibres of the sympathetic system. *J. Physiol. (Lond.)* 25:364–398.
6. Bruce, A.N. (1910). Über die Beziehung der sensiblen Nervenendigungen zum Entzundungsvorgang. *Arch. Exp. Pathol. Pharmakol.* 63:424–433.
7. Bruce, A.N. (1913). Vaso-dilator axon-reflexes. *Q.J. Exp. Physiol.* 6:339–354.
8. Lewis, T. (1927). Mechanism of the flare and of the wheal. (1927) In: *The Blood Vessels of the Human Skin and Their Responses.* (Lewis, T., ed.). Chicago Medical Book Company, Chicago, pp. 67–80.

9. Celander, O., and Folkow, B. (1953). The nature and the distribution of afferent fibres provided with the axon reflex arrangement. *Acta Physiol. Scand. 29*:359–370.

10. Chapman, L.F., Ramos, A.O., Goodell, H., and Wolff, H.G. (1961). Neurohumoral features of afferent fibres in man. *Arch. Neurol. 4*:617–650.

11. Chapman, L.F., and Goodell, H. (1964). The participation of the nervous system in the inflammatory reaction. *Ann. NY Acad. Sci. 161*:21–29.

12. Lynn, B. (1988). Neurogenic inflammation. *Skin Pharmacol. 1*:217–224.

13. Gaylor, J.B. (1934). The intrinsic nervous mechanism of the human lung. *Brain 57*: 143–160.

14. Adami, J.G. (1907). *Inflammation. An Introduction to the Study of Pathology*, 3rd ed., Macmillan, New York.

15. Lundberg, J.M., and Saria, A. (1982). Capsaicin-sensitive vagal neurons involved in control of vascular permeability in rat trachea. *Acta Physiol. Scand. 115*:521–523.

16. Lundberg, J.M., and Saria, A. (1983). Capsaicin-induced desensitization of airway mucosa to cigarette smoke, mechanical and chemical irritants. *Nature 302*:251–253.

17. Lundberg, J.M., Saria, A., Brodin, E., Rosell, S., and Folkers, K. (1983). A substance P antagonist inhibits vagally induced increase in vascular permeability and bronchial smooth muscle contraction in the guinea pig. *Proc. Natl. Acad. Sci. USA 80*:1120–1124.

18. Lundberg, J.M., Martling, C.R., Saria, A., Folkers, K., and Rosell, S. (1983). Cigarette smoke-induced airway oedema due to activation of capsaicin-sensitive vagal afferents and substance P release. *Neuroscience 10*:1361–1368.

19. Lundberg, J.M., Brodin, E., and Saria, A. (1983). Effects and distribution of vagal capsaicin-sensitive substance P neurons with special reference to the trachea and lungs. *Acta Physiol. Scand. 119*:243–252.

20. Umeno, E., McDonald, D.M., and Nadel, J.A. (1990). Hypertonic saline increases vascular permeability in the rat trachea by producing neurogenic inflammation. *J. Clin. Invest. 85*:1905–1908.

21. Sakamoto, T., Elwood, W., Barnes, P.J., and Chung, K.F. (1992). Pharmacological modulation of inhaled sodium metabisulphite-induced airway microvascular leakage and bronchoconstriction in the guinea-pig. *Br. J. Pharmacol. 107*:481–487.

22. Ray, D.W., Hernandez, C., Leff, A.R., Drazen, J.M., and Solway, J. (1989). Tachykinins mediate bronchoconstriction elicited by isocapnic hyperpnea in guinea pigs. *J. Appl. Physiol. 66*:1108–1112.

23. Garland, A., Ray, D.W., Doerschuk, C.M., Alger, L., Eappon, S. Hernandez, C., et al. (1991). Role of tachykinins in hyperpnea-induced bronchovascular hyperpermeability in guinea pigs. *J. Appl. Physiol. 70*:27–35.

24. Lundberg, J.M., Hökfelt, T., Martling, C.R., Saria, A., and Cuello, C. (1984). Substance P–immunoreactive sensory nerves in the lower respiratory tract of various mammals including man. *Cell. Tissue Res. 235*:251–261.

25. Hua, X.Y., Theodorsson-Norheim, E., Brodin, E., Lundberg, J.M., and Hökfelt, T. (1985). Multiple tachykinins (neurokinin A, neuropeptide K and substance P) in capsaicin-sensitive sensory neurons in the guinea-pig. *Regul. Pept. 13*:1–19.

26. Lundberg, J.M., Martling, C.-R., and Hökfelt, T. (1988). Airways, oral cavity and salivary glands: classical transmitters and peptides in sensory and autonomic motor

neurons. In: *Handbook of Chemical Neuroanatomy*. Vol. 6. The Peripheral Nervous System. (Björklund, A., Hökfelt, T., and Owman, C., eds.) Elsevier, Amsterdam, pp. 391–444.

27. Baluk, P., Nadel, J.A., and McDonald, D.M. (1992). Substance P–immunoreactive sensory axons in the rat respiratory tract: a quantitative study of their distribution and role in neurogenic inflammation. *J. Comp. Neurol. 319*:586–598.

28. Kummer, W., Fischer, A., Kurkowski, R., and Heym, C. (1992). The sensory and sympathetic innervation of guinea-pig lung and trachea as studied by retrograde neuronal tracing and double-labelling immunohistochemistry. *Neuroscience 49*: 715–737.

29. Lundberg, J.M., Saria, A., Theodorsson-Norheim, E., Brodin, E., Hua, X.-Y., Martling, C.-R., et al. (1985). Multiple tachykinins in capsaicin-sensitive afferents: occurrence, release and biological effects with special reference to irritation of the airways. In: *Tachykinin Antagonists*. (Håkanson, R., and Sundler, F., eds.) Elsevier North-Holland, New York, pp. 159–169.

30. Saria, A., Martling, C.R., Yan, Z., Theodorsson-Norheim, E., Gamse, R., and Lundberg, J.M. (1988). Release of multiple tachykinins from capsaicin-sensitive sensory nerves in the lung by bradykinin, histamine, dimethylphenyl piperazinium, and vagal nerve stimulation. *Am. Rev. Respir. Dis. 137*:1330–1335.

31. Martins, M.A., Shore, S.A., and Drazen, J.M. (1991). Release of tachykinins by histamine, methacholine, PAF, LTD4, and substance P from guinea pig lungs. *Am. J. Physiol. 261*:L449–L455.

32. Hua, X.Y., and Yaksh, T.L. (1992). Release of calcitonin gene-related peptide and tachykinins from the rat trachea. *Peptides 13*:113–120.

33. Terenghi, G., McGregor, G.P., Bhuttacharji, S., Wharton, J., Bloom, S.R., and Polak, J.M. (1983). Vagal origin of substance P-containing nerves in the guinea pig lung. *Neurosci. Lett. 36*:229–239.

34. Lundberg, J.M., Brodin, E., Hua, X., and Saria, A. (1984). Vascular permeability changes and smooth muscle contraction in relation to capsaicin-sensitive substance P afferents in the guinea-pig. *Acta Physiol. Scand. 120*:217–227.

35. Cadieux, A., Springall, D.R., Mulderry, P.K., Rodrigo, J., Ghatei, M.A., Terenghi, G., et al. (1986). Occurrence, distribution and ontogeny of CGRP immunoreactivity in the rat lower respiratory tract: effect of capsaicin treatment and surgical denervations. *Neuroscience 19*:605–627.

36. Springall, D.R., Cadieux, A., Oliveira, H., Su, H., Royston, D., and Polak, J.M. (1987). Retrograde tracing shows that CGRP-immunoreactive nerves of rat trachea and lung originate from vagal and dorsal root ganglia. *J. Auton. Nerv. Syst. 20*: 155–166.

37. McDonald, D.M., Mitchell, R.A., Gabella, G., and Haskell, A. (1988). Neurogenic inflammation in the rat trachea. II. Identity and distribution of nerves mediating the increase in vascular permeability. *J. Neurocytol. 17*:605–628.

38. Dalsgaard, C.J., and Lundberg, J.M. (1984). Evidence for a spinal afferent innervation of the guinea pig lower respiratory tract as studied by the horseradish peroxidase technique. *Neurosci. Lett. 23*:117–122.

39. Saria, A., Martling, C.R., Dalsgaard, C.J., and Lundberg, J.M. (1985). Evidence

for substance P–immunoreactive spinal afferents that mediate bronchoconstriction. *Acta Physiol. Scand. 125*:407–414.

40. Saria, A., Lundberg, J.M., Skofitsch, G., and Lembeck, F. (1983). Vascular protein linkage in various tissues induced by substance P, capsaicin, bradykinin, serotonin, histamine and by antigen challenge. *Naunyn-Schmiedeberg's Arch. Pharmacol. 324*:212–218.

41. Abelli, L., Maggi, C.A., Rovero, P., Del Bianco, E., Regoli, D., Drapeau, G., et al. (1991). Effect of synthetic tachykinin analogues on airway microvascular leakage in rats and guinea-pigs: evidence for the involvement of NK-1 receptors. *J. Auton. Pharmacol. 11*:267–275.

42. Abelli, L., Nappi, F., Maggi, C.A., Rovero, P., Astolfi, M., Regoli, D., et al. (1991). NK-1 receptors and vascular permeability in rat airways. *Ann. NY Acad. Sci. 632*: 358–359.

43. Delay-Goyet, P., and Lundberg, J.M. (1991). Cigarette smoke-induced airway oedema is blocked by the NK1 antagonist, CP-96,345. *Eur. J. Pharmacol. 2*: 157–158.

44. Delay-Goyet, P., Franco, C.A., Gonsalves, S.F., Clingan, C.A., Lowe, J., and Lundberg, J.M. (1992). CP-96,345 antagonism of NK1 receptors and smoke-induced protein extravasation in relation to its cardiovascular effects. *Eur. J. Pharmacol. 222*:213–218.

45. Lei, Y.H., Barnes, P.J., and Rogers, D.F. (1992). Inhibition of neurogenic plasma exudation in guinea-pig airways by CP-96,345, a new non-peptide NK1 receptor antagonist. *Br. J. Pharmacol. 105*:261–262.

46. Murai, M., Morimoto, H., Maeda, Y., and Fujii, T. (1992). Effects of the tripeptide substance P antagonist, FR113680, on airway constriction and airway edema induced by neurokinins in guinea pigs. *Eur. J. Pharmacol. 217*:23–29.

47. Sakamoto, T., Barnes, P.J., and Chung, K.F. (1993). Effect of CP-96,345, a non-peptide NK1 receptor antagonist, against substance P–induced, bradykinin-induced and allergen-induced airway microvascular leakage and bronchoconstriction in the guinea pig. *Eur. J. Pharmacol. 231*:31–38.

48. Tousignant, C., Chan, C.-C., Guevremont, D., Brideau, C., Hale, J.J., MacCoss, M., et al. (1993). NK$_2$ receptors mediate plasma extravasation in guinea-pig lower airways. *Br. J. Pharmacol. 108*:383–386.

49. Bunnett, N.W., Vigna, S., Bowden, J.J., McDonald, D.M., Fisher, J., Okamoto, A., et al. (1993). Antibodies to the substance P receptor (SPR NK1) and to a chimeric Flag-SPR expressed in mammalian cells. *FASEB J. 7*:A183.

50. Carstairs, J., and Barnes, P. (1986). Autoradiographic mapping of substance P receptors in lung. *Eur. J. Pharmacol. 127*:295–296.

51. Hoover, D., and Hancock, J. (1987). Autoradiographic localization of substance P binding sites in guinea-pig airways. *J. Auton. Nerv. Syst. 19*:171–174.

52. Sertl, K., Wiedermann, C.J., Kowalski, M.L., Hurtado, S., Plutchok, J., Linnoila, I., et al. (1988). Substance P: the relationship between receptor distribution in rat lung and the capacity of substance P to stimulate vascular permeability. *Am. Rev. Respir. Dis. 138*:151–159.

53. O'Flynn, N.M., Helme, R.D., Watkins, D.J., and Burcher, E. (1989). Autoradio-

graphic localization of substance P binding sites in rat footpad skin. *Neurosci. Lett.* *106*:43–48.

54. Nimmo, A.J., Whitaker, E.M., Carstairs, J.R., and Morrison, J.F.B. (1989). The autoradiographic localization of calcitonin gene-related peptide and substance P receptors in human fallopian tube. *Q.J. Exp. Physiol.* *74*:955–958.

55. Black, J., Diment, L., Armour, C., Alouan, L., and Johnson, P. (1990). Distribution of substance P receptors in rabbit airways, functional and autoradiographic studies. *J. Pharmacol. Exp. Ther.* *253*:381–386.

56. Kummer, W., Fischer, A., Preissler, U., Couraud, J.-Y., and Heym, C. (1990). Immunohistochemistry of the guinea-pig trachea using an anti-idiotypic antibody recognizing substance P receptors. *Histochemistry* *93*:541–546.

57. Kummer, W., Fischer, A., Couraud, J.-Y., and Heym, C. (1991). Immunohistochemistry of peptides (substance P and VIP) and peptide receptors in the trachea. *J. Auton. Nerv. Syst.* *33*:121–123.

58. Fischer, A., Kummer, W., Couraud, J.-Y., Adler, D., Branscheid, D., and Heym, C. (1992). Immunohistochemical localization of receptors for vasoactive intestinal peptide and substance P in human trachea. *Lab. Invest.* *67*:387–393.

59. Lundberg, J.M., Franco, C.A., Hua, X., Hökfelt, T., and Fischer, J.A. (1985). Coexistence of substance P and calcitonin gene-related peptide-like immunoreactivities in sensory nerves in relation to cardiovascular and bronchoconstrictor effects of capsaicin. *Eur. J. Pharmacol.* *108*:315–319.

60. Luts, A., Widmark, E., Ekman, R., Waldeck, B., and Sundler, F. (1990). Neuropeptides in guinea-pig trachea: distribution and evidence for the release of CGRP into tracheal lumen. *Peptides* *11*:1211–1216.

61. Terada, M., Iwanaga, T., Takahashiiwanaga, H., Adachi, I., Arakawa, M., and Fujita, T. (1992). Calcitonin gene-related peptide (CGRP)-immunoreactive nerves in the tracheal epithelium of rats: an immunohistochemical study by means of whole mount preparations. *Arch. Histol. Cytol.* *55*:219–233.

62. Baluk, P., Nadel, J.A., and McDonald, D.M. (1993). Calcitonin gene-related peptide (CGRP) in secretory granules of serous cells in the rat tracheal epithelium. *Am. J. Respir. Cell Mol. Biol.* *8*:446–453.

63. Carstairs, J.R. (1987). Distribution of calcitonin gene-related peptide receptors in the lung. *Eur. J. Pharmacol.* *140*:357–358.

64. Mak, J., and Barnes, P.J. (1988). Autoradiographic localization of calcitonin gene-related peptide (CGRP) binding sites in human and guinea-pig lung. *Peptides 9*: 957–963.

65. Barnes, P.J., Baraniuk, J.N., and Belvisi, M.G. (1991). Neuropeptides in the respiratory tract. Part II. *Am. Rev. Respir. Dis.* *144*:1391–1399.

66. Martling, C.R. (1987). Sensory nerves containing tachykinins and CGRP in the lower airways. Functional implications for bronchoconstriction, vasodilatation and protein extravasation. *Acta Physiol. Scand.* *563* (Suppl.):1–57.

67. Martling, C.R., Saria, A., Fischer, J.A., Hökfelt, T., and Lundberg, J.M. (1988). Calcitonin gene-related peptide and the lung: neuronal coexistence with substance P, release by capsaicin and vasodilatory effect. *Regul. Pept.* *20*:125–139.

68. Stjärne, P., Lundblad, L., Änggård, A., Hökfelt, T., and Lundberg, J.M. (1989).

Tachykinins and calcitonin gene-related peptide: co-existence in sensory nerves of the nasal mucosa and effects on blood flow. *Cell. Tissue Res.* 256:439–446.

69. Brokaw, J.J., and White, G.W. (1992). Calcitonin gene-related peptide potentiates substance P–induced plasma extravasation in the rat trachea. *Lung* 170:85–93.

70. Brain, S.D., and Williams, T.J. (1985). Inflammatory oedema induced by synergism between calcitonin gene-related peptide (CGRP) and mediators of increased vascular permeability. *Br. J. Pharmacol.* 86:855–860.

71. Gamse, R., and Saria, A. (1985). Potentiation of tachykinin-induced plasma protein extravasation by calcitonin gene-related peptide. *Eur. J. Pharmacol.* 114:61–66.

72. Raud, J., Lundberg, T., Broddajansen, G., Theodorsson, E., and Hedqvist, P. (1991). Potent anti-inflammatory action of calcitonin gene-related peptide. *Biochem. Biophys. Res. Commun.* 180:1429–1435.

73. Lundblad, L., Saria, A., Lundberg, J.M., and Änggård, A. (1983). Increased vascular permeability in rat nasal mucosa induced by substance P and stimulation of capsaicin-sensitive trigeminal neurons. *Acta Otolaryngol. (Stockh.)* 96:479–484.

74. McDonald, D.M. (1988). Neurogenic inflammation in the rat trachea. I. Changes in venules, leucocytes and epithelial cells. *J. Neurocytol.* 17:583–603.

75. Petersson, G., Bacci, E., McDonald, D.M., and Nadel, J.A. (1993). Neurogenic plasma extravasation in the rat nasal mucosa is potentiated by peptidase inhibitors. *J. Pharmacol. Exp. Ther.* 264:509–514.

76. McDonald, D.M. (1994) Endothelial gaps and permeability of venules of rat tracheas exposed to inflammatory stimuli. *Am. J. Physiol.* (in press).

77. Majno, G., and Palade, G.E. (1961). Studies on inflammation. I. The effect of histamine and serotonin on vascular permeability: an electron microscopic study. *J. Biophys. Biochem. Cytol.* 11:571–604.

78. Clough, G., and Michel, C.C. (1988). The ultrastructure of frog microvessels following perfusion with the ionophore A23187. *Q.J. Exp. Physiol.* 73:123–125.

79. Clough, G., Michel, C.C., and Phillips, M.E. (1988). Inflammatory changes in permeability and ultrastructure of single vessels in the frog mesenteric microcirculation. *J. Physiol. (Lond.)* 395:99–114.

80. Persson, C.G.A., and Erjefält, I.A.L. (1988). Nonneural and neural regulation of plasma exudation in airways. In: *The Airways. Neural Control in Health and Disease.* (Kaliner, M.A., and Barnes, P.J., eds.) Marcel Dekker, New York, pp. 523–549.

81. Erjefält, I., and Persson, C.G.A. (1989). Inflammatory passage of plasma macromolecules into airway wall and lumen. *Pulm. Pharmacol.* 2:93–102.

82. Luts, A., Sundler, F., Erjefält, I., and Persson, C.G.A. (1990). The airway epithelial lining in guinea pigs is intact promptly after the mucosal crossing of a large amount of plasma exudate. *Int. Arch. Allergy Appl. Immunol.* 91:385–388.

83. Persson, C.G.A. (1990). Plasma exudation in tracheobronchial and nasal airways: a mucosal defence mechanism becomes pathogenic in asthma and rhinitis. *Eur. Respir. J. 12* (Suppl.):652s–657s.

84. Persson, C.G.A., Erjefält, I., Alkner, U., Baumgarten, C., Greiff, L., Gustafsson, B., et al. (1991). Plasma exudation as a first line respiratory mucosal defence. *Clin. Exp. Allergy 21*:17–24.

85. Persson, C.G.A., Erjefält, I., Gustafsson, B., and Luts, A. (1990). Subepithelial

hydrostatic pressure may regulate plasma exudation across the mucosa. *Int. Arch. Allergy Appl. Immunol. 92*:148–153.

86. Gustafsson, B.G., and Persson, C.G.A. (1991). Asymmetrical effects of increases in hydrostatic pressure on macromolecular movement across the airway mucosa. A study in guinea-pig tracheal tube preparations. *Clin. Exp. Allergy 21*:121–126.

87. Erjefält, I., Luts, A., and Persson, C.G.A. (1993). Appearance of airway absorption and exudation tracers in guinea pig tracheobronchial lymph nodes. *J. Appl. Physiol. 74*:817–824.

88. Umeno, E., Nadel, J.A., Huang, H.T., and McDonald, D.M. (1989). Inhibition of neutral endopeptidase potentiates neurogenic inflammation in the rat trachea. *J. Appl. Physiol. 66*:2647–2652.

89. Umeno, E., Nadel, J.A., and McDonald, D.M. (1990). Neurogenic inflammation of the rat trachea: fate of neutrophils that adhere to venules. *J. Appl. Physiol. 69*:2131–2136.

90. Katayama, M., Piedimonte, G., Nadel, J.A., and McDonald, D.M. (1993). Peptidase inhibitors reverse steroid-induced suppression of neutrophil adhesion in rat tracheal blood vessels. *Am. J. Physiol. 264*:L316–L322.

91. Helme, R.D., Eglezos, A., and Hosking, C.S. (1987). Substance P induces chemotaxis of neutrophils in normal and capsaicin-treated rats. *Immunol. Cell. Biol. 65*:267–269.

92. Payan, D.G., and Goetzl, E.J. (1987). Substance P receptor-dependent responses of leukocytes in pulmonary inflammation. *Am. Rev. Respir. Dis. 136*:S39–S43.

93. Öhlén, A., Thureson-Klein, Å., Lindbom, L., Persson, M.G., and Hedqvist, P. (1989). Substance P activates leukocytes and platelets in rabbit microvessels. *Blood Vessels 26*:84–94.

94. Perianin, A., Snyderman, R., and Malfroy, B. (1989). Substance P primes human neutrophil activation: a mechanism for neurological regulation of inflammation. *Biochem. Biophys. Res. Commun. 161*:520–524.

95. McDonald, D.M. (1992). Infections intensify neurogenic plasma extravasation in the airway mucosa. *Am. Rev. Respir. Dis. 146*:S40–S44.

96. McDonald, D.M. (1988). Respiratory tract infections increase susceptibility to neurogenic inflammation in the rat trachea. *Am. Rev. Respir. Dis. 137*:1432–1440.

97. Huang, H.T., Haskell, A., and McDonald, D.M. (1989). Changes in epithelial secretory cells and potentiation of neurogenic inflammation in the trachea of rats with respiratory tract infections. *Anat. Embryol. (Berl.) 180*:325–341.

98. McDonald, D.M., Schoeb, T.R., and Lindsey, J.R. (1991). *Mycoplasma pulmonis* infections cause long-lasting potentiation of neurogenic inflammation in the respiratory tract of the rat. *J. Clin. Invest. 87*:787–799.

99. Dusser, D.J., Jacoby, D.B., Djokic, T.D., Rubinstein, I., Borson, D.B., and Nadel, J.A. (1989). Virus induces airway hyperresponsiveness to tachykinins—role of neutral endopeptidase. *J. Appl. Physiol. 67*:1504–1511.

100. Piedimonte, G., Nadel, J.A., Umeno, E., and McDonald, D.M. (1990). Sendai virus infection potentiates neurogenic inflammation in the rat trachea. *J. Appl. Physiol. 68*:754–760.

101. Schoeb, T.R., Davidson, M.K., and Lindsey, J.R. (1982). Intracage ammonia

promotes growth of *Mycoplasma pulmonis* in the respiratory tract of rats. *Infect. Immun. 38*:212–217.

102. Schoeb, T.R., Kervin, K.C., and Lindsey, J.R. (1985). Exacerbation of murine respiratory mycoplasmosis in gnotobiotic F344/N rats by sendai virus infection. *Vet. Pathol. 22*:272–282.

103. Schoeb, T.R., and Lindsey, J.R. (1987). Exacerbation of murine respiratory mycoplasmosis by sialodacryoadenitis virus infection in gnotobiotic F344 rats. *Vet. Pathol. 24*:392–399.

104. Gamse, R. (1982). Capsaicin and nociception in the rat and mouse. Possible role of substance P. *Naunyn-Schmiedeberg's Arch. Pharmacol. 320*:205–216.

105. Keen, P., Tullo, A.B., Blyth, W.A., and Hill, T.J. (1982). Substance P in the mouse cornea: effects of chemical and surgical denervation. *Neurosci. Lett. 29*:231–235.

106. Jancsó, G., Ferencsik, M., Such, G., Király, E., Nagy, A., and Bujdosó, M. (1985). Morphological effects of capsaicin and its analogues in newborn and adult mammals. In: *Tachykinin Antagonists.* (Håkanson, R., and Sundler, F., eds.) Elsevier North-Holland, New York, pp. 35–44.

107. Maggi, C.A., Giuliani, S., Santicioli, P., Abelli, L., Geppetti, P., Somma, V., et al. (1987). Species-related variations in the effects of capsaicin on urinary bladder functions: relation to bladder content of substance P–like immunoreactivity. *Naunyn-Schmiedeberg's Arch. Pharmacol. 336*:546–555.

108. Mantione, C.R., and Rodriguez, R. (1990). A bradykinin (BK)1 receptor antagonist blocks capsaicin-induced ear inflammation in mice. *Br. J. Pharmacol. 99*:516–518.

109. Gábor, M., and Rázga, Z. (1992). Development and inhibition of mouse ear oedema induced with capsaicin. *Agents Actions 36*:83–86.

110. Szolcsányi, J., Sann, H., and Pierau, F.K. (1986). Nociception in pigeons is not impaired by capsaicin. *Pain 27*:247–260.

111. Pierau, F.-K., Sann, H., Harti, G., and Szolcsányi, J. (1988). A possible mechanism for the absence of neurogenic inflammation in birds: Inappropriate release of peptides from small afferent nerves. *Agents Actions 23*:12–13.

112. Coleridge, H.M., Coleridge, J.C.G., and Roberts, A.M. (1983). Rapid shallow breathing evoked by selective stimulation of airway fibres in dogs. *J. Physiol. (Lond.) 340*:415–433.

113. Jancsó, G., Király, E., Such, G., Joó, F., and Nagy, A. (1987). Neurotoxic effect of capsaicin in mammals. *Acta Physiol. Hung. 69*:295–313.

114. Szolcsányi, J. (1987). Selective responsiveness of polymodal nociceptors of the rabbit ear to capsaicin, bradykinin and ultra-violet irradiation. *J. Physiol. (Lond.) 388*:9–23.

115. Maggi, C.H., and Meli, A. (1988). The sensory-efferent function of capsaicin-sensitive sensory neurons. *Gen. Pharmacol. 19*:1–43.

116. Martling, C.R., Matran, R., Alving, K., Lacroix, J.S., and Lundberg, J.M. (1989). Vagal vasodilatory mechanisms in the pig bronchial circulation preferentially involves sensory nerves. *Neurosci. Lett. 30*:306–311.

117. Matran, R., Alving, K., Martling, C.R., Lacroix, J.S., and Lundberg, J.M. (1989). Effects of neuropeptides and capsaicin on tracheobronchial blood flow of the pig. *Acta Physiol. Scand. 135*:335–342.

118. Coleridge, H.M., Coleridge, J.C.G., Green, J.F., and Parsons, G.H. (1992). Pulmonary C-fiber stimulation by capsaicin evokes reflex cholinergic bronchial vasodilation in sheep. *J. Appl. Physiol.* 72:770–778.

119. Stjernschantz, J., Sears, M., and Mishima, H. (1982). Role of substance P in the antidromic vasodilation, neurogenic plasma extravasation and disruption of the blood-aqueous barrier in the rabbit eye. *Naunyn-Schmiedeberg's Arch. Pharmacol.* 321:329–335.

120. Ferrell, W.R., and Russell, N.J. (1986). Extravasation in the knee induced by antidromic stimulation of articular C fibre afferents of the anaesthetized cat. *J. Physiol. (Lond.)* 379:407–416.

121. Lynn, B., and Shakhanbeh, J. (1988). Substance P content of the skin, neurogenic inflammation and numbers of C-fibres following capsaicin application to a cutaneous nerve in the rabbit. *Neuroscience* 24:769–775.

122. Buckley, T.L., Brain, S.D., and Williams, T.J. (1990). Ruthenium red selectively inhibits oedema formation and increased blood flow induced by capsaicin in rabbit skin. *Br. J. Pharmacol.* 99:7–8.

123. Glinsukon, T., Stitmunnaithum, V., Toskulkao, C., Buranawuti, T., and Tangkrisanavinont, V. (1980). Acute toxicity of capsaicin in several animal species. *Toxicon* 18:215–220.

124. Lundberg, J.M., Alving, K., Karlsson, J.A., Matran, R., and Nilsson, G. (1991). Sensory neuropeptide involvement in animal models of airway irritation and of allergen-evoked asthma. *Am. Rev. Respir. Dis.* 143:1429–1431.

125. Barnes, P.J. (1986). Asthma as an axon reflex. *Lancet* 1:242–245.

126. Wolf, G. (1988). Neue Aspekte zur Pathogenese und Therapie der hyperreflektorischen Rhinopathie. *Laryngol. Rhinol. Otol. (Stuttg.)* 67:438–445.

127. Joos, G.F. (1989). The role of sensory neuropeptides in the pathogenesis of bronchial asthma. *Clin. Exp. Allergy* 1:9–13.

128. Barnes, P.J., Belvisi, M.G., and Rogers, D.F. (1990). Modulation of neurogenic inflammation: novel approaches to inflammatory disease. *Trends Pharmacol. Sci.* 11:185–189.

129. Nadel, J.A. (1990). Decreased neutral endopeptidases: possible role in inflammatory diseases of airways. *Lung 168* (Suppl.):123–127.

130. Barnes, P.J. (1991). Sensory nerves, neuropeptides, and asthma. *Ann. NY Acad. Sci.* 629:359–370.

131. Joos, G.F., and Pauwels, R.A. (1991). The bronchoconstrictor effect of sensory neuropeptides in man. *Ann. NY Acad. Sci.* 629:371–382.

132. Nadel, J.A. (1991). Mechanisms of inflammation and potential role in the pathogenesis of asthma. *Allergy Proc.* 12:85–88.

133. Solway, J., and Leff, A.R. (1991). Sensory neuropeptides and airway function. *J. Appl. Physiol.* 71:2077–2087.

134. Barnes, P.J. (1992). Neurogenic inflammation and asthma. *J. Asthma 29*:165–180.

135. Barnes, P.J. (1987). Airway neuropeptides and airway disease. *Ann. Ital. Med. Int.* 2:327–332.

136. Barnes, P.J. (1987). Neuropeptides in human airways: function and clinical implications. *Am. Rev. Respir. Dis. 136*:S77–S83.

137. Casale, T.B. (1987). Neuromechanisms of asthma. *Ann. Allergy 59*:391–398.
138. Joos, G.F.P., Pauwels, R.A.R., and Van Der Straeten, M.E.R.P. (1988). The role of neuropeptides as neurotransmitters of non-adrenergic, non-cholinergic nerves in bronchial asthma. *Bull. Eur. Physiopathol. Respir. 23*:619–637.
139. Baraniuk, J.N., and Kaliner, M.A. (1990). Neuropeptides and nasal secretion. *J. Allergy Clin. Immunol. 86*:620–627.
140. Baraniuk, J.N., and Kaliner, M. (1991). Neuropeptides and nasal secretion. *Am. J. Physiol. 261*:L223–L235.
141. Barnes, P.J. (1991). Neuropeptides and asthma. *Am. Rev. Respir. Dis. 143*:S28–S32.
142. Casale, T.B. (1991). Neuropeptides and the lung. *J. Allergy Clin. Immunol. 88*: 1–14.
143. Barnes, P.J. (1992). Neural mechanisms in asthma. *Br. Med. Bull. 48*:149–168.
144. Chanez, P., Godard, P., Lacoste, J.Y., Bousquet, J., and Michel, F.B. (1992). Médiateurs et neuromédiateurs dans l'asthme. *Presse Med. 21*:259–265.
145. Persson, C.G.A. (1986). Role of plasma exudation in asthmatic airways. *Lancet 2*: 1126–1129.
146. Persson, C.G.A. (1988). Plasma exudation and asthma. *Lung 166*:1–23.
147. Barnes, P.J., Boschetto, P., Rogers, D.F., Belvisi, M., Roberts, N., Chung, K.F., et al. (1990). Effects of treatment on airway microvascular leakage. *Eur. Respir. J. 12* (Suppl.):670s–671s.
148. Chung, K.F., Rogers, D.F., Barnes, P.J., and Evans, T.W. (1990). The role of increased airway microvascular permeability and plasma exudation in asthma. *Eur. Respir. J. 3*:329–337.
149. Persson, C.G.A (1991). Plasma exudation in the airways: mechanisms and function. *Eur. Respir. J. 4*:1268–1274.
150. Yager, D., Shore, S., and Drazen, J.M. (1991). Airway luminal liquid. Sources and role as an amplifier of bronchoconstriction. *Am. Rev. Respir. Dis. 143*:S52–S54.
151. Rogers, D.F., and Evans, T.W. (1992). Plasma exudation and oedema in asthma. *Br. Med. Bull. 48*:120–134.
152. Bascom, R., Kagey, S.A., and Proud, D. (1991). Effect of intranasal capsaicin on symptoms and mediator release. *J. Pharmacol. Exp. Ther. 259*:1323–1327.
153. Rajakulasingam, K., Polosa, R., Lau, L.C.K., Church, M.K., Holgate, S.T., and Howarth, P.H. (1992). Nasal effects of bradykinin and capsaicin: influence on plasma protein leakage and role of sensory neurons. *J. Appl. Physiol. 72*:1418–1424.
154. Collier, J.G., and Fuller, R.W. (1984). Capsaicin inhalation in man and the effects of sodium cromoglycate. *Br. J. Pharmacol. 81*:113–117.
155. Maxwell, D.L., Fuller, R.W., and Dixon, C.M.S. (1987). Ventilatory effects of inhaled capsaicin in man. *Eur. J. Clin. Pharmacol. 31*:715–717.
156. Wolf, G., Loidolt, D., Saria, A., and Gamse, R. (1987). Änderungen des nasalen Volumsstromes nach lokaler Applikation des Neuropeptides Substanz-P und von Capsaicin. *Laryngol. Rhinol. Otol. (Stuttg.) 66*:412–415.
157. Geppetti, P., Fusco, B.M., Marabini, S., Maggi, C.A., Fanciullacci, M., and Sicuteri, F. (1988). Secretion, pain and sneezing induced by the application of capsaicin to the nasal mucosa in man. *Br. J. Pharmacol. 93*:509–514.

158. Stjärne, P., Lundblad, L., Lundberg, J.M., and Änggård, A. (1989). Capsaicin and nicotine-sensitive afferent neurones and nasal secretion in healthy human volunteers and in patients with vasomotor rhinitis. *Br. J. Pharmacol. 96*:693–701.

159. Karlsson, J.A., and Persson, C.G.A. (1989). Novel peripheral neurotransmitters and control of the airways. *Pharmacol. Ther. 43*:397–423.

160. Choudry, N.B., Fuller, R.W., Anderson, N., and Karlsson, J.A. (1990). Separation of cough and reflex bronchoconstriction by inhaled local anaesthetics. *Eur. Respir. J. 3*:579–583.

161. Fuller, R.W. (1990). The human pharmacology of capsaicin. *Arch. Int. Pharmacodyn. Ther. 303*:147–156.

162. Hansson, L., Wollmer, P., Dahlbäck, M., and Karlsson, J.A. (1992). Regional sensitivity of human airways to capsaicin-induced cough. *Am. Rev. Respir. Dis. 145*: 1191–1195.

163. Midgren, B., Hansson, L., Karlsson, J.A., Simonsson, B.G., and Persson, C.G.A. (1992). Capsaicin-induced cough in humans. *Am. Rev. Respir. Dis. 146*:347–351.

164. Choudry, N.B., Studham, J., Harland, D., and Fuller, R.W. (1993). Modulation of capsaicin induced airway reflexes in humans: effect of monoamine oxidase inhibition. *Br. J. Clin. Pharmacol. 35*:184–187.

165. Fuller, R.W., Dixon, C.M., and Barnes, P.J. (1985). Bronchoconstrictor response to inhaled capsaicin in humans. *J. Appl. Physiol. 58*:1080–1084.

166. Fuller, R.W., Karlsson, J.A., Choudry, N.B., and Pride, N.B. (1988). Effect of inhaled and systemic opiates on responses to inhaled capsaicin in humans. *J. Appl. Physiol. 65*:1125–1130.

167. Fuller, R.W., Maxwell, D.L., Dixon, C.M., McGregor, G.P., Barnes, V.F., Bloom, S.R., et al. (1987). Effect of substance P on cardiovascular and respiratory function in subjects. *J. Appl. Physiol. 62*:1473–1479.

168. Devillier, P., Dessanges, J.F., Rakotosihanaka, F., Ghaem, A., Boushey, H.A., Lockhart, A., et al. (1988). Nasal response to substance P and methacholine in subjects with and without allergic rhinitis. *Eur. Respir. J. 1*:356–361.

169. Braunstein, G., Fajac, I., Lacronique, J., and Frossard, N. (1991). Clinical and inflammatory responses to exogenous tachykinins in allergic rhinitis. *Am. Rev. Respir. Dis. 144*:630–635.

170. Joos, G., Pauwels, R., and Van der Straeten, M. (1987). Effect of inhaled substance P and neurokinin A on the airways of normal and asthmatic subjects. *Thorax 42*: 779–783.

171. Cheung, D., Bel, E.H., Den Hartigh, J., Dijkman, J.H., and Sterk, P.J. (1992). The effect of an inhaled neutral endopeptidase inhibitor, thiorphan, on airway responses to neurokinin A in normal humans in vivo. *Am. Rev. Respir. Dis. 145*:1275–1280.

172. Nieber, K., Baumgarten, C., Witzel, A., Rathsack, R., Oehme, P., Brunnee, T., et al. (1991). The possible role of substance P in the allergic reaction, based on two different provocation models. *Int. Arch. Allergy Appl. Immunol. 94*:334–338.

173. Nieber, K., Baumgarten, C.R., Rathsack, R., Furkert, J., Oehme, P., and Kunkel, G. (1992). Substance P and β-endorphin-like immunoreactivity in lavage fluids of subjects with and without allergic asthma. *J. Allergy Clin. Immunol. 90*:646–652.

174. Nieber, K., Baumgarten, C., Rathsack, R., Furkert, J., Laake, E., Müller, S., et al.

(1993). Effect of azelastine on substance P content in bronchoalveolar and nasal lavage fluids of patients with allergic asthma. *Clin. Exp. Allergy 23*:69–71.

175. Tomaki, M., Ichinose, M., Nakajima, N., Miura, M., Yamauchi, H., Inoue, H., et al. (1993). Elevated substance P concentration in sputum after hypertonic saline inhalation in asthma and chronic bronchitis patients. *Am. Rev. Respir. Dis. 147*:A478.

176. Chaen, T., Watanabe, N., Mogi, G., Mori, K., and Takeyama, M. (1993). Substance P and vasoactive intestinal peptide in nasal secretions and plasma from patients with nasal allergy. *Ann. Otol. Rhinol. Laryngol. 102*:16–21.

177. Tønnesen, P., Hindberg, I., Schaffalitzky de Muckadell, O.B., and Mygind, N. (1988) Effect of nasal allergen challenge on serotonin, substance P and vasoactive intestinal peptide in plasma and nasal secretions. *Allergy 43*:310–317.

178. Walker, K.B., Serwonska, M.H., Valone, F.H., Harkonen, W.S., Frick, O.L., Scriven, K.H., et al. (1988). Distinctive patterns of release of neuroendocrine peptides after nasal challenge of allergic subjects with ryegrass antigen. *J. Clin. Immunol. 8*:108–113.

179. Gallego, R., Garciacaballero, T., Roson, E., and Beiras, A. (1990). Neuroendocrine cells of the human lung express substance P-like immunoreactivity. *Acta Anat. 139*:278–282.

180. Ichinose, M., Nakajima, N., Takahashi, T., Yamauchi, H., Inoue, H., and Takishima, T. (1992). Protection against bradykinin-induced bronchoconstriction in asthmatic patients by neurokinin receptor antagonist. *Lancet 340*:1248–1251.

181. Borson, D.B., Brokaw, J.J., Sekizawa, K., McDonald, D.M., and Nadel, J.A. (1989). Neutral endopeptidase and neurogenic inflammation in rats with respiratory infections. *J. Appl. Physiol. 66*:2653–2658.

182. Nadel, J.A. (1990). Neutral endopeptidase modulation of neurogenic inflammation in airways. *Eur. Respir. J. 12* (Suppl):645s–651s.

183. Nadel, J.A., and Borson, D.B. (1991). Modulation of neurogenic inflammation by neutral endopeptidase. *Am. Rev. Respir. Dis. 143*:S33–S36.

184. Lötvall, J.O., Elwood, W., Tokuyama, K., Barnes, P.J., and Chung, K.F. (1991). Differential effects of phosphoramidon on neurokinin A– and substance P–induced airflow obstruction and airway microvascular leakage in guinea-pig *Br. J. Pharmacol. 104*:945–949.

185. Piedimonte, G., McDonald, D.M., and Nadel, J.A. (1991). Neutral endopeptidase and kininase II mediate glucocorticoid inhibition of neurogenic inflammation in the rat trachea. *J. Clin. Invest. 88*:40–44.

186. Nadel, J.A. (1992). Regulation of neurogenic inflammation by neutral endopeptidase. *Am. Rev. Respir. Dis. 145*:S48–S52.

187. Shipp, M.A., Tarr, G.E., Chen, C.Y., Switzer, S.N., Hersh, L.B., Stein, H., et al. (1991). CD10/neutral endopeptidase 24.11 hydrolyzes bombesin-like peptides and regulates the growth of small cell carcinomas of the lung. *Proc. Natl. Acad. Sci. USA 88*:10662–10666.

188. Takaoka, M., Shiragami, K., Fujino, K., Miki, K., Miyake, Y., Yasuda, M., et al. (1991). Phosphoramidon-sensitive endothelin converting enzyme in rat lung. *Biochem. Int. 25*:697–704.

189. Di Maria, G.U., Katayama, M., Borson, D.B., and Nadel, J.A. (1992). Neutral endopeptidase modulates endothelin-1-induced airway smooth muscle contraction in guinea-pig trachea. *Regul. Pept. 39*:137–145.

190. Yamaguchi, T., Kohrogi, H., Kawano, O., Ando, M., and Araki, S. (1992). Neutral endopeptidase inhibitor potentiates endothelin-1-induced airway smooth muscle contraction. *J. Appl. Physiol. 73*:1108–1113.

191. Di Maria, G.U., Bellofiore, S., Pennisi, A., Calgagirone, F., Ciancio, N., and Mistretta, A. (1993). Neutral endopeptidase inhibition increases airway response to inhaled metabisulfite in normal subjects. *Am. Rev. Respir. Dis. 147*:A838.

192. Honda, I., Kohrogi, H., Yamaguchi, T., Ando, M., and Araki, S. (1991). Enkephalinase inhibitor potentiates substance P– and capsaicin-induced bronchial smooth muscle contractions in humans. *Am. Rev. Respir. Dis. 143*:1416–1418.

193. Nichol, G.M., O'Connor, B.J., Lecomte, J.M., Chung, K.F., and Barnes, P.J. (1992). Effect of neutral endopeptidase inhibitor on airway function and bronchial responsiveness in asthmatic subjects. *Eur. J. Clin. Pharmacol. 42*:491–494.

194. Ollerenshaw, S.L., Jarvis, D., Sullivan, C.E., and Woolcock, A.J. (1991). Substance P immunoreactive nerves in airways from asthmatics and nonasthmatics. *Eur. Respir. J. 4*:673–682.

195. Howarth, P.H., Djukanovic, R., Wilson, J.W., Holgate, S.T., Springall, D.R., and Polak, J.M. (1991). Mucosal nerves in endobronchial biopsies in asthma and non-asthma. *Int. Arch. Allergy Appl. Immunol. 94*:330–333.

196. Bowden, J., and Gibbins, I.L. (1992). Relative density of substance P-immunoreactive (SP-IR) nerve fibres in the tracheal epithelium of a range of species. *FASEB J. 6*:A1276.

197. Laitinen, L.A., Laitinen, A., Panula, P.A., Partanen, M., Tervo, K., and Tervo, T. (1983). Immunohistochemical demonstration of substance P in the lower respiratory tract of the rabbit and not of man. *Thorax 38*:531–536.

198. Fick, R., Jr., Metzger, W.J., Richerson, H.B., Zavala, D.C., Moseley, P.L., Schoderbek, W.E., et al. (1987). Increased bronchovascular permeability after allergen exposure in sensitive asthmatics. *J. Appl. Physiol. 63*:1147–1155.

199. Lundblad, L., Änggård, A., and Lundberg, J.M. (1982). Vasodilation in the cat nasal mucosa induced by antidromic trigeminal nerve stimulation. *Acta Physiol. Scand. 115*:517–519.

200. Matran, R., Alving, K., Martling, C.R., Lacroix, J.S., and Lundberg, J.M. (1989). Vagally mediated vasodilatation by motor and sensory nerves in the tracheal and bronchial circulation of the pig. *Acta Physiol. Scand. 135*:29–37.

201. Alving, K. (1991). Airways vasodilatation in the immediate allergic reaction. Involvement of inflammatory mediators and sensory nerves. *Acta Physiol. Scand. 597* (Suppl.):1–64.

202. Piedimonte, G., Hoffman, J.I.E., Husseini, W.K., Hiser, W.L., and Nadel, J.A. (1992). Effect of neuropeptides released from sensory nerves on blood flow in the rat airway microcirculation. *J. Appl. Physiol. 72*:1563–1570.

203. Lundberg, J.M., and Saria, A. (1982). Bronchial smooth muscle contraction induced by stimulation of capsaicin-sensitive sensory neurons. *Acta Physiol. Scand. 116*: 473–476.

204. Lundberg, J.M., Martling, C.R., and Saria, A. (1983). Substance P and capsaicin-induced contraction of human bronchi. *Acta Physiol. Scand. 119*:49–53.

205. Karlsson, J.A., Finney, M.J., Persson, C.G.A., and Post, C. (1984). Substance P antagonists and the role of tachykinins in non-cholinergic bronchoconstriction. *Life Sci. 24*:2681–2691.

206. Kröll, F., Karlsson, J.A., Lundberg, J.M., and Persson, C.G.A. (1990). Capsaicin-induced bronchoconstriction and neuropeptide release in guinea pig perfused lungs. *J. Appl. Physiol. 68*:1679–1687.

207. Grunstein, M.M., Tanaka, D.T., and Grunstein, J.S. (1984). Mechanism of substance P–induced bronchoconstriction in maturing rabbit. *J. Appl. Physiol. 57*: 1238–1246.

208. Tanaka, D.T., and Grunstein, M.M. (1984). Mechanisms of substance P-induced contraction of rabbit airway smooth muscle. *J. Appl. Physiol. 57*:1551–1557.

209. Kuo, H.P., Rohde, J.A.L., Tokuyama, K., Barnes, P.J., and Rogers, D.F. (1990). Capsaicin and sensory neuropeptide stimulation of goblet cell secretion in guinea-pig trachea. *J. Physiol. (Lond.) 431*:629–641.

210. Rogers, D.F., and Dewar, A. (1990). Neural control of airway mucus secretion. *Biomed. Pharmacother. 44*:447–453.

211. Tokuyama, K., Kuo, H.P., Rohde, J.A.L., Barnes, P.J., and Rogers, D.F. (1990). Neural control of goblet cell secretion in guinea pig airways. *Am. J. Physiol. 259*: L108–L115.

212. Baraniuk, J.N., Lundgren, J.D., Okayama, M., Goff, J., Mullol, J., Merida, M., et al. (1991). Substance P and neurokinin A in human nasal mucosa. *Am. J. Respir. Cell Mol. Biol. 4*:228–236.

213. Kondo, M., Tamaoki, J., and Takizawa, T. (1990). Neutral endopeptidase inhibitor potentiates the tachykinin-induced increase in ciliary beat frequency in rabbit trachea. *Am. Rev. Respir. Dis. 142*:403–406.

214. Wong, L.B., Miller, I.F., and Yeates, D.B. (1990). Stimulation of tracheal ciliary beat frequency by capsaicin. *J. Appl. Physiol. 68*:2574–2580.

215. Lindberg, S., and Dolata, J. (1993). NK1 receptors mediate the increase in muco-ciliary activity produced by tachykinins. *Eur. J. Pharmacol. 231*:375–380.

216. Al-Bazzaz, F.J., Kelsey, J.G., and Kaage, W.D. (1985). Substance P stimulation of chloride secretion by canine tracheal mucosa. *Am. Rev. Respir. Dis. 131*:86–89.

217. Rangachari, P.K., McWade, D., and Donoff, B. (1987). Luminal tachykinin receptors on canine tracheal epithelium: functional subtyping. *Regul. Pept. 18*:101–108.

218. Mizoguchi, H., and Hicks, C.R. (1989). Effects of neurokinins on ion transport and sulfated macromolecule release in the isolated ferret trachea. *Exp. Lung Res. 15*: 837–848.

219. Sestini, P., Bienenstock, J., Crowe, S.E., Marshall, J.S., Stead, R.H., Kakuta, Y., et al. (1990). Ion transport in rat tracheal epithelium in vitro: role of capsaicin-sensitive nerves in allergic reactions. *Am. Rev. Respir. Dis. 141*:393–397.

220. Ahlstedt, S., Alving, K., Hesselmar, B., and Olaisson, E. (1986). Enhancement of the bronchial reactivity in immunized rats by neonatal treatment with capsaicin. *Int. Arch. Allergy Appl. Immunol. 80*:262–266.

221. Payan, D.G., and Goetzl, E.J. (1986). Mediation of pulmonary immunity and hypersensitivity by neuropeptides. *Eur. J. Respir. Dis. 144* (Suppl.):77–106.

222. Saria, A. (1988). Neuroimmune interactions in the airways: implications for asthma, allergy, and other inflammatory airway diseases. *Brain Behav. Immun. 2*:318–321.

223. Nilsson, G., Alving, K., Lundberg, J.M., and Ahlstedt, S. (1991). Local immune response and bronchial reactivity in rats after capsaicin treatment. *Allergy 46*:304–311.

224. Bertrand, C., Geppetti, P., Baker, J., Yamawaki, I., and Nadel, J.A. (1993). Role of neurogenic inflammation in antigen-induced vascular extravasation in guinea pig trachea. *J. Immunol. 150*:1479–1485.

225. Spina, D., McKenniff, M.G., Coyle, A.J., Seeds, E.A.M., Tramontana, M., Perretti, F., et al. (1991). Effect of capsaicin on PAF-induced bronchial hyperresponsiveness and pulmonary cell accumulation in the rabbit. *Br. J. Pharmacol. 103*:1268–1274.

226. Riccio, M.M., Manzini, S., and Page, C.P. (1993). The effect of neonatal capsaicin on the development of bronchial hyperresponsiveness in allergic rabbits. *Eur. J. Pharmacol. 232*:89–97.

227. Abraham, W.M., Ahmed, A., Cortes, A., and Delehunt, J.C. (1993). C-fiber desensitization prevents airway hyperresponsiveness to cholinergic and antigenic stimuli after antigen challenge in allergic sheep. *Am. Rev. Respir. Dis. 147*:A478.

228. McDonald, D.M. (1994). Neurogenic inflammation in the airways. In: *Autonomic Control of the Respiratory System.* (Barnes, P., ed.) Harwood Academic Publishers, London (in press).

229. Barnes, P.J. (1991). Neurogenic inflammation in airways. *Int. Arch. Allergy Appl. Immunol. 94*:303–309.

230. Krootila, K., Oksala, O., Zschauer, A. Palkama, A., and Uusitalo, H. (1992). Inhibitory effect of methysergide on calcitonin gene-related peptide-induced vasodilatation and ocular irritative changes in the rabbit. *Br. J. Pharmacol. 106*:404–408.

231. Maggi, C.A. (1992). Therapeutic potential of capsaicin-like molecules: studies in animals and humans. *Life Sci. 51*:1777–1781.

16

Modulation of Neurogenic Inflammation by Peptidases

Jay A. Nadel
University of California at San Francisco, San Francisco, California

I. INTRODUCTION

In addition to the classic antonomic nervous systems (sympathetic and parasympathetic) that innervate airways, a third neural pathway exists in a subpopulation of primary sensory neurons that can be stimulated electrically and that are uniquely sensitive to capsaicin, the pungent substance present in plants of the genus *Capsicum* (Holzer, 1991; Maggi, 1991; Szolcsanyi, 1984). These capsaicin-sensitive neurons are contained in a heterogeneous population of sensory nerves: most of the neurons are C-fiber polymodal nociceptors, some are C-fiber warmth receptors, and some are δ-fiber polymodal nociceptors with cell bodies located in the vagal and trigeminal sensory ganglia and dorsal root ganglia (Holzer, 1991).

When the nerves are stimulated, release of neuropeptides occurs; these peptides include the tachykinins substance P (SP) and neurokinin A (NKA), as well as calcitonin gene-related peptide (CGRP). Local release of these neuropeptides from the sensory nerves results in a series of tissue responses termed "neurogenic inflammation." These responses occur in different organs, including the gastrointestinal and genitourinary tracts, eye, airways, and heart (Holzer, 1991; Maggi, 1991).

A variety of stimuli including hypertonic media (Umeno et al., 1989; Tramontana et al., 1991), irritants such as tobacco smoke (Lundberg and Saria, 1983) and the industrial irritant toluene diisocyanate (Mapp et al., 1991), and mediators such as histamine (Saria et al., 1988), serotonin (Tramontana et al., 1992) and high

K$^+$ (Geppetti et al., 1991a), and prostanoids (Geppetti et al., 1991b) cause the neural release of sensory neuropeptides.

In airways, neurogenic inflammation produces potent responses in a wide variety of cells (Fig. 1). This pathway has intrigued investigators because it mimics responses in inflammatory diseases such as asthma and chronic bronchitis. These responses include stimulation of gland secretion (Gashi et al., 1986; Borson et al., 1987), smooth muscle contraction (Lundberg and Saria, 1982a; Sekizawa et al., 1987a,b; Shore et al., 1988; Djokic et al., 1989a), vasodilation (Piedimonte et al., 1992), vascular extravasation (Lundberg and Saria, 1982b; Umeno et al., 1989), cough (Kohrogi et al., 1988), neutrophil and eosinophil adhesion to endothelium (Umeno et al., 1989; Katayama et al., 1993; Baluk et al., 1993), and immunological and hypersensitity responses in T lymphocytes (Payan et al., 1983), B lymphocytes (Stanisz et al., 1986; Scicchitano et al., 1987), mast cell degranula-

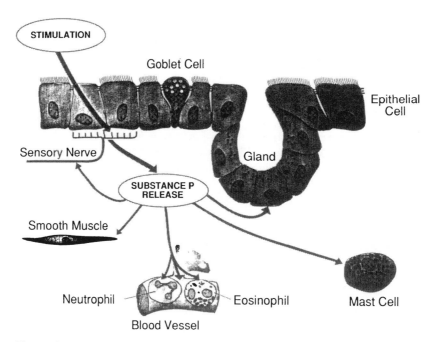

Figure 1   Diagram depicting the events that occur in neurogenic inflammation in airways. mechanical, chemical, and pharmacological stimulation cause sensory nerves to release neuropeptides (here depicted as substance P, but also including neurokinin A and calcitonin gene-related peptide). Released neuropeptides diffuse to target tissues and contract airway smooth muscle, stimulate gland secretion, stimulate mast cells, and have multiple pro-inflammatory effects in blood vessels. For details, see text.

tion (Mazurek et al., 1981), and ion transport (Al-Bazzaz et al., 1985; Mizoguchi and Hicks, 1989).

II. MODIFICATION OF NEUROGENIC INFLAMMATORY RESPONSES

Neurogenic inflammation is generally viewed as protective, a means of responding to noxious stimuli. However, exaggeration of neurogenic inflammation could cause exaggerated inflammatory responses and could thus contribute to the pathogenesis of inflammatory diseases. Various possibilities exist for causing exaggerated responses. Thus, increased synthesis and transport of peptide transmitters have been demonstrated in animal models of arthritis (Lembeck et al., 1981). Increased numbers of receptors for tachykinins could also result in exaggerated responses, and this has been described in chronic inflammatory bowel disease (Manthy et al., 1988). Finally, the actions of peptides are terminated by peptidases that normally cleave and thus inactivate neuropeptides. This would allow higher concentrations of the released peptides to exist in tissues, which would result in exaggerated responses. In this chapter, the role of membrane-bound peptidases in modulating neurogenic inflammation will be discussed. Evidence will be presented that decreased peptidase activity in airways results in exaggerated neurogenic inflammatory responses and that increased peptidase activity decreases neurogenic inflammatory responses.

III. ENZYMES THAT DEGRADE TACHYKININS IN AIRWAYS

In vitro, tachykinins are degraded by various enzymes, including serine proteases (Hanson and Lovenberg, 1980), mast cell chymase (Caughey et al., 1988), calpains (Murachi et al., 1987), neutral endopeptidase (reviewed in Turner, 1987), and angiotensin-converting enzyme (Skidgel et al., 1984). However, only two enzymes are known to play roles in the normal modulation of neurogenic inflammation in vivo: neutral endopeptidase (NEP; also called enkephalinase, EC 3.4.24.11) and angiotensin-converting enzyme (ACE; also called kininase II, dipeptidyl carboxypeptidase, peptidyl dipeptidase A, EC 3.4.15.1). Because NEP plays the predominant role in modulating most airway responses to tachykinins, it will be reviewed in greater detail.

IV. PHYSICAL CHARACTERISTICS OF NEP AND ACE

NEP and ACE are zinc metalloproteases that are variably glycosylated, and the enzymes are anchored to the cell by a single hydrophobic membrane-spanning domain. The catalytic domain faces the extracellular millieu, where it cleaves peptides as they diffuse through tissue. The zinc ion is an essential component of the catalytic site and acts as a target of inhibitors that utilize thiol (thiorphan, captopril), phosphate (phosphoramidon), or carboxyalkyl (enalapril, lisinopril) zinc-coordinating groups.

NEP has been cloned in rats (Malfroy et al., 1987), rabbits (Devault et al., 1987), humans (Malfroy et al., 1988), and mice (Chen et al., 1992). The open reading frame of the human cDNA encodes a 742-amino-acid polypeptide (molecular weight approximately 94 dKa). The N-terminal cytoplasmic domain contains only 27 amino acids. The molecule has a 23-amino-acid hydrophobic signal peptide in its membrane-spanning domain and a large C-terminal extracellular domain containing the catalytic site. There is more than 90% homology between cloned sequences among species, indicating a high degree of interspecies conservation. The human NEP gene is located on chromosome 3 (Tran et al., 1989).

Two ACE isoenzymes have been described. The larger form (150–180 kDa) is distributed broadly throughout the vascular endothelium, the highest concentration being found in kidneys and lungs (Patchett and Cordes, 1985). The smaller form (90–100 kDa) is found in the testis (Velletri, 1985). Human endothelial ACE cDNA encodes a 1306-amino-acid polypeptide. Unlike NEP, ACE is anchored to the cell surface by a hydrophobic region near its C-terminal domain (Soubrier et al., 1988). Endothelial ACE contains two highly symmetrical extracellular domains (Soubrier et al., 1988), each containing putative zinc-binding sequences. In contrast to NEP and testicular ACE, which contain a single extracellular domain, the degree of homology between human and mouse endothelial ACE is lower (83%) (Bernstein et al., 1989) than the homology for NEP.

V. SPECIFICITY OF NEP AND ACE

NEP was first discovered in the brush border epithelium of the kidney (Kerr and Kenny, 1974). Subsequently, a peptidase isolated from rat brain, called "enkephalinase" for its ability to cleave small opioid peptides (Malfroy et al., 1978), and the antigen CD10 (common acute lymphoblastic leukemia, CALLA) (Cossman et al., 1983) were found to be identical to NEP. NEP cleaves internal peptide bonds (hence the name endopeptidase) at the amino side of hydrophobic amino acids, while ACE cleaves C-terminal di- or tripeptides. Thus, NEP cleaves SP at the Gln^6-Phe^7, Phe^7-Phe^8, and Gly^9-Leu^{10} bonds (Matsas et al., 1983; Skidgel et al., 1984). ACE cleaves at the Phe^8-Gly^9 and Gly^9-Leu^{10} bonds (Skidgel et al., 1984). SP 1–9, the principal catabolic fragment generated by enzymatic cleavage of SP, is biologically inactive (Borson et al., 1987; Sekizawa et al., 1987b).

NEP cleaves a variety of substrates of different lengths, including SP, NKA, bradykinin, gastrin-releasing peptide, atrial natriuretic peptide, enkephalins, endothelin, insulin B-chain, neurotensin, and the bacterial chemotactic peptide *N*-formyl-Met-Leu-Phe. NEP cleaves CGRP only slowly (Katayama et al., 1991).

Thus, the selectivity of NEP and ACE depends in part on the ability of the enzyme to cleave and thus inactivate the specific peptide. However, the selectivity also resides in the specific cells where the peptidase is expressed. Because NEP and ACE are bound to cells and their active sites are located on the surface of the cell facing the external millieu, they act on peptides close to the surfaces of the cells to which they are attached.

VI. EVIDENCE FOR ROLE OF PEPTIDASES IN MODULATING NEUROGENIC INFLAMMATION

A. Localization of Peptidases and Sites of Neuropeptide Release from Sensory Nerves

The role of peptidases in modulating neurogenic inflammation depends on the cellular location of the peptidase, because the enzyme is anchored to specific cells and can therefore cleave neuropeptides in close association with the surfaces of the cells on which the peptidase is located. High levels of NEP and ACE activities exist in the lungs (Llorens and Schwartz, 1981; Patchett and Cordes, 1985), and tracheal homogenates contain both activities (Dusser et al., 1988a). Activity assays and immunohistochemical staining have demonstrated the presence of NEP in the tracheal epithelium (Sekizawa et al., 1987a; Borson et al., 1989), tracheal smooth muscle (Sekizawa et al., 1987b), submucosal glands (Nadel, 1992), alveolar epithelium (Johnson et al., 1985), and vascular endothelium (Llorens-Cortes et al., 1992). A similar pattern of immunoreactivity was found in a variety of species, including human, ferret, guinea pig, rat, and dog (Nadel, J.A., unpublished observation).

ACE is concentrated on the luminal surface of the vascular endothelium (Caldwell et al., 1976; Ryan et al., 1976; Johnson et al., 1985).

Thus, because NEP is widely distributed in a variety of cells, it would be anticipated that it would modulate responses in many cell types. On the other hand, ACE, which is predominantly located on endothelial cells, would be anticipated to modulate vascular responses (e.g., leukocyte adhesion, vasomotion, vascular permeability).

In addition to the location of the peptidases, the location of neuropeptide release from sensory nerve terminals is an important determinant of peptide actions. Sensory nerve terminals and effector cells do not possess a specialized unit similar to the neuromuscular junctions between motor nerves and nicotinic receptors on striated muscle cells. Presumably, sensory peptide neurotransmitters diffuse from sites of release to target cells. The distribution of sensory nerve fibers in sensory nerves containing tachykinins has been studied in rodent airways (Baluk et al., 1992): there is a dense network of these nerve fibers beneath the epithelium, and sensory nerve fibers are also present near smooth muscle, submucosal gland cells, and around venules (Baluk et al., 1992). Normal human tissues show a similar pattern of distribution, but SP-containing sensory nerves are less abundant than in rodents (Lundberg et al., 1984; Polak and Bloom, 1982). In disease, this distribution could change.

From the anatomical localization of NEP and the sites of neuropeptide release, it has been suggested (Nadel, 1991, 1992) that NEP modulates neurogenic inflammatory responses both at sites of neuropeptide release (mostly in the epithelium) and at sites of neuropeptide actions. When the airway epithelium is removed in vitro, the responses of airway smooth muscle to substance P (Sekizawa et al., 1991; Devillier et al., 1988) and to capsaicin (Djokic et al., 1989) are increased. However, there is evidence that NEP also modulates neurogenic inflammatory responses at sites of action on target cells; this evidence is reviewed elsewhere (Nadel, 1991, 1992). NEP is located on the surfaces of cells that contain tachykinin receptors, and studies of SP binding suggest that NEP regulates the binding of SP to its receptor by decreasing the amount of SP available to the receptor, without significantly changing the affinity or the number of receptors (Iwamoto et al., 1991). Thus, the active site of NEP faces the interstitial space. Released peptides diffuse through the tissues toward the cell receptor. In the unstirred layer near the cell surface, the enzyme and receptor compete for the peptide. Molecules of peptide are cleaved and inactivated, thus creating a decreased effective concentration of peptide accessible to the receptor. Figure 2 shows this concept diagrammatically. Evidence for this model is provided by studies of isolated airway smooth muscle in a chamber. Addition of SP causes an increase in muscle tension, which reaches a plateau. Addition of an NEP inhibitor to the bath results in a further increase in tension, suggesting that the effective concentration of SP at the muscle receptor has increased (Nadel, J.A., personal observation). Further evidence for competition between NEP and receptors on cells is provided by the fact that responses to SP are smaller in isolated cells engineered to express both NK-1 receptors and NEP than in cells expressing only NK-1 receptors, together with adjacent cells expressing NEP (Okamoto et al., 1993). In different species and in airways of different sizes in a single species, the distribution of sites of neuropeptide release may vary, and these differences could affect the responses of specific cells and the modulation by peptidases.

B. Effects of Selective Inhibitors of Peptidases on Neurogenic Inflammatory Responses

Selective inhibitors of NEP and ACE have been developed, and they have been useful in assessing the role of peptidases in modulating neurogenic inflammation. It was hypothesized that, if NEP modulates neuropeptide actions in airways, then selective inhibition of NEP should potentiate the actions of exogenously delivered and endogenously released tachykinins (Borson et al., 1987). These authors

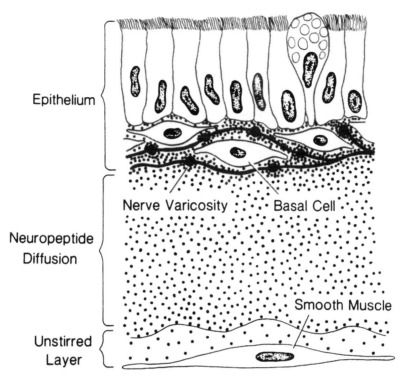

Figure 2 Diagrammatic model hypothesized for mechanisms of neutral endopeptidase modulation of airway smooth muscle responses to substance P (SP) and other tachykinins released from sensory nerves. In large airways, many of the release sites of neuropeptides from the sensory nerves are located in the epithelium in close contact with the basal cells, which contain the enzyme neutral endopeptidase on their surfaces. When the nerves release SP (whose concentration is indicated by the number of solid dots), cleavage and inactivation of the SP reduce the concentration of the peptide near sites of release (indicated by decreased number of dots). The remaining SP diffuses toward target cells (in the example, smooth muscle). Neutral endopeptidase on the surface of the target cell further reduces the concentration of SP in close contact with receptors on the cell surface. (Reproduced with permission from Nadel, 1992b.)

showed that NEP inhibition results in marked potentiation of tachykinin effects on macromolecule secretion in ferret tracheal submucosal glands (Borson et al., 1987). Nadel suggested that peptidase modulation might be a critical determinant of all neurogenic inflammation responses (Nadel, 1991). A variety of studies in various species confirmed this hypothesis. Thus, NEP inhibitors shift to the left the concentration-response curves to tachykinins in ferret (Sekizawa et al., 1987a), guinea pig (Stimler-Gerard, 1987), and human (Black et al., 1988; Naline et al.,

1989) airway smooth muscle. Similar potentiating bronchomotor effects of NEP occur in vivo in guinea pigs (Dusser et al., 1988) and in humans (Cheung et al., 1992). NEP inhibitors also potentiate SP-induced contraction in gut smooth muscle (Djokic et al., 1989b). In awake guinea pigs, cough induced by aerosolized SP is potentiated markedly by NEP inhibitors (Kohrogi et al., 1988; Kohrogi et al., 1989). SP-induced increase in cholinergic neurotransmission is also exaggerated by NEP inhibition (Sekizawa et al., 1987b). The effects of endogenously released tachykinins after electrical stimulation of the vagus nerves or after the administration of capsaicin in airways are also exaggerated by peptidase inhibition. Smooth muscle contraction (Dusser et al., 1988a; Sheppard et al., 1988) and submucosal gland secretion (Haxhiu et al., 1991) in response to inflammation are also potentiated by NEP inhibitors in vivo. The effects are not only potentiated (Fig. 3), but also prolonged (Fig. 4).

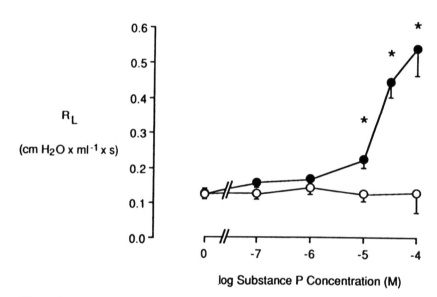

Figure 3 Concentration-response curves to aerosolized substance P (seven breaths) in the absence (open circles) or 15 min after the administration of aerosolized phosphoramidon (10^{-4} M, 90 breaths; filled circles) in anesthetized guinea pigs ($n = 5$ for each condition). Total pulmonary resistance (R_L) is expressed as means ± SEM. *Significant difference between control animals and animals treated with aerosolized phosphoramidon ($p <$ 0.005). Aerosolized substance P alone did not increase R_L at any concentration. However, after administration of phosphoramidon, a neutral endopeptidase inhibitor, aerosolized substance P induced slowly developing and long-lasting increases in R_L that were concentration dependent. (Reproduced with permission from Dusser et al., 1988a.)

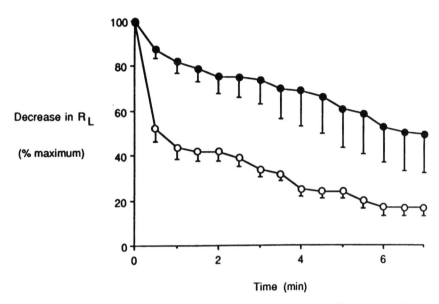

Figure 4   Time course of decrease in total pulmonary resistance (R_L) after maximum bronchoconstrictor response to vagus nerve stimulation (10 V, 3 Hz, 5 msec for 20 sec) either 15 min after administration of aerosolized 0.9% NaCl solution (90 breaths, open circles) or 15 min after administration of aerosolized phosphoramidon (10^{-4} M, 90 breaths, filled circles) in anesthetized guinea pigs ($n = 5$ for each condition). R_L (means $\pm$ SE) is expressed as percent of maximum increase in R_L above baseline value before nerve stimulation. After administration of phosphoramidon, a neutral endopeptidase inhibitor, the increase in R_L was much more prolonged. (Reproduced with permission from Dusser et al., 1988a.)

In glands and airway smooth muscle, ACE inhibitors do not appear to affect neurogenic inflammatory responses. However, tachykinin-induced effects in the vascular bed are modulated by both NEP and ACE. Thus, both NEP and ACE inhibitors potentiate the vascular extravasation induced by sensory nerve stimulation (Umeno et al., 1989) and by injection of SP (Piedimonte et al., 1991). Similarly, neurogenic inflammation induces neutrophil adhesion to endothelium, and both NEP and ACE inhibitors exaggerate this response (Umeno et al., 1989; Katayama et al., 1993). The simultaneous inhibition of NEP and ACE potentiates SP-induced effects more than a maximally effective dose of either inhibitor alone (Piedimonte et al., 1991; Katayama et al., 1993). Tachykinin-induced plasma extravasation is also modulated by NEP in skin (Iwamato et al., 1989).

Capsaicin also causes plasma extravasation in the nose, an effect that is potentiated by NEP inhibition and, to a lesser extent, by ACE inhibition (Petersson et al., 1993). Both SP and capsaicin increase microvascular blood flow in rat airways (Piedimonte et al., 1992) and in rat nasal mucosa (Piedimonte et al., 1993b). The threshold for vasodilation is much less than the dose required to induce vascular extravasation. ACE (Piedimonte et al., 1992) and NEP inhibition (Yamawaki et al., 1993) lower the threshold and prolong the duration of vasodilation in airways induced by SP. The findings are in contrast to isolated pulmonary arteries where the relaxant effect of SP is potentiated by ACE inhibitors but not by NEP inhibitors (Rouissi et al., 1990).

Many stimuli cause inflammatory responses, and peptidase inhibitors have been useful in analyzing the role of neurogenic inflammation in the responses. Thus, inhalation of hypertonic saline aerosol causes plasma extravasation in rats (Umeno et al., 1990). An NEP inhibitor increased this response markedly and capsaicin pretreatment abolished the response. A selective NK-1 receptor antagonist (CP-99,994) inhibited plasma extravasation produced by inhalation of hypertonic saline aerosols (Piedimonte et al., 1993c). Together these results implicate neurogenic inflammation as the cause of hypertonic saline-induced plasma extravasation in rats by releasing tachykinins in airways, acting via NK-1 receptors. Toluene diisocyanate (TDI), a chemical that causes occupational asthma, contracts bronchial smooth muscle in guinea pigs by releasing neuropeptides from sensory nerves. These responses are exaggerated in the presence of NEP inhibitors (Sheppard et al., 1989; Mapp et al., 1991).

VII. POTENTIATION OF NEUROGENIC INFLAMMATION BY STIMULI THAT DECREASE NEUTRAL ENDOPEPTIDASE ACTIVITY

Respiratory viral infections (Jacoby et al., 1988; Dusser et al., 1989a; Borson et al., 1989), cigarette smoke (Dusser et al., 1989b), and the industrial pollutant toluene diisocyanate (Sheppard et al., 1988) all exaggerate neurogenic inflammatory bronchoconstriction, and they decrease NEP activity. In each of these conditions, the fact that pharmacological inhibition of NEP causes little or no further exaggeration of neurogenic inflammatory responses provides further evidence that decreased NEP activity is responsible for the exaggerated responses.

TDI is reported to inhibit NEP activity in airway tissue (Sheppard et al., 1988). TDI can also stimulate the release of tachykinins (Mapp et al., 1991). Sendai virus infection in airways of rats potentiates neurogenic plasma extravasation (Piedimonte et al., 1990), and this exaggerated response is accounted for by decreased NEP activity in airways. Respiratory virus infections also exaggerate the increase in airway blood flow evoked by SP in rats (Yamawaki et al., 1993), and evidence suggests that decreased peptidase activity in the airways is the cause of this exaggerated response.

VIII.   EFFECTS OF INCREASED NEUTRAL ENDOPEPTIDASE ON NEUROGENIC INFLAMMATION

Human NEP has recently been cloned and expressed (Malfroy et al., 1988), and some of its potential therapeutic uses have been evaluated. Aerosolized recombinant human NEP inhibited cough produced by both endogenous and exogenous tachykinins in guinea pigs (Kohrogi et al., 1989). Recombinant human NEP has also been used successfully to prevent neuropeptide responses in other organs. Various tachykinins cause plasma extravasation in guinea pig skin, an effect that is potentiated by NEP inhibitors (Iwamoto et al., 1989). Recombinant human NEP inhibited SP-induced plasma extravasation in the skin in a dose-dependent fashion (Rubinstein et al., 1990).

In the human eye (donors), SP-induced contraction of the iris was potentiated by NEP inhibitors (Anderson et al., 1990). In the rabbit eye, recombinant human NEP prevented SP-induced contraction of the iris (Malfroy et al., 1988). These findings suggest that recombinant NEP might be useful in the treatment of symptoms of diseases involving inflammatory peptides such as SP and other peptides such as bradykinin (Gafford et al., 1983; Dusser et al., 1988) that are cleaved by NEP.

Glucocorticoids are known to suppress inflammation, but the mechanisms are generally unknown. Because these substances are known to up-regulate the synthesis of some peptidases (e.g., Mendelsohn et al., 1982), their effects on neurogenic plasma extravasation were examined. Neutrophil adhesion to post-capillary endothelium is produced in the rat trachea by aerosolized capsaicin (Katayama et al., 1983). Corticosteroids caused dose-dependent suppression of this (NK-1 receptor) neurogenic inflammatory process. When NEP and ACE inhibitors were added together, the corticosteroid suppressive effect was reversed. These results suggest that the suppressive effect of corticosteroids is due to an increase in the activity of peptidases that degrade sensory neuropeptides. Dexamethasone also reduced, in a dose-dependent fashion, the magnitude of plasma extravasation produced in rat airways by neurogenic inflammation, and the evidence suggests that this is due, at least in part, to an increase in NEP activity (Piedimonte et al., 1991). That corticosteroids are capable of up-regulating the expression of NEP has been demonstrated directly in human tracheal epithelial cells (Borson and Gruenert, 1991). Thus, some of the antiinflammatory effects of corticosteroids in disease could be due to up-regulation of peptidases that cleave and inactivate inflammatory peptides such as tachykinins and kinins.

IX.   EVIDENCE THAT NEUROGENIC INFLAMMATION MAY PLAY A ROLE IN ASTHMA AND OTHER DISEASES

Exercise causes bronchoconstriction in asthmatics, and this is believed to be due to airway drying with the resultant increase in airway luminal fluid osmolarity.

Hypertonic saline causes bronchospasm in animals and also increases vascular extravasation in airways. This extravasation is exaggerated by the addition of NEP inhibitors (Umeno et al., 1990). Hypertonic saline-induced extravasation is due to activation of NK-1 receptors (Piedimonte et al., 1993b). Hypertonic saline aerosols also cause increased adhesion of neutrophils and eosinophils to the postcapillary venular endothelium via NK-1 receptor effect (Baluk et al., 1993).

To test the role of neurogenic inflammation allergic airway disease, ovalbumin-sensitized guinea pigs were studied. Inhalation of antigen resulted in vascular extravasation in the trachea, which was present within 5 min and reached a maximum 10 min after exposure. The late effect was potentiated by phosphoramidon, an NEP inhibitor, and was inhibited by CP-96,345, a selective inhibitor of NK-1 receptors (Fig. 5), implicating neurogenic inflammatory pathways in this

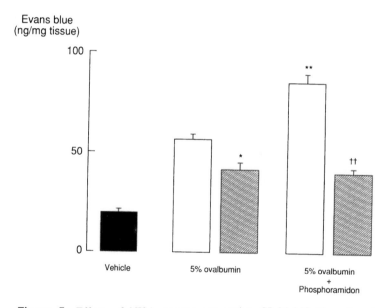

Figure 5 Effects of NK-1 receptor antagonist, CP-96,345 (4 mg/kg, i.v.; hatched columns), on Evans blue extravasation induced in the guinea pig trachea by inhalation of ovalbumin (5% for 2 min; open columns) in absence or presence of phosphoramidon (2.5 mg/kg, i.v. 5 min before the exposure to ovalbumin). Vascular extravasation was evaluated by measuring the amount of Evans blue dye extravasated in the trachea after 10 min. Values are means SEM; $n = 5$/group. **Significantly different from 5% ovalbumin ($p < 0.01$). ††Significantly different from 5% ovalbumin in presence of phosphoramidon ($p < 0.01$). Phosphoramidon, a selective neutral endopeptidase inhibitor, potentiated ovalbumin-induced plasma extravasation, and the CP-96,345, an NK-1 receptor antagonist reduced the effect of ovalbumin markedly. (Reproduced with permission from Bertrand et al., 1993a.)

late response (Bertrand et al., 1993). Bradykinin, a nine-amino-acid peptide, is believed to play a role in the symptoms of allergic rhinitis, common cold, and influenza virus infection (Baumgartner et al., 1986; Carter Barnett et al., 1990; Proud et al., 1990; Regoli and Barabè, 1980; Togias et al., 1985). Bradykinin causes vascular extravasation in the rat nasal mucosa by releasing neuropeptides from sensory nerves, acting via NK-1 receptors (Fig. 6). These effects are exaggerated when NEP and ACE are inhibited. Thus, it is possible that decreased activities of the peptidases (e.g., caused by disease) could exaggerate bradykinin-induced inflammatory responses to the point of causing symptomatic disease. Antigen-induced bronchoconstriction in vitro is also potentiated by NEP inhibition, but only 10 min after challenge, suggesting that neurogenic inflammation plays a similar role in allergic bronchoconstriction (Kohrogi et al., 1991). In sensitized guinea pigs, plasma extravasation induced by antigen is prevented to an equivalent degree by a bradykinin β_2 antagonist and by an NK-1 receptor antago-

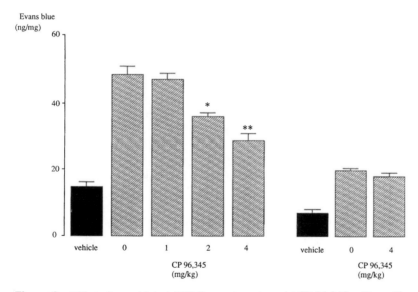

Figure 6   Effect of neurokinin-1 (NK-1) receptor antagonist CP-96,345 on Evans blue dye extravasation in nasal mucosa induced by instillation of 25 nmol bradykinin (left) or by intravenous injection of 1 μmol/kg bradykinin in presence of phosphoramidon and captopril (right). Rats were pretreated with inhibitors and NK-1 receptor antagonist 5 min before exposure to bradykinin. Values are means ± SE; $n = 5$ rats/group. Significantly different from control: *$p < 0.05$, **$p < 0.01$. Pretreatment with CP-96,345, a selective NK-1 receptor antagonist, reduced the increase in extravasation produced by intranasal instillation of bradykinin in a dose-dependent fashion but had no effect of extravasation caused by intravenous bradykinin. (Reproduced with permission from Bertrand et al., 1993b.)

nist. Thus, antigen-antibody response results in kinin generation, which in turn stimulates the release of neuropeptides from sensory nerves. This cascade suggests therapeutic strategies for preventing inflammatory effects in allergic states. Recently, a tachykinin receptor antagonist has been reported to inhibit bronchial responses to inhaled bradykinin aerosols (Ichinose et al., 1992). This is the first study in humans to suggest a role for neurogenic inflammation in airway disease.

Nasal resistance to airflow is increased by local instillation of SP in a dose-dependent fashion in normal subjects and in patients with allergic rhinitis (Lurie et al., in press). NEP inhibition increase SP responses in both normal and allergic subjects. Curiously, ACE inhibition had no effect on the normal subjects. These results suggest that peptidases play a role in modulating nasal responses and that disease may modify these peptidase effects.

As mentioned above, TDI causes occupational asthma. It causes airway inflammation in exposed humans. The asthmatic responses of TDI could be due in part to effects on neurogenic inflammation. Similarly, respiratory viral infections decrease NEP activity and thus magnify responses to neuropeptides (Jacoby et al., 1988; Borson et al., 1989; Piedimonte et al., 1990). One of the mechanisms by which viral infections might trigger asthma attacks is by decreasing peptidase activity and thus magnifying responses to released peptide mediators.

X. SUMMARY

A noncholinergic, nonadrenergic nervous system has been shown to cause a series of inflammatory effects of neuropeptides released from sensory nerves in the airways. These affects are normally limited by peptidases located on cells in the airways that cleave and thus inactivate the peptides. One peptidase, neutral endopeptidase, is located on multiple target cells and near sites of neuropeptide release, and it modulates multiple neurogenic inflammatory responses. Another peptidase, kininase II, exists on endothelial cells and modulates vascular neurogenic responses selectively. Up-and-down regulation of these peptidases has profound effects on neurogenic inflammatory responses. These effects suggest that neurogenic inflammation may play a role in allergic and other inflammatory diseases of airways.

REFERENCES

Al-Bazzaz, F.J., Kelsey, J.G., and Kaage, W.D. (1985). Substance P stimulation of chloride secretion by canine tracheal mucosa. *Am. Rev. Respir. Dis. 131*:86.

Anderson, J.A., Malfroy, B., Richard, N.R., Kullerstrand, L., Lucas, C., and Binder, P.S. (1990). Substance P contracts the human iris sphincter: possible modulation by endogenous enkephalinase. *Regul. Pept. 29*:49.

Baluk, P., Nadel, J.A., and McDonald, D.M. (1992). Substance P–immunoreactive axons in the rat respiratory tract: a quantitative study of their distribution and role in neurogenic inflammation. *J. Comp. Neurol. 319*:586.

Baluk, P., Bertrand, C., Geppetti, P., McDonald, D.M., and Nadel, J.A. (1993). The NK-1 receptor antagonist, CP-96,345, inhibits the adhesion of neutrophils and eosinophils in rat trachea. *Am. Rev. Respir. Dis. 147*:A475.

Baumgartner, C.R., Nicholos, R.C., Naclerio, R.M., Lichtenstein, L.M., Norman, P.S., and Proud, D. (1986). Plasma kallikrein during experimentally induced allergic rhinitis: role in kinin formation and contribution to TAME-esterase activity in nasal secretions. *J. Immunol. 137*:977.

Bernstein, K.E., Martin, B.M., Edwards, A.S., and Bernstein, E.A. (1989). Mouse angiotensin-converting enzyme is a protein composed of two homologous domains. *J. Biol. Chem. 264*:11945.

Bertrand, C., Geppetti, P., Baker, J., Yamawaki, I., and Nadel, J.A. (1993a). Role of neurogenic inflammation in antigen-induced vascular extravasation in guinea pig trachea. *J. Immunol. 150*:1479.

Bertrand, C., Geppetti, P., Baker, J., Petersson, G., Piedimonte, G., and Nadel, J.A. (1993b). Role of peptidases and NK_1 receptors in vascular extravasation induced by bradykinin in rat nasal mucosa. *J. Appl. Physiol. 74*:2456.

Black, J.L., Johnson, P.R.A., and Armour, C.L. (1988). Potentiation of the contractile effects of neuropeptides in human bronchus by an enkephalinase inhibitor. *Pulm. Pharm. 1*:21.

Borson, D.B., and Gruenert, D.C. (1991). Glucocorticoids induce neutral endopeptidase in transformed human tracheal epithelial cell. *Am. J. Physiol. 260*:L83.

Borson, D.B., Corrales, R., Varsano, S., Gold, M., Viro, N., Caughey, G., Ramachandran, J., and Nadel, J.A. (1987). Enkephalinase inhibitors potentiate substance P-induced secretion of $^{35}SO_4$-macromolecules from ferret trachea. *Exp. Lung Res. 12*:21.

Borson, D.B., Brokaw, J.J., Sekizawa, K., McDonald, D.M., and Nadel, J.A. (1989). Neutral endopeptidase and neurogenic inflammation in rats with respiratory infections. *J. Appl. Physiol. 66*:2653.

Caldwell, P.R.B., Seegal, B.C., Hsu, K.C., Das, M., and Soffer, R.L. (1976). Angiotensin-converting enzyme: vascular endothelial localization. *Science 191*:1050.

Carter Barnett, J.K., Cruse, L.W., and Proud, D. (1990). Kinins are generated in nasal secretions during influenza A infections in ferrets. *Am. Rev. Respir. Dis. 142*:162.

Caughey, G.H., Leidig, F., Viro, N.F., and Nadel, J.A. (1988). Substance P and vasoactive intestinal peptide degradation by mast cell tryptase and chymase. *J. Pharmacol. Exp. Ther. 244*:133.

Chen, C.Y., Salles, G., Seldin, M.F., Kister, A.E., Reinherz, E.L., and Shipp, M. (1992). Murine common acute lymphoblastic leukemia antigen (CD10 neutral endopeptidase 24.11). Molecular characterization, chromosomal localization, and modeling of the active sites. *J. Immunol. 148*:2817.

Cheung, D., Bel, E.H., Den Hartigh, J., Dijkman, J.H., and Sterk, P.J. (1992). The effect of inhaled neutral endopeptidase inhibitor, thiorphan, on airway responses to neurokinin A in normal humans in vivo. *Am. Rev. Respir. Dis. 145*:1275.

Cossman, J., Neckers, L.M., Leonard, W.J., and Greene, W.C. (1983). Polymorphonuclear neutrophils express the common acute lymphoblastic leukemia antigen. *J. Exp. Med.* *157*:1064.

Devault, A., Lazure, C., Nault, C., LeMoual, H., Seidah, N.G., Chretien, M., Kahn, P., Powell, J., Mallet, J., Beaumont, A., Roques, B.P., Crine, P., and Boileau, G. (1987). Amino acid sequence of rabbit neutral endopeptidase 24.11 (enkephalinase) deduced from complementary DNA. *EMBO J. 6*:1317.

Devillier, P., Advenier, C., Drapeau, G., Marsac, J., and Regoli, D. (1988). Comparison of the effects of epithelium removal and of an enkephalinase inhibitor on the neurokinin-induced contractions of guinea-pig isolated trachea. *Br. J. Pharmacol. 94*:675.

Djokic, T.D., Nadel, J.A., Dusser, D.J., Sekizawa, K., Graf, P.D., and Borson, D.B. (1989a). Inhibitors of neutral endopeptidase potentiate electrically and capsaicin-induced noncholinergic contraction in guinea pig bronchi. *J. Pharmacol. Exp. Ther. 248*:7.

Djokic, T.D., Sekizawa, K., Borson, D.B., and Nadel, J.A. (1989b). Neutral endopeptidase inhibitors potentiate substance P-induced contraction in gut smooth muscle. *Am. J. Physiol. 256*:G39.

Dusser, D.J., Umeno, E., Graf, P.D., Djokic, T., Borson, D.B., and Nadel, J.A. (1988a). Airway neutral endopeptidase-like enzyme modulates tachykinin-induced broncho-constriction in vivo. *J. Appl. Physiol. 65*:2585.

Dusser, D., Nadel, J., Sekizawa, K., Graf, P., and Borson, D. (1988b). Neutral endopep-tidase and angiotensin converting enzyme inhibitors potentiate kinin-induced contrac-tion of ferret trachea. *J. Pharmacol. Exp. Ther. 244*:531.

Dusser, D.J., Jacoby, D.B., Djokic, T.D., Rubinstein, I., Borson, D.B, and Nadel, J.A. (1989a). Virus induces airway hyperresponsiveness to tachykinins: role of neutral endopeptidase. *J. Appl. Physiol. 67*:1504.

Dusser, D.J., Djokic, T.D., Borson, D.B., and Nadel, J.A. (1989b). Cigarette smoke induces bronchoconstrictor hyperresponsiveness to substance P and inactivates airway neutral endopeptidase in the guinea pig. Possible role of free radicals. *J. Clin. Invest. 84*:900.

Gafford, J.T., Skidgel, R.A., Erdos, E.G., and Hersh, L.B. (1983). Human kidney "enkephalinase," a neutral metalloendopeptidase that cleaves active peptides. *Bio-chemistry 22*:3265.

Gashi, A.A., Borson, D.B., Finkbeiner, W.E., Nadel, J.A., and Basbaum, C. (1986). Neuropeptides degranulate serous cells of ferret tracheal glands. *Am. J. Physiol. 251*:C223.

Geppetti, P., Del Bianco, E., Patacchini, R., Santicioli, P., Maggi, C.A., and Tramontana, M. (1991a). Low pH-induced release of calcitonin gene-related peptide from capsaicin-sensitive sensory nerves: mechanism of action and biological response. *Neuroscience 41*:295.

Geppetti, P., Del Bianco, E., Tramontana, M., Vigano, T., Folco, G.C., Maggi, C.A., Manzini, S., and Fanciullacci, M. (1991b). Arachidonic acid and bradykinin share a common pathway to release neuropeptides from capsaicin-sensitive sensory nerve fibers of the guinea pig heart. *J. Pharmacol. Exp. Ther. 259*:759.

Hanson, G.R., and Lovenberg, W. (1980). Elevation of substance P–like immunoreactivity in the rat central nervous system by protease inhibitors. *J. Neurochem. 35*:1370.

Haxhiu, M.A., Tseng, H.C., and Davis, B. (1988). Thiorphan enhances baseline and substance P induced tracheal gland secretion in dogs. *Clin. Res. 36*:506A.

Holzer, P. (1991). Capsaicin: cellular targets, mechanisms of actions, and selectivity for thin sensory neurons. *Pharmacol. Rev. 43*:143.

Ichinose, M., Nakajima, N., Takahashi, T., Yamauchi, H., Inoue, H., and Takishima, T. (1992). Protection against bradykinin-induced bronchoconstriction in asthmatic patients by neurokinin receptor antagonist. *Lancet 340*:1248.

Iwamoto, I., Ueki, I., and Nadel, J.A. (1988). Effect of neutral endopeptidase inhibitors in ^{3}H-substance P binding in rat ileum. *Neuropeptides 11*:185.

Iwamoto, I., Ueki, I.F., Borson, D.B., and Nadel, J.A. (1989). Neutral endopeptidase modulates tachykinin-induced increase in vascular permeability in guinea pig skin. *Int. Arch. Allergy Appl. Immunol. 88*:288.

Jacoby, D.B., Tamaoki, J., Borson, D.B., and Nadel, J.A. (1988b). Influenza infection causes airway hyperresponsiveness by decreasing enkephalinase. *J. Appl. Physiol. 64*:2653.

Johnson, A.R., Ashton, J., Schulz, W.W., and Erdos, E.G. (1985). Neutral metalloendopeptidase in human lung tissue and cultured cells. *Am. Rev. Respir. Dis. 132*:564.

Katayama, M., Nadel, J.A., Bunnett, N.W., Di Maria, G.U., Haxhiu, M., and Borson, D.B. (1991). Catabolism of calcitonin gene-related peptide and substance P by neutral endopeptidase. *Peptides 12*:563.

Katayama, M., Nadel, J.A., Piedimonte, G., and McDonald, D.M. (1993). Peptidase inhibitors reverse steroid-induced suppression of neutrophil adhesion in rat tracheal blood vessels. *Am. J. Physiol. 264*:L316.

Kerr, M.A., and Kenny, A.J. (1974). The purification and specificity of a neutral endopeptidase from rabbit kidney brush border. *Biochem. J. 137*:477.

Kohrogi, H., Graf, P.D., Sekizawa, K., Borson, D.B., and Nadel, J.A. (1988). Neutral endopeptidase inhibitors potentiate substance P- and capsaicin-induced cough in awake guinea pigs. *J. Clin. Invest. 82*:2063.

Kohrogi, H., Nadel, J.A., Malfroy, B., Gorman, C., Bridenbaugh, R., Patton, J.S., and Borson, D.B. (1989). Recombinant human enkephalinase (neutral endopeptidase) prevents cough induced by tachykinins in awake guinea pigs. *J. Clin. Invest. 84*:781.

Kohrogi, H., Yamaguchi, T., Kawano, O., Honda, I., Ando, M., and Araki, S. (1991). Inhibition of neutral endopeptidase potentiates bronchial contraction induced by immune response in guinea pigs in vitro. *Am. Rev. Respir. Dis. 144*:636.

Lembeck, F., Donnerer, J., and Copaert, F.C. (1981). Increase in substance P primary afferent nerves during chronic inflammation. *Neuropeptides 1*:175.

Llorens-Cortes, C., Huang, H., Vicart, P., Gasc, G.-M., Paulin, D., and Corvol, P. (1992). Identification and characterization of neutral endopeptidase in endothelial cells from venous and arterial origin. *J. Biol. Chem. 267*:14012.

Lundberg, J.M., and Saria, A. (1982a). Bronchial smooth muscle contraction induced by stimulation of capsaicin-sensitive sensory neurons. *Acta Physiol. Scand. 116*:473.

Lundberg, J.M., and Saria, A. (1982b). Capsaicin-sensitive vagal neurons involved in control of vascular permeability in rat trachea. *Acta Physiol. Scand. 115*:521.

Lundberg, J.M., and Saria, A. (1983). Capsaicin-induced desensitization of airway mucosa to cigarette smoke, mechanical and chemical irritants. *Nature 302*:251.

Lurie, A., Nadel, J.A., Roisman, G., Siney, H., and Dusser, D.J. (1993). Role of neutral endopeptidase and kininase II on substance P-induced increase in nasal obstruction in patients with allergic rhinitis. *Am. Rev. Respir. Dis.* (in press).

Malfroy, B., Swerts, J.P., Guyon, A., Roques, B.P., and Schwartz, J.C. (1978). High-affinity enkephalin-degrading peptidase in brain is increased after morphine. *Nature 276*:523.

Malfroy, B., Schofield, P.R., Kuang, W.J., Seeburg, P.H., Mason, A.J., and Henzel, W.J. (1987). Molecular cloning and amino acid sequence of rat neutral endopeptidase. *Biochem. Biophys. Res. Commun. 144*:59.

Malfroy, B., Kuang, W.J., Seeburg, P.H., Mason, A.J., and Schofield, P.R. (1988). Molecular cloning and amino acid sequence of human enkephalinase (neutral endopeptidase). *FEBS Lett. 229*:206.

Malfroy, B., Liggit, D., McCaber, J., Baughman, R., Kado-Fong, H., Mulholland, K., Bridenbaugh, R., and Anderson, J. (1990). Administration of recombinant enkephalinase (neutral endopeptidase) prevents capsaicin-induced miosis in the rabbit eye in vivo. *J. Pharmacol. Exp. Ther. 252*:462.

Mantyh, C.R., Gates, T.S., Zimmermann, R.P., Welton, M.L., Passaro, E.P., Vigna, S.R., Maggio, J.E., Kruger, L., and Mantyh, P.W. (1988). Receptor binding sites for substance P but not substance K or neuromedin K are expressed in high concentrations by arterioles, venules and lymph nodules in surgical specimens obtained from patients with ulcerative colitis and Crohn disease. *Proc. Natl. Acad. Sci. USA 85*:3235.

Mapp, C.E., Boniotti, A., Graf, P.D., Chitano, P., Fabbri, L.M., and Nadel, J.A. (1991a). Bronchial smooth muscle responses evoked by toluene diisocyanate are inhibited by ruthenium red and by indomethacin. *Eur. J. Pharmacol. 200*:73.

Mapp, C.E., Graf, P.D., Boniotti, A., and Nadel, J.A. (1991b). Toluene diisocyanate contracts guinea pig bronchial smooth muscle by activating capsaicin-sensitive sensory nerves. *J. Pharmacol. Exp. Ther. 256*:1082.

Matsas, R., Fulcher, I.S., Kenny, A.J., and Turner, A.J. (1983). Substance P and (Leu)-enkephalin are hydrolyzed by an enzyme in pig caudate synaptic membranes that is identical with the endopeptidase of kidney microvilli. *Proc. Natl. Acad. Sci. USA 80*:3111.

Mazurek, N., Pecht, I., Teichburg, V.I., and Blumberg, S. (1981). The role of the N-terminal tetrapeptide in the histamine-releasing action of substance P. *Neuropharmacology 20*:1025.

Mendelsohn, F.A.O., Lloyd, C.J., Kachel, C., and Funder, J.W. (1982). Induction of glucocorticoids of angiotensin converting enzyme production from bovine endothelial cells in culture and rat lung in vivo. *J. Clin. Invest. 70*:684.

Mizoguchi, H., and Hicks, C.R. (1989). Effects of neurokinins on ion transport and sulfated macromolecule release in the isolated ferret trachea. *Exp. Lung Res. 15*:837.

Murachi, T., Hatanaka, M., and Hakamoto, T. (1987). Calpains and neuropeptide metabolism. In: *Neuropeptides and Their Peptidases.* (Turner, A.J., ed.). Ellis Horwood, Chichester, England, pp. 202–228.

Nadel, J.A. (1991). Neutral endopeptidase modulates neurogenic inflammation. *Eur. J. Respir. 4*:745.

Nadel, J.A., and Borson, D.B. (1991). Modulation of neurogenic inflammation by neutral endopeptidase. *Am. Rev. Respir. Dis. 143*:S33.

Nadel, J.A. (1992a). Membrane-bound peptidases: endocrine, paracrine, and autocrine effects. *Am. J. Respir. Cell Mol. Biol. 7*:469.

Nadel, J.A. (1992b). Regulation of neurogenic inflammation by neutral endopeptidase. *Am. Rev. Respir. Dis. 145*:S48.

Naline, E., Devillier, P., Drapeau, G., et al. (1989). Characterization of neurokinin effects and receptor selectivity in human isolated bronchi. *Am. Rev. Respir. Dis. 140*:679.

Okamoto, A., Payan, D.G., and Bunnett, N.W. (1993). Interactions between neutral endopeptidase (NEP, EC 3.4.24.11) and the substance P receptor (SPR, NK-1) expressed in mammalian cells. *Gastroenterology 104*:A560.

Patchett, A.A., and Cordes, E.H. (1985). The design and properties of N-carboxy-alkyldipeptide inhibitors of angiotensin converting enzyme, *Adv. Enzym. 57*:1.

Payan, D.G, Brewster, D.R., and Goetzl, E.J. (1983). Specific stimulation of human T-lymphocyte by substance P. *J. Immunol. 131*:1613.

Petersson, G., Bacci, E., McDonald, D.M., and Nadel, J.A. (1992). Neurogenic plasma extravasation in the rat nasal mucosa is potentiated by peptidase inhibitors. *J. Pharmacol. Exp. Ther. 264*:509.

Piedimonte, G., Nadel, J.A., Umeno, E., and McDonald, D.M. (1990). Sendai virus infection potentiates neurogenic inflammation in the rat trachea. *J. Appl. Physiol. 68*:754.

Piedimonte, G., McDonald, D.M., and Nadel, J.A. (1991). Neutral endopeptidase and kininase II mediate glucocorticoid inhibition of neurogenic inflammation in the rat trachea. *J. Clin. Invest. 88*:40.

Piedimonte, G., Hoffman, J.I.E., Husseini, W.K., Hiser, W.L., and Nadel, J.A. (1992). Effect of neuropeptides released from sensory nerves on blood flow in the rat airway microcirculation. *J. Appl. Physiol. 72*:1563.

Piedimonte, G., Hoffman, J.I.E., Husseini, W.K., Snider, R.M., Desai, M.C., and Nadel, J.A. (1993a). NK_1 receptors mediate neurogenic inflammatory increase in blood flow in rat airways. *J. Appl. Physiol. 74*:2462.

Piedimonte, G., Hoffman, J.I.E., Husseini, W.K., Bertrand, C., Snider, R., Desai, M.C., Petersson, G., and Nadel, J.A. (1993b). Neurogenic vasodilation in the rat nasal mucosa involves neurokinin₁ tachykinin receptors. *J. Pharmacol. Exp. Ther. 265*:36.

Piedimonte, G., Bertrand, C., Geppetti, P., Snider, R.M., Desai, M.C., and Nadel, J.A. (1993c). A new NK_1 receptor antagonist (CP-99,994) prevents the increase in tracheal vascular permeability produced by hypertonic saline. *J. Pharmacol. Exp. Ther. 266*:270.

Polak, J.M., and Bloom, S.R. (1982). Regulatory peptides and neuron specific enolase in the respiratory tract of man and other mammals. *Exp. Lung Res. 3*:313.

Proud, D., Naclerio, R.M., Gwaltney, J.M., and Hardley, J.O. (1990). Kinins are generated in nasal secretions during natural rhino-virus colds. *J. Infect. Dis. 161*:120.

Regoli, D., and Barabè, J. (1980). Pharmacology of bradykinin and related kinins. *Pharmacol. Rev. 32*:1.

Rouissi, N., Nantel, F., Drapeau, G., Rhaleb, N.E., Dion, S., and Regoli, D. (1990). Inhibitors of peptidases: how they influence the biological activities of substance P, neurokinins, kinins, and angiotensin in isolated vessels. *Pharmacology 40*:185.

Rubinstein, I., Iwamoto, I., Ueki, I.F., Borson, D.B., and Nadel, J.A. (1990). Recombinant neutral endopeptidase attenuates substance P-induced plasma extravasation in the guinea pig skin. *Int. Arch. Allergy Appl. Immunol. 91*:232.

Ryan, U.S., Ryan, J.W., Whitaker, C., and Chiu, A. (1976). Localization of angiotensin converting enzyme (kininase II). II. Immunocytochemistry and immunofluorescence. *Tissue Cell 8*:125.

Saria, A., Martling, C.R., Yan, Z., et al. (1988). Release of multiple tachykinins from capsaicin-sensitive sensory nerves in the lung by bradykinin, dimethylphenyl piperazinium, and vagal nerve stimulation. *Am. Rev. Respir. Dis. 137*:1330.

Scicchitano, R., Bienenstock, J., and Stanisz, A.M. (1987). In vivo immunomodulation by substance P. *Fed. Proc. 46*:781.

Sekizawa, K., Tamaoki, J., Graf, P.D., Basbaum, C., Borson, D.B., and Nadel, J.A. (1987a). Enkephalinase inhibitor potentiates mammalian tachykinin-induced contraction in ferret trachea. *J. Pharmacol. Exp. Ther. 243*:1211.

Sekizawa, K., Tamaoki, J., Nadel, J.A., and Borson, D.B. (1987b). Enkephalinase inhibitor potentiates substance P- and electrically induced contraction in ferret trachea. *J. Appl. Physiol. 63*:1401.

Sheppard, D., Thompson, J.E., Scypinski, L., Dusser, D., Nadel, J.A., and Borson, D.B. (1988). Toluene diisocyanate increases airway responsiveness to substance P and decreases airway enkephalinase. *J. Clin. Invest. 81*:1111.

Shore, S.A., Stimler-Gerard, N.P., Coats, S.R., and Drazen, J.M. (1988). Substance P-induced bronchoconstriction in the guinea pig: enhancement by inhibitors of neutral metalloendopeptidase and angiotensin-converting enzyme. *Am. Rev. Respir. Dis. 137*:331.

Skidgel, R.A. (1992). Bradykinin-degrading enzymes—structure, function, distribution, and potential roles in cardiovascular pharmacology. *J. Cardiovasc. Pharmacol. 20*:S4.

Skidgel, R.A., Engelbrecht, A., Johnson, A.R., and Erdos, E.G. (1984). Hydrolysis of substance P and neurotensin by converting enzyme and neutral endoproteinase. *Peptides 5*:769.

Soubrier, F., Alhenc-Gelas, F., Hubert, C., Allegrini, J., John, M., Tregear, G., and Corvol, P. (1988). Two putative active centers in human angiotensin I-converting enzyme revealed by molecular cloning. *Proc. Natl. Acad. Sci. USA 85*:9386.

Stanisz, A.M., Befus, A.D., and Bienenstock, J. (1986). Differential effects of vasoactive intestinal peptide, substance P and somatostatin on immunoglobulin synthesis and proliferation by lymphocytes from Peyer's patches, mesenteric lymph nodes, and spleen. *J. Immunol. 132*:152.

Stimler-Gerard, N.P. (1987). Neutral endopeptidase-like enzyme controls the contractile activity of substance P in guinea pig lung. *J. Clin. Invest. 79*:1819.

Togias, A.G., Naclerio, R.M., Proud, D., Fish, J.E., Adkinson, N.F., Kagey-Sobotka, A., Norman, P.S., and Lichtenstein, L.M. (1985). Nasal challenge with cold, dry air results in release of inflammatory mediators: possible mast cell involvement. *J. Clin. Invest. 76*:1375.

Tramontana, M., Cecconi, R., Del Bianco, E., Santicioli, P., Maggi, C.A., Alessandri, M., and Geppetti, P. (1991). Hypertonic media produce Ca^{2+}-dependent release of calcitonin gene-related peptide from capsaicin-sensitive nerve fibres in the rat urinary bladder. *Neurosci. Lett. 124*:79.

Tramontana, M., Giuliani, S., Del Bianco, E., Lecci, A., Maggi, C.A., Evangelista, S., and Geppetti, P. (1992). Effects of capsaicin and $5HT_3$ receptor antagonists on 5-HT-evoked release of calcitonin gene-related peptide in the guinea-pig heart. *Br. J. Pharmacol. 108*:431.

Tran, P.R., Willard, H.F., and Letarte, M. (1989). The common acute lymphoblastic leukemia antigen (neutral endopeptidase 3.4.24.11) gene is located on human chromosome 3. *Cancer Genet. Cytogenet. 42*:129.

Turner, A.J. (1987). Endopeptidase-24.11. In: *Neuropeptides and Their Peptidases*. (Turner, A.J., ed.). Ellis Horwood, Chichester, England, pp. 183–201.

Umeno, E., Nadel, J.A., Huang, H.T., and McDonald, D.M. (1989). Inhibition of neutral endopeptidase potentiates neurogenic inflammation in the rat trachea. *J. Appl. Physiol. 66*:2647.

Umeno, E., McDonald, D.M., and Nadel, J.A. (1990). Hypertonic saline increases vascular permeability in the rat trachea by producing neurogenic inflammation. *J. Clin. Invest. 85*:1905.

Velletri, P.A. (1985). Testicular angiotensin I-converting enzyme. *Life Sci. 36*:1597.

Yamawaki, I., Geppetti, P., Bertrand, C., Chan, B., Massion, P., Piedimonte, G., and Nadel, J.A. (1993). Sendai virus infection potentiates the increase in airway blood flow induced by substance P in rat. *Am. Rev. Respir. Dis. 147*:A476.

17

Control of Airway Vascular Beds

John Widdicombe

St. George's Hospital Medical School, London, England

I. INTRODUCTION

The vascular beds of the nose, trachea, and bronchi show some structural similarities but also considerable differences. This is not surprising in view of the anatomy of these structures. The nose has a rigid, bony case, and distention of the vascular bed, possibly with edema and exudate, will encroach on the nasal cavity and tend to block it. The tracheal cartilages have some rigidity, but are partly flexible; changes in mucosal thickness will have little effect on airflow resistance, especially since the thickness of the mucosa is small compared with the diameter of the lumen. For the bronchi, however, thickening of the mucosa due to vascular congestion will encroach on the airway lumen and will tend to cause appreciable obstruction. For the trachea and bronchi, the interaction between smooth muscle contraction and mucosal thickness is important, since both can influence the airway caliber whereas for the nose there is no equivalent of the lower airway smooth muscle. In addition to these differences in structure of the various parts of the respiratory tract, there are striking species differences for the vascular beds, especially for the lower airways. Some of these will be mentioned later.

Nearly all vascular studies have been applied to the nose and to the tracheobronchial tree, and there are very few observations on the vasculature of the pharynx and larynx. Here the whole structure of the respiratory tract is very different, the mucosa having a stratified epithelium and the caliber of the lumen being profoundly affected by the contraction of striated muscles in the airway walls. Although neuropeptide-containing nerves have been described here (Albegger et al., 1991), there has been little functional research; the impression is given that changes in vascular congestion and resultant mucosal thickness are of little

373

aerodynamic importance compared with the dramatic luminal changes that can be caused by contraction of the striated muscles.

The three components of the nervous system that exist throughout the respiratory tract can all act on the blood vessels there. These are the parasympathetic vasodilator, the sympathetic vasoconstrictor, and the sensory systems; the last mediate axon reflexes and neurogenic inflammation (see Chapter 15). All three systems act by the release of neuropeptides, and the two motor systems also act by release of conventional neurotransmitters.

Control of the airway vascular beds, including by neuropeptides, has not only pharmacological and physiological interest, but clinical significance. Thus both allergic rhinitis (hay fever) and nonallergic (vasomotor) rhinitis involve considerable vasodilatation and vascular congestion, which lead to nasal blockage. For the lower airways there is now abundant evidence that part of the primary pathology of asthma is an inflammatory change in the mucosa, which includes vasodilatation, edema, and exudate. For the bronchi and bronchioles this can contribute considerably to obstruction.

II. STRUCTURE OF VASCULAR BEDS

A. Capillaries

A common feature of the vascular anatomy of all parts of the airways is a subepithelial capillary plexus (Widdicombe, 1990, 1993). This is supplied by arteries and arterioles that penetrate the mucosa, and for the bronchi also the smooth muscle, radially to reach the subepithelium. In all species studied the capillary network in the nose is abundant and lies within about 15–30 μm of the epithelium (Cauna, 1992; Widdicombe, 1993). This is also true for the lower airways of most species, but an exception is the rabbit that has only a sparse capillary network (Hughes, 1965; A. Robson and J.G. Widdicombe, unpublished results). The capillary network can be seen under the light and electron microscopes and can be displayed by making plastic casts of the vasculature (Hill et al., 1989; Laitinen et al., 1989). It would provide a supply of nutrients and other blood-borne agents to the epithelium and glands and would also constitute a variable barrier to the diffusion of drugs and mediators from the lumen and epithelium through to deeper tissues. As for other capillary systems, there is no evidence that it is under nervous control, and there seem to be no studies on whether neuropeptides directly affect capillary structure and function.

The capillaries in the subepithelial plexuses of the nose are fenestrated, whereas in most species those in the lower airways are nonfenestrated (Watanabe and Watanabe, 1980; Cauna, 1982; Widdicombe, 1993). Exceptions seem to be rodents such as guinea pigs and rats, where the capillaries are usually fenestrated, and for all species at sites such as the neuroepithelial bodies [NEBs; paracrine structures containing bioactive peptides (Sorokin and Hoyt, 1990)] and bronchus-associated lymphoid tissue (BALT), where again the capillaries are fenestrated. In humans the lower airway capillaries are nonfenestrated, although in asthma fenestrations develop (Laitinen and Laitinen, 1991, 1992). The function of fenestrations is thought to be to aid the transport of water (but not of plasma proteins). A large water transudation might be appropriate for the nose with its air-conditioning role, and for the lower airways in conditions such as asthma where the epithelium is damaged and water evaporation may be increased.

B. Postcapillary Venules

The capillary network drains into postcapillary venules, which, unlike the capillaries, have smooth muscle in their walls as a discontinuous sheet. These vessels are responsible for vascular transudation in inflammatory conditions (McDonald, 1992; Persson, 1987, 1992). The endothelial cells contract and open gaps between them through which water and plasma constituents including albumin can flow. This in turn leads to exudation into the airway lumen by opening the tight junctions between the epithelial cells. As described elsewhere (see Chapter 15), this is an important site of action of sensory neuropeptides.

C. Capacitance Vessels (Sinuses)

The nasal mucosa has an exceptionally well-developed system of capacitance vessels or sinuses (Cauna and Hinderer, 1969; Cauna, 1970; Widdicombe, 1990; Baraniuk et al., 1990, 1991). These have smooth muscle in their walls and drain through throttle or cushion vessels. When they distend, the mucosa thickens and causes nasal blockage, with the opposite effect when they collapse. In theory their distention could be controlled by several factors: constriction of the throttle vessels, relaxation of the sinus wall smooth muscle, and opening of the entry vessels to the capacitance network would all lead to sinus distention. However, little is known about the relative importance of these three mechanisms and how they are affected by mediators including neuropeptides. Although there are assertions that arteriovenous anastomoses (AVAs) open directly into the capacitance system, so that dilatation of the AVAs would distend the sinuses, there is no functional and little histological support for this view. In the nose the vascular capacitance system is most prominent in the anterior respiratory regions, especially over the turbinates and septum, whereas further caudally in the olfactory region of the nose it is less prominent. The main physiological role of the capacitance system is probably to control the patency of the nasal lumen and thus to affect the air-conditioning and filtration powers of the nose.

The lower airways also possess a vascular capacitance system, but this is less well developed than in the nose and shows striking species differences (Hill et al., 1989; Hughes, 1965; Sobin et al., 1963). It is most prominent in the trachea but is also present in the bronchi, becoming less conspicuous as the airways become smaller. For the sheep and rabbit in the trachea the sinuses form an interrupted sheet between the subepithelial capillary network and the deeper tissues such as cartilage and smooth muscle. For the bronchi the network is most conspicuous at points of branching, at least for the sheep (Hill et al., 1989). Some species such as rodents have only a scanty capacitance system, whereas in humans the system is intermediate in density. For the sheep the sinuses have no smooth muscle in their walls, but only occasional pericytes outside. In both rabbits and humans there is a clear smooth muscle layer. Measurements of tracheal mucosal thickness in the sheep show that congestion can almost double mucosal thickness (Corfield et al., 1991), and presumably this is largely due to distention of the sinus vessels. Whether the same applies to the bronchi has not been determined.

D. Anastomoses

The nasal mucosa has an abundant system of AVAs, especially in the anterior respiratory part. They have been demonstrated both physiologically and anatomically and can take up to 50% of the nasal blood flow (Anggard, 1974; Malm, 1974). For the tracheobronchial tree no AVAs have been demonstrated histologically. Functional studies show that none can be detected in the sheep (Corfield et al., 1993), although they are present in the rabbit and open in response to vasodilator agents such as methacholine (Robson et al., 1993).

The bronchial vasculature has a complex system of anastomoses with the pulmonary blood vessels, and probably 75% of the bronchial blood flow drains into the pulmonary vascular bed. The remainder enters the systemic circulation via the azygos and hemiazygos veins (Matthay, 1992; Widdicombe, 1993). Thus the bronchial blood flow, unlike that for the trachea and nose, will be considerably influenced by the changes in the pulmonary vascular bed.

III. PARASYMPATHETIC MOTOR NEUROPEPTIDES

Vasoactive intestinal polypeptide (VIP), the parasympathetic neuropeptide most frequently studied, is found both in the vidian nerve supplying the nose and in the vagal supply to the lower airways. Peptide histidine isoleucine (PHI) and its human variant peptide histidine methionine (PHM) may coexist and be more important in some species. In general, the vascular actions of the three peptides seem to be similar and they come from a common molecular precursor. They coexist in parasympathetic nerves with acetylcholine (ACh) and probably nitric oxide (NO).

A. Histology

1. The Nose

The neurohistology is described elsewhere (see Chapter 1). Immunohistochemical methods show that VIP and PHI occur in many nerve fibers in the nasal mucosa, and the two peptides probably coexist. They are especially prominent in the anterior respiratory mucosa, where both the vasculature and the submucosal glands are also better developed. Fibers containing these neuropeptides lie around small blood vessels and glands and beneath the epithelium (Uddman and Sundler, 1986, 1987; Uddman et al., 1987; Albegger et al., 1991). They have been described in humans. There are also many nerves containing VIP/PHI in the parasympathetic motor ganglia to the nose (sphenopalatine), suggesting that there may be local reflex modulation at this site. Since the VIP is present in nerves that also stain for acetylcholinesterase, VIP and ACh probably coexist in the same neurons.

Galanin has also been identified by immunohistochemical methods in the nasal mucosa, with the same distribution as for VIP and ACh (Cheung et al., 1985). However, the evidence for galanin as a parasympathetic neuropeptide is far less secure than that for VIP, and galanin may also occur in sensory nerves.

2. The Tracheobronchial Tree

Here the general arrangement of VIP/PHI-containing nerves is very similar to that for the nose, with a clear association with blood vessels and submucosal glands as well as with airway smooth muscle. The ganglia in the tracheobronchial mucosa also contain VIP in both fibers and cells (Uddman et al., 1978). However, few VIP/PHI-containing nerves are found in the bronchi and smaller airways, at least for the pig and probably the cat (Martling et al., 1990). This observation correlates with functional studies indicating that parasympathetic neuropeptides have a more powerful action on the vasculature of the trachea than on that of the bronchi. In small mammals VIP-containing nerves seem to be less common near vascular smooth muscle and more conspicuous around submucosal glands (Springall et al., 1988).

Galanin has a similar distribution to VIP/PHI in many species, as shown by immunohistological methods (Cheung et al., 1985; Springall et al., 1988). It is found around blood vessels and submucosal glands and in the airway smooth muscle, with the highest concentration in the main bronchi. Its presence in the bronchi, with the virtual absence of VIP/PHI at this site in the pig, could point to a functional difference for the neuropeptides, which, however, has not been tested.

B. Pharmacology and Physiology

1. The Nose

There have been many studies of perfused vasculatures of the nose, especially for the dog, pig, and cat. Most neuropeptides have been tested by vascular injection. This shows that VIP is a powerful vasodilator and long-lasting compared with ACh (Lung et al., 1984; Anggard et al., 1979; Uddman et al., 1980; Stjarne, 1991) (Fig. 1). Other parasympathetic neuropeptides, such as PHI, PHM, and galanin, do not seem to have been tested. Vascular perfusions do not allow determination of which individual blood vessels are affected, a problem to be discussed later.

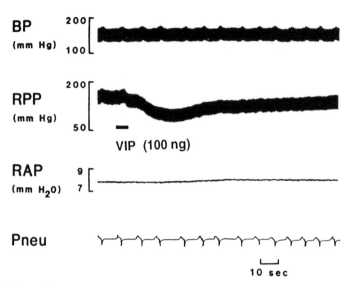

Figure 1 Effects on nasal arterial perfusion pressure in a dog after intra-arterial injection of VIP (100 ng injected at signal bar). (Traces from above down) Systemic arterial blood pressure (BP); right nasal arterial perfusion pressure (RPP); right nasal airway pressure (RAP); trachea airflow (Pneu). Note the vasodilatation without increase in airways resistance. (From Lung et al., 1984.)

However, VIP increased mucosal volume only slightly in the pig, together with a large decrease in total vasculature resistance; laser-Doppler measurements, possibly reflecting flow in subepithelial capillaries, were decreased (Stjarne, 1991) (Fig. 2). Thus it seems likely that at least the arterioles, the main site of vascular resistance, are dilated together with other vascular adjustments. Other neuropeptides decreased total vascular resistance but with different actions on mucosal volume and laser-Doppler signal; thus they must have different actions on different components of the vascular bed. In the dog, simultaneous measurements of nasal airflow resistance and vascular resistance during the administration of VIP suggest that the vasodilatation is accompanied by very little mucosal congestion and thickening, since airflow resistance increases only slightly (Lung et al., 1984) (Fig. 3). This is in contrast with other vasodilators such as bradykinin and histamine. It is possible that VIP opens AVAs, thus increasing nasal blood flow, but without greatly distending nasal capacitance vessels.

A strong vasodilatation is caused when parasympathetic nerves to the nose of cats are stimulated, and this response is atropine-resistant (Anggard et al., 1979). There is a simultaneous release of VIP into the venous outflow of the nose

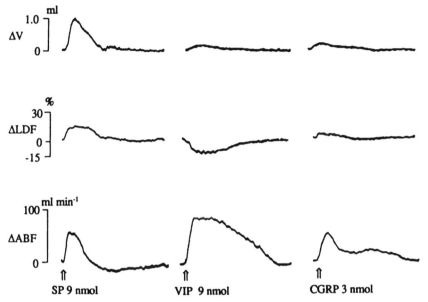

Figure 2 Typical recordings of changes (Δ) in nasal mucosal volume (V), laser-Doppler signal (LDF), and nasal arterial blood flow (ABF) upon infusion of substance P (SP), vasoactive intestinal polypeptide (VIP), and calcitonin gene-related peptide (CGRP) in a pig. All agents caused increased blood flow with various changes in laser-Doppler signal and in mucosal volume. (From Stjarne, 1991.)

(Lundberg et al., 1981). These experiments suggest that VIP is the effective vasodilator from parasympathetic nerves under physiological conditions.

2. The Tracheobronchial Tree

In the dog and pig, both VIP and PHI are powerful tracheal vasodilators, as shown by vascular perfusion experiments (Laitinen et al., 1987b; Salonen et al., 1988; Matran, 1991; Matran et al., 1989b) (Fig. 4). VIP is 10 times more active than PHI in the dog. The bronchial vasculature also dilates in response to exogenous VIP, but far less than does that of the trachea, at least for the pig (Matran, 1991; Matran et al., 1989a). These results support the histological evidence mentioned earlier.

When the parasympathetic (vagal) nerves to the tracheobronchial tree are electrically stimulated, there is dilatation of both the tracheal and bronchial vasculatures (Laitinen et al., 1987b; Martling et al., 1985; Matran et al., 1989a,b) (Figs. 5, 6, and 7). This could be due to three mechanisms: release of ACh from vagal motoneurons, release of VIP/PHI from the same motoneurons, and release of sensory neuropeptides by antidromic activation of sensory nerves. Prevention

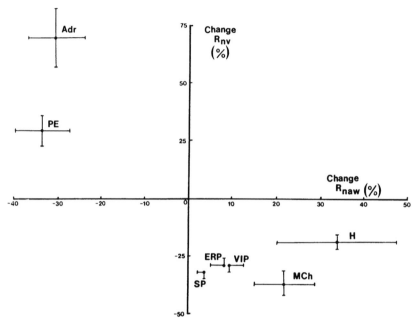

Figure 3 Comparison of the effects of various mediators and drugs injected into the perfused nasal vasculature of the dog. (Ordinate) Change in vascular resistance (R_{nv}); (abscissa) change in nasal airflow resistance (R_{naw}). Note that the neuropeptides substance P (SP), vasoactive intestinal polypeptide (VIP), and eloidsin-related peptide (ERP) cause similar decreases in vascular resistance compared with other vasodilators such as histamine (H) and methacholine (MCh), but far smaller increases in nasal airflow resistance. PE, phenylephine; Adr, adrenaline. Mean values, with bars indicating SEM. (Modified from Lung et al., 1984.)

of the first by giving atropine approximately halves the vasodilatation (Figs. 5, 6, and 7). Administration of hexamethonium, which would block the motor release of VIP/PHI but retain the release of sensory neuropeptides, again cuts the response by more than half. Thus all three mechanisms seem to exist (Laitinen et al., 1987c). The atropine-sensitive and -insensitive components are stimulus-frequency-dependent, and this result is consistent with the view that high-frequency nerve impulses release predominantly VIP-like transmitters whereas low-frequency stimulation is more effective in releasing ACh.

Neuropeptides such as VIP, administered into the vascular perfusate of the dog trachea, thicken the mucosa, as do various inflammatory mediators (Laitinen et al., 1986) (Fig. 8). However, as for the nose, VIP has a weaker action on thickness compared with vasodilatation than do mediators such as histamine and bradykinin.

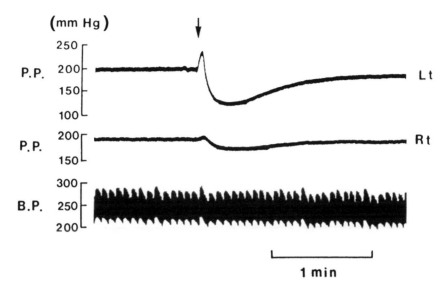

Figure 4 Responses of right (Rt) and left (Lt) perfused tracheal vascular beds of a dog to close-arterial injection of VIP (60 nmol, at arrow) on the left side. (Traces from above down) Left arterial perfusion pressure (P.P.); right arterial perfusion pressure; systemic arterial blood pressure (BP). The transient rise in left arterial P.P. immediately after the injection is a pressure artifact due to the injection. Subsequently, both perfusion pressures fall, with a large effect on the ipsilateral side (43% decrease) and a smaller response on the contralateral side (12% decrease). There are insignificant changes in systemic arterial blood pressure. (From L.A. Laitinen et al., 1987b.)

Galanin has been little studied on the tracheobronchial vasculature physiologically, but at least for the dog it seems to be completely inactive in concentrations that, for other neuropeptides, can cause a considerable vascular response (Salonen et al., 1988).

IV. SYMPATHETIC MOTOR NEUROPEPTIDES

Sympathetic postganglionic fibers to many tissues contain neuropeptide Y (NPY) as the cotransmitter with noradrenaline (NA) (see Chapter 7). NPY seems to be a long-lasting vasoconstrictor and can enhance and prolong the action of NA.

A. Histology

1. The Nose

Immunohistochemistry shows that nerves containing NPY are abundant in the nasal mucosa, especially around blood vessels, including resistance vessels,

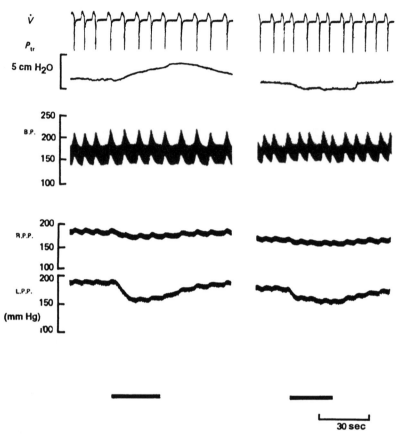

Figure 5 Electrical stimulation (5 V, 1 msec, 10 Hz) of the peripheral end of the cut left superior laryngeal nerve in a spontaneously breathing dog before (left) and after (right) administration of atropine (1 mg/kg i.v.). The tracheal vasculatures were perfused on both sides. $\dot{V}$, airflow; P_{tr}, tracheal pressure; B.P., blood pressure; R.P.P., right perfusion pressure; L.P.P., left perfusion pressure. Before administration of atropine, stimulation causes falls in both perfusion pressures, greater on the ipsilateral side, an increase in tracheal pressure, and no change in blood pressure. After administration of atropine, the falls in perfusion pressure are smaller, and there is a decrease in tracheal pressure, effects presumably medicated by neuropeptides. (From Laitinen et al., 1987c.)

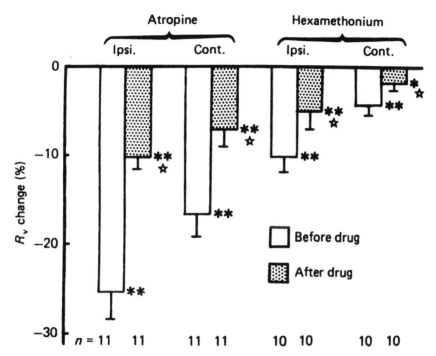

Figure 6 Changes in tracheal vascular resistance (R_v) in anesthetized dogs with bilateral tracheal perfusions. Electrical stimulation (10 V, 1 msec, 10 Hz) of the peripheral end of one superior laryngeal nerve, both being cut, causes vasodilatation on both sides (open columns), the larger effect being ipsilateral. After administration of atropine (0.5–1.0 mg/kg i.v.), the responses are about halved (stippled columns, left). After administration of the ganglionic blocking agent hexamethonium (0.5 mg/kg i.v.) in addition to atropine, the responses are again reduced but there is a residual vasodilatation (stippled columns, right), presumably due to sensory nerve stimulation. The latter two effects are presumably mediated by neuropeptides. Ipsi., ipsilateral; Cont., contralateral. *$p < 0.05$; **$p < 0.01$ for change compared with zero effect. ☆< 0.05 for response after drug compared to response before drug. (From Laitinen et al., 1987b.)

capacitance vessels, and AVAs (Baraniuk et al., 1990, 1991; Uddman et al., 1987; Lacroix, 1989). Surgical denervation (sympathectomy) or depletion of neurotransmitters by 6-hydroxydopamine or reserpine leads to disappearance in the mucosa of both NPY and NA. Although NPY is seen in nerves close to all mucosal blood vessels except capillaries, it is especially frequent in the walls of arterioles and AVAs, and receptor-binding studies in humans show an abundance of NPY receptors at these latter sites (Baraniuk et al., 1991). Both nerves and receptors are more frequent in the respiratory mucosa than in the olfactory mucosa, possibly

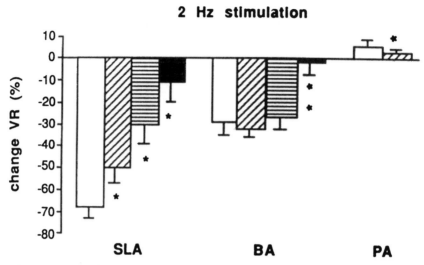

Figure 7 Changes in resistance (VR) in the vascular beds supplied by the superior laryngeal (SLA), bronchial (BA), and pulmonary (PA) arteries caused by vagal nerve stimulation at 2 Hz over 2 min, in controls (open bars) and after successive pretreatment with atropine (diagonal bars), the ganglionic blocking drug chlorisondamine (horizontal bars), and capsaicin (solid bars). Responses after atropine are presumably due to release of neuropeptides. Data are given as means ± SEM and expressed as percentages of basal conditions during a control period immediately preceding the stimulations. *$p < 0.05$; **$p < 0.001$ with the Quade test. (From Matran, 1991.)

because the vascular bed and submucosal glands have a denser distribution at the former site.

2. The Tracheobronchial Tree

Nerves containing NPY can be identified in the mucosa of the lower respiratory tract by immunohistochemical methods (Lundberg et al., 1989; Martling et al., 1990). They are especially prominent around blood vessels, particularly arterioles, but are also found in the airway smooth muscle. There are a few in the region of submucosal glands and the airway epithelium. They are particularly frequent in the guinea pig but are rather uncommon in humans and the pig (Springall et al., 1988). Comparison of the trachea with the smaller bronchi suggests that the former contains 15 times more fibers than the latter in the pig (Martling et al., 1990). NPY almost certainly coexists in the same nerves with NA, as for the nose and other vascular sites. Around glands and in the smooth muscle, the rather sparse fibers contain not only these two neurotransmitters, but some VIP/PHI. NPY is also found in sympathetic ganglia in the lower airways, especially those associated with vasomotor nerve fibers; VIP is found in both cell bodies and nerves. Some NPY

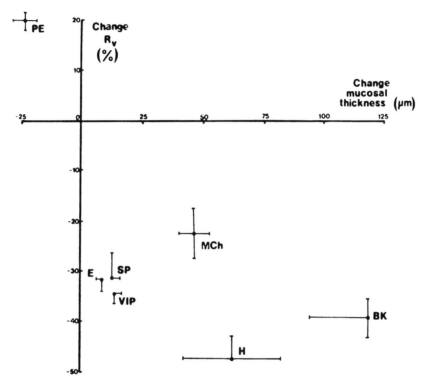

Figure 8 Comparison of the effects of various mediators and drugs injected into the perfused tracheal vasculature of the dog. (Ordinate) Change in vascular resistance (R_v); (abscissa) change in tracheal mucosal thickness (μm). Note that the neuropeptides substance P (SP), vasoactive intestinal polypeptide (VIP), and eloidsin-related peptide (E) cause similar decreases in vascular resistance compared with other vasodilators such as histamine (H), bradykinin (BK), and methacholine (MCh), but with smaller increases in mucosal thickness. PE, phenylephrine. (From Laitinen et al., 1986.)

has also been detected in vagal ganglia (Uddman and Sundler, 1987; Springall et al., 1987). As with the nose, surgical (sympathectomy) and chemical (with 6-hydroxydopamine or reserpine) interventions lead to depletion of both NA and NPY in the tracheobronchial tree.

B. Pharmacology and Physiology

1. The Nose

Injection of NPY into the perfused nasal vascular bed increases the nasal vascular resistance and decreases nasal mucosal volume, in other words causes nasal decongestion (Lacroix, 1989) (Fig. 9). NPY causes a vasoconstriction with a

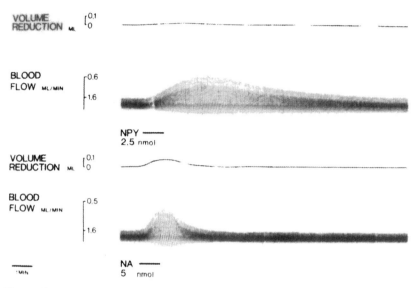

Figure 9   Effects of intra-arterial infusions of NPY (2.5 nmol) and NA (5 nmol) on cat nasal mucosal blood flow (ml/min) and volume of the nasal cavity (volume reduction in ml). NPY induces a slowly developing, long-lasting vasoconstriction compared to NA. On the other hand, NA induces a more pronounced volume reduction relative to NPY. (From Lundblad et al., 1987.)

longer latency and a far longer action than does NA and is not influenced by α-adrenoceptor antagonists. Stimulation of sympathetic nerves to the nose causes strong vasoconstriction and a fall in airflow resistance, and these responses are decreased but not eliminated by α-adrenoceptor antagonists (Lacroix et al., 1988a,b). Stimulation of the sympathetic nerves also increases the output of NPY into venous effluent, a response not prevented by α-adrenoceptor antagonists. All these results indicate that NPY is a functional nasal vasoconstrictor agent, in cooperation with NA. NPY is released into the venous effluent of the nose (Lundblad et al., 1987; Lacroix et al., 1988a,b). With regard to frequency-dependent responses, NA is released more at low-frequency stimulation and NPY at higher frequencies (Lundblad et al., 1987; Lacroix et al., 1988a). Thus the results support colocalization of NPY and NA, with a frequency-dependent release mechanism. Although NPY seems effectively to constrict both nasal resistance and capacitance vessels and probably AVAs, the exact activities of NPY and its physiological actions at the different sites have not been determined.

2. Tracheobronchial Tree

As for the nose, injection of NPY into the arterial perfusate of the tracheal and bronchial vasculatures increases vascular resistance (Salonen et al., 1988; Pernow, 1988; Martling et al., 1990; Matran et al., 1990) (Fig. 10). Stimulation of sympathetic nerves to the lower airways causes vasoconstriction and tissue depletion of NA and NPY, and the mediator effects depend on stimulus frequency; NA is preferentially released by low frequencies while NPY is released more at high frequencies. The actions of NPY on the capacitance vessels of the tracheobronchial tree do not seem to have been studied.

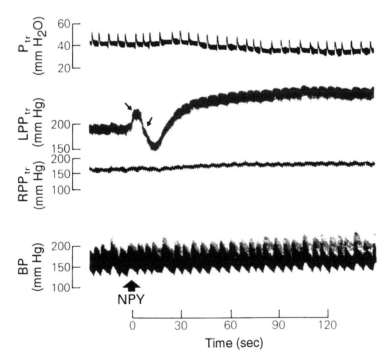

Figure 10   Tracheal vascular response to neuropeptide Y (NPY, 2 nmol) injected into the arterial catheter supplying the left side of the perfused tracheal vascular bed of a dog. A long-lasting increase in left arterial perfusion pressure (LPP_{tr}) is seen after two transient artifacts (small arrows); the initial increase in perfusion pressure is due to injection, and the subsequent decrease is caused by lowered viscosity of the blood due to the saline. There are also a small increase in right arterial perfusion pressure (RPP_{tr}) and a decrease in intratracheal pressure (P_{tr}; not seen in two other dogs) but no change in systemic arterial blood pressure (BP). (From Salonen et al., 1988.)

V.   SENSORY NEUROPEPTIDES

The neuropeptides released from sensory nerves in the airway mucosa are powerfully vasodilator, and this response is part of the mechanism of neurogenic inflammation. When activated, the same nerves can also set up central reflexes than may include mucosal vasodilatation mediated by the parasympathetic and sympathetic nervous systems (Widdicombe, 1986, 1993). The nerves probably have nonmyelinated (C-fiber) fibers, and their terminals probably ramify in the mucosa and penetrate into the epithelium (Uddman et al., 1987; Lundberg, 1990). Their central reflex actions include not only mucosal vasodilatation, but mucus secretion, sneezing, and other changes in pattern of breathing (Widdicombe, 1986). The neuropeptides most frequently studied include substance P (SP), neurokinin A (NKA), and calcitonin gene-related peptide (CGRP). Some species may also have neurokinin B (NKB) and galanin. Gastrin-releasing peptide (GRP), cholecystokinin, and somatostatin have also been found in presumed sensory nerves in the nasal mucosa.

A.   Histology

1.   The Nose

The most common neuropeptide in the nasal mucosa, identified by immunohisto-chemistry, seems to be SP. This is found especially around blood vessels and submucosal glands, but also within and below the respiratory epithelium (Uddman and Sundler, 1986; Uddman et al., 1987; Springall et al., 1988; Lundberg, 1990; Albegger et al., 1991). Cats have a particularly copious supply of SP-containing nerves, although it seems to be present in most species. SP-containing cell bodies and fibers are seen in the trigeminal ganglion (Lundblad, 1984), supporting the idea that the sensory nervous receptors may be involved in local reflexes with modulation of activity via the ganglia. CGRP, NKA, and galanin have also been identified in the nasal mucosa, including human nasal mucosa (Baraniuk et al., 1990). Their distribution seems to be rather similar to that of SP. Depletion of the sensory neuropeptides in the nose can be produced by administering large doses of capsaicin, supporting the view that the peptides are in fact in sensory nerves.

2.   The Tracheobronchial Tree

The distribution of sensory neuropeptides here is rather similar to that in the nose. While SP and NKA are found in nerves particularly close to blood vessels, glands, and in the epithelium, the smooth muscle is less well supplied (Uddman et al., 1987; Lundberg et al., 1987, 1988; Martling et al., 1990). These two sensory neuropeptides are also found in the airway ganglia and the nodose (vagal) ganglia (Martling et al., 1990), as well as in sympathetic ganglia and dorsal roots (Springall et al., 1987). The last observation suggests that there could be important sensory pathways to the central nervous system in the sympathetic nerves as well as in the more frequently studied vagal supply to the airways. SP and NKA are especially abundant in the tracheobronchial nerves of rodents and less common in cats and humans (Springall et al., 1988).

CGRP in general has a similar distribution to SP and NKA (Martling et al., 1990). However, it is particularly conspicuous in neuroepithelial bodies, which are groups of paracrine cells found especially in some species (Martling et al., 1988; McBride et al., 1990; Shimosegawa and Said, 1991). There seems to be little CGRP in human epithelium. In the rat, CGRP is abundant in the vagal jugular ganglia, and SP more in the vagal nodose ganglia. In the cat, CGRP is present in the stellate sympathetic ganglia (Lindh et al., 1987). Like SP, it is also found in dorsal root ganglia (Springall et al., 1988). Thus although in general CGRP has a similar localization to SP and NKA, there are some differences that may point to functional diversions. The results also point to the potential importance of sensory neuropeptides in complex local sympathetic and central reflex interactions, as well as in the more studied axon reflexes of neurogenic inflammation.

As for the nose, other neuropeptides have been identified in the tracheobronchial mucosa, whose function is uncertain. These include GRP, eledoisin-like peptide, and pituitary adenylate cyclase-activating peptide (PACAP) (Lundberg et al., 1988; Uddman et al., 1991).

B. Pharmacology and Physiology

1. The Nose

All the main sensory neuropeptides cause vasodilatation in the nasal mucosa, a conclusion based on perfusion studies with direct injection of the peptides into the vascular beds (Fig. 2). These experiments can only easily be done in the dog and pig (Lundberg et al., 1984; Matran et al., 1989a; Stjarne, 1991), but indirect support of this evidence has been obtained for small animals such as the guinea pig and rat (Lundberg and Saria, 1982; Lundblad, 1984). The coincident extravasation of Evans-blue-labeled albumin into the mucosal interstitium has been especially well studied in these small rodents. The effect is thought to be due to opening of intercellular gaps between the endothelial cells of the postcapillary venules (see Chapter 15). Antidromic stimulation of sensory nerves to the airway mucosa mimics these vascular and interstitial changes presumably by release of sensory neuropeptides; CGRP-like material can be detected in the venous effluent blood. In the dog, nasal vasodilatation due to SP is associated with little increase in nasal airflow resistance, suggesting that the capacitance vessels are not greatly dilated (Lung et al., 1984) (Fig. 3).

The destruction of sensory nerves by large doses of capsaicin, or their degeneration after nerve section, leaves vascular beds on which neuropeptides are still effective in causing vasodilatation and transudation of plasma proteins; however, irritation by chemicals supplied locally no longer sets up neurogenic inflammation (Lundblad et al., 1985). The results support the view that nasal sensory nerves mediate axon reflex neurogenic inflammation. Similar experiments are difficult to conduct in humans, but capsaicin desensitization has been attempted and may destroy sensory nerves and block neurogenic inflammation in patients with rhinitis (Wolfe, 1988).

2. The Tracheobronchial Tree

When the vasculature of the trachea or bronchi of large mammals is perfused, and neuropeptides are injected into the perfusate, all the main sensory neuropeptides mentioned above cause vasodilatation (Laitinen et al., 1987a,b; Salonen et al., 1988; Martling et al., 1990). For the dog tracheal vasculature, the order of potency

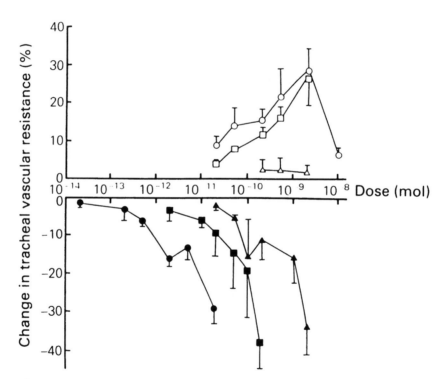

Figure 11   Dose-response curves showing the effects of some neuropeptides on tracheal vascular resistance in dogs. Five to six doses of each neuropeptide (only three doses of galanin) were injected into the arterial catheters supplying the perfused tracheal vascular bed: (○) bombesin; (□) neuropeptide Y; (△) galanin; (●) neurokinin A; (■) calcitonin gene-related peptide; (▲) peptide histidine isoleucine. The values presented are mean percentage changes from control values in three to four separate dogs; vertical lines indicate SEM. (From Salonen et al., 1988.)

is NKA > CGRP > SP (Fig. 11). For the dog bronchial artery in vitro, the order is different: SP > NKA > CGRP (McCormack et al., 1989). However, these methods of determining potency may not be relevant to the situation in vivo, since the concentrations of peptides released endogenously from sensory nerves and acting on airway vascular smooth muscle are unknown. As with VIP, SP causes relatively less mucosal thickening than do mediators such as histamine and bradykinin, for equivalent decreases in vascular resistance (Laitinen et al., 1986) (Fig. 8). This may indicate that SP has only a weak action on the capacitance system.

Antidromic electrical stimulation of parasympathetic nerves to the lower airways causes tracheal and bronchial vasodilatation, as already mentioned (Laitinen et al., 1987c; Martling et al., 1985; Matran et al., 1989b). In the presence of atropine to block cholinergic actions and of hexamethonium to block ganglionic transmission in motor nerves, there is a residual vasodilatation presumably due to antidromic release of sensory neuropeptides. This is the dominant vasodilator mechanism for the bronchi in the pig, in which species parasympathetic motor nerve fibers with release of VIP seem to be rather inactive.

Sensory neuropeptides can be released by locally applied capsaicin and by various irritants such as cigarette smoke, which presumably act on the sensory terminals in the epithelium (Lundberg et al., 1985, 1990). The release of the neuropeptides causes vasodilatation and plasma transudation into the interstitium, as a component of neurogenic inflammation. These responses are abolished by administration of large doses of capsaicin that deplete the nerves of neuropeptides or by the degeneration of sensory nerves due to nerve section.

VI. CONCLUSIONS

It is clear that all three components of the nervous system in the airways, parasympathetic, sympathetic, and sensory, can release neuropeptides that have strong actions on the airway vasculatures. These responses can be elicited by local application of neuropeptides and by stimulation of various nervous pathways.

However, several features of the control of the airway vasculatures require much further research. First, we know little about the interaction of the three components of the nervous system or their relative potencies on the vascular beds. For example, with physiological nervous control, the balance of parasympathetic and sympathetic motor activities may be crucial. Second, there is little quantitative analysis of the local interplay between the conventional neurotransmitters, such as ACh and NA, and the more recently studied neuropeptides. Third, even if only the sensory neuropeptides are considered, so many of them are present in these nerves and so little quantitative information is available about either the amounts released or their activities on blood vessels that little can be said about their relative importance. The availability of good neuropeptide antagonists would be valuable in this respect. Finally, nearly all studies of the effects of neuropeptides on the airways vasculature have been with rather coarse methods. For example, measurement of total vascular resistance with a perfused airway mucosa says nothing about whether a neuropeptide is acting on arterioles, sinus smooth muscle, or AVAs. Similarly, the measurement of Evans-blue-labeled albumin transudation via post-capillary venules isolates only one component of the vascular bed and says little about whether the agent may be acting on vessels other than the postcapillary venules.

The importance of the neuropeptides in controlling the airway vasculature has been very convincingly established, but the quantitative aspects of their actions, and the way in which multiple neurotransmitters may interact on different components of the vascular bed, present a complex and unexplored area. Finally, it should be mentioned that nearly all the relevant studies have been on experimental animals, and there is a great deficiency in our knowledge of the importance of neuropeptides acting on the mucosal vascular bed of humans.

REFERENCES

Albegger, K., Hauser-Kronberger, C.E., Saria, A., Graf, A.H., Bernatzky, G., and Hacker, G.W. (1991). Regulatory peptides and general neuroendocrine markers in the human nasal mucosa, the soft palate and the larynx. *Acta Otolaryngol. 111*:373–378.

Anggard, A. (1974). Capillary and shunt blood flow in the nasal mucosa of the cat. *Acta Otolaryngol. 78*:418–422.

Anggard, A., Lundberg, J.M., Hokfelt, T., Nilsson, G., Fahrenkrug, J., and Said, S. (1979). Innervation of the cat nasal mucosa with special reference to relation between pep-tidergic and cholinergic neurons. *Acta Physiol. Scand. 473*:50.

Baraniuk, J.N., Lundgren, J.D., Goff, J., Mullol, J., Castellino, S., Merida, M., Shelhamer, J.H., and Kaliner, M. (1990). Calcitonin gene-related peptide (CGRP) in human nasal mucosa. *Am. J. Physiol. 258*:L81–L88.

Baraniuk, J.N., Lundgren, J.D., Okayama, M., Goff, J., Mullol, J., Merida, M., Shelhamer, J.H., and Kaliner, M.A. (1991). Substance P and neurokinin A in human nasal mucosa. *Am. J. Respir. Cell Mol. Biol. 4*:228–236.

Cauna, N. (1970). Electron microscopy of the nasal vascular bed and its nerve supply. *Ann. Otol. Rhinol. Laryngol. 79*:443–450.

Cauna, N. (1982). Blood and nerve supply of the nasal lining. In: *The Nose: Upper Airway Physiology and the Atmospheric Environment.* (Proctor, D.E., and Andersen, I.B., eds.) Elsevier Biomedical, Amsterdam, pp. 45–69.

Cauna, N., and Hinderer, K.H. (1969). Fine structure of blood vessels of the human nasal respiratory mucosa. *Ann. Otol. Rhinol. Laryngol. 78*:865–879.

Cheung, A., Polak, J.M., Bauer, F.E., Cadieux, A., Christofides, N.D., Springall, D.R., and Bloom, S.R. (1985). Distribution of galanin immunoreactivity in the respiratory tract of pig, guinea-pig, rat and dog. *Thorax 40*:889–896.

Corfield, D.R., Hanafi, Z., Webber, S.E., and Widdicombe, J.G. (1991). Changes in mucosal thickness and blood flow in sheep. *J. Appl. Physiol. 71*:1282–1288.

Corfield, D.R., Webber, S.E., and Widdicombe, J.G. (1993). Distribution of blood flow in the perfused tracheas of sheep: a search for arteriovenous anastomoses. *J. Appl. Physiol. 74*: 1856–1861.

Hill, P., Goulding, D., Webber, S.E., and Widdicombe, J.G. (1989). Blood sinuses in the submucosa of the large airways of the sheep. *J. Anat. 162*:235–247.

Hughes, T. (1965). Microcirculation of the tracheobronchial tree. *Nature (Lond.) 206*: 425–426.

Lacroix, J.S. (1989). Adrenergic and non-adrenergic mechanisms in sympathetic vascular control of the nasal mucosa. *Acta Physiol. Scand. 136*:1–63.

Lacroix, J.S., Stjarne, P., Anggard, A., and Lundberg, J.M. (1988a). Sympathetic vascular control of the pig nasal mucosa. 1. Increased resistance and capacitance vessel responses upon stimulation with irregular bursts compared to continuous impulses. *Acta Physiol. Scand. 132*:83–90.

Lacroix, J.S., Stjarne, P., Anggard, A., and Lundberg, J.M. (1988b). Sympathetic vascular control of pig nasal mucosa. 2. Reserpine-resistant, non-adrenergic nervous responses in relation to neuropeptide Y and ATP. *Acta Physiol. Scand. 133*:183–197.

Laitinen, L. (1986). Detailed analysis of the neural elements in human airways. In: *Neural Regulation of the Airways in Health and Disease*. (Kaliner, M., and Barnes, P.J., eds.) Marcel Dekker, New York, pp. 35–36.

Laitinen, L.A., and Laitinen, A. (1991). Overview of the pathology of asthma. In: *Pharmacology of Asthma, Handbook of Experimental Pharmacology*. (Page, C.P., and Barnes, P.J., eds.) Springer-Verlag, New York, pp. 1–25.

Laitinen, L.A., and Laitinen, A. (1992). The bronchial circulation. Histology and electron microscopy. In: *The Bronchial Circulation*. (Butler, J., ed.) Marcel Dekker, New York, pp. 79–98.

Laitinen, L.A., Robinson, N.P., Laitinen, A., and Widdicombe, J.G. (1986). Relationship between tracheal mucosal thickness and vascular resistance in dogs. *J. Appl. Physiol. 61*:2186–2194.

Laitinen, L.A., Laitinen, A., and Widdicombe, J.G. (1987a). Effects of inflammatory and other mediators on airway vascular beds. *Am. Rev. Respir. Dis. 135*:S67–S70.

Laitinen, L.A., Laitinen, M.A., and Widdicombe, J.G. (1987b). Dose-related effects of pharmacological mediators on tracheal vascular resistance in dogs. *Br. J. Pharmacol. 92*:703–709.

Laitinen, L.A., Laitinen, M.V.A., and Widdicombe, J.G. (1987c). Parasympathetic nervous control of tracheal vascular resistance in the dog. *J. Physiol. 385*:135–146.

Laitinen, A., Laitinen, L.A., Moss, R., and Widdicombe, J.G. (1989). Organisation and structure of the tracheal and bronchial blood vessels in the dog. *J. Anat. 165*:133–140.

Lindh, B., Lundberg, J.M., Hokfelt, T., Elfvin, L.-G., Fahrenkrug, J., and Fisher, J. (1987). Co-existence of CGRP- and VIP-like immunoreactivities in a population of neurons in the cat stellate ganglion. *Acta Physiol. Scand. 131*:475–476.

Lundberg, J.M. (1990). Peptide and classical transmitter mechanisms in the autonomic nervous system. *Arch. Int. Pharmacodyn. 303*:9–19.

Lundberg, J.M., and Saria, A. (1982). Capsaicin-sensitive vagal neurons are involved in control of vascular permeability in cat trachea. *Acta Physiol. Scand. 115*:521–523.

Lundberg, J.M., Anggard, A., Enson, T., Fahrenkrug, J., and Hokfelt, T. (1981) Vasoactive intestinal polypeptide and cholinergic mechanisms in cat nasal mucosa. Studies on choline acetyltransferase and release of vasoactive intestinal polypeptide. *Proc. Natl. Acad. Sci. USA 78*:5255–5259.

Lundberg, J.M., Brodin, E., Hua, X.-Y., and Saria, A. (1984). Vascular permeability changes and smooth muscle contraction in relation to capsaicin sensitive substance P afferents in the guinea-pig. *Acta Physiol. Scand. 120*:217–227.

Lundberg, J.M., Franco-Cereoda, A., Huang, X.-Y., Hokfelt, T., and Fischer, J.A. (1985). Coexistence of substance P and calcitonin gene related peptide-like immunoreactivities in sensory nerves in relation to cardiovascular and bronchoconstrictor effects of capsaicin. *Eur. J. Pharmacol. 108*:315–219.

Lundberg, J.M., Saria, A., Lundblad, L. Anggard, A., Martling, C.-R., Theodorsson-Norheim, E., Stjarne, P., and Hokfelt, T. (1987). Bioactive peptides in capsaicin sensitive C-fibre afferents of the airways: functional and pathophysiological implications. In: *The Airways: Neural Control in Health and Disease.* (Kaliner, M.A., and Barnes, P.J., eds.) Marcel Dekker, New York, pp. 417–445.

Lundberg, J.M., Martling, C.R., and Hokfelt, T. (1988). Airways, oral cavity and salivary glands: classical transmitters and peptides on sensory and autonomic motor neurons. In: *Handbook of Chemical Neuroanatomy. The Peripheral Neurons System.* (Bjonkund, A., Hokfelt, T., and Owman, C., eds.) Elsevier, Amsterdam, pp. 391–444.

Lundberg, J.M., Pernow, J., and Lacroix, J.S. (1989). Neuropeptide Y: sympathetic con-transmitter and modulation? *News Physiol. Sci. 4*:13–17.

Lundberg, J.M., Alving, K., Lacroix, J.S., and Matran, R. (1990). Local and central reflex mechanisms in the neural control of airway microcirculation. *Eur. Respir. J. 3*:624a–628s.

Lundblad, L. (1984). Protective reflexes and vascular beds in the nasal mucosa elicited by activation of capsaicin sensitive substance P immunoreactive trigeminal neurons. *Acta Physiol. Scand. 539*:1–45.

Lundblad, L., Brodin, E., Lundberg, J.M., and Anggard, A. (1985). Effects of nasal capsaicin pretreatment and cryosurgery on sneezing reflexes, neurogenic plasma extravasation, sensory and sympathetic neurons. *Acta Otolaryngol. 100*:117–127.

Lundblad, L., Anggard, A., Saria, A., and Lundberg, J.M. (1987). Neuropeptide Y and non-adrenergic sympathetic vascular control of the cat nasal mucosa. *J. Auton. Nerv. Syst. 20*:189–197.

Lung, M.A., Phipps, R.C., Wang, J.C.C., and Widdicombe, J.G. (1984). Control of nasal vasculature and airflow resistance in the dog. *J. Physiol. 349*:535–551.

Malm, L. (1974). Response of resistance and capacitance vessels in feline nasal mucosa to vasoactive agents. *Acta Otolaryngol. 78*:90–97.

Martling, C.R., Anggard, A., and Lundberg, J.M. (1985). Noncholinergic vasodilation in the tracheobronchial tree of the cat induced by vagal nerve stimulation. *Acta Physiol. Scand. 125*:343–346.

Martling, C.R., Saria, A., Fischer, J.A., Hokfelt, T., and Lundberg, J.M. (1988). Calcitonin gene-related peptide and the lung: Neuronal coexistence with substance P releases by capsaicin and vasodilatory effect. *Regul. Pept. 20*:125–139.

Martling, C.R., Matran, R., Alving, K., Hokfelt, T., and Lundberg, J.M. (1990). Innervation of lower airways and neuropeptide effects on bronchial and vascular tone in the pig. *Cell. Tissue Res. 260*:223–233.

Matran, R. (1991). Neural control of lower airway vasculature. *Acta Physiol. Scand. 142*: 1–54.

Matran, R., Alving, K., Martling, C.R., Lacroix, J.S., and Lundberg, J.M. (1989a). Effects of neuropeptides and capsaicin on tracheobronchial blood flow of the pig. *Acta Physiol. Scand. 135*:335–342.

Matran, R., Alving, K., Martling, C.R., Lacroix, J.S., and Lundberg, J.M. (1989b). Vagally mediated vasodilatation by motor and sensory nerves in the tracheal and bronchial circulation of the pig. *Acta Physiol. Scand. 135*:29–37.

Matran, R., Franco-Cereceda, A., Alving, K., and Lundberg, J.M. (1990). Sympathetic control of tracheal, bronchial and pulmonary circulation in the pig. *Eur. J. Pharmacol. 183*:1570.

Matthay, M.A. (1992). The bronchial and systemic circulations in lung and pleural fluid and protein balance. In: *The Bronchial Circulation*. (Butler, J., ed.) Marcel Dekker, New York, pp. 389–415.

McBride, J.T., Springall, D.R., Winter, R.J.D., and Polak, J.M. (1990). Quantitative immunocytochemistry shows calcitonin gene-related peptide-like immunoreactivity in lung neuroendocrine cells is increased by chronic hypoxia in rat. *Am. J. Respir. Cell Mol. Biol. 3*:587–593.

McDonald, D.M. (1992). Infections intensify neurogenic plasma leakage in the airway mucosa. *Am. Rev. Respir. Dis. 146*:540–544.

Pernow, J. (1988). Co-release and functional interactions of neuropeptide Y and noradrenaline in peripheral sympathetic vascular control. *Acta Physiol. Scand. 133*(Suppl. 568):1–56.

Persson, C.G. (1987). Leakage of macromolecules from the tracheobronchial microcirculation. *Am. Rev. Respir. Dis. 135*:S71–S75.

Persson, C.G. (1992). Plasma exudation from tracheobronchial microvessels in health and disease. In: *The Bronchial Circulation*. (Butler, J., ed.) Marcel Dekker, New York, pp. 443–473.

Robson, A.G., Skasick, A., Webber, S.E., and Widdicombe, J.G. (1993). Measurement of airway blood flow using radiolabelled microspheres in the anaesthetized rabbit. *Am. Rev. Respir. Dis. 147*:A657.

Salonen, R.O., Webber, S.E., and Widdicombe, J.G. (1988). Effects of neuropeptides and capsaicin on the canine tracheal vasculature in vivo. *Br. J. Pharmacol. 95*:1262–1270.

Shimosegawa, T., and Said, S.I. (1991). Pulmonary calcitonin gene-related peptide immunoreactivity: nerve-endocrine cell interrelationships. *Am. J. Respir. Cell Mol. Biol. 4*:126–134.

Sobin, S.S., Frasher, W.G., Tremer, H.M., and Hadley, G.G. (1963). The microcirculation of the tracheal mucosa. *Angiology 14*:165–170.

Sorokin, S.P., and Hoyt, R.F., Jr. (1990). On the supposed function of neuroepithelial bodies in adult mammalian lungs. *News Physiol. Sci. 5*:89–95.

Springall, D.R., Cadieux, A., Oliveira, H., Su, H., Royston, D., and Polak, J.M. (1987). Retrograde tracing shows that CGRP-immunoreactive nerves of rat trachea and lung originate from vagal and dorsal root ganglia. *J. Auton. Nerv. Syst. 20*:155–166.

Springall, D.R., Bloom, S.R., and Polak, J.M. (1988). Distribution, nature, and origin of peptide containing nerves in mammalian airways. In: *The Airways*. (Kaliner, M.A., and Barnes, P.J., eds.) Marcel Dekker, New York, pp. 299–342.

Stjarne, P. (1991). Sensory and motor reflex control of nasal mucosal blood flow and secretion: clinical implications in non-allergic nasal hyperreactivity. *Acta Physiol. Scand. 142*:1–64.

Uddman, R., and Sundler, F. (1986). Innervation of the upper airways. In: *Clinics in Chest Medicine, Vol. 7. The Upper Airways*. (Widdicombe, J.H., ed.) W.B. Saunders, Philadelphia, pp. 201–209.

Uddman, R., and Sundler, F. (1987). Neuropeptides in the airways: a review. *Am. Rev. Respir. Dis. 136*:S3–S8.

Uddman, R., Alumets, J., Densert, O., Hakanson, R., and Sandler, F. (1978). Occurrence and distribution of VIP nerves in the nasal mucosa and tracheobronchial wall. *Acta Otolaryngol. 86*:443–448.

Uddman, R., Malm, L., and Sundler, F. (1980). Effects of vasoactive intestinal polypeptide on resistance and capacitance vessels in the nasal mucosa. *Acta Otolaryngol. 90*:304–308.

Uddman, R., Anggard, A., and Widdicombe, J.G. (1987). Nerves and neurotransmitters in the nose. In: *Allergic and Vasomotor Rhinitis: Pathophysiological Aspects*. (Mygind, N., and Pipkorn, U., eds.) Munksgaard, Copenhagen, pp. 50–62.

Uddman, R., Lutz, A., Atrimura, A., and Sundler, F. (1991). Pituitary adenylate cyclase-activating peptide (PACAP) a new vasoactive intestinal polypeptide (VIP) like peptide in the respiratory tract. *Cell. Tissue Res. 265*:197–200.

Watanabe, K., and Watanabe, I. (1980). The ultrastructural characteristics of the capillary walls in human nasal mucosa. *Rhinology 18*:183–195.

Widdicombe, J.G. (1986). Sensory innervation of the lungs and airways. In: *Progress in Brain Research, Vol. 67. Visceral Sensation*. (Cervero, F., and Morrison, J.F.B., eds.) Elsevier, Amsterdam, pp. 49–64.

Widdicombe, J.G. (1990). Comparison between the vascular beds of upper and lower airways. *Eur. Respir. J. 3*:564–571.

Widdicombe, J.G. (1993). Microvascular anatomy of the airways. In: *Asthma and Rhinitis*. (Busse, W., and Holgate, S.T., eds.) Blackwell Scientific Publications, Boston (in press).

Wolfe, G. (1988). Neue Aspekte zur Pathogenese und Therapie der hyperflektorischen Rhinopathie. *Laryngol. Rhinol. Otol. 67*:438–445.

DISCUSSION

Baraniuk: What are the relative effects of capsaicin on nasal airflow resistance and blood flow?

Widdicombe: We have not looked at the effect of capsaicin on tracheal mucosal thickness, but in the nose the biphasic response in vascular resistance—constriction followed by congestion—is not accompanied by appreciable changes in airflow resistance.

McDonald: How long does the increase in vascular permeability last? The increased vascular permeability produced by mediators such as bradykinin, 5-HT, histamine, tachykinins, and PAF lasts only a few minutes. Allergens can cause a more prolonged change in permeability but the mechanism, probably involving neutrophils, is different from that of the inflammatory mediators.

Widdicombe: With single injections of mediators into the tracheal vasculature the vasodilatation usually lasts only 2–5 min, but the changes in mucosal thickness may last for up to 20 min or more. This may suggest that edema is occurring and taking far longer to resolve.

Solway: Could the epithelial shrinkage in response to mucosal hypertonicity mechanically stimulate sensory nerve endings within the epithelium and thereby recruit water-replenishing mechanisms that include neuropeptide release?

Widdicombe: Both in the larynx and in the tracheobronchial tree, the sensory receptors respond to hypertonic solutions, as shown by Boushey, Coleridge, and others. Hypertonic solutions also cause airway vasodilatation and mucus secretion; so they may evoke neurogenic inflammation.

Kaliner: In a model being studied by Togias (at John Hopkins), human volunteers breathe cold dry air to possibly cause hypertonic challenge. Secretions formed are rich in glandular products. Indeed, atropine prevents this response, suggesting that cholinergic reflexes participate in the response to hyperosmolar challenge.

18

Basophil and Mast Cell Activation: Neuropeptides and Nerves

John Bienenstock
McMaster University, Hamilton, Ontario, Canada

I. INTRODUCTION

For a number of years, my colleagues and I have been studying the relationships and communication pathways between mast cells and nerves. Most of these studies have been performed in experimental animals, especially rodents (Undem, 1990). Some of the morphological associations have been also demonstrated in the human (Stead et al., 1989). The primary tissues we have looked at have been in the gastrointestinal tract, but some of our studies have also been performed in the respiratory tract (Sestini et al., 1989, 1990). Our observations have led us to believe that there are meaningful and significant communications, directly or indirectly, between mast cells as representative cells of the immune system and the nervous system. This communication occurs in a bidirectional mode and has been demonstrated in vivo, ex vivo, and in vitro (Sestini et al., 1990).

II. MORPHOLOGICAL OBSERVATIONS

Some of our initial observations were in the form of morphological and morphometric studies in which we demonstrated, at both the light and electron microscopic levels, close associations and even apposition between mast cells and nerves in the gastrointestinal tract (Stead et al., 1987). These were found in the lamina propria of the small and large intestines in rats and in mice and subsequently in humans (Stead et al., 1989). Mast cells were seen closely apposed to enteric nerves throughout their length. These were particularly revealed on exact

399

transverse sectioning of villi. The nerves that were in apposition seemed mostly to contain substance P (SP) or calcitonin gene-related peptide (CGRP). These associations were far more frequent than expected by chance alone. Arizono et al. (1990) have more recently confirmed these associations and extended them to cells other than mast cells. We ourselves have shown that similar conclusions can be drawn for eosinophils in the gastrointestinal tract (Quinonez et al., 1986). Others have shown such mast cell–nerve associations in relation to either boutons or nerve terminals in the rat intestinal tract (Newsom et al., 1983) or an association between mast cells and nerves in the mesentery, liver, skin, and diaphragm (Heine and Forster, 1975; Skofitsch et al., 1985; Dimlich, 1984).

III. SUBSTANCE P

In this chapter, I will discuss some of our work relating especially to substance P, since the subject given to me was mast cell and basophil activation—presumably in the context of neuropeptides in the respiratory tract. Mast cells were shown some time ago to be degranulated by substance P, but the dose-response curves obtained suggested that relatively high concentrations were needed (Johnson and Erdös, 1973; Shanahan et al., 1985). For example, in excess of 10^{-5} M was needed to get 50% release from rat peritoneal mast cells. Our own attempts to characterize secretagogues for secretion of histamine from rat intestinal mucosal mast cells showed that SP was the only neuropeptide of many tested that was able to release histamine at all from these cells, again at relatively high concentrations (Shanahan et al., 1985).

Much has been written about SP and its potential mechanism of action in several different systems (Mousli et al., 1990). For example, SP has multiple effects on cells of the immune system, and some of these occur through classical neuropeptide receptors, which have been classified as NK-1, -2, and -3. It is known that most of the neurological effects of SP acting in the central nervous system occur through NK-1 receptors as a result of the binding of the C-terminus. As summarized in Table 1, the two ends of the molecule appear to have different activities, and therefore it may not be surprising to learn that the highly charged N-terminus of this undecapeptide neurotransmitter appears to act in a number of substantive biological systems, including mast cell degranulation, without any evidence for the presence of SP receptors expressed on the cells (Mousli, 1990). It is thought that SP might interact directly with G proteins on the cell surface, with a subsequent signal transduction mechanism distinct from that associated with SP binding to its natural receptors. The exact pathway for this has not been worked out.

Human basophils do not appear to be degranulated by substance P, even at high concentrations (Lowmann et al., 1988). Human skin and lung mast cells, however, do appear to respond similarly in this respect to rat peritoneal mast cells (Church et al., 1991).

Table 1 Effects of Substance P

C-terminus: -Phe-Phe-Gly-Leu-Met-NH$_2$	N-terminus: -Arg-Pro-Lys-Pro-Gln-Gln
Most CNS, smooth muscle, glands	Catecholamine release by rat adrenal cells
Human T-lymphocyte proliferation	Histamine release by rat peritoneal, human
Rat lymph node response	skin mast cells
Chemiluminescence increase in human	Increased phagocytosis by mouse
monocytes	macrophages and human polymorphs
Human monocyte interleukin secretion	

Source: From Mousli et al. (1990).

IV. USSING CHAMBER EXPERIMENTS

There are various ways of investigating the hypothesis that mast cells may play an important role in normal physiological mechanisms. The system of antigen-induced intestinal epithelial cell chloride ion secretion has been employed by a number of individuals. First to use this system to show these associations were Baird and Cuthbert (1987), who demonstrated that antigen would cause chloride ion secretion in the sensitized guinea pig intestine as evidenced by electrophysiological changes measured in the classical Ussing chamber. Dr. Mary Perdue, one of my current colleagues, introduced us to this system after she too had demonstrated such changes in the sensitized rat intestinal epithelium, which were also due to nerves, as evidenced by the abolition of the short-circuit-current generation by tetrodotoxin, a sodium pump and nerve-specific toxin (Perdue and Bienenstock, 1991). We ourselves, together with Dr. Perdue and colleagues (Crowe and Perdue, 1992), have shown that antigen, mast cell, and nerve-dependent systems similar to those described by her and others in the intestine also exist in the sensitized rat trachea (Sestini et al., 1990).

In brief summary, Table 2 outlines these findings and indicates by pharmacological means that an antigen–mast cell–substance P–containing nerve-dependent

Table 2 Ussing Chamber Experiments

N-phenylanthranillic acid (chloride channel blockers)
Doxantrazole (mast cell stabilizer)
Diphenhydramine (H-1 antagonist)
Tetrodotoxin (neurotoxin)
Capsaicin (depletes substance P–containing nerves)

Immediate responses to luminal antigen occur in this model of intestinal and tracheal hypersensitivity.

system in part regulates normal homeostasis in the rat intestinal and tracheal epithelium.

The proof that this system was indeed mast cell dependent came subsequently from work done by Perdue et al. (1991) in w/w^v mice, which are mast cell deficient owing to the lack of C-*kit*. In this animal, it is possible to reconstitute the mast cell population by the syngeneic transfer of bone marrow from a +/+ littermate or to reconstitute locally with cultured mast cells. In these experiments, the w/w^v had low antigen-dependent generation of short-circuit changes but still retained some changes in the order of 30% of the +/+ control. The full extent of response was restored in the bone-marrow-reconstituted animal, indicating that this was primarily a mast-cell-dependent phenomenon. Similar low levels of antigen-dependent short-circuit-current generation were found in the mast-cell-deficient sl/sl^d counterpart to the w/w^v strain. In this latter case, the deficiency is due to the c-*kit* ligand itself. Therefore, this effect is reconstituted by replacement of the deficient c-*kit* ligand (stem cell factor).

Important additional results in the former study were the effects of electrical field stimulation in which enteric nerves were activated, which showed marked reduction in the mast-cell-deficient animal compared to the +/+ control. The full effect was restored in the reconstituted animal, indicating that mast cell–nerve communication was bidirectional. Similar ex vivo experiments by Bani-Sacchi et al. (1986) have shown a mast cell response to field stimulation in the intestine, which was atropine sensitive.

Dr. Helen Cooke and her associates (Javed et al., 1992) have recently shown, with a sensitized intestinal model in the Ussing chamber, that acetylcholine was released simultaneously in parallel to the generation of the short-circuit current, thus clearly proving that antigen causes the release of a specific neurotransmitter in the intestine. Weinreich and colleagues (Weinreich et al., 1992; Undem et al., 1990) have similarly shown antigen- and cell-dependent long-term potentiation of synaptic transmission in autonomic ganglia. Wood (1993) has been able to dissect this system further, and his work indicates very complex histamine-dependent H1, 2, and 3 effects at both the prejunctional and postjunctional levels.

V. ELECTRICAL STIMULATION OF NERVES

In electrical stimulation of the nerves of the respiratory tract, McDonald (1987) has failed to show mast cell degranulation. Similarly, mast cell degranulation was only found in the skin after long-term antidromic electrical stimulation of the saphenous nerve in the rat by Kowalski and Kaliner (1988). There is, however, a significant earlier literature that does indicate the likely role of mast cells in antidromic electrical stimulation of the rat saphenous nerve, which stands in contradiction to this more recent work.

In experiments again involving the rat, Dimitriadou et al. (1990) have shown by electrical stimulation of the trigeminal ganglion that mast cells on the ipsilateral side in the dura mater and tongue showed granular changes immediately upon stimulation, and that the mast cells were eventually degranulated, but only on the ipsilateral side after prolonged stimulation.

Perhaps some of the differences that have been seen and noted relate to the different types of electrical stimulation that have been used. An alternative explanation may come from the interesting results reported by Miura et al. (1990). These workers showed in *Ascaris*-infected cats that inhalation of *Ascaris* antigen caused an increase in pulmonary resistance and elevated levels of histamine in the blood. If the animals were pretreated with propranolol and atropine, and both vagus nerves were electrically stimulated prior to or at the time of the inhalational challenge with *Ascaris* antigen, both the histamine increases previously seen, as well as the pulmonary resistance increases, were abrogated. The only interpretation of these observations is that a nonadrenergic, noncholinergic (NANC) inhibitory nervous system exists that is capable of directly or indirectly inhibiting antigen-induced mast cell degranulation. This important observation has not yet been confirmed. However, it could be one of the reasons why differing results are found by different workers, using different systems and modes of electrical stimulation. It is possible that inhibitory systems are variously at play, and that electrical stimulation may promote them, thus actively inhibiting degranulation, depending on the type of current level and duration of stimulation.

VI. PAVLOVIAN CONDITIONING

Because we and others had evidence that mast cells were in communication with peripheral nerves, we wondered if it would be possible to cause degranulation via effects of the central nervous system. There is considerable literature suggesting that this is indeed the case in various allergic phenomena. Russell et al. (1984) have shown in a classic conditioning model that guinea pigs could be induced to release histamine upon exposure to the conditioning stimulus. Since it was unclear whether this was due to mast cell degranulation, we decided to design experiments to test the hypothesis more directly (MacQueen et al., 1989). We conditioned rats that had been infected with *Nippostrongylus brasiliensis* to promote various arms of the allergic reaction. The animals were then conditioned by exposing them to loud noises and flashing lights at the same time as they received an injection of antigen that caused them to suffer mild anaphylaxis. After three rounds of conditioning, the animals were rested and then challenged either with the conditioning stimulus alone or coupled with injection of saline. Various groups of animals acted as the obvious positive or negative controls. The conditioning stimulus alone was able, in this set of experiments, to elicit mast cell degranulation, as evidenced by the release of an enzyme marker found only in rat intestinal mucosal mast cells. We concluded that the brain, either directly or indirectly, was able to cause degranulation of peripheral mast cells. Although we would prefer to explain this as a result of direct communication between mast cells and nerves, there is no direct evidence for or against this hypothesis.

VII. COCULTURE OF MAST CELLS AND NERVES

We ourselves have undertaken extensive efforts to determine what the effects of coculture of mast cells of various types and nerves would be (Blennerhassett et al., 1991, 1992; Stead et al., 1990). We have used both dorsal root and superior cervical ganglia either as explants or as dissociated nerves and have cultured these in low concentrations of nerve growth factor and then have carefully placed 1000 or so mast cells at a point distal to the growth cones of the extending neurites. Our findings can be summarized as follows (Table 3). Superior cervical ganglion neurites tend to associate with mast cells and form (at least in culture) permanent (up to 120 hr) membrane associations with the growth cones of neurites. Neurites that make contact with mast cells do so, tend to branch after contact, and carry on their extension, leaving the association with the mast cell intact. In the electron microscope, there appeared to be less than 20 nm in separation between nerve and mast cell contact. Rat basophil leukemia cells in similar situations were prevented from division and differentiated into much more mature-looking cells. There were significant electrical physiological changes in the mast cells associated with these nerves.

VIII. PATCH CLAMP STUDIES

Since we thought that SP may be involved from our morphological studies with enteric nerve mast cell associations, we decided to approach this from the point of view of patch clamping (Janiszewski et al., 1992). Accordingly, we set up baseline studies similar to those carried out by Neher and Sakmann (1992). We used primarily rat peritoneal mast cells, freshly isolated, and we tested the effects of SP on electrical recordings taken in the whole cell mode (Januszewski et al., submitted). At 5 μM, degranulation accompanied by the expected electrical changes occurs. We found that the time to an electrical effect from the point of

Table 3 Coculture of Mast Cells and Nerves (Superior Cervical Ganglia)

Selective association (neurotropic and neurotrophic effects)
Contact made and maintained (<20 nm) for up to 120 hr
Mast cells cease to divide (rat basophilic leukemia cells) and differentiate
Association causes electrophysiological changes

application of SP varied according to the concentration, so that 102 sec elapsed before the effect was seen at 10^{-6} M. At 2×10^{-8} M, the elapsed time was 594 sec and at 2×10^{-12} M, it was 802 sec. Accordingly, we began to look at the effects of low concentrations of SP. In brief, at low concentrations of SP, for example 5 pM, only a few transient electrical impulses were induced. However, remarkably, a second application of 5 pM caused 50% of the cells tested to degranulate, but invariably caused a significant elevation of the electrical current generated.

Thus, we have demonstrated a priming event, in mast cells, by a first very low dose of SP followed by a second challenge at the same subthreshold level. The time elapsed to degranulation in the repeated 5 pM experiments was in excess of 15 min. Thus, this may account for some of the findings already referred to earlier in the in vivo electrical stimulation experiments, in which prolonged stimulation appears to be necessary for degranulation to occur. The extent to which mast cells have been exposed or primed by neurotransmitters prior to antigen challenge, whether due to previous nervous stimulation or not, is therefore important to help us understand the physiological significance of nerve mast cell interactions.

It is perhaps interesting at this point to record that in the experiments conducted in the last few years on priming (most of these have been with basophils), a number of important observations have been made. A variety of hemopoietic growth factors can prime human basophils for subsequent synergistic or exaggerated responses to agonists such as C 5a (Bischoff and Dahinden, 1992a; Bischoff et al., 1990). These priming agents have been IL-3, IL-5, GM-CSF, and most recently in human mast cells, with human recombinant stem cell factor, the c-*kit* ligand. In addition, Dahinden, who with his associates has done much of the work in this area, has shown that nerve growth factor also has the same priming effect on human basophils (Bischoff and Dahinden, 1992b).

IX. NERVE GROWTH AND NERVE GROWTH FACTOR

These data appear to fit the model that is slowly developing in terms of mast cell–nerve associations. However, at this stage a brief description of nerve growth factor (NGF) is in order. It is an essential growth factor for a variety of nerves, including myelinated nerves in the peripheral nervous system, as well as for some in the central nervous system (Levi-Montalcini, 1987). It is especially important for the development of the sympathetic nervous system since anti-NGF treatment in early life abrogates the development of this system. The generation of NGF in myelinated nerves is dependent on the synthesis of IL-1 by macrophages that either are already present or are brought in as a result of injury (Lindholm et al., 1987).

IL-1 causes up-regulation of messenger RNA for NGF itself as well as for the NGF receptor. In this regard, it is perhaps even more interesting that IL-1 or NGF will together, or separately, induce the message for preprotachykinin in the superior cervical ganglion, with subsequent SP expression, synthesis, and transport down the efferent nerves (Hart et al., 1991; Lindsay and Harmer, 1989). NGF is made not only by cells of the nervous system, but also by many different glandular cells and by fibroblasts and keratinocytes (Levi-Montalcini, 1987; Davies et al., 1987; Tron et al., 1990).

In careful morphometric experiments we have carried out in the course of inflammation using a parasite (*N. brasiliensis*), we have shown that significant changes occur in the nerves of the intestine (Tomioka et al., 1993). Long after the inflammation has gone and the normal mast cell hyperplasia that is induced by the parasite has disappeared, significant increases in nerve density were seen in the enteric villi. Many additional changes were observed, including an increase in the number of small nerve profiles, which we interpreted to indicate nerve sprouting. Thus, in this model of inflammation, extensive restructuring of the enteric nervous system occurred, and we feel that it is likely that this is a regular accompaniment to inflammatory events, whenever they occur, in all tissues in the body.

NGF injected into neonates causes extensive mast cell hyperplasia in the skin, spleen, and intestine (Aloe and Levi-Montalcini, 1977). We ourselves have extended these original observations and have shown a 64-fold increase in the numbers of mast cells found per unit area in the spleen following NGF injections into the neonate (Tomioka et al., 1993).

We have also shown that it is likely that the NGF-dependent connective tissue mast cell hyperplasia occurs as a result of the direct effect of NGF on mast cells causing degranulation (Marshall et al., 1990). The likely release of cytokines from mast cells causes mast cell hyperplasia since mast cell degranulation products will produce the same end result. The effects of NGF on intestinal mast cell hyperplasia may be different, or at least occur through different mechanisms, since this was not caused directly by mast cell degranulation products. It is quite possible that this occurs through a T-cell-mediated effect, since in examining the effect of NGF on human hemopoietic colony growth, Matsuda, Denburg, and myself showed that this may well have occurred as a result of T-cell activation (Matsuda et al., 1988). NGF has many other significant nonneuronal effects (Table 4), which

Table 4 Nonneuronal Effects of Nerve Growth Factor

Enhanced wound healing, phagocytosis, macrophage activation
Promotes Con A–induced lymphocyte mitogenesis, MLR, human B-cell differentiation with selective synthesis of IgG4
Potent mast cell degranulator
Induces extensive mastopoiesis in neonatal rats and mice
Synergizes with GM-CSF and IL-5 to promote murine and human hemopoietic colony growth

may be interesting in terms of allergic inflammation but are not the subject of this chapter.

X. SUMMARY

Mast cells clearly can communicate with nerves. This has been shown by ourselves and others, in a variety of systems. These associations occur at a morphological level, and physiological interactions have been shown in vivo, ex vivo, as well as in vitro. It is likely that electrical nervous stimulation has direct effects on mast cells. These effects may not always be those of degranulation, but local secretion through differential effects on cytokine production as opposed to the release of granular products may be occurring. It is important to note that inhibitory nervous system pathways of a NANC nature exist and may inhibit mast cell degranulation so that electrical stimulation may inhibit mast cell degranulation. The priming effect of neuropeptides on mast cells is interesting and may help explain these relationships on an efferent arm. The effects of NGF on mast cell growth and the relationship of mast cells to nerve growth itself are interesting and may be important since structural remodeling of the nervous system occurs upon inflammation, and event likely to be seen in other forms of inflammation in other tissues as well.

Although we believe that the interaction between mast cells and the nervous system is important and occurs in a bidirectional mode, we believe that the mast cell is only a model for the interactions that occur between the nervous and immune systems.

ACKNOWLEDGMENTS

The author gratefully acknowledges the many major contributions of his colleagues to the work described in this chapter, especially Drs. Blennerhassett, Janiszewski, Marshall, Perdue, Stanisz, and Stead. Special acknowledgments go to the Medical Research Council of Canada, the Crohn's and Colitis Foundation of Canada, and AB Draco (Astra group) for support of the research described.

REFERENCES

Aloe, L., and Levi-Montalcini, R. (1977). Mast cells increase in tissues of neonatal rats injected with the nerve growth factor. *Brain Res. 133*:358–366.

Arizono, N., Matsuda, S., Hattori, T., Kojima, Y., Maeda, T., and Galli, S.J. (1990). Anatomical variation in mast cell nerve associations in the rat small intestine, heart, lung, and skin: similarities of distances between neural processes and mast cells, eosinophils, or plasma cells in the jejunal lamina propria. *Lab. Invest. 62*:626–634.

Baird, A.G., and Cuthbert, A.W. (1987). Neuronal involvement in type I hypersensitivity reactions in gut epithelia. *Br. J. Pharmacol.* *92*:647–655.

Bani-Sacchi, T., Barattini, M., Bianchi, S., et al. (1986). The release of histamine by parasympathetic stimulation in guinea-pig auricle and rat ileum. *J. Physiol.* *371*:29–43.

Bienenstock, J. (1992). Cellular communications networks: implications for our understanding of gastrointestinal physiology. In: *Neuro-immuno-physiology of the Gastrointestinal Mucosa, Implications for Inflammatory Diseases.* (Stead, R., Perdue, M.H., Cooke, H., Powell, D.W., and Barrett, K.E., eds.). New York Academy of Sciences, New York, pp. 1–9.

Bienenstock, J. (1993). Neuroimmune interactions in the regulation of mucosal immunity. In: *Immunophysiology of the Gut*, Vol. 12. Academic Press, New York, pp. 171–181.

Bienenstock, J., MacQueen, G., Sestini, P., Marshall, J.S., Stead, R.H., and Perdue, M.H. (1991). Mast cell/nerve interactions in vitro and in vivo. *Am. Rev. Respir. Dis.* *143*:S55–S58.

Bischoff, S.C., and Dahinden, C.A. (1992a). c-*kit* ligand: a unique potentiator of mediator release by human lung mast cells. *J. Exp. Med.* *175*:237–244.

Bischoff, S.C., and Dahinden, C.A. (1992b). Effect of nerve growth factor on the release of inflammatory mediators by mature human basophils. *Blood* *79*:2662–2669.

Bischoff, S.C., de Weck, A.L., and Dahinden, C.A. (1990). Interleukin 3 and granulocyte/macrophage-colony-stimulating factor render human basophils responsive to low concentrations of complement component C3a. *Proc. Natl. Acad. Sci. USA* *87*:6813–6817.

Blennerhassett, M.G., Tomioka, M., and Bienenstock, J. (1991). Formation of contacts between mast cells and sympathetic neurons in vitro. *Cell. Tissue Res.* *265*:121–128.

Blennerhassett, M.G., Janiszewski, J., and Bienenstock, J. (1992). Sympathetic nerve contact alters membrane resistance of cells of the RBL-2H3 mucosal mast cell line. *Am. J. Respir. Cell Mol. Biol.* *6*:504–509.

Church, M.K., El-Lati, S., and Caulfield, J.P. (1991). Neuropeptide-induced secretion from human skin mast cells. *Int. Arch. Allergy Appl. Immunol.* *94*:310–318.

Crowe, S.E., and Perdue, M.H. (1992). Gastrointestinal food hypersensitivity: basic mechanisms of pathophysiology. *Gastroenterology* *103*:1075–1095.

Davies, A.M., Bandtlow, C., Hermann, R., Korsching, S., Rohrer, H., and Thoenen, H. (1987). Timing and site of nerve growth factor synthesis in developing skin in relation to innervation and expression of the receptor. *Nature* *326*:353–358.

Dimitriadou, V., Buzzi, M.G., Moskowitz, M.A., and Theoharides, T.C. (1991). Trigeminal sensory fiber stimulation induces morphological changes reflecting secretion in rat dura mater mast cells. *Neuroscience* *44*:97–112.

Dimlich, R.V.W. (1984). Electron microscopic examination of mast cells and nerve processes in hepatic portal areas of the rat. In: *Proceedings of the 42nd Annual Meeting of the Electron Microscopy Society of America*, pp. 230–231.

Foreman, J.C., and Jordan, C.C. (1984). Neurogenic inflammation. *Trends Pharmacol. Sci.* *5*:116–119.

Hart, R.P., Shadiack, A.M., and Jonakait, G.M. (1991). Substance P gene expression is regulated by interleukin-1 in cultured sympathetic ganglia. *J. Neurosci. Res.* *29*:282–291.

Heine, H., and Forster, F.J. (1975). Relationship between mast cells and preterminal nerve fibers. *Z. Mikrosk-anat. Forsch.* *89*:934–937.

Jancso, N., Jancso-Gabor, A., and Szolcsanyi, J. (1967). Direct evidence for neurogenic inflammation and its prevention by denervation and by pretreatment with capsaicin. *Br. J. Pharmacol. 31*:138–151.

Janiszewski, J., Bienenstock, J., and Blennerhassett, M.G. (1992). Substance P induces whole-cell current transients in RBL-2H3 cells. *Am. J. Physiol. 263*:C736–C742.

Javed, N.H., Wang, Y.Z., and Cooke, H.J. (1992). Neuroimmune interactions—role for cholinergic neurons in intestinal anaphylaxis. *Am. J. Physiol. 263*:G847–G852.

Johnson, A.R., and Erdös, E.G. (1973). Release of histamine from mast cells by vasoactive peptides. *Proc. Soc. Exp. Biol. Med. 142*:1252.

Kowalski, M.L., and Kaliner, M.A. (1988). Neurogenic inflammation vascular permeability, mast cells. *J. Immunol. 140*:3905–3911.

Levi-Montalcini, R. (1987). The nerve growth factor: thirty-five years later. *EMBO J. 6*: 1145–1154.

Lindholm, D., Heumann, R., Meyer, M., and Thoenen, H. (1987). Interleukin-1 regulates synthesis of nerve growth factor in non-neuronal cells of rat sciatic nerve. *Nature 330*:658–659.

Lindsay, R.M., and Harmar, A.J. (1989). Nerve growth factor regulates expression of neuropeptide genes in adult sensory neurons. *Nature 337*:362–364.

Lowmann, M.A., Rees, P.H., Benyon, R.C., and Church, M.K. (1988). Human mast cell heterogeneity: histamine release from mast cells dispersed from skin, lung, adenoids, tonsils and intestinal mucosa in response to IgE-dependent and nonimmunological stimuli. *J. Allergy Clin. Immunol. 81*:590–597.

MacQueen, G., Siegel, S., Marshall, J., Perdue, M.H., and Bienenstock, J. (1989). Pavlovian conditioning of rat mucosal mast cells to secrete rat mast cell protease II. *Science 243*:83–85.

Marshall, J.S., Stead, R.H., McSharry, C., Nielsen, L., and Bienenstock, J. (1990). The role of mast cell degranulation products in mast cell hyperplasis. I. Mechanism of action of nerve growth factor. *J. Immunol. 144*:1886–1892.

Matsuda, H., Coughlin, M.D., Bienenstock, J., and Denburg, J.A. (1988). Nerve growth factor promotes human hemopoietic colony growth and differentiation. *Proc. Natl. Acad. Sci. USA 850*:6508–6512.

McDonald, D.M. (1987). Neurogenic inflammation in the respiratory tract: actions of sensory nerve mediators on blood vessels and epithelium of the airway mucosa. *Am. Rev. Respir. Dis. 136* (Suppl.):65–72.

Miura, M., Inoue, H., Ichinose, M., Kimura, K., Katsumata, U., and Takishima, T. (1990). Effect of nonadrenergic noncholinergic inhibitory nerve stimulation on the allergic reaction in cat airways. *Am. Rev. Respir. Dis. 141*:29–32.

Mousli, M., Bueb, J.L., Bronner, C., Rouot, B., and Landry, Y. (1990). G protein activation: a receptor-independent mode of action for cationic amphiphillic neuropeptides and venom peptides. *Trends Pharmacol. Sci. 11*:358–362.

Neher, E., and Sakmann, B. (1992). The patch clamp technique. *Sci. Am. 266*:28–35.

Newsom, B., Dahlstrom, A., Enerback, L.L., and Ahlman, H. (1983). Suggestive evidence for a direct innervation of mucosal mast cells. An electron microscope study. *Neuroscience 10*:565–570.

Perdue, M.H., and Bienenstock, J. (1991). Immunophysiology of the gut. *Curr. Opin. Gastroenterol. 7*:421–431.

Perdue, M.H., Masson, S., Wershil, B.K., and Galli, S. (1991). Role of mast cells in ion transport abnormalities associated with intestinal anaphylaxis: correction of the diminished secretory response in genetically mast cell-deficient W/W^v mice by bone marrow transplantation. *J. Clin. Invest. 87*:687–693.

Quinonez, G., Stead, R.H., Simon, G.T., and Bienenstock, J. (1986). Close relationship of eosinophils with post ganglionic axons of the rat. 49th Annual Meeting of the Ontario Association of Pathologists, Alton, Ontario, Canada.

Russell, M., Dark, K.A., Cummings, R.W., et al. (1984). Learned histamine release. *Science 225*:733.

Sestini, P., Dolovich, M., Vancheri, C., Stead, R.H., Marshall, J.S., Perdue, M., Gauldie, J., and Bienenstock, J. (1989). Antigen-induced lung solute clearance in rats is dependent on capsaicin-sensitive nerves. *Am. Rev. Respir. Dis. 139*:401–406.

Sestini, P., Bienenstock, J., Crowe, S.E., Marshall, J.S., Stead, R.H., Kahuta, Y., and Perdue, M.H. (1990). Ion transport in rat tracheal epithelium in vitro. Role of capsaicin-sensitive nerves in allergic reactions. *Am. Rev. Respir. Dis. 141*:393–397.

Shanahan, F., Denburg, J.A., Fox, J., Bienenstock, J., and Befus, D. (1985). Mast cell heterogeneity: effects of neuroenteric peptides on histamine release. *J. Immunol. 135*: 1331–1337.

Skofitsch, G., Savitt, S.M., and Jacobwitz, D.M. (1985). Suggestive evidence for a functional unit between mast cells and substance P fibers in rat diaphragm and mesentery. *Histochemistry 82*:5–8.

Stead, R.H., and Bienenstock, J. (1990). Cellular interactions between the immune and peripheral nervous systems. A normal role for mast cells? In: *Cell to Cell Interaction*. (Burger, M.M., Sordat, B., and Zinkernagel, R.M., eds.). Karger A.G., Basel, pp. 170–187.

Stead, R.H., Tomioka, M., Quinonez, G., Simon, G.T., Felten, S.Y., and Bienenstock, J. (1987). Intestinal mucosal mast cells in normal and nematode-infected rat intestines are in intimate contact with peptidergic nerves. *Proc. Natl. Acad. Sci. USA 4*:2975–2979.

Stead, R.H., Dixon, M.F., Bramwell, N.H., Riddell, R.H., and Bienenstock, J. (1989): Mast cells are closely apposed to nerves in the human gastrointestinal mucosa. *Gastroenterology 97*:575–585.

Stead, R.H., Perdue, M.H., Blennerhassett, M.G., Kakuta, Y., Sestini, P., and Bienenstock, J. (1990). The innervation of mast cells. In: *The Neuroendocrine-immune Network*. (Freier, S., ed.) CRC Press, Boca Raton, FL, pp. 19–37.

Stead, R.H., Kosecka-Janiszewska, U., Oestreicher, A.B., Dixon, M.F., and Bienenstock, J. (1991). Remodeling of B-50 (GAP-43)- and NSE-immunoreactive mucosal nerves in the intestines of rats infected with *Nippostronglyus brasiliensis*. J. Neurosci. 11:3809–3821.

Tomioka, M., Stead, R.H., Marshall, J., McSharry, C., Nielsen, L., Hamil, R.W., Coughlin, M.D, and Bienenstock, J. (1993). Nerve growth factor stimulates thymus-independent growth of connective tissue and intestinal mucosal mast cells in neonatal rats (in preparation).

Tron, V.A., Coughlin, M.D., Jang, D.E., Stanisz, J., and Sauder, N. (1990). Expression and modulation of nerve growth factor in murine keratinocytes (PAM 212). *J. Clin. Invest. 85*:1085–1089.

Undem, B.J., Hubbard, W.C., Christian, E.P., and Weinreich, D. (1990). Mast cells in the guinea pig superior cervical ganglion: a functional and histological assessment. *J. Auton. Nerv. Syst. 30*:75–88.

Weinreich, D., and Undem, B.J. (1987). Immunological regulation of synaptic transmission in isolated guinea pig autonomous ganglia. *J. Clin. Invest. 79*:1529–1532.

Weinreich, D., Undem, B.J., and Leal-Cardoso, J.H. (1992). Functional effects of mast cell activation in sympathetic ganglia. In: *Neuro-immuno-physiology of the Gastrointestinal Mucosa. Implications for Inflammatory Diseases*. (Stead, R.H., Perdue, M.H., Cooke, H., Powell, D.W., and Barrett, K.E., eds.). New York Academy of Sciences, New York, pp. 293–308.

Wood, J.D. (1993). Enteric neuroimmune interactions. In: *Immunophysiology of the Gut* (Walker, W.A., Harmatz, P.R., and Wershil, B.K., eds.). Academic Press, New York, pp. 207–227.

DISCUSSION

Polak: I was interested in your findings of hyperplastic nerves in an animal model of inflammatory bowel disease. In humans the same observations apply: hyperplastic and disorganized nerves are commonly found in inflammatory bowel disease.

Bienenstock: The work showing neuronal hyperplasia in Crohn's disease and ulcerative colitis is convincing. We are convinced that mast cells are not the only cell types involved or associated with nerves. Lymphocytes and other cells participating in immune processes may also associate with nerves.

Erdös: About 20 years ago A.R. Johnson and I described the release of histamine from rat peritoneal mast cells by substance P. We related this activity to the positively charged amino acids in the peptide. Of other peptides, kallidin was more effective than bradykinin; however, the nonpeptide polistes kinin was a potent histamine releaser, approaching 48/80 in effectiveness. In more recent experiments, human skin mast cells have been shown to be more reactive to substance P than peritoneal mast cells.

Bienenstock: Thank you for pointing out the regional differences in mast cell responsiveness. The varied responses to substance P are also seen in responses to NGF.

McDonald: What is the best evidence of nerve-mediated mast cell degranulation and does it occur in the respiratory tract?

Bienenstock: This is still a controversial area. Dimitriadov et al. showed that mast cell granule changes occurred after trigeminal nerve electrical stimulation. Field stimulation causes mast-cell-dependent intestinal epithelial chloride ion secretion, as shown by Perdue, Wershall, and Galli. Miura showed that bilateral vagal stimulation in the sensitized cat inhibited histamine release and increased pulmonary resistance with antigen exposure. There is no current evidence that nerve stimulation in the airways causes mast cell degranulation.

Solway: Which cell types secrete nerve growth factor?

Bienenstock: Cells of the nervous system, fibroblasts, keritinocytes, and glandular cells, especially salivary gland cells.

Goetzl: Does substance P activate basophils?

Bienenstock: Substance P does not affect basophils.

Goetzl: What permits such as exaggerated response to substance P by mast cells in your patch-clamp studies?

Bienenstock: It appears to involve a calcium-activated chloride channel, inhibitable by NPPB.

Kaliner: There are two relevant experiments I'd like to share: First, Kowalski et al., in published work failed to demonstrate that neurogenic inflammation of rat skin or lung involved mast cells. Second, Metcalfe and his colleagues at the National Institutes of Health have shown that mast cells begin to synthesize cytokines in response to substance P stimulation, in the absence of histamine release. Thus, exposure to neuropeptides might facilitate mast cell participation in late-phase allergic reactions without causing immediate degranulation.

Bienenstock: Nerve stimulation clearly affects mast cells, while prolonged electrical stimulation is required for degranulation. We need to sort out the effects of the NANC before we can determine all the inhibitor and stimulatory actions of nerve-derived factors on mast cells.

19

Inositol Trisphosphate and Smooth Muscle Function

R. F. Coburn, H. Matsumoto, and C. B. Baron
University of Pennsylvania School of Medicine, Philadelphia, Pennsylvania

I. INTRODUCTION

This chapter reviews selected areas in the field of inositol 1,4,5-trisphosphate [$Ins(1,4,5)P_3$] signal transduction in airway smooth muscle, smooth muscle, and other cell types. $Ins(1,4,5)P_3$ signal transduction has not been previously reviewed for smooth muscle at the time of this writing. Of course we will stress issues that relate to our research in this area, which is largely on airway smooth muscle. It is clear that there are multiple signal transduction systems that control smooth muscle function and may all relate to one another. We make no attempt to discuss in any depth all these areas. We will also not cover recent developments regarding the established or putative inositol phospholipid signal transduction mechanisms involving diacylglycerol (Berridge, 1993), phosphatidic acid (Bocckino et al., 1991), or inositol tetrakisphosphate (Irvine and Moor, 1986).

There are a number of recent reviews of the general area of $Ins(1,4,5)P_3$ signal transduction (Gilman, 1987; Fain, 1990; Joseph and Williamson, 1989; Downes and Carter, 1990; Rodbell, 1991; Rooney and Thomas, 1991; Berridge, 1993).

Two apparently independent limbs of inositol phospholipid cascades that produce $Ins(1,4,5)P_3$ have been described: a limb that exerts control on cellular function and a limb activated by growth factors and other mediators, which controls gene expression and regulation (Berridge, 1993) (Fig. 1). Our review is concerned primarily with inositol phospholipid metabolism and $Ins(1,4,5)P_3$ production linked to function. A challenge for the future is to determine how smooth muscle cells are remodeled during disease and the significance of this

413

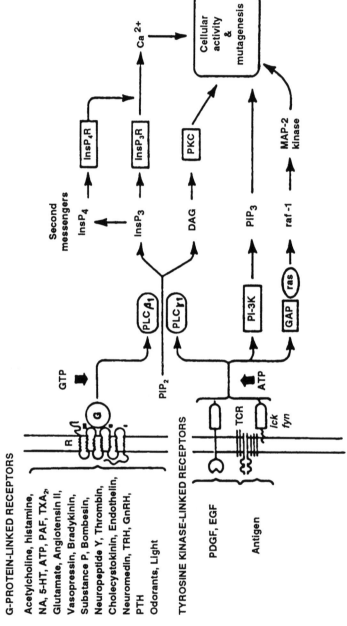

Figure 1 Two major receptor-mediated pathways for stimulating the formation of Ins(1,4,5)P₃. PLC-β isoforms are involved in PIP₂ hydrolysis that is driven by receptors that control cellular function, and PLC-γ isoforms are involved in PIP₂ hydrolysis that directs signals to the nucleus. (From Berridge, 1993. Published with permission.)

remodeling. At present, we note the provocative findings that a single inflammatory mediator can activate inositol phospholipid signal transduction systems that release Ca^{2+} and control function and signal transduction systems that induce proliferation (Panettieri et al., 1990) and that a growth factor can cause smooth muscle contraction (Berk et al., 1986). Remodeled, hypersensitive smooth muscle cells are capable of producing higher rates of $Ins(1,4,5)P_3$ formation (Salari et al., 1992).

II. RECEPTOR-INOSITOL-PHOSPHOLIPID-SPECIFIC PHOSPHOLIPASE C COUPLING

Most receptors that are linked to G proteins have seven membrane-spanning domains and cytosolic and extracellular domains. For referral to studies of muscarinic receptor subtypes, the reader is referred to publications by Bonner et al. (1987), Maeda et al., (1988), Buckley et al. (1989), and Bonner (1992). Similar data have been obtained for other receptor subtypes. Agonist binding evokes conformational changes responsible for activation of G proteins (Gillman, 1987; Spiegel et al., 1991; Rodbell, 1992). G proteins have been linked to β isomers of inositol-phospholipid-specific phospholipase C (PLC) (Roffel et al., 1990; Yang et al., 1991; Lee et al., 1991; Katz and Simon, 1992). The "wiring" between receptor subtypes, G proteins, and effector proteins is complex. Binding of an agonist to a single receptor subtype gives divergent signals in that multiple G proteins and effector mechanisms can be activated (Ashkenazi et al., 1987; Peralta et al., 1987; Luchese et al., 1990; Fargin et al., 1991; Yang et al., 1991). Multiple different receptor subtypes can activate PLC-mediated PIP_2 hydrolysis (Chilvers et al., 1989; Murray and Kotlikoff, 1991; Baron et al., 1993; Berridge, 1993) using different G proteins (Gerwins and Fredholm, 1992; Katz and Simon, 1991). α or $\beta\gamma$ G-protein subunits can bind to effector proteins; the same subunit can activate different effectors (Berridge, 1993). Technology now allows reconstitution systems that include a single muscarinic receptor subtype and a single G protein and PLC isoform (Berstein et al., 1992; Lee et al., 1992). There is evidence that G-protein function can be modulated by protein-kinase-C-mediated phosphorylations (Sagi-Eisenberg, 1989).

Little is known about diversity of G proteins in airway or other smooth muscle. The approach of deducing the presence of G proteins using pertussis toxin has been applied to smooth muscle (Marc et al., 1988; Gu et al., 1991).

At the present time at least seven established, immunologically distinct inositol-phospholipid-specific PLC isoforms with different molecular weights and tissue distributions have been reported (Low et al., 1986; Rhee et al., 1989; Fain, 1990; Homma et al., 1993). These have been classified into three different families, β, γ, and δ (Rhee et al., 1989). Catalytic properties are similar with the different isoforms, each hydrolyzing PI, PIP, and PIP_2, with PIP_2 hydrolysis occurring at the lowest [Ca^{2+}]. Most PLC-γ and PLC-δ has been recovered in the cytosol of various cells; PLC-β, at least in the brain, was identified in particulate fractions (Lee et al., 1992). The same cell types may have multiple PLC isoforms (Homma et al., 1989).

In many tissues, PLC-β isoforms are involved in receptor-Gq activation of PIP$_2$ hydrolysis and Ins(1,4,5)P$_3$ production directed at the Ins(1,4,5)P$_3$-receptor-channel, controlling cellular function (Rhee et al., 1989; Katz and Simon, 1992; Berridge, 1993; Park et al., 1993). There can be specificity in interactions of different Gq family members with different PLC-β isoforms (Lee et al., 1992). Both α and β-γ G-protein subunits can activate PLC-β isoforms (Katz and Simon, 1992). Site-specific mutagenesis studies of brain PLC-β have delineated the G-protein-binding sites (C-terminal end of the molecule); the site of enzyme activity also has been partly characterized (Wu et al., 1993).

PLC-γ isoforms are involved in transducing signals directed to the nucleus. PLC-γ is not considered to be activated by G proteins. Rather, PLC-γ is linked to receptors that have inherent tyrosine kinase activity, and PLC-γ is activated by a tyrosine-kinase-mediated phosphorylation (Rhee et al., 1989; Meisenhelder et al., 1989; Goldschmidt-Clermont et al., 1991). PLC-γ isoforms are associated with proteins located at the boundary between the cytoskeleton and plasma membrane (Forscher, 1989; Grondin et al., 1991). Binding to profilin (Beck and Keen, 1991; Goldschmidt-Clermont et al., 1991) may be necessary for PLC-γ activity to be expressed. Translocation of PLC-γ may occur during activation (Moriarty et al., 1988; Constantinescu and Popescu, 1991; Payrastre et al., 1991).

PLC-δ isoforms have been identified in multiple tissues (Rhee, 1989; Kato et al., 1992; Park et al., 1993; Rebecchi et al., 1993). The mechanisms of activation of these isoforms are still uncertain. However, PLC-δ isoforms are reported to be Ca^{2+} sensitive and activated by G proteins (Kato et al., 1992; Park et al., 1993).

PLC-γ and PLC-δ isoforms have been identified in a number of smooth muscles (Zhou et al., 1993; LaBelle and Gu, 1993; Kato et al., 1992; Homma et al., 1993). In only one of these studies (Zhou et al., 1993) was a PLC-β isomer present in detectable quantities. In vascular smooth muscle, PLC-γ2 was shown to be involved in signal transduction evoked by platelet-derived growth factor (Homma et al., 1993). At present, the identification of smooth muscle PLC isomers that catalyze PIP$_2$ hydrolysis linked to development of muscle force is uncertain. Data from other tissues suggest that a β isomer is involved; however, the lack of detection of β isomers in most smooth muscles studied to data suggests that the δ isomer may be involved in receptor-PLC-Ins(1,4,5)P$_3$-induced Ca^{2+} release. Because of this uncertainty, in the subsequent sections of this chapter, which are entirely involved in inositol phospholipid transduction directed at cellular function, we use the general term "PLC."

It is not clear if the same or different PIP$_2$ pools are hydrolyzed by different PLC isoforms present in the same cell or if Ins(1,4,5)P$_3$ produced by PIP$_2$ hydrolysis by different isoforms have separate functions. It is not understood why the two limbs of PLC signal transduction cascades (Berridge, 1993) are organized when PLC in each limb has the same function.

PLC activity as determined by $Ins(1,4,5)P_3$ formation measurements is activated by smooth muscle stretch (Kulik et al., 1991), by agents that open K^+ channels (Challis et al., 1992), and may be modulated by the extent of Ca^{2+} loading of the sarcoplasmic reticulum (SR) (Berman and Goldman, 1992). None of these phenomena are understood, but they emphasize possible complex cellular control mechanisms.

Although there are reports of cAMP-mediated inhibition of $Ins(1,4,5)P_3$ formation in many tissues, including airway muscle (Madison and Brown, 1988; Abdel-Latiff, 1988; Hall et al., 1989), this was probably not mediated by phosphorylation of PLC-β isoforms by cAMP-dependent protein kinases (Kim et al., 1989). These findings may be due to cAMP-dependent, protein-kinase-mediated phosphorylation of adrenergic receptors (Limas and Limas, 1985; Leeb-Lundberg et al., 1987). Activation of protein kinase C by phorbol esters caused a decrease in agonist-activated $Ins(1,4,5)P_3$ formation in many tissues. In airway smooth muscle, phorbol esters completely inhibited $Ins(1,4,5)P_3$ formation evoked by carbachol (Baba et al., 1989). The mechanism for this effect is not completely understood.

III. INOSITOL PHOSPHOLIPID METABOLISM

The principal inositol phospholipids involved in the "PI cycle" are phosphatidylinositol (PI), phosphatidylinositol 4-phosphate (PIP), and phosphatidylinositol 4,5-bisphosphate (PIP_2) (Berridge, 1993). With receptor activation of a PLC, the first event is considered to be PIP_2 hydrolysis to $Ins(1,4,5)P_3$ and diacylglycerol (DAG). In many cells this is associated with rapid large decreases in PIP_2 contents. The cycle is completed with onset of PI resynthesis in the endoplasmic reticulum (ER) and transport to, and insertion into, the plasma membrane (Wolf, 1990; Berridge, 1993).

The above schema has been modified with the discovery of 3-kinases that convert PI to PI(3)P, PI(4)P to $PI(3,4)P_2$ and $PI(4,5)P_2$ to $PI(3,4,5)P_3$ (Auger et al., 1989; Carpenter and Cantley, 1990; Hawkins et al., 1992).

The state of knowledge of PI 4-kinase and PI(4)P 5-kinase isomers has been reviewed (Ling et al., 1989; Carpenter and Cantley, 1990; Pike, 1992; Devecha et al., 1992). Both PI 4-kinases and PI(4)P 5-kinases have been recovered in the cytoplasm and in cytoskeleton and plasma membrane preparations (Dale, 1985; Payrastre et al., 1991; Pike, 1992).

Various investigators have postulated that PI(4)P 5-kinase may exert control of $Ins(1,4,5)P_3$ formation under the condition where PI resynthesis is occurring, which is replacing PIP_2 hydrolyzed by PLC. PI(4)P 5-kinase may be regulated by G proteins (Smith and Chang, 1989), by phosphatidic acid concentrations (Moritz et al., 1992), by tyrosine kinase (Gaudette and Holub, 1990), by translocation to cytoskeleton (Eng and Lo, 1990; Dale, 1985; Payrastre et al., 1991) or receptor (Cochet et al., 1991), and/or by changes in [Ca^{2+}]i (Mesaeli et al., 1992). Evidence that PI 4-kinase activities can be increased by tyrosine kinase phosphorylation and modulated by cAMP-dependent protein kinase (PKC) has been reviewed by Pike (1992).

Although PLC activation is the major control point for receptor-activated inositol phospholipid metabolism linked to cell function, control is also exerted by processes that control PI resynthesis in the ER. The reactions involved in PI resynthesis are illustrated in Figure 2. Mechanisms for control of PI resynthesis and insertion into the plasma membrane remain obscure; there is some evidence that increases in [Ca^{2+}]i are involved (Takanawa and Egawa, 1977; Eberhard and Holz, 1991).

There are several methods for study of receptor-activated inositol phospholipid metabolism in intact cells: (1) measurements of changes in pool sizes; (2) determinations of rates of incorporation of radioactive D-myo-inositol into various lipids, which gives an estimation of flux from PI to PIP to PIP$_2$ (Baron et al., 1989b). In canine trachealis muscle (CTM) and swine trachealis muscle (STM), there seem to be at least two phases of activation of inositol phospholipid metabolism (Baron et al., 1989b). The early phase is characterized by PIP$_2$ hydrolysis and decreases in inositol phospholipid pool sizes (Baron et al., 1989b, 1993; Chilvers et al., 1991). This phase occurs prior to the onset of PI resynthesis and occurs during the time of development of force. Estimations of PIP$_2$ hydrolysis rates during maximal carbachol stimulation were made from rates of decrease in PIP$_2$ + PI, which gave a value of 0.42 ± 0.001 nmol × 100 nmol total lipid Pi^{-1} × min^{-1} (Baron et al., 1989b). The second phase, which occurs at a time when force is maintained, is characterized by PI resynthesis under conditions where pool sizes of PI, PIP, and PIP$_2$ remain nearly constant. Under this condition, flux rate in inositol phospholipids, and Ins(1,4,5)P$_3$ formation rate, becomes limited by the rate of PI resynthesis. Specific radioactivities in PI, PIP, and PIP$_2$ increase steeply in parallel suggesting over the slow time frame that the kinase reactions are in near equilibrium (Fig. 3). Flux rates during this phase, computed from incorporation of myo-inositol radioactivity into PI, PIP, and PIP$_2$, averaged 0.14 ± 0.01 nmol × 100 nmol total lipid Pi^{-1} × min^{-1}, a value 20 times that determined in unstimulated tissue and less than that which occurred during the force development phase of contraction (Baron et al., 1989b). We computed that 50% of PI, PIP, and PIP$_2$ turned over every 10, 0.8, and 2.0 min, respectively, as a result of high flux rates and small pools utilized in receptor-evoked activation. These data suggest the importance of inositol phospholipid metabolism in providing Ins(1,4,5)P$_3$ (and DAG) during force development and during maintained force.

There are very large changes in total tissue inositol phospholipid contents in receptor-activated cells, which must reflect large changes in plasma membrane contents of these lipids. The rapid turnover times of inositol phospholipids in

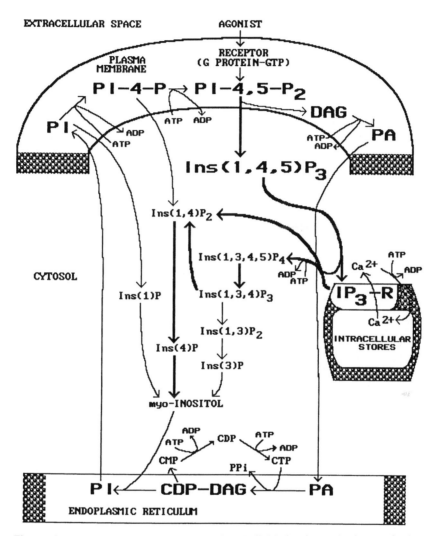

Figure 2 Major components of inositol phospholipid signal transduction mechanisms during muscarinic activation of trachealis muscle. Heavy lines indicate inositol phosphate degradative pathways found in STM (Baron et al., 1992b). Phosphatidic acid (PA) is shown to be formed within the plasma membrane because of our unpublished data of receptor-evoked increases in PA in this cellular fraction. Light lines indicate the possibility of PLC-mediated PI and PIP hydrolysis

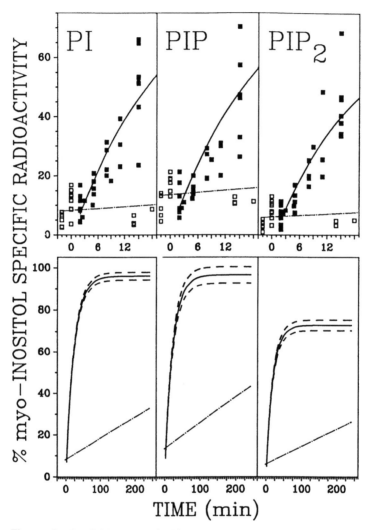

Figure 3 Parallel increases in PI-, PIP-, and PIP$_2$-specific radioactivities in canine trachealis muscle cells following carbachol (5.5 μM) administration. All data were taken from Baron et al. (1989b, 1993). (Upper graph) Rapid changes occurring following carbachol stimulation (filled squares and solid line (exponential fit)). Open squares and dashed-dotted line (linear fit) indicate unstimulated data over the same time period. (Lower graph) Exponential fit ± SEM (dashed lines), over a longer time period, and unstimulated linear fit. All lines shown were obtained fitting all of the data points: stimulated n = 56–58; unstimulated, n = 29–30.

maximally stimulated CTM and STM indicate that pool sizes can change rapidly. An example of this is seen after a rapid inhibition of PLC, effected by atropine, in carbachol-contracted STM (Baron et al., 1992b). Although PLC was rapidly inhibited, inhibition of PI resynthesis was delayed for several minutes, which resulted in a rapid increase in PIP_2 and PIP contents and a slower increase in PI content. Whether these changes can exert effects on plasma membrane proteins that control cell function is still not known. Changes in PIP pool size may control the smooth muscle sarcolemmal Ca^{2+}-ATPase (Vrolix et al., 1988; Verbist et al., 1991) or alter actin assembly (Pike, 1992).

More information about the organization of plasma membrane receptor–G protein–PLC "complexes" and PIP_2 pools needs to be obtained to be able to understand states where cells are stimulated simultaneously by different agonists, each of which activates a PLC isoform. It is possible that PIP_2 pools involved in receptor activation are compartmentalized to specific receptor subtypes (Vickers and Mustard, 1986; Kirk et al., 1989; Monaco et al., 1990). In STM, however, a common PIP pool was shown to be involved in carbachol and serotonin activation (Baron et al., 1993). Thus, despite evidence that G-protein targets are limited to microdomains in the plasma membrane (reviewed in Baron et al., 1993), activated PLC had access to a pool of PIP_2 that was not limited to individual receptor–G protein microdomains. These results introduced the possibility that there could be competition for a common PIP_2 pool during simultaneous serotonin and carbachol activation pathways in this muscle.

IV. INS(1,4,5)P$_3$ DEGRADATION

$Ins(1,4,5)P_3$ can be degraded via two pathways: 3-kinases forming $Ins(1,3,4,5)P_4$ and 5-phosphatases degrading the compound to $Ins(1,4)P_2$ (Williamson et al., 1988; Shears, 1989). Cascades of phosphatases involved in degradation of $Ins(1,4)P_2$ and $Ins(1,3,4,5)P_4$ are shown in Figure 2.

From studies of rapid relaxations of carbachol-contracted STM by atropine, it was established that 5-phosphatase is the rate-limiting reaction in this muscle for degradation of $Ins(1,4,5)P_3$ (Baron et al., 1992b). 5-Phosphatases in nonmuscle cells are primarily membrane bound (perhaps plasma membrane bound); smaller activities are found in cytosol (reviewed by Shears, 1989). There are multiple isoforms of this enzyme with K_ms of approximately 20 μM, a value considerably larger than a likely $[Ins(1,4,5)P_3]i$ for most cells, even during receptor stimulation. 5-Phosphatases can use either $Ins(1,4,5)P_3$ or $Ins(1,3,4,5)P_4$ as substrates. Many $Ins(1,4,5)P_3$ 5-phosphatases show augmented activity following phorbol ester treatment. An $Ins(1,4,5)P_3$ 5-phosphatase studied in a vascular smooth muscle showed increased activity as $[Ca^{2+}]i$ was varied from 0.2 to 1 μM (Sasaguri et al., 1985). Thus, it is possible that modulation of 5-phosphatase activities via other second-messenger systems can exert control of $[Ins(1,4,5)P_3]$. This has not been demonstrated experimentally in intact cells.

$Ins(1,4,5)P_3$ 3-kinase is another enzyme whose activity can be modulated. Increases in $[Ca^{2+}]i$ augmented 3-kinase activity in a calmodulin-dependent manner, and this enzyme may also be activated by PKC and PKC-mediated phosphorylations (Shears, 1989).

Patterns of inositol phosphates observed during smooth muscle activation can vary considerably in different muscles. This indicates that activities of enzymes in the inositol phosphate degradative pathway can vary in different muscles. One of the most striking examples was the almost complete absence of $Ins(1,3,4)P_3$ and only small $Ins(1,3,4,5)P_4$ contents in stimulated vascular smooth muscle (Gu et al., 1991; Matsumato et al, 1993), whereas $Ins(1,3,5)P_3$ and $Ins(1,3,4,5)P_4$ contents rose significantly in activated STM (Chilvers et al., 1990; Baron et al., 1992b).

Li^+ is commonly used to enhance the yield of inositol phosphates. However, since only some of the degradative enzymes are inhibited (Shears, 1989), the products can vary widely from those studies in which Li_+ is not used. This can occur not only by direct inhibition of certain phosphatases, but by end-product inhibition of other enzymes. The effect of Li^+ on smooth muscle function has not been examined and is only known for its effect on brain inositol phospholipid metabolism.

Isolated inositol-phospholipid-specific PLC can utilize PI and PIP as well as PIP_2 as substrates (Wilson et al., 1985; Majerus et al., 1990); however, PIP_2 hydrolysis may occur preferentially, at least under conditions of physiological $[Ca^{2+}]i$ (Berridge, 1993; Joseph and Williamson, 1989; Batty and Nahorski, 1992). In vascular smooth muscle, PI hydrolysis did not occur during early phases of adrenergic-receptor-evoked contraction. After 20 min of stimulation, there may have been some PI hydrolysis, based on the finding that Ins(1)P was formed and that $Ins(1,4)P_2$ was converted to Ins(4)P, not Ins(1)P, in a permeabilized muscle (LaBelle et al., 1992).

Small quantities of IP_5 and IP_6 have been recovered in some tissues. The function of these compounds is unknown, and so far, there is no evidence of receptor-evoked changes in tissue contents of these compounds (Shears, 1989). Cyclic IP_3 has been postulated to function as a second messenger (Majerus et al., 1990). There is no report of the presence of this compound in airway or other smooth muscles.

V. INS(1,4,5)P₃-INDUCED RELEASE OF Ca²⁺ FROM INTRACELLULAR STORES

Following the initial demonstration that $Ins(1,4,5)P_3$ releases Ca^{2+} from intracellular stores of pancreatic islet cells (Streb et al., 1983), there have been measurements of this phenomenon in many other tissues, including smooth muscle cells (Suematsu et al., 1984; Somlyo et al., 1985; Chopra et al., 1991). $Ins(1,4,5)P_3$ formed as a result of PIP_2 hydrolysis catalyzed by a PLC isomer is thought to diffuse to an $Ins(1,4,5)P_3$-sensitive Ca^{2+} store. Recent progress in this field includes (1) the isolation and partial characterization of $Ins(1,4,5)P_3$-receptor protein and a related protein, the ryanodine receptor; (2) partial characterization of intracellular Ca^{2+} stores that are released in receptor-activated cells; and (3) the biology of Ca^{2+} release from various intracellular Ca^{2+} stores.

$Ins(1,4,5)P_3$-sensitive Ca^{2+} stores can be conceptualized as organelles (perhaps modified ER, such as sarcoplasmic reticulum, or calciosomes) (Joseph and Williamson, 1989; Rooney and Thomas, 1991) that have lipid membranes into which $Ins(1,4,5)P_3$-receptor-channel proteins and Ca^{2+}-ATPases are inserted and that contain in their lumen a Ca^{2+}-binding protein such as calsequestrin. (As will be discussed below, the lipid membrane of these organelles may also contain a ryanodine receptor.) Existing evidence suggests $Ins(1,4,5)P_3$ binds to a receptor-channel protein evoking an allosteric change, which increases the probability of the channel being operated in an open state thereby allowing Ca^{2+} to move down its electrochemical gradient into the cytosol (Berridge, 1993).

An $Ins(1,4,5)P_3$ receptor [$Ins(1,4,5)P_3$-R] was isolated from rat cerebellar Purkinje cells by Supattapone et al. (1988), purified and cloned by Furachi et al. (1989) and by Mignery et al. (1990), and studied by reconstitution into lipid vesicles or bilayers (Supattapone et al., 1988; Ferris et al., 1989; Ehrlich and Watras, 1990). The monomer has typical membrane-spanning domains in the C-terminal region. The $Ins(1,4,5)P_3$ binding site is located on the large N-terminal, which projects into cytoplasm and is far from channel domains. Tetrameric assembly is required for function. Although still controversial (Joseph and Williamson, 1989), the assembled tetramer may bind one $Ins(1,4,5)P_3$ molecule per monomer. A steep $Ins(1,4,5)P_3$ concentrations–binding relationship was demonstrated (Iino and Endo, 1992). Sequential $Ins(1,4,5)P_3$ binding to the tetramer complex showed positive cooperativity (Meyer et al., 1988). Early studies indicated that $Ins(1,4,5)P_3$-R transports Ca^{2+} via a channel, rather than via a transporter (Joseph et al., 1989). The finding of different single-unit conductances of reconstituted $Ins(1,4,5)P_3$-Rs into lipid bilayers [i.e., >100 pS for Purkinje cells vs. <20 pS for the protein isolated from vascular smooth muscle (Ehrlich and Watrus, 1990)] suggests there is a family of $Ins(1,4,5)P_3$-Rs with different isomers in different cell types, or perhaps within the same cell. $Ins(1,4,5)P_3$-evoked increase in the probability of channel opening persisted as long as $Ins(1,4,5)P_3$ was present (Supattapone et al., 1988). Channel opening probability exhibited a biphasic Ca^{2+} sensitivity over physiological [Ca^{2+}] (Iino, 1990). Low concentrations of Ca^{2+} potentiate $Ins(1,4,5)P_3$-evoked Ca^{2+} release with a maximum at 300 μM (positive feedback) and higher concentrations inhibit it (Yao and Parker, 1992). This characteristic may explain the rapid initial $Ins(1,4,5)P_3$-evoked Ca^{2+}

release, which some have described as "all or none" (Parker and Ivorra, 1991; Bootman et al., 1991), and a slower second phase of Ca^{2+} release (Missiaen et al., 1992). $Ins(1,4,5)P_3$-induced Ca^{2+} release was dependent on K^+ counter transport into SR (Joseph et al., 1989). In brain microsomes, $Ins(1,4,5)P_3$-evoked Ca^{2+} release was inhibited by the K^+ channel inhibitor TEA (Shah and Pant, 1988), $Ins(1,4,5)P_3$ affinities for binding (K_d) vary from 0.1 to 0.6 nM in different tissues (Joseph and Williamson, 1989). There may be multiple conformational states with different $Ins(1,4,5)P_3$-binding affinities (Rouxel et al., 1992; Mohr et al., 1993). On the basis of results evoked by caged $Ins(1,4,5)P_3$ injection into nonmuscle cells, $Ins(1,4,5)P_3$-sensitive Ca^{2+} stores have been modeled as a collection of independent stores; as $Ins(1,4,5)P_3$ concentration increased, recruitment of $Ins(1,4,5)P_3$-sensitive stores occurred (Parker and Ivorra, 1990). Within the same cell there was heterogeneity regarding the $Ins(1,4,5)P_3$ sensitivity of different Ca^{2+} stores (Bootman et al., 1991; Chopra et al., 1991).

The state of Ca^{2+} loading of $Ins(1,4,5)P_3$-sensitive stores was critical for their function. Depletion of stores resulted in a decrease in sensitivity to $Ins(1,4,5)P_3$ (Missiaen et al., 1991). The relationship of Ca^{2+}-ATPase-mediated refilling of stores to $[Ca^{2+}]$ was found to vary in different stores (Bian et al., 1991). Thus, the characteristics of Ca^{2+}-ATPases exerted control on the sensitivity of $Ins(1,4,5)P_3$ binding to $Ins(1,4,5)P_3$-R. Filling of intracellular Ca^{2+} stores using extracellular Ca^{2+} may involve uptake via plasma membrane channels and utilize a mechanism that is bypassing increases in $[Ca^{2+}]i$ [the capacitative model of Putney (1986)]. In smooth muscle, filling of the SR was inhibited by agents that block voltage-gated Ca^{2+} channels (L channels) (Bourreau et al., 1991; Low et al., 1993) as predicted by the capacitance model. Ca^{2+} released from $Ins(1,4,5)P_3$-sensitive stores could be transferred to $Ins(1,4,5)P_3$-insensitive stores (Nilsoon et al., 1987; Ogden, 1988). There may be a cytoplasmic protein that modulates $Ins(1,4,5)P_3$-R function (Danoff et al., 1988; Hershey et al., 1993). In purified cerebellar cells, cAMP-dependent, protein-kinase-mediated phosphorylation resulted in an increase in the half-maximal $Ins(1,4,5)P_3$ concentration required to induce Ca^{2+} release (Supattapone et al., 1989).

Heparin and caffeine have been useful in studying $Ins(1,4,5)P_3$-sensitive intracellular Ca^{2+} stores. Heparin binds to $Ins(1,4,5)P_3$-R and is an inhibitor of $Ins(1,4,5)P_3$-mediated Ca^{2+} release (Kobayashi et al., 1989; Chopra et al., 1989; Iino, 1990). Limitations in the use of heparin in studies of $Ins(1,4,5)P_3$ function have been documented (Challis et al., 1991). Caffeine has been shown to inhibit rat cerebellar Purkinje cell $Ins(1,4,5)P_3$-R mediated Ca^{2+} release (Brown et al., 1992), perhaps by inhibiting positive cooperativity during progressive binding to the $Ins(1,4,5)P_3$ tetramer (Meyer et al., 1988).

In smooth muscle, and other cells, all of the Ca^{2+} released from internal stores could not be explained entirely on the basis of $Ins(1,4,5)P_3$-triggered release. In most cells 20–40% of the nonmitochondrial releasable intracellular Ca^{2+} is

Ins(1,4,5)P_3 sensitive (Berridge, 1993). Exceptions include the rat insulinoma cells, where about 90% of the nonmitochondrial releasable store was Ins(1,4,5)P_3 sensitive (Prentki et al., 1985), and cultured airway smooth muscle, where almost the entire releasable Ca^{2+} store was sensitive to Ins(1,4,5)P_3 (Chopra et al., 1991).

In addition to Ins(1,4,5)P_3-sensitive stores, two other types of intracellular Ca^{2+} stores have been identified: ryanodine-sensitive stores and GTP-sensitive stores. Ryanodine-receptor channel (RYR) (so called because ryanodine binds to this protein) is a Ca^{2+}-sensitive, voltage-dependent channel investigated most intensely in cardiac and skeletal muscle SR (Hymel et al., 1988; Zorzato et al, 1990). RYRs have been found in other tissues, including smooth muscle (Hwang and Van Breemen, 1987; Hisayama and Takayanagi, 1988; Chopra et al., 1991). The RYR molecular structure shows 70% homology with the structure of Ins(1,4,5)P_3-R. The C-terminal region forms the Ca^{2+} channel and the protein functions as a tetramer. As with Ins(1,4,5)P_3-Rs, there may be a heterogeneous family of RYRs in different cells. Caffeine binds to RYR and increases the probability of the Ca^{2+} channel open state. The Ca^{2+} effect, which occurs over physiological concentrations, increases the opening state probability. These findings suggest there may be two different mechanisms that trigger RYR-containing organelles to release Ca^{2+}: (1) membrane depolarization (if the organelles have junctions with plasma membrane); (2) Ca^{2+} entering the cell via plasma membrane channels, or released from intracellular Ca^{2+} stores, induces Ca^{2+} release from RYR-containing stores.

With reconstituted RYR prepared from aortic smooth muscle (Herrmann-Frank et al., 1991), the probability of channel opening was increased by μM Ca^{2+}, and by caffeine, as occurred with skeletal and cardiac RYR. In the majority of preparations, however, voltage dependence was undetectable. Multiple conduction states were present. The unitary conductance in 100 mM Ca^{2+} was 100 pS and in 250 mM K^+ it was 360 pS.

In some cell types, including smooth muscle, a GTP-binding protein is involved in release of Ca^{2+} from intracellular stores (Gill et al., 1986; Cheuh and Gill, 1986; Kobayashi et al., 1986; Thomas, 1988). The significance of GTP-induced Ca^{2+} release is not clear. A hypothesis based on studies performed on nonmuscle cells (Ghosh et al., 1989) is that a GTP-binding protein is involved in refilling Ins(1,4,5)P_3-sensitive stores using Ca^{2+} located within Ins(1,4,5)P_3-insensitive stores.

In smooth muscle, junctional and central SR operate to release Ca^{2+} into the cytoplasm (Somlyo et al., 1988). In the guinea pig portal vein, a maximal concentration of norepinephrine released about 50% of Ca^{2+} present in central SR (Somlyo et al., 1988). Agonist-evoked intracellular Ca^{2+} release could be explained by SR Ca^{2+} release.

The relative importance of Ins(1,4,5)P_3-mediated SR Ca^{2+} release versus Ca^{2+} release from SR that contains RYR, and from other possible intracellular Ca^{2+}

stores, is still not precisely defined. Ca^{2+} can be released in chemically skinned smooth muscle cells by $Ins(1,4,5)P_3$, caffeine, ryanodine, and GTP or related nucleotides (Kobayashi et al., 1986; Matsumoto et al., 1990; Chopra et al., 1991; Bian et al., 1991; Missiaen et al., 1991; Yamazawa et al., 1992). The different types of intracellular Ca^{2+} stores cannot be quantified because Ca^{2+} released by one compound could release Ca^{2+} from other types of stores via Ca^{2+}-induced Ca^{2+} release or be taken up by other Ca^{2+} stores. $Ins(1,4,5)P_3$ is specific for $Ins(1,4,5)P_3$-R-containing organelles; caffeine can release all Ca^{2+} from RYR-containing organelles, but also release Ca^{2+} from $Ins(1,4,5)P_3$-R containing organelles (Brown et al., 1992). Smooth muscles studied to date contain both $Ins(1,4,5)P_3$-ryanodine- and caffeine-releasable stores, but there are marked differences in the characteristics of these stores in different muscles. In cultured airway smooth muscle cells, the entire releasable intracellular Ca^{2+} store was $Ins(1,4,5)P_3$ sensitive (Chopra et al., 1991) (See Fig. 4). $Ins(1,4,5)P_3$-sensitive stores could be divided on the basis of their sensitivity to $Ins(1,4,5)P_3$, GTPγS, caffeine, or ryanodine. The Ca^{2+} store most sensitive to $Ins(1,4,5)P_3$ was not released by GTPγS, ryanodine, or caffeine. Although results of reconstitution experiments, discussed above, suggested that Ca^{2+} stores released by ryanodine

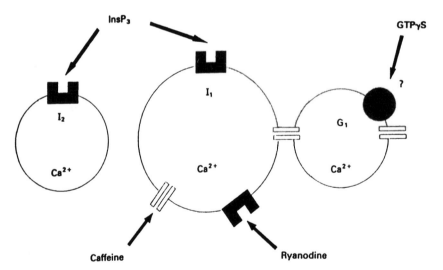

Figure 4 Intracellular Ca^{2+} stores in trachealis muscle as determined by Chopra et al. (1991). Two compartments are shown: an $Ins(1,4,5)P_3$-sensitive store that was emptied by low $Ins(1,4,5)P_3$ concentrations and not influenced by prior administration of ryanodine, GTP, or caffeine, and an $Ins(1,4,5)P_3$-sensitive store emptied by higher $[Ins(1,4,5)P_3]$, and also emptied by caffeine or ryanodine, and refilled by a GTP-sensitive mechanism. (From Chopra et al., 1991. Published with permission.)

participate in Ca^{2+}-induced Ca^{2+} release, this is not always the case in some smooth muscle (Missiaen et al., 1991). The dual action of caffeine was documented in smooth muscle, where this agent released Ca^{2+} from both ryanodine- and $Ins(1,4,5)P_3$-sensitive stores (Yamazawa et al., 1992). In one vascular smooth muscle, agonist-releasable Ca^{2+} stores were distinct from caffeine-sensitive stores (Matsumoto et al., 1990). In another vascular smooth muscle, agonist-releasable stores seemed to be identical to caffeine-sensitive stores (Kobayashi et al., 1986). Caffeine- and ryanodine-sensitive Ca^{2+} stores may be more important than $Ins(1,4,5)P_3$-sensitive stores in fast intestinal smooth muscles (Yamazawa et al., 1992). In slow smooth muscles, $Ins(1,4,5)P_3$-R-containing stores may have primary importance (Matsumoto et al., 1990; Bian et al., 1991; Chopra et al., 1991). There may be differences in affinities of Ca^{2+}-ATPases for Ca^{2+} in different smooth muscle organelles. In one study, $Ins(1,4,5)P_3$-sensitive pools could be loaded at a lower $[Ca^{2+}]$ than required for loading of $Ins(1,4,5)P_3$-insensitive pools (Missiaen et al., 1991). In another study, $Ins(1,4,5)P_3$-sensitive stores were more sensitive to thapsigargin than were $Ins(1,4,5)P_3$-insensitive stores (Bian et al., 1991).

In two studies, inhibitors of $Ins(1,4,5)P_3$-mediated release were utilized to assess the relative importance of $Ins(1,4,5)P_3$-sensitive pools. Kitazawa et al. (1989) determined that heparin completely inhibited receptor-evoked release of Ca^{2+} from intracellular stores. Baba et al. (1989) inhibited carbachol-evoked $Ins(1,4,5)P_3$ formation in STM by treating the muscle with phorbol dibutyrate. Under this condition, carbachol-evoked force was almost completely inhibited by verapamil, or low $[Ca^{2+}]$ in the bathing solution, suggesting $Ins(1,4,5)P_3$-evoked Ca^{2+} release was entirely responsible for intracellular Ca^{2+} release. The findings in these two studies suggest that agonist-evoked intracellular Ca^{2+} release is dependent on $Ins(1,4,5)P_3$ formation.

VI. INS(1,4,5)P₃- AND INS(1,3,4,5)P₄-INDUCED PLASMA MEMBRANE Ca²⁺ EXCHANGE

$Ins(1,4,5)P_3$ has been found to stimulate Ca^{2+} influx in T lymphocytes (Kuno and Gardener, 1987) and in mast cells (Penner et al., 1988). High-affinity, specific $Ins(1,4,5)P_3$ binding to membranes or plasma membranes has been reported in a number of different cells (Theibert et al., 1987, 1990; Challis et al., 1991; Khan et al., 1992; Fujimoto et al., 1992). Little is known about $Ins(1,4,5)P_3$-mediated Ca^{2+} influx in airway and other smooth muscles.

$Ins(1,3,4,5)$ tetrakisphosphate was shown to mediate Ca^{2+} influx into oocytes (Irvine and Moore, 1986; Irvine, 1990). High-affinity, specific $Ins(1,3,4,5)P_4$ binding sties are present in some cells (Challis et al., 1991; Mouillac et al., 1992). Whether $Ins(1,3,4,5)P_4$ is a second messenger is controversial (Snyder et al., 1988). Both $Ins(1,4,5)P_3$ and $Ins(1,3,4,5)P_4$ may be required for Ca^{2+} influx

(Morris et al., 1987; Downes and Carter, 1990). $Ins(1,3,4,5)P_4$ in combination with changes in $[Ca^{2+}]i$ regulated histamine-induced Ca^{2+} influx in DDT,MF-2 smooth muscle cells (Molleman et al., 1991). Increases in $Ins(1,3,4,5)P_4$ and $Ins(1,3,4)P_3$ contents during carbachol stimulation, discussed earlier, suggest there may be a role of these compounds in this tissue (Baron et al., 1992a), in contrast to vascular smooth muscles studied to date, where $Ins(1,3,4,5)P_4$ contents are very low and do not increase with receptor activation (Matsumoto et al., 1993; Gu et al., 1991).

VII. $INS(1,4,5)P_3$ AND INOSITOL PHOSPHOLIPID METABOLISM IN AIRWAY SMOOTH MUSCLE

We will now discuss some unresolved issues regarding roles of the inositol phospholipid transduction system in smooth muscle and in airway smooth muscle.

Issue 1: The importance of the inositol phospholipid transduction system in airway smooth muscle.

Most studies of inositol phospholipid and phosphate metabolism have utilized supramaximal agonist concentrations. As discussed earlier, the evidence that signals generated by receptor-activated inositol phospholipid metabolism control smooth muscle function under conditions of maximal agonist concentrations includes: (1) data obtained in many nonmuscle cells; (2) activation of inositol phospholipid metabolism and $Ins(1,4,5)P_3$ formation; (3) increases in contents of $Ins(1,4,5)P_3$ during activation (in some muscles); (4) studies of isolated smooth muscle $Ins(1,4,5)P_3$-R; (5) $Ins(1,4,5)P_3$ or caged $Ins(1,4,5)P_3$ release intracellular Ca^{2+} in smooth muscle; (6) effects of inhibitors of receptor-activated PIP_2 hydrolysis on muscle force development; effects of inhibitors of $Ins(1,4,5)P_3$ binding to $Ins(1,4,5)P_3$-Rs on force or Ca^{2+} release; (7) evidence that intracellular Ca^{2+} activates force development. Gaps in the story are a result of: (1) the lack, at the present time, of a method to determine free $[Ins(1,4,5)P_3]$; (2) uncertainty about roles of $Ins(1,4,5)P_3$-sensitive Ca^{2+} stores; (3) uncertainty about $Ins(1,4,5)P_3$ concentrations required to release Ca^{2+} from $Ins(1,4,5)P_3$-sensitive Ca^{2+} stores; (4) no method at present to precisely separate force triggered by membrane depolarization versus force triggered by $Ins(1,4,5)P_3$-induced Ca^{2+} release; (5) uncertainty about mechanisms involved in the so-called latch state force and possible role of $Ins(1,4,5)P_3$-induced Ca^{2+} during this phase of contraction.

The importance of $Ins(1,4,5)P_3$ formation in submaximally activated smooth muscle has not been adequately quantified. In CTM, acetylcholine (ACh)-evoked force closely paralleled membrane potential change at low [ACh], but not at high [Ach] (Farley and Miles, 1977). In addition, at low [ACh], force was sensitive to inhibitors of gated Ca^{2+} channels and removal of Ca^{2+} from the bathing solution (Farley and Miles, 1978) and insensitive to these perturbations at high [ACh] (Coburn, 1979). These findings suggested that release of Ca^{2+} from intracellular stores (and pharmacomechanical coupling) is not important at low [ACh]. In our opinion, these data need to be reinterpreted since we now know that these stores can be fed via L Ca^{2+} channels, and that the $Ins(1,4,5)P_3$ sensitivity of these stores is decreased when the stores are partly depleted (see above). Thus, these findings may not potently argue against a role of inositol phospholipid signal transduction during submaximal contractions.

So far, there are only a few published studies of effects of graded concentrations of agonists on inositol phospholipid metabolism in smooth muscle, and the role of $Ins(1,4,5)P_3$ formation during submaximal agonist contractions is still controversial (Grandordy et al., 1986). Incorporation of ^{32}P into PI in rabbit aorta correlated well with force under conditions of varying α_1-adrenergic activation (Villalboos-Molina et al., 1982). Most data in the literature indicate that submaximal agonist concentrations evoke only small increases in inositol phosphate production and that there is a large reserve in inositol phosphate formation (Meurs et al., 1988; Pijuan et al., 1993; Al-Hassani et al., 1993) even at agonist concentrations that produce near-maximal force. Study of the significance of such a large reserve of $Ins(1,4,5)P_3$ formation rate is indicated. All existing studies have utilized isometric force measurements; studies should be performed using isotonic force measurements and should study force transients, looking for a physiological role of very large $Ins(1,4,5)P_3$ formation rates.

Issue 2: The state of Ins(1,4,5)P$_3$ in unstimulated smooth muscle.

The $Ins(1,4,5)P_3$ content in unstimulated STM, computed from radioactivity measurements, is equivalent to about 3 μM assuming the compound is dissolved in all cellular water (Baron et al., 1992a). This concentration is severalfold higher than the $[Ins(1,4,5)P_3]$ required to release Ca^{2+} from $Ins(1,4,5)P_3$ reconstituted in lipid bilayers (Joseph and Williamson, 1989). This high concentration is surprising because under unstimulated conditions turnover of inositol phospholipids is extremely low (Baron et al., 1989b), suggesting $Ins(1,4,5)P_3$ is not being formed from labeled PIP_2. In unstimulated STM, the large $Ins(1,4,5)P_3$ pool does not have access to 3-kinases or 5-phosphatases or to $Ins(1,4,5)P_3$-sensitive SR (Baron et al., 1992a). High unstimulated $Ins(1,4,5)P_3$ contents have been previously reported for several nonmuscle cells [referred to in Joseph and Williamson (1989), Shears (1989), and Baron et al. (1992a)]. Similar data have been reported using $Ins(1,4,5)P_3$-binding protein analyses or, as performed in our laboratory, measuring $Ins(1,4,5)P_3$ radioactivity in equilibrium-labeled muscle.

The finding of a high $Ins(1,4,5)P_3$ content in unstimulated muscle is probably not an artifact of the extraction procedure. We were able to deplete the unstimulated store after a contraction-relaxation cycle (Baron et al., 1992a; Matsumoto et al., 1993) and by incubating the tissue in low $[Ca^{2+}]$ buffer for 10 min (unpublished data). The high unstimulated $Ins(1,4,5)P_3$ content could be explained by compartmentalization or binding of this compound or if $Ins(1,4,5)P_3$-R and 5-phosphatases and 3-kinases are not activated in unstimulated muscle despite a high [Ins(1,4,5)P$_3$]. We will discuss the issue of Ins(1,4,5)P$_3$ binding later. Uncertainty regarding the meaning of high unstimulated Ins(1,4,5)P$_3$ contents adds uncertainty to the meaning of Ins(1,4,5)P$_3$ contents determined in stimulated tissue.

Issue 3: Is the onset of Ins(1,4,5)P$_3$ formation rapid enough to explain rapid development of force in smooth muscle?

In most smooth muscles, including airway smooth muscle, the initial rapid rise in [Ca^{2+}]i is driven by release from internal Ca^{2+} stores (Himpens and Somlyo, 1988; Al-Hassani et al., 1993; Kajita and Yamaguchi, 1993). In a slow smooth muscle, release of Ca^{2+} from internal stores should have an onset within approximately 400 msec following agonist binding to receptors (Miller-Hance et al., 1988). For Ins(1,4,5)P$_3$ to function during force development in intact muscle requires that receptor-PLC activation and diffusion of formed Ins(1,4,5)P$_3$ to Ins(1,4,5)P$_3$-sensitive Ca^{2+} stores occur within this time period. Walker et al. (1987) measured the timing of a permeabilized guinea pig portal vein contraction evoked by laser pulse photolysis of caged Ins(1,4,5)P$_3$; the timing of contraction was also measured following a laser pulse photolysis of caged phenylephrine. Considering the delay between increases in [Ca^{2+}]i and onset of force (Miller-Hance et al., 1988) and temperature corrections, the data of Walker are consistent with rapid Ins(1,4,5)P$_3$-mediated force development in smooth muscle.

The next step is to determine whether Ins(1,4,5)P$_3$ is formed rapidly enough and whether Ins(1,4,5)P$_3$ concentrations at the site of Ins(1,4,5)P$_3$-sensitive Ca^{2+} stores are sufficient for Ca^{2+} release. We need to know whether Ca^{2+} stores containing Ins(1,4,5)P$_3$-R are located in junctional SR and affinities of Ins(1,4,5)P$_3$-R for Ins(1,4,5)P$_3$ in intact muscle. The diffusion coefficient for Ins(1,4,5)P$_3$ within cytosol suggests a long range of messenger action (Hagelberg and Allan, 1990; Allbritton et al., 1992).

An alternative mechanism for release of Ca^{2+} during force development involves RYR Ca^{2+} stores. Possible scenarios are: (1) junctional SR-containing RYR, which are characterized by Ca^{2+}-induced Ca^{2+} release mechanisms, are activated by Ca^{2+} influx across the plasma membrane via gated Ca^{2+} channels, or (2) electrical communications occur between junctional SR and plasma membrane (Somlyo et al., 1988) so that Ca^{2+} is released via voltage-gated RYR-containing SR. Since agonist-activated trachealis muscle shows membrane depolarization that parallels force development (Coburn and Yamaguchi, 1977; Farley and Miles, 1977; Coburn, 1979), either of the above scenarios is possible. Interactions between Ca^{2+} entering and Ca^{2+} released from Ins(1,4,5)P$_3$-triggered stores seem likely, perhaps by a mechanism whereby entering Ca^{2+} augments Ins(1,4,5)P$_3$-evoked Ca^{2+} release by increasing cooperative Ins(1,4,5)P$_3$ binding to Ins(1,4,5)P$_3$-R.

In STM, carbachol-induced depolarization was completely inhibited by bathing the muscle in a low [Na$^+$]-containing solution (Baba et al., 1989). (The mechanism for low [Na$^+$] inhibition of carbachol-induced depolarization is unknown.) Despite the absence of membrane depolarization, carbachol-induced force was rapid and unchanged. This finding argues against the existence of a mechanism whereby Ca^{2+} influx via a gated channel operates to release internal Ca^{2+} stores. The results of this experiment also suggests the importance of $Ins(1,4,5)P_3$-induced Ca^{2+} release during force development. In another experiment, also discussed above, STM was pretreated with phorbol esters, which entirely inhibited carbachol-evoked $Ins(1,4,5)P_3$ formation. Under this condition the rate of development of force was markedly decreased. Although phorbol esters have multiple effects on airway muscle, including effects on plasma membrane ion channels (Baba et al., 1989), and sensitization of contractile apparatus to Ca^{2+} (Itoh et al., 1988), data are consistent with a role of $Ins(1,4,5)P_3$ in force development.

An obvious approach to the question whether $Ins(1,4,5)P_3$ pool sizes increase rapidly enough to drive force development is to determine the timing of receptor-evoked increases in $Ins(1,4,5)P_3$ content. As indicated above, many cell types show transient increases in $Ins(1,4,5)P_3$ content following agonist activation. Shears (1989) summarized data obtained from many different cultured cells that indicate a long-lasting two- to threefold increase in $Ins(1,4,5)P_3$ content during agonist-evoked activation. In bovine trachealis muscle (Chilvers et al., 1989) and in vascular smooth muscles (Gu et al., 1991; Matsumoto et al., 1993; Pijuan et al., 1993), transient increases in $Ins(1,4,5)P_3$ content have been reported. So far increases have not been recorded at times following agonist administration associated with increases in [Ca]i. In STM, carbachol- and field-stimulation-evoked force development occurred without measurable increases in $Ins(1,4,5)P_3$ content (Baron et al., 1989a, 1992b) (even at times less than 200 msec following onset of field stimulation or carbachol administration). However, there was a rapid increase in the contents of $Ins(1,4,5)P_3$ by-products, suggesting that $Ins(1,4,5)P_3$ was formed but that the degradation rate was nearly equal to the formation rate. If formation of $Ins(1,4)P_2$ is an index of free $[Ins(1,4,5)P_3]$, we could conclude that free $[Ins(1,4,5)P_3]$ increased early during development of force. As discussed earlier, PIP_2 contents decreased rapidly during development of force; our calculation of $Ins(1,4,5)P_3$ formation rate from this decrease, given above, suggested a role of $Ins(1,4,5)P_3$ formation in force development mechanisms.

Issue 4: Does $Ins(1,4,5)P_3$ participate in maintenance of force and relaxation of maintained force?

There is evidence that $Ins(1,4,5)P_3$ is not important during smooth-muscle-maintained force. This evidence includes findings that agonist-evoked $Ins(1,4,5)P_3$ content increases are phasic in some smooth muscles (Chilvers et al., 1990; Gu et al., 1991) and that force is dependent on extracellular $[Ca^{2+}]$. However, in contrast to this evidence, changes in inositol phospholipid contents associated with activation persisted during smooth muscle maintained force (Baron et al., 1984, 1989b; Hashimato et al., 1985; Chilvers et al., 1991). Rates of increases in total inositol phosphates (with Li^+ present) remained elevated while force was maintained (Grandordy et al., 1986, 1988; Hall, Donaldson and Hill, 1989; Chilvers et al., 1990; Gu et al., 1991). With STM, $Ins(1,4,5)P_3$ content remained unchanged during the maintained phase of carbachol-induced force; however, inositol phospholipid metabolic flux remained markedly elevated, and contents of inositol phosphate by-products continued to rapidly increase (Baron et al., 1992a,b). These data indicated that the rate of $Ins(1,4,5)P_3$ formation was still markedly elevated, suggesting $Ins(1,4,5)P_3$ played a role during maintained force. This concept was supported by the finding that following inhibition of $Ins(1,4,5)P_3$ formation using phorbol esters in CTM, force was phasic, not maintained (Baba et al., 1989).

The finding that there was a rapid decrease in $Ins(1,4,5)P_3$ and $Ins(1,4)P_2$ contents during atropine-evoked relaxations and that these changes preceded onset of relaxation gave additional evidence that $Ins(1,4,5)P_3$ may play a role during maintained force.

The finding that $Ins(1,4,5)P_3$ and $Ins(1,4)P_2$ contents decreased rapidly following atropine-evoked relaxations of carbachol-contracted STM also suggests that decreases in free $Ins(1,4,5)P_3$ concentrations may participate in the mechanisms of relaxation (Baron et al., 1992b). These decreases preceded decreases in force. There is abundant evidence that rapid smooth muscle relaxations are driven by decreases in $[Ca^{2+}]i$ (Himpens and Somlyo, 1988; Anwer et al., 1989; Gunst and Bandyopadayax, 1989; Driska et al., 1989; DeFeo and Morgan, 1989), and it is likely that multiple mechanisms are involved, including: turnoff of $Ins(1,4,5)P_3$-induced release of Ca^{2+}, turnoff of Ca^{2+} influx into the cell, and Ca^{2+} scavenger mechanisms involving Ca^{2+}-ATPases.

Issue 5: Is Ins(1,4,5)P₃ bound in swine trachealis muscle?

We have postulated that some of the above data can be explained if $Ins(1,4,5)P_3$ is present in both free and bound states. This could explain the high $Ins(1,4,5)P_3$ content in unstimulated muscle, which does not have access to activated 5-phosphatases or 3-kinases or to $Ins(1,4,5)P_3$-R. Bound $Ins(1,4,5)P_3$ which could be rapidly liberated during muscle activation would also explain our findings for carbachol-activated STM where there was no increase in total $Ins(1,4,5)P_3$ content, but $Ins(1,4,5)P_3$ gained access to degrading enzymes. Findings during relaxation gave the strongest evidence for $Ins(1,4,5)P_3$ binding because the total content was depleted to levels about 60% of the unstimulated store. Thus, there may an unloading and loading cycle for bound $Ins(1,4,5)P_3$. Our postulate, shown in Figure 5, is that the bound store is utilized for rapid increases in free $Ins(1,4,5)P_3$ during force development and that following relaxation the store is depleted. The model is based on the assumption that $Ins(1,4)P_2$ content reflects free $Ins(1,4,5)P_3$.

The location of a putative bound $Ins(1,4,5)P_3$ store is unknown. Cytosolic $Ins(1,4,5)P_3$-binding proteins have been reported (Koppitz et al., 1986; Kanematsu et al., 1992) as well as membrane $Ins(1,4,5)P_3$-R binding sites, discussed above.

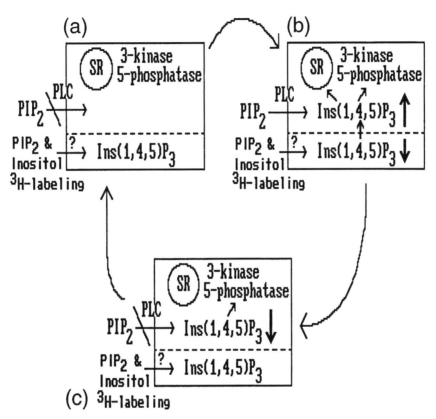

Figure 5 Schema showing proposed explanations for changes in inositol phosphate concentrations during the unstimulated state, carbachol-evoked force development, maintained force, and atropine-induced relaxations. (a) Unstimulated muscle where PLC is not activated and most or all of $Ins(1,4,5)P_3$ is sequestered. Arrows entering the box indicate formation and/or labeling of $Ins(1,4,5)P_3$ and diagonal line indicates a pathway operating at a low basal level (in the absence of agonist stimulation). (b) Stimulation with carbachol where $Ins(1,4,5)P_3$ formation appears to increase in the absence of any change in total tissue $Ins(1,4,5)P_3$ content, with a rapid shift of sequestered $Ins(1,4,5)P_3$ into the nonsequestered compartment, along with formation of new $Ins(1,4,5)P_3$ via PLC activation. Arrows entering the box are as in (a); thin arrows within the box indicate movement of $Ins(1,4,5)P_3$ and accessibility to $Ins(1,4,5)P_3$-sensitive SR and degrading enzymes; thick arrows indicate changes in $Ins(1,4,5)P_3$ content in the compartments. (c) During atropine-induced relaxations, there is a rapid hydrolysis of $Ins(1,4,5)P_3$ in the nonsequestered compartment with a decrease in total tissue $Ins(1,4,5)P_3$ content. Arrows are the same as in (b) and the diagonal line indicates a blocked pathway. (From Baron et al., 1992a.)

ACKNOWLEDGMENTS

The authors acknowledge the assistance of Dr. Edward LaBelle in preparing the section on PLC isoforms.

REFERENCES

Abdel-Latiff, A.A. (1991). Biochemical and functional interactions between the inositol 1,4,5-trisphosphate-Ca^{2+} and cyclic AMP signalling systems in smooth muscle. *Cell Signal 3*:371.

Al-Hassani, M.M., Garcia, J.G.N., and Gunst, S.J. (1993). Differences in Ca^{2+} mobilization by muscarinic agonists in tracheal smooth muscle. *Am. J. Physiol. 264*:L53.

Allbritton, N.L., Meyer, T., and Stryer, L. (1992). Range of messenger action of calcium ion and inositol 1,4,5-trisphosphate. *Science 258*:1813.

Anwer, K., Hovington, J.A., and Sanborn, B.M. (1989). Antagonism of contractants and relaxants at the level of intracellular calcium and phosphoinositide turnover in the rat uterus. *Endocrinology 124*:2995.

Ashkenazi, A., Winslow, J.W., Peralta, E.G., Peterson, G.L., Schimerlik, M.E., Capon, D.J., and Ramachandran, J. (1987). A M2 muscarinic receptor subtype coupled to both adenylylcyclase and phosphoinositide turnover. *Science 238*:672.

Auger, K.R., Serunian, L.A., Soltoff, S.P., Libby, P., and Cantley, L.C. (1989). PDGF-dependent tyrosine phosphorylation stimulates production of novel polyphosphoinositides in intact cells. *Cell 57*:167.

Baba, K., Baron, C.B., and Coburn, R.F. (1989). Phorbol ester effects on coupling mechanisms during cholinergic contraction of swine tracheal smooth muscle. *J. Physiol. (Lond.) 412*:23.

Baron, C.B., Cunningham, M., Strauss, J.F., and Coburn, R.F. (1984). Pharmacomechanical coupling in smooth muscle may involve phosphatidylinositol metabolism. *Proc. Natl. Acad. Sci. USA 81*:6899.

Baron, C.B., Baba, K., Person, C.R., and Coburn, R.F. (1989a). Timing of inositol phosphate increases in field-stimulated pig trachealis smooth muscle strips. *FASEB J. 3*:A1297.

Baron, C.B., Pring, M., and Coburn, R.F. (1989b). Inositol lipid turnover and compartmentation in canine trachealis smooth muscle. *Am. J. Physiol. 256*:C375.

Baron, C.B., Pompeo, J.N., and Azim, S. (1992a). Inositol 1,4,5-trisphosphate compartmentalization in tracheal smooth muscle. *Arch. Biochem. Biophys. 292*:382.

Baron, C.B., Pompeo, J.N., and Coburn, R.F. (1992b). Inositol 1,4,5-trisphosphate, inositide flux rates and pool sizes during smooth muscle relaxation. *Am. J. Physiol. 262*:L100.

Baron, C.B., Pompeo, J., Blackman, D.R., and Coburn, R.F. (1993). Common phosphatidylinositol 4,5-bisphosphate pools are involved in carbachol and serotonin activation of tracheal smooth muscle. *J. Pharmacol. Exp. Ther. 266*:8–15.

Batty, I.H., and Nahorski, S.R. (1992). Analysis of [^{3}H]inositol phosphate formation and metabolism in cerebral-cortical slices. *Biochem. J. 288*:807.

Beck, K.A., and Keen, J.H. (1991). Interaction of phosphoinositide cycle intermediates with the plasma membrane-associated clathrin assembly AP-2. *J. Biol. Chem. 266*:4442.

Berk, B.C., Alexander, R.W., Brock, T.A., Gimbrone, M.A., and Webb, R.C. (1986). Vasoconstriction: a new activity for platelet-derived growth factor. *Science 232*:87.

Berman, D.M., and Goldman, W.F. (1992). Stored calcium modulates inositol phosphate synthesis in cultured smooth muscle cells. *Am. J. Physiol. 263*:C535.

Berridge, M.J. (1993). Inositol trisphosphate and calcium signalling. *Nature 361*:315.

Berstein, G., Blank, J.L., Smrcka, A.V., Higashijima, T., Sternweis, P.C., Exton, J.H., and Ross, E.M. (1992). Reconstitution of agonist-stimulated phosphatidylinositol 4,5-bisphosphate hydrolysis using purified m 1 muscarinic receptor, $G_{q/11}$ and phospholipase C_{-B1}. *J. Biol. Chem. 267*:8081.

Bian, J., Ghosh, T.K., Wang, J., and Gill, D.L. (1991). Identification of intracellular calcium pools. Selective modification by thapsigargin. *J. Biol. Chem. 266*:8801.

Bocckino, S.B., Wilson, P.B., and Exton, J.H. (1991). Phosphatidate-dependent protein phosphorylation. *Proc. Natl. Acad. Sci. USA 88*:6210.

Bonner, T.I. (1992). Domains of muscarinic acetylcholine receptors that confer specificity of G protein coupling. *Trends Pharmacol. Sci. 13*:48.

Bonner, T.I., Buckley, N.J., Young, A.C., and Brann, M.R. (1987). Identification of family of muscarinic acetylcholine receptor genes. *Science 237*:527.

Bootman, M.D., Berridge, M.J., and Taylor, C.W. (1991). All-or-nothing Ca^{2+} mobilization from the intracellular stores of single histamine-stimulated HeLa cells. *J. Physiol. (Lond.) 450*:163.

Bourreau, J.P., Abela, A.P., Kwan, C.Y., and Daniel, E.E. (1991). Acetylcholine Ca^{2+} stores refilling directly involves a dihydropyridine-sensitive channel in dog trachea. *Am. J. Physiol. 261*:C497.

Brooksbank, C.E., Hutchings, A., Butcher, G.W., Irvine, R.F., and Divecha, N. (1993). Monoclonal antibodies to phosphatidylinositol 4-phosphate 5-kinase: distribution and intracellular localization of the C-isoform. *Biochem. J. 291*:77.

Brown, G.R., Sayers, L.G., Kirk, C.J., Michell, R.H., and Michelangeli, F. (1992). The opening of the inositol 1,4,5-trisphosphate-sensitive Ca^{2+} channel in rat cerebellum is inhibited by caffeine. *Biochem. J. 282*:309.

Buckley, N.J., Bonner, T.I., Buckley, C.M., and Brann, M.R. (1989). Antagonist binding properties of five cloned muscarinic receptors expressed in CHO-K1 cells. *Mol. Pharmacol. 35*:469.

Carpenter, C.L., and Cantley, L.C. (1990). Phosphoinositide kinases. *Biochemistry 29*:11147.

Challis, R.A.J., Safrany, S.T., Potter, B.V.L., and Nahorski, S.R. (1991). Intracellular recognition sites for inositol 1,4,5-trisphosphate and inositol 1,3,4,5-tetrakisphosphate. *Pharmacol. Soc. Trans. 19*.

Challis, R.A.J., Patel, N., Adams, D., and Arch, J.R.S. (1992). Inhibitory action of the potassium channel opener BRL 38227 on agonist-stimulated phosphoinositide metabolism in bovine tracheal smooth muscle. *Biochem. Pharmacol. 43*:17.

Cheuh, S.H., and Gill, D.L. (1986). Inositol 1,4,5-trisphosphate and guanine nucleotides activate calcium release from endoplasmic reticulum via distinct mechanisms. *J. Biol. Chem. 261*:13883.

Chilvers, E.R., Barnes, P.J., and Nahorski, S.R. (1989). Characterization of agonist-stimulated incorporation of myo-3[H] inositol into inositol phospholipids and [³H]inositol phosphate formation in tracheal smooth muscle. *Biochem. J. 262*:739.

Chilvers, E.R., Batty, I.H., Barnes, P.J., and Nahorski, S.R. (1990). Formation of inositol polyphosphates in airway smooth muscle after muscarinic receptor stimulation. *J. Pharmacol. Exp. Ther. 252*:786.

Chilvers, E.R., Batty, I.H., Challis, R.A., Barnes, P.J., and Nahorski, S.R. (1991). Determination of mass changes in phosphatidylinositol 4,5-bisphosphate and evidence for agonist-stimulated metabolism of inositol 1,4,5-trisphosphate in airway smooth muscle. *Biochem. J. 275*:373.

Chopra, L.C., Twort, C.H.C., Ward, J.P.T., and Cameron, I.R. (1989). Effects of heparin on inositol 1,4,5-trisphosphate and guanosine triphosphate induced calcium release in cultured smooth muscle trachea. *Biochem. Biophys. Res. Commun. 163*:262.

Chopra, L.C., Twort, C.H.C., Cameron, I.R., and Ward, J.P.T. (1991). Inositol 1,4,5-trisphosphate- and guanosine 5-0-(3-thiotriphosphate)-induced Ca²⁺ release in culture airway smooth muscle. *Br. J. Pharmacol. 104*:901.

Coburn, R.F. (1979). Electromechanical coupling in canine trachealis muscle: acetylcholine contractions. *Am. J. Physiol. 236*:C177.

Coburn, R.F., and Yamaguchi, T. (1977). Membrane potential-dependent and -independent tension in the canine tracheal muscle. *J. Pharmacol. Exp. Ther. 201*:276.

Cochet, C., Filhol, O., Payrastre, B., Hunter, T., and Gill, G. (1991). Interaction between the epidermal growth factor receptor on phosphoinositide kinases. *J. Biol. Chem. 266*:637.

Constantinescu, St., N., and Popescu, L.M. (1991). Topological regulation of cell-membrane phosphoinositidase C. *Biochem. Biophys. Res. Commun. 178*:773.

Dale, E. (1985). Phosphatidylinositol 4-phosphate kinase is associated with the membrane skeleton in human erythrocytes. *Biochem. Biophys. Res. Commun. 133*:189.

Danoff, S.K., Supattapone, S., and Snyder, S.H. (1988). Characterization of a membrane protein from brain mediating the inhibition of inositol 1,4,5-trisphosphate receptor binding by calcium. *Biochem. J. 254*:701.

DeFeo, T.T., and Morgan, K.G. (1989). Calcium force coupling mechanisms during vasodilation-induced relaxations of ferret aorta. *J. Physiol. (Lond.) 412*:123.

Devecha, N., Brooksbank, E.L., and Irvine, R.F. (1992). Purification and characterization of phosphatidylinositol 4-phosphate 5-kinases. *Biochem. J. 288*:637.

Downes, C.P., and Carter, A.N. (1990). Inositol lipids and phosphates. *Curr. Opin. Cell Biol. 2*:185.

Driska, S.T., Stein, P.G., and Porter, R. (1989). Myosin dephosphorylation during rapid relaxation of hog carotid artery smooth muscle. *Am. J. Physiol. 256*:C315.

Eberhard, D.A., and Holz, R.W. (1991). Calcium promotes the accumulation of polyphosphoinositides in intact and permeabilized bovine adrenal chromaffin cells. *Cell. Mol. Neurobiol. 11*:357.

Ehrlich, B.W., and Watras, J. (1990). Inositol 1,4,5-trisphosphate activates a channel from smooth muscle sarcoplasmic reticulum. *Nature 336*:583.

Eng, S.P., and Lo, C.S. (1990). Mastoparan increases membrane bound phosphatidylinositol kinase and phosphatidylinositol 4-monophosphate kinase activities in Madin-Darby canine kidney cells. *Life Sci. 46*:273.

Fain, J.N. (1990). Regulation of phosphoinositide-specific phospholipase C. *Biochim. Biophys. Acta 1053*:81.

Fargin, A., Yamamoto, K., Cotecchia, S., Goldsmith, P.K., Spiegel, A.M., Lapetina, E.G., Caron, M.G., and Lefkowtiz, R.J. (1991). Dual coupling of the cloned 5-HT$_{1A}$ receptor to both adenylyl cyclase and phospholipase C. *Cell-Signal. 3*:547.

Farley, J.M., and Miles, P.R. (1977). Role of depolarization in acetylcholine-induced contractions of dog trachealis muscle. *J. Pharmacol. Exp. Ther. 201*:199.

Farley, J.M., and Miles, P.R. (1978). The sources of calcium for acetylcholine-induced contractions of dog tracheal smooth muscle. *J. Pharmacol. Exp. Ther. 207*:340.

Ferris, C.B., Huganir, R.L., Supattapone, S., and Snyder, S.H. (1989). Purified inositol 1,4,5-trisphosphate receptor mediates calcium flux in reconstituted lipid vesicles. *Nature 342*:87.

Forscher, P. (1989). Calcium and polyphosphoinositide control of cytoskeletal dynamics. *Trends Neurosci. 12*:468.

Fujimoto, T., Nakade, S., Miyawaki, A., Mikoshiba, K., and Ogawa, K. (1992). Localization of inositol 1,4,5-trisphosphate receptor-like protein in plasmalemmal caveolae. *J. Cell Biol. 119*:1057.

Furuichi, T., Yoshikawa, S., Miyawaki, A., Wada, K., Maeda, N., and Mikoshiba, K. (1989). Primary structure and functional expression of the inositol 1,4,5-trisphosphate-binding protein P400. *Nature 342*:32.

Gaudette, D.C., and Holub, B.J. (1990). Effect of genistein, a tyrosine kinase inhibitor, on U46619-induced phosphoinositide phosphorylations in human platelets. *Biochem. Biophys. Res. Commun. 170*:238.

Gerwins, P., and Fredholm, B.B. (1992). Stimulation of adenosine A$_1$ receptors and bradykinin receptors, which act via different G proteins, synergistically raises inositol 1,4,5-trisphosphate and intracellular free calcium in DDT$_1$ MF-2 smooth muscle cells. *Proc. Natl. Acad. Sci. USA 89*:7330.

Ghosh, T.K., Mullaney, J.M., Tarazi, F.I., and Gill, D.L. (1989). GTP-activated communication between distinct inositol 1,4,5-trisphosphate-sensitive and insensitive calcium pools. *Nature 340*:236.

Gill, D.L., Ueda, T., Chueh, S.-H., and Noel, M.W. (1986). Ca^{2+} release from endoplasmic reticulum is mediated by a guanine nucleotide regulatory mechanism. *Nature 320*:461.

Gilman, A.G. (1987). G proteins: transducers of receptor-operated signals. *Annu. Rev. Biochem. 56*:615.

Goldman, W.F., Wier, W.G., and Blaustein, M.P. (1989). Effects of activation on distribution of Ca^{2+} in single arterial smooth muscle cells. *Circ. Res. 64*:1019.

Goldschmidt-Clermont, P.J., Kim, J.W., Machesky, L.M., Rhee, S.G., and Pollard, T.D. (1991). Regulation of phospholipase C-gamma 1 by profilin and tyrosine phosphorylation. *Science 251*:1231.

Grandordy, B.M., Cuss, F.M., Sampson, A.S., Palmer, J.B., and Barnes, P.J. (1986). Phosphatidylinositol response to cholinergic agonists in airway smooth muscle: relationship to contraction and muscarinic receptor occupancy. *J. Pharmacol. Exp. Ther. 238*:273.

Grandordy, B.M., Frossard, N., Rhoden, K.J., and Barnes, P.J. (1988). Tachykinin-induced phosphoinositide breakdown in airway smooth muscle and epithelium: relationship to contraction. *Mol. Pharmacol. 33*:515.

Grondin, P., Plaintive, M., Sultan, C., Breton, M., Mauco, G., and Chap, H. (1991).

Interaction of pp60c-src, phospholipase C. inositol-lipid, and diacylglycerol kinases with the cytoskeletons of thrombin-stimulated platelets. *J. Biol. Chem. 266*:25705.

Gu, H., Martin, H., Barsotti, R.J., and LaBelle, E.F. (1991). Rapid increase in inositol phosphate levels in norepinephrine-stimulated vascular smooth muscles. *Am. J. Physiol. 261*:C17.

Gunst, S.J. (1989). Effects of muscle length and load on intracellular Ca in tracheal smooth muscle. *Am. J. Physiol. 256*:C807.

Gunst, S.J., and Bandyopadayax, S. (1989). Contractile force and intracellular Ca^{2+} during relaxation of canine tracheal smooth muscle. *Am. J. Physiol. 257*:C355.

Hagelberg, C., and Allan, D. (1990). Restricted diffusion of integral membrane proteins and polyphosphoinositides leads to their depletion in microvesicles released from human erythrocytes. *Biochem. J. 271*:831.

Hall, I.P., Donaldson, J., and Hill, S.J. (1989). Inhibition of histamine-stimulated inositol phospholipid hydrolysis by agents which increase cyclic AMP levels in bovine tracheal smooth muscle. *Br. J. Pharmacol. 97*:603.

Hashimato, T., Hirata, M., and Ito, Y. (1985). A role of inositol 1,4,5-trisphosphate in the initiation of agonist-induced contractions of dog tracheal smooth muscle. *Br. J. Pharmacol. 86*:191.

Hawkins, P.T., Jackson, T.R., and Stephens, L.R. (1992). Platelet-derived growth factor stimulates synthesis of PtdIns $(3,4,5)P_3$ by activating a PtdIns$(4,5)P_2$ 3 OH kinase. *Nature (Lond.) 358*:157.

Herrmann-Frank, A., Darling, E., and Meissner, G. (1991). Functional characterization of the Ca^{2+}-gated Ca^{2+} release channel of vascular smooth muscle sarcoplasmic reticulum. *Pflügers Arch. 418*:353.

Hershey, P.E.C., Pessah, I.N., and Mohr, F.C. (1993). Regulation of inositol 1,4,5-trisphosphate receptors in rat basophilic leukemia cells. II. Modulation of the receptor in permeabilized cells by the cytosolic compartment. *Biochim. Biophys. Acta 1147*:115.

Himpens, B., and Somlyo, A.P. (1988). Free-calcium and force transients during depolarization and pharmacomechanical coupling in guinea-pig smooth muscle. *J. Physiol (Lond.) 395*:507.

Hisayama, T., and Takayanagi, I. (1988). Ryanodine: its possible mechanisms of action in the caffeine-sensitive calcium store of smooth muscle. *Pflügers Arch. 412*:376.

Homma, Y., Takenawa, T., Emori, Y., Sorimachi, H., and Suzuki, K. (1989). Tissue and cell type-specific expression of mRNA's for four types of inositol phospholipid-specific phospholipase C. *Biochem. Biophys. Res. Commun. 164*:406.

Homma, Y., Sakamoto, T., Sunoda, M., Aoki, M., and Takenawa, T. (1993). Evidence for involvement of phospholipase C-gamma 2 in signal transduction of platelet-derived growth factor in vascular smooth muscle cells. *Biochem. J. 290*:649.

Hwang, K.S., and Van Breemen, C. (1987). Ryanodine modulation of 45Ca efflux and tension in rabbit aortic smooth muscle. *Pflügers Arch. 408*:343.

Hymel, L., Inui, M., Fleisher, S., and Schindler, H. (1988). Purified ryanodine receptor of skeletal muscle SR forms a Ca^{2+}-activated oligomeric Ca^{2+} channel in planar bilayers. *Proc. Natl. Acad. Sci. USA 85*:441.

Iino, M. (1990). Biphasic Ca^{2+} dependence of inositol 1,4,5-trisphosphate-induced Ca release in smooth muscle cells of the guinea pig taenia caeci. *J. Gen. Physiol. 95*:1103.

Iino, M., and Endo, M. (1992). Calcium-dependent immediate feedback control of inositol 1,4,5-trisphosphate-induced Ca^{2+} release. *Nature 360*:76.

Irvine, R.F. (1990). Quantal Ca^{2+} release and the control of Ca^{2+} entry by inositol phosphates—a possible mechanism. *FEBS Lett. 263*:5.

Irvine, R.F., and Moor, R.M. (1986). Micro-injection of inositol 1,3,4,5-tetrakisphosphate activates sea urchin eggs by a mechanism dependent on external Ca^{2+}. *Biochem. J. 240*:917.

Itoh, Y., Kubota, Y., and Kuriyama, H. (1988). Effects of phorbol ester on acetylcholine-induced Ca^{2+} mobilization and contraction in the porcine coronary artery. *J. Physiol. 397*:401.

Joseph, S.K., and Williamson, J.R. (1986). Characteristics of inositol trisphosphate-mediated Ca^{2+} release from permeabilized hepatocytes. *J. Biol. Chem. 261*:14658.

Joseph, S.K., and Williamson, J.R. (1989). Inositol polyphosphates and intracellular calcium release. *Arch. Biochem. Biophys. 273*:1.

Joseph, S.K., Rice, H.L., and Williamson, J.R. (1989). The effect of external calcium and pH on inositol trisphosphate-mediated calcium release from cerebellum microsomal fractions. *Biochem. J. 258*:261.

Kajita, J., and Yamaguchi, H. (1993). Calcium mobilization by muscarinic cholinergic stimulation in bovine single airway smooth muscle. *Am. J. Physiol. 264*:L496.

Kanematsu, T., Takeye, H., Watanabe, Y., Ozaki, S., Yoshida, M., Koga, T., Iwanaga, S., and Hirata, M. (1992). Putative inositol 1,4,5-trisphosphate binding proteins in rat brain cytosol. *J. Biol. Chem. 267*:6518.

Kato, H., Fukami, K., Shibasaki, F., Homma, Y., and Takenawa, T. (1992). Enhancement of phospholipase C-δ1 activity in the aortas of spontaneously hypertensive rats. *J. Biol. Chem. 267*:6483.

Katz, A., Wu, D., and Simon, M.I. (1992). Subunits beta gamma of heterotrimeric G proteins activate beta 2 isoforms of phospholipase C. *Nature 360*:686.

Khan, A.A., Steiner, J.P., and Snyder, S.H. (1992). Plasma membrane inositol 1,4,5-trisphosphate receptor of lymphocytes: selective enrichment in sialic acid and unique binding specificity. *Proc. Natl. Acad. Sci. USA 89*:2849.

Kim, U.H., Kim, J.W., and Rhee, S.G. (1989). Phosphorylation of phospholipase C-gamma by cAMP-dependent protein kinase. *J. Biol. Chem. 264*:20167.

Kirk, C.J., Hunt, P.A., and Michell, R.H. (1989). Do cells contain discrete pools of inositol lipids that are coupled to receptor activation? *Biochem. Soc. Trans. 17*:978.

Kitazawa, T., Kobayashi, S., Horiuti, K., Somlyo, A.V., and Somlyo, A.P. (1989). Receptor-coupled permeabilized smooth muscle. Role of the phosphatidylinositol cascade, G-proteins, and modulation of the contractile response to Ca^{2+}. *J. Biol. Chem. 264*:5339.

Kobayashi, S., Kanaide, H., and Nakamura, M. (1986). Complete overlap of caffeine- and K depolarization-sensitive cellular calcium storage site in cultured rat arterial smooth muscle cells. *J. Biol. Chem. 261*:15709.

Kobayashi, S., Kitazawa, T., Somlyo, A.V., and Somlyo, A.P. (1989). Cytosolic heparin inhibits muscarinic and alpha adrenergic Ca release in smooth muscle. *J. Biol. Chem. 264*:17997.

Koppitz, B., Vogel, F., and Mayr, G.W. (1986). Mammalian aldolases are isomer-selective high-affinity inositol polyphosphate binders. *Eur. J. Biochem. 161*:421.

Kulik, T.J., Bialecki, R.A., Colucci, W.S., Rothman, A., Glennon, E.T., and Underwood, R.H. (1991). Stretch increases inositol trisphosphate and inositol tetrakisphosphate in cultured pulmonary vascular smooth muscle cells. *Biochem. Biophys. Res. Commun.* *180*:982.

Kuno, M., and Gardner, P. (1987). Ion channels activated by inositol 1,4,5-trisphosphate in plasma membrane of human T-lymphocytes. *Nature 326*:301.

LaBelle, E.F., and Gu, H. (1993). Isoforms of phospholipase C in rat tail artery. *FASEB J.* *7*:A1263.

LaBelle, E.F., Gu, H., and Trajkovic, S. (1992). Norepinephrine stimulates the direct breakdown of phosphatidylinositol in rat tail artery. *J. Cell. Physiol. 153*:234.

Lee, C.H., Park, D., Wu, D., Rhee, S.G., and Simon, M.I. (1992). Members of the Gq alpha subunit gene family activate phospholipase C beta isozymes. *J. Biol. Chem.* *267*:16044.

Leeb-Lundberg, L.M., Cotecchia, F.S., DiBlasi, A., Caron, M.G., and Lefkowitz, R.J. (1987). Regulation of adrenergic receptor function by phosphorylation: agonist-promoted desensitization and phosphorylation of alpha-1-adrenergic receptors coupled to inositol phospholipid metabolism in DDT, MF-2 smooth muscle cells. *J. Biol. Chem.* *262*:3098.

Limas, C.J., and Limas, C. (1985). Phorbol esters and diacylglycerol-mediated desensitization of cardiac beta-adrenergic receptors. *Circ. Res. 57*:443.

Ling, L.E., Schulz, J.T., and Cantley, L.C. (1989). Characterization and purification of membrane-associated phosphatidylinositol 4-phosphate kinase from human red blood cells. *J. Biochem. 264*:5080.

Low, A.M., Darby, P.J., Kwan, C.Y., and Daniel, E.E. (1993). Effects of thapsigargin and ryanodine on vascular contractility: cross-talk between sarcoplasmic reticulum and plasmalemma. *Eur. J. Pharmacol. 230*:53.

Low, M.G., Carroll, R.C., and Cox, A.C. (1986). Characterization of multiple forms of phosphoinositide-specific phospholipase C purified from human platelets. *Biochem. J.* *237*:139.

Lucchesi, P.A., Scheid, C.R., Romano, F.D., Kargacin, M.E., Mullikin-Kilpatrick, D., Yamaguchi, H., and Honeyman, T.W. (1990). Ligand binding and G protein coupling of muscarinic receptors in airway smooth muscle. *Am. J. Physiol. 258*:C730.

Madison, J.M., and Brown, J.K. (1988). Differential inhibitory effects of forskolin, isoproterenol and dibutyryl cyclic adenosine monophosphate on phosphoinositide hydrolysis in canine tracheal smooth muscle. *J. Clin. Invest. 82*:1462.

Maeda, A., Kubo, T., Mishina, M., and Numa, S. (1988). Tissue distribution of mRNAs encoding muscarinic acetylcholine receptor subtypes. *FEBS Lett. 239*:339.

Majerus, P.W., Ross, T.S., Cunningham, T.W., Caldwell, K.K., Jefferson, A.B., and Bansal, V.S. (1990). Recent insights in phosphatidylinositol signaling. *Cell 63*:459.

Marc, S., Leiber, D., and Harbon, S. (1988). Fluoroaluminates mimic muscarinic- and oxytocin-receptor-mediated generation of inositol phosphates and contraction in the intact guinea-pig myometrium. Role for a pertussis/cholera-toxin-insensitive G protein. *Biochem. J. 255*:705.

Matsumoto, T., Kanaide, H., Shogakiuchi, Y., and Nakamura, M. (1990). Characteristics of the histamine-sensitive calcium store in vascular smooth muscle. Comparison with norepinephrine- or caffeine-sensitive stores. *J. Biol. Chem. 265*:5610.

Matsumoto, H., Baron, C.B., and Coburn, R.F. (1993). Inositol tris-phosphate and smooth muscle oxidative metabolism contraction coupling. *FASEB J. 7*:A332.

Meisenhelder, J., Sho, P.G., Rhee, S.G., and Hunter, T. (1989). Phospholipase C-gamma is a substrate for the PDGF and EGF receptor protein tyrosine kinases in vivo and in vitro. *Cell 57*:1109.

Mesaeli, N., Lamers, J.M.J., and Panagia, V. (1992). Phosphoinositide kinases in rat heart sarcolemma: biochemical properties and regulation by calcium. *Mol. Cell. Biochem. 117*:181.

Meurs, H., Roffel, A.F., Postema, J.B., Timmermans, A., Elzinga, C.R.S., Kauffman, H.F., and Zaagsma, J. (1988). Evidence for a direct relationship between phospho-inositide metabolism and airway smooth muscle contraction induced by muscarinic agonists. *Eur. J. Pharmacol. 156*:271.

Meyer, T., Holowka, D., and Stryer, L. (1988). Highly cooperative opening of calcium channels by inositol 1,4,5-trisphosphate. *Science 240*:655.

Mignery, G.A., Newton, C.L., Archer, B.T., and Sudhof, T.C. (1990). Structure and expression of the rat inositol 1,4,5-trisphosphate receptor. *J. Biol. Chem. 265*:12679.

Miller-Hance, W.C., Miller, J.R., Wells, J.N., Stull, J.T., and Kamm, K.E. (1988). Biochemical events associated with activation of smooth muscle contraction. *J. Biol. Chem. 263*:13979.

Missiaen, L., De Smedt, H., Droogmans, G., Declerck, I., Plessers, L., and Casteels, R. (1991). Uptake characteristics of the $InsP_3$-sensitive and -insensitive Ca^{2+} pools of porcine aortic smooth muscle cells: different Ca^{2+} sensitivity of the Ca^{2+}-uptake mechanism. *Biochem. Biophys. Res. Commun. 174*:1183.

Missiaen, L., Taylor, C.W., and Berridge, M.J. (1992). Luminal Ca^{2+} promoting sponta-neous Ca^{2+} release from inositol trisphosphate-sensitive stores in rat hepatocytes. *J. Physiol. (Lond.) 455*:623.

Mohr, C.F., Hershey, P.E.C., Zimanyi, I., and Pessah, I.N. (1993). Regulation of inositol 1,4,5-trisphosphate receptors in rat basophilic leukemia cells. I. Multiple conforma-tional states of the receptor in a microsomal preparation. *Biochim. Biophys. Acta 1147*:105.

Molleman, A., Hoiting, B., Duin, M., van dan Akker, J., Nelemans, A., and Den Hertog, A. (1991). Potassium channels regulated by inositol 1,3,4,5-tetrakisphosphate and internal calcium in DDT_1, MF-2 smooth muscle cells. *J. Biol. Chem. 266*:5658.

Monaco, M.E., Attinasi, M., and Koreh, K. (1990). Effect of dual agonists on phospho-inositide pools in WRK-1 cells. *Biochem. J. 269*:633.

Moriarty, T.M., Gillo, B., Carty, D.J., Premont, R.T., Landau, E.M., and Iyengar, R. (1988). Beta, gamma subunits of GTP-binding proteins inhibit muscarinic receptor stimulation of phospholipase C. *Proc. Natl. Acad. Sci. USA 85*:8865.

Moritz, A., De Graan, P.N., Gispen, W.H., and Wirtz, K.W. (1992). Phosphatidic acid is a specific activator of phosphatidylinositol-4-phosphate kinase. *J. Biol. Chem. 267*:7207.

Morris, A.P., Gallacher, D.V., Irvine, R.F., and Peterson, O.H. (1987). Synergism of inositol trisphosphate and tetrakisphosphate in activating Ca-dependent K channels. *Nature 330*:653.

Mouillac, B., Devilliers, G., Jard, S., and Guillon, G. (1992). Pharmacological characteriz-ation of inositol 1,4,5-trisphosphate binding sites: relation to Ca^{2+} release. *Eur. J. Pharmacol. 225*:179.

Murray, R.K., and Kotlikoff, M.K. (1991). Receptor-activated calcium influx in human airway smooth muscle cells. *J. Physiol. 435*:123.

Nilsoon, T., Arkhammar, P., Hallberg, A., Hellman, B., and Bergren, P.O. (1987). Characterization of the inositol 1,4,5-trisphosphate-induced Ca^{2+} release in pancreatic beta-cells. *Biochem. J. 248*:329.

Ogden, D. (1988). Cell physiology. Answer in a flash [news]. *Nature 336*:16.

Panettieri, R.A., Yadvish, P.A., Kelly, A.M., Rubinstein, N.A., and Kotlikoff, M.I. (1990). Histamine stimulates proliferation of airway smooth muscle and induces c-*fos* expression. *Am. J. Physiol. 259*:L365.

Park, D., Deok-Young, J., Lee, C.W., Kee, K.H., and Rhee, S.G. (1993). Activation of phospholipase C isozymes by G protein beta-gamma subunits. *J. Biol. Chem. 268*:4573.

Parker, C., and Ivorra, I. (1990). Localized all-or-none calcium liberation by inositol trisphosphate. *Science 250*:977.

Parker, I., and Ivorra, I. (1991). Caffeine inhibits inositol trisphosphate-mediated liberation of intracellular calcium in *Xenopus* oocytes. *J. Physiol. (Lond.) 433*:229.

Payrastre, B., van Bergen en Henegouwen, P.M.P., Breton, M., den Hartigh, J.C., Plantavid, M., Verkleij, A.J., and Boonstra, J. (1991). Phosphoinositide kinase, diacyl-glycerol kinase, and phospholipase C activities associated in the cytoskeleton: effect of epidermal growth factor. *J. Cell Biol. 115*:121.

Penner, R., Matthews, G., and Neher, E. (1988). Regulation of calcium influx by second messengers in rat mast cells. *Nature 334*:499.

Peralta, E.G., Ashkenazi, A., Winslow, J.W., Ramachandran, J.R., and Capon, D.J. (1987). Differential regulation of PI hydrolysis and adenylate cyclase by muscarinic receptor subtypes. *Nature 334*:434.

Pijuan, V., Sukholutskaya, I., Kerrick, W.G., Lam, M., van Breemen, C., and Litosch, I. (1993). Rapid stimulation of Ins(1,4,5)P$_3$ production in rat aorta by NE: correlation with contractile state. *Am. J. Physiol. 264*:H126.

Pike, L.J. (1992). Phosphatidylinositol 4-kinases and the role of polyphosphoinositides in cellular regulation. *Endocr. Rev. 13*:692.

Popescu, L.M., Hinescu, M.E., Musat, S., Ionescu, M., and Pistritzu, F. (1986). Inositol trisphosphate and the contraction of vascular smooth muscle cells. *Eur. J. Pharmacol. 123*:167.

Prentki, M., Corkey, B.E., and Matschinsky, F.M. (1985). Inositol 1,4,5-trisphosphate and the endoplasmic reticulum Ca^{2+} cycle of a rat insulinoma cell line. *J. Biol. Chem. 260*:9184.

Putney, J.W. (1986). A model for receptor-regulated calcium entry. *Cell Calcium 7*:1.

Rebecchi, M.J., Eberhard, T., Delaney, T., Ali, S., and Bittman, R. (1993). Hydrolysis of short acyl chain inositol lipids by phospholipase C-δ1. *J. Biol. Chem. 268*:1735.

Rhee, S.G., Suh, P.G., Ryu, S.H., and Lee, S.Y. (1989). Studies of inositol phospholipid-specific phospholipase C. *Science 244*:546.

Rodbell, M. (1992). The role of GTP-binding proteins in signal transduction: from the sublimely simple to the conceptually complex. *Curr. Topics Cell Regul. 32*:1.

Roffel, A.E., Meurs, H., Elzinga, C.R.S., and Zaagsma, J. (1990). Characterization of the muscarinic receptor subtype involved in phosphoinositide metabolism in bovine tracheal smooth muscle. *Br. J. Pharmacol. 99*:293.

Rooney, T.A., and Thomas, A.P. (1991). Organization of intracellular calcium signals generated by inositol lipid-dependent hormones. *Pharmacol. Ther. 49*:223.

Rouxel, F.P., Hilly, M., and Mauger, J. (1992). Characterization of a rapidly dissociating inositol 1,4,5-trisphosphate-binding site in liver membranes. *J. Biol. Chem. 267*:10017.

Sagi-Eisenberg, R. (1989). GTP-binding proteins as possible targets for protein kinase-C action. *Trends Biochem. Sci. 14*:355.

Salari, H., Yeung, M., Howard, S., and Schellenberg, R.R. (1992). Increased contraction and inositol phosphate formation of tracheal smooth muscle from hyperresponsive guinea pigs. *J. Allergy Clin. Immunol. 90*:918.

Sasaguri, T., Hirata, M., and Kuriyama, H. (1985). Dependence on Ca^{2+} on the activities of phosphatidylinositol 4-5 bisphosphate phosphodiesterase and inositol 1,4,5-trisphosphate phosphatase in smooth muscles of the porcine coronary artery. *Biochem. J. 231*:497.

Shah, J., and Pant, H.C. (1988). Potassium channel blockers inhibit inositol trisphosphate-induced Ca^{2+} release in the microsome fractions isolated from the rat brain. *Biochem. J. 250*:617.

Shears, S.B. (1989). Metabolism of the inositol phosphates produced upon receptor activation. *Biochem. J. 260*:313.

Smith, C.D., and Chang, J. (1989). Regulation of brain phosphatidylinositol-4 phosphate kinase by GTP analogues. A potential role for guanine nucleotide regulatory proteins. *J. Biol. Chem. 264*:3206.

Snyder, P.M., Krause, K.H., and Welsh, M.J. (1988). Inositol trisphosphate isomers but not inositol 1,3,4,5-tetrakisphosphate induced calcium influx in Xenopus laevis oocytes. *J. Biol. Chem. 263*:11048.

Somlyo, A.V., Bond, M. Somlyo, A.P., and Scarpa, P. (1985). Inositol trisphosphate-induced calcium release and contraction in vascular smooth muscle. *Proc. Natl. Acad. Sci. USA 82*:5231.

Somlyo, A.P., Somlyo, A.V., Bond, M., Broderick, R., Goldman, Y.E., Shuman, H., Walker, J.W., and Trentham, D.R. (1987). Calcium and magnesium movements in cells and the role of inositol trisphosphate in muscle. *Soc. Gen. Physiol. Ser. 42*:77.

Somlyo, A.P., Walker, J.W., Goldman, Y.E., Trentham, D.R., Kobayashi, S., Kitazawa, T., and Somlyo, A.V. (1988). Inositol trisphosphate, calcium and muscle contraction. *Phil. Trans. R. Soc. Lond. B 320*:399.

Spiegel, A.J., Backlund, P.S., Jr., Butrynski, J.E., Jones, T.L., and Simonds, W.F. (1991). The G protein connection: molecular basis of membrane association. *Trends Biochem. Sci. 16*:338.

Streb, H., Irvine, R.F., Berridge, M.J., and Schulz, I. (1983). Release of Ca^{2+} from a non mitochondrial intracellular store in pancreatic acinar cells by inositol 1,4,5-trisphosphate. *Nature 306*:67.

Suematsu, E., Hirata, M., Hashimoto, T., and Kuriyama, H. (1984). Inositol 1,4,5-trisphosphate releases Ca^{2+} from intracellular store sites in skinned single cells of porcine coronary artery. *Biochem. Biophys. Res. Commun. 120*:481.

Supattapone, S., Worley, P.F., Baraban, J.M., and Snyder, S.H. (1988). Solubilization, purification and characterization of an inositol trisphosphate receptor. *J. Biol. Chem. 263*:1530.

Supattapone, S., Danoff, S., Thiebert, A., Joseph, S.K., Steiner, J., and Snyder, S.H. (1989). Cyclic AMP-dependent phosphorylation of a brain inositol trisphosphate receptor decreases its release of calcium. *Proc. Natl. Acad. Sci. USA 85*:8747.

Takazawa, K., Passaveiro, H., Dumont, J.E., and Erneux, C. (1989). Purification of bovine brain inositol 1,4,5-triphosphate 3-kinase. Identification of the enzyme by sodium dodecyl sulphate/polyacrylamide-gel electrophoresis. *Biochem. J. 261*:483.

Takenawa, T., and Egawa, K. (1977). CDP-diglyceride inositol transferase from rat liver. Purification and properties. *J. Biol. Chem. 252*:5429

Theibert, A.B., Supattapone, S., Worley, P.E., Baraban, J.M., Meek, J.L., and Snyder, S.H. (1987). Demonstration of inositol 1,3,4,5-tetrakisphosphate receptor binding. *Biochem. Biophys. Res. Commun. 148*:1283.

Theibert, A.B., Supattapone, S., Ferris, C.D., Danoff, S.K., Evans, R.K., and Snyder, S.H. (1990). Solubilization and separation of inositol 1,3,4,5-tetrakisphosphate- and inositol 1,4,5-trisphosphate-binding proteins and metabolizing enzymes in rat brain. *Biochem. J. 267*:441.

Thomas, A.P. (1988). Enhancement of the inositol 1,4,5-trisphosphate-releasable Ca^{2+} pool by GTP in permeabilized hepatocytes. *J. Biol. Chem. 263*:2704.

Verbist, J., Gadella, T.W.J., Jr., Raeymaekers, L., Wuytack, F., Wirtz, K.W.A., and Casteels, R. (1991). Phosphoinositide-protein interactions of the plasma-membrane Ca^{2+}-transport ATPase as revealed by fluorescence energy transfer. *Biochim. Biophys. Acta 1063*:1.

Vickers, J.D., and Mustard, J.F. (1986). The phosphoinositides exist in multiple metabolic pools in rabbit platelets. *Biochem. J. 238*:411.

Villalboos-Molina, R., Uc, M., Hong, E., and Garcia-Sainz, A. (1982). Correlation between phosphatidylinositol labeling and contraction in rabbit aorta. Effect of alpha-1 adrenergic activation. *J. Pharmacol. Exp. Ther. 222*:258.

Vrolix, M.L., Raeymaekers, F., Wuytack, F., Hoffman, F., and Casteels, R. (1988). Cyclic GMP-dependent protein kinase stimulates the plasmallemmal Ca^{2+} pump of smooth muscle via phosphorylation of phosphatidylinositol. *Biochem. J. 255*:855.

Walker, J.W., Somlyo, A.V., Goldman, Y.E., Somlyo, A.P., and Trentham, D.R. (1987). Kinetics of smooth and skeletal muscle activation by laser pulse photolysis of caged inositol 1,4,5-trisphosphate. *Nature 327*:249.

Williamson, J.R., Hansen, C.A., Johanson, R.A., Coll, K.E., and Williamson, M. (1988). Formation and metabolism of inositol phosphates: the inositol tris/tetrakisphosphate pathway. *Adv. Exp. Med. Biol. 232*:183.

Wilson, D.B., Neufeld, E.J., and Majerus, P.W. (1985). Phosphoinositide interconversion in thrombin-stimulated human platelets. *J. Biol. Chem. 260*:1046.

Wolf, R.A. (1990). Synthesis, transfer and phosphorylation of phosphoinositides in cardiac membranes. *Am. J. Physiol. 259*:C987.

Wu, D.J.H., Katz, A., and Simon, M.I. (1993). Identification of critical regions on phospholipase C-beta-1 required to activation by G-proteins. *J. Biol. Chem. 268*:3704.

Yamazawa, T., Iino, M., and Endo, M. (1992). Presence of functionally different compartments of the Ca^{2+} store in single intestinal smooth muscle cells. *FEBS Lett. 301*:181.

Yang, C.M., Chou, S., and Sung, T. (1991). Muscarinic receptor subtypes coupled to generation of different second messengers in isolated tracheal smooth muscle cells. *Br. J. Pharmacol. 104*:613.

Yao, Y., and Parker, I. (1992). Potentiation of inositol trisphosphate-induced Ca^{2+} mobilization by xenopus oocytes by cytosolic Ca^{2+}. *J. Physiol. (Lond.) 458*:319.

Zhou, C.J., Akhtar, R.A., and Abdel-Latif, A.A. (1993). Purification and characterization of phosphoinositide-specific phospholipase C from bovine iris sphincter smooth muscle. *Biochem. J. 289*:401.

Zorzato, F., Fujii, J., Otsu, K., Phillips, M., Green, N.M., Lai, F.A., Meissner, G., and MacLennan, D.H. (1990). Molecular cloning of cDNA encoding human and rabbit forms of the Ca^{2+} release channel (ryanodine receptor) of skeletal muscle sarcoplasmic reticulum. *J. Biol. Chem. 265*:2244.

DISCUSSION

Boucher: What controls the rate of Ca^{2+} influx across the plasma membrane in smooth muscle? Does the Putney "capacitative" model apply here, or is there a unique signaling pathway?

Coburn: Ca^{2+} influx to airway smooth muscle cells probably occurs via L-type Ca^{2+} channels. There is evidence for the "capacitative" mechanism in some smooth muscles.

Nadel: In diseases such as asthma, the airways narrow exclusively in response to many bronchoconstrictor mediators. This implies more shortening of the airway smooth muscle. How does isometric tension relate to the muscle shortening and how does shortening relate to disease?

Coburn: In disease, many systems may be altered; possibilities include a change in myosin, phosphoinositide C isoforms, or changes in coupling mechanisms. To understand smooth muscle remodeling is a challenge for the future.

Elde: Please restate the evidence against a ryanodine-sensitive Ca^{2+} release pathway in smooth muscle force generation.

Coburn: We assume this store is released by increased Ca^{2+} (i.e., Ca^{2+}-induced Ca^{2+} release). But development of force is not changed under conditions where depolarization is inhibited or intracellular Ca^{2+} is low.

20

Neuropeptides and Airway Macromolecule Secretion

Jens D. Lundgren
Hvidovre Hospital, University of Copenhagen, Copenhagen, Denmark

James H. Shelhamer
Clinical Center, National Institutes of Health, Bethesda, Maryland

I. INTRODUCTION

Excessive airway secretion may be a common feature of a variety of diseases characterized by airways inflammation. These excessive secretions may play an important role in the airway obstruction noted in a variety of obstructive diseases, including chronic bronchitis, asthma, and cystic fibrosis. Respiratory secretions (mucus) consist of locally produced and serum proteins, glycoproteins, proteoglycans, and lipids (Table 1). Macromolecules in these secretions include the glycoconjugates, proteoglycans, and mucus glycoproteins (MGP). It is the MGP that at least in part gives mucus its characteristic viscosity and elasticity. Presumably much of this viscosity and elasticity is due to interaction between mucin molecules. It is also possible that interactions between mucins and other proteins, glycoproteins, or cell products may contribute to these characteristics. The depth of the mucus layer and the amount of mucin present may be critical to the ability of the mucociliary apparatus to clear airway mucus.

II. SOURCES OF AIRWAY MACROMOLECULES

Submucosal gland mucus cells and the goblet cells of the respiratory epithelium are the sources of MGP secreted into the airways. The goblet cells are mucin-secreting cells located in the pseudostratified columnar epithelium that lines the

Table 1 Airway Mucus
Constituents

Electrolytes
Proteoglycans
Mucin glycoproteins
Locally produced proteins
   Lactoferrin
   Lysozyme
   Secretory IgA
   Secretory leukoprotease inhibitor
   Other protease inhibitors
Serum proteins
Lipids
Cells and cell products
Entrapped materials

conducting airways. The submucosal glands are complex tubuloacinar glands that consist of a glandular duct to which are connected multiple acinar structures that contain both mucus and serous cells, so designated for their staining characteristics with light microscopy. The mucus cells produce and secrete mucin. The serous cells produce and secrete lactoferrin, lysozyme, proteoglycans, protease inhibitor, and other products. Myoepithelial cells may be seen around the tubuloacinar gland structure. In the proximal airway in humans, the submucosal gland cells outnumber the goblet cells by about 40:1. Beyond the cartilagenous airways, submucosal glands are absent, but goblet cells may extend to the smaller bronchi. Thus a discussion of the effect of neuropeptides on airway macromolecule secretion must, where possible, distinguish among the specific type of secretory cell as the source of the secretion and, if possible, the type of glycoconjugate measured.

III. METHODS OF STUDY OF AIRWAY MACROMOLECULES

A variety of experimental approaches have been utilized for the study of the effect of neuropeptides on airway secretion. Each experimental approach has advantages and disadvantages; therefore, the data resulting from these different approaches must be interpreted in light of the experimental design. Examples of these designs (not intended to be a complete list) are presented in Table 2. Some of these approaches have included in vivo studies of feline tracheal secretion (Peatfield et al., 1983), histological examination of experimental animal trachea (Gashi et al., 1986; Tokuyama, 1990), studies of experimental animal tracheal tissue in organ culture (Borson et al., 1984; Lundgren et al., 1987, 1990), of human

Table 2 Neuropeptides and Airway Secretion: Experimental Approaches

Species	Parameter	Example reference
Feline		
In vivo	RGC release	Peatfield, 1983
Organ culture	RGC release	Lundgren, 1989
Isolated glands	RGC release	Shimura, 1987
Canine		
Organ culture	RGC release	Coles, 1984
Ferret		
Ussing chamber	RGC release	Borson, 1987
Tracheal tissue	Histology	Gashi et al., 1986
Trachea	Immunohistochemistry	Lazarus, 1986
Guinea pig	Histology	Kuo, 1990
Human bronchus	Fucose, hexose,	
Organ culture	protein release	Rogers, 1989
Organ culture	RGC release	Coles, 1981
Human nasal mucosa		
Organ culture	RGC, lactoferrin release	Baraniuk, 1990
In vivo	RGC, lysozyme	Baraniuk, 1992b

RGC = respiratory glycoconjugate. See text for details.

bronchus in organ culture (Coles et al., 1981; Rogers et al., 1989), of isolated submucosal glands from experimental animals (Shimura et al., 1988), and of human nasal mucosa in vivo (Baraniuk et al., 1992b).

IV. NEUROPEPTIDES AND AIRWAY MACROMOLECULAR SECRETION

It has been clear for years that the autonomic nervous system exerts a profound control over airway submucosal gland secretion. In 1983, Peatfield and Richardson observed that mucus release following vagal nerve stimulation was not completely inhibited by atropine or by atropine, phentolamine, and propranolol. Borson, using studies in isolated ferret tracheal segments and measuring $^{35}SO_4$ labeled macromolecules as a measure of glandular secretion, produced similar findings (Borson et al., 1984). Thus it was postulated that a nonadrenergic, noncholinergic nervous system was capable of stimulating submucosal gland secretion. Neuropeptides that have been noted in airway nerves may be localized to sensory nerves or may colocalize with mediators of the cholinergic or the adrenergic nervous system. The neuropeptides that have been identified in airway sensory nerves include substance P, neurokinin A, neurokinin B, and calcitonin gene-related peptide (CGRP). Gastrin-releasing peptide (GRP) may also be contained in sensory nerves (Baraniuk et al., 1990; Gawin et al., 1993). Vasoactive intestinal peptide (VIP) and VIP analogs are localized at least in part to cholinergic nerves, while neuropeptide Y (NPY) is localized to adrenergic nerves. The effect of each of these peptides on airway macromolecule secretion will be addressed. Among the neuropeptides, substance P is by far the most extensively studied and will be dealt with most extensively in this chapter.

A. Substance P

Substance P is an undecapeptide that has been found in sensory nerves in the mucosa and submucosa. It is distributed in nerves throughout the airway mucosa from smooth muscle to the superficial epithelium in experimental animal and human airway tissue (Lundberg et al., 1984). Substance P–immunoreactive nerves are also found in parasympathetic ganglia, suggesting that it may modulate parasympathetic ganglionic reflexes (Lundberg et al., 1984). Substance P receptors have been identified by autoradiography in the airway smooth muscle, vascular endothelium, submucosal glands, and superficial epithelium (Baraniuk et al., 1990c). That substance P is a potent mucus secretagogue has been reported in studies of a variety of species including dogs (Coles et al., 1984), ferrets (Gashi et al., 1986; Borson et al., 1987; Webber, 1989), cats (Shimura et al., 1987; Lundgren et al., 1989), and humans (Rogers et al., 1989). Table 3 lists the ranges of doses of substance P required for stimulation of airway secretion in various experimental models. Substance P may exert its secretagogue effect via activation of NK-1

Table 3 Effect of Substance P on Mucus Secretion

Species	Parameter	Minimal effective conc. of SP (nM/L)	Ref.
Canine	RGC release	0.1	Coles, 1984
Ferret	RGC release	10	Borson, 1987
	Serous cell degranulation	1000	Gashi et al., 1986
Cat	RGC release from explants	1	Lundgren, 1989
	RGC release from isolated glands	100	Shimura, 1987
	Gland contraction	0.0001	Shimura, 1987
Guinea pig	Tracheal goblet cell histology	0.003	Kuo, 1990
Human	Fucose, hexose, protein release	10	Rogers, 1989

RGC = respiratory glycoconjugate. See text for details.
Source: Adapted with permission from Lundgren and Shelhamer (1990).

receptors (Lundgren et al., 1989; Gentry, 1991; Barnes et al., 1991). The stimulation of glandular airway macromolecule release may occur in two ways. First, it may cause contraction of myoepithelial cells resulting in expulsion of mucus contained in the glandular duct (Shimura et al., 1987). Second, substance P may bind to receptors on glandular cells to cause secretory cell exocytosis (Gashi et al., 1986). Goblet cells of the guinea pig airway also are sensitive to substance P stimulation. Morphometric studies of guinea pig airway have documented that substance P more than neurokinin A stimulates goblet cell secretion (presumably via a NK-1 receptor) (Kuo et al., 1990).

B. Gastrin-Releasing Peptide

GRP is a 27-amino-acid peptide that shares sequence homology with bombesin, the 14-amino-acid amphibian peptide. In the respiratory mucosa of the upper airway, GRP is found in sensory nerve fibers in or near arteries, veins, and submucosal glands (Baraniuk et al., 1990c). Stimulation of sensory nerves of the nasal mucosa of guinea pigs results in the release of GRP into nasal secretions, implying GRP is localized to capsaicin-sensitive sensory nerves. GRP is also found in neuroendocrine cells in the superficial epithelium and alveolar macrophages and has been reported in nasal trigeminal neurons and in feline tracheal ganglion (Lundgren et al., 1990; Uddman et al., 1984; Baraniuk et al., 1990c). The COOH-terminal portion of the molecule binds to and activates GRP receptors (Lundgren et al., 1990; Gawin et al., 1993). GRP receptors have been localized by autoradiography to the superficial respiratory epithelium and to submucosal glands in feline trachea and in human nasal and bronchial mucosa (Baraniuk et al., 1990c, 1992; Lundgren et al., 1990). In in vitro studies of human nasal mucosa and feline tracheal mucosa, GRP stimulates secretion of respiratory glycoconjugates and of the serous cell marker lactoferrin (Lundgren et al., 1990; Baraniuk et al., 1990c). In vivo, GRP stimulates the release of mucus and serous cell products from the nasal mucosa of guinea pigs and from the nasal mucosa of normal human volunteers (Gawin et al., 1993; Baraniuk et al., 1992b).

C. Vasoactive Intestinal Peptide

VIP is a 28-amino-acid peptide that is among the neuropeptides found in relatively high concentration in the nasal and bronchial mucosa (Baraniuk et al., 1990e). VIP has been localized primarily to cholinergic neurons in the respiratory mucosa (Laitinen, 1985; Baraniuk et al., 1990e). Thus, the distribution of VIP-containing neurons in the airway mucosa includes a high concentration around the submucosal gland acini. VIP-binding sites in the airway have been identified in vascular and airway smooth muscle, in arterial vessels, and in epithelial cells and submucosal glands (Carstairs and Barnes, 1986; Baraniuk et al., 1990e). The reported effect of VIP on airway secretion has been variable and may depend on the species and study design. Some of the studies of the effect of VIP on airway macromolecular secretion are listed in Table 4. VIP stimulates ^{35}S-labeled respiratory glycoconjugates from ferret trachea (Peatfield et al., 1983). Further, VIP has been reported to modulate ferret airway macromolecule secretion induced by other secretagogues. VIP treatment may inhibit cholinergic-induced secretion while it augments adrenergic-induced secretion (Webber and Widdicombe, 1987). In feline tracheal submucosal glands, VIP potentiates cholinergic-stimulated secretion of airway glycoconjugates (Shimura et al., 1988). In the human airway, the results are less clear. In human nasal mucosa explants, VIP stimulates serous cell secretion, as evidenced by an increase in the release of the serous cell marker lactoferrin (Baraniuk et al., 1990e). Furthermore, consistent with the effect reported by Shimura on ferret tracheal submucosal glands, the effect of VIP on nasal mucosal gland serous cell secretion is additive to that of methacholine (Baraniuk et al., 1990e). In human bronchial organ culture, VIP has been reported to inhibit secretion of respiratory glycoconjugates (Coles et al., 1981). VIP may also have an effect on airway secretion by its stimulation of epithelial cell chloride transport (Carstairs and Barnes, 1986; Natanson et al., 1983).

D. Neuropeptide

NPY is a 36-amino-acid peptide that localizes to adrenergic nerves. Thus in the airway mucosa the distribution of NPY is around blood vessels and submucosal glands (Baraniuk et al., 1990a; Barnes et al., 1991). The presence of NPY in parasympathetic ganglia suggests that it may modulate cholinergic stimulation of secretion (Barnes et al., 1991). NPY has been reported to enhance cholinergically and adrenergically mediated secretion from ferret trachea (Webber, 1988). NPY has no direct macromolecule secretion from human nasal mucosa in vitro (Baraniuk et al., 1990a).

Table 4 VIP Effect on Airway Secretion

Species	Condition	VIP effect	Ref.
Ferret	Trachea in vitro	Stimulation	Peatfield, 1983
Ferret	Chol stimulation	Inhibition	Webber, 1987
	Adre stimulation	Augmentation	
Cat	Chol stimulation	Augmentation	Shimura, 1988
Human	Bronchial explants	Inhibition	Coles, 1981
Human	Nasal mucosa, chol. stim	S. cell secretion, additive with	Baraniuk, 1990

Chol = cholinergic; Adre = adrenergic; Stim = stimulation; S. cell = serous cell.

E.   Calcitonin Gene-Related Peptide

CGRP colocalizes with substance P in sensory nerves in the airway (Barnes et al., 1991a). Therefore, the distribution of CGRP-containing neurons is throughout the respiratory mucosa. CGRP receptors have not been clearly demonstrated by autoradiographic methods on airway secretory cells. CGRP has been reported to have a weak stimulatory effect on serous cell and goblet cell secretion from ferret trachea in vitro (Webber et al., 1991). It has also been reported to be a weak stimulant of goblet cell exocytosis in the airway of guinea pigs challenged in vivo (Kuo et al., 1990). In in vitro studies of human nasal mucosa, CGRP produced no measurable increase in markers for serous cell or mucus cell secretion (Baraniuk et al., 1990d).

The effect on airway macromolecule secretion induced by a variety of neuropeptides clearly associated with airway sensory nerves, with parasympathetic nerves, and with adrenergic nerves and the location of receptors for those peptides in a variety of experimental animals and in humans are summarized in Table 5.

F.   Other Peptides

1.   Endothelins

Endothelin-1 is a recently recognized peptide produced by endothelial and epithelial cells. In the respiratory mucosa, it is produced by vascular endothelial cells and by serous cells of the submucosal glands and probably by superficial epithelial cells. Endothelin-1 binding sites are located on the smooth muscle of vessel walls, on bronchial smooth muscle, and around submucosal glands. High-affinity binding of endothelin-1 to epithelial cells of the superficial epithelium has also been described (Wu et al., 1993). Endothelin-1 stimulates macromolecule secretion in vitro from isolated feline submucosal glands (Shimura et al., 1992). Endothelin-1, more than endothelin-2 or endothelin-3, stimulates glandular secretion of both mucus cell products and serous cell products of submucosal glands in the human nasal and bronchial mucosa (Mullol et al., 1993; Johnson, 1991). Because endothelin-1 is also capable of stimulating eicosanoid generation in the respiratory mucosa, it may also be capable of producing an indirect eicosanoid-mediated effect on macromolecule secretion (Wu et al., 1990, 1993).

2.   Enkephalins

Endorphin peptides have been detected in the lung tissue of experimental animals and humans. Similarly, binding sites for enkephalins have been detected in the airway. Met-enkephalin immunoreactivity has been described in nerves in the airway mucosa (Gibbins et al., 1987). Dynorphin A is capable of stimulating airway macromolecule secretion from feline airways in vitro (Lundgren et al., 1987). This secretory effect was blocked by the opioid antagonist naloxone,

Table 5 Neuropeptide Effect on Mucus Secretion

	Species	Peptide	Physiological activity	Receptor location SE	Receptor location SG	Nervous distribution
Afferent	Human	SP	++	+	+	SG
		NKA	+			
		CGRP	0			
		GRP	+	+	+	SG
	Feline	SP	++	+	+ (NK-1)	SG
		NKA	+			
		CGRP	0			
		GRP	++	+	+	SG
	Canine	SP	++			
	Ferret	SP	++			
		CGRP	+			
	Guinea pig	SP	++			
		GRP	++			
		CGRP	+			
Parasympathetic	Human	VIP	+/−	+	+	SG
	Feline explants	VIP	0			
	Feline glands	VIP	++			
	Ferret	VIP	++			
	Canine	VIP	++	+	+	SG
Sympathetic	Human	NPY	0			
Endothelin	Human		++	+	+	
	Feline		++			

SE = superficial epithelium; SG = submucosal glands. See text for details and references.
Source: Adapted from Larivée, P., Levine, S., Rieves, D., and Shelhamer, J. (1994). Airway inflammation and mucous hypersecretion. In: Takishima, T., and Shimura, S., eds. *Airway Secretion*. Marcel Dekker, New York, 1994, and Johnson, C., Larivée, P., and Shelhamer, J. (1993). Epithelial cells: regulation of mucous secretion. In: Busse, W., and Holgate, S., eds. *Asthma and Rhinitis*. Blackwell Scientific Publications, Cambridge, MA.

suggesting that the effect is via a kappa-opioid receptor. Conversely, in a study of the release of protein and fucose from human bronchial explants, morphine was reported to inhibit capsaicin-stimulated release of these markers for respiratory macromolecules, whereas naloxone augmented the capsaicin-stimulated release of these markers. Therefore, an inhibitory role of the μ receptor on nonadrenergic, noncholinergic–mediated airway secretion in human bronchial explants has been postulated (Rogers et al., 1989).

V. NEUROPEPTIDES, INFLAMMATORY CELLS, AND AIRWAY SECRETION

A variety of inflammatory cells and their products have been shown to cause airway macromolecular secretion. These immune and inflammatory cells include mucosal mast cells, macrophages, neutrophils and eosinophils. On activation, mast cells may release histaine, chymase, leukotrienes, platelet-activating factor, and cytokines such as tumor necrosis factor, all of which have been shown to cause increased airway macromolecule secretion (Lundgren et al., 1990; Larivee et al., 1993; Levine et al., 1993). Macrophages release eicosanoids, platelet-activating factor, and macrophage mucous secretagogue (Larivee et al., 1993). Neutrophils may release elastase, cathepsin G, eicosanoids, platelet-activating factor, and oxygen species, all of which have been identified as stimulants of airway macromolecule secretion during the inflammatory response (Larivee et al., 1993). Eosinophils release the airway secretagogues eosinophil cationic protein, platelet-activating factor, eicosanoids, and oxygen species (Lundgren et al., 1991; Larivee et al., 1993). The tachykinins have been demonstrated to have effects on each of these inflammatory cell types, although the importance of these interactions in the induction of airway hypersecretion is uncertain, especially in humans.

Substance P causes the degranulation of mast cells from rat connective tissue, peritoneum, intestinal mucosa, and lung (Ali et al., 1986; Stanisz et al., 1987). Further, the bronchoconstrictor effect of the tachykinins substance P and neurokinin A may be due in part to tachykinin-induced mast cell degranulation (Joos et al., 1988). Similarly, vagus nerve stimulation on a canine model leads to increased pulmonary mast cell release of histamine during antigen challenge, suggesting the possibility of sensory nerve modulation of mast cell mediator release (Leff, 1986). While substance P causes histamine release from human skin mast cells, this effect has not been demonstrated in human lung (Lowman et al., 1988; Ali et al., 1986). Substance P and neurokinin A have been reported to activate alveolar macrophages, probably via a NK-2 receptor (Brunelleschi et al., 1990). Furthermore, substance P stimulates human monocyte chemotaxis and induces the release of proinflammatory cytokines interleukin-1, tumor necrosis factor-alpha, and interleukin-6 from human peripheral blood mononuclear cells (McGillis et al., 1987; Lotz et al., 1988). Direct and indirect effects of the tachykinins on neutrophil function have been reported. An increase in neutrophils in venules of the airway is characteristic of neurogenic inflammation of the rat trachea, and the number of neutrophils is increased by pretreatment of the animals with a neutral endopeptidase inhibitor (Umeno et al., 1989). Studies in mice have suggested that substance P–induced tissue infiltration of neutrophils is dependent on mast cell degranulation, implying an indirect effect of substance P on neutrophil chemotaxis (Yano et al., 1989). However, there is evidence in rabbits and in humans of substance P binding to receptors on neutrophils with induction of chemotaxis and lysozyme secretion (in rabbit neutrophils) or chemotaxis and superoxide genera-tion (in human neutrophils) (Marasco et al., 1981; Kolasinski et al., 1992). Simi-larly, substance P has been reported to induce degranulation of eosinophils although it is uncertain that a receptor-mediated mechanism is involved (Kroegel et al., 1990). If the effects of the tachykinins on inflammatory cells tend to be "proinflammatory," the reported effects of VIP on inflammatory cells tend to be "anti-inflammatory." VIP is reported to inhibit antigen-induced histamine release from peripheral lung (Undem et al., 1983) and to reduce acute inflammatory responses in the lung (Foda et al., 1988; Said, 1990). While there is species-to-species variability in the responsiveness of inflammatory cells to peptide stimulation, it seems possible that some effects of neuropeptides on airway macromolecule secretion might be indirect and mediated by inflammatory cells and their products.

VI. NEUTRAL ENDOPEPTIDASE

Neuropeptides in the airways may be degraded by a variety of enzymes, including angiotensin-converting enzyme and neutral endopeptidase (3.4.24.11) (NEP). NEP is produced by the airway epithelial cells and may play an important role in the regulation of airway responses involving neuropeptides. Inhibition of NEP by thiorphan or phosphoramidon potentiates the responses of airway cells to exog-enous substance P, NKA, electric field stimulation, or capsaicin (Borson et al., 1987; Rogers et al., 1989). A variety of conditions may inactivate NEP or decrease the production of NEP by epithelial cells. Inactivation of NEP in the airway may be due to ozone, cigarette smoke, or toluene diisocyanates (Yeadon et al., 1990; Dusser et al., 1989; Sheppard et al., 1988). NEP production by epithelial cells may be reduced by viral infection (Jacoby et al., 1988; McDonald, 1987). Thus, it seems likely that some inflammatory states may be associated with unopposed neurogenic inflammation because part of the inflammatory process resulted in decreased epithelial cell production of NEP or in inactivation of NEP.

VII. ROLE OF NEUROPEPTIDES IN AIRWAY DISEASE

It is possible that the pathophysiology of a variety of airway diseases associated with hypersecretion of respiratory mucus may relate in part to neurogenic inflammation. Sensory nerve stimulation may result in airway macromolecule secretion in two ways. First, sensory nerve stimulation may activate a cholinergi-cally mediated transspinal reflex, resulting in stimulation of submucosal gland secretion. Second, sensory nerve stimulation may result in an axon reflex release of neuropeptides from neurons near secretory cells in the submucosa or in the superficial epithelium. In addition to these two mechanisms by which macro-molecule secretion may be stimulated, it is possible that the loss of an inhibitory influence such as VIP may also be involved in disease processes. These processes may include allergic rhinitis, asthma, and chronic bronchitis.

A. Rhinitis

A role of neuropeptides and of neurogenic inflammation in the pathogenesis of allergic rhinitis has been postulated (Baraniuk and Kaliner, 1991). These authors have speculated that antigen-induced nasal mast cell degranulation results in the release of mast cell products such as histamine, leukotrienes, and mast cell chymase and in the production of bradykinin. Both histamine and bradykinin may stimulate sensory nerves, resulting in axon responses including epithelial cell secretion and glandular secretion. Other inflammatory cells, such as eosinophils, may release products such as major basic protein which may damage superficial epithelial cells, resulting in decreased production of neutral endopepticase and in unopposed effects of sensory nerve peptides. Tachykinins may also act on mast cells, eosinophils, and other cells to increase the release of cell products (such as histamine or eosinophil cationic protein) that may act directly on secretory epithelial cells to stimulate secretion of macromolecules. Sensory nerve stimulation may also result in parasympathetic reflex-mediated stimulation of submucosal gland secretion.

B. Asthma

Neuropeptide participation in the airway inflammation and hypersecretion of asthma has also been postulated (Barnes, 1987; Barnes et al., 1991a). Damage to airway epithelium from chemicals, irritants, or inflammatory cell products such as eosinophil major basic protein or infectious agents such as respiratory viruses may expose sensory fibers that penetrate the basement membrane of the superficial epithelial surface. Sensory nerves may be activated by irritants, bradykinin, or other stimuli, resulting in axon reflex-mediated release of substance P and other neuropeptides with the subsequent stimulation of superficial epithelial cell and submucosal gland secretion. Stimulation of these sensory neurons may also result in reflex cholinergic stimulation of submucosal gland secretion. Substance P may also stimulate mast cell degranulation and attract inflammatory cells (Barnes, 1987). Conversely, a portion of the ongoing inflammatory process in the airways may be the absence or deficiency of inhibitory influences such as might be provided by VIP. Ollerenshaw, in a study of VIP immunoreactivity in the airway mucosa of normal persons and patients with severe asthma, found a relative absence of VIP in the asthmatic airways, suggesting the loss of a negative modulating influence on neurogenic inflammation (Ollerenshaw et al., 1989).

C. Chronic Bronchitis

Irritants such as cigarette smoke or ozone may have three effects that may propagate airway inflammation and the hypersecretion of airway mucus characteristic of chronic bronchitis. First, cigarette smoke and ozone may inactivate NEP (Sheppard et al., 1988; Yeadon et al., 1990). Second, these irritants may directly activate sensory neurons in the airway. Third, these irritants may damage epithelial cells, resulting in the production of eicosanoids and other chemoattractants and in the exposure of sensory neurons in the superficial epithelium. Activation of sensory neurons may result in axon reflex release of neuropeptides such as substance P, which may stimulate airway macromolecule release from superficial epithelial cells and from submucosal glands. Activation of the same sensory neurons may also result in reflex cholinergic stimulation of submucosal gland secretion. There is some preliminary evidence for a role of substance P in the airway hypersecretion of chronic bronchitis. First, substance P is detectable in the airway secretions induced by hypertonic saline from patients with chronic bronchitis or from patients with asthma but not from normals (Tomaki et al., 1993). Second, a substance P receptor antagonist in preliminary studies appears to be capable of reducing the expectorated sputum volumes of patients with chronic bronchitis (Ichinose et al., 1993). Finally, it seems possible that airway hypersecretion in patients with chronic bronchitis is in part due to a decreased response of the secretory apparatus to VIP. Coles observed that while VIP appeared to be capable of inhibiting baseline and methacholine-stimulated respiratory macromolecule secretion in human airway explants, it did not inhibit baseline or stimulated secretion from explants from patients with chronic bronchitis (Coles et al., 1981). One interpretation of these data is that the absence of modulation with VIP may lead to enhanced responses to reflex stimuli.

Although it is possible to postulate mechanisms by which neuropeptides are integrally involved, directly or indirectly, in airway hypersecretion in a variety of disease states characterized by airway inflammation, a clear role for neurogenic inflammation has not been proven in the hypersecretion associated with any of these states. Additional work in humans with these problems will be required before these mechanisms are clearly implicated in these human diseases, and therapeutic strategies are devised.

REFERENCES

Ali, H., Leung, K., Pearce, F., Hayes, N., and Foreman, J. (1986). Comparison of the histamine-releasing action of substance P on mast cells and basophils from different species and tissues. *Int. Arch. Allergy Appl. Immunol. 79*:413–418.

Baluk, P., Nadel, J., and McDonald, D. (1993). Calcitonin gene-related peptide in secretory granules of serous cells in the rat tracheal epithelium. *Am. J. Respir. Cell Mol. Biol. 8*: 446–453.

Baraniuk, J., and Kaliner, M. (1991). Neuropeptides and nasal secretion. *Am. J. Physiol. 261*:L223–L235.

Baraniuk, J., Castellino, S., Lundgren, J., Goff, J., Merida, M., Shelhamer, J., and Kaliner, M. (1990a). Neuropeptide Y (NPY) in human nasal mucosa. *Am. J. Respir. Cell Mol. Biol. 3*:165–173.

Baraniuk, J., Lundgren, J., Gawin, A., Muzuguchi, H., Peden, D., Shelhamer, J., and Kaliner, M. (1990b). Bradykinin receptor distribution in human nasal mucosa, and analysis of secretory responses in vivo and in vitro. *Am. Rev. Respir. Dis. 141*:706–714.

Baraniuk, J., Lundgren, J., Goff, J., Peden, D., Merida, M., Shelhamer, M., and Kaliner, M. (1990c). Gastrin-releasing peptide in human nasal mucosa. *J. Clin. Invest. 85*:998–1005.

Baraniuk, J., Lundgren, J., Goff, J., Mullol, J., Castellino, S., Merida, M., Shelhamer, J., and Kaliner, M. (1990d). Calcitonin gene-related peptide in human nasal mucosa. *Am. J. Physiol. 258*:L81–L88.

Baraniuk, J., Lundgren, J., Okayama, M., Mullol, J., Merida, M., Shelhamer, J., and Kaliner, M. (1990e). Vasoactive intestinal peptide (VIP) in human nasal mucosa. *J. Clin. Invest. 86*:825–831.

Baraniuk, J., and Kaliner, M. (1991). Neuropeptides and nasal secretion. *Am. J. Physiol. 261*:L223–235.

Baraniuk, J., Lundgren, J., Shelhamer, J., and Kaliner, M. (1992a). Gastrin-releasing peptide binding sites in human tracheobronchial mucosa. *Neuropeptides 21*:81–84.

Baraniuk, J., Silver, P., Lundgren, J., Cole, P., Kaliner, M., and Barnes, P. (1992b). Bombesin stimulates human nasal mucous and serous cell secretion in vivo. *Am. J. Physiol. 262*:L48–L52.

Barnes, P. (1987). Neuropeptides in human airways: function and clinical implications. *Am. Rev. Respir. Dis. 136*:S77–S83.

Barnes, P., Baraniuk, J., and Belvisi, M. (1991a). Neuropeptides in the respiratory tract. *Am. Rev. Respir. Dis. 144*:1187–1198.

Barnes, P., Baraniuk, J., and Belvisi, M. (1991b). Neuropeptides in the respiratory tract. *Am. Rev. Respir. Dis. 144*:1391–1399.

Borson, D., Charlin, M., Gold, B., and Nadel, J. (1984). Neural regulation of $^{35}SO_4$-macromolecule secretion from tracheal glands of ferrets. *J. Appl. Physiol. 57*:457–466.

Borson, D., Corrales, R., Varsano, S., Gold, M., Viro, N., Gaughey, G., Ramachandran, J., and Nadel, J. (1987). Enkephalinase inhibitors potentiate substance P-induced secretion of $^{35}SO_4$-macromolecules from ferret trachea. *Exp. Lung. Res. 12*:21–36.

Brunelleschi, S., Vanni, L., Ledda, F., Giotti, A., Maggi, C., and Fantozzi, R. (1990). Tachikinins activate guinea-pig alveolar macrophages: involvement of NK_2 and NK_1 receptors. *Br. J. Pharmacol. 100*:417–420.

Carstairs, J., and Barnes, P. (1986). Visualization of vasoactive intestinal peptide receptors in human and guinea pig lung. *J. Pharmacol. Exp. Ther. 239*:249–255.

Coles, S., Said, S., and Reid, L. (1981). Inhibition by vasoactive intestinal peptide of glycoconjugate and lysozyme secretion by human airways in vitro. *Am. Rev. Respir. Dis. 124*:531–536.

Coles, S., Neill, K., and Reid, L. (1984). Potent stimulation of glycoprotein secretion in canine trachea by substance P. *J. Appl. Physiol. 57*:1323–1327.

Dusser, D., Djokic, T., Borson, D., and Nadel, J. (1989). Cigarette smoke induces bronchoconstrictor hyperresponsiveness to substance P and inactivates airway neutral endopeptidase in the guinea pig. *J. Clin. Invest. 84*:900–906.

Foda, H., Iwanaga, T., Lui, L., and Said, S. (1988). Vasoactive intestinal peptide protects against HCl-induced pulmonary edema in rats. *Ann. NY Acad. Sci. 527*:633.

Gashi, A.A., Borson, D.B., Finkbeiner, W.E., Nadel, J.A., Basbaum, C.B. (1986).

Neuropeptides degranulate serous cells of ferret tracheal glands. *Am. J. Physiol. 251*:C223–229.

Gawin, A., Baraniuk, J., Lundgren, J., and Kaliner, M. (1993). Effects of gastrin-releasing peptide and analogues on guinea pig mucosal secretion. *Am. J. Physiol. 264*:L345–350.

Gentry, S. (1991). Tachykinin receptors mediating airway macromolecular secretion. *Life Sci. 48*:1609–1618.

Gibbins, I., Furness, J., and Costa, M. (1987). Pathway specific patterns of the coexistence of substance P, calcitonin gene-related peptide, cholecystokinin and dynorphin in neurons of the dorsal root ganglia of the guinea-pig. *Cell Tissue Res. 248*:417–437.

Ichinose, M., Katsumata, U., Kikuchi, R., Fudushima, T., Ishii, M., Inoue, C., Shirato, K., and Takishima, T. (1993). Effect of tachikinin receptor antagonist on chronic bronchitis patients. *Am. Rev. Respir. Dis. 147*:A318.

Ingenito, E., Pliss, L., Martins, M., and Ingram, R. (1991). Effects of capsaicin on mechanical, cellular and mediator responses to antigen in sensitized guinea pigs. *Am. Rev. Respir. Dis. 143*:572–577.

Jacoby, D., Tamaoki, J., Borson, D., and Nadel, J. (1988). Influenza infection causes airway hyperresponsiveness by decreasing enkephalinase. *J. Appl. Physiol. 64*:2653–2658.

Joos, G., Pauwels, R., and Van Der Straeten, M. (1988). The mechanism of tachykinin induced bronchoconstriction in the rat. *Am. Rev. Respir. Dis. 137*:1038–1044.

Kolasinski, S., Haines, K., Siegel, E., Cronstein, B., and Abramson, S. (1992). Neuropeptides and inflammation: a somatostatin analog as a selective antagonist of neutrophil activation by substance P. *Arthritis Rheum. 35*:369–375.

Kroegel, C., Giembycz, M., and Barnes, P. (1990). Characterization of eosinophil cell activation by peptides. *J. Immunol. 145*:2581–2587.

Kuo, H., Rohde, J., Tokuyama, K., Barnes, P., and Rogers, D. (1990). Capsaicin and sensory neuropeptide stimulation of goblet cell secretion in guinea-pig trachea. *J. Physiol. 431*:629–641.

Laitinen, L.A., Heino, M., Laitinen, A., Kava, T., and Haahtela, T. (1985). Damage of the airway epithelium and bronchial reactivity in patients with asthma. *Am. Rev. Respir. Dis. 131*:599–606.

Larivée, P., Levine, S., Rieves, R., and Shelhamer, J. (1993). Airway inflammation and mucous hypersecretion. In: *Airway Secretion: Physiological Bases for the Control of Mucous Hypersecretion.* (Takishima, T., and Shimura, S., eds.). Marcel Dekker, New York, 1994.

Lazarus, S., Basbaum, C., Parnes, P., and Gold, W. (1986). cAMP immunocytochemistry provides evidence of functional VIP receptors in trachea. *Am. J. Physiol. 251*:C115–C119.

Leff, A.R., Stimler, N.P., Munoz, N.M., Shioya, T., Tallet, J., and Dame, C. (1986). Augmentation of respiratory mast cell secretion of histamine caused by vagus nerve stimulation during antigen challenge. *J. Immunol. 136*:1066–1073.

Levine, S., Logun, C., Larivee, P., and Shelhamer, J. (1993). TNF-alpha induces secretion of respiratory mucous glycoprotein from human airways in vitro. *Am. Rev. Respir. Dis. 147*:1011A.

Lotz, M., Vaughan, J., and Carson, D. (1988). Effect of neuropeptides on production of inflammatory cytokines by human monocytes. *Science 241*:1218–1221.

Lowman, M., Benyon, C., and Chruch, M. (1988). Characterization of neuropeptide-induced histamine release from human dispersed skin mast acells. *Br. J. Pharmacol. 95*: 121–130.

Lundberg, J., Hokfelt, T., Martling, C., Saria, A., and Cuello, C. (1984). Substance P–immunoreactive sensory nerves in the lower respiratory tract of various mammals including man. *Cell Tissue Res. 235*:251–261.

Lundgren, J., and Shelhamer, J. (1990). Pathogenesis of airway mucus hypersecretion. *J. Allergy Clin. Immunol. 85*:399–419.

Lundgren, J., Kaliner, M., Logun, C., and Shelhamer, J. (1987). The effects of endorphins on mucous glycoprotein secretion from feline airways in vitro. *Exp. Lung Res. 12*:303–309.

Lundgren, J., Wiedermann, C., Logun, C., Plutchok, J., Kaliner, M., and Shelhamer, J. (1989). Substance P mediated secretion of respiratory glycoconjugate from feline airways in vitro. *Exp. Lung Res. 15*:17–29.

Lundgren, J., Ostrowski, N., Baraniuk, J., Kaliner, M., and Shelhamer, J. (1990). Gastrin-releasing peptide stimulates glycoconjugate release from feline tracheal explants. *Am. J. Physiol. 258*:L68–L74.

Lundgren, J., Davey, R., Lundgren, B., Mullol, J., Marom, Z., Logun, C., Baraniuk, J., Kaliner, M., and Shelhamer, J. (1991). Eosinophils and mucus airway secretion: Eosinophil cationic protein stimulates and major basic protein inhibits secretion from airway organ culture. *J. Allergy Clin. Immunol. 87*:689–698.

Marasco, W., Showell, H., and Becker, E. (1981). Substance P binds to the formylpeptide chemotaxis receptor on the rabbit neutrophil. *Biochem. Biophys. Res. Commun. 99*: 1065–1072.

McDonald, D. (1987). Neurogenic inflammation in the respiratory tract: actions of sensory nerve mediators on blood vessels and epithelium of the airway mucosa. *Am. Rev. Respir. Dis. 136*:S65–S71.

McGillis, J., Organist, M., and Payan, D. (1987). Substance P and immunoregulation. *Fed. Proc. 46*:196–199.

Mullol, J., Chowdhury, B., White, M., Ohkubo, K., Rieves, R., Baraniuk, J., Hausfeld, J., Shelhamer, J., and Kaliner, M. (1993). Endothelin in human nasal mucosa. *Am. J. Respir. Cell Mol. Biol. 8*:393–402.

Natanson, I., Widdicombe, J., and Barnes, P. (1983). Effect of vasoactive intestinal peptide on ion transport across dog tracheal epithelium. *J. Appl. Physiol. 55*:1844–1848.

Ollerenshaw, S., Jarvis, D., Woolcock, A., Sullivan, C., and Scheibner, T. (1989). Absence of immunoreactive vasoactive intestinal polypeptide in tissue from lungs of patients with asthma. *N. Engl. J. Med. 320*:1244–1248.

Peatfield, A., and Richardson, P. (1983a). The action of dust in the airways on secretion into the trachea of the cat. *J. Physiol. (Lond.) 342*:327–334.

Peatfield, A., and Richardson, P. (1983b). Evidence for non-cholinergic, non-adrenergic nervous control of mucus secretion into the cat trachea. *J. Physiol. (Lond.) 342*:335–345.

Peatfield, A., Barnes, P., Bratcher, C., Nadel, J., and Davis, B. (1983). Vasoactive intestinal peptide stimulates tracheal submucosal gland secretion in ferrets. *Am. Rev. Respir. Dis. 128*:89–93.

Rogers, D., and Barnes, P. (1989). Opioid inhibition of neurally mediated mucus secretion in human bronchi. *Lancet 1*:930–932.

Rogers, D., Aursudkij, B., and Barnes, P. (1989). Effects of tachykinins on mucus secretion in human bronchi in vitro. *Eur. J. Pharmacol. 174*:283–286.

Said, S. (1990). Neuropeptides as modulators of injury and inflammation. *Life Sci. 47*: PL19–PL20.

Sestini, P., Bieninstock, J., Crowe, S., Mashall, J., Stead, R., Kakuta, Y., and Perdue, M. (1990). Ion transport in rat tracheal epithelium in vitro. *Am. Rev. Respir. Dis. 141*:393–397.

Sheppard, D., Thompson, J., Scypinski, L., Dusser, D., Nadel, J., and Borson, B. (1988). Toluene diisocyanate increases airway responsiveness to substance P and decreases airway neutral endopeptidase. *J. Clin. Invest. 81*:1111–1115.

Shimura, S., Sasaki, T., Okayama, H., Sasaki, H., and Takashima, T. (1987). Effect of substance P on mucus secretion of isolated submucosal gland from feline trachea. *J. Appl. Physiol. 63*:646–653.

Shimura, S., Sasaki, T., Ekeda, K., Sasaki, H., and Takishima, T. (1988). VIP augments cholinergic-induced glycoconjugate secretion in tracheal submucosal glands. *J. Appl. Physiol. 65*:2537–2544.

Shimura, S., Ishihara, H., Satoh, M., Matsuda, T., Nagaki, N., Sasaki, H., and Takishima, T. (1992). Endothelin regulation of mucus glycoprotein secretion from feline submucosal glands. *Am. J. Physiol. 262*:L208–L213.

Stanisz, A., Scicchitano, R., Stead, R., Matsuda, H., Tomioka, M., Denburg, J., and Bienenstock, J. (1987). Neuropeptides and immunity. *Am. Rev. Respir. Dis. 136*:S48–S51.

Tomaki, M., Ichinose, M., Nakajima, N., Miura, M., Yamauchi, H., Inoue, H., and Shirato, K. (1993). Elevated substance P concentration in sputum after hypertonic saline inhalation in asthma and chronic bronchitis patients. *Am. Rev. Respir. Dis. 147*:A478.

Uddman, R., Moghimzadeh, E., and Sundler, F. (1984). Occurrence and distribution of GRP-immunoreactive nerve fibres in the respiratory tract. *Arch. Otorhinolaryngol. 239*: 145–151.

Umeno, E., Nadel, J., Huang, H., and McDonald, D. (1989). Inhibition of neutral endopeptidase potentiates neurogenic inflammation in the rat trachea. *J. Clin. Invest. 66*: 2647–2652.

Umeno, E., Mcdonald, D., and Nadel, J. (1990). Hypertonic saline increases vascular permeability in the rat trachea by producing neuorogenic inflammation. *J. Clin. Invest. 85*:1905–1908.

Undem, B., Disk, E., and Bucner, C. (1983). Inhibition by vasoactive intestinal peptide of antigen-induced histamine release from guinea-pig minced lung. *Eur. J. Pharmacol. 88*: 247–250.

Webber, S. (1988). The effects of peptide histidine isoleucine and neuropeptide Y on mucous volume output from the ferret trachea. *Br. J. Pharmacol. 95*:49–54.

Webber, S. (1989). Receptors mediating the effects of substance P and neurokinin A on mucus secretion and smooth muscle tone of the ferret trachea: potentiation by an enkephalinase inhibitor. *Br. J. Pharmacol. 98*:1197–1206.

Webber, S., and Widdicombe, J. (1987). The effect of vasoactive intestinal peptide on smooth muscle tone and mucous volume output from fetter trachea. *Br. J. Pharmacol. 91*:139–148.

Webber, S., Lim, J., and Widdicombe, J. (1991). The effects of calcitonin gene-related peptide on submucosal gland secretion and epithelial albumin transport in the ferret trachea in vitro. *Br. J. Pharmacol. 120*:79–84.

Wu, T., Mullol, J., Rieves, R., Logun, C., Hausfield, J., Kaliner, M., and Shelhamer, J. (1990). Endothelin-1 stimulates eicosanoid production in cultured human nasal mucosa. *Am. J. Respir. Cell Mol. Biol. 6*:168–174.

Wu, T., Rieves, R., Larivee, P., Logun, C., Lawrence, M., and Shelhamer, J. (1993).

Production of eicosanoids in response to endothelin-1 and identification of specific endothelin-1 binding sites in airway epithelial wells. *Am. J. Respir. Cell Mol. Biol.* 8: 282–290.

Yano, H., Wershil, B., Arizono, N., and Galli, S. (1989). Substance P–induced augmentation of cutaneous vascular permeability and granulocyte infiltration in mice is mast cell dependent. *J. Clin. Invest.* 84:1276–1286.

Yeadon, M., Wilkinson, D., and Payan, A. (1990). Ozone induces bronchial hyperreactivity to inhaled substance P by functional inhibition of enkephalinase (abstract). *Br. J. Pharmacol.* 99:91.

DISCUSSION

Peterson: There is a discrepancy between some of the in vitro studies and in vivo observations. When substance P is sprayed onto the human nose in vivo, no increase in the amount of secretions is noted, but you showed that incubation of nasal tissue in vitro in substance P caused increased glandular secretion. Is there an explanation for these discrepancies?

Kaliner: Was any attempt to inhibit the actions of NEP included in the study?

Baraniuk: Using the weight of collected secretions may be an insensitive measure of mucus secretion. Lavages with measurement of specific proteins in the fluid might be a far more sensitive way to assess mucus secretion. Braustein demonstrated that substance P caused increases in albumin secretion reflecting increased vascular permeability, while I showed that nasal bombesin induced increased lysozyme and mucus secretion. In both cases, the in vivo actions were consistent with the distribution of receptors and the actions that were proposed based on in vitro studies.

Goetzl: You suggest that both alpha and beta agonists might stimulate secretions in the airways, and Dr. Mak showed information suggesting that you are correct. Can you comment further?

Shelhamer: Both alpha and beta agonists have been reported to have an effect on macromolecular transport, but the beta actions are more variable. Our results definitely show alpha-agonist-induced secretions, but have failed to confirm a beta action. Raphael examined the effects of alpha and beta agonists in vitro and in vivo on human nasal secretions and showed a minimal in vivo effect on alpha agonists and no effect of beta agonists at all, in either system.

Nadel: Using micropipettes, we have shown that both alpha and beta agonists cause secretion of fluid in animal models. Beta agonists produce a more viscous secretion, and histological studies show a selective mucus cell secretion. Alpha agonists selectively degranulate serous cells.

Widdicombe: Clinicians have the impression that atropine completely blocks secretions, and this observation suggests that cholinergic efferents are paramount in controlling secretions. Are these beliefs valid?

Nadel: During bronchoscopy, the procedure causes secretions due to stimulation of sensory nerves located most densely at the bifurcations. These responses are due to cholinergic reflexes. Moreover, using the micropipette technique, it can be shown that secretion can continue after atropine treatment, but that reflex secretions are totally impaired.

Kaliner: Part of the problem is that the currently available anticholinergic compound used clinically is not as potent as atropine, and therefore, the clinical data are still not clearly understood.

Bienenstock: Did you say that endothelin binds to neuroendocrine but not epithelial cells?

Shelhamer: We have not studied endothelin receptors on neuroendocrine cells.

Linnoila: How would you rank the importance of neuropeptides in secretions in intact humans? Which is the most important?

Shelhamer: Of all the secretagogues, serine proteases are the most potent. I would rank cholinergic agonists as the next most important. Of the neuropeptides, substance P probably ranks below cholinergic agonists.

Linnoila: In normal subjects, cholinergic reflexes are likely the predominant stimuli, based on the extensive drying action of atropine.

Nadel: Much of pathological secretion is "pus." What is the modern definition of pus?

Shelhamer: Pus is probably reasonably defined as purulent secretions and that indicates the presence of many granulocytes.

Kaliner: After a rhinovirus nasal infection, secretions are composed of products of vascular permeability, as well as granulocytes. Analysis of lavages on days 2–4 of a rhinovirus upper respiratory infection demonstrates that albumin and IgG make up more than 50% of the proteins, instead of the usual 15%.

Nadel: In considering neutrophil infiltration, the state of the neutrophil itself may be critical. While destroying bacteria is certainly a primary action of the neutrophil, another may be the release of proteases into the tissue, with profound effects on the tissue itself. These actions might include glandular secretion.

Berger: Couldn't lysozyme found in secretions reflect neutrophil degranulation, rather than glandular secretion?

Shelhamer: We have measured mucus secretion by an ELISA directed to the mucus glycoprotein, as well as with lactoferrin secretion.

21

ATP as Neurally Released Modulator of Human Airway Ion Transport

**Richard C. Boucher, Jr., Michael R. Knowles,
Eduardo Lazarowski, Sarah J. Mason, and M. Jackson Stutts**
University of North Carolina at Chapel Hill, Chapel Hill, North Carolina

I. INTRODUCTION

The volume and composition of airway surface liquids, "secretions," are controlled by the aggregate ion transport activities of pulmonary epithelia. Although it is likely that major differences in patterns of ion transport exist within pulmonary regions, the dominant basal mode of ion transport is Na^+ absorption in proximal airway epithelia, which also exhibit the potential for a limited capacity to secrete Cl^- ions. A simplified diagram of the mechanisms that support human airway epithelial ion transport is presented in Figure 1. As depicted in this schema, Na^+ is actively transported from the airway lumen to the blood via the actions of oubain-sensitive Na^+-K^+-ATPase that creates a favorable electrochemical gradient for Na^+ to enter the cell through an apical membrane, amiloride-sensitive Na^+ channel. If this channel is blocked by amiloride, proximal human airway epithelia (defined as cells from nasal and third- to sixth-generation bronchial regions) can secrete Cl^- utilizing a Na^+-K^+-$2Cl^-$ cotransport system in the basolateral membrane to move Cl^- into the cell in series with at least two different types of Cl^- channels on the apical membrane for Cl^- exit into the lumen. The best-characterized Cl^- channel is the one coded for by the CF gene, the "CFTR Cl^- channel" (Cl^-_{CFTR}), which may be the channel that mediates most basal Cl^- conductance. In addition, there is an alternative Cl^- channel (Cl^-_a), which is likely regulated by intracellular Ca^{2+} and extracellular triphosphate nucleotides.

Because the volume and composition of airway surface liquids may change in

AIRWAY LUMEN

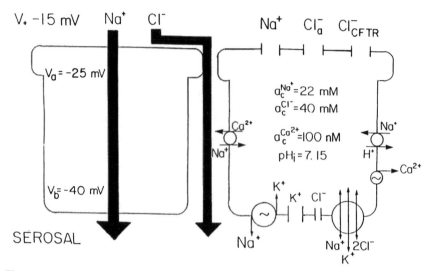

Figure 1 Model depicting ion transport elements in human airway epithelial cell. Two types of Cl⁻ channels [the cystic fibrosis transmembrane regulator (Cl^-_{CFTR}) and the alternative, Ca^{2+}-regulated channel (Cl^-_a)] and an amiloride-sensitive Na^+ channel are shown on the apical membrane. The basolateral membrane expresses a Na^+-K^+-ATPase and a Na^+-K^+-$2Cl^-$ cotransporter. Vectors depict magnitude and routes of Na^+ and Cl^- flow under physiological conditions.

a rapid fashion, the requirements for acute regulation of ion transport rates exist. In this chapter, we will explore the role that ATP, perhaps released in combination with other neuropeptides from afferent nerves located within epithelial lateral intracellular spaces and possibly other sources, may play in this function. In addition, because the effects of ATP are so diverse and profound, the investigation of various epithelial receptor-effector interactions will be described, and the consequences for normal mucosal function and potential therapeutic implications will be described.

II. THE EFFECTS OF ATP ON HUMAN AIRWAY EPITHELIAL ION TRANSPORT

We initially characterized the effects of exogenously added extracellular ATP on ion transport activities of human airway epithelia. Primary cultures of human nasal epithelia were studied in Ussing chambers under basal (Na^+ absorbing) and Cl^--secreting (amiloride pretreated) modes. A comparison of apical and baso-

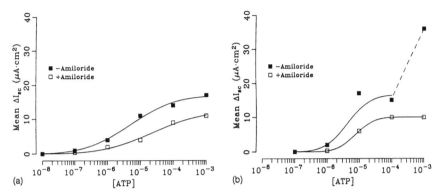

Figure 2 Dose-effect relationship for luminal (a) and basolateral additions (b) of ATP and short-circuit current (I_{sc}) in cultures of human nasal epithelia with or without amiloride pretreatment.

lateral additions of ATP under basal versus amiloride-treated conditions revealed that the effects on Na^+ absorption are greater than on Cl^- secretion (Fig. 2). These data suggest that, depending on the activation state of the apical membrane Na^+ channel ($\pm$ amiloride treatment), ATP can accelerate Na^+ absorption or Cl^- secretion (Mason et al., 1991). This pattern is similar to that observed for peptide agonists, e.g., bradykinin (Clarke et al., 1992a). The concept that ATP can accelerate Na^+ absorption, particularly when presented to the basolateral membrane, may have important consequences for the composition and clearance of airway secretions. As shown in Figure 3, the consequence of accelerating Na^+ and Cl^- absorption from the airway surface liquid compartment is to reduce the quantity of liquid on airway surfaces, concentrate mucins on airway surfaces, and slow mucociliary clearance. Thus, it is likely that both acute and possibly chronic release of this type of mediator may be associated with desiccation of secretions and consequently may be treated best with agents that restore water balance by slowing Na^+ transport, e.g., amiloride.

III. ATP REGULATION OF Cl⁻ SECRETION

Because of the abnormalities in Cl^- secretion associated with lung disease, particularly cystic fibrosis (CF), a major amount of work has been devoted to understanding the mechanisms by which extracellular triphosphate nucleotides (ATP or UTP) can regulate Cl^- secretion in human airway epithelial cells. The major questions that have been investigated are: (1) the pharmacological characterization of the extracellular receptors that recognize ATP/UTP; (2) the intracellular

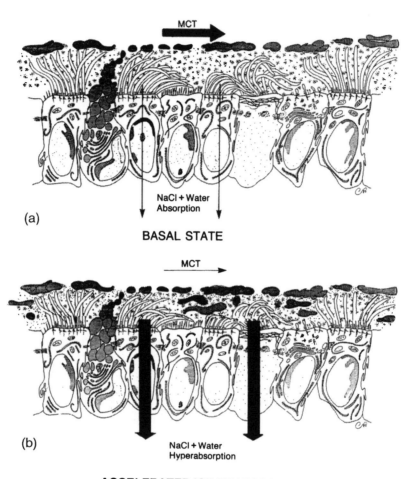

Figure 3 Schema of anatomy of airway surface liquids and proximal airway epithelia under conditions of normal electrolyte transport (a) and accelerated electrolyte transport (b). Vectors depict rates of mucociliary transport (MCT).

signal transduction systems that are activated by these receptors; and (3) the identity of the Cl^- channel(s) that are activated by ATP/UTP. As shown in Figure 4, the apical membrane of the human airway epithelial cells expresses a diverse armamentarium of receptors and effectors. In brief, it is likely that airway epithelial cells exhibit three types of extracellular receptors that recognize triphosphate nucleotides or triphosphate nucleotide breakdown products. The simplest and best characterized are the A_2 adenosine receptors, which recognize the

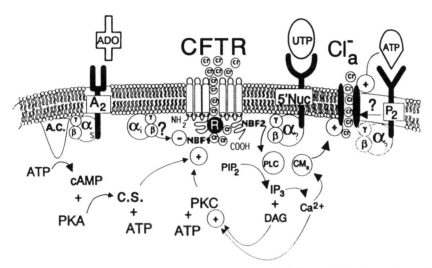

Figure 4 Receptors and channels expressed in the apical (lumen facing) membrane of human airway epithelial cells. Adenosine (A_2) and two types of triphosphate nucleotide receptors (P_2 and 5'-nucleotide) and two Cl^- channel types are shown. Intracellular transduction systems include G proteins, adenylyl cyclase (AC), and phospholipase C (PLC).

nucleoside breakdown product of ATP, adenosine. A_2 adenosine receptors are linked to adenylyl cyclase via G proteins in both positive (G_s) and negative (G_i) fashions (Lazarowski et al., 1992). However, the more recently recognized and potentially more effective receptors with regard to regulation of Cl^- secretion are the so-called P_2 receptors, which recognize the triphosphate nucleotides. These receptors have typically been termed purinoceptors, but the equipotency of pyrimidine triphosphate nucleotides (UTP) with the purine triphosphate nucleotides (ATP) for some receptors has led to a revision of this nomenclature (Brown et al., 1991). For purposes of this review, we will term receptors that recognize ATP but not UTP P_2 receptors and receptors that recognize UTP as well as ATP 5'-nucleotide receptors. There is evidence for both types of receptors on human airway epithelial cells. In addition, the two potential effector Cl^- channels, the CFTR Cl^- channel and the alternative Cl^- channel (Cl_a^-), are also depicted. It is highly likely that the CFTR Cl^- channel is regulated both by activation of protein kinase A (PKA) by cAMP and by protein kinase C (PKC) in response to DAG and raised intracellular Ca^{2+} (Ca_i^{2+}). In contrast, the alternative channel appears to be regulated by intracellular Ca^{2+}, which may be elevated by interactions of 5'-nucleotide receptors with G proteins, phospholipase C (PLC), and the consequent generation of IP_3 and intracellular Ca^{2+} mobilization. In addition, there are data that suggest that the P_2 receptor may be more directly coupled by mechanisms that require no soluble intracellular messengers to the alternative Cl^- channel (Stutts et al., 1992).

The interactions between triphosphate nucleotides and receptors to regulate Cl^- channels are complex and may depend on the sidedness of application of ATP/ UTP. For example, the application of ATP to the basolateral membrane of the cell appears to primarily induce Cl^- secretion via Ca^{2+}-mediated activation of basolateral membrane K^+ channels (Fig. 5). This response leads to intracellular hyperpolarization and an increase of the electrochemical driving force for Cl^- flow through the alternative Cl^- channel. In contrast, application of ATP to the luminal membrane appears to activate the apical membrane Cl^- channels with relatively little effect on the basolateral membrane K^+ channels. The precise mechanism for this segregation of distinct transduction pathways is unknown. Indeed, ATP regulation of the apical membrane is more complex than depicted in Figure 5 because of the presence of at least two different channels in the luminal membrane. As shown in Figure 6, it is possible that ATP or UTP, interacting with a single $5'$-nucleotide receptor in normal cells, can activate both channel types. The alternative Cl^- channel may be activated directly by increases in intracellular Ca_i^{2+} or direct mechanisms whereby the CFTR Cl^- channel is activated by the elevations in Ca^{2+} and diacylglycerol (DAG) via PKC-dependent mechanisms.

These analyses of signal transduction pathways hinge on the notion that there

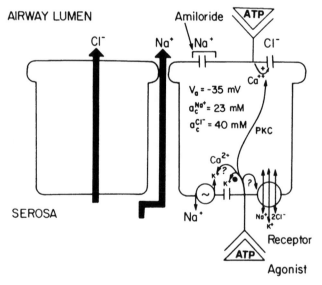

Figure 5 Models describing modes of activation of Cl^- secretion by ATP added selectively to the basolateral or apical membrane.

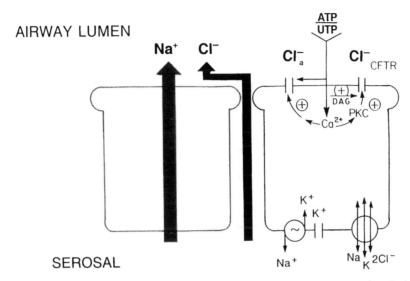

Figure 6 Model depicting mechanisms by which ATP/UTP can regulate both Cl⁻ channel types.

are formally two distinct Cl⁻ channels in the apical membrane, the CFTR Cl⁻ channel and an alternative Cl⁻ channel. The first clues that this was so came from studies of additivity of maximal concentration of agonists in primary cultures of normal human nasal epithelial cells. As shown in Figure 7, ATP and forskolin are about equieffective in inducing Cl⁻ secretion in human nasal epithelium. As would be expected for two independent pathways, the combination of forskolin and ATP is additive. However, the formal proof that there is a second channel that can serve as an alternative to the CFTR Cl⁻ channel present in the apical membrane came from studies of gene-targeted mice. In these animals, the murine CFTR gene was disrupted by homologous recombination techniques, which led to a complete disabling of the murine CF gene (Snouwaert et al., 1992). The result is that these mice produce no normal CFTR messenger RNA or CFTR protein. Cultured airway cells from normal mice and CF mice were compared with respect to their ability to respond to activation via cAMP-mediated (forskolin) and ATP-mediated pathways (Clarke et al., 1992b). As shown in Figure 8, normal mice respond to forskolin with an increase in Cl⁻ secretion whereas CF mice do not. These data are consistent with the absence of the CFTR Cl⁻ channel in the CF mice. In contrast, the CF mice respond equally well to the addition of luminal ATP with Cl⁻ secretion, formally demonstrating that Cl⁻ secretion across the luminal membrane triggered by ATP is mediated by an alternative channel. Therefore, although the molecular identity of the alternative Cl⁻ channel has not been established, it is clear that it is a polypeptide distinct from CFTR.

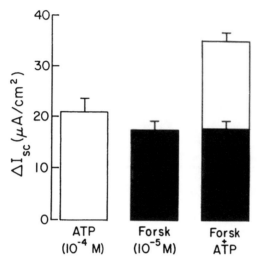

Figure 7 Effect on Cl⁻ secretion of single and cumulative additions of ATP (10^{-4} M) and forskolin (10^{-5} M) to cultures of human airway epithelia.

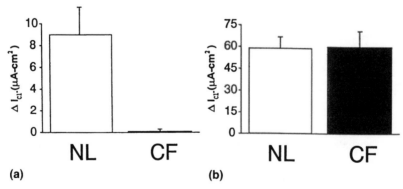

(a)   (b)

Figure 8 Comparison of response of airway epithelia from normal and CF mice to forskolin (10^{-5} M) (a) or forskolin (10^{-5} M) (b).

IV. USE OF EXOGENOUS TRIPHOSPHATE NUCLEOTIDES AS THERAPEUTIC AGENTS TO INCREASE MUCOCILIARY CLEARANCE

Triphosphate nucleotides, including UTP, potentially via their actions on salt and water secretion, may normalize the water content of airway surface liquids and promote clearance. The actions of these compounds to effect these maneuvers are shown in Figure 9. As is shown, the addition of extracellular triphosphate nucleotides to airway surfaces may have profound effects on salt and water metabolism. It is unclear at present whether UTP or ATP alone will promote Cl⁻ secretion, or whether they may require the coadministration of a Na⁺ channel blocker, e.g., amiloride, to effect this type of activity. However, it has become clear that extracellular triphosphate nucleotides have other actions that may be beneficial with regard to increasing the mucociliary clearance. Several investigators have observed that ATP dramatically increases ciliary beat frequency and promotes clearance.

There are, however, some potential disadvantages to using these compounds to increase mucociliary clearance. As shown in Figure 9, extracellular triphosphate nucleotides may release mucin granules from goblet cells (Lethem et al., 1993). This response may or may not have an acute beneficial effect on mucociliary clearance, and the possible long-term effects of chronic triphosphate nucleotide addition to airway surfaces on goblet cell numbers are unknown. In addition, as

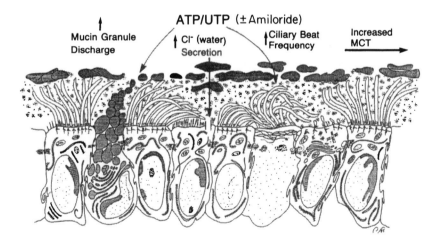

Figure 9 Effects of exogenous (aerosolized) triphosphate nucleotides on elements of the mucociliary clearance. Actions on Cl⁻ secretion, ciliary beat frequency, and goblet cell secretion are shown.

AIRWAY LUMEN

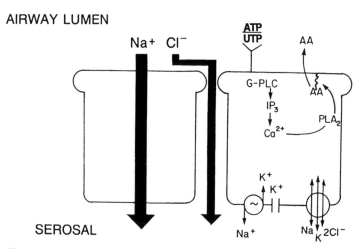

Figure 10 Mechanism for purinoceptor-mediated release of arachidonic acid (AA) from human airway epithelial cells. AA is depicted within the cell as associated with plasma membrane lipids and is the target for cleavage of a Ca^{2+}-activated phospholipase A_2 (PLA$_2$).

shown in Figure 10, it has recently been shown that purinoceptors can activate the release of arachidonic acid from cells (Lazarowski et al., 1993). This activity appears to reflect an ATP/UTP-mediated increase in intracellular Ca^{2+} that induces the translocation of a cytosolic phospholipase A_2 to the plasma membrane, where it catalyzes the release of arachidonic acid from membrane stores. Again, the long-term effects of chronic release of arachidonic acid and its metabolism to potential inflammatory agents are not known.

In summary, it appears that extracellular ATP has dramatic actions on all facets of mucosal physiology. To be investigated in the future are the sites of ATP release, including nerve terminals that may interdigitate the latter aspects of airway epithelial cells. Further research is needed to molecularly characterize these receptors and to take advantage of subtypes of receptors for therapeutic development of selective analogs that may safely increase clearance of materials or perturb any of the actions of ATP on mucus secretion, ciliary beat frequency, or salt and water metabolism, in a controllable fashion.

REFERENCES

Brown, H.A., Lazarowski, E.R., Boucher, R.C., and Harden, T.K. (1991). Evidence that UTP and ATP regulate phospholipase C through a common extracellular 5′-nucleotide receptor in human airway epithelial cells. *Mol. Pharmacol. 40*:648.

Clarke, L.L., Grubb, B.R., Gabriel, S.E., Smithies, O., Koller, B.H., and Boucher, R.C. (1992a). Defective epithelial chloride transport in a gene targeted mouse model of cystic fibrosis. *Science 257*:1125.

Clarke, L.L., Paradiso, A.M., Mason, S.J., and Boucher, R.C. (1992b). Effects of bradykinin on Na^+ and Cl^- transport in human nasal epithelium. *Am. J. Physiol. 262*:C644.

Lazarowski, E.R., Mason, S.J., Clarke, L.L., Harden, T.K., and Boucher, R.C. (1992). Adenosine receptors on human airway epithelia and their relationship to chloride secretion. *Br. J. Pharmacol. 106*:774.

Lazarowski, E.R., Boucher, R.C., and Harden, T.K. (1993). Calcium-dependent release of arachidonic acid in response to purinergic receptor activation in airway epithelium. *Am. J. Physiol.* (in press).

Lethem, M.I., Dowell, M.L., Van Scott, M., Yankaskas, J.R., Egan, T., Boucher, R.C., and Davis, C.W. (1993). Nucleotide regulation of goblet cells in human airway epithelium, in vitro. *Am. J. Respir. Cell Mol. Biol. 9*:315–322.

Mason, S.J., Paradiso, A.M., and Boucher, R.C. (1991). Regulation of transepithelial ion transport and intracellular calcium by extracellular adenosine triphosphate in human normal and cystic fibrosis airway epithelium. *Br. J. Pharmacol. 103*:1649.

Snouwaert, J., Brigman, K., Latour, A.M., Malouf, N.N., Boucher, R.C., Smithies, O., and Koller, B. H. (1992). An animal model for CF made by gene targeting. *Science 257*:1083.

Stutts, M.J., Chinet, Th.C., Mason, S.J., Fullton, J.M., Clarke, L.L., and Boucher, R.C. (1992). Regulation of chloride channels in normal and cystic fibrosis airway epithelial cells by extracellular ATP. *Proc. Natl. Acad. Sci. USA 89*:1621–1625.

DISCUSSION

McDonald: Do sensory nerve stimuli, such as capsaicin, evoke goblet cell secretion from your explants of human airway epithelium?

Boucher: The idea of exposing human airway preparations to electrical field stimulation or capsaicin as a test of neural regulation is excellent.

Linnoila: Have you studied the role of Clara cells in airway salt-water balance?

Boucher: We have studied Clara cells from the rabbit in culture with respect to salt and water transport; these cells are, if anything, more absorptive than proximal airway cells. We have also demonstrated that these cells respond to ATP but not to tachykinins.

Said: Does adenosine have the same effect as ATP?

Boucher: The relative importance of ATP and adenosine varies with the tissue. Adenosine, acting via P_1 receptors, may dominate in gut epithelium. However, triphosphate nucleotides (ATP and UTP) clearly exert their actions via P_2 receptors in the airways, whereas the P_1-receptor effects are minor.

Solway: How does the epithelium respond to mucosal hypertonicity, as might be induced by inhalation of hypertonic saline or conditions of increased respiratory water loss?

Boucher: The responses of human airway epithelia to hypertonic solutions are complex. The nasal epithelial cell has an apical membrane water permeability that exceeds the water permeability of the basolateral membrane, which is almost unique. This configuration allows the epithelial cell to act as an osmometer to sense airway surface liquid tonicity. The cellular responses to this stress are (1) to slow electrolyte absorption and (2) to tighten the tight junctions. We observed that the major response of the epithelium is to signal the microvasculature to dilate.

Kaliner: What concentration of ATP is needed to degranulate goblet cells?

Boucher: The normal ATP concentration appears high (10^{-4} M). However, the "real" ligand is ATP^{4-}, which is in solution at about 100-fold lower concentrations than MG ATP^{2-}. The question about the physiological role relates how much release occurs into the very small volume on airway surfaces on the confined lateral spaces between cells.

Kaliner: Could you comment on the effects of airway stress from noxious substances such as ammonia on goblet cell secretion?

Boucher: We believe that environmental stress is a common trigger for goblet cell secretion. It is possible that environmental stress will release ATP into the lumen. If so, this would be an attractive mechanism to unify a lot of disparate observations.

22

Effects of Neuropeptides on Neurotransmission in the Airways

Maria G. Belvisi
National Heart and Lung Institute, London, England

I. INTRODUCTION

A wide range of peptides have been localized in the respiratory tract of many species of animals including humans. Many of these peptides are localized to nerves in the airways (Table 1). In fact, neuropeptides have been demonstrated in sensory, parasympathetic, and sympathetic neurons in the respiratory tract where in many cases they are colocalized with classical neurotransmitters (Barnes et al., 1991). Neuropeptides have been shown to have effects on mucus secretion, bronchial blood flow, vascular permeability, and bronchomotor tone, and more recently, a neuromodulatory role has been proposed for neuropeptides (Table 2) (Barnes et al., 1990).

II. INTERACTIONS WITH CHOLINERGIC NERVES

The predominant control of human and animal airways is exerted by cholinergic nerves (Barnes, 1987). Stimulation of parasympathetic pathways causes bronchoconstriction, mucus secretion, and bronchial vasodilatation.

A. Tachykinins

The first report that suggested that tachykinins may have a neuromodulatory role in the peripheral nervous system was in the guinea pig myenteric plexus where substance P (SP) was found to evoke the release of acetylcholine (ACh) in a concentration-dependent manner (Yau and Youther, 1982). More recently, it has

477

Table 1 Neuromodulatory Peptides in the Airways

Peptide	Nerve type
Vasoactive intestinal peptide	Parasympathetic
Galanin	
Substance P	Afferent
Neurokinin A	
Neuropeptide tyrosine	Sympathetic
Somatostatin	Afferent/uncertain
Enkephalin	
Cholecystokinin octapeptide	

Table 2 Effect of Exogenous Neuropeptides on Neurotransmission in the Airways

Peptide	Species	Cholinergic neurotransmission	e-NANC neurotransmission
Substance P	Guinea pig	↑	?
	Rabbit	↑	
	Human	No effect	
Neurokinin A	Guinea pig	↑	?
	Rabbit	↑	
	Human	No effect	
Neuropeptide Y	Guinea pig	↓	↓
	Human	No effect	
Opioids	Guinea pig	↓	↓
	Dog	↓	
	Human	↓	
Vasoactive intestinal peptide	Guinea pig	↓	↓
	Cat	↓	
	Ferret (<1 nM)	↑	
	Ferret (>1 nM)	↓	
Galanin	Guinea pig	No effect	↓
	Human	No effect	
Somatostatin	Ferret	↑	No effect
	Guinea pig		No effect
Cholecystokinin octapeptide	Guinea pig	No effect	?
	Human	No effect	

been suggested that tachykinins may play an important role in modulating cholinergic neurotransmission in airway smooth muscle on the basis of immuno-histochemical data (Dey et al., 1991) and functional studies.

1. Rabbit

Exogenous tachykinins have been shown previously to facilitate cholinergic neurotransmission in airway smooth muscle. In rabbit isolated trachea, SP potenti-ated, in a concentration-dependent manner, contractile responses evoked by cholinergic nerve stimulation by a postganglionic, prejunctional mechanism (Tanaka et al., 1986; Armour et al., 1991). SP also induced bronchoconstriction in rabbit trachea in vitro that was significantly inhibited by atropine, suggesting that SP evokes ACh release from cholinergic nerve terminals (Tanaka et al., 1986). In a recent study, experiments were performed using selective tachykinin agonists and antagonists to elucidate the receptor(s) involved in the prejunctional modula-tion of cholinergic neurotransmission in rabbit bronchi. The results obtained suggested the presence of both NK-1 and NK-2 receptors on cholinergic nerves in rabbit bronchi (Belvisi et al., 1994) (Fig. 1). In addition, the full agonist activity of MDL 28, 564 and the rank order of potency of the NK-2 receptor antagonists MEN 10, 376, L659, 877, and R 396 indicates that the NK-2 receptors mediating

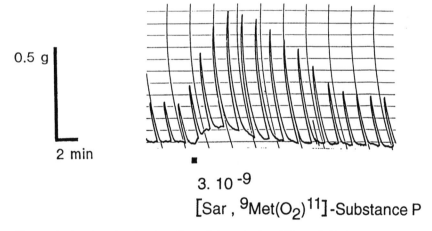

0.5 g

2 min

$3. 10^{-9}$

$[Sar, {}^{9}Met(O_2)^{11}]$-Substance P

Figure 1 Tracing showing the effect of the selective NK-1 tachykinin receptor agonist [Sar[9]]SP sulphone (3 nM) on cholinergic responses to EFS (60 V, 0.4 msec, 2 Hz for 10 sec every min) in rabbit bronchi. [Sar[9]]SP sulfone (3 nM) was used at a concentration that approximated the EC_{50} value for enhancement of responses. (Adapted from Belvisi et al., 1994.)

facilitation of ACh release in rabbit bronchi belong to the same subtype that mediates contraction of the endothelium-deprived rabbit pulmonary artery (termed NK-2A) (Maggi et al., 1990).

2. Guinea Pig

Exogenous tachykinins also potentiate cholinergic neurotransmission at pre- and postganglionic nerve terminals in guinea pig trachea (Hall et al., 1989; Watson et al., 1993a). The tachykinin receptor mediating this facilitatory effect appears to be of the NK-1-receptor subtype (Belvisi et al., 1994; Watson et al., 1993a).

Facilitatory effects of tachykinins on cholinergic neurotransmission may have physiological relevance as there has been some suggestion that endogenous tachykinins facilitate cholinergic contractile responses in airway smooth muscle. The metallopeptidase neutral endopeptidase 24.11 is a major enzyme involved in the breakdown of tachykinins (Erdös and Skidgel, 1989). Inhibition of this enzyme by phosphoramidon (an inhibitor of neutral endopeptidase) (NEP) would be expected to augment the actions of endogenously released tachykinins. In guinea pig trachea, phosphoramidon facilitates contractile responses evoked by preganglionic vagal nerve stimulation (PGS), and not by transmural stimulation of the trachea, in a concentration-dependent manner, and this effect is blocked by capsaicin pretreatment (Watson et al., 1993a). In addition, the facilitatory effect of phosphoramidon was also blocked by the NK-1 receptor antagonist GR71251, and the antagonist, when applied to the tissue during nerve stimulation, caused a small but significant inhibition of the responses evoked by PGS while having no effect on transmural stimulation (TMS). These results suggest that there is release of endogenous tachykinins during pre- but not postganglionic nerve stimulation in guinea pig trachea, suggesting that there are facilitatory tachykinin receptors (probably of the NK-1-receptor subtype) at the level of the parasympathetic ganglia (Watson et al., 1993a). In contrast, Aizawa et al. (1990) have reported that phosphoramidon increases contractions to EFS in guinea pig trachea without changing responses to exogenous ACh.

In addition, capsaicin pretreatment, which depletes sensory nerves of tachykinins, results in a significant reduction in cholinergic contractile responses to both electrical field stimulation (EFS) in vitro and vagal nerve stimulation in vivo in guinea pig airways (Stretton et al., 1992a), suggesting a role for endogenous tachykinins in the facilitation of cholinergic neurotransmission via a postganglionic mechanism. In addition, capsaicin, at a subthreshold concentration, acutely releases tachykinins that enhance cholinergic responses to EFS in guinea pig trachea in vitro, again suggesting endogenous tachykinin-induced, postganglionic modulation of cholinergic responses (Aizawa et al., 1990).

Therefore, in summary, the experiments using capsaicin suggest that tachykinins can modulate contractile responses to EFS, i.e., via a postganglionic mechanism, whereas the experiments using the NK-1-receptor antagonist suggest that modulation only occurs preganglionically. The obvious explanation is that capsaicin has more far-reaching effects (with the possibility of certain nonspecific effects) than the NK-1 antagonist and apart from anything it will release/deplete SP and NKA, which can act at other tachykinin receptors, e.g., NK-2 and NK-3.

3. Ferret

Sekizawa and colleagues (1987) have demonstrated that inhibitors of NEP increase the contractions of ferret trachea evoked by EFS in vitro.

4. Human

Finally, in human bronchial rings none of the selective tachykinin receptor agonists had any effect on cholinergic neurotransmission (Belvisi et al., 1994). In fact, it has been shown previously that NKA produces potentiation of the contractile response to EFS in human bronchi, but only in the presence of K^+ channel blockade (Black et al., 1990). This points to a neuromodulatory role for NKA in human airways, only in situations where the K^+ channel activity is decreased.

The results described above suggest that tachykinins may play an important role in modulating cholinergic neurotransmission in animal airways with no demonstrable effect on human airways. However, this does not rule out a role for endogenous tachykinins in the modulation of cholinergic neurotransmission in human airways. Another consideration is that while this may not be important under normal conditions, the system may be active in disease. For example, if K^+ channels were impaired in disease, then modulatory effects of tachykinins on cholinergic neurotransmission might become evident. In fact, previous studies have demonstrated that NEP is inactivated by viral infection (Jacoby et al., 1988).

B. Opioids

Opioid agonists inhibit stimulus-evoked release of neurotransmitters such as ACh from the central and peripheral nervous systems. Leu-enkephalin selectively inhibits ACh release in the rat corpus striatum (Mulder et al., 1984), and in the peripheral nervous system, morphine inhibits cholinergic neurotransmission in guinea pig ileum (Paton, 1957) and rabbit heart (Kosterlitz and Taylor, 1959).

1. Bronchoconstriction

Opioids have been shown to inhibit cholinergic neurotransmission, at low frequencies of stimulation, in canine airways via a postganglionic, prejunctional mechanism of action (Russell and Simons, 1985). In guinea pig airways cholinergic neurotransmission is similarly inhibited by agonists selective for μ and δ opioid receptors in a frequency-dependent manner (Belvisi et al., 1990). The mechanism of action appeared to be prejunctional and postganglionic as the μ-opioid agonist [D-Ala^2-NMePhe^4-Gly-ol^5] (DAMGO) had no effect on contractile responses to exogenously administered ACh. More recently, experiments in which ACh release was determined, by measurement of [³H]-release from guinea pig tracheal strips preincubated with [³H]-choline while undergoing EFS, demonstrated that opioids (μ and δ agonists) do inhibit ACh release evoked by EFS (Belvisi et al., 1993a). Interestingly, the inhibitory effect of DAMGO on cholinergic neurotransmission is markedly reduced in animals pretreated with capsaicin, suggesting that μ-opioid agonists are acting partly by inhibiting the release of tachykinins from sensory nerves in the airways that facilitate cholinergic neurotransmission (Belvisi et al., 1990).

In addition, DAMGO inhibited cholinergic contractile responses to EFS in a concentration- and frequency-dependent manner in human airway smooth muscle. In fact, at low frequencies there was more than 85% inhibition of the cholinergic response (Belvisi et al., 1992c) (Fig. 2). The mechanism of action appeared to be prejunctional as DAMGO had no effect on contractile responses to ACh (Belvisi

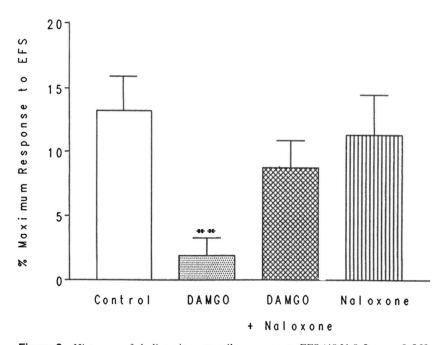

Figure 2　Histogram of cholinergic contractile responses to EFS (40 V, 0.5 msec, 0.5 Hz for 15 sec) in human bronchi in vitro in the presence of DAMGO (1 μM), DAMGO (1 μM) and naloxone (10 μM), and naloxone (10 μM) alone. Contractile responses were expressed as a percentage of the maximum response obtained to EFS in each preparation. Mean values are shown; vertical bars indicate SE. Responses were compared using Student's t-test for unpaired data. $**p < 0.01$.

et al., 1992), suggesting that DAMGO inhibits ACh release. The contractile responses evoked by EFS in vitro under these conditions were not affected by the ganglion blocker hexamethonium (Rhoden et al., 1988), so the mechanism of action of opioids must be postganglionic. Recently, experiments have demonstrated that μ-opioid agonists inhibit ACh release evoked by EFS in human tracheal strips (Belvisi et al., 1993a) (Figs. 3 and 4). In the experiments described, the opioid receptor antagonist naloxone completely blocks opioid-mediated modulatory effects but has no effect on cholinergic neurotransmission when given alone, indicating that endogenous opioids do not modulate release of ACh under the experimental conditions used (Figs. 3 and 4). However, the endogenous opioid Met-enkephalin-Arg[6]-Gly[7]-Leu[8] has been described in guinea pig and rat airways (Shimosegawa et al., 1990), and it is therefore possible that it may act as a neuromodulator if selectively released in the airways. It may be that enkephalins are released in inflammatory conditions from cells such as lymphocytes that are known to be present at sites of inflammation in the airway.

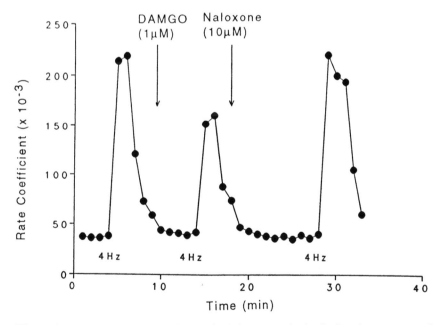

Figure 3 Profile of ACh release from a single human tracheal strip. Results are expressed as rate coefficient ($\times 10^{-3}$), which is a measure of the fractional [³H]-release, plotted against time (min). EFS (40 V, 0.5 msec, 4 Hz for 1 min) produced an increase in [³H]-release, which was inhibited by the μ-opioid agonist DAMGO (1 μM). The inhibitory effect was reversed by the opioid antagonist naloxone.

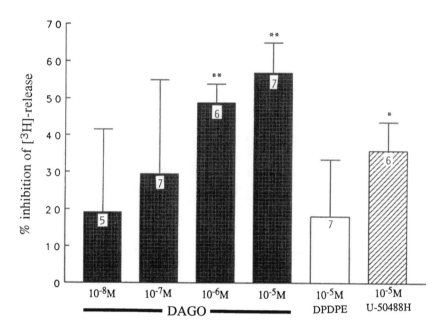

Figure 4   Concentration-dependent inhibition of ACh release evoked by EFS (40 V, 0.5 msec, 4 Hz for 1 min) in human trachea by the μ-opioid receptor agonist (Tyr-(D-Ala)-Gly-(N-Me-Phe)-Gly-ol) (DAGO) (10 nM–10 μM) and the k-opioid receptor agonist [trans-3,4-dichloro-N-methyl-N-(2-(1-pyrrolidinyl)-cyclohexyl)-benzeneacetamine] U-50, 488H) (10 μM) but not the δ-opioid receptor agonist [Tyr-(D-Pen)-Gly-Phe-(D-Pen)] (DPDPE) (10 μM). Mean values are shown; vertical bars indicate SEM. Responses compared using Student's *t*-test for paired data with each tissue serving as its own control. *$p < 0.05$; **$p < 0.01$.

2. *Mucus Secretion*

Opioids can inhibit the mucus discharge from airway goblet cells that is stimulated by cigarette smoke exposure (Kuo et al., 1992). In this system the cholinergic component of the response is inhibited by a μ- and δ-opioid receptor agonists as opposed to the nonadrenergic, noncholinergic (NANC) component of the response, which seems to be inhibited by opioids acting at μ-opioid receptors (Kuo et al., 1992).

C.   **Vasoactive Intestinal Peptide**

Vasoactive intestinal peptide (VIP) is a 28-amino-acid peptide originally isolated from lung extracts (Said et al., 1969). VIP immunoreactivity has been localized to ganglion cells in the trachea and bronchi and to nerve fibers around blood vessels and exocrine glands and within the tracheobronchial smooth muscle layer (Dey et al., 1981; Uddman et al., 1984). It has been suggested that ACh and VIP may coexist and therefore may be coreleased in several organs. For example, in cat submandibular gland VIP is coreleased with ACh and, in fact, enhances the affinity of muscarinic receptors to ACh (Lundberg et al., 1982). It has also been suggested that ACh and VIP may coexist in the same nerve terminals in the airways (Laitinen et al., 1985).

1. Exogenously Added VIP

VIP modulates cholinergic neurotransmission in the guinea pig trachea (Ellis and Farmer, 1989a; Stretton et al., 1991). This modulation may occur via a combination of prejunctional and postjunctional mechanisms because the inhibition of the neural cholinergic response was greater than the inhibition by VIP of the equivalent response elicited by exogenous ACh. VIP also seems to modulate cholinergic neurotransmission in guinea pig airways at the level of the airway parasympathetic ganglia (Martin et al., 1990). This may be physiologically relevant, since VIP-immunoreactive neurons are present in local ganglia in airways (Dey et al., 1981). The nature of the postjunctional effects of VIP is dependent on the concentration of VIP. At low doses, VIP increased the sensitivity of guinea pig trachea to methacholine presumably owing to an increase in the affinity of the muscarinic receptors (Ellis and Farmer, 1989a), and in cat submandibular gland, VIP enhanced postjunctional muscarinic receptor affinity to ACh (Lundberg et al., 1982). At high concentrations VIP produced inhibition of contractile responses to ACh (Ellis and Farmer, 1989a; Stretton et al., 1991), presumably owing to functional antagonism since VIP at these concentrations is a bronchodilator. VIP at low concentrations has a dual effect: it enhances contractions to muscarinic agents, but inhibits cholinergic neural responses to EFS. However, the inhibitory effect is dominant, suggesting that VIP inhibits the release of ACh from cholinergic nerve endings.

In feline airways, VIP inhibits cholinergic nerve-induced contraction and reduces the amplitude of excitatory junction potentials (EJPs) at low concentrations that do not affect the response to exogenous ACh (Ito and Hakoda, 1990).

In contrast to the above findings, in ferret trachea in vitro low concentrations of VIP (<1 nM) appear to enhance cholinergic neurotransmission, whereas higher concentrations are inhibitory (Sekizawa et al., 1988).

In summary, it appears that VIP may act as a bronchodilator or to inhibit ACh release in the airways, and thereby VIP may have a dual mechanism for combating cholinergic bronchoconstriction.

2. Endogenously Released VIP

In guinea pig trachea there is a prominent nonadrenergic (i-NANC) neural bronchodilator response, which is evoked in vitro to EFS (Coburn and Tomita,

1973; Coleman and Levy, 1974; Taylor et al., 1984; Ellis and Farmer, 1989b,c) and in vivo by electrical stimulation of vagus nerves (Chesrown et al., 1980). The role of endogenously released VIP is uncertain, since there are no potent and selective antagonists available. However, in guinea pig trachea, α-chymotrypsin, a proteolytic enzyme that degrades VIP, inhibits the i-NANC neural response to EFS by approximately 35%, and incubation with VIP antisera markedly reduces the response, suggesting that VIP and related peptides, released during EFS of guinea pig trachea, may partly mediate i-NANC relaxation responses (Ellis and Farmer, 1989b,c; Tucker et al., 1990; Li and Rand, 1991).

α-Chymotrypsin enhances cholinergic contractile responses to EFS in guinea pig trachea by between 31 and 38% with no effect on contractile responses to exogenous ACh, suggesting that endogenous VIP may modulate cholinergic neurotransmission via a prejunctional action on cholinergic nerves to inhibit ACh release or by functional antagonism of ACh at the level of the airway smooth muscle, or both (Belvisi et al., 1993b; Brave et al., 1991). The percentage enhancement of cholinergic responses produced by α-chymotrypsin could be related to the percentage inhibition of the i-NANC neural response produced by α-chymotrypsin [i.e., Ellis and Farmer (1989c) found that the i-NANC neural response was inhibited by approximately 35%, Tucker et al. (1990) by 41.5%, and Li and Rand (1991) by 36.6%]. Interestingly, Watson et al. (1993b) have demonstrated that although α-chymotrypsin facilitated contractions of the guinea pig trachea, in a vagally innervated tracheal tube preparation, induced by TMS, there was no effect on contractile responses induced by PGS. These results suggest that stimulation of preganglionic vagal nerve fibers does not cause release of VIP, but VIP can be released in these preparations by stimulation of intrinsic nerves via TMS. Therefore, it may be that the preganglionic inhibitory NANC fibers enter the trachea from a source other than the main vagal nerve trunks.

However, to directly establish whether VIP acts prejunctionally to inhibit transmitter release from cholinergic nerves, it would be necessary to measure the effect of VIP on ACh release.

α-Chymotrypsin, at a concentration shown to inhibit VIP-induced neural responses in guinea pig trachea (Ellis and Farmer, 1989c; Tucker et al., 1990; Li and Rand, 1991) and abolish responses to exogenously applied VIP in human trachea (Belvisi et al., 1992a,b), had no effect on cholinergic contractile responses to EFS in human airways. This suggests that endogenous VIP does not modulate cholinergic neurotransmission in human trachea, exposing a marked difference from the guinea pig airways where endogenous VIP has been shown to inhibit cholinergic contractile responses (Belvisi et al., 1993b). This is somewhat surprising as it has been demonstrated that there are large numbers of VIP-immunoreactive nerves in human airway smooth muscle (Laitinen et al., 1985); however, it may be that VIP is more involved in pulmonary vasodilation than in bronchodilation. Alternatively, the absence of effects of endogenously released VIP from human airways could be due to the release of tryptase, which has been shown to inhibit the bronchodilator action of VIP in airways in vitro (Caughey, 1989), from activated mast cells in human airways.

D. Neuropeptide Tyrosine

Neuropeptide tyrosine (NPY), a 36-amino-acid peptide, is present within the mammalian respiratory tract, where it is believed to be localized to adrenergic nerve fibers innervating blood vessels and bronchial smooth muscle (Sheppard et al., 1984). However, NPY has also been described in cholinergic nerves (Luts and Sundler, 1989). NPY may function to autoregulate sympathetic nerve function by inhibiting the release of noradrenaline (Potter et al., 1988), but this has not yet been demonstrated in the airways.

NPY inhibits cholinergic neurotransmission in guinea pig airways probably via an effect on ACh release, as it reduces cholinergic nerve contractile responses but has no effect on the contractile response to exogenous ACh (Stretton and Barnes, 1988; Grundemar et al., 1988; Matran et al., 1989). This inhibitory effect is more pronounced at low frequencies of stimulation. Furthermore, the inhibitory effect is unaffected by adrenergic blockade, suggesting that NPY acts at a prejunctional receptor on postganglionic cholinergic nerves. NPY appears to act on at least two subtypes of receptors. Y_2 receptors are presynaptic and are thought to inhibit adenylyl cyclase (Wahlestedt et al., 1986). It has been demonstrated that NPY inhibits hippocampal neurons via a Y_2 receptor (Colmers et al., 1991), and it is possible that the same receptor subtype is involved in NPY-induced inhibition of cholinergic neurotransmission in guinea pig airways.

In human airways in vitro, NPY has no effect on cholinergic neurotransmission (C. D. Stretton and P. J. Barnes, unpublished data).

E. Other Neuropeptides

1. Galanin

Galanin, which seems to be present in airway cholinergic nerves (Cheung et al., 1985), has no effect on cholinergic neurotransmission in human or guinea pig airways (C. D. Stretton and P. J. Barnes, unpublished data).

2. Somatostatin

Somatostatin facilitates cholinergic neurotransmission in ferret airways in vitro (Sekizawa et al., 1989) possibly via a prejunctional effect on receptors located on cholinergic nerves.

3. Cholecystokinin

Cholecystokinin octapeptide (CCK) is a potent constrictor of guinea pig and human airways in vitro but has no effect on contractile responses to EFS or to PGS, suggesting that CCK has no effect on cholinergic neurotransmission in the airways (Stretton and Barnes, 1989).

III. INTERACTIONS OF NEUROPEPTIDES WITH NANC NERVES

Neural control of the airways is complex, and in addition to afferent, cholinergic and adrenergic nerves, there are NANC mechanisms whose effects are mediated by the release of neuropeptides probably from classical autonomic nerves.

Electrical stimulation of guinea pig bronchi and distal trachea in vitro and vagus nerve stimulation in vivo evokes an atropine-sensitive bronchoconstrictor response (Barnes et al., 1991). This bronchoconstrictor response has been termed excitatory NANC (e-NANC) and is probably due to the retrograde release of tachykinins from a certain population of sensory nerves (capsaicin-sensitive). Other NANC responses that have been described in the airways include vascular permeability, secretion of mucus, and the regulation of airway blood flow. While it has proved difficult to demonstrate e-NANC neural responses in human airways, these mechanisms may become more evident in diseases such as asthma where there may be less degradation of released tachykinins (Barnes et al., 1991). In a similar manner as described earlier, the release of tachykinins from sensory nerves in the airways can be modulated by other factors, including a variety of neuropeptides, acting at prejunctional receptors (Fig. 5).

A. Opioids

Opioid receptors appear to be present on capsaicin-sensitive sensory nerves (Laduron, 1984), which contain tachykinins including neurokinin A (NKA) and SP. Opioid agonists inhibit stimulus-evoked release of SP from the rat trigeminal nucleus in vitro (Jessell and Iversen, 1977). Since then, many studies have concentrated on the peripheral modulatory actions of opioids on responses evoked by e-NANC nerve stimulation and the release of tachykinins.

1. Bronchoconstriction

NANC bronchoconstriction evoked by EFS in vitro and vagal stimulation in vivo is due to the release of tachykinins from sensory nerves in the airways (Grundstrom et al., 1981; Lundberg et al., 1983). Opioids inhibit e-NANC constrictor responses in guinea pig bronchi in vitro in a concentration-dependent manner (Frossard and Barnes, 1987; Bartho et al., 1987; Belvisi et al., 1990; Kamikawa and Shimo, 1990). DAMGO, a selective μ-opioid receptor agonist, inhibited the e-NANC response with a maximum inhibition of approximately 60%, an effect that was antagonized by the opioid receptor-selective antagonist naloxone. A selective δ-opioid receptor agonist [D-Pen^{2,5}]enkephalin (DPDPE) inhibited the e-NANC

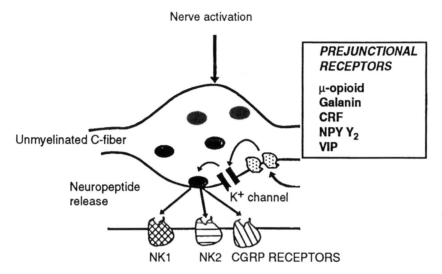

Figure 5 Several prejunctional receptors appear to modulate the release of peptides from airway sensory nerves; opioids and neuropeptide Y (NPY) may inhibit release by opening a common K^+ channel. It is not known how corticotropin-releasing factor (CRP), vasoactive intestinal peptide (VIP), and galanin inhibit NANC neurotransmission prejunctionally. (Adapted from Barnes et al., 1990.)

response but was less potent than DAMGO, and the magnitude of the inhibition achieved was reduced (Belvisi et al., 1990). In addition, a selective k-opioid agonist was ineffective. DAMGO had no direct effect on smooth muscle tone and did not affect SP-induced contractions. Therefore, its effects are likely to be mediated via μ- and possibly δ-opioid receptors on sensory nerves that can modulate the release of NKA and/or SP. To firmly establish which opioid receptor subtypes are involved in the prejunctional modulation of NANC bronchoconstriction in guinea pig airways, it would be necessary to repeat these studies in the presence of selective opioid receptor antagonists that have been developed in the last few years. The same pattern of events is seen in vivo where selective μ- and δ-opioid receptor agonists inhibit vagally induced NANC bronchoconstriction in the guinea pig (Belvisi et al., 1988). However, in contrast to the in vitro data where DAMGO produced 60% inhibition of the e-NANC response in vivo, there was complete inhibition of the NANC vagally induced bronchoconstriction.

2. Vascular Permeability

Increased permeability of the microvasculature to plasma proteins is characteristic of the inflammatory response and may be involved in diseases such as asthma. Opioids have been shown to inhibit neurogenic plasma exudation in skin by a prejunctional mechanism of action, so experiments were performed to investigate whether a similar phenomenon occurs in the airways.

One method that can be employed when measuring plasma extravasation in the airways in animal models is to use Evans blue dye as a marker of plasma leakage. Stimulation of the vagus nerves in anesthetized guinea pigs markedly increases the leakage of dye in trachea and main bronchi by between 300 and 600% (Belvisi et al., 1989). Leakage is not reduced by ganglionic blockade and is not atropine-sensitive but is abolished by pretreatment with capsaicin (which depletes sensory nerves of tachykinins) or an NK-1 antagonist, so this suggests that SP release mediates this neurogenic response (Barnes et al., 1991). Morphine inhibited this neurogenic plasma extravasation in the airways in a dose-dependent manner (Belvisi et al., 1989). The inhibition was reversed by the opioid receptor antagonist naloxone, which suggests that the effect is mediated by opioid receptors. Morphine had no effect on SP-induced leakage, suggesting that morphine acts pre-junctionally, on opioid receptors located on sensory nerves, to inhibit the release of SP.

3. Mucus Secretion

Opioids also inhibit mucus secretion stimulated by capsaicin in human bronchi, in Ussing chambers, in vitro (Rogers and Barnes, 1989). Cigarette smoke can stimulate mucus discharge from airway goblet cells in the guinea pig, which may be inhibited by opioids (Kuo et al., 1992). The NANC component of the cigarette-smoke-induced goblet cell secretion is inhibited by DAMGO, indicating the involvement of μ-opioid receptors in the inhibition of release of tachykinins from sensory nerves.

4. Cough

When investigating whether an agent has any effect on cough, it is difficult to demonstrate whether, if it is effective, it is acting centrally or peripherally. However, it is likely that opioids may have an inhibitory effect on sensory nerve activation as well as on the release of tachykinins from sensory nerves. Thus, opioids may inhibit sensory nerve activation and thereby suppress cough in conscious guinea pigs (Adcock et al., 1988). Interestingly, the peripherally acting μ-opioid agonist BW 443C is effective in this respect, indicating that opioids may well act peripherally to inhibit sensory nerve activation.

B. Vasoactive Intestinal Peptide

1. Exogenously Added VIP

VIP inhibits e-NANC contractile responses in guinea pig bronchi in a concentration-dependent manner, but at higher concentrations (<0.1 μM) it also inhibits contractile responses to SP. These results suggest that at low concentrations exogenous VIP inhibits the release of tachykinins from airway sensory nerves

(Stretton et al., 1991). Thus, any increase in VIP degradation may therefore enhance neurogenic inflammation in the airways.

2. Endogenously Released VIP

Interestingly, α-chymotrypsin, which degrades VIP, has no effect on e-NANC constrictor responses evoked by EFS in guinea pig bronchi (Belvisi et al., 1993b), although exogenous VIP inhibits e-NANC responses in guinea pig bronchi at concentrations that have no effect on contractile responses to exogenous SP (Stretton et al., 1991). One explanation for the apparent difference between the modulatory effects of exogenous and endogenous VIP on e-NANC responses could be related to the innervation of the tracheobronchial tree of the guinea pig. Guinea pig tracheal but not bronchial smooth muscle exhibits i-NANC relaxation responses to EFS (Grundstrom et al., 1981), so α-chymotrypsin may have no effect on e-NANC neural responses in guinea pig bronchi as VIP may not be released in these airways. Alternatively, α-chymotrypsin potentiated NANC bronchoconstriction induced by both capsaicin and vagal stimulation in the guinea pig in vivo (Lei et al., 1993). These results suggest that, at least in vivo where one is probably stimulating the trachea and bronchi simultaneously, endogenous VIP is involved in the regulation of neurogenic bronchoconstriction.

C. Neuropeptide Tyrosine

NPY also modulates e-NANC constrictor responses in guinea pig airways in vitro and in vivo (Stretton et al., 1990; Giuliani et al., 1989a; Matran et al., 1989; Grundemar et al., 1990). This inhibitory action of NPY is unaffected by α-adrenoceptor blockade, and NPY had no effect on bronchoconstrictor responses evoked by exogenous SP. These results suggest that the inhibitory action of NPY is likely to be mediated directly on its prejunctional receptor on sensory nerves. The NPY receptor mediating this effect appears to be of the Y_2 subtype because the COOH-terminal fragment NPY-(13-36) has a similar inhibitory potency but the full sequence peptide produced a more complete inhibition (Grundemar et al., 1990). Finally, the modulatory effect of NPY appears to be greater in main bronchi than in hilar bronchi, and this may relate to the distribution of adrenergic nerves (Stretton et al., 1990). In summary, adrenergic nerves may therefore modulate NANC bronchoconstriction in vivo and in vitro via NPY receptors in addition to α_2-adrenoceptors as previously described (Grundstrom et al., 1985).

D. Other Neuropeptides

1. Galanin

It has been suggested that galanin may be stored in capsaicin-sensitive sensory nerves (Ju et al., 1987). In addition, galanin has an inhibitory action on e-NANC bronchoconstrictor responses in guinea pig bronchi in vitro (Giuliani et al.,

1989b). Therefore, galanin may have a role in autoregulation of the release of tachykinins from airway sensory nerves.

2. Somatostatin

Somatostatin has no effect on e-NANC neurotransmission in the airways (Stretton and Barnes, unpublished data) but does have an inhibitory effect on neurogenic microvascular leakage (another e-NANC phenomenon) in the rat footpad (Lembeck et al., 1982).

3. Corticotropin-Releasing Factor

Corticotropin-releasing factor inhibits neurogenic, NANC, microvascular leakage induced in the rat trachea (Wei et al., 1987).

IV. MECHANISMS OF PREJUNCTIONAL NEUROMODULATION

A. Mechanisms of Facilitation

As mentioned earlier, it has been demonstrated that tachykinins may facilitate cholinergic neurotransmission in the rabbit and guinea pig bronchi via activation of NK-1 receptors (and also NK-2 receptors in the rabbit). Tachykinins mediate contraction of airway smooth muscle by stimulating phosphoinositide hydrolysis and increasing the formation of inositol (1,4,5) triphosphate, which releases calcium ions from their intracellular stores in airway smooth muscle, and diglyceride, which activates protein kinase C (PKC) (Grandordy et al., 1988). However, the cell-signaling mechanisms involved in neuromodulation are unknown. A possible mechanism is that activation of receptors on cholinergic nerves could lead to phosphoinositide hydrolysis, with subsequent Ca^{2+} mobilization and activation of PKC. PKC stimulation with phorbol diesters may have effects on neurotransmission via phosphorylation of a specific 87-kDa protein (a synapsin) that is prominent in nerve terminals (Wang et al., 1989). However, the role of this protein in the modulation of neurotransmission is unknown.

B. Mechanisms of Inhibition

The presynaptic inhibitory modulation of neurotransmitter release is thought to involve receptor-mediated regulation of ion channels in the nerve endings (Miller, 1990). In the central nervous system, some agonists inhibit neuronal function by opening a common potassium (K^+) channel (North et al., 1987). The presence of several types of K^+ channels on nerve cells has been reported, including voltage-sensitive K^+ channels (K_v^+) (Mackinnon and Yellen, 1990), large-conductance calcium (Ca^{2+})-activated K^+ channels (K_{Ca}^+) (Reinhart et al., 1989), small-conductance K_{Ca}^+ channels (Blatz and Magleby, 1986), and ATP-sensitive K^+

channels (K_{ATP}^+) (Schmid-Antomarchi et al., 1990). Charybdotoxin (ChTX), a scorpion venom toxin, selectively blocks large-conductance K_{Ca}^+ (Giminez-Gallego et al., 1988). ChTX reverses and reduces the prejunctional modulation of cholinergic responses by NPY and a μ-opioid in guinea pig airways, and ChTX also attenuates a μ-opioid-induced inhibitory modulation of cholinergic responses in human airways (Miura et al., 1992). Since ChTX does not block any other ion channels than K^+ channels, these results support a hypothesis that ChTX-sensitive K^+ channels, presumably K_{Ca}^+, are involved in the prejunctional inhibition of airway cholinergic neurotransmission by agonist acting on presynaptic receptors in guinea pig and human airways. A similar mechanism appears to operate in NANC nerves. ChTX completely blocks the modulatory action of NPY and opioids on e-NANC responses in guinea pig bronchi. On the other hand, apamin, which inhibits a small-conductance K_{Ca}^+, is without effect (Stretton et al., 1992b).

Recently, it has been suggested that ChTX may not be a selective K^+ channel blocker in nerve cells and that it may block different types of K^+ channels, e.g., Ca^{2+}-insensitive voltage-dependent K^+ channels (K_v^+) (Schweitz et al., 1989). Iberiotoxin (IbTX), a homologous polypeptide to ChTX, purified from the venom of the scorpion *Buthus tamulus*, has been shown to be highly selective for K_{Ca}^+ and without effect on ChTX-sensitive K_v^+ (Galves et al., 1990). Another toxin, noxiustoxin (NxTX), has now been recognized as a potent blocker for ChTX-sensitive K_v^+ in human T lymphocytes (Leonard et al., 1992). Recent data suggest that IbTX can reverse the μ-opioid-induced inhibition of NANC responses in guinea pig bronchi whereas NxTX was without effect, suggesting that large-conductance K_{Ca}^+ are involved in the μ-opioid-induced prejunctional modulation of e-NANC neurotransmission (Miura et al., 1993). Interestingly, in the presence of IbTX, NANC constrictor responses were significantly potentiated by a lower (10 nM) concentration of the μ-opioid receptor agonist DAMGO (this concentration of DAMGO was inhibitory in the absence of the toxin).

These results suggest that large-conductance K_{Ca}^+, but not ChTX-sensitive K_v^+, play a role in the μ-opioid-induced inhibition of NANC neural responses and that the blockade of these channels reveals an excitatory effect of opioids on airway sensory nerves. Therefore, if large-conductance K_{Ca}^+ are impaired, for example in disease, this could induce a potentiation of neurogenic inflammation, since opioids may be endogenous inhibitory modulators of sensory nerves.

V. CONCLUSIONS

Many examples of modulation of neurotransmission in the airways have been discussed in this chapter, but there is still no information about whether these neuromodulatory mechanisms are invoked in pathophysiological conditions. One area that has been highlighted here is the differences in neuromodulation between species, and many experiments are still needed in human airways. Another factor is that the innervation of the airway changes dramatically depending on the airway level investigated, and this is especially true for VIP.

In the years to come, the development of different techniques and investigational tools will enable us to elucidate the complex neural networks that control airway function. For example, the use of cultured neurons and electrophysiological techniques together with the development of potent and selective receptor agonists and antagonists for neuropeptides may lead to a greater understanding of neuromodulatory processes in the airways.

REFERENCES

Adcock, J.J., Schneider, C., and Smith, T.W. (1988). Effects of codeine, morphine and a novel opioid pentapeptide BW 443C on cough, nociception and ventilation in the unanesthetised guinea-pig. *Br. J. Pharmacol. 93*:93–100.

Aizawa, H., Miyazaki, N., Inoue, H., Ikeda, T., and Shigematsu, N. (1990). Effect of endogenous tachykinins on neuro-effector transmission of vagal nerve in guinea-pig tracheal tissue. *Respiration 57*:338–342.

Armour, C.L., Johnson, P.R.A., and Black, J.L. (1991). Nedocromil sodium inhibits substance P–induced potentiation of cholinergic neural responses in the isolated innervated rabbit trachea. *J. Auton. Pharmacol. 11*:167–172.

Barnes, P.J. (1987). Cholinergic control of airway smooth muscle. *Am. Rev. Respir. Dis. 136*:S42–S45.

Barnes, P.J., Belvisi, M.G., and Rogers, D.F. (1990). Modulation of neurogenic inflammation: novel approaches to inflammatory disease. *Trends Pharmacol. Sci. 11*:185–189.

Barnes, P.J., Baraniuk, J.N., and Belvisi, M.G. (1991). Neuropeptides in the respiratory tract. *Am. Rev. Respir. Dis. 144*:1187–1198 (part 1); 1391–1399 (part 2).

Bartho, L., Amann, R., Saria, A., Szolcsanyi, J., and Lembeck, F. (1987). Peripheral effects of opioid drugs in capsaicin-sensitive neurones of the guinea-pig bronchus and rabbit ear. *Naunyn-Schmiedeberg's Arch. Pharmacol. 336*:316–320.

Belvisi, M.G., Chung, K.F., Jackson, D.M., and Barnes, P.J. (1988). Opioid modulation of non-cholinergic neural bronchoconstriction in guinea-pig in vivo. *Br. J. Pharmacol. 95*:413–418.

Belvisi, M.G., Rogers, D.F., and Barnes, P.J. (1989). Neurogenic plasma extravasation: inhibition by morphine in guinea-pig airways in vivo. *J. Appl. Physiol. 66*:268–272.

Belvisi, M.G., Stretton, C.D., and Barnes, P.J. (1990). Modulation of cholinergic neurotransmission in guinea-pig airways by opioids. *Br. J. Pharmacol. 100*:131–137.

Belvisi, M.G., Stretton, C.D., and Barnes, P.J. (1991). Nitric oxide as an endogenous modulator of cholinergic neurotransmission. *Eur. J. Pharmacol. 198*:219–221.

Belvisi, M.G., Stretton, C.D., Yacoub, M. H., and Barnes, P. J. (1992a). Nitric oxide is the endogenous neurotransmitter of bronchodilator nerves in humans. *Eur. J. Pharmacol. 210*:221–222.

Belvisi, M.G., Stretton, C.D., Miura, M., Verleden, G.M., Tadjkarimi, S., Yacoub, M.H., and Barnes, P.J. (1992b). Inhibitory NANC nerves in human tracheal smooth muscle: a quest for the neurotransmitter. *J. Appl. Physiol. 73*:2505–2510.

Belvisi, M.G., Stretton, C.D., Verleden, G.M., Ledingham, S.J.L., Yacoub, M.H., and Barnes, P.J. (1992c). Inhibition of cholinergic neurotransmission in human airways by opioids. *J. Appl. Physiol. 72*:1096–1100.

Belvisi, M.G., Ward, J.K., Patel, H.J., Tadjkarimi, S., Yacoub, M.H., and Barnes, P.J. (1993a). μ-Opioids inhibit electrically evoked acetylcholine release in human and guinea-pig trachea. *Am. Rev. Respir. Dis. 147*:A502.

Belvisi, M.G., Miura, M., Stretton, C.D., and Barnes, P.J. (1993b). Endogenous vasoactive intestinal peptide and nitric oxide modulate cholinergic neurotransmission in guinea-pig trachea. *Eur. J. Pharmacol. 231*:97–102.

Belvisi, M.G., Patacchini, R., Barnes, P.J., and Maggi, C.A. (1994). Facilitatory effects of selective agonists for tachykinin receptors on cholinergic neurotransmission: evidence for species differences. *Br. J. Pharmacol. 111*:103–110.

Black, J.L., Johnson, P.R.A., Alouan, L., and Armour, C.L. (1990). Neurokinin A with K^+ channel blockade potentiates contraction to electrical stimulation in human bronchus. *Eur. J. Pharmacol. 180*:311–317.

Blatz, A.L., and Magleby, K.L. (1986). Single apamin-blocked Ca^{++}-activated K^+ channel of small conductance in cultured rat skeletal muscle. *Nature 323*:718–720.

Brave, S.R., Hobbs, A.J., Gibson, A., and Tucker, J.F. (1991). The influence of L-N^G-nitroarginine on field stimulation induced contractions and acetylcholine release in guinea-pig isolated tracheal smooth muscle. *Biochem. Biophys. Res. Commun. 179*: 1017–1022.

Caughey, G.H. (1989). Roles of mast cell tryptase and chymase in airway function. *Am. J. Physiol. 257 (Lung Cell. Mol. Physiol. 1)*:L39–L46.

Chesrown, S.E., Venugoplan, C.S., Gold, W.M., and Drazen, J.M. (1980). In vivo demonstration of non-adrenergic inhibitory innervation of the guinea-pig trachea. *J. Clin. Invest. 65*:314–320.

Cheung, A., Polak, J.M., Bauer, F.E., Christofides, N.D., Cadieux, A., Springall, D.R., and Bloom, S.R. (1985). The distribution of galanin immunoreactivity in the respiratory tract of pig, guinea-pig, rat, and dog. *Thorax 40*:889–896.

Coburn, R.F., and Tomita, T. (1973). Evidence for non-adrenergic inhibitory nerves in the guinea-pig trachealis muscle. *Am. J. Physiol. 224*:1072–1080.

Coleman, R.A., and Levy, G.P. (1974). A non-adrenergic inhibitory nervous pathway in guinea-pig trachea. *Br. J. Pharmacol. 52*:167–174.

Colmers, W.F., Klapstein, G.J., Fournier, A., St. Pierre, S., and Treherne, K.A. (1991). Presynaptic inhibition by neuropeptide Y in rat hippocampal slices in vitro is mediated by a Y_2-receptor. *Br. J. Pharmacol. 102*:41–44.

Dey, R.D., Shannon, W.A., and Said, S.I. (1981). Localisation of VIP-immunoreactive nerves in airways and pulmonary vessels of dogs, cats and human subjects. *Cell Tissue Res. 220*:231–238.

Dey, R.D., Altemus, J.B., and Michalkiewicz, M. (1991). Distribution of vasoactive intestinal peptide and substance P-containing nerves originating from neurons of airway ganglia in cat bronchi. *J. Comp. Neurol. 304*:330–340.

Ellis, J.L., and Farmer, S.G. (1989a). Modulation of cholinergic neurotransmission by vasoactive intestinal peptide and histidine isoleucine in guinea-pig tracheal smooth muscle. *Pulmon. Pharmacol. 2*:107–112.

Ellis, J.L., and Farmer, S.G. (1989b). The effect of VIP antagonists, and VIP and PHI antisera on non-adrenergic, non-cholinergic relaxations of tracheal smooth muscle. *Br. J. Pharmacol. 96*:513–520.

Ellis, J.L., and Farmer, S.G. (1989c). Effect of peptidases on non-adrenergic non-cholinergic inhibitory responses of tracheal smooth muscle: a comparison with effects on VIP and PHI-induced relaxation. *Br. J. Pharmacol. 96*:521–526.

Erdös, E.G., and Skidgel, R.A. (1989). Neutral endopeptidase 24.11 (enkephalinase) and regulatory peptide hormones. *FASEB J. 3*:145–151.

Frossard, N., and Barnes, P.J. (1987). μ-Opioid receptors modulate non-cholinergic constrictor nerves in guinea-pig airways. *Eur. J. Pharmacol. 141*:519–521.

Galves, A., Gimenez-Gallego, G., Reuben, J.P., Roy-Constantin, L., Feigenbaum, P., Kaczorowski, G.M., and Garcia, M.L. (1990). Purification and characterisation of a unique, potent, peptidyl probe for the high conductance calcium-activated potassium channel from venom of the scorpion *Buthus tamulus*. *J. Biol. Chem. 265*:11083–11090.

Giminez-Gallego, G., Navia, M.A., Reuben, J.P., Katz, G.M., Kaczorowski, G.M., and Garcia, M.L. (1988). Purification, sequence and model structure of charybdotoxin, a potent selective inhibitor of calcium-activated potassium channels. *Proc. Natl. Acad. Sci. USA 85*:3329–3333.

Giuliani, S., Maggi, C.A., and Meli, A. (1989a). Prejunctional modulatory action of neuropeptide Y on peripheral terminals of capsaicin-sensitive sensory nerves. *Br. J. Pharmacol. 98*:407–412.

Giuliani, S., Amann, R., Papini, M., Maggi, C.A., and Meli, A. (1989b). Modulatory action of galanin on responses due to antidromic activation of peripheral terminals of capsaicin-sensitive sensory nerves. *Eur. J. Pharmacol. 163*:91–96.

Grandordy, B.M., Frossard, N., Rohden, K.J., and Barnes, P.J. (1988). Tachykinin-induced phosphoinositide breakdown in airway smooth muscle and epithelium: relationship to contraction. *Mol. Pharmacol. 33*:515–519.

Grundemar, L., Widmark, E., Waldeck, B., and Hakanson, R. (1988). Neuropeptide Y: prejunctional inhibition of vagally induced contraction in the guinea-pig trachea. *Regul. Pept. 23*:309–314.

Grundemar, L., Grundstrom, N., Johansson, I.G.M., Andersson, R.G.G., and Hakanson, R. (1990). Supression by neuropeptide Y of capsaicin-sensitive sensory nerve mediated contraction in guinea-pig airways. *Br. J. Pharmacol. 99*:473–476.

Grundstrom, N., and Andersson, R.G.G. (1985). In vivo demonstration of α_2-adrenoceptor mediated inhibition of the excitatory noncholinergic neurotransmission in guinea-pig airways. *Naunyn-Schmiedeberg's Arch. Pharmacol. 328*:236–240.

Grundstrom, N., Andersson, R.G.G., and Wikberg, J.E.S. (1981). Pharmacological characterisation of the autonomous innervation of guinea-pig tracheobronchial smooth muscle. *Acta Pharmacol. Scand. 49*:150–157.

Hall, A.K., Barnes, P.J., Meldrum, L.A., and Maclagan, J. (1989). Facilitation by tachykinins of neurotransmission in guinea-pig pulmonary parasympathetic nerves. *Br. J. Pharmacol. 97*:274–280.

Ito, Y., and Hakoda, H. (1990). Modulation of cholinergic neurotransmission by VIP, VIP-antiserum and VIP antagonists in dog and cat trachea: VIP plays a role of "double braking" in bronchoconstriction. *Agents Actions 31*(Suppl):197–203.

Jacoby, D.B., Tamaoki, J., Borson, D.B., and Nadel, J.A. (1988). Influenza infection causes hyper-responsiveness by decreasing enkephalinase. *J. Appl. Physiol. 64*:2653–2658.

Jessell, T.M., and Iversen, L.L. (1977). Opiate analgesics inhibit substance P release from trigeminal nucleus. *Nature 268*:549–551.

Ju, G., Hokfelt, T., Brodin, E., Fahrenkrug, J., Fischer, J.A., Frey, P., Elde, R. P., and Brown, J.C. (1987). Primary sensory neurons of the rat showing calcitonin gene-related peptide immunoreactivity and their relation to substance P-, somatostatin, galanin-. vasoactive intestinal polypeptide- and cholecystokinin-immunoreactive ganglion cells. *Cell Tissue Res. 247*:417–431.

Kamikawa, Y., and Shimo, Y. (1990). Morphine and opioid peptides selectively inhibit the non-cholinergically mediated neurogenic contraction of guinea-pig isolated bronchial muscle. *J. Pharmacol. 42*:214–216.

Kosterlitz, H.W., and Taylor, D.W. (1959). The effect of morphine on vagal inhibition of the heart. *Br. J. Pharmacol. Chemother. 14*:209–214.

Kuo, H-P., Rohde, J.A.L., Barnes, P.J., and Rogers, D.F. (1992). Opioid inhibition of neurogenic goblet cell secretion: differential effects on cigarette smoke, capsaicin and electrically induced responses in guinea-pig trachea in vitro. *Br. J. Pharmacol. 105*: 361–366.

Laduron, P.M. (1984). Axonal transport of opiate receptors in capsaicin-sensitive neurones. *Brain Res. 294*:157–160.

Laitinen, A., Partanen, M., Hervonen, A., Pelto-Huikko, M., and Laitinen, L.A. (1985). VIP like immunoreactive nerves in human respiratory tract. *Histochemistry 82*: 313–319.

Lei, Y-H., Barnes, P.J., and Rogers, D.F. (1993). Regulation of NANC neural broncho-constriction in vivo in the guinea-pig: involvement of nitric oxide, vasoactive intestinal peptide and soluble guanylyl cyclase. *Br. J. Pharmacol. 108*:228–235.

Lembeck, F., Donnerer, J., and Bartho, L. (1982). Inhibition of neurogenic vasodilation and plasma extravasation by substance P antagonists, somatostatin and [D-Met^2, Pro^5]-enkephalinamide. *Eur. J. Pharmacol. 85*:171–176.

Leonard, R.J., Garcia, M.L., Slaughter, R.S., and Reuben, J.P. (1992). Selective blockers of voltage-gated K$^+$ channels depolarise human T lymphocytes: mechanism of the anti-proliferative effect of charybdotoxin. *Proc. Natl. Acad. Sci. USA 89*:10094–10098.

Li, C.G., and Rand, M.J. (1991). Evidence that part of the NANC relaxant response of guinea-pig trachea to electrical field stimulation is mediated by nitric oxide. *Br. J. Pharmacol. 102*:91–94.

Lundberg, J.M., Hedlung, B., and Bartai, T. (1982). Vasoactive intestinal peptide enhances muscarinic ligand binding in cat submandibular salivary gland. *Nature 295*:147–149.

Lundberg, J.M., Saria, A., Brodin, E., Rosell, S., and Folkers, K. (1983). A substance P antagonist inhibits vagally-induced increase in vascular permeability and bronchial smooth muscle contraction in the guinea-pig. *Proc. Natl. Acad. Sci. USA 80*:1120–1124.

Luts, A., and Sundler, F. (1989). Peptide containing nerve fibres in the respiratory tract of the ferret. *Cell Tissue Res. 258*:259–267.

Mackinnon, R., and Yellen, G. (1990). Mutation affecting TEA blockade and ion permea-tion in voltage-activated K$^+$ channels. *Science 250*:276–279.

Maggi, C.A., Patacchini, R., Guiliani, S., Rovero, P., Dion, S., Regoli, D., Giachetti, A., and Meli, A. (1990). Competitive antagonists discriminate between NK$_2$ receptor subtypes. *Br. J. Pharmacol. 100*:588–592.

Martin, J.G., Wang, A., Zacour, M., and Biggs, D.F. (1990). The effects of vasoactive intestinal polypeptide on cholinergic neurotransmission in an isolated innervated guinea-pig tracheal preparation. *Respir. Physiol. 79*:111–122.

Matran, R., Martling, C-R., and Lundberg, J.M. (1989). Inhibition of cholinergic and nonadrenergic, non-cholinergic bronchoconstriction in the guinea-pig mediated by neuropeptide Y and α_2-adrenoceptors and opiate receptors. *Eur. J. Pharmacol. 163*: 15–23.

Miller, R.J. (1990). Receptor-mediated regulation of calcium channels and neurotransmitter release. *FASEB J. 4*:3291–3299.

Miura, M., Belvisi, M.G., Stretton, C.D., Yacoub, M.H., and Barnes, P.J. (1992). Role of K$^+$ channels in the modulation of cholinergic neural responses in guinea-pig and human airways. *J. Physiol. 445*:1–15.

Miura, M., Belvisi, M.G., Ward, J.K., and Barnes, P.J. (1993). Role of Ca^{2+}-activated K$^+$ channels in opioid-induced pre-junctional modulation of airway sensory nerves. *Am. Rev. Respir. Dis. 147*:A815.

Mulder, A.H., Wardeh, G., Hogenboom, F., and Frank-Huyzen, A.L. (1984). Kappa and delta opioid receptor agonists differentially inhibit striatal dopamine and acetylcholine release. *Nature 308*:278–280.

North, R.A., Williams, J.T., Surprenant, A., and Christie, M.J. (1987). μ and δ receptors belong to a family of receptors that are coupled to potassium channels. *Proc. Natl. Acad. Sci. USA 84*:5487–5491.

Paton, W.D.M. (1957). The action of morphine and related substances on contraction and on acetylcholine output of coaxially stimulated guinea-pig ileum. *Br. J. Pharmacol. Chemother. 12*:119–127.

Potter, E.K. (1988). Neuropeptide Y as an autonomic neurotransmitter. *Pharmacol. Ther. 37*:251–273.

Reinhart, P.H., Chung, S., and Levitan, I.B. (1989). A family of calcium-dependent potassium channels from rat brain. *Neuron 2*:1031–1041.

Rhoden, K.J., Meldrum, L.A., and Barnes, P.J. (1988). Inhibition of cholinergic neurotransmission in human airways by β_2-adrenoceptors. *J. Appl. Physiol. 65*:700–705.

Rogers, D.F., and Barnes, P.J. (1989). Opioid inhibition of neurally mediated mucus secretion in human bronchi. *Lancet 1*:930–932.

Russell, J.A., and Simons, E.J., Modulation of cholinergic neurotransmission in airways by enkephalin. *J. Appl. Physiol. 58*:853–858.

Said, S.I., and Mutt, V. (1969). Long acting vasodilator peptide from lung tissue. *Nature 224*:699–700.

Schmid-Antomarchi, H., Amoroso, S., Fosset, M., & Lazdunski, M. (1990). K$^+$ channel openersactivate sulfonylurea-sensitive K$^+$ channels and block neurosecretion. *Proc. Natl. Acad. Sci. USA 87*:3489–3492.

Schweitz, H., Stansfield, C.E., Bidard, J.-N., Fagni, L., Meas, P., and Lazdunski, M. (1989). Charybdotoxin blocks dendrotoxin-sensitive voltage-activated K$^+$ channels. *FEBS Lett. 250*:519–522.

Sekizawa, K., Tamaoki, J., Nadel, J.A., and Borson, D.B. (1987). Enkephalinase inhibitor potentiates substance P and electrically induced contraction in ferret trachea. *J. Appl. Physiol. 63*:1401–1405.

Sekizawa, K., Tamaoki, J., Graf, P., and Nadel, J.A. (1988). Modulation of cholinergic transmission by vasoactive intestinal peptide in ferret trachea. *J. Appl. Physiol. 69*:2433–2437.

Sekizawa, K., Graf, P.D., and Nadel, J.A. (1989). Somatostatin potentiates cholinergic neurotransmission in ferret trachea. *J. Appl. Physiol. 67*:2397–2400.

Sheppard, M.N., Polak, J.M., Allen, J.M., and Bloom, S.R. (1984). Neuropeptide tyrosine (NPY): a newly discovered peptide is present in the mammalian respiratory tract. *Thorax 39*:326–330.

Shimosegawa, T., Foda, H.D., and Said, S.I. (1990). [Met]enkephain-Arg^6-Gly^7-Leu^8-immunoreactive nerves in guinea-pig and rat lungs: distribution, origin, and coexistence with vasoactive intestinal polypeptide immunoreactivity. *Neuroscience 36*:737–750.

Stretton, C.D., and Barnes, P.J. (1988). Modulation of cholinergic neurotransmission in guinea-pig trachea by neuropeptide Y. *Br. J. Pharmacol. 93*:672–678.

Stretton, C.D., and Barnes, P.J. (1989). Cholecysokinin-octapeptide constricts guinea-pig and human airways. *Br. J. Pharmacol. 97*:675–682.

Stretton, C.D., Belvisi, M.G., and Barnes, P.J. (1990). Neuropeptide Y modulates non-adrenergic, non-cholinergic neural bronchoconstriction in vivo and in vitro. *Neuropeptides 17*:163–170.

Stretton, C.D., Belvisi, M.G., and Barnes, P.J. (1991). Modulation of neural broncho-constrictor responses in the guinea-pig respiratory tract by vasoactive intestinal peptide. *Neuropeptides 18*:149–157.

Stretton, C.D., Belvisi, M.G., and Barnes, P.J. (1992a). The effect of sensory nerve depletion on cholinergic neurotransmission in guinea-pig airways. *J. Pharmacol. Exp. Ther. 260*:1073–1080.

Stretton, C.D., Miura, M., Belvisi, M.G., and Barnes, P.J. (1992b). Calcium-activated potassium channels mediate pre-junctional inhibition of peripheral sensory nerves. *Proc. Natl. Acad. Sci. USA 89*:1325–1329.

Tanaka, D.T., and Grundstein, N.M. (1986). Effect of substance P on neurally-mediated contraction of rabbit airway smooth muscle. *J. Appl. Physiol. 60*:458–463.

Taylor, J.F., Pare, P.D., and Schellenburg, R.R. (1984). Cholinergic and non-adrenergic mechanisms in human and guinea-pig airways. *J. Appl. Physiol. 56*:958–965.

Tucker, J.F., Brave, S.R., Charalambous, L., Hobbs, A.J., and Gibson, A. (1990). L-N^G-nitroarginine inhibits non-adrenergic, non-cholinergic relaxations of guinea-pig isolated tracheal smooth muscle. *Br. J. Pharmacol. 100*:663–664.

Uddman, R., Alumets, J., Densert, O., Hakansson, R., and Sundler, F. (1984). Occurrence and distribution of VIP nerves in the nasal mucosa and tracheobronchial wall. *Acta. Otolaryngol. 86*:443–448.

Wahlestedt, D., Yanaihara, N., and Hakanson, R. (1986). Evidence for pre- and post-junctional receptors for neuropeptide Y and related peptide. *Regul. Pept. 13*: 307–318.

Wang, J.K.T., Walnas, S.T., Sihra, T.S., Aderem, A., and Greengard, P. (1989). Phospho-rylation and associated translocation of the 87KDa protein, a major protein kinase C substrate in isolated nerve terminals. *Proc. Natl. Acad. Sci. USA 86*:2253–2256.

Watson, N., Maclagan, J., and Barnes, P.J. (1993a). Endogenous tachykinins facilitate transmission through parasympathetic ganglia in guinea-pig trachea. *Br. J. Pharmacol. 109*:751–759.

Watson, N., Maclagan, J., and Barnes, P.J. (1993b). Vagal control of guinea-pig tracheal smooth muscle: lack of involvement of VIP or nitric oxide. *J. Appl. Physiol. 74*:1964–1971.

Wei, E.T., and Kiang, J.C. (1987). Inhibition of protein exudation from the trachea by corticotropin releasing factor. *Eur. J. Pharmacol. 140*:63–67.

Yau, W.M., and Youther, M.L. (1982). Direct evidence for a release of acetylcholine from the myenteric plexus of guinea-pig small intestine by substance P. *Eur. J. Pharmacol. 81*: 665–668.

DISCUSSION

Bienenstock: In your VIP studies, do different depolarization frequencies stimulate different populations of neurons?

Belvisi: We cannot be sure if different populations of neurons are affected by VIP. The inhibitory effect of VIP is easier to demonstrate with responses evoked by lower-stimulation frequencies as there is less signal to combat.

Coburn: The frequency of stimulation may be of importance since the neurotransmitters released may be regulated by the frequency of depolarization.

Baraniuk: I wonder if the frequency of depolarization could affect the ability of autoreceptors to regulate preganglionic or postganglionic neurons.

Lundberg: This could be tested with charybdotoxin. Have you observed any bronchoconstricting effects of charybdotoxin that may indicate that a postjunctional facilitation can counteract the prejunctional inhibitory effects of autoreceptor (e.g., NPY) stimulation on cholinergic transmission?

Belvisi: Yes, higher concentrations of charybdotoxin increase bronchial tone.

Kummer: Do you have evidence for an intracellular link between neuropeptide receptors and K^+ channels in cholinergic terminals?

Belvisi: It is unclear if the inhibitory NPY, opioid, and other autoreceptors act via cyclic nucleotides, such as cAMP and protein kinase A, or via G proteins that directly activate K^+ channels and bypass cyclic nucleotide mechanisms.

23

Neuropeptides and Asthma

Peter J. Barnes

National Heart and Lung Institute, London, England

I. INTRODUCTION

Many neuropeptides are localized to sensory, parasympathetic, and sympathetic neurons in the respiratory tract (Table 1) (Uddman et al., 1993; Barnes et al., 1991). These peptides have potent effects on bronchomotor tone, airway secretions, the bronchial circulation, and inflammatory and immune cells. Although the precise physiological roles of each peptide are not yet understood, some clues are provided by their localization and functional effects. Many of the inflammatory and functional effects of neuropeptides are relevant to asthma, and there is compelling evidence for the involvement of neuropeptides in the pathophysiology and symptomatology of asthma (Barnes, 1991a). The purpose of this chapter is to discuss effects of airway neuropeptides that are relevant to the pathophysiology of asthma and whether this might lead to new therapeutic approaches in the future.

A. Experimental Approaches

Several approaches have been used to investigate the role of neuropeptides in asthma. The effects of exogenous neuropeptides on various target cells relevant to asthma in vitro and their effects on airway function in vivo have been widely studied in animals and humans (Barnes et al., 1991). This approach is valuable in revealing the potential effects of a particular neuropeptide, but it is not possible to know exactly what the local concentration of a particular peptide might be. Furthermore, there are striking differences between species. Even data in normal human airways may not be relevant to the situation in the diseased airway, where there might be alterations in neuropeptide receptor expression and metabolic breakdown.

501

Table 1 Neuropeptides in the Respiratory Tract

Peptide	Localization
Vasoactive intestinal peptide Peptide histidine isoleucine/methionine Peptide histidine valine-42 Helodermin PACAP-27 Galanin	Parasympathetic (afferent)
Substance P Neurokinin A Neuropeptide K Calcitonin gene-related peptide Gastrin-releasing peptide	Afferent
Neuropeptide Y	Sympathetic
Somatostatin Enkephalin Cholecystokinin octapeptide	Afferent/uncertain

A more informative approach is to investigate the action of specific blockers or enhancers, or to study depletion of the relevant peptide since this can reveal the role of the endogenous neuropeptide. Again, it is possible that the disease state may alter the synthesis, release, or metabolism of a particular peptide or its receptors and therefore produce changes in the effects of blocking drugs. It is only recently that potent specific neuropeptide receptor blockers have become available, and these will prove to be important tools in the investigation of the role of neuropeptides in disease.

Several animal models of asthma have been investigated, but none of these closely mimic the chronic eosinophilic inflammation characteristic of asthma and they have been poorly predictive of drugs that will have clinical efficacy (Smith, 1989). The only certain way to evaluate the role of neuropeptides in asthma is to study the effect of specific antagonists or inhibitors in patients with the disease. Specific neuropeptide antagonists suitable for clinical use are now under development, and studies are already underway in asthma. Again, there may be pitfalls in this approach, as it is usual practice to select patients with mild asthma for such studies. It is possible that neuropeptides are relevant only in certain types of asthma or in more severe and intractable disease. Furthermore, it may be difficult to evaluate the effects of neuropeptides on airway function in clinical studies if their main action is on mucosal inflammation, mucus secretion, or airway blood flow, since techniques to evaluate these responses are difficult to use in patients.

B. Neuropeptide Interactions

In this chapter each neuropeptide is considered separately, but it is important to recognize that neuropeptides act as cotransmitters of classical autonomic nerves and that each peptide may have interactions with other nerves, resulting in complex effects on a tissue. An abnormality in one neuropeptide component may therefore have effects on the release and effects of other neuropeptides and autonomic neurotransmitters. For example, vasoactive intestinal peptide (VIP) (from parasympathetic nerves) and neuropeptide Y (NPY) (from sympathetic nerves) may have an inhibitory effect on the release of acetylcholine from parasympathetic nerves and neuropeptides from sensory nerves (Ellis and Farmer, 1989b; Martin et al., 1990; Stretton et al., 1990a, 1991; Hakoda and Ito, 1990; Stretton and Barnes, 1988), and many other such interactions have been reported (Barnes, 1992a). Depletion of neuropeptides from sensory nerves with capsaicin may markedly reduce cholinergic neurotransmission in guinea pigs (Martling et al., 1984; Stretton et al., 1992), but at the same time may enhance inhibitory nonadrenergic, noncholinergic (i-NANC) responses (Stretton et al., 1990b). Another possible interaction is that neuropeptides may affect the expression of autonomic receptors, either by influencing the intracellular pathways activated by the receptor or even by regulating the gene expression of receptors.

II. VIP AND RELATED PEPTIDES

VIP-immunoreactive nerves are widely distributed throughout the respiratory tract in humans (Uddman and Sundler, 1987), and there is also evidence for the presence of closely related peptides peptide histidine methionine (PHM) and pituitary adenylate cyclase activating peptide-27 (PACAP-27) (Lundberg et al., 1984; Uddman and Sundler, 1987). VIP may be localized to parasympathetic and sensory nerves (Dey et al., 1991).

A. Airway Effects

VIP is a potent relaxant of human bronchi in vitro but has little effect on peripheral airways (Palmer et al., 1986b). This is consistent with autoradiographic mapping studies, which show that VIP receptors are expressed in airway smooth muscle of proximal but not distal human airways (Carstairs and Barnes, 1986). This suggests that VIP, released from parasympathetic nerves in proximal airways, may act as an endogenous bronchodilator and may counteract cholinergic bronchoconstriction. VIP acts as a functional antagonist by increasing cyclic AMP concentrations in airway smooth muscle, but also inhibits the release of acetylcholine from airway cholinergic nerves at a ganglionic and postganglionic level via prejunctional receptors (Martin et al., 1990; Ellis and Farmer, 1989b; Stretton et al., 1991; Hakoda and Ito, 1990). Although VIP is a potent bronchodilator after intravenous administration in cats (Diamond and O'Donnell, 1980), it has no effect on airway function in normal human subjects, despite profound vascular effects that limit the dose that can be administered (Palmer et al., 1986). VIP is approximately 10-fold more potent as a vasodilator than a bronchodilator in vitro (Greenberg et al., 1987), and this is reflected by a higher density of VIP receptors in pulmonary vascular compared with airway smooth muscle (Carstairs and Barnes, 1986). In asthmatic patients, inhaled VIP has no bronchodilator effect, although a β-adrenergic agonist in the same subjects is markedly effective (Barnes and Dixon, 1984). Inhaled VIP has a small protective effect against the bronchoconstrictor effect of histamine (Barnes and Dixon, 1984) but has no effect against exercise-induced bronchoconstriction (Bungaard et al., 1983). This lack of potency of inhaled VIP may be explained by the epithelium since this possesses proteolytic enzymes and may present a barrier to diffusion.

It is likely that VIP is more important as a regulator of airway blood flow than airway smooth muscle tone. VIP increases airway blood flow in dogs and pigs and is more potent on tracheal than on bronchial vessels (Widdicombe, 1990; Matran et al., 1989a). There is convincing evidence that VIP is a mediator of NANC vasodilatation in trachea, whereas in more peripheral airways other neuropeptides are involved (Matran et al., 1989b). Since VIP is likely to have a greater effect on bronchial vessels than on airway smooth muscle, it may provide a mechanism for increasing blood flow to contracted smooth muscle. Thus, if VIP is released from cholinergic nerves, it may improve muscular perfusion during cholinergic contraction. Perhaps the apparent protective effect of inhaled VIP against histamine-induced bronchoconstriction in human subjects (Barnes and Dixon, 1984), despite a lack of effect on bronchomotor tone, may be explained by an increase in bronchial blood flow, which would more rapidly remove inhaled histamine from sites of deposition in the airways.

VIP also stimulates mucus secretion, measured by ^{35}S-labeled glycoprotein secretion, in ferret airway in vitro (Peatfield et al., 1983), and there is a high density of VIP receptors in human airway submucosal glands (Carstairs and Barnes, 1986). VIP has a surprising inhibitory effect on glycoprotein secretion from human tracheal explants (Coles et al., 1981), but the effects of VIP on mucus secretion may be complex and may depend on the drive to gland secretion. Mucus secretion stimulated by cholinergic agonists is inhibited in ferret trachea but stimulated in cat trachea, whereas secretion stimulated with the α-adrenergic agonist phenylephrine is augmented (Webber and Widdicombe, 1987; Shimura et al., 1988). VIP is a potent stimulant of chloride ion transport and therefore water secretion in dog tracheal epithelium (Nathanson et al., 1983), suggesting that VIP may be a regulator of airway water secretion and therefore mucociliary clearance. The high density of VIP receptors on epithelial cells of human airways suggests that VIP may regulate ion transport and other epithelial functions in human airways (Carstairs and Barnes, 1986).

VIP inhibits release of mediators from pulmonary mast cells (Undem et al., 1983) and may have several other anti-inflammatory actions in airways (Said, 1991). VIP may interact with T lymphocytes and has the potential to act as a local immunomodulator in airways (O'Dorisio et al., 1989).

B. VIP as i-NANC Neurotransmitter

Several lines of evidence implicate VIP as a neurotransmitter of i-NANC nerves in airways, but this is species dependent (Lammers et al., 1992). In guinea pig trachea, α-chymotrypsin, which degrades VIP, blocks responses to exogenous VIP and results in a reduction in i-NANC response by about 50% (Ellis and Farmer, 1989c), and antiserum to VIP also reduces i-NANC responses (Ellis and Farmer, 1989a). Furthermore, a cyclic-AMP-selective phosphodiesterase inhibitor enhances the i-NANC response in guinea pig trachea, suggesting that the neurotransmitter increases cyclic AMP; this would be consistent with the effect of VIP (Rhoden and Barnes, 1990). In human airways in vitro, α-chymotrypsin, under conditions that completely block the bronchodilator response to exogenous VIP, has no effect on the pronounced i-NANC response, strongly suggesting that neither VIP nor related peptides (PACAP-27, PHM) that are also susceptible to degradation by α-chymotrypsin are involved in the i-NANC response in human airways (Belvisi et al., 1992) (Fig. 1). Similar findings have been reported in feline airways (Altiere et al., 1985).

In guinea pig airways it is apparent that VIP and related peptide account for only about half of the i-NANC response, and recent studies have demonstrated that nitric oxide accounts for the remaining response (Li and Rand, 1991; Tucker et al., 1990). In human and feline airways, however, nitric oxide (NO) appears to account for all of the i-NANC response (Belvisi et al., 1992b; Fisher et al., 1993), indicating that VIP is likely to play little role in regulating airway smooth muscle tone.

It is probably misleading to think of i-NANC nerves as a discrete bronchodilator pathway; it is more likely that it functions as a braking mechanism to cholinergic bronchoconstriction, particularly as both transmitters may be co-released from parasympathetic nerves in the airways. In guinea pig trachea α-chymotrypsin increases cholinergic nerve-induced bronchoconstriction (presumably by removing the braking action of endogenously released VIP) (Belvisi et al., 1993a), but this is not the case in human airways (Ward et al., 1993). However, NO acts as a modulator of cholinergic bronchoconstrictor in both species (Belvisi et al., 1991a; Ward et al., 1993).

C. Role in Asthma?

The role of VIP in the pathophysiology of asthma is far from certain in the absence of potent specific antagonists. Indeed, it is not clear whether VIP has beneficial or

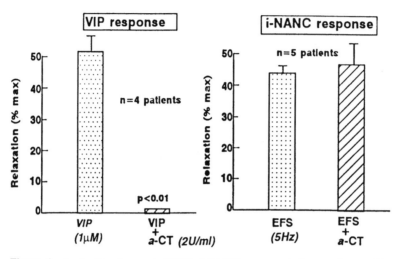

Figure 1 Lack of involvement of VIP in i-NANC responses in human airways. Vasoactive intestinal peptide (VIP)-induced relaxation of human airways in vitro is blocked by incubation with α-chymotrypsin (α-CT), but inhibitory nonadrenergic, noncholinergic (i-NANC) neural bronchodilatation of a similar degree is not affected. EFS: 40 V, 0.5 msec, 30 sec, 5 Hz. (Data from Belvisi et al., 1992.)

deleterious effects, since its vasodilator and mucus secretory effects predominate over any effect on airway smooth muscle in the airways, particularly in peripheral airways. A striking absence of VIP-immunoreactive nerves has been described in the lungs of patients with asthma in tissues largely obtained at postmortem (Ollerenshaw et al., 1989). The loss of VIP immunoreactivity from all tissues including pulmonary vessels is so complete that it seems unlikely to represent a fundamental absence of VIP-immunoreactive nerves in asthma. More likely is the possibility that enzymes, such as mast cell tryptase, are released from inflammatory cells in asthma and that these rapidly degrade VIP when sections are cut (Barnes, 1989b). Biopsies taken from patients with mild asthma suggest that VIP-immunoreactive nerves appear normal in asthma (Howarth et al., 1991) and the amount of VIP mRNA is normal in asthmatic lungs (Adcock et al., 1991). Nor is there any abnormality in the distribution of VIP receptors in the airways of asthmatic patients (Sharma and Jeffery, 1990). VIP antibodies, which would neutralize the effects of VIP, have also been described in the plasma of asthmatic patients, but as they are found just as often in nonasthmatic patients, their significance is doubtful (Paul et al., 1989). While it seems unlikely that there would be any primary abnormality in VIP innervation in the airways of patients with asthma, it is possible that a secondary abnormality may arise as a result of the inflammatory process in the airway.

Mast cell tryptase is particularly active in degrading VIP (Caughey, 1989) and is known to be elevated in asthmatic airways (Wenzel et al., 1988). Inhibition of tryptase potentiates the bronchodilator response to VIP in human airways in vitro (Tam et al., 1990), and mast cell tryptase reverses the relaxation of airways induced by VIP (Franconi et al., 1989) and markedly increases the in vitro responsiveness of canine airways (Sekizawa et al., 1989). Tryptase released from mast cells in the asthmatic airway may then more rapidly degrade VIP and related peptides released from airway cholinergic nerves. This would remove a "brake" from cholinergic nerves and lead to exaggerated cholinergic reflex bronchoconstriction. This may also have the effect of increasing inflammatory responses in the airway, since VIP has anti-inflammatory actions. Our own data in human airways argue against a role for VIP in the regulation of airway tone and suggest that this peptide is more important as a vasoregulator.

Whether i-NANC responses are impaired in asthma is not yet certain. In vitro the i-NANC response is not reduced compared to responses in normal airways (Belvisi et al., 1993b). In patients with mild asthma, no evidence for an impaired NANC bronchodilator reflex has been observed (Michoud et al., 1988; Lammers et al., 1989). However, this does not preclude a defect in more severe asthmatics, in whom the degree of airway inflammation may be greater. In sensitized guinea pigs exposed to allergen, a reduction in i-NANC responses has been reported (Miura et al., 1990). This is presumably due to the release of enzymes or reactive oxygen species from inflammatory cells in the airways. However, as discussed above, the contribution of VIP to i-NANC responses in human airways is not established (although it is possible that it is released with certain neural activation patterns not mimicked by electrical field stimulation in vitro), and increased degradation of this peptide in asthma may have a relatively minor effect on airway tone.

D. VIP-Related Peptides

Several other peptides have now been identified in the mammalian nervous system that are similar in structure and effect to VIP. Peptide histidine isoleucine (PHI) and PHM have a marked structural similarity to VIP, with 50% amino acid sequence homology. PHI and PHM are encoded by the same gene as VIP, and both peptides are synthesized in the same prohormone (Tatemoto, 1984). It is therefore not surprising to find that PHI has a similar immunocytochemical distribution in lung to VIP (Christofides et al., 1984). PHI has similar effects to VIP and may activate the same receptor, although it is less potent as a vasodilator and equipotent as a bronchodilator. Peptide histidine valine (PHV-42) is an N-terminally extended precursor of VIP that also has bronchodilator action (Yiangou et al., 1987).

Helodermin is a 35-amino-acid peptide of similar structure to VIP that has been isolated from the salivary gland venom of the Gila monster lizard. Helodermin-like immunoreactivity has been localized to airway nerves and has similar effects to VIP, but has a longer duration of action. Helodermin is a potent relaxant of airway smooth muscle in vitro, and helodermin-like immunoreactivity has been found in trachea (Foda et al., 1990). Helodermin appears to activate a high-affinity form of the VIP receptor (Robberecht et al., 1988).

PACAP, a 38-amino-acid peptide isolated from sheep hypothalamus, and PACAP-27, a truncated fragment, have marked sequence homology with VIP and have been demonstrated in the peripheral nervous system (Miyata et al., 1990). PACAP-like immunoreactivity has a similar distribution to VIP in airways of several species and may be localized to cholinergic and also to capsaicin-sensitive afferent nerves (Cardell et al., 1991). PACAP-like immunoreactivity is particularly prominent in human airways (Uddman et al., 1993). The effects of PACAP-27 are similar to those of VIP, and there appears to be a particularly high density of PACAP receptors in lung tissue (which appear to be distinct from VIP receptors) (Gottschall et al., 1990).

The reason for the coexistence of so many similar peptides with similar effects is not clear, and until specific antagonists are developed, it will be difficult to elucidate.

III. TACHYKININS

Substance P (SP) and neurokinin A (NKA), but not neurokinin B, are localized to sensory nerves in the airways of several species. SP-immunoreactive nerves are abundant in rodent airways, but are very sparse in human airways (Martling et al., 1987; Laitinen et al., 1992; Komatsu et al., 1991). Rapid enzymatic degradation of SP in airways, and the fact that SP concentrations may decrease with age and possibly after cigarette smoking, could explain the difficulty in demonstrating this peptide in some studies. SP-immunoreactive nerves in the airway are found beneath and within the airway epithelium, around blood vessels, and, to a lesser extent, within airway smooth muscle. SP-immunoreactive nerve fibers also innervate parasympathetic ganglia, suggesting a sensory input that may modulate ganglionic transmission and result in ganglionic reflexes.

SP in the airways is localized predominantly to capsaicin-sensitive unmyelinated nerves in the airways, but chronic administration of capsaicin only partly depletes the lung of tachykinins, indicating the presence of a population of capsaicin-resistant SP-immunoreactive nerves, as in the gastrointestinal tract (Lundberg et al., 1987; Day et al., 1991). Similar capsaicin denervation studies are not possible in human airways, but after extrinsic denervation by heart-lung transplantation, there appears to be a loss of SP-immunoreactive nerves in the submucosa (Springall et al., 1990).

A. Effects on Airways

Tachykinins have many different effects on the airways that may be relevant to asthma, and these effects are mediated via NK-1 receptors (preferentially activated by SP) and NK-2 receptors (activated by NKA) (Nakanishi, 1991). Tachykinins constrict smooth muscle of human airways in vitro via NK-2 receptors (Naline et al., 1989; Advenier et al., 1992). The contractile response to NKA is significantly greater in small human bronchi than in more proximal airways, indicating that tachykinins may have a more important constrictor effect on peripheral airways (Frossard and Barnes, 1988), whereas cholinergic constriction tends to be more pronounced in proximal airways. This is consistent with the autoradiographic distribution of tachykinin receptors, which are distributed to small and large airways. In vivo SP does not cause bronchoconstriction or cough, either by intravenous infusion (Fuller et al., 1987c; Evans et al., 1988) or by inhalation (Fuller et al., 1987c; Joos et al., 1987), whereas NKA causes bronchoconstriction both after intravenous administration (Evans et al., 1988) and after inhalation in asthmatic subjects (Joos et al., 1987). Mechanical removal of airway epithelium potentiates the bronchoconstrictor response to tachykinins (Frossard et al., 1989; Devillier et al., 1988a), largely because the ectoenzyme neutral endopeptidase (NEP), which is a key enzyme in the degradation of tachykinins in airways, is strongly expressed on epithelial cells.

SP stimulates mucus secretion from submucosal glands in human airways in vitro (Rogers et al., 1989) and is a potent stimulant to goblet cell secretion in guinea pig airways (Kuo et al., 1990). Indeed, SP is likely to mediate the increase in goblet cell discharge after vagus nerve stimulation and exposure to cigarette smoke (Tokuyama et al., 1990; Kuo et al., 1992b).

Stimulation of the vague nerve in rodents causes microvascular leakage, which is prevented by prior treatment with capsaicin or by a tachykinin antagonist, indicating that release of tachykinins from sensory nerves mediates this effect. Among the tachykinins, SP is most potent at causing leakage in guinea pig airways (Rogers et al., 1988), and NK-1 receptors have been localized to postcapillary venules in the airway submucosa (Sertl et al., 1988). Inhaled SP also causes microvascular leakage in guinea pigs, and its effect on the microvasculature is more marked than its effect on airway smooth muscle (Lotvall et al., 1990a). It is difficult to measure airway microvascular leakage in human airways, but SP causes a wheal in human skin when injected intradermally, indicating the capacity to cause microvascular leak in human postcapillary venules; NKA is less potent, indicating that an NK-1 receptor mediates this effect (Fuller et al., 1987a).

Tachykinins have potent effects on airway blood flow. Indeed, the effect of tachykinins on airway blood flow may be the most important physiological and pathophysiological role of tachykinins in airways. In canine and porcine trachea, both SP and NKA cause a marked increase in blood flow (Salonen et al., 1988b;

Matran et al., 1989b). Tachykinins also dilate canine bronchial vessels in vitro, probably via an endothelium-dependent mechanism (McCormack et al., 1989b). Tachykinins also regulate bronchial blood flow in the pig; stimulation of the vagus nerve causes a vasodilatation mediated by the release of sensory neuropeptides, and it is likely that calcitonin gene-related peptide (CGRP) as well as tachykinins are involved (Matran et al., 1989b).

Tachykinins may also interact with inflammatory and immune cells (McGillis et al., 1987; Daniele et al., 1992), although whether this is of pathophysiological significance remains to be determined. SP degranulates certain types of mast cell, such as those in human skin, although this is not mediated via a tachykinin receptor (Lowman et al., 1988). There is no evidence that tachykinins degranulate lung mast cells (Ali et al., 1986). SP has a degranulating effect on eosinophils (Kroegel et al., 1990); again, the degranulation is related to high concentrations of peptide and, as for mast cells, is not mediated via a tachykinin receptor. At lower concentrations tachykinins have been reported to enhance eosinophil chemotaxis. Tachykinins may activate alveolar macrophages (Brunelleschi et al., 1990) and monocytes to release inflammatory cytokines, such as IL-6 (Lotz et al., 1988). Tachykinins and vagus nerve stimulation also cause transient vascular adhesion of neutrophils in the airway circulation (Umeno et al., 1989). In the skin SP induces an infiltration with neutrophils, which appears to be dependent on degranulation of dermal mast cells (Matsuda et al., 1989), but this may not be relevant in the airways for the reasons discussed above.

Tachykinins may also be involved in airway fibrosis. Both SP and NKA stimulate proliferation of human lung fibroblasts and NKA also stimulates chemotaxis (Harrison et al., 1992). Studies with selective tachykinin receptor antagonists indicate that both NK-1 and NK-2 receptors may be involved.

In guinea pig trachea, tachykinins also potentiate cholinergic neurotransmission at postganglionic nerve terminals, and an NK-2 receptor appears to be involved (Hall et al., 1989). There is also potentiation at the ganglionic level (Undem et al., 1991; Watson et al., 1993), which appears to be mediated via a NK-1 receptor (Watson et al., 1993). Endogenous tachykinins may also facilitate cholinergic neurotransmission since capsaicin pretreatment results in a significant reduction in cholinergic neural responses both in vitro and in vivo (Martling et al., 1984; Stretton et al., 1989). However, in human airways there is no evidence for a facilitatory effect on cholinergic neurotransmission (Belvisi et al., 1994), although such an effect has been reported in the presence of potassium channel blockers (Black et al., 1990).

B. Metabolism

Tachykinins are subject to degradation by at least two enzymes, angiotensin-coverting enzyme (ACE) and NEP (Nadel, 1991). ACE is predominantly localized to vascular endothelial cells and therefore breaks down intravascular peptides. ACE inhibitors, such as captopril, enhance bronchoconstriction due to intravenous SP (Shore et al., 1988; Martins et al., 1990), but not inhaled SP (Lotvall et al., 1990b). NKA is not a good substrate for ACE, however. NEP appears to be the most important enzyme for the breakdown of tachykinins in tissues. Inhibition of NEP by phosphoramidon or thiorphan markedly potentiates bronchoconstriction in vitro in animal (Sekizawa et al., 1987) and human airways (Black et al., 1988) and after inhalation in vivo (Lotvall et al., 1990b). NEP inhibition also potentiates mucus secretion in response to tachykinins in human airways (Rogers et al., 1989). NEP inhibition enhances e-NANC and capsaicin-induced bronchoconstriction, due to the release of tachykinins from airways sensory nerves (Djokic et al., 1989; Frossard et al., 1989).

The activity of NEP in the airways appears to be important in determining the effects of tachykinins; any factors that inhibit the enzyme or its expression may be associated with increased effects of exogenously or endogenously released tachykinins. Several of the stimuli known to induce bronchoconstrictor responses in asthmatic patients have been found to reduce the activity of airway NEP (Nadel, 1991).

IV. CALCITONIN GENE-RELATED PEPTIDE

CGRP-immunoreactive nerves are abundant in the respiratory tract of several species. CGRP is costored and colocalized with SP in afferent nerves (Martling, 1987). CGRP has been extracted from and is localized to human airways (Palmer et al., 1987; Komatsu et al., 1991). CGRP-immunoreactive nerve fibers appear to be more abundant than SP-immunoreactive fibers, possibly because CGRP has greater stability and is also present in some nerves that do not contain SP. CGRP is found in trigeminal, nodose-jugular, and dorsal root ganglia (Uddman et al., 1985) and has also been detected in neuroendocrine cells of the lower airways.

CGRP is a potent vasodilator, which has long-lasting effects. CGRP is an effective dilator of human pulmonary vessels in vitro and acts directly on receptors on vascular smooth muscle (McCormack et al., 1989). It also potently dilates bronchial vessels in vitro (McCormack et al., 1989) and produces a marked and long-lasting increase in airway blood flow in anesthetized dogs (Salonen et al., 1988a) and conscious sheep in vivo (Parsons et al., 1992). Receptor mapping studies have demonstrated that CGRP receptors are localized predominantly to bronchial vessels rather than to smooth muscle or epithelium in human airways (Mak and Barnes, 1988). It is possible that CGRP may be the predominant mediator of arterial vasodilatation and increased blood flow in response to sensory nerve stimulation in the bronchi (Matran et al., 1989b). CGRP may be an important mediator of airway hyperemia in asthma.

By contrast, CGRP has no direct effect of airway microvascular leak (Rogers et al., 1988). In the skin, CGRP potentiates the leakage produced by SP, presumably by increasing the blood delivery to the sites of plasma extravasation in the postcapillary venules (Khalil et al., 1988). This does not occur in guinea pig airways when CGRP and SP are coadministered, possibly because blood flow in the airways is already high (Rogers et al., 1988), although an increased leakage response has been reported in rat airways (Brockaw and White, 1992). It is possible that potentiation of leak may occur when the two peptides are released together from sensory nerves.

CGRP causes constriction of human bronchi in vitro (Palmer et al., 1987). This is surprising since CGRP normally activates adenylyl cyclase, an event usually associated with bronchodilatation. Receptor mapping studies suggest few, if any, CGRP receptors over airway smooth muscle in human or guinea pig airways, and this suggests that the paradoxical bronchoconstrictor response reported in human airways may be mediated indirectly. In guinea pig airways, CGRP has no consistent effect on tone (Martling et al., 1988).

CGRP has a weak inhibitory effect on cholinergically stimulated mucus secretion in ferret trachea (Webber et al., 1991) and on goblet cell discharge in guinea pig airways (Kuo et al., 1990). This is probably related to the low density of CGRP receptors on mucus secretory cells, but does not preclude the possibility that CGRP might increase mucus secretion in vivo by increasing blood flow to submucosal glands.

CGRP injection into human skin causes a persistent flare, but biopsies have revealed an infiltration of eosinophils (Pietrowski and Foreman, 1986). CGRP itself does not appear to be chemotactic for eosinophils, but proteolytic fragments of the peptide are active (Haynes and Manley, 1988), suggesting that CGRP released into the tissues may lead to eosinophilic infiltration.

CGRP inhibits the proliferative response of T lymphocytes to mitogens, and specific receptors have been demonstrated on these cells (Umeda and Arisawa, 1989). CGRP also inhibits macrophage secretion and the capacity of macrophages to activate T lymphocytes (Nong et al., 1989). This suggests that CGRP has potential anti-inflammatory actions in the airways.

V. NEUROGENIC INFLAMMATION IN ANIMAL MODELS OF ASTHMA

Pain, heat, redness, and swelling are the cardinal signs of inflammation. Sensory nerves may be involved in the generation of each of these signs. There is now considerable evidence that sensory nerves participate in inflammatory responses. This "neurogenic inflammation" is due to the antidromic release of neuropeptides from nociceptive nerves or C-fibers via an axon reflex. The phenomenon is well documented in several organs, including skin, eye, gastrointestinal tract, and bladder (Maggi and Meli, 1988). There is also increasing evidence that neurogenic inflammation occurs in the respiratory tract (Barnes, 1990; McDonald, 1987; Solway and Leff, 1991), and it is possible that it may contribute to the inflammatory response in asthma (Barnes, 1986, 1991b) (Fig. 2).

There are several lines of evidence that neurogenic inflammation may be important in animal models that may have relevance to asthma. These models have usually been in rodents where tachykinin effects are pronounced and may not be predictive of the role of tachykinins in human airways, however.

Four main experimental approaches have been used to assess the role of sensory neuropeptides in animal models of asthma; these include studies of depletion with capsaicin depletion, enhancement with inhibitors of NEP, tachykinin receptor antagonists, and inhibitors of sensory neuropeptide release.

A. Capsaicin Depletion Studies

Capsaicin pretreatment to deplete neuropeptides from C-fibers, either in neonatal animals (which results in degeneration of C-fibers) or in acute treatment of adult animals (resulting in depletion of sensory neuropeptides). In rat trachea, capsaicin

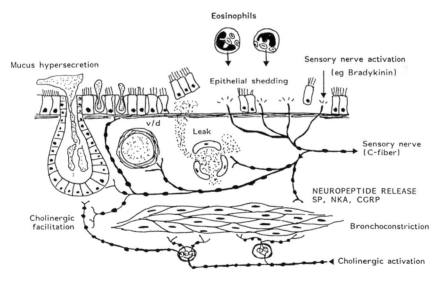

Figure 2 Neurogenic inflammation in asthma. Possible neurogenic inflammation in asthmatic airways via retrograde release of peptides from sensory nerves via an axon reflex. Substance P (SP) causes vasodilatation, plasma exudation, and mucus secretion, whereas neurokinin A (NKA) causes bronchoconstriction and enhanced cholinergic reflexes and calcitonin gene-related peptide (CGRP) vasodilatation.

pretreatment inhibits the microvascular leakage induced by irritant gases, such as cigarette smoke (Lundberg and Saria, 1983), and inhibits goblet cell discharge and microvascular leak induced by cigarette smoke in guinea pigs (Kuo et al., 1992b). Capsaicin-sensitive nerves may also contribute to the bronchoconstriction and microvascular leak induced by isocapnic hyperventilation (Ray et al., 1989), hypocapnia (Reynolds and McEvoy, 1989), inhaled sodium metabisulfite (Sakamoto et al., 1992), and nebulized hypertonic saline (Umeno et al., 1990) and toluene diisocyanate (Thompson et al., 1987) in rodents. In guinea pigs capsaicin pretreatment has little or no effect on the acute bronchoconstrictor or plasma exudation response to allergen inhalation in sensitized animals (Lötvall et al., 1991). Administration of capsaicin increases airway responsiveness in guinea pigs to cholinergic agonists, and this effect is prevented by prior treatment with capsaicin, suggesting that capsaicin-sensitive nerves release products that increase airway responsiveness (Hsiug et al., 1992). In pigs capsaicin pretreatment inhibits the vasodilator response to allergen (which may be mediated by the release of CGRP) (Alving et al., 1988). In allergic sheep capsaicin pretreatment prevents the airway hyperresponsiveness to both allergen and cholinergic agonists (Abraham et al., 1993). In a model of chronic allergen exposure in guinea pigs, capsaicin pretreatment results in complete inhibition of airway hyperresponsiveness, without any change in the eosinophil inflammatory response (Matsuse et al., 1991). In rabbits neonatal capsaicin treatment inhibits the airway hyperresponsiveness associated with neonatal allergen sensitization, although this does not appear to be associated with any change in content of sensory neuropeptides in lung tissue (Riccio et al., 1993). This suggests that capsaicin-sensitive nerves may play a role in chronic inflammatory responses to allergen.

There has been speculation that mast cells in the airways might be influenced by capsaicin-sensitive nerves. Histological studies have demonstrated a close proximity between mast cells and sensory nerves in airways (Bienenstock et al., 1988). There is also evidence that antidromic stimulation of the vagus nerve leads to mast cell mediator release in canine airways (Leff et al., 1982). Furthermore, allergen exposure has effects on ion transport in guinea pig airways that are dependent on capsaicin-sensitive nerves (Sestini et al., 1990).

B.   Inhibition of Neuropeptide Metabolism

The activity of NEP may be an important determinant of the extent of neurogenic inflammation in airways, and inhibition of NEP in the rodent by thiorphan or phosphoramidon has been shown to enhance neurogenic inflammation in various rodent models. NEP is not specific to tachykinins and is also involved in the metabolism of other bronchoactive peptides, including kinins and endothelins. Certain virus infections enhance e-NANC responses in guinea pigs (Saban et al., 1987), and *Mycoplasma* infection enhances neurogenic microvascular leakage in rats (McDonald, 1987), an effect that is mediated by inhibition of NEP activity. Influenza virus infection of ferret trachea in vitro and of guinea pigs in vivo inhibits the activity of epithelial NEP and markedly enhances the bronchoconstrictor responses to tachykinins (Jacoby et al., 1988). Similarly, Sendai virus infection potentiates neurogenic inflammation in rat trachea (Piedimonte et al., 1990). This may explain why respiratory tract virus infections are so deleterious to patients with asthma. Hypertonic saline also impairs epithelial NEP function, leading to exaggerated tachykinin responses (Umeno et al., 1990), and cigarette smoke exposure has a similar effect, which can be explained by an oxidizing effect on the enzyme (Dusser et al., 1989). Toluene diisocyanate, albeit at rather high doses, also reduces NEP activity, and this may be a mechanism contributing to the airway hyperresponsiveness that may follow exposure to this chemical (Sheppard et al., 1988). Thus, many of the agents that lead to exacerbations of asthma appear to reduce the activity of NEP at the airway surface, thus leading to exaggerated responses to tachykinins (and other peptides) and so to increased airway inflammation.

C. Inhibition of Sensory Neuropeptide Effects

Specific antagonists of tachykinin receptors have now been developed and provide a more specific tool to investigate the role of tachykinins in animal models. Several highly potent and stable peptide and nonpeptide tachykinin antagonists have recently been developed that are highly selective for either NK-1 or NK-2 receptors (Watling, 1992). The NK-1 receptor antagonist CP 96,345 is able to block the plasma exudation response to vagus nerve stimulation and to cigarette smoke in guinea pig airways (Lei et al., 1992; Delay-Goyet and Lundberg, 1991), without affecting the bronchoconstrictor response, which is blocked by the NK-2 antagonist SR 48,968 (Advenier et al., 1992). Similar results have been obtained with the very potent NK-1 selective antagonist FK 888 (Hirayama et al., 1993). CP 96,345 also blocks hyperpnea- and bradykinin-induced plasma exudation in guinea pigs (Garland et al., 1992; Sakamoto et al., 1993), but has no effect on the acute plasma exudation induced by allergen in sensitized animals (Sakamoto et al., 1993). These specific antagonists are useful new tools in probing the involvement of tachykinins in disease and will be invaluable in clinical studies in the future.

D. Inhibition of Sensory Neuropeptide Release

Several agonists act on prejunctional receptors on airway sensory nerves to inhibit the release of neuropeptides and neurogenic inflammation (Barnes et al., 1990), as discussed in Chapter 22. Opioids are the most effective inhibitory agonists, acting via prejunctional μ-receptors, and have been shown to inhibit cigarette-smoke-induced discharge from goblet cells in guinea pig airways in vivo (Kuo et al., 1990) and to inhibit ozone-induced hyperreactivity in guinea pigs, which appears to be mediated via sensory nerves (Yeadon et al., 1992). Several other agonists are also effective and may act by opening a common calcium-activated large conductance potassium channel in sensory nerves (Stretton et al., 1992). Openers of other potassium channels, which achieve the same hyperpolarization of the sensory nerve, are also effective in blocking neurogenic inflammation in rodents (Ichinose and Barnes, 1990) and have been shown to block cigarette-smoke-induced goblet cell secretion in guinea pigs (Kuo et al., 1992).

VI. NEUROGENIC INFLAMMATION IN ASTHMA?

Although it was proposed several years ago that neurogenic inflammation and peptides released from sensory nerves might be important as an amplifying mechanism in asthmatic inflammation (Barnes, 1986), there is little evidence to date to support this idea, despite the extensive work in rodent models (Table 2). This is partly because it has proved difficult to apply the same approaches to human volunteers.

A. Sensory Neuropeptides in Human Airways

In comparison with rodent airways, SP- and CGRP-immunoreactive nerves are sparse in human airways. Quantitative studies indicate that SP-immunoreactive fibers constitute only 1% of the total number of intraepithelial fibers, whereas in guinea pig they make up 60% of the fibers (Bowden and Gibbins, 1992a). This raises the possibility that sensory nerves in humans may contain some unidentified transmitter that may be involved in neurogenic inflammation. Chronic inflamma-

Table 2 Neurogenic Inflammation in Asthma

Evidence in favor
    Increased SP-immunoreactive nerves in bronchial biopsies
    Increased SP in bronchoalveolar lavage
    Bronchoconstrictor response to NKA increased in asthma
    Increased NK-1-receptor expression
    Activation of airway sensory nerves by bradykinin
    Reduced bronchoconstrictor response to bradykinin with tachykinin antagonist
Evidence against
    SP-immunoreactive nerves not increased in bronchial biopsies
    Only tansient airway response to inhaled capsaicin (and not increased in asthma)
    No effect of neutral endopeptidase inhibitors on airway function
    No effect of sensory nerve modulation (opioids, local anesthetics)
    No effect of tachykinin antagonists in asthma?

tion may lead to changes in the pattern of innervation, through the release of neurotrophic factors from inflammatory cells. Thus in chronic arthritis and inflammatory bowel disease, there is an increase in the density of SP-immunoreactive nerves (Levine et al., 1986; Holzer, 1988). A striking increase in SP-like immunoreactive nerves has been reported in the airway of patients with fatal asthma (Ollerenshaw et al., 1991). This increased density of nerves is particularly noticeable in the submucosa. Whether this apparent increase is due to proliferation of sensory nerves or to increased synthesis of tachykinins has not yet been established. Recently, elevated concentrations of SP in bronchoalveolar lavage of patients with asthma have been reported, with a further rise after allergen challenge (Nieber et al., 1992), suggesting that there may be an increase in SP in the airways of asthmatic patients. Similarly, SP has been detected in the sputum of asthmatic patients after hypertonic saline inhalation (Tomaki et al., 1993).

Cultured sensory neurons are stimulated by nerve growth factor (NGF), which markedly increases the transcription of preprotachykinin A gene, the major precursor peptide for tachykinins (Lindsay and Harmar, 1989). Similarly, adjuvant-induced inflammation in rat spinal cord increases the gene expression of PPT-A (Minami et al., 1989). Since NGF may be released from several types of inflammatory cell, it is possible that this could lead to increased tachykinin synthesis and increased nerve growth. Several other neurotrophic factors have also recently been identified. However, bronchial biopsies of mild asthmatic patients have not revealed any evidence of increased SP-immunoreactive nerves (Howarth et al., 1991). This may indicate that the increased innervation (Ollerenshaw et al., 1991) may be a feature of either prolonged or severe asthma and indicates the need for more studies.

B. Sensory Nerve Activation

Sensory nerves may be activated in airway disease. In asthmatic airways the epithelium is often shed, thereby exposing sensory nerve endings. Sensory nerves in asthmatic airways may be "hyperalgesic" as a result of exposure to inflammatory mediators such as prostaglandins and certain cytokines (such as IL-1β and TNFα) (Cunha et al., 1992). Hyperalgesic nerves may then be activated more readily by other mediators, such as kinins.

Capsaicin induces bronchoconstriction and plasma exudation in guinea pigs (Lundberg et al., 1987) and increases airway blood flow in pigs (Alving et al., 1988). In humans, capsaicin inhalation causes cough and a *transient* bronchoconstriction, which is inhibited by cholinergic blockade and is probably due to a laryngeal reflex (Fuller et al., 1985; Midgren et al., 1992). This suggests that neuropeptide release does not occur in human airways, although it is possible that insufficient capsaicin reaches the lower respiratory tract because the dose is limited by coughing. In patients with asthma, there is no evidence that capsaicin induces a greater degree of bronchoconstriction than in normal individuals (Fuller et al., 1985).

Bradykinin is a potent bronchoconstrictor in asthmatic patients and also induces coughing and a sensation of chest tightness, which closely mimics a naturally occurring asthma attack (Barnes, 1992b; Fuller et al., 1987b). Yet it is a weak constrictor of human airways in vitro, suggesting that its potent constrictor effect is mediated indirectly. Bradykinin is a potent activator of bronchial C-fibers in dogs (Kaufman et al., 1980) and releases sensory neuropeptides from perfused rodent lungs (Saria et al., 1988). In guinea pigs, bradykinin instilled into the airways causes bronchoconstriction, which is reduced significantly by a cholinergic antagonist (as in asthmatic patients (Fuller et al., 1987b)) and also by capsaicin pretreatment (Ichinose et al., 1990b). The plasma leakage induced by inhaled bradykinin is inhibited by an NK-1 antagonist (Sakamoto et al., 1993). This indicates that bradykinin activates sensory nerves in the airways and that part of the airway response is mediated by release of constrictor peptides from capsaicin-sensitive nerves. In asthmatic patients, an inhaled, nonselective tachykinin antagonist, FK 224, has recently been shown to reduce the bronchoconstrictor response to inhaled bradykinin and also to block the cough response in those subjects who coughed in response to bradykinin (Ichinose et al., 1992).

C. Studies with NEP Inhibitors

In rodents, inhibition of NEP with thiorphan or phosphoramidon results in striking potentiation of tachykinin- and sensory-nerve-induced effects and has been used as an approach to explore the potential for neurogenic inflammation in disease (Nadel, 1991). Intravenous acetorphan, which is hydrolyzed to thiorphan, was administered to asthmatic subjects, and although there was potentiation of the wheal-and-flare response to intradermal SP, there was no effect on baseline airway caliber or on bronchoconstriction induced by a "neurogenic" trigger sodium metabisulfite (Nichol et al., 1992). The lack of effect could be due to inadequate inhibition of NEP in the airways, particularly at the level of the epithelium. Nebulized thiorphan has been shown to potentiate the bronchoconstrictor response to inhaled NKA in normal and asthmatic subjects (Cheung et al., 1992a,b), but there was no effect on baseline lung function in asthmatic patients (Cheung et al., 1992b), indicating that there is unlikely to be any basal release of tachykinins. NEP is strongly expressed in the human airway (Baraniuk et al., 1991), but there is no evidence based on immunocytochemical staining or in situ hybridization that it is defective in asthmatic airways (Baraniuk, J., and Barnes, P.J., unpublished), and the fact that after inhaled thiorphan the bronchoconstrictor response to inhaled NKA is further enhanced in asthmatic subjects provides supportive functional data that NEP function may not be impaired, at least in mild asthma (Cheung et al., 1992b). Of course, it is possible that NEP may become dysfunctional after viral infections or exposure to oxidants and thus contribute to asthma exacerbations.

D. Studies with Tachykinins

Inhaled or i.v. SP infusions have no significant effect on airway function in normal or asthmatic volunteers (Fuller et al., 1987c; Evans et al., 1988; Joos et al., 1987), although NKA causes bronchoconstriction (Evans et al., 1988; Joos et al., 1987). The apparent lack of response to SP may be because changes mediated via NK-1 receptors, such as increased mucus secretion, increased airway blood flow, or increased plasma exudation, cannot easily be measured in patients. In inflammatory bowel disease, there is evidence for a marked up-regulation of tachykinin receptors, particularly in the vasculature, suggesting that chronic inflammation may lead to changes in tachykinin receptor expression (Mantyh et al., 1988; Mantyh, 1991). In patients with allergic rhinitis, an increased vascular response to nasally applied SP is observed (Devillier et al., 1988b). There is evidence that NK-1-receptor gene expression may be increased in the lungs of asthmatic patients (Adcock et al., 1993). This might be due to increased transcription in response to activation of transcription factors, such as AP-1, which are activated in human lung by cytokines such as TNFα (Adcock et al., 1992). A consensus sequence both for AP-1 and for glucocorticoid receptor binding has been identified upstream of the NK-1-receptor gene (Nakanishi, 1991). Corticosteroids conversely reduce NK-1-receptor gene expression (Ihara and Nakanishi, 1990) and reduce NK-1-receptor mRNA in asthmatic lungs (Adcock et al., 1993).

E. Modulation of Neurogenic Inflammation

Apart from tachykinin receptor antagonists, neurogenic inflammation may be modulated by either preventing activation of sensory nerves or preventing the release of neuropeptides (Fig. 3). Both approaches may be tried in asthmatic patients, using currently available drugs, although these approaches are not as specific as tachykinin antagonists, as the drugs used have additional effects.

Activation of sensory nerves may be inhibited by local anesthetics, but it has proved to be very difficult to achieve adequate local anesthesia of the respiratory tract. Inhalation of local anesthetics, such as lidocaine, has not been found to have consistent inhibitory effects on various airway challenges and indeed may even promote bronchoconstriction in some patients with asthma (McAlpine and Thomson, 1989). This paradoxical bronchoconstriction may be due to the greater anesthesia of laryngeal afferents, which are linked to a tonic nonadrenergic bronchodilator reflex (Lammers et al., 1988, 1989). Other drugs may inhibit the activation of airway sensory nerves. Cromolyn sodium and nedocromil sodium may have direct effects on airway C-fibers (Dixon et al., 1979; Jackson et al., 1989), and this might contribute to their antiasthma effect. Nedocromil sodium is

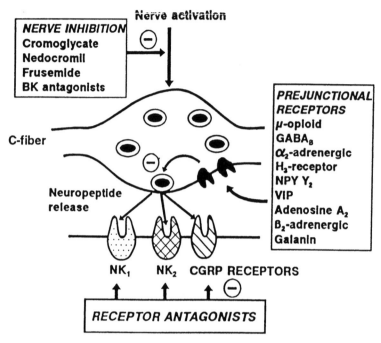

Figure 3 Modulation of neurogenic inflammation. There are several ways in which neurogenic inflammation may be modulated in asthma. These include antagonists of neuropeptide receptors, inhibition of neuropeptide release from sensory nerves, and inhibition of nerve activation.

highly effective against bradykinin-induced and sulfur-dioxide-induced broncho-constriction in asthmatic patients (Dixon et al., 1979, 1987), which are believed to be mediated by activation of sensory nerves in the airways. In addition, nedocromil sodium, and to a much lesser extent cromolyn sodium, inhibit the e-NANC neural bronchoconstriction due to tachykinin release from sensory nerves in guinea pig bronchi in vitro, indicating an effect on release of sensory neuropeptides as well as on activation (Verleden et al., 1991). The loop diuretic furosemide (frusemide), given by nebulization, behaves in a similar fashion to nedocromil sodium and inhibits metabisulfite-induced bronchoconstriction in asthmatic patients (Nichol et al., 1990) and also e-NANC and cholinergic broncho-constriction in guinea pig airways in vitro (Elwood et al., 1991). In addition, nebulized furosemide also inhibits certain types of cough (Ventresca et al., 1990), providing further evidence for an effect on sensory nerves.

Many drugs act on prejunctional receptors to inhibit the release of neuropep-tides, as discussed above. Opioids are the most effective inhibitors, but an inhaled μ-opioid agonist, the pentapeptide BW 443C, was found to be ineffective in inhibiting metabisulfite-induced bronchoconstriction, which is believed to act via neural mechanisms (O'Connor et al., 1991). One problem with BW 443C is that it may be degraded by NEP in the airway epithelium and therefore may not reach a high enough concentration in the vicinity of the airway sensory nerves. Another agent that has a prejunctional modulatory effect in guinea pigs is the H_3-receptor agonist α-methyl histamine (Ichinose et al., 1990a). However, inhalation of α-methyl histamine had no effect on either resting tone or metabisulfite-induced bronchoconstriction in asthmatic patients (O'Connor et al., 1993b).

F. Future Directions

The imminent availability of potent nonpeptide tachykinin antagonists will make it much easier to evaluate the role of neurogenic inflammation in asthma. Current evidence suggests that while sensory nerves may be important in mediating symptoms of asthma (cough and chest tightness) and may be hyperalgesic in asthma (Fig. 4), the contribution of sensory nerves to inflammation in asthma may

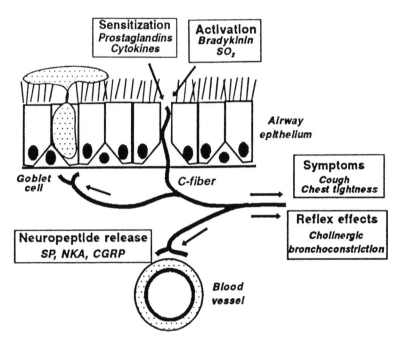

Figure 4 Airway sensory nerves in asthma. Sensory nerves in the airways may be sensitized and activated in asthma, leading to symptoms and reflex effects, but also to neurogenic inflammation via the retrograde release of neuropeptides.

not be very great, at least in patients with mild disease who are suitable for study. It is possible that the disease process may change the efficacy of neuropeptides, however, by either increasing neuropeptide synthesis, reducing degradation, or increasing receptor expression. This is demonstrated dramatically in rodents with viral or bacterial infections that have a markedly exaggerated neurogenic inflammatory response (McDonald et al., 1991). It is possible that in more severe asthma this becomes relatively more important or there may be certain types of asthma (e.g, brittle asthma, steroid-resistant asthma) where neurogenic inflammation is more important. In evaluating the effects of procedures designed to increase or decrease neurogenic inflammation in the airways, it is important to recognize that this is usually judged by effects on airway caliber, which is largely determined by airway smooth muscle contraction. It is more likely that sensory neuropeptides exert a much greater effect on blood vessels and mucus-secreting cells than on airway smooth muscle, but we do not have useful techniques to evaluate these responses.

VII. OTHER NEUROPEPTIDES

Several other peptides have been localized to airway nerves, but even less it known of their role in asthma than that of the peptides discussed previously.

A. Neuropeptide Y

NPY is localized to adrenergic nerves, but is also colocalized with VIP in some species. In heart-lung transplant recipients, there is an apparent increase in NPY-like immunoreactive nerves, suggesting that there may normally be some descending inhibitory influence to the expression of this peptide (Springall et al., 1990). In rodents, depletion of sensory neuropeptides with capsaicin is associated with an increase in adrenergic nerves, indicating that there may be a reciprocal interaction between sensory and adrenergic innervation in lung (van Ranst and Lauweryns, 1990). NPY may also be found within parasympathetic ganglia, where it coexists with VIP since sympathectomy does not completely deplete NPY. There is a population of NPY- and VIP-immunoreactive fibers in guinea pig trachea that are sympathetic in origin but do not contain norepinephrine (Bowden and Gibbins, 1992b).

NPY is most important in the regulation of airway blood flow. NPY causes a long-lasting reduction in tracheal blood flow in anesthetized dogs (Salonen et al., 1988a) but has no direct effect on canine bronchial vessels in vitro (McCormack et al., 1989b), suggesting a preferential effect on resistance vessels in the airway. NPY may constrict resistance vessels, reducing mucosal blood flow and reducing microvascular leak through the reduction in the perfusion of permeable postcapillary venules. NPY has no direct effect on airway smooth muscle of guinea pig

(Stretton and Barnes, 1988) but may cause bronchoconstriction via release of prostaglandins (Cadieux et al., 1989). NPY has a modulatory effect on cholinergic transmission of postganglionic cholinergic nerves (Stretton and Barnes, 1988). This appears to be a direct effect on prejunctional NPY-receptors, rather than secondary to any effect on α-adrenoceptors. NPY also has a modulatory effect on e-NANC bronchoconstricton both in vitro and in vivo, and this effect is surprisingly long-lasting (Stretton and Barnes, 1988; Matran et al., 1989c). NPY has no direct effect on secretion from ferret airways, although it has complex effects on stimulated secretion. NPY enhances both cholinergic and adrenergic stimulation of mucus secretion, but inhibits stimulated serous cell secretion (Webber, 1988).

The role of NPY in asthma is unknown. By reducing airway blood flow (and possibly airway microvascular leak), together with its modulatory action of cholinergic and sensory nerves, it may play a beneficial role in down-regulating inflammatory effects. Whether NPY is defective in asthma is not known.

B. Gastrin-Releasing Peptide

Gastrin-releasing peptide (GRP)/bombesin-immunoreactive nerves are present in the lower respiratory tract of several animal species, including humans, and are probably localized to sensory nerves (Uddman et al., 1984). GRP and bombesin-like peptides may play important roles in lung maturation and epithelial differentiation. Bombesin is a potent bronchoconstrictor in guinea pigs in vivo (Belvisi et al., 1991b). However, in vitro it has no effect on either proximal airways or lung strips, indicating that it produces bronchoconstriction indirectly, although the mechanism is not yet clear. Bombesin has a constrictor effect on airway vessels (Salonen et al., 1988a). GRP and bombesin are potent stimulants of airway mucus secretion in human and cat airways in vitro (Baraniuk et al., 1990; Lundgren et al., 1990). The role of GRP in asthma is unknown.

C. Cholecystokinin

Cholecystokinin octapeptide (CCK_8) has been identified in low concentration in lungs and airways of several species (Ghatei et al., 1982). CCK_8 is a potent constrictor of guinea pig and human airways in vitro (Stretton and Barnes, 1989). The bronchoconstrictor response is potentiated by epithelial removal and by phosphoramidon, suggesting that it is degraded by epithelial NEP. The bronchoconstrictor effect of CCK_8 is also potentiated in guinea pigs sensitized and exposed to inhaled allergen, possibly because allergen exposure reduces epithelial NEP function. CCK_8 acts directly on airway smooth muscle and is potently inhibited by the specific CCK antagonist L363,851, indicating that CCK_A receptors (peripheral type) are involved. CCK_8 has no apparent effect on cholinergic neurotransmission either at the level of parasympathetic ganglia or at postganglionic nerve terminals. Although few CCK-immunoreactive nerves are present in airways, it may still have a significant effect on airway tone if these particular neural fibers are activated selectively.

D. Somatostatin

Somatostatin has been localized to some afferent nerves (Jansco et al., 1981), but the concentration detectable in lung is low (Ghatei et al., 1982; Uddman and Sundler, 1987). Somatostatin has no direct action on airway smooth muscle in vitro but appears to potentiate cholinergic neurotransmission in ferret airways (Sekizawa et al., 1989b). Although somatostatin has a modulatory effect on neurogenic inflammation in the rat footpad (Lembeck et al., 1982), no modulation of e-NANC nerves in airways is apparent (Stretton, C.D., and Barnes, P.J., unpublished observations).

E. Galanin

Galanin is widely distributed in the respiratory tract innervation of several species. It is colocalized with VIP in cholinergic nerves of airways and is present in parasympathetic ganglia (Dey et al., 1990; Cheung et al., 1985). It is also colocalized with SP/CGRP in sensory nerves and dorsal root, nodose, and trigeminal ganglia (Uddman and Sundler, 1987). Galanin has no direct effect on airway tone in guinea pigs but modulates e-NANC neurotransmission (Guiliani et al., 1989). It has no effect on airway blood flow in dogs (Salonen et al., 1988a), and its physiological role in airways remains a mystery.

F. Enkephalins

Leucine-enkephalin has been localized to neuroendocrine cells in airways (Cutz, 1982), and [Met]enkephalin-Arg^6-Gly^7-Leu^8-immunoreactive nerves have been described in guinea pig and rat lungs (Jansco et al., 1981; Gibbins et al., 1987), with a similar distribution to VIP (Shimosegawa et al., 1990). The anatomical origins and functional roles of the endogenous opioids are not clear since the opioid antagonist naloxone has no effect on neurally mediated airway effects (Belvisi et al., 1988, 1989). However, it is possible that these opioid pathways may be selectively activated from brainstem centers under certain conditions. Exogenous opioids potently modulate neuropeptide release from sensory nerves in airways via μ-opioid receptors (Frossard and Barnes, 1987; Belvisi et al., 1988, 1989).

VIII. ROLE OF NEUROPEPTIDES IN ASTHMA

The presence of so many neuropeptides in the respiratory tract raises questions about their physiological role. It is now appreciated that many of these peptides are cotransmitters in classic autonomic nerves and may be regarded as modulators of autonomic effects, perhaps acting to "fine-tune" airway functions (Barnes, 1989a) and to modulate the release of other neurotransmitters (Barnes, 1992a). Although much of the research on neuropeptides in the airways has previously concentrated on their effects on airway smooth muscle (Barnes, 1988), it is now clear that the most potent effects of many of the relevant peptides are on airway vasculature and secretions, and that neuropeptides may have an important role in regulating the mucosal surface of the airways. Another important area that is very relevant to asthma is whether neuropeptides influence the immune system, in particular the immune cells involved in asthma (Daniele et al., 1992). There is likely to be increasing research in the area of neuroimmune interaction, and in some species there is already evidence for neuropeptide innervation of bronchus-associated lymphoid tissue (Nohr and Weihe, 1991). The possibility that inflammatory cells, such as macrophages, lymphocytes, and eosinophils, may themselves produce neuropeptide-like peptides under certain conditions is also an important area of future research (Daniele et al., 1992).

The lack of understanding of the physiological role of individual peptides is largely due to the lack of specific antagonists that can be given safely to humans. Several selective neuropeptide receptor antagonists are now under development, which may soon be available for clinical studies.

A. Therapeutic Prospects

Although there has been optimism that inhibitors or mimics of neuropeptides might have therapeutic application in asthma, it is unlikely that such drugs would have a major advantage over existing agents (Barnes, 1992c). However, it is also possible that such drugs may have value in the treatment of other inflammatory airway diseases, such as chronic obstructive pulmonary disease (COPD), cystic fibrosis, and bronchiectasis, and in the treatment of cough.

VIP is a potent relaxant of human airways in vitro but is ineffective by inhalation, presumably because of metabolism by airway enzymes. A stable derivative could be developed to overcome this drawback, but it is most unlikely that such an agonist could be more effective than a β-agonist. A VIP analog would have the disadvantage of increased side effects due to vasodilatation and might not be effective in relaxing peripheral airways, since VIP receptor are not expressed on airway smooth muscle of small airways, whereas β-receptors are distributed throughout the respiratory tract.

Inhibition of neurogenic inflammation is probably achieved most effectively with tachykinin antagonists, particularly NK-1-receptor antagonists. Preliminary results suggest that tachykinin antagonists have no significant effect in controlling clinical asthma, presumably because other inflammatory mediators are still exerting their effects. By contrast, tachykinin antagonists appear to improve the symptoms of patients with COPD (Ichinose et al., 1993). Further clinical studies over prolonged periods will now be needed before the role of tachykinin antagonists in asthma becomes clear.

Although neurogenic inflammation may not be important in the majority of asthmatic patients, it is likely that sensory nerves may play an important role in the symptomatology of asthma, and therefore drugs that reduce airway hyperalgesia or inhibit airway sensory nerve activation may have therapeutic effects. Bradykinin is likely to be an important activator of airway sensory nerves in asthma, and potent bradykinin antagonists are now in clinical development.

IX. CONCLUSIONS

Many neuropeptides have now been localized to the respiratory tract, and almost certainly more will be discovered. These peptides often have potent actions on airway and vascular tone and on lung secretions, but the presence of so many peptides raises questions about their physiological role. Unique combinations of peptides are colocalized and coreleased from the various subpopulations of sensory, parasympathetic, and sympathetic nerve fibers. The neuropeptides may produce synergistic and/or antagonist events at both pre- and postsynaptic neurons and on any surrounding target cells that possess the appropriate spectrum of peptide receptors. In this way, neuropeptides may act as subtle regulators of tissue activities under physiological conditions. However, in inflammatory diseases such as asthma, they may have important pathogenetic roles. Alterations of degrading enzymes such as neutral endopeptidase may result in unopposed actions of proinflammatory neuropeptides. Until specific antagonists have been developed, it will be difficult to evaluate the precise roles of each of the neuropeptides in disease. It is certainly possible that pharmacological agents that interact with neuropeptides by affecting their release, metabolism, or receptors may be developed in the future with therapeutic potential.

REFERENCES

Abraham, W.M., Ahmed, A., Cortes, A., and Delehunt, J.C. (1993). C-fiber desensitization prevents hyperresponsiveness to cholinergic and antigenic stimuli after antigen challenge in allergic sheep. *Am. Rev. Respir. Dis. 147*:A478.

Adcock, I.M., Belvisi, M.G., Stretton, C.D., Yacoub, M., and Barnes, P.J. (1991). Inhibitory NANC responses and VIP mRNA levels in asthmatic and denervated human lung. *Am. Rev. Respir. Dis. 143*:A355.

Adcock, I.M., Shirasaki, H., Yacoub, M., and Barnes, P.J. (1992). Effects of steroids on transcription factors in human lung. *Am. Rev. Respir. Dis. 145*:A834.

Adcock, I.M., Peters, M., Gelder, C.M., Shirasaki, H., Brown, C.R., and Barnes, P.J.

(1993). Increased tachykinin receptor gene expression in asthmatic lungs and its modulation by steroids. *J. Mol. Endocrinol. 11*:1–7.

Advenier, C., Naline, E., Toty, L., et al. (1992). Effects on the isolated human bronchus of SR 48968, a potent and selective nonpeptide antagonist of the neurokinin A (NK$_2$) receptors. *Am. Rev. Respir. Dis. 146*:1177.

Ali, H., Leung, K.B.I., Pearce, F.L., Hayes, N.A., and Foreman, J.C. (1986). Comparison of histamine releasing activity of substance P on mast cells and basophils from different species and tissues. *Int. Arch. Allergy 79*:121.

Altiere, R.J., Szarek, J.L., and Diamond, L. (1985). Neurally mediated nonadrenergic relaxation in cat airways occurs independent of cholinergic mechanisms. *J. Pharmacol. Exp. Ther. 234*:590.

Alving, K., Matran, R., Lacroix, J.S., and Lundberg, J.M. (1988). Allergen challenge induces vasodilation in pig bronchial circulation via a capsaicin sensitive mechanism. *Acta Physiol. Scand. 134*:571.

Baraniuk, J.N., Lundgren, J.D., Goff, J., et al. (1990). Gastrin releasing peptide (GRP) in human nasal mucosa. *J. Clin. Invest. 85*:998.

Baraniuk, J.N., Mak, J., Letarte, M., Davis, R., Twort, C., and Barnes, P.J. (1991). Neutral endopeptidase mRNA expression in airways. *Am. Rev. Respir. Dis. 143*:A40.

Barnes, P.J. (1986). Asthma as an axon reflex. *Lancet 1*:242.

Barnes, P.J. (1988). Neuropeptides and airway smooth muscle. *Pharmacol. Ther. 36*:119.

Barnes, P.J. (1989a). Airway neuropeptides: roles in fine tuning and in disease. *News Physiol. Sci. 4*:116.

Barnes, P.J. (1989b). Vasoactive intestinal peptide and asthma. *N. Engl. J. Med. 321*:1128.

Barnes, P.J. (1990). Neurogenic inflammation in airways and its modulation. *Arch. Int. Pharmacodyn. 303*:67.

Barnes, P.J. (1991a). Neuropeptides and asthma. *Am. Rev. Respir. Dis. 143*:S28.

Barnes, P.J. (1991b). Sensory nerves, neuropeptides and asthma. *Ann. NY Acad. Sci. 629*:359.

Barnes, P.J. (1992a). Modulation of neurotransmission in airways. *Physiol. Rev. 72*:699.

Barnes, P.J. (1992b). Bradykinin and asthma. *Thorax 47*:979.

Barnes, P.J. (1992c). New drugs for asthma. *Eur. Respir. J. 5*:1126.

Barnes, P.J., and Dixon, C.M.S. (1984). The effect of inhaled vasoactive intestinal peptide on bronchial reactivity to histamine in man. *Am. Rev. Respir. Dis. 130*:162.

Barnes, P.J., Belvisi, M.G., and Rogers, D.F. (1990). Modulation of neurogenic inflammation: novel approaches to inflammatory diseases. *Trends Pharmacol. Sci. 11*:185.

Barnes, P.J., Baraniuk, J., and Belvisi, M.G. (1991). Neuropeptides in the respiratory tract. *Am. Rev. Respir. Dis. 144*:1187.

Belvisi, M.G., Chung, K.F., Jackson, D.M., and Barnes, P.J. (1988). Opioid modulation of non-cholinergic neural bronchoconstriction in guinea-pig in in vivo. *Br. J. Pharmacol. 95*:413.

Belvisi, M.G., Rogers, D.F., and Barnes, P.J. (1989). Neurogenic plasma extravasation: inhibition by morphine in guinea pig airways in vivo. *J. Appl. Physiol. 66*:268.

Belvisi, M.G., Stretton, C.D., and Barnes, P.J. (1991a). Nitric oxide as an endogenous modulator of cholinergic neurotransmission in guinea pig airways. *Eur. J. Pharmacol. 198*:219.

Belvisi, M.G., Stretton, C.D., and Barnes, P.J. (1991b). Bombesin-induced broncho-constriction in the guinea pig: mode of action. *J. Pharmacol. Exp. Ther. 258*:36.

Belvisi, M.G., Stretton, C.D., Miura, M., et al. (1992a). Inhibitory NANC nerves in human tracheal smooth muscle: a quest for the neurotransmitter. *J. Appl. Physiol. 73*:2505.

Belvisi, M.G., Stretton, C.D., and Barnes, P.J. (1992b). Nitric oxide is the endogenous neurotransmitter of bronchodilator nerves in human airways. *Eur. J. Pharmacol. 210*:221.

Belvisi, M.G., Miura, M., Stretton, C.D., and Barnes, P.J. (1993a). Endogenous vasoactive intestinal peptide and nitric oxide modulate cholinergic neurotransmission in guinea pig trachea. *Eur. J. Pharmacol. 231*:97.

Belvisi, M.G., Ward, J.K., Tadjarimi, S., Yacoub, M.H., and Barnes, P.J. (1993b). Inhibitory NANC nerves in human airways: differences in disease and after extrinsic denervation. *Am. Rev. Respir. Dis. 147*:A286.

Belvisi, M.G., Patacchini, R., Barnes, P.J., and Maggi, C.A. (1994). Facilitatory effects of selective agonists for tachykinin receptors on cholinergic neurotransmission. *Br. J. Pharmacol. 111*:103–110.

Bienenstock, J., Perdue, M., Blennerhassett, M., et al. (1988). Inflammatory cells and epithelium: mast cell/nerve interactions in lung in vitro and in vivo. *Am. Rev. Respir. Dis. 138*:S31.

Black, J.L., Johnson, P.R.A., and Armour, C.L. (1988). Potentiation of the contractile effects of neuropeptides in human bronchus by an enkephalinase inhibitor. *Pulm. Pharmacol. 1*:21.

Black, J.L., Johnson, P.R., Alouvan, L., and Armour, C.L. (1990). Neurokinin A with K$^+$ channel blockade potentiates contraction to electrical stimulation in human bronchus. *Eur. J. Pharmacol. 180*:311.

Bowden, J., and Gibbins, I.L. (1992a). Relative density of substance P-immunoreactive nerve fibres in the tracheal epithelium of a range of species. *FASEB J. 6*:A1276.

Bowden, J.J., and Gibbins, I.L. (1992b). Vasoactive intestinal peptide and neuropeptide Y coexist in non-noradrenergic sympathetic neurons to guinea pig trachea. *J. Auton. Nerv. System. 38*:1.

Brockaw, J.J., and White, G.W. (1992). Calcitonin gene-related peptide potentiates sub-stance P–induced plasma extravasation in the rat trachea. *Lung. 170*:89.

Brunelleschi, S., Vanni, L., Ledda, F., Giotti, A., Maggi, C.A., and Fantozzi, R. (1990). Tachykinins activate guinea pig alveolar macrophages: involvement of NK2 and NK1 receptors. *Br. J. Pharmacol. 100*:417.

Bungaard, A., Enehjelm, S.D., and Aggestrop, S. (1983). Pretreatment of exercise-induced asthma with inhaled vasoactive intestinal peptide. *Eur. J. Respir. Dis. 64*:427.

Cadieux, A, Benchekroun, M.T., St. Pierre, S., and Fournier, A. (1989). Bronchoconstric-tive action of neuropeptide Y (NPY) on isolated guinea pig airways. *Neuropeptides 13*:215.

Cardell, L-O., Uddman, R., Luts, A., and Sundler, F. (1991). Pituitary adenylate cyclase activating peptide (PACAP) in guinea-pig lung: distribution and dilatory effects *Regul. Pept. 36*:379.

Carstairs, J.R., and Barnes, P.J. (1986). Visualization of vasoactive intestinal peptide receptors in human and guinea pig lung. *J. Pharmacol. Exp. Ther. 239*:249.

Caughey, G.H. (1989). Roles of mast cell tryptase and chymase in airway function. *Am. J. Physiol. 257*:L39.

Cheung, A., Polak, J.M., Bauer, F.E., et al. (1985). The distribution of galanin immuno-reactivity in the respiratory tact of pig, guinea pig, rat, and dog. *Thorax 40*:889.

Cheung, D., Bel, E.H., den Hartigh, J., Dijkman, J.H., and Sterk, P.J. (1992a). An effect of an inhaled neutral endopeptidase inhibitor, thiorphan, on airway responses to neuro-kinin A in normal humans in vivo. *Am. Rev. Respir. Dis. 145*:1275.

Cheung, D., Timmers, M.C., Bel, E.H., den Hartigh, J., Dijuman, J.H., and Sterk, P.J. (1992b). An isolated neutral endopeptidase inhibitor, thiorphan, enhances airway narrowing to neurokin A in asthmatic subjects in vivo. *Am. Rev. Respir. Dis. 195*: A682.

Christofides, N.O., Yiangou, Y., Piper, P.J., et al. (1984). Distribution of peptide histidine isoleucine in the mammalian respiratory tract and some aspects of its pharmacology. *Endocrinology 115*:1958.

Coles, S.J., Said, S.I., and Reid, L.M. (1981). Inhibition by vasoactive intestinal peptide of glycoconjugate and lysozyme secretion by human airways in vitro. *Am. Rev. Respir. Dis. 124*:531.

Cunha, F.Q., Poole, S., Lorenzetti, B.B., and Ferreira, S.H. (1992). The pivotal role of tumour necrosis factor α in the development of inflammatory hyperalgesia. *Br. J. Pharmacol. 107*:660.

Cutz, E. (1982). Neuroendocrine cells of the lung—an overview of morphological charac-teristics and development. *Exp. Lung Res. 3*:185.

Daniele, R.P., Barnes, P.J., Goetzl, E.J., et al. (1992). Neuroimmune interactions in the lung. *Am. Rev. Respir. Dis. 145*:1230.

Delay-Goyet, P., and Lundberg, J.M. (1991). Cigarette smoke-induced airway oedema is blocked by the NK_1-antagonist CP-96,345. *Eur. J. Pharmacol. 203*:157.

Devillier, P., Advenier, C., Drapeau, G., Marsac, J., and Regoli, D. (1988a). Comparison of the effects of epithelium removal and of an enkephalinase inhibitor on the neurokinin-induced contractions of guinea pig isolated trachea. *Br. J. Pharmacol. 94*:675.

Devillier, P., Dessanges, J.F., Rakotashanaka, F., Ghaem, A., Boushey, H.A., and Lockhart, A. (1988b). Nasal response to substance P. and methacholine with and without allergic rhinitis. *Eur. Respir. J. 1*:356.

Dey, R.D., Mitchell, H.W., and Coburn, R.F. (1990). Organization and development of peptide-containing neurons in the airways. *Am. J. Respir. Cell Mol. Biol. 3*:187.

Dey, R.D., Altemus, J.B., and Michalkiewicz, M. (1991). Distribution of vasoactive intestinal peptide- and substance P–containing nerves originating from neurons of airway ganglia in cat bronchi. *J. Comp. Neurol. 304*:330.

Diamond, L., and O'Dennell, M. (1980). A nonadrenergic vagal inhibitory pathway to feline airways. *Science 208*:185.

Dixon, C.M.S., Fuller, R.W., and Barnes, P.J. (1987). The effect of nedocromil sodium on sulphur dioxide induced bronchoconstriction. *Thorax 42*:462.

Dixon, N., Jackson, D.M., and Richards, I.M. (1979). The effect of sodium cromogly-cate on lung irritant receptors and left ventricular receptors in anasthetized dogs. *Br. J. Pharmacol. 67*:569.

Djokic, T.D., Nadel, J.A., Dusser, D.J., Sekizawa, K., Graf, P.D., and Borson, D.B. (1989).

Inhibitors of neutral endopeptidase potentiate electrically and capsaicin-induced non-cholinergic contraction in guinea pig bronchi. *J. Pharmacol. Exp. Ther. 248*:7.

Dusser, D.J., Djoric, T.D., Borson, D.B., and Nadel, J.A. (1989). Cigarette smoke induces bronchoconstrictor hyperresponsiveness to substance P and inactivates airway neutral endopeptidase in the guinea pig. *J. Clin. Invest. 84*:900.

Ellis, J.L., and Farmer, S.G. (1989a). The effects of vasoactive intestinal peptide (VIP) antagonists, and VIP and peptide histidine isoleucine antisera on nonadrenergic, noncholinergic relaxations of tracheal smooth muscle. *Br. J. Pharmacol. 96*:513.

Ellis, J.L., and Farmer, S.G. (1989b). Modulation of cholinergic neurotransmission by vasoactive intestinal peptide and peptide histidine isoleucine in guinea pig tracheal smooth muscle. *Pulm. Pharmacol. 2*:107.

Ellis, J.L., and Farmer, S.G. (1989c). Effects of peptidases on nonadrenergic, noncholinergic inhibitory responses of tracheal smooth muscle: A comparison with effects on VIP- and PHI-induced relaxation. *Br. J. Pharmacol. 96*:521.

Elwood, W., Lötvall, J.O., Barnes, P.J., and Chung, K.F. (1991). Loop diuretics inhibit cholinergic and non-cholinergic nerves in guinea pig airways. *Am. Rev. Respir. Dis. 143*:1340.

Evans, T.W., Dixon, C.M., Clarke, B., Conradson, T.B., and Barnes, P.J. (1988). Comparison of neurokinin A and substance P on cardiovascular and airway function in man. *Br. J. Pharmacol. 25*:273.

Fisher, J.T., Anderson, J.W., and Waldron, M.A. (1993). Nonadrenergic noncholinergic neurotransmitter of feline trachealis: VIP or nitric oxide? *J. Appl. Physiol. 74*:31.

Foda, H.D., Higuchi, J., and Said, S.I. (1990). Helodermin, a VIP-like peptide, is a potent long-acting pulmonary vasodilator. *Am. Rev. Respir. Dis. 141*:A486.

Franconi, G., Graf, P., Lazarus, S., Nadel, J., and Caughey, G. (1989). Mast cell tryptase and chymase reverse airway smooth muscle relaxation induced by vasoactive intestinal peptide in ferret. *J. Pharmacol. Exp. Ther. 248*:947.

Frossard, N., and Barnes, P.J. (1987). μ-Opioid receptors modulate non-cholinergic constrictor nerves in guinea-pig airways. *Eur. J. Pharmacol. 141*:519.

Frossard, N., and Barnes, P.J. (1988). Effects of tachykinins on small human airways and the influence of thiorphan. *Am. Rev. Respir. Dis. 137*:195A.

Frossard, N., Rhoden, K. J., and Barnes, P.J. (1989). Influence of epithelium on guinea pig airway responses to tachykinins: role of endopeptidase and cyclooxygenase. *J. Pharmacol. Exp. Ther. 248*:292.

Fuller, R.W., Dixon, C.M.S., and Barnes, P.J. (1985). The bronchoconstrictor response to inhaled capsaicin in humans. *J. Appl. Physiol. 85*:1080.

Fuller, R.W., Conradson, T-B., Dixon, C.M.S., Crossman, D.C., and Barnes, P.J. (1987a). Sensory neuropeptide effects in human skin. *Br. J. Pharmacol. 92*:781.

Fuller, R.W., Dixon, C.M.S., Cuss, F.M.C., and Barnes, P.J. (1987b). Bradykinin-induced bronchoconstriction in man: mode of action. *Am. Rev. Respir. Dis. 135*:176.

Fuller, R.W., Maxwell, D.L., Dixon, C.M.S., et al. (1987c). The effects of substance P on cardiovascular and respiratory function in human subjects. *J. Appl. Physiol. 62*:1473.

Garland, A., Jordan, J.E., Kao, R., et al. (1992). Neurokinin-1 receptor blockade with (±)CP-96,345 inhibits hyperpnea-induced bronchoconstriction in guinea pigs. *Am. Rev. Respir. Dis. 145*:A45.

Ghatei, M.A., Sheppard, M., O'Shaunessy, D.J., et al. (1982). Regulatory peptides in the mammalian respiratory tract. *Endocrinology 111*:1248.

Gibbins, I.L., Furness, J.B., and Costa, M. (1987). Pathway specific patterns of coexistence of substance P, calcitonin gene related peptide, cholecystokinin, and dynorphin in neurons of the dorsal root ganglion of the guinea pig. *Cell. Tissue Res. 248*:417.

Gottschall, P.E., Tatsumo, I., Miyata, A., and Arimura, A. (1990). Characterization and distribution of binding sites for the hypothalamic peptide pituitary adenylate cyclase activating polypeptide. *Endocrinology 127*:272.

Greenberg, B., Rhoden, K., and Barnes, P.J. (1987). Relaxant effects of vasoactive intestinal peptide and peptide histidine isoleucine in human and bovine pulmonary arteries. *Blood Vessels 24*:45.

Guiliani, S., Amann, R., Papini, A.M., Maggi, C.A., and Meli, A. (1989). Modulatory action of galanin on responses due to antidromic activation of peripheral terminals of capsaicin sensitive sensory nerves. *Eur. J. Pharmacol. 163*:91.

Hakoda, H, and Ito, Y. (1990). Modulation of cholinergic neurotransmission by the peptide VIP, VIP antiserum and VIP antagonists in dog and cat trachea. *J. Physiol. 428*:133.

Hall, A.K., Barnes, P.J., Meldrum, L.A., and Maclagan, J. (1989). Facilitation by tachykinins of neurotransmission in guinea-pig pulmonary parasympathetic nerves. *Br. J. Pharmacol. 97*:274.

Harrison, N.K., Dawes, K.E., Barnes, P.J., Laurent, G.J., and Chung, K.F. (1992). Effects of neurokinin A, substance P and vasoactive intestinal peptide on human lung fibroblast proliferation and chemotaxis. *Am. Rev. Respir. Dis. 145*:A681.

Haynes, L.W., and Manley, C. (1988). Chemotactic response of guinea pig polymorphonucleocytes in vivo to rat calcitonin gene related peptide and proteolytic fragments. *J. Physiol. 43*:79P.

Hirayama, Y., Lei, Y.H., Barnes, P.J., and Rogers, D.F. (1993). Effects of two novel tachykinin antagonists FK 224 and FK 888 on neurogenic plasma exudation, bronchoconstriction and systemic hypotension in guinea pigs in vivo. *Br. J. Pharmacol. 108*:844.

Holzer, P. (1988). Local effector functions of capsaicin-sensitive sensory nerve endings: involvement of tachykinins, calcitonin gene related peptide, and other neuropeptides. *Neuroscience 24*:739.

Howarth, P.H., Djukanovic, R., Wilson, J.W., Holgate, S.T., Springall, D.R., and Polak, J.M. (1991). Mucosal nerves in endobronchial biopsies in asthma and non-asthma. *Int. Arch. Allergy Appl. Immunol. 94*:330.

Hsiug, T-R., Garland, A., Ray, D.W., Hershenson, M.B., Leff, A.R., and Solway, J. (1992). Endogenous sensory neuropeptide release enhances non specific airway responsiveness in guinea pigs. *Am. Rev. Respir. Dis. 146*:148.

Ichinose, M., and Barnes, P.J. (1990). A potassium channel activator modulates both noncholinergic and cholinergic neurotransmission in guinea pig airways. *J. Pharmacol. Exp. Ther. 252*:1207.

Ichinose, M., Belvisi, M.G., and Barnes, P.J. (1990a). Histamine H_3-receptors inhibit neurogenic microvascular leakage in airways. *J. Appl. Physiol. 68*:21.

Ichinose, M., Belvisi, M.G., and Barnes, P.J. (1990b). Bradykinin-induced bronchoconstriction in guinea-pig in vivo: role of neural mechanisms. *J. Pharmacol. Exp. Ther. 253*:1207.

Ichinose, M., Nakajima, N., Takahashi, T., Yamauchi, H., Inoue, H., and Takishima, T. (1992). Protection against bradykinin-induced bronchoconstriction in asthmatic patients by a neurokinin receptor antagonist. *Lancet 340*:1248.

Ichinose, M., Katsumata, U., Kikuchi, R., et al., (1993). Effect of tachykinin receptor antagonist on chronic bronchitis patients. *Am. Rev. Respir. Dis. 147*:A318.

Ihara, H., and Nakanishi, S. (1990). Selective inhibition of expression of the substance P receptor mRNA in pancreatic acinar AR42J cells by glucocorticoids. *J. Biol. Chem. 36*:22,441.

Jackson, D.M., Norris, A.A., and Eady, R.P. (1989). Nedocromil sodium and sensory nerves in the dog lung. *Pulm. Pharmacol. 2*:179.

Jacoby, D.B., Tamaoki, J., Borson, D.B., and Nadel, J.A. (1988). Influenza infection increases airway smooth muscle responsiveness to substance P in ferrets by decreasing enkephalinase. *J. Appl. Physiol. 64*:2653.

Jancso, G., Hökfelt, T., Lundberg, J.M., et al. (1981). Immunohistochemical studies on the effect of capsaicin on spinal and medullary peptide and monoamine neurons using antisera to substance P, gastrin/CCK, somatostatin, VIP, enkephalin, neurotensin, and 5-hydroxytryptamine. *J. Neurocytol. 10*:963.

Joos, G., Pauwels, R., and van der Straeten, M.E. (1987). Effect of inhaled substance P and neurokinin A in the airways of normal and asthmatic subjects. *Thorax 42*:779.

Kaufman, M.P., Coleridge, H.M., Coleridge, J.C.G., and Baker, D.G. (1980). Bradykinin stimulates afferent vagal C-fibres in intrapulmonary airways of dogs. *J. Appl. Physiol. 48*:511.

Khalil, Z., Andrews, P.V., and Helme, R.D. (1988). VIP modulates substance P induced plasma extravasation in vivo. *Eur. J. Pharmacol. 151*:281.

Komatsu, T., Yamamoto, M., Shimokata, K., and Nagura, H. (1991). Distribution of substance-P-immunoreactive and calcitonin gene related peptide-immunoreactive nerves in normal human lungs. *Int. Arch. Allergy Appl. Immunol. 95*:23.

Kroegel, C., Giembycz, M.A., and Barnes, P.J. (1990). Characterization of eosinophil activation by peptides. Differential effects of substance P, mellitin, and f-met-leu-phe. *J. Immunol. 145*:2581.

Kuo, H.P., Rohde, J.A.L., Barnes, P.J., and Rogers, D.F. (1990a). Morphine inhibition of cigarette smoke induced goblet cell secretion in guinea pig trachea in vivo. *Respir Med. 84*:425.

Kuo, H-P., Rohde, J.A.L., Tokuyama, K., Barnes, P.J., and Rogers, D.F. (1990b). Capsaicin and sensory neuropeptide stimulation of goblet cell secretion in guinea pig trachea. *J. Physiol. 431*:629.

Kuo, H-P., Rohde, J.A.L., Barnes, P.J., and Rogers, D.F. (1992a). K^+ channel activator inhibition of neurogenic goblet cell secretion in guinea pig trachea. *Eur. J. Pharmacol. 221*:385.

Kuo, H-P., Barnes, P.J., and Rogers, D.F. (1992b). Cigarette smoke-induced airway goblet cell secretion: dose dependent differential nerve activation. *Am. J. Physiol. 7*:L161.

Laitinen, L.A., Laitinen, A., and Haahtela, T. (1992). A comparative study of the effects of an inhaled corticosteroid, budesonide, and of a β_2-agonist, terbutaline, on airway inflammation in newly diagnosed asthma. *J. Allergy Clin. Immunol. 90*:32.

Lammers, J-W.J., Minette, P., McCusker, M., Chung, K.F., and Barnes, P.J. (1988).

Nonadrenergic bronchodilator mechanisms in normal human subjects in vivo. *J. Appl. Physiol. 64*:1817.

Lammers, J-W.J., Minette, P., McCusker, M., Chung, K.F., and Barnes, P.J. (1989). Capsaicin-induced bronchodilatation in mild asthmatic subjects: possible role of non-adrenergic inhibitory system. *J. Appl. Physiol. 67*:856.

Lammers, J.W.J., Barnes, P.J., and Chung, K.F. (1992). Non-adrenergic, non-cholinergic airway inhibitory nerves. *Eur. Respir. J. 5*:239.

Leff, A.R., Stimler, N.P., Munoz, N.M., Shioya, T., Talley, J., and Dame, C. (1982). Augmentation of respiratory mast cell secretion of histamine caused by vagal nerve stimulation during antigen challenge. *J. Immunol. 136*:1066.

Lei, Y-H., Barnes, P.J., and Rogers, D.F. (1992). Inhibition of neurogenic plasma exudation in guinea pig airways by CP-96,345, a new non-peptide NK_1-receptor antagonist. *Br. J. Pharmacol. 105*:261.

Lembeck, F., Donnerer, J., and Bartho, L. (1982). Inhibition of neurogenic vasodilation and plasma extravasation by substance P antagonists, somatostatin and [D-Met2, Pro5]-enkephalinamide. *Eur. J. Pharmacol. 85*:171.

Levine, J.D., Dardick, S.J., Roizan, M.F, Helms, C., and Basbaum, A.I. (1986). Contribution of sensory afferents and sympathetic efferents to joint injury in experimental arthritis. *J. Neurosci. 6*:3423.

Li, C.G., and Rand, M.J. (1991). Evidence that part of the NANC relaxant response of guinea-pig trachea to electrical field stimulation is mediated by nitric oxide. *Br. J. Pharmacol. 102*:91.

Lindsay, R.M., and Harmar, A.J. (1989). Nerve growth factor regulates expression of neuropeptide genes in sensory neurons. *Nature 337*:362.

Lötvall, J.O., Lemen, R.J., Hui, K.P., Barnes, P.J., and Chung, K.F. (1990a). Airflow obstruction after substance P. aerosol: contribution of airway and pulmonary edema. *J. Appl. Physiol. 69*:1473.

Lötvall, J.O., Skoogh, B-E., Barnes, P.J., and Chung, K.F. (1990b). Effects of aerosolized substance P on lung resistance in guinea pigs: a comparison between inhibition of neutral endopeptidase and angiotensin-converting enzyme. *Br. J. Pharmacol. 100*:69.

Lötvall, J.O., Hui, K.P., Löfdahl, C.-G., Barnes, P.J., and Chung, K.F. (1991). Capsaicin pretreatment does not inhibit allergen-induced airway microvascular leakage in guinea pig. *Allergy 46*:105.

Lotz, M., Vaughn, J.H., and Carson, D.M. (1988). Effect of neuropeptides on production of inflammatory cytokines by human monocytes. *Science 241*:1218.

Lowman, M.A., Benyon, R.C., and Church, M.K. (1988). Characterization of neuropeptide-induced histamine release from human dispersed skin mast cells. *Br. J. Pharmacol. 95*:121.

Lundberg, J.M., and Saria, A. (1983). Capsaicin-induced desensitization of the airway mucosa to cigarette smoke, mechanical and chemical irritants. *Nature 302*:251.

Lundberg, J.M., Fahrenkrug, J., Hokfelt, T., et al. (1984). Coexistence of peptide histidine isoleucine (PHI) and VIP in nerves regulating blood flow and bronchial smooth muscle tone in various mammals including man. *Peptides 5*:593.

Lundberg, J.M., Saria, A., Lundblad, L., et al. (1987). Bioactive peptides in capsaicin-sensitive C-fiber afferents of the airways: functional and pathophysiological implications. In: *The Airways: Neural Control in Health and Disease* (Kaliner, M.A., and Barnes, P.J., eds.) Marcel Decker, New York, p. 417.

Lundgren, J.D., Ostrowski, N., Baraniuk, J.N., Shelhamer, J.H., and Kaliner, M.A. (1990). Gastrin releasing peptide stimulates glycoconjugate release from feline tracheal explants. *Am. J. Physiol. 258*:L68.

Maggi, C.A., and Meli, A. (1988). The sensory efferent function of capsaicin sensitive sensory nerves. *Gen. Pharmacol. 19*:1.

Mak, J.C.W., and Barnes, P.J. (1988). Autoradiographic localization of calcitonin gene-related peptide binding sites in human and guinea pig lung. *Peptides 9*:957.

Mantyh, P.W. (1991). Substance P and the inflammatory and immune response. *Ann. NY Acad. Sci. 632*:263.

Mantyh, P.W., Gates, T.S., Zimmerman R.P., et al. (1988). Receptor binding sites for substance P but not substance K or neuromedin K are expressed in high concentrations by arterioles, venules and lymph nodes in surgical specimens obtained from patients with ulcerative colitis and Crohns disease. *Proc. Natl. Acad. Sci. USA 85*:3235.

Martin, J.G., Wang, A., Zacour, M., and Biggs, D.F. (1990). The effects of vasoactive intestinal polypeptide on cholinergic neurotransmission in isolated innervated guinea pig tracheal preparations. *Respir. Physiol. 79*:111.

Martins, M.A., Shore, S.A., Gerard, N.P., Gerald, C., and Drazen, J.M. (1990). Peptidase modulation of the pulmonary effects of tachykinins in tracheal superfused guinea pig lungs. *J. Clin. Invest. 85*:170.

Martling, C.R. (1987). Sensory nerves containing tachykinins and CGRP in the lower airways: functional implications for bronchoconstriction, vasodilation, and protein extavasation. *Acta Physiol. Scand.* (Suppl. 563):1.

Martling, C., Saria, A., Andersson, P., and Lundberg, J.M. (1984). Capsaicin pretreatment inhibits vagal cholinergic and noncholinergic control of pulmonary mechanisms in guinea pig. *Naunyn-Schmiedeberg Arch. Pharmacol. 325*:343.

Martling, C.R., Theodorsson-Norheim, E., and Lundberg, J.M. (1987). Occurrence and effects of multiple tachykinins: substance P, neurokinin A, and neuropeptide K in human lower airways. *Life Sci. 40*:1633.

Martling, C.R., Saria, A., Fischer, J.A., Hokfelt, T., and Lundberg, J.M. (1988). Calcitonin gene related peptide and the lung: neuronal coexistence and vasodilatory effect. *Regul. Pept. 20*:125.

Matran, R., Alving, K., Martling, C-R., Lacroix, J.S., and Lundberg, J.M. (1989a). Vagally mediated vasodilatation by motor and sensory nerves in the tracheal and bronchial circulation of the pig. *Acta Physiol. Scand. 135*:29.

Matran, R., Alving, K., Martling, C.R., Lacroix, J.S., and Lundberg, J.M. (1989b). Effects of neuropeptides and capsaicin on tracheobronchial blood flow in the pig. *Acta Physiol. Scand. 135*:335.

Matran, R., Martling, C.-R., and Lundberg, J.M. (1989c). Inhibition of cholinergic and nonadrenergic, noncholinergic bronchoconstriction in the guinea-pig mediated by neuropeptide Y and alpha$_2$-adrenoceptors and opiate receptors. *Eur. J. Pharmacol. 163*:15.

Matsuda, H., Kawkita, K., Kiso, Y., Nakano, T., and Kitamura, Y. (1989). Substance P induces granulocyte infiltration through degranulation of mast cells. *J. Immunol. 142*:927.

Matsuse, T., Thomson, R.J., Chen, X-R., Salari, H., and Schellenberg, R.R. (1991). Capsaicin inhibits airway hyperresponsiveness, but not airway lipoxygenase activity nor eosinophilia following repeated aerosolized antigen in guinea pigs. *Am. Rev. Respir. Dis. 144*:368.

McAlpine, L.G., and Thomson, N.C. (1989). Lidocaine-induced bronchoconstriction in asthmatic patients. Relation to histamine airway responsiveness and effect of preservative. *Chest 96*:1012.

McCormack, D.G., Mak, J.C.W., Coupe, M.O., and Barnes, P.J. (1989a). Calcitonin gene-related peptide vasodilation of human pulmonary vessels: receptor mapping and functional studies. *J. Appl. Physiol. 67*:1265.

McCormack, D.G., Salonen, R.O., and Barnes, P.J. (1989b). Effect of sensory neuropeptides on canine bronchial and pulmonary vessels in vitro. *Life Sci. 45*:2405.

McDonald, D.M. (1987). Neurogenic inflammation in the respiratory tract: actions of sensory nerve mediators on blood vessels and epithelium of the airway mucosa. *Am. Rev. Respir. Dis. 136*:S65.

McDonald, D.M., Schoeb, T.R., and Lindsey, J.R. (1991). Mycoplasma pulmonis infections cause long-lasting protection of neurogenic inflammation in the respiratory tract of the rat. *J. Clin. Invest. 87*:787.

McGillis, J.P., Organist, M.L., and Payan, D.G. (1987). Substance P and immunoregulation. *Fed. Proc. 14*:120.

Michoud, M.-C., Jeanneret-Grosjean, A., Cohen, A., and Amyot, R. (1988). Reflex decrease of histamine-induced bronchoconstriction after laryngeal stimulation in asthmatic patients. *Am. Rev. Respir. Dis. 138*:1548.

Midgren, B., Hansson, L., Karlsson, J.A., Simonsson, B.G., and Persson, C.G.A. (1992). Capsaicin-induced cough in humans. *Am. Rev. Respir. Dis. 146*:347.

Minami, M., Kuraishi, Y., Kawamura, M., Yamaguchi, T., Masu, Y., and Nakanishi, S. (1989). Enhancement of preprotachykinin A gene expression by adjuvant-induced inflammation in the rat spinal cord: possible inducement of substance P-containing spinal neurons in nociceptor. *Neurosci. Lett. 98*:105.

Miura, M., Noue, H., Ichinose, M., Kimura, K., Matsumata, U., and Takishima, T. (1990). Effect of nonadrenergic, nocholinergic inhibitory nerve stimulation on the allergic reaction in cat airways. *Am. Rev. Respir. Dis. 141*:29.

Miyata, A., Jiang, L., Dahl, R.D., et al. (1990). Isolation of a neuropeptide corresponding to the N-terminal 27 residues of the pituitary adenylate cyclase activating polypeptide with 38 residues (PACAP38). *Biochem. Biophys. Res. Commun. 170*:643.

Nadel, J.A. (1991). Neutral endopeptidase modulates neurogenic inflammation. *Eur. Respir. J. 4*:745.

Nakanishi, S. (1991). Mammalian tachykinin receptors. *Annu. Rev. Neurosci. 14*:123.

Naline, E., Devillier, P., Drapeau, G., et al. (1989). Characterization of neurokinin effects on receptor selectivity in human isolated bronchi. *Am. Rev. Respir. Dis. 140*:679.

Nathanson, I., Widdicombe, J.H., and Barnes, P.J. (1983). Effect of vasoactive intestinal peptide on ion transport across dog tracheal epithelium. *J. Appl. Physiol. 55*:1844.

Nichol, G.M., Alton, E.W.F.W., Nix, A., Geddes, D.M., Chung, K.F., and Barnes, P.J. (1990). Effect of inhaled furosemide on metabisulfite- and methacholine induced bronchoconstriction and nasal potential difference in asthmatic subjects. *Am. Rev. Respir. Dis. 142*:567.

Nichol, G.M., O'Connor, B.J., Le Compte, J.M., Chung, K.F., and Barnes, P.J. (1992). Effect of neutral endopeptidase inhibitor on airway function and bronchial responsiveness in asthmatic subjects. *Eur. J. Clin. Pharmacol. 42*:495.

Nieber, K., Baumgarten, C.R., Rathsack, R., Furkert, J., Oehame, P., and Kunkel, G. (1992). Substance P and β-endorphin-like immunoreactivity in lavage fluids of subjects with and without asthma. *J. Allergy Clin. Immunol. 90*:646.

Nohr, D., and Weihe, E. (1991). The neuroimmune link in the bronchus-associated lymphoid tissue (BALT) of cat and rat: peptides and neural markers. *Brain Behav. Immun. 5*:84.

Nong, Y.H., Titus, R.G., Riberio, J.M., and Remold, H.G. (1989). Peptides encoded by the calcitonin gene inhibit macrophage function. *J. Immunol. 143*:45.

O'Connor, B.J., Chen-Wordsell, M., Barnes, P.J., and Chung, K.F. (1993a). Effect of an inhaled opioid peptide on airway responses to sodium metabisulphite in asthma. *Br. J. Clin. Pharmacol.* (in press).

O'Connor, B.J., Lecomte, J.M., and Barnes, P.J. (1993b). Effect of an inhaled H_3-receptor agonist on airway responses to sodium metabisulphite in asthma. *Br. J. Clin. Pharmacol. 35*:55.

O'Dorisio, M.S., Shannaon, B.T., Fleshman, D.J., and Campolito, L.B. (1989). Identification of high affinity receptors for vasoactive intestinal peptide on human lymphocytes of B cell lineage. *J. Immunol. 142*:3533.

Ollerenshaw, S., Jarvis, D., Woolcock, A., Sullivan, C., and Scheibner, T. (1989). Absence of immunoreactive vasoactive intestinal polypeptide in tissue from the lungs of patients with asthma. *N. Engl. J. Med. 320*:1244

Ollerenshaw, S.L, Jarvis, D., Sullivan, C.E., and Woolcock, A.J. (1991). Substance P immunoreactive nerves in airways from asthmatics and non-asthmatics. *Eur. Respir. J. 4*:673.

Palmer, J.B.D., Cuss, F.M.C., Warren, J.B., and Barnes, P.J. (1986a). The effect of infused vasoactive intestinal peptide on airway function in normal subjects. *Thorax 41*:663.

Palmer, J.B.D., Cuss, F.M.C., and Barnes, P.J. (1986b). VIP and PHM and their role in nonadrenergic inhibitory responses in isolated human airways. *J. Appl. Physiol. 61*:1322.

Palmer, J.B.D., Cuss, F.M.C., Mulderry, P.K., et al. (1987). Calcitonin gene-related peptide is localized to human airway nerves and potently constricts human airway smooth muscle. *Br. J. Pharmacol. 91*:95.

Parsons, G.H., Nichol, G.M., Barnes, P.J., and Chung, K.F. (1992). Peptide mediator effects on bronchial blood velocity and lung resistance in conscious sheep. *J. Appl. Physiol.*

Paul, S., Said, S.I., Thompson, A.B., et al. (1989). Characterization of autoantibodies to vasoactive intestinal peptide in asthma. *J. Neuroimmunol. 23*:133.

Peatfield, A.C., Barnes, P.J., Bratcher, C., Nadel, J.A., and Davis, B. (1983). Vasoactive intestinal peptide stimulates tracheal submucosal gland secretion in ferret. *Am. Rev. Respir. Dis. 128*:89.

Piedimonte, G., Nadel, J.A., Umeno, E., and McDonald, D.M. (1990). Sendai virus infection potentiates neurogenic inflammation in the rat trachea. *J. Appl. Physiol. 68*:754.

Pietrowski, W., and Foreman, J.C. (1986). Some effects of calcitonin gene related peptide in human skin and on histamine release. *Br. J. Dermatol. 114*:37.

Ray, D.W., Hernandez, C, Leff, A.R., Drazen, J.M., and Solway, J. (1989). Tachykinins mediate bronchoconstriction elicited by isocapnic hyperpnea in guinea pigs. *J. Appl. Physiol. 66*:1108.

Reynolds, A.M., and McEvoy, R.D. (1989). Tachykinins mediate hypocapnia-induced bronchoconstriction in guinea pigs. *J. Appl. Physiol. 67*:2454.

Rhoden, K.J., and Barnes, P.J. (1990). Potentiation of non-adrenergic non-cholinergic relaxation in guinea pig airways by a cAMP phosphodiesterase inhibitor. *J. Pharmacol. Exp. Ther. 282*:396.

Riccio, M.M., Manzini, S., and Page, C.P. (1993). The effect of neonatal capsaicin in the development of bronchial hyperresponsiveness in allergic rabbits. *Eur. J. Pharmacol. 232*:89.

Robberecht, P., Waelbroeck, M., deNeef, P., Camus, J.C., Coy, D.H., and Christophe, J. (1988). Pharmacological characterization of VIP receptors in human lung membranes. *Peptides 9*:339.

Rogers, D.F., Belvisi, M.G., Aursudkij, B., Evans, T.W., and Barnes, P.J. (1988). Effects and interactions of sensory neuropeptides on airway microvascular leakage in guinea pigs. *Br. J. Pharmacol. 95*:1109.

Rogers, D.F., Aursudkij, B., and Barnes, P.J. (1989). Effects of tachykinins on mucus secretion on human bronchi in vitro. *Eur. J. Pharmacol. 174*:283.

Saban, R., Dick, E.C., Fishlever, R.I., and Buckner, C.K. (1987). Enhancement of parainfluenze 3 infection of contractile responses to substance P and capsaicin in airway smooth muscle from guinea pig. *Am. Rev. Respir. Dis. 136*:586.

Said, S.I. (1991). VIP as a modulator of lung inflammation and airway constriction. *Am. Rev. Respir. Dis. 143*:S22.

Sakamoto, T., Elwood, W., Barnes, P.J., and Chung, K.F. (1992). Pharmacological modulation of inhaled metabisulphite-induced airway microvascular leakage and bronchoconstriction in guinea pig. *Br. J. Pharmacol. 107*:481.

Sakamoto, T., Barnes, P.J., and Chung, K.F. (1993). Effect of CP-96,345, a non-peptide NK$_1$-receptor antagonist, against substance P–, bradykinin-, and allergen-induced airway microvascular leaks and bronchoconstriction in the guinea pig. *Eur. J. Pharmacol. 231*:31.

Salonen, R.O., Webber, S.E., and Widdicombe, J.G. (1988a). Effects of neuropeptides and capsaicin on the canine tracheal vasculature in vivo. *Br. J. Pharmacol. 95*:1262.

Salonen, R.O., Webber, S.E., and Widdicombe, J.G. (1988b). Effects of neuropeptides and capsaicin on the canine tracheal vasculature in vivo. *Br. J. Pharmacol. 95*:1262.

Saria, A., Martling, C.R., Yan, Z., Theodorsson-Norheim, E., Gamse, R., and Lundberg, J.M. (1988). Release of multiple tachykinins from capsaicin-sensitive nerves in the lung by bradykinin, histamine, dimethylphenylpiperainium, and vagal nerve stimulation. *Am. Rev. Respir. Dis. 137*:1330.

Sekizawa, K., Tamaoki, J., Graf, P.D., Basbaum, C.B., Borson, D.B., and Nadel, J.A. (1987). Enkephalinase inhibitors potentiate mammalian tachykinin-induced contraction in ferret trachea. *J. Pharmacol. Exp. Ther. 243*:1211.

Sekizawa, K., Laughey, G.H., Lazarus, S.C., Gold, W.M., and Nadel, J.A. (1989a). Mast cell tryptase causes airway smooth muscle hyperresponsiveness in dogs. *J. Clin. Invest.* *83*:175.

Sekizawa, K, Graf, P.D., and Nadel, J.A. (1989b). Somatostatin potentiates cholinergic neurotransmission in ferret trachea. *J. Appl. Physiol. 67*:2397.

Sertl, K., Wiedermann, C.J., Kowalski, M.L., et al. (1988). Substance P: The relationship between receptor distribution in rat lung and the capacity of substance P to stimulate vascular permeability. *Am. Rev. Respir. Dis. 138*:151.

Sestini, P., Bienenstock, J., Crowe, S.E., et al. (1990). Ion transport in rat tracheal ganglion in vitro. Role of capsaicin-sensitive nerves in allergic reactions. *Am. Rev. Respir. Dis. 141*:393.

Sharma, R.K., and Jeffery, P.K. (1990). Airway VIP receptor number is reduced in cystic fibrosis but not asthma. *Am. Rev. Respir. Dis. 141*:A726.

Sheppard, D., Thompson, J.E., Scypinski, L., Dusser, D.J., Nadel, J.A., and Borson, D.B. (1988). Toluene diisocyanate increases airway responsiveness to substance P and decreases airway neutral endopeptidase. *J. Clin. Invest. 81*:1111.

Shimosegawa, T., Foda, H.D., and Said, S.I. (1990). [Met]enkephalin-Arg^6-Gly^7-Leu^8-immunoreactive nerves in guinea pig and rat lungs: distribution, origin, and coexistence with vasoactive intestinal polypeptide immunoreactivity. *Neuroscience 36*:737.

Shimura, S., Sasaki, T., Ekeda, K., Sasaki, H., and Takishima, T. (1988). VIP augments cholinergic-induced glycoconjugate secretion in tracheal submucosal glands. *J. Appl. Physiol. 65*:2537.

Shore, S.A., Stimler-Gerard, N.P., Coats, S.R., and Drazen, J.M. (1988). Substance P induced bronchoconstriction in guinea pig. Enhancement by inhibitors of neutral metalloendopeptidase and angiotensin converting enzyme. *Am. Rev. Respir. Dis. 137*:331.

Smith, H. (1989). Animal models of asthma. *Pulm. Pharmacol. 2*:59.

Solway, J., and Leff, A.R. (1991). Sensory neuropeptides and airway function. *J. Appl. Physiol. 71*:2077.

Springall, D.R., Polak, J.M., Howard, L., et al. (1990). Persistence of intrinsic neurones and possible phenotypic changes after extrinsic denervation of human respiratory tract by heart-lung transplantation. *Am. Rev. Respir. Dis. 141*:1538.

Stretton, C.D., and Barnes, P.J. (1988). Modulation of cholinergic neurotransmission in guinea pig trachea by neuropeptide Y. *Br. J. Pharmacol. 93*:672.

Stretton, C.D., and Barnes, P.J. (1989). Cholecystokinin octapeptide constricts guinea-pig and human airways. *Br. J. Pharmacol. 97*:675.

Stretton, C.D., Belvisi, M.G., and Barnes, P.J. (1989). The effect of sensory nerve depletion on cholinergic neurotransmission in guinea pig airways. *Br. J. Pharmacol. 98*:782P.

Stretton, C.D., Belvisi, M.G., and Barnes, P. J. (1990a). Neuropeptide Y modulates non-adrenergic non-cholinergic neural bronchoconstriction in vivo and in vitro. *Neuropeptides 17*:163.

Stretton, C.D., Belvisi, M.G., and Barnes, P.J. (1990b). Sensory nerve depletion potentiates inhibitory NANC nerves in guinea pig airways. *Eur. J. Pharmacol. 184*:333.

Stretton, C.D., Belvisi, M.G., and Barnes, P.J. (1991). Modulation of neural broncho-constrictor responses in the guinea pig respiratory tract by vasoactive intestinal peptide. *Neuropeptides 18*:149.

Stretton, C.D., Miura, M., Belvisi, M.G., and Barnes, P.J. (1992a). Calcium-activated potassium channels mediate prejunctional inhibition of peripheral sensory nerves. *Proc. Natl. Acad. Sci. USA 89*:1325.

Stretton, C.D., Belvisi, M.G., and Barnes, P.J. (1992b). The effect of sensory nerve depletion on cholinergic neurotransmission in guinea pig airways. *J. Pharmacol. Exp. Ther. 260*:1073.

Tam, E.K., Franconi, G.M., Nadel, J.A., and Caughey, G.H. (1990). Protease inhibitors potentiate smooth muscle relaxation induced by vasoactive intestinal peptide in isolated human bronchi. *Am. J. Respir. Cell Mol. Biol. 2*:449.

Tatemoto, K. (1984). PHI—a new brain-gut peptide. *Peptides 5*:151.

Thompson, J.E., Scypinski, L.A., Gordon, T., and Sheppard, D. (1987). Tachykinins mediate the acute increase in airway responsiveness by toluene diisocyanate in guinea-pigs. *Am. Rev. Respir. Dis. 136*:43.

Tokuyama, K., Kuo, H-P., Rohde, J.A.L., Barnes, P.J., and Rogers, D.F. (1990). Neural control of goblet cell secretion in guinea pig airways. *Am. J. Physiol. 259*:L108.

Tomaki, M., Ichinose, M., Nakajima, N., et al. (1993). Elevated substance P concentration in sputum after hypertonic saline inhalation in asthma and chronic bronchitis patients. *Am. Rev. Respir. Dis. 147*:A478.

Tucker, J.F., Brave, S.R., Charalambons, L., Hobbs, A.J., and Gibson, A. (1990). L-N^G-nitro arginine inhibits non-adrenergic, noncholinergic relaxations of guinea pig isolated tracheal smooth muscle. *Br. J. Pharmacol. 100*:663.

Uddman, R., and Sundler, F. (1987). Neuropeptides in the airways: a review. *Am. Rev. Respir. Dis. 136*:S3.

Uddman, R., Moghimzadeh, E., and Sundler, F. (1984). Occurrence and distribution of GRP-immunoreactive nerve fibres in the respiratory tract. *Arch. Otorhinolaryngol. 239*:145.

Uddman, R., Luts, A., and Sundler, F. (1985). Occurrence and distribution of calcitonin gene related peptide in the mammalian respiratory tract and middle ear. *Cell Tissue Res. 214*:551.

Uddman, R., Hakanson, R., Luts, A., and Sundler, F. (1993). Distribution of neuropeptides in airways. In: *Autonomic Control of the Respiratory System* (Barnes, P.J., ed.) Harvard Academic, London (in press).

Umeda, Y., and Arisawa, H. (1989). Characterization of the calcitonin gene related peptide receptor in mouse T lymphocytes. *Neuropeptides 14*:237.

Umeno, E., Nadel, J.A., Huang, H.T., and McDonald, D.M. (1989). Inhibition of neutral endopeptidase potentiates neurogenic inflammation in the rat trachea. *J. Appl. Physiol. 66*:2647.

Umeno, E., McDonald, D.M., and Nadel, J.A. (1990). Hypertonic saline increases vascular permeability in the rat trachea by producing neurogenic inflammation. *J. Clin. Invest. 85*:1905.

Undem, B.J., Dick, E.C., and Buckner, C.K. (1983). Inhibition by vasoactive intestinal peptide of antigen-induced histamine release from guinea pig minced lung. *Eur. J. Pharmacol. 88*:247.

Undem, B.J., Myers, A.C., Barthlow, H., and Weinreich, D. (1991). Vagal innervation of guinea pig bronchial smooth muscle. *J. Appl. Physiol. 69*:1336.

van Ranst, L., and Lauweryns, J.M. (1990). Effects of long-term sensory vs. sympathetic denervation of the distribution of calcitonin gene-related peptide and tyrosine hydroxylase immunoreactivity in the rat lung. *J. Neuroimmunol.* *29*:131.

Ventresca, G.P., Nichol, G.M., Barnes, P.J., and Chung, K.F. (1990). Inhaled furosemide inhibits cough induced by low chloride content solutions but not by capsaicin. *Am. Rev. Respir. Dis.* *142*:143.

Verleden, G.M., Belvisi, M.G., Stretton, C.D., and Barnes, P.J. (1991). Nedocromil sodium modulates non-adrenergic non-cholinergic bronchoconstrictor nerves in guinea-pig airways in vitro. *Am. Rev. Respir. Dis.* *143*:114.

Ward, J.K., Belvisi, M.G., Fox, A.J., et al. (1993). Modulation of cholinergic neural bronchoconstriction by endogenous nitric oxide and vasoactive intestinal peptide in human airways in vitro. *J. Clin. Invest.* (in press).

Watling, K.J. (1992). Non peptide antagonists heralded a new era in tachykinin research. *Trends Pharmacol. Sci.* *13*:266.

Watson, N., Maclagan, J., and Barnes, P.J. (1993). Endogenous tachykinins facilitate transmission through parasympathetic ganglia in guinea-pig trachea. *Br. J. Pharmacol.* (in press).

Webber, S.E. (1988). The effects of peptide histidine isoleucine and neuropeptide Y on mucous volume output from ferret trachea. *Br. J. Pharmacol.* *55*:40.

Webber, S.E., and Widdicombe, J.G. (1987). The effect of vasoactive intestinal peptide on smooth muscle tone and mucous volume output from ferret trachea. *Br. J. Pharmacol.* *91*:139.

Webber, S.G., Lim, J.C.S., and Widdicombe, J.G. (1991). The effects of calcitonin gene related peptide on submucosal gland secretion and epithelial albumin transport on ferret trachea in vitro. *Br. J. Pharmacol.* *102*:79.

Wenzel, S.E., Fowler, A.A., and Schwartz, L.B. (1988). Activation of pulmonary mast cells by bronchoalveolar allergen challenge. In vivo release of histamine and tryptase in atopic subjects with and without asthma. *Am. Rev. Respir. Dis.* *137*:1002.

Widdicombe, J.G. (1990). The NANC system and airway vasculature. *Arch. Int. Pharmacodyn.* *303*:83.

Yeadon, M., Wilkinson, D., Darley-Usmar, V., O'Leary, V.J., and Payne, A.N. (1992). Mechanisms contributing to ozone-induced bronchial hyperreactivity in guinea pigs. *Pulm. Pharmacol.* *5*:39.

Yiangou, Y., DiMarzo, V., Spokes, R.A., Panico, M., Morris, H.R., and Bloom, S.R. (1987). Isolation, characterization, and pharmacological actions of peptide histidine valine 42, a novel prepro-vasoactive intestinal peptide derived peptide. *J. Biol. Chem.* *262*:14010.

DISCUSSION

Erdös: Is it known where VIP is cleaved by chymotrypsin?

Barnes: The amino acid sequence for VIP is well known. The N-terminal end is rich in acidic amino acids, and there is a decapeptide in the middle containing five positively charged amino acids (Lys, Arg). Thus, even when chymotrypsin cleaves VIP, these sequences may act as histamine releasers. This may be the reason why even cleaved VIP has activity in guinea pigs.

McDonald: Considering the paucity of SP-containing nerves in the airways and the putative increase in NK-1 receptors in the airways of asthmatics, is it possible that an as-yet-unidentified peptide might be released from sensory nerves in human airways and act on NK-1 receptors in human airway?

Barnes: I am interested in your comments on the need to investigate the sensory component of asthma. Clearly, the sensory neurotransmitters, SP and CGRP, are more evident in animals than in humans. However, sensory nerves are abundant, as reflected in staining with the marker PGP. Thus, it is possible that a sensory neuropeptide other than SP or CGRP might be important in humans.

Nadel: You have concluded that neurogenic inflammation (via tachykinins) is not important in humans. Yet, you showed evidence that SP-containing nerves exist in asthma and that an NK-1-receptor antagonist reduced symptoms of asthma. These findings support a role for neurogenic inflammation in asthma. Any comments?

Barnes: The role of tachykinins in asthma will only be known when potent and specific antagonists have been studied in asthmatic patients.

24

Inflammatory Mechanisms in Cystic Fibrosis Lung Disease

Melvin Berger

Case Western Reserve University School of Medicine and Rainbow Babies and Children's Hospital, Cleveland, Ohio

I. INTRODUCTION

Very little has been written about the role of neuropeptides in cystic fibrosis (CF). In the past, when the manifestations of CF in different organ systems were difficult to understand and interrelate, it was proposed that abnormalities in secretion of or responsiveness to vasoactive intestinal peptide, substance P, or other neuropeptides might account for many effects associated with CF (Said and Heinz-Erian, 1988). In recent years, the basic defects in ion permeability have become much better defined, particularly since the CF gene and its product have been identified, and a primary role for abnormalities in neuropeptide signaling seem less likely. The most serious complication of CF is chronic lung disease, which remains the major cause of mortality. This lung disease is typified by a continuous neutrophil-dominated inflammatory response to chronic infection with *Pseudomonas aeruginosa*. The factors that predispose the CF patient to infection with this particular organism are not clear, but it is increasingly obvious that the inability to eradicate this organism leads to a vicious cycle of neutrophil influx, inflammatory damage, and continued infection. This inflammatory process can thus serve as a model for effects of neutrophils in other situations, which may be more limited in space and/or time, but in which similar pathophysiology is likely to occur.

In recent years, many observations related to CF have been unified by the elucidation of the characteristic ion transport abnormalities and the discovery of the defective gene. In a stunning triumph of "reverse genetics" or "positional cloning," the defective gene and its product, which has been designated cystic fibrosis transmembrane regulator (CFTR) were identified by Collins, Tsui, Riordan, and their co-workers in Michigan and Toronto in 1989 (Rommens et al., 1989; Riordan et al., 1989; Kerem et al., 1989). A recent volume presents an extensive comprehensive review of CF research (Davis, 1993), so in this chapter I will only provide a brief overview of recent discoveries and unanswered questions, to bring our current understanding of the mechanisms of inflammation in the CF lung into focus.

II. STRUCTURE AND FUNCTION OF CFTR

The gene for CFTR is some 250,000 base pairs in length and resides on chromosome 7, region q31. It is a complex gene with 27 exons. The product is a protein of 1480 amino acids with a predicted M_r of 170 kDa, which can be divided into two symmetrical halves each of which consists of six membrane-spanning regions and an intracellular nucleotide binding fold. In addition, there is a single large globular domain with many potential phosphorylation sites, which has been designated the "R," or regulatory, domain. With the exception of this R domain, CFTR appears to be homologous to a family of proteins that contain similar nucleotide-binding regions called "ATP-binding cassettes" and are hence referred to as "ABC" proteins. These proteins all contain similar membrane-spanning regions that are visualized as forming a cylindrical transmembrane channel, accounting for the functions of other members of this ABC family, which include a variety of bacterial sugar and amino acid transporters as well as the mammalian multidrug-resistant or "P" glycoprotein, which is believed to pump drugs out of cells (Hyde et al., 1990). When speculative models of the CFTR structure are visualized in comparison to these molecules, the R domain hangs below the transmembrane channel like a ball on a chain, as if movement of the ball in or out of the end of the cylinder formed by the 12 membrane-spanning regions would close or open a channel. It is postulated that the state of occupancy of the nucleotide-binding folds determines the state of phosphorylation of the R domain, thus determining the activity of the putative channel (see below), correlating with the observation that the most common CF mutation, deletion of phenylalanine at amino acid position 508 (ΔF508), is in one of (the first) of these nucleotide-binding folds. Further evidence consistent with this model of the role of the R domain in regulating ion channel function of CFTR is provided from experiments by Davis et al., which show that overexpression of the R domain in normal cells decreases their Cl^- permeability and that expression of forms of the CFTR gene that lack the R domain is associated with greater permeability than native forms (Davis, 1993).

It has been shown that expression of wild-type CFTR confers characteristic Cl^- permeability properties (see below) to cells that lack this conductance (Kartner et al., 1991), and that insertion of purified CFTR into artificial lipid membranes also confers similar characteristic permeabilities (Bear et al., 1992). Insertion of wild-type CFTR also corrects ion transport abnormalities in CF cell lines (Drumm et al., 1990; Rich, 1990). In contrast, expression of mutant forms of CFTR fails to confer the same properties. These results are all consistent with the hypothesis that CFTR itself is a Cl^- channel that is defective in CF cells, although it is still possible that CFTR regulates the function or expression of other channels (as well).

The sequence of CFTR predicts two potential glycosylation sites in one of the exposed extracellular loops. On Western blots, forms of CFTR as large as 165,000 daltons have been identified, suggesting that it is glycosylated in vivo. Treatment with N-glycanase reduces these to 145,000 daltons, confirming that the increased molecular mass is in fact due to sugars. Recent studies suggest that glycosylation is intimately related to processing and translocation of newly synthesized CFTR within the cell. When CFTR is expressed in COS cells, the protein itself has $M_r = 130,000$ daltons. A "core glycosylated" form with $M_r = 135,000$ daltons is found in the endoplasmic reticulum, and a mature or fully glycosylated form with $M_r = 165,000$ daltons is subsequently found in the Golgi and plasma membrane. When many mutant forms of CFTR, including the most common $\Delta F508$ form, are expressed in these tissue culture cells, little of the mature form is made, and the incompletely glycosylated product fails to reach the plasma membrane (Cheng et al., 1990). This is not true for all CFTR mutants and may be partly related to the marked overexpression of the gene in this in vitro model system. The in vivo situation is unclear since both wild-type and $\Delta F508$ mutant CFTR are difficult to identify immunochemically in the apical membranes of secretory cells in vivo (Marino et al., 1991), and characteristic Cl^- conductances (see below) have been observed in some situations where CFTR antigen is undetectable.

III. ION TRANSPORT DEFECTS IN CF

It had long been suspected that the primary defect in CF would be related to an ion transport defect because of the well-recognized, characteristic increase in NaCl concentration of CF sweat that is used as a major diagnostic criterion. Indeed, studies of the physiology of the sweat gland led to the first documentation that CF involved a basic defect in Cl^- permeability (reviewed in Quinton and Reddy, 1993). In more recent years, attention has been focused more on Cl^- permeability and other ion transport characteristics of airway cells, since the major morbidity in CF is due to the lung disease. It is now clear that airway and most other cells have several different types of ion channels that have distinct electrical properties and are regulated in different ways (Anderson and Welsh, 1991; Cliff and Frizzell, 1990; Guggino, 1993). Airway cells, in particular, increase Cl^- secretion in response to a wide variety of stimuli, which employ several different intracellular messengers: increased cAMP, activation of protein kinase A, activation of protein kinase C, and increased free intracellular Ca^{2+}. CF cells have been shown to respond normally to increased intracellular Ca^{2+} and to extracellular ATP, but they fail to respond to stimuli such as β-adrenergic agents that act through cAMP, to cAMP itself, to agents that activate protein kinase A, or to the catalytic subunit of protein A kinase itself. It is now believed that the Ca^{2+}-regulated Cl^- conductance is due to a channel other than CFTR and that the failure to increase Cl^- permeability in response to cAMP is the most characteristic abnormality caused by mutations in CFTR. Increased Cl^- conductance in response to cAMP, measured either electrically or by decreasing fluorescence of Cl^- sensitive indicators such as SPQ, is now the trait most commonly used in different experimental systems in vitro to determine whether active CFTR is being expressed and/or if the CF defect is being expressed (or corrected as the case may be) in vitro (Welsh et al., 1993). Evidence that this abnormality is also expressed in the airway in vivo is provided by measurements of the transepithelial electrical potential, whose magnitude is reduced in CF (Knowles et al., 1981, 1983). The Cl^- permeability defect due to CFTR may not account for all of the transport abnormalities in CF, however, since CF airways in vivo as well as cultured cells and tissues also have been found to have increased Na^+ reabsorption, which has not yet been explained. This may be especially significant since CFTR is not abundantly expressed in the lung, and the increased Na^+ reabsorption may be more important than the decreased Cl^- secretion in determining the amount of water in the secretions.

IV. ABNORMALITIES IN SECRETIONS

It is widely assumed that the problems with CF secretions relate primarily to decreased hydration caused by impaired secretion of water or excess reabsorption, which in turn is secondary to impaired secretion of Cl^- or increased reabsorption of Na^+, since in most secretions water follows salt passively. These alterations in the salt and water concentration are felt to lead to precipitation of mucins and other macromolecules, which in turn block the secretory duct of the gland or the lumen in which the secretions collect. However, other hypotheses have also been advanced to account for specific structural and/or chemical abnormalities that have been reported in CF secretions. First, it has been reported that zymogen and mucin granules contain ion channels in their membranes that are activated at the time of secretion (Gasser et al., 1988). Decreased chloride content or other ionic changes resulting from abnormalities (i.e., mutant CFTR) in these channels could lead to abnormal initial unfolding and hydration of mucins, leading to persistent differences in their aqueous structure. Second, it has been reported that CF mucins have decreased sialylation and increased sulfation (Cheng et al., 1989). Barasch and co-workers have postulated that decreased Cl^- conductance of intracellular organelles such as the Golgi complex might be caused by abnormal CFTR and might impair their acidification (Barasch et al., 1991). Deficient acidification would shift the relative activity of different enzymes with different pH optima, resulting in abnormalities in the postsynthetic modification of mucins and other macromolecular glycoconjugates.

V. RELATION TO PATHOLOGY

Although many organs are affected in CF, with current therapy most of the morbidity and mortality is due to the lung disease. Outside of the lung, there is little evidence that inflammation plays a major role in the pathology of CF or that inflammatory processes are regulated abnormally. The increase in NaCl content of sweat is a classic pathognomic feature of CF, yet there is little evidence of gross pathology in the eccrine sweat gland (Tomashefski et al., 1993). Abnormalities in exocrine secretion more clearly lead to gross structural changes. The reduced water and Cl⁻ content of CF pancreatic juice leads to plugging of pancreatic ducts and secondary destruction of glands, producing another of the major characteristics of CF, but there is only minimal inflammation (Tomashefski et al., 1993). Analogous obstructive phenomena result in structural and functional alterations in other tissues as well, but these are rarely accompanied by inflammatory changes. In the intestine, the abnormal secretions may result in meconium ileus at birth or other obstructive phenomena in later life. Plugging of the tubules of the vas deferens generally leads to obliteration even before birth, resulting in sterility in men with CF. Similarly, the decreased water content of uterine mucus may lead to formation of a cervical plug that may decrease fertility in the woman as well. Obstruction of the ducts of submucosal glands in the lung may antedate infection as the earliest change observable in that organ, but the predominance of lung destruction follows infection, as discussed below. In all the above situations, plausible relations between the basic defect in CFTR, the corresponding changes in the secretions, and the pathology have been suggested, although many questions still remain. The ultimate destruction of the lung is unique in involving infection as a critical etiological element.

VI. INFECTION IN THE LUNG IN CF

Chronic infection of the airways remains the most important clinical problem in CF. Although many organisms may contribute to this, a few bacteria—*Staphylococcus aureus*, *Hemophilus influenzae*, and especially, a unique mucoid form of *Pseudomonas aeruginosa*—stand out as the major pathogens. In recent years, *Pseudomonas cepacia* has also been a problem in some areas. Point prevalence studies show that approximately 25% of patients below the age of 5 are infected with *S. aureus*, 15–20% with *H. influenzae*, and 20–30% with *P. aeruginosa* (reviewed in Konstan and Berger, 1993). With time, nearly all patients become infected with *P. aeruginosa*, and the percentages of older patients with *S. aureus* and *H. influenza* fall somewhat. Initially, culture results may be only intermittently positive for *P. aeruginosa*, but most patients then reach a stage in which the *P. aeruginosa* can no longer be eradicated from the airways and it is subsequently continuously present on culture. For most patients, the transition to this phase of their illness is associated with the beginning of a progressive downhill course in which a vicious cycle of infection, inflammation, and obstruction culminates in death.

The clinical course of lung disease is quite variable, even in patients with the same genotype (i.e., homozygous for ΔF508), although a few genotypes do seem to correlate with milder disease. Differences in the timing, frequency, and/or severity of exposure to infections and/or other environmental insults are almost certainly major determinants of any individual patient's course. Differences in the individuals' immune and inflammatory responses to infection and in the aggressiveness of therapy, as well as compliance, all contribute to the variability of the clinical course. Cough is the most frequent pulmonary symptom, although some infants and toddlers may present with wheezing and/or recurrent bronchiolitis. The onset of pulmonary symptoms is generally insidious, although occasionally, an infant may present with severe staphylococcal pneumonia. Most patients experience a slowly progressive course punctuated by periodic exacerbations of their chronic bronchitis. These may be due to increased numbers of bacteria, acquisition of new organisms, and/or increased invasion. The latter, in particular, may follow airway epithelial damage due to intercurrent viral infections or other insults. These exacerbations are generally associated with increased cough and sputum production, shortness of breath and decreased exercise tolerance, and rarely, fever and/or other systemic signs of infection. Wheezing generally accompanies CF lung disease and may result from narrowing of airways by inspissated mucopurulent secretions, inflammation and mucosal edema, and/or bronchospasm. As the disease progresses, these signs and symptoms increase in frequency and severity, and systemic effects such as poor weight gain, cachexia, and general malaise become prominent. Physical findings of chronic hyperinflation; continuously present crackles, rhonchi, and patchy air exchange; and digital clubbing become readily apparent. Pulmonary function tests demonstrate severely limited expiratory flow and marked air trapping, and chest X-rays show pronounced hyperinflation, peribronchial thickening, and impaction, with bronchiectasis in nearly all patients. As pulmonary disease continues, hemoptysis may occur due to erosions of airway walls adjacent to engorged bronchial vessels. Pneumothorax and cor pulmonale are also common.

Current treatment for CF emphasizes pancreatic enzyme replacement and good nutrition, relief of airway obstruction, and control of infection (Fiel, 1993). Treatments to relieve airway obstruction include bronchodilators, mucolytics, and various techniques of postural drainage. Efforts to control infection include immunization against influenza, measles, and other viruses; antibiotics; and efforts to limit exposure to *P. cepacia*. The treatments may help slow the inevitable progression of the lung disease but have not been able to halt it, and the mean life expectancy in most centers is only 20–30 years. In many centers, lung and/or heart-lung transplantation has been employed to prolong life, but this approach is obviously limited by its own complications, costs, and the lack of availability of donor organs. Newer approaches to therapy include attempts to diminish the products of inflammation (see below) with DNAase (Shak et al., 1990; Aitken et al., 1992) and antiproteases; anti-inflammatories to diminish the inflammatory response itself; agents designed to increase salt and water content of airway secretions by non-CFTR secretory pathways including amiloride (Knowles et al., 1990b) and UTP or ATP (Knowles et al., 1991); and finally, gene therapy.

VII. RELATION OF INFLAMMATION AND INFECTION

As noted above, infection of the CF airways occurs at an early age and rapidly becomes permanently established. The processes of infection and inflammation seem intimately related, and there is no evidence for primary inflammation in the absence of infection. Indeed, other organs such as the vas deferens and pancreas may become obliterated or severely damaged without evidence of an external inflammatory response. The inflammatory lung disease in CF is centered in the airways rather than the alveoli or interstitium, again suggesting the importance of externally acquired infection in its development. These processes become locally invasive, crossing the epithelium and going into the submucosa, then attacking the airway wall and supporting structures. The earliest pathological changes are hyperplasia of goblet cells and hypertrophy of mucous glands, which have been found been in infants (Bedrossian et al., 1976). Epithelial changes begin with denudation, loss of cilia, and squamous metaplasia. Recent evidence, showing that even infants who are too young to expectorate sputum are frequently positive for *S. aureus*, *H. influenza*, and *P. aeruginosa* when cultured bronchoscopically (Armstrong et al., 1993), suggests that infection and/or the inflammatory response may play a role in these early changes. With time, mucopurulent plugging of the airways occurs in all patients and is accompanied by acute and chronic infiltrates in the mucosa and submucosa. The invasion of the inflammatory process from the airway lumen into the wall causes ulceration and microabscess formation, which lead to progressive bronchiectasis.

Organized infiltrates are also found in nearly all patients but consolidation is not a major problem early in the course. What is most striking about the lung pathology in CF is that it continues to be dominated by a neutrophil-rich acute-type picture. Organization of the exudate and/or granuloma formation is not common. Immunohistopathological studies by Baltimore et al. (1989) confirm that *P. aeruginosa* antigens are found predominantly in an endobronchiolar distribution, closely associated with intense inflammation and obliterative changes in airways ≤1 mm in diameter. Although the organisms were seen to a lesser extent in alveolar spaces, they were not commonly identified in interstitial areas. This correlates with the clinical observations that rapidly progressive pneumonia is rare and bacteremia and sepsis almost unheard of except in terminally ill CF patients. It thus seems clear that the primary pathological process is an intense inflammatory response to a relatively noninvasive pathogen, rather than a deeply invasive, necrotizing infection in which severe tissue damage is produced by organisms invading across tissue planes and secreting extracellular toxins, as might be seen in neutropenic cancer patients. Similarly, the formation of granulomas containing small numbers of organisms that are resistant to killing by host phagocytes, as might be found in tuberculosis or histoplasmosis, does not seem to play a major role in CF. As will be explained below, this peculiar pattern of inflammatory disease in CF is most likely related to the unique characteristics of the major infecting pathogen, mucoid *P. aeruginosa*, and the interplay of the host inflammatory response with this infection.

Despite the inability of CF patients to eradicate *P. aeruginosa* or other organisms from the lungs, there is no increased incidence of infection outside of the respiratory tract and no evidence of any generalized disorder of regulation of inflammatory reactions in other parts of the body. Eradication of bacterial infection is usually accomplished by the phagocytic defense system, consisting of the phagocytes (neutrophils and macrophages) and opsonins (antibodies and complement). Despite extensive study in many laboratories, no consistent primary defect in chemotactic or phagocytic function has ever been identified (Santos and Hill, 1984). In fact, the predominance of neutrophils in CF sputum and bronchoalveolar lavage confirm that neutrophil migration is normal in vivo. The lack of metastatic infection and the lack of circulating immune complexes (except in very advanced cases) are consistent with the normal phagocytosis by CF alveolar macrophages of organisms other than *Pseudomonas* and with normal results in reticuloendothelial system clearance assays (Mantzouranis et al., 1988), suggesting that macrophage function is normal in vivo. Despite the fact that the gene for CFTR is expressed in lymphoblasts (Chen et al., 1989), no apparent functional consequences of CF mutations in lymphocytic cells are detectable. In vitro assays of lymphocyte proliferation in response to mitogens and standard test antigens are generally normal in CF patients, and the lack of increased susceptibility to protozoans or other opportunistic pathogens suggests that cell-mediated immunity is intact in vivo as well. Although there have been some reports of diminished lymphocyte responses to *Pseudomonas* antigens in advanced cases (Sorensen et al., 1991) and of imbalances among the IgG subclasses (Moss et al., 1987), these most likely are the result of prolonged LPS and/or antigenic stimulus rather than primary defects. In the absence of any systemic immune defect or phagocytic abnormality, it seems likely that local factors in the lung must predispose CF patients to the particular infections to which they are susceptible and/or must prevent eradication of such infections. Each of these possibilities will be discussed below.

VIII. SUSCEPTIBILITY OF CF PATIENTS TO INFECTION WITH MUCOID *PSEUDOMONAS*

The link between mutations in the CF gene, its biochemical consequences, and the increased susceptibility to lung infection, particularly with mucoid forms of *P. aeruginosa*, remains one of the most important unanswered questions in CF research. Clearly, the abnormalities in mucus secretions, which probably relate directly to the basic defect, must play a role. The underhydration of secretions and/or specific chemical alterations in mucins lead to inspissation of secretions in the ducts of various exocrine glands as well as in the lumens into which their secretions flow. This mechanism probably accounts for obstruction and obliteration in the vas deferens, pancreas, biliary tree, and even the intestine. A similar process occurs in the lung, but since this organ is open to the outside environment, airway obstruction and impaired mucus clearance lead to infection. Increased mucus secretion and airway obstruction do not by themselves explain the increased susceptibility to infection or to the selection of the particular flora unique to CF. Mucus hypersecretion and airway obstruction also occur in asthma, yet infection is usually absent in that disease. Impaired mucociliary clearance in primary ciliary dyskinesis syndromes results in chronic infection of the airways, but *P. aeruginosa* is not a predominant problem. Similarly, primary antibody deficiency syndromes are associated with chronic and recurrent sinopulmonary infection, but *P. aeruginosa* does not emerge as a predominant pathogen: *H. influenzae*, *S. pneumoniae*, and other encapsulated organisms are more often involved. Thus, there are striking differences between CF and other chronic pulmonary diseases associated with infection, most prominently the relatively restricted bacterial flora and the tendency of the infection to remain predominately endobronchial. Therefore, other factors in the CF lung must contribute to the increased susceptibility to a restricted spectrum of pathogens.

Although *S. aureus* and *H. influenzae* are commonly found in CF patients, there is nothing unique about CF isolates of these organisms, and they are both found frequently in other conditions that predispose patients to acute and chronic airway infection. Both organisms may be found in the upper airways and pharynx of normal children and adults, and the frequency with which they infect the lower airways may result from their presence at these sites when obstruction or epithelial damage occurs from other causes. Most CF patients have chronic sinusitis, and *S. aureus* as well as *H. Influenzae* are common causes of sinusitis in CF as well as non-CF patients. Sinusitis increases the burden of organisms in the upper airway and serves as a reservoir that may cause recurrent descending infection of the lower airways, particularly if the latter frequently become partly obstructed. It is

thus possible that lung infection with these organisms does not reflect a specifically increased susceptibility in CF per se, although it is possible that alterations
in the chemical structure of mucins and/or cell surfaces could increase the
adherence of these organisms.

The most distinctive finding in CF, and the one that is most clearly associated
with the progressive downhill course, is the chronic endobronchial infection with
mucoid strains of *P. aeruginosa*. Although *P. aeruginosa* may colonize the
oropharynx of non-CF patients, and this organism is widely prevalent in the
environment, lower airway infection with this organism is unusual in immunocompetent hosts. *P. aeruginosa* is also commonly present in the sinuses of CF patients,
yet this organism is not characteristic of sinusitis in non-CF patients, and even if it
does occur, it is rarely of the mucoid type. As discussed for the other organisms
above, an increased burden of *P. aeruginosa* in the sinuses may serve as a reservoir
for descending infection into the lower airways. Descending infections with *S.
aureus* and *P. aeruginosa* are seen in CF patients after lung transplantation, but
their incidence in this setting is apparently similar to that in non-CF transplant
recipients. Interestingly, and in contrast to the CF lung before transplantation,
these organisms can be eradicated from the lungs of posttransplant CF patients
with appropriate antibiotic therapy (Smyth et al., 1991).

Three major hypotheses have been advanced to account for the specific
predilection of CF patients to infection with the mucoid type of pseudomonas.
These are: (1) selection by the recurrent and/or chronic treatment of *Hemophilus*
and *Staphylococcus* infections with antibiotics to which *Pseudomonas* are not
sensitive, (2) a specific increase in affinity of CF cells and/or mucins for
Pseudomonas adhesins, and (3) an undefined, but unique characteristic of the CF
airway epithelial microenvironment or airway lining fluid that induces the mucoid
phenotype. Each of these will be discussed below.

The source from which *P. aeruginosa* is acquired remains controversial. The
organism is ubiquitous in the environment and may be acquired randomly by any
patient, as is frequently the case with debilitated or immunosuppressed individuals. However, some investigators have suggested that cross-infection is the major
route of initial acquisition of *Pseudomonas* (reviewed by Koch and Høiby, 1993).
Cross-infection could occur by direct contact with siblings or other CF patients, or
indirectly from contaminated hospital or home environments or from equipment
used by other patients. Studies in Denmark suggested that *P. aeruginosa* was more
prevalent and was acquired earlier in patients treated in CF centers than in those
not attending these centers (Pedersen, S.S., et al., 1986). Cohorting of infected
patients and segregation of uninfected patients has resulted in a decrease in the
incidence of new infections (Koch and Høiby, 1993). Several observations argue
against the role of selective pressure of antibiotics as a major contributor to
Pseudomonas infection in CF. Most important, many patients, even infants,
already harbor *Pseudomonas* at the time of initial diagnosis and treatment (May

et al., 1972; Stern et al., 1977; Abman et al., 1991). By the use of bronchoscopy to obtain cultures from young infants who do not expectorate, it is becoming clear that the initial prevalence of *P. aeruginosa* in extremely young patients identified by genetic screening is much higher than formerly suspected (Armstrong, 1993). These patients have not had prolonged courses of antibiotics, weakening the argument that selective pressure is caused by antibiotic treatment. Conversely, patients with other causes of chronic lung infection and/or bronchiectasis may be on antibiotics almost continuously for decades before they acquire *Pseudomonas*, if they ever do, and even then, mucoid types are extremely rare.

As in any other infection, establishment of *Pseudomonas* in the CF airway must start with adherence of the organisms to the epithelial surface. This may be facilitated if the organisms adhered to large amounts of uncleared mucins in CF (Vishwanath and Rampal, 1984, 1985), but despite careful assays, no enhanced specific binding of *P. aeruginosa* to CF mucins has been demonstrated (Sajjan et al., 1992). It has also been proposed that *Pseudomonas* may bind preferentially to airway cells that have been damaged by viruses or prior infection with *S. aureus* and/or *H. influenzae*. In particular, it has been shown that neutrophil elastase and other proteases may remove fibronectin from the surface of epithelial cells and expose receptors for *Pseudomonas* pili or other adhesins (Woods et al., 1981; Suter, 1988; Plotowski et al., 1991). These types of damage also occur in other lung diseases that do not favor infection with *P. aeruginosa*, and thus do not seem likely to explain the problem with this organism in CF. One attractive hypothesis that would explain the specificity of this association is suggested by the observations of Krivan et al. (1988) that asialogangliosides may serve as high-affinity binding sites for *P. aeruginosa* and the reports that CF mucins (and perhaps cell surface glycolipids as well as other glycoconjugates) are undersialylated (Cheng et al., 1989). Besides underhydration and other abnormalities in the viscoelastic properties of CF secretions, such specific alterations in their chemical structure could serve as the basis for the specificity of the association of one particular type of organism with this disease. The hypothesis has been advanced by Barasch and colleagues (1991) that mutations in CFTR cause decreased Cl^- permeability in intracellular organelles as well as the apical membrane and that this might result in decreased acidification of intracellular compartments like the Golgi apparatus. Decreased acidification might in turn shift the relative activity of different enzymes involved in postsynthetic modification of mucins, glycoproteins, and other complex carbohydrates, since competing transferases have different pH optima. This could readily explain why mucins (and, by implication, cell surface glycoconjugates as well) in CF are undersialylated and oversulfated and ties the latter observations together with the basic defect. Recent work by Zar et al. (1993) suggests increased binding of *P. aeruginosa* to CF nasal epithelial cells as compared to cells from normal donors, but the differences are modest at best, and there is much overlap between the two groups.

Since a major factor in the chronicity of the *Pseudomonas* infection in CF is likely to be the shift to the mucoid phenotype, it is possible that alterations in the airway lining fluid or other changes in the CF microenvironment that encourage this conversion account for the predilection of CF patients to chronic infection with this organism (May et al., 1991). Recent studies suggest that mucoid *P. aeruginosa* grow as a biofilm that physically obstructs phagocytosis (Anwar et al., 1992). These results are consistent with earlier electron microscopic studies by Lam et al. (1980) showing that in mucoid colonies, individual organisms are separated by copious amounts of amorphous material, and it is easy to visualize this as interfering with phagocytosis and stabilizing the organisms as a microcolony on the airway wall. In addition, because of its anionic nature, the mucoid exopolysaccharide (MEP) may serve to bind antibiotics (Slack and Nichols, 1981; Bayer et al., 1991) and/or antibodies and complement components, again restricting their access to the organisms deep within the mucoid colony. CF patients are initially infected with classic, nonmucoid strains of *P. aeruginosa*. Animal studies have shown that the shift to mucoidy can occur with prolonged persistent infection, even in otherwise normal hosts (Woods et al., 1991). The shift to mucoidy has also been induced in vitro under dehydrating conditions (Roychoudhury et al., 1991; May et al., 1991). Mucoid strains are rarely recovered from non-CF individuals, and mucoid phenotypes have occasionally been reported in other gram-negative bacteria recovered from CF patients. It has also been observed frequently that mucoid *P. aeruginosa* isolates from CF patients rapidly revert to the nonmucoid phenotype in the laboratory. It seems likely, therefore, that the specific chemical milieu in the CF lung accelerates or induces the shift to the mucoid phenotype, but the actual constituents or factors that are responsible remain undefined.

Although they are presented as three competing hypotheses for the sake of argument, it is certainly possible that each of these mechanisms contributes to the overall predilection of CF patients to mucoid *P. aeruginosa* infection. The large controlled study of early chronic antibiotic (Keflex) therapy now underway in the United States should shed light on whether this will decrease the subsequent infection with *P. aeruginosa* by decreasing the initial damage from *Staphylococcus* or *Hemophilus* infections, or whether it will increase acquisition of *P. aeruginosa* by its selective pressure. It is certainly possible that a minor degree of increased adherence of *P. aeruginosa* to CF mucins or epithelial cell surfaces promotes persistence of this organism, and continuous growth of the organisms in the CF milieu leads to the shift to mucoidy. *P. aeruginosa* infection in CF does not become rapidly invasive as it does in patients with neutropenia or neutrophil defects, but the shift to mucoidy usually leads to a state in which the CF patient can no longer eradicate the organism from the lung. MEP not only increases retention of the organisms in the airway and impairs phagocytosis and penetration of antibiotics, but it is highly antigenic and very likely serves as a major inflammatory stimulus that contributes to the inability of the host's homeostatic mechanisms to control the vicious cycle of inflammation and persistent infection that in the end destroys the patient's lungs.

IX. THE INFLAMMATORY RESPONSE IN THE CF LUNG

Since there is no apparent systemic immune defect in CF or disorder of regulation of inflammatory responses outside of the lung, my colleagues and I turned to bronchoalveolar lavage (BAL) as a tool to study the pathophysiology of the interaction of infection and inflammation in situ. This technique offers the advantage of sampling the milieu in the airways even in patients who are young and/or have only mild disease, and who consequently do not produce much sputum. In these patients, and with experienced bronchoscopists who have established good relationships with them, the procedure is quite benign, and serial studies are possible. Normally, alveolar macrophages are the predominant phagocytes recovered in BAL, with neutrophils generally accounting for less than 3% of the cells (Reynolds and Newball, 1974). In contrast, even in patients with mild CF, neutrophils predominate (Table 1). Recent studies show similar changes even in infants who were diagnosed by genetic screening and who have had little or no clinical evidence of lung problems (Khan, 1993). As the lung disease progresses, the number of neutrophils increases progressively and the sputum becomes markedly purulent and increasingly tenacious. Barton et al. (1976) showed that

Table 1 Bronchoalveolar Lavage (BAL) Findings in Stable CF Patients with Mild Lung Disease ($FEV_1 = 78 \pm 18\%$ of Predicted, $n = 20$) versus Non-CF Controls ($n = 29$)

	CF	Non-CF	CF/Non-CF ratio
1. Neutrophils ($\times 10^6$/ml)	46 ± 60 $(1-222)$	0.1 ± 0.2 $(0-0.6)$	460
Neutrophils/total cells (%)	58 ± 23 $(12-95)$	3 ± 3 $(0-9)$	
2. Active elastase (μM)	2.3 ± 3.8 $(0-14.1)$	0.005 ± 0.03 $(0-0.1)$	460
3. Total α_1-PI (μM)	2.7 ± 2.4 $(0.3-8.0)$	0.8 ± 0.6 $(0-2.8)$	3

Values expressed as mean $\pm$ SD (range) in respiratory epithelial lining fluid (ELF). ELF in recovered BAL fluid was calculated from simultaneous determination of urea concentration in BAL and serum. Elastase was assayed by hydrolysis of MeO-Suc-Ala-Ala-Pro-Val-pNa, which measures active elastase, above and beyond that which may be present in an inactive, complexed form. CF patients were aged 12–36 years; non-CF controls were healthy adults aged 18–36 years.

Source: Adapted from Konstan and Berger (1993), with permission of the copyright holder.

leukocytes were the major source of the greatly increased amount of DNA found in CF sputum. It has been more recently shown that improvements in pulmonary function after intensive inpatient antibiotic and pulmonary toilet "cleanouts" were associated with decreased density of *P. aeruginosa* and decreased content of DNA in the sputum, implying decreased neutrophil infiltration (Smith et al., 1988). This correlation strengthens the etiological relationship between the *Pseudomonas* infection, neutrophilic response, and decreased pulmonary function. Interestingly, this DNA makes a major contribution to the viscosity of the sputum in CF, and reducing the length of the DNA strands by the use of recombinant human DNase allows increased clearance of sputum and decreases airway obstruction, resulting in improvement in pulmonary function (Shak et al., 1990; Aitken et al., 1992).

Although animal models and the rapidly invasive nature of *Pseudomonas* infections in non-CF hosts suggest that extracellular virulence factors play an important role in *Pseudomonas* infections in those settings, they may be less important in CF. Pseudomonads are capable of secreting a number of factors such as exotoxin A, alkaline protease, and elastase as well as low-molecular-weight siderophores and phenazine pigments. The enzymes and toxins may be important early in the establishment of *Pseudomonas* in the airway by directly inhibiting phagocytosis, but they are unlikely to play much of a role in the chronic stage of the infection or in the progressive lung damage that ensues, since extensive necrosis is not a prominent feature of CF lung disease, and because they induce neutralizing antibodies that are found at high titer in most CF patients (Klinger et al., 1978; Hollsing et al., 1987). The pigments and siderophores may avoid immunological neutralization because of their low molecular weights. Phenazines may contribute to persistence of *Pseudomonas* by inhibiting lymphocyte responses (Nutman et al., 1987), thus depriving macrophages of cytokines such as γ-interferon that enhance their phagocytic and bactericidal activity. These pigments have also been shown to inhibit ciliary activity (Wilson et al., 1987). The siderophores may enable the bacteria to obtain iron despite the presence of host iron-binding proteins and may thus favor their growth (Sokol and Woods, 1984). In addition, the latter compounds may increase the toxicity of reactive oxygen species formed by macrophages and neutrophils in the inflammatory milieu (Coffman et al., 1990). Since there is not yet an animal model that mimics the initial colonization with *P. aeruginosa* and the conversion to chronic infection, the importance of any of these factors in vivo remains difficult to evaluate.

As discussed above, it is likely that the production of the extracellular alginate "slime" that gives CF strains of *P. aeruginosa* their characteristic phenotype makes a major contribution to the pathogenicity of chronic infection with this organism by protecting microcolonies in vivo and allowing their survival. When classical strains of *P. aeruginosa* shift to the mucoid phenotype, they generally decrease production of extracellular enzymes and toxins and decrease production of the long O-polysaccharide side chains on their LPS (Hancock et al., 1983). This shift from smooth to rough types of LPS is associated with a marked increase in the susceptibility to the bactericidal activity of serum in vitro, but despite the fact that antibodies and active complement have been found in CF sputum and lung fluid, the CF host is still unable to eradicate these organisms. In addition to its effects in restricting access of phagocytes, antibiotics, and host proteins to the individual organisms deep within the colonies and stabilizing these microcolonies in the airway, MEP may play an important role in the elaboration of lung damage in CF because it is highly antigenic. Although much of the antibody CF patients form against MEP may be ineffective in opsonophagocytic killing of organisms in suspension culture (Pier et al., 1987), these antibodies may still contribute to the formation of immune complexes in situ, which are likely to stimulate the inflammatory response and intensify the local tissue damage. Regardless of how bacterial factors and the host-bacterial interactions lead to the establishment of chronic *Pseudomonas* infection and conversion to the mucoid phenotype (reviewed by May et al., 1991), this is clearly a critical event that does not occur with other pathogens even in CF, and it is responsible for turning the inflammatory response into a self-perpetuating process that escapes from homeostatic control. This unchecked inflammatory response leads to a vicious cycle in which the products of the continued neutrophil influx actually impair clearance of the bacteria and eventually destroy the lung (Fig. 1).

X. NEUTROPHIL CHEMOATTRACTANTS IN THE CF LUNG

Several potent neutrophil attractants have been documented to be present in CF lung fluids, including the following:

LTB$_4$ (from PMN and AM)
C5a and other peptide fragments
Bacterial products
Cytokines: IL-8, TNF

Bacterial products that resemble the prototypic synthetic chemoattractant N-formyl-Met-Leu-Phe are almost certainly present, and although experiments with *P. aeruginosa* culture filtrates have not been reported per se, products of another gram-negative organism, *Escherichia coli*, are markedly chemotactic in vitro and rapidly recruit neutrophils into the lung in animal models. Bacterial LPS and/or other products are also probably important inducers of chemotactic cytokines from alveolar macrophages as well as epithelial cells (see below). Available evidence showing that *P. aeruginosa* in CF sputum are coated with immunoglobulin and complement (Hann and Holsclaw, 1976) and that soluble C3c in sputum correlates with *P. aeruginosa* infection (Schiøtz et al., 1979) strongly suggests that complement is activated in situ. Fick et al. (1986) have shown that the potent chemo-

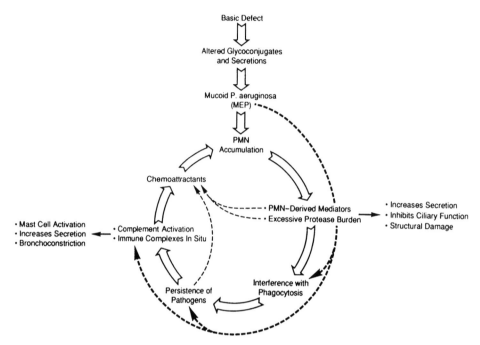

Figure 1 The inflammatory response in the CF lung as a vicious cycle initiated by *Pseudomonas* infection. Conversion to the mucoid phenotype with production of mucoid exopolysaccharide (MEP) by *P. aeruginosa* is critical because it physically impedes access of phagocytes to the bacteria. Persistence of bacteria leads to chronic neutrophil (PMN) influx by contributing bacterial chemoattractants per se as well as antigens that activate complement and generate C5a. In addition, LTB$_4$ produced by the PMNs themselves and IL-8, which is induced by PMN elastase, also contribute to the PMN influx. The PMNs bring an excessive protease burden to the lung, which causes structural damage and impairs phagocytic killing of the organisms, exacerbating the vicious cycle. (Reprinted from Konstan and Berger, 1993, with permission.)

attractant C5a is present in CF lung fluids. Furthermore, these workers have also shown that CF sputum contains proteases that can liberate chemotactic fragments from C5 in vitro. In addition, LTB4, which can be produced by both macrophages and neutrophils (Martin et al., 1984), has been shown to be the major eicosanoid present in CF BAL and sputum (Cromwell et al., 1981; Zakrewski et al., 1987; Konstan and Walenga, 1993). The increasing recovery of LTB4 upon incubation of CF sputum samples in vitro suggests that neutrophils that have migrated into the airways may be major sources of this chemoattractant in vivo. This is one example of how the neutrophils themselves can establish and perpetuate a positive feedback cycle that could amplify other chemotactic signals (Fig. 1).

Even though the macrophages in CF BAL show little or no phagocytic activity (Thomassen et al., 1980), they may still be potent producers of LTB4 and have been clearly shown to be actively producing IL-8 by immunofluorescence of cells fixed immediately upon removal from the lung by BAL (Bonfield et al., 1993). In addition to LPS, other bacterial products, antibody-antigen complexes, opsonized bacteria, and activated complement fragments, including C5a, are all present in the CF lung and have all been shown to be potent stimuli for the production of IL-8 and other neutrophil-attracting and -activating cytokines in vitro. IL-8 has also been shown to be produced by respiratory epithelial cells in response to soluble products of *P. aeruginosa* (Massion et al., 1993) and in response to neutrophil elastase itself (Nakamura, 1992), although it is not clear how much the cytokines produced by epithelial cells add to the large amounts that are likely produced by the macrophages.

In addition to IL-8 (Fick et al., 1991), other cytokines, including IL-1, tumor necrosis factor/cachectin (TNF-α), and IL-6, have all been found at elevated concentrations in CF BAL, although reports of cytokines in CF plasma remain controversial (Bonfield et al., 1993; Suter et al., 1989; Wilmott et al., 1990, 1991; Balcom et al., 1991; Brown et al., 1990). TNF has been demonstrated to enhance neutrophil oxidative responses and other activities that may be beneficial or deleterious depending on the setting (Hostoffer et al., 1993). In addition to their effects on neutrophils and other cells of the immune system, both TNF and IL-1 can induce fever and other systemic effects and both can induce production of IL-6 and IL-8 by macrophages. IL-1 is important in stimulating acute-phase responses and muscle protein catabolism. TNF causes alterations in lipoprotein metabolism and may be responsible for the cachexia that accompanies many chronic diseases (Cerami and Beutler, 1988). It is thus widely speculated that in addition to stimulating the local inflammatory response in the lung, IL-1 and TNF may contribute to the poor weight gain and other systemic manifestations of CF. Since IL-6, in addition to other activities, stimulates antibody production by B cells, it may be important in the hyperimmunoglobulinemia found in most CF patients. Assessing the in vivo role of any individual cytokine in a process as complicated as the inflammatory response in the CF lung is difficult at best. Cytokines have short half-lives, and circulating concentrations or activities may not be representative of local effects. Many binding factors and receptor antagonists are produced at the same time as the cytokines themselves and may be induced by the same stimuli. For example, we have found high levels of IL-1 receptor antagonist in CF BAL (Bonfield et al., 1993), and it is tempting to speculate that these may reduce the systemic response to the cytokine itself, accounting for the lack of fever, but that the cytokine's effects may predominate in the local microenvironment. Because of the complexity of cytokine networks, it is difficult to assess their true importance in CF lung disease, but it does seem likely that at least IL-8 and TNF are major contributors to the overall chemotactic and neutrophil-activating activity in the CF lung. Thus, many potent neutrophil attractants are produced in the inflammatory milieu of the CF lung, and the neutrophils themselves contribute as well—both directly, by producing LTB4, and indirectly, by the effects of elastase in inducing IL-8 (see below). It has been difficult to single out any one chemoattractant as "the most important," and it similarly seems unlikely that any single type of antagonist will be effective at significantly reducing the neutrophil influx into the lung.

XI. Neutrophil-Mediated Lung Damage in CF

Besides furthering the inflammatory response by both direct and indirect contributions to the total chemoattractant activity (see above), neutrophils may directly injure the lung by producing highly reactive oxygen metabolites and by releasing lysosomal enzymes and proteases (Weiss, 1989). Following stimulation, neutrophils and macrophages assemble a multienzyme complex that reduces molecular oxygen to produce several highly reactive and very toxic species, including singlet oxygen, superoxide, hydroxyl radical, and hydrogen peroxide (Babior, 1978; Hopkins et al., 1992). Together with the neutrophil enzyme myeloperoxidase, hydrogen peroxide can react with halides such as Cl^- to produce hypochlorous acid (bleach), which has potent microbicidal activity but may be less toxic to normal tissues than the oxygen radicals per se. In the context of host defense, the active oxidative products are generally considered important in killing bacteria intracellularly, and the proteases and lysosomal enzymes are believed to be necessary to break down and digest the bacterial remnants. However, these same oxidative products and enzymes may act extracellularly at inflammatory sites, damaging the surrounding tissues and altering the inflammatory milieu (Weiss, 1989).

Many stimuli that are chemotactic for neutrophils also cause secretion of granular contents (Weismann et al., 1980; Gallin, 1984). Enzymes such as collagenase and gelatinase that are released during neutrophil activation may have a normal role in facilitating movement of the neutrophil through basement membranes and tissue planes, while proteins such as lactoferrin, lysozyme, and defensins may exert direct bactericidal effects. The neutrophil has thus been described as a "secretory organ of inflammation" (Weismann et al., 1980; Gallin, 1984). It is obvious, however, that excess enzymes and/or oxidative products can digest connective tissue and damage normal tissues (Weiss, 1989). When the surface receptors and adhesion proteins of neutrophils and macrophages are engaged and phagocytosis is initiated, some of the oxidative metabolites and granular enzymes may escape—either before the incipient phagocytic vacuole is sealed around the ingested material, or if this seal is incomplete. When the target of phagocytosis is too large to be ingested and/or if the neutrophils' receptors are engaged by antibody and/or complement fragments bound to tissue surfaces, phagocytosis may be "frustrated." In this situation, the products of the oxidative respiratory burst and the granular enzymes are released into the interface between the neutrophil and the surface, rather than into a phagocytic vacuole (Wright and Silverstein, 1984; Rice and Weiss, 1990). This may occur when the neutrophils attack a microcolony or biofilm of *P. aeruginosa* whose surface is coated with antibodies to the MEP. Obviously, this could be an important mechanism of tissue damage, while organisms deep in the colony remain unaffected. Once a neutrophil has left the vascular space and migrates into the tissues, it does not return to the circulation. Recent observations by Tosi et al. showing that respiratory epithelial cells express ICAM-1 and other neutrophil adhesion molecules in response to treatment with TNF and other stimuli in vitro (Tosi et al., 1988) and confirmation that expression of this molecule is up-regulated on CF epithelia in vivo (Tosi et al., 1992) suggest that adhesion molecules on the epithelial cells may hold the neutrophils in place and retard their removal by mucociliary clearance. Thus, many neutrophils probably die and decay in situ, leaving their entire content of proteases and digestive enzymes to contribute to tissue damage and dysfunction at the inflammatory site.

Although host protease inhibitors should limit the injury caused by proteolytic enzymes, once the inhibitory capacity of these inhibitors has been exceeded, further activity is unimpeded and the damage becomes extensive because in CF, the *Pseudomonas* infection that provides the primary inflammatory stimulus persists and the neutrophil influx continues. Evidence from BAL studies confirms that the concentration of α_1-antitrypsin in the BAL of mild CF patients is severalfold higher than in normal subjects (Table 1), but the numbers of neutrophils in the same specimens are elevated several-hundred-fold. There is a corresponding several-hundred-fold excess of elastase and other neutrophil proteases, which clearly exceed the inhibitory capacity of the inhibitors. Tissue macrophages are larger than neutrophils and have a greater lifespan in situ, so it is tempting to speculate that a given amount of phagocytosis by macrophages might cause less tissue damage than the equivalent amount of ingestion by neutrophils. Studies from our own and other laboratories, however, have shown that alveolar macrophages express relatively few complement receptors and that their overall phagocytosis of fully opsonized bacteria does not match that of the neutrophils.

XII. NEUTROPHIL PROTEASES IN THE LUNG IN CF

A considerable body of evidence suggests that neutrophil-derived proteases play major roles not only in damaging lung tissue in CF, but also in perpetuating the inflammatory response. It is well established that unchecked neutrophil elastase is the major cause of emphysema in patients with α_1-antitrypsin deficiency. An analogous mechanism may account for the emphysema occasionally seen in CF. However, since the airways are the primary locus of neutrophil accumulation in CF, the effect of the uninhibited proteases is primarily on the airway rather than the alveolar wall. Thus, proteolytic destruction of structural proteins leads to bronchiectasis in CF rather than alveolar destruction as in α_1-antitrypsin deficiency. In vitro studies have shown that incubation of human bronchial epithelial cell monolayers with concentrations of elastase found in CF BAL leads to loss of tight junctions (Chung et al., 1991), and this has also been found in biopsy specimens. This in turn could lead to increased access of neutrophil products and inflammatory mediators to the submucosa, increasing their ability to damage the structural framework of the airway and to further stimulate the inflammatory response.

Many studies have documented active proteases in CF sputum and BAL (Table 2). Bruce et al. (1985) provided direct evidence for proteolytic destruction of lung tissue in vivo in CF by demonstrating increased urinary excretion of desmosines, cross-linked amino acid degradation products of elastin, with higher concentrations correlating with increased severity of lung disease. These investigators also demonstrated disruption and exfoliation of elastin fibers in areas of inflammation in situ, providing direct evidence that proteases released as part of the inflammatory response could generate substantial lung injury in CF.

The lung is normally protected against proteolytic damage by α_1-protease inhibitor (α_1-PI, α_1-antitrypsin), secretory leukocyte protease inhibitor (SLPI, formerly termed bronchial mucus protease inhibitor), and, to a lesser extent, α_2-macroglobulin. The latter may play some role at inflammatory sites, partic-

Table 2 Adverse Effects of Proteases in the CF Lung

Adverse effect	Ref.
On epithelium	
1. Structural damage	Bruce et al., 1985
2. Increased secretion and induction of secretory metaplasia	Snider et al., 1984
	Sommerhoff et al., 1990
	Schuster et al., 1992
3. Inhibit ciliary beating	Tegner et al., 1979
4. Induction of IL-8	Nakamura et al., 1992
On opsonophagocytic defense	
1. Cleavage of IgG	
Production of inhibitory fragments	Fick et al., 1984
Inhibition of cell activation by immune complexes	Döring et al., 1986
2. Inactivation of C3	Suter et al., 1984
3. Cleavage of C5 and generation of chemoattractants	Fick et al., 1986
4. Cleavage and inactivation of fibronectin	Suter et al., 1988
5. Cleavage of C3b receptor (CR1) from neutrophils	Berger et al., 1989
6. Cleavage of C3bi from pseudomonas	Tosi et al., 1990

Source: Reprinted from Konstan and Berger (1993), with permission of the publisher.

ularly if there is increased vascular permeability, but in general, its high molecular weight (725,000 daltons) limits its penetration into tissues and its efficacy at inflammatory sites. This protein probably exerts its most important effects intravascularly, limiting the activation of proteases in the coagulation and complement cascades and promoting their clearance from the circulation by macrophages of the RES. Local synthesis of α_2-macroglobulin by alveolar macrophages and/or penetration into tissues is difficult to assess. SLPI is produced and secreted by airway mucosal cells and is the predominant inhibitor in the conducting airways (Tegner, 1978; Ohlsson, 1980). α_1-PI (M_r = 45,000 daltons) diffuses from the blood and predominates in smaller airways and alveoli.

Both SLPI and α_1-PI can be inactivated by *Pseudomonas* proteases (Morihara, 1979; Johnson et al., 1982), which may limit their efficacy in CF, although the extent to which this occurs in vivo is difficult to assess (Tournier et al., 1985). α_1-PI can also be inactivated by the reactive oxygen species released during neutrophil activation, since its active site contains a methionine residue that is critical for activity. Oxidation of this methionine reduces the affinity of α_1-PI for neutrophil elastase by three orders of magnitude (Carp and Janoff, 1980; Janoff, 1988), as has been demonstrated with a variety of chemical oxidants in vitro as well as with activated neutrophils, monocytes, and alveolar macrophages (Carp and Janoff, 1980). This type of inactivation can also be produced by cell-free systems containing myeloperoxidase, H_2O_2, and Cl^-, which takes on particular significance since free myeloperoxidase is readily demonstrable in CF sputum (Goldstein and Döring, 1986). Inactivated α_1-PI can then be cleaved by a variety of proteases, as can α_1-PI complexed to serine proteases it has inhibited. Since there is a tremendous excess of proteases over inhibitors in the CF lung, it is not surprising that most of the α_1-PI in CF BAL or sputum is in the form of low-molecular-weight cleavage fragments rather than high-molecular-weight complexes (Goldstein and Döring, 1986). In most studies of free proteases in CF BAL and/or sputum, the majority of the activity is inhibitable by inactivators of serine proteases such as phenylmethyl sulfonyl fluoride, soybean trypsin inhibitor, and exogenous α_1-PI, rather than by chelators such as EDTA or phenanthrolene, or by the relative specific inhibitor of *Pseudomonas* elastase, phosphoramidon (reviewed in Konstan and Berger, 1993). This strongly suggests that neutrophil elastase and cathepsin G are the major enzymes present, since macrophage and *Pseudomonas* elastases are metalloenzymes (Mandl et al., 1962; Banda and Werb, 1981). Several studies have shown correlations between protease activity in CF sputum or BAL with the clinical condition of the patient and/or the occurrence of exacerbations, and increased release of proteases during exacerbations has also been suggested by reports that the serum concentration of elastase–α_1-PI complexes correlates with the activity of the lung infection (Hollsing et al., 1987). A particularly disconcerting recent observation about CF is that even in the first 6 months of life, there are significant increases in the amount of elastase bound to

α_1-PI and excess free elastase, which are readily detectable in BAL of CF infants identified by neonatal screening (Khan et al., 1993). In addition to direct structural damage, neutrophil elastase and other proteases that are thus present chronically in CF cause functional changes that exacerbate the abnormalities in secretion, contribute to the excessive inflammatory response, and interfere with local host defense mechanisms, as will be explained below.

XIII. FUNCTIONAL EFFECTS OF NEUTROPHIL PROTEASES

Although administration of high doses of human neutrophil extracts or purified elastase to animals causes pulmonary hemorrhage, lower doses cause less acute damage but can cause dramatic secretory metaplasia (Snider et al., 1984). Elastase inhibitors such as SLPI block these effects, confirming that are they due to enzymatic activity per se (Lucey et al., 1990). Somerhof et al. (1990) have shown that human neutrophil elastase and cathepsin G are extremely potent secreta-gogues for cultured bovine tracheal serous gland cells, releasing almost 50-fold more high-molecular-weight mucin glycoproteins than optimal doses of hista-mine. These results have more recently been extended to human cells as well (Schuster et al., 1992), and Klinger et al. (1984) have shown that *Pseudomonas* enzymes also have similar effects. Although the mechanisms by which the proteases induce secretion and/or secretory metaplasia are unknown, the results suggest that uninhibited neutrophil elastase may cause similar effects in CF both acutely and chronically. In light of the abnormalities in CF secretions, it seems likely that any increased secretory stimulus would increase the amount of mucus plugging and airway obstruction. It has also been shown that purulent sputum and purified elastase inhibit the beating of respiratory cilia in vitro and that these effects can be inhibited by α_1-PI (Tegner et al., 1979; Smallman et al., 1984). Together these effects of elastase may compound the problem and seriously exacerbate the problems of abnormal secretions and decreased mucociliary clear-ance in CF.

Another important functional change caused by elastase is the induction of IL-8 synthesis by respiratory epithelial cells (Nakamura et al., 1992). High concentra-tions of this potent chemotactic cytokine have been reported in CF BAL by several investigators (Fick et al., 1991; Bonfield et al., 1993; Khan et al., 1993), and CF BAL induces IL-8 production by bronchial epithelial cell lines in vitro (Nakamura et al., 1992), although it is not clear how much of the IL-8 is produced by epithelial cells versus alveolar macrophages. The induction of chemoattractant synthesis by neutrophil elastase provides yet another example of how the inflammatory process may become a vicious cycle, which feeds on itself and escapes from control.

As noted above, no systemic immunological or phagocytic defect has ever been documented in CF, and it thus seems likely that local factors in the lung must interfere with eradication of the infection. Uninhibited proteases in the inflammatory milieu seem likely candidates for causing such local interference, because in the lung milieu their activity greatly exceeds the activities of their inhibitors, but elsewhere in the body the normal excess inhibitory capacity is maintained (Goldstein and Döring, 1986; Hollsing et al., 1987). Studies from several laboratories have documented that soluble proteins important in host defense against bacteria may be cleaved and inactivated by proteases in CF BAL or sputum, as summarized in Table 2. In most cases, the proteolytic activity is inhibitable by serine protease inhibitors, suggesting that elastase and, to a lesser extent, cathepsins are responsible. Optimal phagocytosis requires deposition of opsonins on the bacteria that can interact with specific surface receptors on the phagocytes and that the latter, neutrophils and macrophages, must be activated to ingest and kill the bound bacteria. The opsonins include fibronectin, antibodies, and complement fragments derived from C3, all of which have been shown to be inactivated by proteases in CF sputum or BAL, reducing their efficacy in vivo or even producing inhibitory fragments that may block antigenic sites on the bacteria or occupy receptors on phagocytes. The effects of cleavage of fibronectin may further complicate the situation in CF because removal of epithelial cell surface fibronectin has been shown to increase the adherence of gram-negative bacteria to the cell surfaces (Woods et al., 1981; Abraham et al., 1983). Our own work has also focused on the effects of proteases on cell surfaces, but in the context of opsonophagocytosis. These studies have shown that neutrophils in CF BAL lack the C3b receptor, and that a serine protease in the BAL as well as purified elastase can efficiently remove this receptor from the cell surface, disabling one of the two major complement receptors and markedly decreasing the bactericidal activity of the cells (Berger et al., 1989). Studies by Tosi et al. (1990) showed that elastase also had effects on the opsonized bacteria, removing C3bi, the ligand for the other major complement receptor, further decreasing the bactericidal activity by creating an "opsonin-receptor mismatch."

Since the concentration of elastase necessary for the latter effect is less than one-tenth the amount necessary to remove C3b receptors from neutrophils, the observation that the neutrophils in fresh BAL are already lacking C3b receptors suggests that concentrations of elastase commonly present in the lung fluid of CF patients virtually eliminate the activity of the complement system in vivo. Studies by McElvaney et al. (1991) have shown that administration of α_1-PI in vitro or in vivo can reverse these effects and eliminate the inhibitory effects of CF BAL on neutrophil killing of *Pseudomonas*. Comparison of the maximal phagocytic activity of alveolar macrophages and neutrophils for *Pseudomonas* show that by far, the greatest enhancement of activity is when complement is included with neutrophils (Berger, 1990). Thus, elimination of complement function in vivo is likely to greatly reduce the overall local phagocytic activity. Macrophage function may be further compromised by proteases since optimal intracellular killing by macrophages depends on activation of these cells by protein lymphokines produced by activated T cells. *Pseudomonas* proteases have been shown to be capable of cleaving lymphocyte surface molecules including CD4 (Pedersen et al., 1987) and can inhibit lymphocyte activation and natural killer cell activity (Pedersen and Kharazmi, 1987). It is likely that excess host proteases can also mediate similar effects in the local environment. Inhibition of lymphocyte activation would consequently decrease γ-interferon production and alter production of other immunoregulatory cytokines as well. In addition, *Pseudomonas* alkaline protease and elastase degrade and inactivate γ-interferon (Horvat and Parmely, 1988; Horvat et al., 1989), suggesting that human elastase may have similar effects. Thus, neutrophil elastase is likely to exert a number of effects in the CF lung that specifically exacerbate the major morbidity of this disease by increasing secretions, attracting more neutrophils, and interfering with eradication of the chronic infection. Thus, the neutrophil's own products create a vicious cycle that leads to severe inflammatory damage to the lung and is ultimately responsible for the death of the patient.

XIV. NEUROPEPTIDES IN CF

There have been few specific studies of functions of neuropeptides and/or of nonadrenergic, noncholinergic nerves in CF. Before our current understanding of the mechanism of decreased Cl^- permeability in many CF tissues was so greatly advanced by the discovery of the gene and description of the likely structure and function of CFTR, Said and Heinz-Erian (1987) postulated that a deficiency of VIP or VIP-secreting nerves could account for many of the manifestations of CF. This was based on the widespread innervation of exocrine glands with VIP-secreting nerves and the fact that VIP stimulates secretion whose characteristics (i.e., increased Cl^- movement into sweat, increased water content of exocrine secretions, and increased pancreatic HCO_3^- secretion) seem opposite abnormalities in these secretions found in CF patients. Said and his colleagues also reported increased plasma VIP levels in CF patients, ruling out a defect in VIP synthesis, but noted decreased VIP-immunoreactive nerves in the sweat glands of CF patients (Heinz-Erian et al., 1985). They thus postulated that these two observations might be connected if the VIP was depleted from the nerves due to exaggerated postsynaptic release as an attempt to reverse the CF abnormalities or overcome a possible receptor dysfunction. Recent studies by Rogers et al. (1993) have shown that CF bronchial tissue secretes less mucus in response to substance P than normal bronchial tissue, and these differences were also found in response to methacholine and terbutaline. The decreased response to agonists that act through three different receptors suggests a defect in a common intermediate signaling step or in the end response itself, so this information does not shed much light on the presence or function of neuropeptide receptors per se.

In the normal airway, neutral endopeptidase is believed to modulate neurogenic neutrophil influx and inflammation by degrading substance P and/or other transmitters (Umeno et al., 1989). The excess proteases in the chronically infected CF lung may thus influence neuropeptide signaling in at least three ways: (1) cleavage of the neuropeptides themselves; (2) cleavage of the neutral endopeptidase; or (3) cleavage of receptors for the neuropeptides on epithelial, secretory, and/or inflammatory cells. The balance between these effects in the inflammatory milieu of the CF lung has not been defined, but this may well determine whether the neuropeptides play any role in the chronic phase of this disease. The lack of abnormalities in inflammatory responses elsewhere in the body and the likelihood that the reported abnormalities in neuropeptide-containing nerves and/or in the responsiveness of CF tissues may be secondary to other pathophysiological phenomena in CF suggest that the neuropeptides probably do not play much of a role in the initial establishment of CF lung disease either.

XV. SUMMARY

Little is known about the functions of neuropeptides in CF. Because VIP-secreting nerves are found in association with many exocrine glands as well as eccrine sweat glands, the hypothesis was advanced that abnormalities of neuropeptide secretion or responsiveness might account for the manifestations of CF in many different organs. However, the recent discovery of the CF gene and the predicted functional model of its product CFTR probably explain most of the ion transport abnormalities in CF. Most of the morbidity and mortality in CF in the present era is due to the chronic lung disease. This in turn is intimately related to persistent infection with an unusual mucoid type of *P. aeruginosa*. It is not clear why this particular organism is so specifically associated with CF, but several hypotheses have been proposed. There is no unusual incidence of infection outside the respiratory tract in CF, and there is no suggestion of dysregulated inflammatory reactions in other parts of the body. It thus seems unlikely that any generalized disorder of neurogenic inflammation or of chemotactic neuropeptides is involved. In the absence of any systemic immunological or phagocytic defect, it seems clear that local factors in the lung must prevent eradication of this organism. Studies of BAL cells and fluid suggest that once established, the chronic neutrophil-dominated inflammatory response becomes a vicious cycle that escapes from homeostatic control and continues to stimulate itself while damaging the lung and interfering with local host defense mechanisms. Although the discovery of the gene and rapid progress toward gene therapy hold great promise for CF patients, current therapeutic trials focus on ameliorating the process in the lungs. These include the use of aerosolized amiloride and purine nucleotides to increase secretion of water and facilitate mucociliary clearance, anti-inflammatory drugs to decrease the penetration of neutrophils into the lung, and DNase and protease inhibitors to diminish the deleterious effects of neutrophil products.

REFERENCES

Abman, S.H., Ogle, J.W., Harbeck, R.J., Butler-Simon, N., Hammond, K.B., and Accurso, F.J. (1991). Early bacteriologic, immunologic, and clinical courses of young infants with cystic fibrosis identified by neonatal screening. *J. Pediatr. 119*:211–217.

Abraham, S.N., Beachey, E.H., and Simpson, W.A. (1983). Adherence of *Streptococcus pyogenes, Escherichia coli,* and *Pseudomonas aeruginosa* to fibronectin-coated and uncoated epithelial cells. *Infect. Immun. 41*:1261–1268.

Aitken, M., Burke, W., McDonald, G., et al. (1992). Recombinant human DNase inhalation in normal and CF subjects: a phase I study. *JAMA 267*:1947–1951.

Anderson, M.P., and Welsh, M.J. (1991). Calcium and cAMP activate different chloride channels in the apical membrane of normal and cystic fibrosis epithelia. *Proc. Natl. Acad. Sci. USA 88*:6003–6007.

Anwar, H., Strap, J.L., and Costerton, J.W. (1992). Susceptibility of biofilm cells of *Pseudomonas aeruginosa* to bactericidal actions of whole blood and serum. *FEMS Microbiol. Lett. 71*:235–241.

Armstrong, D.S., Grimwood, K., Carzino, R., Carlin, J., Olinsky, A., and Phelan, P.D. (1993). Bacterial colonization of the lungs in infants with cystic fibrosis diagnosed by neonatal screening. *Am. Rev. Respir. Dis. 147*:A463.

Babior, B.M. (1978). Oxygen-dependent microbial killing by phagocytes. *N. Engl. J. Med. 298*:659-668, 721–725.

Balcom, D.F., Kushmerick, P.S., Casey, S.C., Sherry, B.L., Smith, A.L., and Wilson, C.B. (1991). Cytokines and clinical status in CF patients. *Pediatr. Pulmonol.* (Suppl. 6):302.

Baltimore, R.S., Christie, C.D.C., and Smith, G.J. (1989). Immunohistopathologic localization of *Pseudomonas aeruginosa* in lungs from patients with cystic fibrosis: Implications for the pathogenesis of progressive lung deterioration. *Am. Rev. Respir. Dis. 140*:1650–1661.

Banda, M.J., and Werb, Z. (1981). Mouse macrophage elastase: purification and characterization as a metalloproteinase. *Biochem. J. 589*:685–687.

Barasch, J., Kiss, B., Prince, A., Saiman, L., Gruenert, K., and Al-Awqati, Q. (1991). Defective acidification of intracellular organelles in cystic fibrosis. *Nature 352*:70–73.

Barton, A.D., Ryder, K., Lourenco, R.V., Dralle, W., and Weiss, S.G. (1976). Inflammatory reaction and airway damage in cystic fibrosis. *J. Lab. Clin. Med. 88*:423–426.

Bayer, A.S., Speert, D.P., Park, S., Tu, J., Witt, M., Nast, C.C., and Norman, D.C. (1991). Functional role of mucoid exopolysaccharide (alginate) in antibiotic-induced and polymorphonuclear leukocyte-mediated killing of *Pseudomonas aeruginosa. Infect. Immunol. 59*(1):302–308.

Bear, C.E., Li, C., Kartner, N., Bridges, R.J., Jensen, T.J., Ramjeesingh, M., and Riordan, J.R. (1992). Purification and functional reconstitution of the cystic fibrosis transmembrane conductance regulator (CFTR). *Cell 68*:809–818.

Bedrossian, C.W.M., Greenberg, S.D., Singer, D.B., Hansen, J.J., and Rosenberg, H.S. (1976). The lung in cystic fibrosis: A quantitative study including prevalence of pathologic findings among different age groups. *Hum. Pathol. 7*:195–204.

Berger, M., Sorensen, R.U., Tosi, M.F., Dearborn, D.G., and Döring, G. (1989). Complement receptor expression on neutrophils at an inflammatory site, the *Pseudomonas*-infected lung in cystic fibrosis. *J. Clin. Invest. 84*:1302–1313.

Berger, M., Norvell, T., Tosi, M, Konstan, M., and Schrieber, J. (1990). Lack of complement receptor expression by alveolar macrophages (AM) correlates with poor enhancement of phagocytosis of *P. aeruginosa* by complement (C). *Pediatr. Res.* 27:154A.

Bonfield, T.L., Ghnaim, H.A., Panusk, J.R., Konstan, M., and Berger, M. (1993). Cytokines in cystic fibrosis BAL. *Pediatr. Pulmonol.* (Suppl. 9):287.

Brown, M.A., Scuderi, P., Clarke McIntosh, J., Pfeffer, K., and Parsons, P. (1990). Relation between tumor necrosis factor-α and granulocyte elastase-α1-proteinase inhibitory complexes in the plasma of patients with cystic fibrosis. *Am. Rev. Respir. Dis.* 140:984–985.

Bruce, M.C., Poncz, L., Klinger, J.D, Stern, R.C., Tomashefski, J.F. Jr., and Dearborn, D.G. (1985). Biochemical and pathologic evidence for proteolytic destruction of lung connective tissue in cystic fibrosis. *Am. Rev. Respir. Dis.* 132:529–535.

Carp. H., and Janoff, A. (1980). Potential mediator of inflammation: Phagocyte-derived oxidants suppress the elastase-inhibitory capacity of alpha-1-proteinase inhibitor in vitro. *J. Clin. Invest.* 66:987–995.

Cerami, A., and Beutler, B. (1988). The role of cachectin/TNF in endotoxic shock and cachexia. *Immunol. Today* 9(1):28–31.

Chen, J.H., Schulman, H., and Gardner, P. (1989). A cAMP-regulated chloride channel in lymphocytes that is affected in cystic fibrosis. *Science* 243:657–660.

Cheng, P.W., Boat, T.F., Cranfill, K., Yankaskas, J.R., and Boucher, R.C. (1989). Increased sulfation of glycoconjugates by cultured nasal epithelial cells from patients with cystic fibrosis. *J. Clin. Invest.* 84:68–72.

Cheng, S.H., Gregory, R.J., Marshall, J., Paul, S., Sauza, D.W., White, G.A., O'Riordan, C.R., and Smith, A.E. (1990). Defective intracellular transport and processing CFTR is the molecular basis of most cystic fibrosis. *Cell* 63:827–834.

Chung, Y., Kercsmar, C.M., and Davis, P. B. (1991). Ferret tracheal epithelial cells grown in vitro are resistant to lethal injury by activated neutrophils. *Am. J. Respir. Cell Mol. Biol.* 5:125–132.

Cliff, W.H., and Frizzell, R.A., (1990). Separate Cl⁻ conductances activated by cAMP and Ca²⁺ in Cl⁻-secreting epithelial cells. *Proc. Natl. Acad. Sci. USA* 87:4956–4960.

Coffman, T.J., Cox, C.D., Edeker, B.L., and Britigan, B.E. (1990). Possible role of bacterial siderophores in inflammation. *J. Clin. Invest.* 86:1030–1037.

Cromwell, O., Walport, M.J., Morris, H.R., Taylor, G.W., Hodson, M.E., Batten, J., and Kay, A.B. (1981). Identification of leukotrienes D and B in sputum from cystic fibrosis patients. *Lancet* 2:164–165.

Davis, P.B., ed. (1993). *Cystic Fibrosis.* Lung Biology in Health and Diseases Series (C. Lenfant, exec. ed.), Vol. 64. Marcel Dekker, New York.

Döring, G., Goldstein, W., Botzenhart, K., Kharazmi, A., Schiotz, P. O., Høiby, N., and Dasgupta, M. (1986). Elastase from polymorphonuclear leucocytes: a regulatory enzyme in immune complex disease. *Clin. Exp. Immunol.* 64:597–605.

Drumm, M.L., Pope, H.A., Cliff, W.H., Rommens, J.M., Marvin, S.A., Tsui, L.-C., Collins, F.C. Frizzell, R.A., and Wilson, J.M. (1990). Correction of the cystic fibrosis defect in vitro by retrovirus-mediated gene transfer. *Cell* 62:1277–1233.

Fick R.B., Jr., Nagel, G.P., Squier, S.U., Wood, R.E., Gee, J.B.L., and Reynolds, H.Y. (1984). Proteins of the cystic fibrosis respiratory tract. Fragmented immunoglobulin G opsonic antibody causing defective opsonophagocytosis. *J. Clin. Invest.* 74:236–248.

Fick, R.B., Jr., Robbins, R.A., Squier, S.U., Schoderbek, W.E., and Russ, W.D. (1986). Complement activation in cystic fibrosis respiratory fluids: in vivo and in vitro generation of C5a and chemotactic activity. *Pediatr. Res. 20*:1258–1268.

Fick, R.B., Standiford, T.J., Kunkel, S.L., and Streiter, R.M. (1991). Interleukin-8 (IL-8) and neutrophil accumulation in the inflammatory airways disease of cystic fibrosis (CF). *Clin. Res. 39*:292A.

Fiel, S.B. (1993). Clinical management of pulmonary disease in cystic fibrosis. *Lancet 341*:1070–1074.

Gallin, J.I. (1984). Neutrophil specific granules: A fuse that ignites the inflammatory response. *Clin. Res. 32*:320–328.

Gasser, K.W., DiDomenico, J., and Hopfer, U. (1988). Secretagogues activate chloride transport pathways in pancreatic zymogen granules. *Am. J. Physiol. 254*:G93–G99.

Goldstein, W., and Döring, G. (1986). Lysosomal enzymes from polymorphonuclear leukocytes and proteinase inhibitors in patients with cystic fibrosis. *Am. Rev. Respir. Dis. 134*:49–56.

Guggino, W.D. Outwardly rectifying chloride channels and CF: a divorce and remarriage. (1993). *J. Bioenerg. Biomembr. 25*:27–35.

Hancock, R.E.W., Mutharia, L.M., Chan, L., Darveau, R.P., Speert, D.P., and Pier, G.B. (1983). *Pseudomonas aeruginosa* isolates from patients with cystic fibrosis: a class of serum-sensitive, nontypable strains deficient in lipopolysaccharide O side chains. *Infect. Immun. 42*(1):170–177.

Hann, S., and Holsclaw, D.S. (1976). Interactions of *Pseudomonas aeruginosa* with immunoglobulins and complement in sputum. *Infect. Immun. 14*:114–117.

Heinz-Erian, P., Dey, R.D., Flux, M., and Said, S.I. (1985). Deficiency of vasoactive intestinal peptide innervation in the sweat glands of cystic fibrosis patients. *Science 229*:1407–1408.

Hollsing, A.E., Granstrom, M., Vasil, M.L., Wretlind, B., and Strandvik, B. (1987). Prospective study of serum antibodies to *Pseudomonas aeruginosa* exoproteins in cystic fibrosis. *J. Clin. Microbiol. 25*:1868–1874.

Hopkins, P.J., Bemiller, L.S., and Curnutte, J.T. (1992). Chronic granulomatous diseases: diagnosis and classification at the molecular level. *Clin. Lab Med. 12*:277–304.

Horvat, R.T., and Parmely, M.J. (1988). *Pseudomonas aeruginosa* alkaline protease degrades human gamma interferon and inhibits its bioactivity. *Infect. Immun. 56*:2925–2932.

Horvat, R.T., Clabaugh, M., Duval-Jobe, C., and Parmely, M.J. (1989). Inactivation of human gamma interferon by *Pseudomonas aeruginosa* proteases: elastase augments the effects of alkaline protease despite the presence of α-2-macroglobulin. *Infect Immun. 57*:1668–1674.

Hostoffer, R.W., Krukovets, I., and Berger, M. (1993). The regulation of FcαR on neutrophils by TNFα and its role in cystic fibrosis. *Pediatr. Res. 33*:154A:907.

Hyde, S.C., Emsley, P., Hartshorn, M.J., Mimmack, M.M., Gileadia, U., Pearce, S.R., Gallagher, M.P., Gill, D.R., Hubbard, R.E., and Higgins, C.F. (1990). Structural model of ATP-binding proteins associated with cystic fibrosis, multidrug resistance and bacterial transport. *Nature 346*:362–365.

Janoff, A. (1988). Emphysema: proteinase-antiproteinase imbalance. In: *Inflammation:*

Basic Principles and Clinical Correlates. (Goldstein, I.M., and Snyderman, R., eds.) Raven Press, New York, pp. 803–814.

Johnson, D.A., Carter-Hamm, B., and Dralle, W.M. (1982). Inactivation of human bronchial mucosal proteinase inhibitor by *Pseudomonas aeruginosa* elastase. *Am. Rev. Respir. Dis. 126*:1070–1073.

Kartner, N., Hanrahan, J.W., Jensen, T.J., Naismith, A.L., Sun, S., Ackerley, C.A., Reyes, E.F., Tsui, L.-C., Rommens, J.M., Bear, C.E., and Riordan, J.R. (1991). Expression of the cystic fibrosis gene in non-epithelial invertebrate cells produces a regulated anion conductance. *Cell 64*:681–691.

Kerem, B.-S., Rommens, J.M., Buchanan, J.A. (1989). Identification of the cystic fibrosis gene: genetic analysis. *Science 245*:1073–1080.

Khan, T.Z., Wagener, J.S., Riches, D.W.H., and Accurso, F.J. (1993). Increased interleukin-8 levels and gene expression by pulmonary macrophages in bronchoalveolar lavage fluid from infants with cystic fibrosis. *Am. Rev. Respir. Dis. 147*:A463.

Klinger, J.D., Straus, D.C., Hilton, C.B., and Bass, J.A. (1978). Antibodies to proteases and exotoxin A of *Pseudomonas aeruginosa* in patients with cystic fibrosis: demonstration by radioimmunoassay. *J. Infect. Dis. 138*:49–58.

Klinger, J. D., Tandler, B., Liedtke, C.M., and Boat, T.F. (1984). Proteinases of *Pseudomonas aeruginosa* evoke mucin release by tracheal epithelium. *J. Clin. Invest. 74*:1669–1678.

Knowles, M., Boucher, R., and Gatzy, J.T. (1981). Increased bioelectric potential difference across respiratory epithelia in cystic fibrosis. *N. Engl. J. Med. 305*:1489–1495.

Knowles, M.R., Stutts, M.J., Spock, A., Fisher, N., Gatzy, J.T., and Boucher, R.C. (1990a). Abnormal ion permeation through cystic fibrosis respiratory epithelium. *Science 221*:1067–1070.

Knowles, M.R., Church, M.L., Waltner, W.E., Yankaskas, J.R., Gilligan, P., King, M., Edwards, L.J., Helms, R.W., and Boucher, R.C. (1990b). A pilot study of aerosolized amiloride for the treatment of lung diseases in cystic fibrosis. *N. Engl. J. Med. 322*:1189–1194.

Knowles, M.R., Clark, L.L., and Boucher, R.C. (1991). Activation by extracellular nucleotides of chloride secretion in the airway epithelium of patients with CF. *N. Engl. J. Med. 325*:533–538.

Koch, C., and Høiby, N. (1993). Pathogenesis of cystic fibrosis. *Lancet 341*:1065–1069.

Konstan, M.W., and Berger, M. (1993). Infection and inflammation of the lung in cystic fibrosis. In: *Cystic Fibrosis.* (Davis, P.B., ed.) Marcel Dekker, New York, pp. 219–276.

Konstan, M.W., and Walenga, R. (1993). Leukotriene B$_4$ is markedly elevated in the epithelial lining fluid of patients with cystic fibrosis. *Am. Rev. Respir. Dis. 148*:896–901.

Krivan, H.C., Ginsburg, V., and Roberts, D.D. (1988). *Pseudomonas aeruginosa* and Pseudomonas cepacia isolated from cystic fibrosis patients bind specifically to gangliotetraosylceramide (asialo GM1) and gangliotriaosylceramide (asialo GM2). *Arch. Biochem. Biophys. 260*(1):493–496.

Lam, J., Chan, R., Lam, K., and Costerton, J.W. (1980). Production of mucoid microcolonies by *Pseudomonas aeruginosa* within infected lungs in cystic fibrosis. *Infect. Immun. 28*:546–556.

Lucey, E.C., Stone, P.J., Ciccolella, D.E., Breuer, R., Christensen, T.G., Thompson, R.C., and Snider, G.L. (1990). Recombinant human secretory leukocyte-protease inhibitor: in vitro properties, and amelioration of human neutrophil elastase-induced emphysema and secretory cell metaplasia in the hamster. *J. Lab. Clin. Med. 115*:224–232.

Mandl, I., Keller, S., and Cohen, B. (1962). Microbial elastases: a comparative study. *Proc. Soc. Exp. Biol. Med. 109*:923–925.

Marino, C.R., Matovcik, L.M., Gorelick, F.S., and Cohn, J.A. (1991). Localization of the cystic fibrosis transmembrane conductance regulator in pancreas. *J. Clin. Invest. 88*:712–716.

Martin, T.R., Altman, L.C., Albert, R.K., and Henderson, W.R. (1984). Leukotriene B4 production by the human alveolar macrophage: a potential mechanism for amplifying inflammation in the lung. *Am. Rev. Respir. Dis. 129*:106–111.

Massion, P.P., Inoue, H., Richman-Eisenstat, J.B.Y., Jorens, P.G., Housset, B., and Nadel, J.A. (1993). *Pseudomonas aeruginosa* is chemotactic for neutrophils by stimulating interleukin-8 (IL-8) synthesis and secretion in human airway epithelial cells. *Clin. Res. 41*:138A.

May, J.R., Herrick, N.C., and Thompson, D. (1972). Bacterial infection in cystic fibrosis. *Arch. Dis. Child 47*:908–913.

May, T.B., Shinabarger, D., Maharaj, R., Kato, J., Chu, L., Devault, J.D., Roychoudhury, S., Zielinski, N.A., Berry, A., Rothmel, R.K., Misra, T.K., and Chakrabarty, A.M. (1991). Alginate synthesis by *Pseudomonas aeruginosa*: a key pathogenic factor in chronic pulmonary infections of cystic fibrosis patients. *Clin. Microbiol. Rev. 4*(4): 191–206.

McElvaney, N.G., Hubbard, R.C., Birrer, P., Chernick, M.S., Caplan, D.B., Frank, M.M., and Crystal, R.G. (1991). Aerosol α_1-antitrypsin treatment for cystic fibrosis. *Lancet 337*:392–394.

McElvaney, N.G., Nakamura, H., Birrer, P., Hébert, Wong, W.L., Alphonso, M., Baker, J.B., Catalano, M.A., and Crystal, R.G. (1992). Modulation of airway inflammation in cystic fibrosis. *J. Clin. Invest. 90*:1296–1301.

Morihara, K., Tsuzuki, H., and Oda, K. (1979). Protease and elastase of *Pseudomonas aeruginosa*: Inactivation of human plasma alpha 1 - proteinase inhibitor. *Infect. Immun. 24*:188–193.

Moss, R.B., and Lewiston, N.J. (1980). Immune complexes and humoral response to *Pseudomonas aeruginosa* in cystic fibrosis. *Am. Rev. Respir. Dis. 121*:23–29.

Moss, R.B., Hsu, Y.P., Van Eede, P.H., Van Leeuwen, A.M., Lewiston, N.J., and De Lange, G. (1987). Altered antibody isotype in cystic fibrosis: Impaired natural antibody response to polysaccharide antigens. *Pediatr. Res. 22*:708–713.

Nakamura, H., Yoshimura, K., McElvaney, N.G., Crystal, R.G. (1992). Neutrophil elastase in respiratory epithelial lining fluid of individuals with cystic fibrosis induces interleukin-8 gene expression in human bronchial epithelial cell line. *J. Clin. Invest. 89*:1478–1484.

Nutman, J., Berger, M., Chase, P.A., Dearborn, D.G., Miller, K.M., Waller, R.L., and Sorensen, R.U. (1987). Studies on the mechanism of T-cell inhibition by *Pseudomonas aeruginosa* phenazine pigment pyocyanine. *J. Immunol. 138*:3481–3487.

Ohlsson, K. (1980). Interactions between granulocyte proteases and protease inhibitors in the lung. *Bull. Eur. Physiopathol. Respir. 16*(Suppl.):209–222.

Pedersen, B.K., and Kharazmi, A. (1987). Inhibition of human natural killer cell activity by *Pseudomonas aeruginosa* alkaline protease and elastase. *Infect. Immun. 55*: 986–989.

Pedersen, S.S., Koch, C., Høiby, N., and Rosendal, K. (1986). An epidemic spread of multiresistant *Pseudomonas aeruginosa* in a cystic fibrosis centre. *J. Antimicrob. Chemother. 17*:505–516.

Pedersen, B.K., Kharazmi, A., Theander, T.G., Odum, N., Andersen, V., and Bendtzen, K. (1987). Selective modulation of the CD4 molecular complex by *Pseudomonas aeruginosa* alkaline protease and elastase. *Scand. J. Immunol. 26*:91–94.

Pfeffer, K., Huecksteadt, T., Hoidal, J. (1991). Tumor necrosis factor: *in vitro* production and expression in macrophages from CF patients. *Pediatr. Pulmonol.* (Suppl. 6):303.

Pier, G.B., Saunders, J.M., Ames, P., Edwards, M.S., Auerback, H., Goldfarb, J., Speert, D.P., and Hurwitch, S. (1987). Opsonophagocytic killing antibody to *Pseudomonas aeruginosa* mucoid exopolysaccharide in older noncolonized patients with cystic fibrosis. *N. Engl. J. Med. 317*:793–798.

Plotkowski, M.C., Chevillard, M., Pierrot, D., Altemayer, D., Zahm, J.M., Colliot, G., and Puchelle, E. (1991). Differential adhesion of *Pseudomonas aeruginosa* to human respiratory epithelial cells in primary culture. *J. Clin. Invest. 87*:2018–2028.

Quinton, P.M., and Reddy, M.M. (1993). The sweat gland. In: *Cystic Fibrosis*. (Davis, P.B., ed.) Marcel Dekker, New York, pp. 137–156.

Reynolds, H.Y., and Newball, H.H. (1974). Analysis of proteins and respiratory cells obtained from human lungs by bronchial lavage. *J. Lab. Clin. Med. 84*:559–573.

Rice, W.G., and Weiss, S.J. (1990). Regulation of proteolysis at the neutrophil-substrate interface by secretory leukoprotease inhibitor. *Science 249*:178–181.

Rich, D.P., Anderson, M.P., Gregory, R.J., Cheng, S., Paul, S., Jefferson, D.M., McCann, J.D., Klinger, K.W., Smith, A.E., and Welsh, M.J. (1990). Expression of cystic fibrosis transmembrane conductance regulator corrects defective chloride channel regulation in cystic fibrosis airway epithelial cells. *Nature 347*:358–363.

Riordan, J.R., Rommens, J.M., Kerem, B., Alon, N., Rozmahel, R., Grzelczak, Z., Zielenski, J., Lok, S., Plavsic, N., Chou, J.L., Drumm, M.L., Iannuzzi, M.C., Collins, F.S., and Tsui, L.-C. (1989). Identification of the cystic fibrosis gene: cloning and characterization of complementary DNA. *Science 245*:1066–1073.

Rogers, D.F., Alton, E.W.F.W., Dewar, A., Lethem, M.I., and Barnes, P.J. (1993) Impaired stimulus-evoked mucus secretion in cystic fibrosis bronchi. *Exp. Lung Res. 19*:37–53.

Rommens, J.M., Iannuzzi, M.C., Kerem, B., Drumm, M.L., Helmer, G., Dean, M., Rozmahel, R., Cole, J.L., Kennedy, D., Hidaka, N., Zsiga, M., Buchwald, M., Riordan, J.R., Tsui, L.C., and Collins, F.S. (1989). Identification of the cystic fibrosis gene: Chromosome walking and jumping. *Science 245*:1059–1065.

Roychoudhury, S., Zielinski, N.A., DeVault, J.D., Kato, J., et al. (1991). *Pseudomonas aeruginosa* infection in cystic fibrosis: biosynthesis of aliginate as a virulence factor. *Antibiot. Chemother. 44*:63–67.

Said, S.L., and Heinz-Erian, P. (1988). VIP and exocrine function: possible role in cystic fibrosis. In: *Cellular and Molecular Basis of Cystic Fibrosis*. (Mastella, G., and Quinton, P.M., eds.) San Francisco Press, San Francisco, pp. 355–361.

Sajjan, U., Reisman, J., Doig, P., Irvin, R.T., Forstner, G., and Forstner, J. (1992). Binding of nonmucoid *Pseudomonas aeruginosa* to normal human intestinal mucin and respiratory mucin from patients with cystic fibrosis. *J. Clin. Invest.* 89:657–665.

Santos, J.I., and Hill, H.R. (1984). Neutrophil function in cystic fibrosis. In: *Immunological Aspects of Cystic Fibrosis*. (Shapira, E., and Wilson, G.B., eds.) Boca Raton, FL, CRC Press, pp. 29–37.

Schiøtz, P.O., Sorensen, H., and Høiby, N. (1979). Activated complement in the sputum from patients with cystic fibrosis. *Acta. Pathol. Microbiol. Scand.* 87(Sect. C):1–5.

Schuster, A., Ueki, I., and Nadel, J.A. (1992). Neutrophil elastase stimulates tracheal submucosal gland secretion that is inhibited by ICI 200,355. *Am. J. Physiol.* 262: L86–91.

Shak, S., Capon, D.J., Hellmiss, R., Marsters, S.A., and Baker, C.L. (1990). Recombinant human DNase I reduces the viscosity of cystic fibrosis sputum. *Proc. Natl. Acad. Sci. USA* 87:9188–9192.

Slack, M.P.E., and Nichols, W.W. (1981). The penetration of antibiotics through sodium alginate and through the exopolysaccharide of a mucoid strain of *Pseudomonas aeruginosa*. *Lancet* 2:502–503.

Smallman, L.A., Hill, S.L., and Stockley, R.A. (1984). Reduction of ciliary beat frequency in vitro by sputum from patients with bronchiectasis: A serine proteinase effect. *Thorax* 39:663–667.

Smith, A.L., Redding, G., Doershuk, C., Goldmann, D., Gore, E., Hilman, B., Marks, M., Moss, R., Ramsey, B., Rubio, T., Schwartz, R.H., Thomassen, M.J., Williams-Warren, J., Weber, A., Wilmott, R.W., Wilson, H.D., and Yogev, R. (1988). Sputum changes associated with therapy for endobronchial exacerbation in cystic fibrosis. *J. Pediatr.* 112:547–554.

Smyth, R.L., Higenbottam, T., Scott, J., and Wallwork, J. (1991). Cystic fibrosis. 5. The current state of lung transplantation for cystic fibrosis. *Thorax* 46:213–216.

Snider, G.L., Lucey, E.C., Christensen, T.G., Stone, P.J., Calore, J.D., Catanese, A., and Franzblau, C. (1984). Emphysema and bronchial secretory cell metaplasia induced in hamsters by human neutrophil products. *Am. Rev. Respir. Dis.* 129:155–160.

Sokol, P.A., and Woods, D.E. (1984). Relationship of iron and extracellular virulence factors to *Pseudomonas aeruginosa* lung infections. *J. Med. Microbiol.* 18:125–133.

Sommerhoff, C.P., Nadel, J.A., Basbaum, C.B., and Caughey, G.H. (1990). Neutrophil elastase and cathepsin G stimulate secretion from cultured bovine airway gland serous cells. *J. Clin. Invest.* 85:682–689.

Sorensen, R.U., Waller, R.L., and Klinger, J.D. (1991). Cystic fibrosis. Infection and immunity to *Pseudomonas*. *Clin. Rev. Allergy* 9(1–2):47–74.

Stern, R.C., Boat, T.F., Doershuk, C.F., Tucker, A.S., Miller, R.B., and Matthews, L.W. (1977). Cystic fibrosis diagnosed after age 13: twenty-five teenage and adult patients including three asymptomatic men. *Ann. Intern. Med.* 87:188–191.

Suter, S., Schaad, U.B., Roux, L., Nydegger, U.E., and Waldvogel, F.A. (1984). Granulocyte neutral proteases and *Pseudomonas* elastase as possible causes of airway damage in patients with cystic fibrosis. *J. Infect. Dis.* 149:523–531.

Suter, S., Schaad, U.B., Morgenthaler, J.J., Chevallier, I., and Schnebli, H.P. (1988). Fibronectin-cleaving activity in bronchial secretions of patients with cystic fibrosis. *J. Infect. Dis.* 158:89–100.

Suter, S., Schaad, U.B., Roux-Lombard, P., Girardin, E., Grau, G., and Dayer, J.M. (1989). Relation between tumor necrosis factor-alpha and granulocyte elastase-alpha 1-proteinase inhibitor complexes in the plasma of patients with cystic fibrosis. *Am. Rev. Respir. Dis. 140*:1640–1644.

Tegner, H. (1978). Quantitation of human granulocyte protease inhibitors in non-purulent bronchial lavage fluids. *Acta. Otolaryngol. 85*:282–289.

Tegner, H., Ohlsson, K., Toremalm, N.G., and von Mecklenburg, C. (1979). Effect of human leukocyte enzymes on tracheal mucosa and its mucociliary activity. *Rhinology 17*:199–206.

Thomassen, M.J., Demko, C.A., Wood, R.E., Tandler, B., Dearborn, D.G., Boxerbaum, B., and Kuchenbrod, P.J. (1980). Ultrastructure and function of alveolar macrophages from cystic fibrosis patients. *Pediatr. Res. 14*:715–721.

Tomashefski, J.F., Abramowsky, C.R., and Dahms, B.B. (1993). The pathology of cystic fibrosis. In: *Cystic Fibrosis*. (Davis, P.B., ed.) Marcel Dekker, New York, pp. 435–481.

Tosi, M.F., Zakem, H., and Berger, M. (1990). Neutrophil elastase cleaves C3bi on opsonized *Pseudomonas* as well as CR1 on neutrophils to create a functionally important opsonin receptor mismatch. *J. Clin. Invest. 86*:300–308.

Tosi, M.F., Stark, J.M., Smith, C.W., Hamedani, A., Gruenert, D.C., and Infeld, M.D. (1992). Induction of ICAM-1 expression on human airway epithelial cells by inflammatory cytokines: effects on neutrophil-epithelial cell adhesion. *Am. J. Respir. Cell Mol. Biol. 7*:214–221.

Tournier, J.M., Jacquot, J., Puchelle, E., and Bieth, J.G. (1985). Evidence that *Pseudomonas aeruginosa* elastase does not inactivate the bronchial inhibitor in the presence of leukocyte elastase: studies with cystic fibrosis sputum and with pure proteins. *Am. Rev. Respir. Dis. 132*:524–528.

Umeno, E., Nadel, J.A., Huang, H., and McDonald, D.M. (1989). Inhibition of neutral endopeptidase potentiates neurogenic inflammation in the rat trachea. *J. Appl. Physiol. 66*:2647–2652.

Vishwanath, S., and Ramphal, R. (1984). Adherence of *Pseudomonas aeruginosa* to human tracheobronchial mucin. *Infect. Immun. 45*(1):197–202.

Vishwanath, S., and Ramphal, R. (1985). Tracheobronchial mucin receptor for *Pseudomonas aeruginosa*: predominance of amino sugars in binding sites. *Infect. Immun. 48*(2):331–335.

Weiss, S.J. (1989). Tissue destruction by neutrophils. *N. Engl. J. Med. 320*(6):365–376.

Weissman, G., Smolen, J.E., and Korchak, H.M. (1980). Release of inflammatory mediators from stimulated neutrophils. *N. Engl. J. Med. 303*:27–34.

Welsh, M.J., Anderson, M.P., Rich, D.P., Ostedgarrd, L.S., Gregory, R.J., Cheng, S.H., and Smith, A. (1993). Cystic fibrosis, CFTR, and abnormal electrolyte transport. In: *Cystic Fibrosis* (Davis, P.B., ed.) Marcel Dekker, New York, pp. 29–52.

Wilmott, R.W., Kassab, J.T., Kilian, P.L., Benjamin, W.R., Douglas, S.D., and Wood, R.E. (1990). Increased levels of interleukin-1 in bronchoalveolar washings from children with bacterial pulmonary infections. *Am. Rev. Respir. Dis. 142*:365–368.

Wilmott, R.W., Kociela, K., and Frenzke, M. (1991). Relationship of peripheral blood cytokine concentrations and cytokine production to clinical status in cystic fibrosis (CF). *Pediatr. Pulmonol.* (Suppl. 6):303.

Wilson, R., Pitt, T., Taylor, G., Watson, D., MacDermot, J., Sykes, D., Roberts, D., and Cole, P. (1987). Pyocyanin and 1-Hydroxyphenazine produced by *Pseudomonas aeruginosa* inhibit the beating of human respiratory cilia in vitro. *J. Clin. Invest.* *79*:221–229.

Woods, D.E., Straus, D.C., Johanson, W.G., Jr., and Bass, J.A. (1981). Role of fibronectin in the prevention of adherence of *Pseudomonas aeruginosa* to buccal cells. *J. Infect. Dis.* *143*:784–790.

Woods, D.E., Sokol, P.A., Bryan, L.E., Storey, D.G., Mattingly, S.J., Vogel, H.J., and Ceri, H. (1991). In vivo regulation of virulence in *Pseudomonas aeruginosa* associated with genetic rearrangement. *J. Infect. Dis. 163*:143–149.

Wright, S.D., and Silverstein, S.C. (1984). Phagocytosing macrophages exclude proteins from the zones of contact with opsonized targets. *Nature 309*:359–361.

Zakrzewski, J.T., Barnes, N.C., Costello, J.F., and Piper, P.J.(1987). Lipid mediators in cystic fibrosis and chronic obstructive pulmonary disease. *Am. Rev. Respir. Dis. 136*:779–782.

Zar, H., Saiman, L., Quittell, L., and Prince, A. (1993). Correlation between CF genotype and *P. aeruginosa* adherence to respiratory epithelial cells. *Pediatr. Res. 33*:389A.

DISCUSSION

Goetzl: You suggest that disordered phagocytosis of multicellular mucoid aggregates of pseudomonal organisms is a primary mechanism for the continued activation of neutrophils. What is the role of immune complexes or immunoglobulins?

Berger: Immune responses are normal in CF patients with high titres of antipseudomonal antibodies, hypergammaglobulinemia, and normal antibody and lymphocyte proliferation responses to test antigens. Some studies show a conversion to IgG_2 and IgG_4 antipseudomonal antibodies that have reduced phagocytic activity. This may be due to the prolonged antigenic stimulus.

25

Allergic Rhinitis

Michael A. Kaliner
Institute for Asthma and Allergy, Washington, D.C.

I. INTRODUCTION

The prevalence of allergic rhinitis is staggering; it is the most common immunological disease and also the most common chronic disease experienced by humans. About 10–17% of the American population experiences allergic rhinitis (Broder et al., 1974). Most patients develop symptomatic allergic rhinitis during childhood or young adulthood, although at least 30% of patients develop symptoms beyond age 30. Patients have been seen in their eighth decade who have developed new allergies. It is estimated that at least 25–30 million Americans have allergic rhinitis (Druce and Kaliner, 1988). The disease is chronic and requires frequent physician visits. Astoundingly, allergic rhinitis accounts for 2.5% for all physician visits, for all diseases. It is estimated that another 0.5% of all visits is for the purpose of immunotherapy.

II. CLINICAL PRESENTATION OF ALLERGIC RHINITIS

Clues to the allergic nature of the disease, and to the underlying causes, can be ascertained by the periodicity of the symptoms, precipitating events or exposures, and what helps the patient treat the symptoms. Suspicions generated by the history are strengthened by the signs of allergic rhinitis: swollen conjunctiva, swollen and boggy nasal mucosa with clear watery secretions, and the presence of extra creases and folds below the eyes (caused by the edema and rubbing) as well as vascular pooling over the checks giving an appearance described as allergic shiners. Skin testing is an invaluable confirmatory (not diagnostic) tool useful in helping establish the precise identity of aeroallergens causing the disease complex.

577

The pattern of the disease is determined in part by the spectrum of sensitivities exhibited by the patient. Thus, the most common symptom complex involves seasonal allergies, worse in the spring and fall of each year. These seasonal exacerbations correspond to the pollinating seasons of the trees (early spring), grasses (late spring and early summer), and weeds, especially ragweed (late summer through early fall). Conversely, year-around allergens such as dust mites, animal emanations, cockroaches, and household molds can cause perennial symptoms.

A. IgE Production

The development of sensitivity to an offending allergen requires the synthesis of IgE antibodies directed at the antigenic epitope. IgE production is carefully regulated and requires the recognition of the allergen, development of IgE-producing plasma cells, and the synthesis and secretion of IgE. All of these events are regulated by T-lymphocyte-derived factors, with IL-4 increasing IgE production and interferon-γ suppressing it. The capacity to produce IgE in increased amounts is genetically controlled, and therefore, allergies run in families.

B. Mast Cells

Once sufficient exposure to allergen has resulted in increased IgE production with sensitization of mast cells, subsequent exposure to allergen can elicit an allergic response. In humans, the mast cell is found in the loose connective tissue of all organs, especially around blood vessels, nerves, and lymphatics. Mast cells are most abundant in the skin, the upper and lower respiratory tract, and the gastrointestinal and reproductive mucosa (Metcalfe et al., 1981). In the nasal mucosa, most mast cells are found in the superficial 200 μm, generally clustered just beneath the basement membrane and in the epithelium. The majority of mast cells in the nasal mucosa contain both chymase and tryptase and are known as MC_{tc} (Igarashi et al., 1993). The number of mast cells increases during the allergy season, and the number of intraepithelial mast cells may increase dramatically. The intraepithelial mast cells are predominantly of the MC_t type, containing trypsin but not chymase (Igarashi et al., 1993). In electron micrographs of resting nasal mast cells, the ubiquitous, densely stained, secretory granules are roughly spherical, and most are 0.2–0.5 μm in diameter. Most granules comprise a dense, amorphous matrix material with embedded or interspersed crystalline constituents in the form of scrolls, gratings, or lattices (Friedman and Kaliner, 1985). Scrolls are found most commonly in cells undergoing degranulation. It should be noted that, even in experimentally unstimulated nasal tissue, variable numbers of degranulating mast cells are present. In unstimulated nasal epithelium, one-third of mast cells may be degranulated, and the percentage progressively increases from the lamina propria to the mucosal surface (Friedman and Kaliner, 1985).

Early events following IgE-mediated stimulation of human lung mast cells include granule swelling with loss of stainable matrix, an increase in the proportion of granules demonstrating a scroll pattern, and the appearance of electron-dense clumps in the granules. Granules progressively become more rope-like, with eventual solubilization of granule contents. In nasal and lung mast cells, extruded granules are never observed intact in the external environment of the degranulated nasal mast cell. Over the course of a few minutes, the degranulating mast cell may discharge the solubilized contents from 1000 secretory granules into the surrounding interstitial fluid (Friedman et al., 1986). Mast cells are found near superficial postcapillary venules (which respond with increased vascular permeability), sensory nerves (which respond by initiating the sensation of itching and elicit the sneeze reflex), and glands (which respond by secretion).

C. Mast Cell Mediators

The mechanism by which these responses are elicited involves the actions of the mediators of allergy, some of which are enumerated in Table 1 (Kaliner and Lemanske, 1992). The pattern of allergic response in the nasal mucosa usually exhibits one of two syndromes: episodic sneezing, itching, and rhinorrhea, and/or nasal congestion. These two syndromes may overlap, although most patients will describe one or the other pattern as their primary problem. Associated symptoms include palatal and eye itching, conjunctival swelling, postnasal drip, coughing, and the irrepressible need to sneeze.

The symptoms of allergic rhinitis are caused by the following processes (Table 2) (Druce and Kaliner, 1989): (1) Vascular permeability causes edema fluid to collect in the mucosa causing nasal congestion and also contributing to the secretions in the nasal lumen. Permeability (Table 3) is caused by the actions of vasoactive amines including histamine (through its H1 receptor), prostaglandins

Table 1 Mediators of Allergy (Selected for Relevance to Allergic Rhinitis)

Histamine
Chemotactic factors
Kininogenase
Prostaglandins, Hete's, leukotrienes
Prostaglandin-generating factor
Bradykinin
Platelet-activating factor
Tryptase and chymase
Inflammatory factor

Table 2 Allergic Rhinitis—Major Symptoms and Putative Mediators

Symptom	Mediator
Pruritus	Histamine (H1 receptor), prostaglandins
Swelling	Histamine (H1 receptor) eicosanoids, kinins, PAF, several neuropeptides
Sneezing	Histamine (H1 receptor), eicosanoids
Mucus secretion	Histamine (reflex through H1 receptor), eicosanoids, PAF, kinins, CGRP, ET-1, SP, VIP
Nasal irritability	Inflammatory factors, eicosanoids, chemotactic factors, cytokines

and leukotrienes, bradykinin, platelet-activating factor (PAF), and secondarily by the release of neuropeptides on the endothelial cells in the superficial post-capillary venules (Persson et al., 1991; Raphael et al., 1991). (2) Mucus secretion contributes important molecules serving host-defense functions to the secretions (Kaliner, 1991). Glands are stimulated directly by prostaglandins, and reflexly by histamine, which acts by causing the discharge from nerves of acetylcholine and several neuropeptides (Raphael et al., 1989b). (3) Neural reflexes stimulated by histamine (through its H1 receptor) and possibly by other mast cell mediators cause the pruritus and sneezing reflexes, as well as glandular secretion (Raphael et al., 1988b). (4) Late-phase reactions (LPR) contribute to the edema and irritability of the nose. LPR are caused by inflammatory factors released from mast cells and include chemotactic factors, PAF, eicosanoids, and a number of inflammatory cytokines (which are also synthesized and released by lymphocytes) (Lemanske and Kaliner, 1993). These factors lead to the expression of adhesion molecules, the attraction of inflammatory cells, and the infiltration of the mucosa with neutrophils, eosinophils, lymphocytes, mast cells, and basophils. Biopsies of the nasal mucosa after allergen challenge reveal the following changes: increased

Table 3 Mediators Thought to Cause Microvascular Permeability in Allergic Reactions

Mast cell mediators	Neuropeptides
Histamine	Substance P
Bradykinin	Neurokinin A
Leukotrienes	Calcitonin gene-related peptide
Several prostaglandins	
Chymase	
Platelet-activating factor	
Reactive oxygen species	

expression of the receptor for IL-2, increased numbers of eosinophils, increased mast cells in the epithelium, and mast cell degranulation (Bentley et al., 1992; Igarashi et al., in press). This inflammation plays a major role in the increased irritability of the nose characteristically seen during the allergy season.

In general, allergic reactions occur at the interface between humans and their environments. These surfaces are endowed with a rapidly deployable host-defense mechanism, the production of a copious supply of mucus, for which the mucous membrane was named. Many of the symptoms of allergic reactions are due to the glandular secretion, vascular permeability, edema, and vasodilatation, which are really expressions of this primitive host-defense mechanism (Kaliner, 1991).

III. NASAL SECRETIONS

Nasal airway secretions and their constituent proteins derive from epithelial cells (including goblet cells), submucosal glands (including both serous and mucous cells), blood vessels, and secretory cells resident in the mucosa (including plasma cells, mast cells, lymphocytes, and fibroblasts). Respiratory secretions consist of a mixture of mucous glycoproteins, glandular products, and plasma proteins (Table 4). Baseline, resting secretions include the following major proteins: albumin

Table 4 Constituents of Nasal Secretions

Mucous cell products
   Mucous glycoproteins
Serous cell products
   Lactoferrin
   Lysozyme
   Secretory IgA and secretory component
   Neutral endopeptidase
   Aminopeptidase
   Peroxidase
   Secretory leukoprotease inhibitor
Plasma proteins
   Albumin
   Immunoglobulins-IgG, IgA (monomeric, 7S), IgM, IgE
   Carboxypeptidase N
   Angiotensin-converting enzyme
   Kallikrein
Indeterminate sources
   Calcitonin gene-related peptide
   Urea
   Uric acid

Table 5 Roles and Functions of Human Nasal
Secretions

Protective functions
   Antioxidant (uric acid)
   Humidification
   Lubrication
   Waterproofing
   Insulation
   Provide proper medium for ciliary actions
Barrier functions
   Macromolecular sieve
   Entrapment of microorganisms and particulates
   Transport media for elimination of entrapped materials
Host-defense functions
   Extracellular source of IgA/IgG
   Extracellular site for multiple enzyme actions
   Antimicrobial functions
      Lysozyme
      Lactoferrin
      IgA/IgG
      Rapid deployment of multiple plasma proteins

(representing about 15% of total protein), IgG (2–4%), secretory IgA (15%), lactoferrin (2–4%), lysozyme (15–30%), nonsecretory IgA (about 1%), IgM (<1%), secretory leukoprotease inhibitor (3%), and mucous glycoproteins (about 10–15%) (Raphael et al., 1988b, 1989b; Meredith, 1989; Lee, 1993). The functions of these secretory proteins are summarized in Table 5.

IV. NASAL AIRWAY HYPERREACTIVITY

Increasing reactivity to inhaled allergens is one characteristic of allergic rhinitis. In other words, during the course of a pollen season, it requires less and less pollen exposure to elicit symptoms. This phenomenon was observed by Connell many years ago and became known as "allergic priming" (Connell, 1969). The underlying causes of airway hyperreactivity are not clear as yet, but it is clear that LPR significantly contributes to increases in reactivity (Lemanske and Kaliner, 1993; Kaliner, 1987). LPR denotes the inflammatory response noted some hours after allergen challenge, which may be manifest by airflow obstruction and the presence of increased infiltrating cells including eosinophils, neutrophils, basophils, and lymphocytes. Nasal allergen challenge, under specific experimental conditions, may result in both an immediate and late-phase-allergic response (Naclerio et al.,

1983, 1985). Repeated allergen challenge increases both specific and nonspecific nasal reactivity (Iliopoulos et al., 1990). The mechanism by which LPR actually changes airway reactivity is under investigation. One recent study in a monkey model of asthma suggested that mast cell activation and elaboration of tumor necrosis factor (TNF) led to the generation of adhesion molecules, which then facilitated eosinophil infiltration and the generation of increased airway reactivity. Pretreatment with antibodies directed at TNF prevented the response (Wegner et al., 1990). Compelling arguments, however, can also be raised for a role for neutral endopeptidase, increased or decreased amounts of neuropeptides, and specific actions of mediators derived from mast cells, lymphocytes, or eosinophils. Current work on nasal airway hyperreactivity has indicated that allergic rhinitis subjects always have increased cholinergic responsiveness and that antigen causes increased reactivity to histamine and other vasoactive amines (White, in press).

V. NASAL NEUROPEPTIDES

There is compelling evidence for the role that parasympathetic innervation plays both in nasal health and in disease (Kaliner, 1992). Extensive work, reviewed by Baraniuk elsewhere in this book, indicates that neuropeptides might also contribute to the state of the nasal mucosa. Peptide neurotransmitters can be found in the nasal mucosa and include substance P (SP), neurokinin A (NKA), calcitonin gene-related peptide (CGRP), vasoactive intestinal peptide (VIP), neuropeptide Y (NPY), and gastrin-releasing peptide (GRP) or bombesin. These neuropeptides are found in sensory nerves (SP, NKA, CGRP, GRP), in cholinergic efferents (VIP), and in α-adrenergic nerves (NPY) (Baraniuk and Kaliner, 1991). Stimulation of the autonomic nerves results in the release of classical neurotransmitter plus peptide neurotransmitter. The exact role of the neuropeptides in nasal physiology is still conjectural, but our data suggest that GRP and VIP are important glandular secretagogues, that NPY is a potent vasoconstrictor, that CGRP acts as a vasodilator as does VIP, and that SP and NKA are mucus secretagogues but likely act primarily on blood vessels as dilators and to cause vascular permeability. We have recently found that endothelin-1 (ET-1) is also synthesized in the mucosa by gland cells an endothelial cells. ET-1 is found in secretions and acts on submucosal glands to induce both serous and mucous cell secretion.

The neuropeptides are degraded by four enzymes: neutral endopeptidase (NEP), aminopeptidase M (AP-M), carboxypeptidase (CP), and angiotensin-converting enzyme (ACE). Each of these enzymes can be found in secretions and in the nasal mucosa. Both NEP and AP-M are actually made in the serous cells of the submucosal glands and are secreted along with other gland products. By contrast, both CP and ACE are plasma proteins that make their way into secretions along with other plasma proteins during the process of vascular permeability. NEP is the major metabolizing enzyme for SP, NKA, ET-1, VIP, and other peptides.

NEP in the mucosa, a 90-kDa protein, is found in the cytoplasmic fraction of mucosal homogenates, and the message for NEP protein is found in the serous cell and endothelial cell (Ohkuba et al., 1993).

A. Neuropeptide Secretion

Adding a "cocktail" of inhibitors to nasal secretions (thiorphan, captopril, aprotinin) permits recovery of neuropeptides secreted in vivo in response to antigen. In subjects challenged with antigen to which they were sensitive, SP, CGRP, and VIP levels in nasal washings all increased significantly. The neuropeptides were secreted within 3 min of allergen challenge, but peaked after 13 min. Histamine challenge elicited a significant increase in VIP but not CGRP or SP release, suggesting that histamine stimulated a cholinergic reflex response associated with the selective release of VIP (Mossiman et al., 1993). Thus, neuropeptides are secreted during allergic reactions, and histamine might be responsible for VIP release. The underlying mechanism for SP and CGRP release is not known, but we suspect that other mediators besides histamine act to cause sensory nerve discharge of SP and CGRP.

B. Vasomotor Rhinitis

The other experiment suggesting that neuropeptides might play a role in rhinitis involves the treatment of vasomotor rhinitis (VMR). VMR is a common disease recognized by the nasal congestion and secretion elicited by environmental factors such as cold weather, high humidity, strong smells, exposure to alcohol, or emotional states. There is no satisfactory treatment of this disease at this time. However, several uncontrolled experiments suggested that topical exposure to capsaicin, the essence of red peppers, could dramatically reduce the symptoms of VMR. Capsaicin usually acts by causing sensory nerves to secrete neuropeptides and causing the tissue to become "desensitized" to subsequent exposures to neuropeptides (Stjarne et al., 1991; Lacroix et al., 1991). If reproduced in controlled studies, these data would suggest that subjects with VMR either secrete excessive neuropeptides or are unduly responsive to the actions of neuropeptides. Moreover, this observation would be the first instance where neuropeptides are thought to be responsible for a disease.

The symptoms of allergic rhinitis are therefore caused by both acute and chronic events (Table 2). Therapy is aimed at these two targets, reversing the acute events and preventing the late responses that cause the chronic inflammatory changes. Although allergic rhinitis is not life-threatening, it causes an unbelievable amount of morbidity, affecting the lives of nearly one-fifth of the population. It has often been said that allergic rhinitis is a minor problem, unless you suffer from it yourself! Only an affected individual is aware of the distraction caused by episodes of sneezing or the irrepressible need to sneeze; the bother of chronic rhinorrhea (and what to do with the soggy tissues generated); the unbearable itching of eyes, nose, and palate; and the fatigue elicited by these recurrent symptoms, which may occur daily for months. Physicians need to be aware that we now understand this disease quite well, that excellent short-term and long-term therapeutic approaches are now available, and that the disease can be controlled quite effectively. It is a rare circumstance where therapeutic advances can make the life of one out of five people more comfortable! Certainly all physicians should be made aware of these advances.

VI. CONCLUSIONS

Allergic rhinitis is the most common immunological disease experienced by humans and accounts for as many as 3% of all office visits for all diseases in the United States. Intense research over the past two decades has uncovered the underlying mechanisms responsible for the symptomatology and appropriate treatment for the disease. Appreciation of the inflammatory component of the late phase of the nasal allergic reaction has allowed us to gain insights into airway hyperactivity and its proper treatment. Thus, allergic rhinitis can be diagnosed and treated effectively and great improvement in the life-styles of nearly 20% of the inhabitants of the temperate regions of earth can be enhanced.

REFERENCES

Baraniuk, J.N., and Kaliner, M. (1991). Neuropeptides and nasal secretions. *Am. J. Physiol. Lung. Cell. Mol. Physiol. 261*:L223.

Bentley, A.M., Jacobson, M.R., Cumberworth, V., Barkans, J.R., Moqbel, R., Schwartz, L.B., Irani, A.A., Kay, A.B., and Durham, S.R. (1992). Immunohistology of the nasal mucosa in season allergic rhinitis: increases in activated eosinophils and epithelial mast cells. *J. Allergy Clin. Immunol. 89*:877.

Broder, I., Barlow, P.P., and Hortin, R.J.M. (1974). The epidemiology of asthma and hay fever in a total community, Tecumseh, Michigan. *J. Allergy Clin. Immunol. 54*:100.

Connell, J.T. (1969). Quantitative intranasal pollen challenges. III. The priming effect of allergic rhinitis. *J. Allergy 43*:33–44.

Druce, H., and Kaliner, M. (1988). Allergic rhinitis. *JAMA 259*:260.

Druce, H.M., and Kaliner, M.A. (1989). Allergic rhinitis. In: *Current Therapy of Respiratory Disease-3*. (Cherniack, R.M., ed.) B.C. Decker, Toronto, p. 16.

Friedman, M.M., and Kaliner, M.A. (1985). In situ degranulation of human nasal mucosal mast cells: ultrastructural features and cell-cell interactions. *J. Allergy Clin. Immunol. 76*:70.

Friedman, M.M., Metcalfe, D.D., and Kaliner, M. (1986). Electron microscopic comparison of human nasal and lung mast cell degranulation. In: *Mast Cell Differentiation and Heterogeneity*. (Befus. A.D., and Bienenstock, J., eds.) Raven Press, New York, p. 367.

Igarashi, Y., Kaliner, M.A., Hausfeld, J.N., Irani, A.M.A., Schwartz, L.B., and White, M.V. (1993). Quantification of inflammatory cells in the nasal mucosa. *J. Allergy Clin. Immunol.* *91*:1082.

Igarashi, Y., Goldrich, M.S., Kaliner, M.A., Hausfeld, J.N., Irani, A.M.M., Schwartz, L.B., and White, M.V. (in press). Quantification of inflammatory cells in the nasal mucosa of allergic rhinitis and normals. *J. Allergy Clin. Immunol.*

Iliopoulos, O., Proud, D., Adkinson, N.F., Norman, P.S., Kagey-Sobotka, A., Lichtenstein, L.M., and Naclerio, R.M. (1990). Relationship between the early, late, and rechallenge reaction to nasal challenge with antigen: observations on the role of inflammatory mediators and cells. *J. Allergy Clin. Immunol.* *86*:851.

Kaliner, M. (1987). The late phase reaction and its clinical implications. *Hosp. Pract.* *22*:73.

Kaliner, M.A. (1991). Human respiratory secretions and host defense. *Am. Rev. Respir. Dis.* *144*:S52.

Kaliner, M.A. (1992). The physiology and pathophysiology of the parasympathetic nervous system in nasal disease: an overview. *J. Allergy Clin. Immunol.* *90*:1044.

Kaliner, M.A., and Lemanske, R.F. (1992). Rhinitis and asthma. *JAMA 268*:2807.

LaCroix, J.S., Buvelot, J.M., Polla, B.S., and Lundberg, J.M. (1991). Improvement of symptoms of non-allergic rhinitis by local treatment with capsaicin. *Clin. Exp. Allergy* *21*:595.

Lee, C.H., Igarashi, Y., Hohman, R.J., Kaulbach, H., White, M.V., and Kaliner, M.A. (1993). Distribution of secretory leukoprotease inhibitor in the human nasal airway. *Am. Rev. Respir. Dis.* *147*:710.

Lemanske, R.F., and Kaliner, M. (1993). Late phase allergic reactions. In: *Allergy Principles and Practice*, 4th ed. (Middleton, E., Reed, C.E., Ellis, E.F., Adkinson, N.F., Jr., Yunginger, J.W., and Busse, W.W., eds.) C.V. Mosby, St. Louis, p. 320.

Meredith, S.D., Raphael, G.D., Baraniuk, J.N., Banks, S.M., and Kaliner, M.A. (1989). The pathogenesis of rhinitis. III. The control of IgG secretion. *J. Allergy Clin. Immunol.* *84*:920.

Metcalfe, D.D., Kaliner, M.A., and Donlon, M.A. (1981). The mast cell. *CRC Crit. Rev. Immunol.* *3*:23.

Mosimann, B.L., White, M.V., Hohman, R.J., Goldrich, M.S., Kaulbach, H.C., and Kaliner, M.A. (1993). Substance P, calcitonin gene related peptide and vasoactive intestinal peptide increase in nasal secretions after allergen challenge in atopic patients. *J. Allergy Clin. Immunol.* *92*:95–104.

Naclerio, R.M., Meier, H.L., Kagey-Sobotka, A., Adkinson, N.F., Meyers, D.A., Norman, P.S., and Lichtenstein, L.M. (1983). Mediator release after nasal airway challenge with allergen. *Am. Rev. Respir. Dis.* *128*:597.

Naclerio, R.M., Proud, D., Togias, A., Kagey-Sobotka, A., Adkinson, N.F., Meyers, D.A., Plaut, M., Norman, P.S., and Lichtenstein, L.M. (1985). Inflammatory mediators in late antigen-induced rhinitis. *N. Engl. J. Med. 31*365.

Ohkuba, K., Baraniuk, J., Hohman, R.J., Kaulbach, H.C., Hausfield, J.N., Merida, M., and Kaliner, M.A. (1993). Human nasal mucosal neutral endopeptidase (NEP): location, quantitation and secretion. *Am. J. Respir. Cell Mol. Biol.* *9*:557–567.

Persson, C.G.A., Ejjefalt, I., Alkner, U., Baumgarten, C., Grieff, L., Gustafasson, B., Luts, A., Pipkorn, U., Sundler, F., Svensson, C., and Wollmer, P. (1991). Plasma exudation as a first line mucosal defense. *Clin. Exp. Allergy 21*:17.

Raphael, G.D., Meredith, S.D., Baraniuk, J.N., and Kaliner, M.A. (1988a). Nasal reflexes. *Am. J. Rhinol.* 2:109.

Raphael, G.D., Druce, H.M., Baraniuk, J.N., and Kaliner, M.A. (1988b). Pathophysiology of rhinitis. 1. Assessment of the sources of protein in methacholine-induced nasal secretions. *Am. Rev. Respir. Dis. 138*:413.

Raphael, G.D., Jeney, E.V., Baraniuk, J.N., Kim, I., Meredith, S.D., and Kaliner, M.A. (1989a). The pathophysiology of rhinitis: lactoferrin and lysozyme in nasal secretions. *J. Clin. Invest. 84*:1528.

Raphael, G.D., Meredith, S.D., Baraniuk, J.N., Druce, H.M., Banks, S.M., and Kaliner, M.A. (1989b). The pathophysiology of rhinitis. II. Assessment of the sources of protein in histamine-induced nasal secretions. *Am. Rev. Respir. Dis. 139*:791.

Raphael, G.D., Baraniuk, J.N., and Kaliner, M.A. (1991). How the nose runs and why. *J. Allergy Clin. Immunol. 87*:457.

Stjarne, P., Lundblad, L., Angaard, A., and Lundberg, J.M. (1991). Local capsaicin treatment of the nasal mucosa reduces symptoms in patients with non-allergic nasal hyperreactivity. *Am. J. Rhinol. 5*:145.

Wegner, C.D., Gundel, R.H., Reilly, P., Haynes, N., Letts, L.G., and Rothlein, R. (1990). Intercellular adhesion molecule-1 (ICAM-1) in the pathogenesis of asthma. *Science 247*:456.

White, M.V. (in press). Nasal cholinergic responsiveness in atopic subjects studied out of season. *J. Allergy Clin. Immunol.*

White, M.V., and Kaliner, M.A. (1990). Mast cells and asthma. In: *Asthma: Its Pathology and Treatment.* (Kaliner, M.A., Persson, C., and Barnes, P., eds.) Marcel Dekker, New York, p. 409.

DISCUSSION

Nadel: Tachykinins potentiate cholinergic neurotransmission in guinea pig trachea and bronchus. Could the effects of atropine in your nasal studies be due to tachykinin-mediated effects on cholinergic neurotransmission?

Kaliner: Since atropine was used, a muscarinic mechanism was inhibited. We cannot be sure of the site of atropine's effects in vivo.

Kunkel: Local bradykinin generation could also play a role in neurogenic inflammation by stimulating sensory nerves and substance P release.

McDonald: What is the source of the SP and CGRP in nasal secretion in your studies?

Kaliner: My guess is that the SP and CGRP are released from nociceptive nerves and carried by bulk movement of fluid across the epithelial barrier into the nasal lumen. VIP is released by histamine, presumably from parasympathetic nerves.

Goetzl: We have also demonstrated CGRP release after allergen challenge. Prolonged vasodilatation is prolonged after CGRP nasal provocation. This is a major difference from the transient effects of other peptides.

26

Neuropeptides in Small Cell Lung Carcinoma

Terry W. Moody

Biomarkers and Prevention Research Branch, National Cancer Institute, National Institutes of Health, Rockville, Maryland

I. INTRODUCTION

Small cell lung carcinoma (SCLC) is a neuroendocrine tumor that skills approximately 35,000 patients in the United States annually (Minna et al., 1992). Traditionally it is treated with chemotherapy, but usually relapse occurs. The tumor is initially localized to the lung and then metastasizes to the bone, brain, liver, and lymph nodes. Median survival time is approximately 1 year.

Our knowledge of SCLC greatly increased with the advent of cell lines derived from patient biopsy specimens. In the past decade hundreds of continuous cell lines have been cultured. SCLC cells have dense-core neurosecretory granules reminiscent of neuronal synaptosomes. High levels of dopa decarboxylase (DDC) are found in SCLC cells, suggesting that catecholamines can be synthesized (Carney et al., 1985). Because peptides are colocalized with central nervous system neurotransmitters, we investigated the peptide content of SCLC cells. High levels of bombesin/gastrin-releasing peptide (BN/GRP) were detected by radioimmunoassay (Moody et al., 1981). Because SCLC cells also contain immunoreactive neurotensin (NT), they can synthesize multiple neuropeptides (Moody et al., 1985b).

Subsequent studies showed that BN/GRP could be released from SCLC cells into the conditioned medium by vasoactive intestinal peptide (VIP), which elevates the intracellular cAMP (Korman et al., 1986). Also, BN/GRP was detected in the plasma of SCLC patients and was elevated in patients with extensive disease.

The secreted peptide can bind with high affinity to cell surface receptors for BN/GRP that are present on SCLC cells (Moody et al., 1985a). Additional studies showed that high-affinity binding sites were present for NT and VIP (Allen et al., 1988; Shaffer et al., 1987). Thus SCLC cells have multiple receptors for neuropeptides.

Because SCLC cells have both peptides and receptors, they can function in an autocrine manner. A monoclonal antibody (MAb) that neutralizes BN/GRP (2A11) inhibits SCLC growth in vitro and in vivo. MAb 2A11, similar to BN/GRP receptors, recognizes the C-terminal of BN (Cuttitta et al., 1985). Thus when BN is bound to 2A11, it is unable to bind with high affinity to cell surface receptors. As a result, 2A11 neutralizes secreted BN/GRP so that it is unable to bind to receptors. Subsequent studies showed that exogenous BN stimulated SCLC growth (Carney et al., 1987). In this chapter, SCLC peptides will be discussed.

II. GRP AND NMB-LIKE PEPTIDES

BN is a 14-amino-acid peptide initially isolated from frog skin (Table 1). The mammalian equivalent of BN is GRP, a 27-amino-acid peptide that has nine of the 10 same C-terminal amino acids as does BN. Neuromedin B is a structural analog that has seven of the 10 same C-terminal amino acids as does BN. NMB and GRP are the two main members of the BN family of peptides.

Both GRP and NMB have been cloned. The genes for GRP and NMB are located on chromosomes 15 and 18, respectively. By Northern analysis, the predominant mRNA for GRP and NMB is comprised of 0.9 and 0.8 kb, respectively (Cardona et al., 1991; Giaccone et al., 1992). Both peptides are derived from high-molecular-weight precursor proteins. NMB is synthesized as a 114-amino-acid preproNMB (Fig. 1). A 24-amino-acid fragment at the N-terminal is cleaved by signal proteases to form a 90-amino-acid proNMB. ProNMB is cleaved by a trypsin-like enzyme to form NMB-32-glycyl-lysyl-lysine (NMB-32GKK). NMB-32GKK is cleaved by carboxypeptidase B-like enzymes to NMB-32G, and NMB-32G is metabolized by peptidyl-α-amidating monooxygenase (PAM) to NMB-32, which has an amidated C-terminal. NMB-32, which is biologically active, may be metabolized by trypsin-like enzymes to NMB, which

Table 1 Structure of BN-like Peptides

GRP	Ala-Pro-Val-Ser-Val-Gly-Gly-Gly-Thr-Val-Leu-Ala-Lys-Met-Tyr-Pro-Arg-Gly-Asn-His-Trp-Ala-Val-Gly-His-Leu-Met-NH$_2$
BN	Pyr-Gln-Arg-Leu-Gly-Asn-Gln-Trp-Ala-Val-Gly-His-Leu-Met-NH$_2$
NMB	Gly-Asn-Leu-Trp-Ala-Thr-Gly-His-Phe-Met-NH$_2$

Sequence homologies relative to BN are underlined.

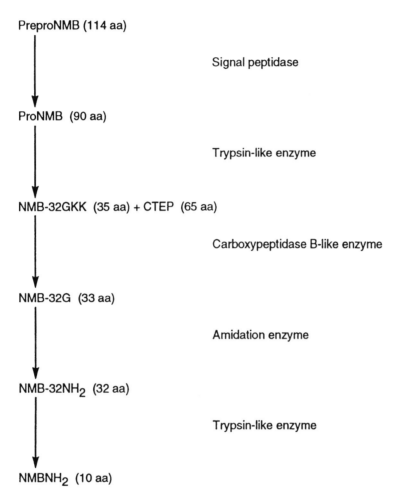

Figure 1 Posttranslational processing of NMB. The peptide intermediates for NMB are indicated. Also, the enzymes involved in the processing are shown. Standard amino acid abbreviations are used (G indicates glycine and K indicates lysine).

has 10 amino acids and is biologically active. NMB may be inactivated by enkephalinase to NMB[1-7] and NMB[8-10]. Similarly, GRP type I is derived from a 148-amino-acid preproGRP, and similar enzymes are utilized in its processing.

Because an amidated peptide is required for biological activity, the precursor proteins for GRP and NMB are inactive (Mervic et al., 1991). Thus it is possible to regulate biological activity using inhibitors that regulate enzyme activity. In the developing lung, GRP peptides and mRNA are transiently expressed at postnatal week 16 in the human especially in neuroendocrine cells (Spindel et al., 1987). GRP may play a role in the development of the fetal lung; however, in the normal adult lung, GRP is rarely expressed.

In the developing and adult rat lung, the levels of GRP are approximately 0.08 and 0.02 pmol/mg protein. In contrast, in SCLC cells the GRP levels are as high as 18 pmol/mg protein. Therefore, the levels of GRP are markedly elevated in the malignant lung relative to fetal or adult lung. Secretion studies have indicated that those agents that elevate cAMP, such as VIP, secretin, or corticotropin-releasing factor (CRF), increase the secretion rate of GRP from SCLC cells approximately twofold. Routinely 100–1000 pM concentrations of GRP are present in conditioned medium exposed to SCLC cells. In the plasma of normal patients, the basal level is 10–50 pM; however, in patients with extensive SCLC, the GRP concentration can be as high as 1 nM. This can be transiently elevated by secretin infusion in patients. Also, in patients with respiratory distress syndrome (e.g., smokers), high levels of BN-like peptides (50 nM) are present in bronchoalvelar lavage fluid (BAL) as well as the urine (Aguayo et al., 1992). The chemical nature of this BN-like peptide in BAL remains to be determined.

Secretion of GRP results in increased GRP gene expression as new peptide must be made. In this regard we have found that agents that elevate the cAMP, such as forskolin, increase the 0.9-kb GRP mRNA (Draoui et al., 1993). Also, agents that stimulate protein kinase C, such as phorbol ester, increase GRP gene expression.

SCLC cells have high-affinity binding sites for NMB and GRP (Moody et al., 1985a, 1992). Cell line NCI-H345 bound both (^{125}I-Tyr^0)NMB and (^{125}I-Tyr^4)BN (Kd = 1 and 2 nM, respectively) to a single class of sites (B_{max} = 800 and 1500/cell, respectively). The specificity of binding was very different. Specific (^{125}I-Tyr^0)NMB binding was inhibited with high affinity by NMB and NMB-32, moderate affinity by BN, GRP, (D-Phe^6)BN$^{6-13}$methyl ester (ME), (Psi^{13,14}, Leu^{14})BN (Psi-13,14), and (D-Arg^1, D-Pro^2, D-Trp^{7,9}, Leu^{11})substance P ((APTTL)SP), and low affinity by GRP$^{1-16}$ (Table 2) Staley et al., 1991; Mahmoud, 1991). In contrast, specific (^{125}I-Tyr^4)BN binding was inhibited with high affinity by BN, GRP, (D-Phe^6)BN$^{6-13}$ME, and Psi-13,14 (IC$_{50}$ = 1–30), moderate affinity by NMB, NMB-32, and (APTTL)SP (IC$_{50}$ = 100–1500 nM), and low affinity by GRP$^{1-16}$ (IC$_{50}$ > 10,000 nM). These data indicate that the NMB receptor prefers NMB relative to GRP or BN whereas the GRP receptor prefers BN or GRP relative to NMB. Also, the GRP receptor prefers (D-Phe^6)BN$^{6-13}$)ME or Psi-13,14 relative to (APTTL)SP. The NMB receptor has no preference for (APTTL)SP, (D-Phe^6)BN$^{6-13}$, or Psi-13,14. Therefore, the specificity of binding for the GRP and NMB receptor is distinct.

The GRP and NMB receptors were recently cloned. The NMB receptor is comprised of 390 amino acid residues whereas the GRP receptor contains 384 amino acid residues (Battey et al., 1991; Spindel et al., 1990; Wada et al., 1991).

Table 2 Specificity of Binding to SCLC Cells

Peptide	(^{125}I-Tyr^0)NMB	(^{125}I-Tyr^4)BN
NMB	1	100
NMB-32	2	200
BN	100	1
GRP	50	5
(D-Phe^6)BN$^{6-13}$ME	700	5
(Psi^{13,14}, Leu^{14})BN	1,500	30
(APTTL)SP	1,000	1,000
GRP$^{1-16}$	>10,000	>10,000

The ability to inhibit specific binding to NCI-H345 (IC$_{50}$, nM) was determined using 0.1 nM (^{125}I-Tyr^0)NMB or (^{125}I-Tyr^4)BN. The mean value of three determinations is indicated.

Each receptor has seven hydrophobic domains, an extracellular N-terminal, and an intracellular C-terminal (Fig. 2). In addition, there are three extracellular loops, which may participate in ligand binding, and three intracellular loops, which may participate in interaction with G proteins. There is 56% sequence homology between the NMB and GRP receptor especially in the seven hydrophobic membrane-spanning domains. The NMB receptor has a 41-amino-acid N-terminal, which contains potential glycosylation sites such as serine and threonine amino acid residues. Transmembrane domain 1 contains 24 amino acid residues and, like domains 2–7, is enriched in hydrophobic amino acid residues such as leucine, valine, and isoleucine. The first extracellular loop contains 18 amino acids and contains hydrophilic amino acids such as aspartate. Similarly, the third intracellular loop contains 31 amino acids and hydrophilic amino acids such as lysine. The C-terminal contains 64 amino acid residues and many serine and threonine amino acids. It remains to be determined whether these amino acids are phosphorylated after exposure to NMB resulting in receptor desensitization. The NMB receptor binding site is currently being defined using receptor chimeras and site-directed mutagenesis techniques.

GRP and NMB receptors interact with a G protein stimulating phosphatidyl inositol turnover. As a result, phosphatidylinositol (PI) diphosphate (PIP$_2$) is metabolized by phospholipase C to inositol 1,4,5-trisphosphate (IP$_3$) and diacylglycerol (DAG). The IP$_3$ may interact with receptors in the endoplasmic reticulum resulting in the opening of Ca^{2+} channels. As a result, 10 nM BN elevates the cytosolic Ca^{2+} in SCLC cell line NCI-H345 (Fig. 3A). NCI-H345 also has NMB receptors that cause PI turnover as shown in Figure 3B, 10 nM NMB elevates cytosolic Ca^{2+}. The SP antagonist (APTTL)SP (10 μM) has no effect on cytosolic Ca^{2+} but blocks the increase in cytosolic Ca^{2+} caused by BN (Fig. 3C). Sur-

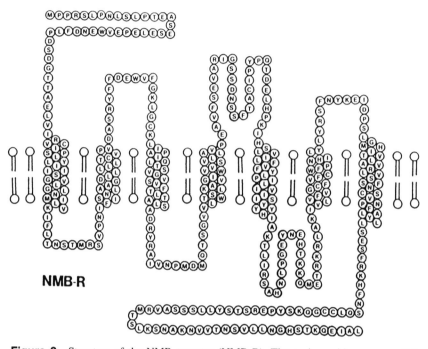

NMB-R

Figure 2 Structure of the NMB receptor (NMB-R). The amino acid sequence of the NMB-R (390 amino acid) is shown using standard single-letter abbreviation. (Derived from Wada et al., 1991.)

prisingly, (APTTL)SP also blocks the increase in cytosolic Ca^{2+} caused by 10 nM NMB (Fig. 3D). Because the SP antagonist has little structural homology with BN or NMB, it is possible that it alters the receptor conformation and impairs receptor–G protein interaction. In contrast, (D-Phe[6])BN[6-13] methylester or Psi-13,14 (1 μM) inhibited the increase in cystosolic Ca^{2+} caused by 10 nM BN but not 10 nM NMB. These data indicate that (D-Phe[6])BN[6-13] methylester and Psi-13,14 are specific GRP receptor antagonists whereas (APTTL)SP is a more broad-spectrum antagonist.

BN (10 nM) causes a transient increase in the cytosolic Ca^{2+}. The intracellular Ca^{2+} levels are increased from 150 to 190 nM after 1 min. The Ca^{2+} levels then slowly decrease and return to basal levels after 3 min possibly owing to receptor desensitization and/or down-regulation. The Ca^{2+} levels in NCI-H345 are not depleted because subsequent addition of another stimulus such as 10 nM NMB or 1 μg/ml ionomycin causes a strong Ca^{2+} response. Also, within 5 min after addition of 10 nM BN, protein kinase C is translocated from the cytosol to the membrane. BN may elevate the DAG levels, resulting in protein kinase C (PKC)

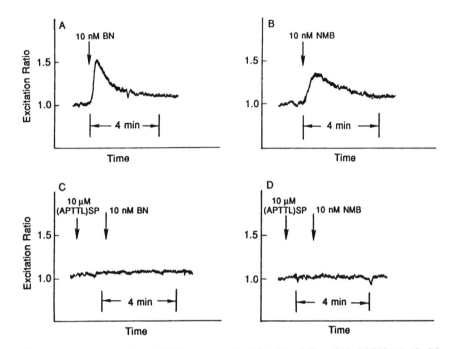

Figure 3 Effect of BN and NMB on cytosolic Ca^{2+}. The ability of 10 nM BN (A), 10 nM NMB (B), 10 nM BN + 10 μM (APTTL)SP (C), and 10 nM NMB + 10 μM (APTTL)SP (D) to alter cytosolic Ca^{2+} in Fura-2-AM-loaded NCI-H345 cells is indicated.

activation. Also, BN causes a transient increase in the protooncogene c-*fos*. NCI-H345 c-*fos* mRNA levels are maximal 30–60 min after the addition of 10 nM BN, followed by a slow decline so that basal values are reached after 4 hr. c-*fos* may form a heterodimer with c-*jun*, resulting in activation of AP-1 sites and altered SCLC gene expression. The effects of NMB on PKC and oncogene expression remain to be determined.

BN and NMB stimulate the proliferation of SCLC cells in vitro. Psi-13,14 (1 μM) inhibits the increase in NCI-H345 colony caused by 10 nM BN and also reduces basal colony formation. Thus Psi-13,14 inhibits the clonal growth caused by exogenous or endogenous BN. Psi-13,14 had little effect on the clonal growth caused by 10 nM NMB. In contrast, (APTTL)SP inhibited the increase in clonal growth caused by 10 nM BN or 10 nM NMB (Bepler et al., 1988). These data suggest that (APTTL)SP is a more broad-spectrum antagonist than is Psi-13,14.

Previously, BN was demonstrated to stimulate SCLC proliferation in vivo (Alexander et al., 1988). Psi-13,14 (10 μg) injected daily into nude mice slowed tumor growth by approximately 50% (Mahmoud et al., 1991). The effects of

Psi-13,14 were reversible in that if peptide administration was discontinued, the tumor rapidly regrew. In contrast, (APTTL)SP (10 μg) did not inhibit xenograft formation, but administration of mg quantities of SP antagonists daily did slow xenograft formation (Sethi and Rozengurt, 1992). These data indicated that Psi-13,14 is approximately 2 orders of magnitude more potent than is (APTTL)SP in vivo. Currently the biodistribution and pharmacokinetics of Psi-13,14 are being investigated in nude mice (Davis et al., 1992).

A summary of the signal transduction mechanisms for GRP is shown in Figure 4. BN and GRP bind with high affinity to the GRP receptor (78 kDa), stimulating PI turnover. PLC catalyzes the formation of IP_3, which releases Ca^{2+} from the endoplasmic reticulum, and DAG, which causes the translocation of PKC (80 kDa); PKC can also be activated by phorbol-12-myristate-13-acetate (PMA). PKC phosphorylates various substrates, resulting in increased c-*fos* mRNA. The c-*fos* in turn may form heterodimers with c-*jun*, resulting in activation of AP-1 regulatory sequences in the 5′ region upstream from genes. This may result in increased SCLC proliferation.

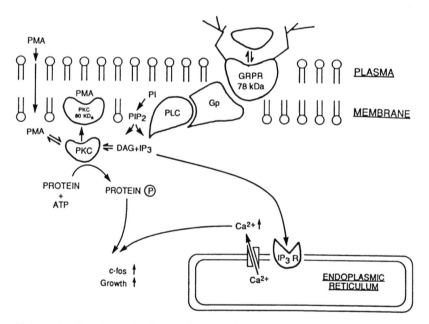

Figure 4 Signal transduction mechanism of BN. BN or GRP binds to cell surface receptors (78 kDal) and interacts with a guanine-nucleotide-binding protein (Gp). Protein lipase C (PLC) is stimulated, resulting in PI turnover. The DAG released activates protein kinase C (80 kDal), and the IP_3 released caused Ca^{2+} release from the endoplasmic reticulum. This ultimately stimulates c-*fos* mRNA and proliferation.

III. VIP/PACAP

^{125}I-VIP binds with high affinity to almost all lung cancer cell lines studied (Lee et al., 1990). VIP is a 28-amino-acid peptide with an amidated C-terminal, which elevates cAMP levels in SCLC cells. The cAMP may activate protein kinase A, resulting in phosphorylation of proteins such as synapsin I. This causes exocytosis of SCLC granules, resulting in increased GRP secretion. The GRP may then bind to receptors stimulating SCLC growth.

VIP stimulates SCLC clonal growth and VIPhybrid (VIPhyb) inhibits SCLC growth. VIPhyb is a VIP antagonist that has the same C-terminal 22 amino acid residues as does VIP but a different N-terminal. As a result, VIP and the structurally related pituitary adenylate cyclase activating peptide (PACAP) inhibit specific ^{125}I-VIP binding with high affinity (IC_{50} = 10 and 20 nM, respectively) where VIPhyb inhibits binding with moderate affinity (IC_{50} = 0.7 μM) (Moody et al., 1993a). VIP (10 nM) elevates cAMP 10-fold in NCI-H345 cells whereas the increase in cAMP caused by VIP is reversed by 10 μM VIPhyb. Also, VIP stimulates SCLC growth whereas VIPhyb inhibits the number of SCLC colonies. VIP immunoreactivity is present in SCLC cells (Gozes et al., 1993). Also, VIP mRNA (2.1 kb) is present in lung cancer cells. Because VIP-like peptides and receptors are present in SCLC cells, VIP may function as a SCLC autocrine growth factor.

Recent data indicate that ^{125}I-PACAP-27 binds with high affinity to SCLC cells (Moody et al., 1993b). Specific ^{125}I-PACAP binding to SCLC cells is inhibited with high affinity by PACAP (IC_{50} = 10 nM) but not VIP (IC_{50} = 1000 nM). Also, PACAP elevates the cAMP and cytosolic Ca^{2+} levels whereas VIP elevates cAMP but has no effect on cytosolic Ca^{2+}. Because PACAP receptors may cause PI turnover and elevate cAMP, both PKC and protein kinase A will be stimulated. PACAP strongly increases SCLC proliferation. These data indicate that SCLC cells have both VIP and PACAP receptors.

IV. NEUROTENSIN

Approximately 50% of the SCLC cell lines examined (Moody et al., 1985b) have immunoreactive NT. NT, a 13-amino-acid peptide, is secreted into conditioned medium. Also, high-affinity binding sites for NT (Allen et al., 1988) are present (Kd = 2 nM) on SCLC cell line NCI-H209 (B_{max} = 2300/cell). Specific ^{125}I-NT binding is inhibited with high affinity by NT[8-13] (IC_{50} = 10 nM) but not NT[1-8]. NT and NT[8-13] elevate the cytosolic Ca^{2+} from 150 to 270 nM. Figure 5 shows that the increase in cytosolic Ca^{2+} caused by NT is not reversed by (APTTL)SP. Similarly, Psi-13,14 had no effect on the increase in cytosolic Ca^{2+} caused by NT. These data indicate that SCLC cells have NT receptors that are distinct from those of GRP or NMB. Also, NT stimulated the clonal growth of SCLC cells (Davis et al., 1989).

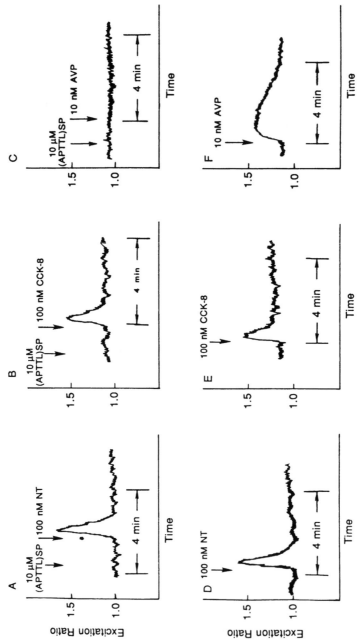

Figure 5 Effect of AVP, CCK-8, and NT on cytosolic Ca^{2+}. The ability of 100 nM NT (D), 100 nM CCK-8 (E), and 10 nM AVP (F) to elevate cytosolic Ca^{2+} in Fura-2-AM-loaded NCI-H345 cells is indicated. (APTTL)SP (10 μM) had no effect on the ability of 100 nM NT (A) or 100 nM CCK-8 (B) to elevate cytosolic Ca^{2+} but blocked the effect of 10 nM AVP (C).

Because NT and its receptors are present in SCLC cells, it may function as a SCLC autocrine growth factor.

V. CHOLECYSTOKININ/GASTRIN

^{125}I-CCK-8 binds with high affinity (Kd = 2 nM) to a single class of sites (B_{max} = 1700/cell) (Yoder and Moody, 1987). Specific ^{125}I-CCK-8 binding was inhibited with high affinity by nonsulfated CCK-8, gastrin-I, and L-365,260 (IC_{50} = 5, 10, and 0.2 nM, respectively) but not L-364,718 (IC_{50} = 0.5 μM). CCK-8 or gastrin-I (10 nM) elevated the cytosolic Ca^{2+} in NCI-H345 cells from 150 to 180 nM (Staley et al., 1989), and the increase in cytosolic Ca^{2+} was blocked by 10 nM L-365,260 but not L-364,718. L-365,260 did not block the increase in cytosolic Ca^{2+} caused by BN, NMB, NT, or AVP (Staley et al., 1990). Also, Figure 5 shows that (APTTL)SP does not impair the ability of CCK-8 to elevate cytosolic Ca^{2+}. CCK stimulated the clonal growth of SCLC cells (Sethi and Rozengurt, 1991).

VI. VASOPRESSIN/OXYTOCIN

Vasopressin (AVP) receptors of the V1 type are present on SCLC cells (Hong and Moody, 1991), and AVP, a nine-amino-acid peptide, stimulates the clonal growth of SCLC cells (Sethi and Rozengurt, 1991). AVP (10 nM) or oxytocin (1 μM) elevated the cytosolic Ca^{2+} in a concentration-dependent manner and is approximately 100-fold more potent than oxytocin. The increase in cytosolic Ca^{2+} caused by 10 nM AVP is reversed by (β-mercapto-β,β-cyclopentamethylene propionic acid)1,O-MeTyr^2, Arg^8)vasopressin, a V1 antagonist, or (APTTL)SP, a substance P antagonist (Fig. 5). Because (APTTL)SP blocks GRP, NMB, and AVP receptors, it is a broad-spectrum SCLC peptide receptor antagonist, and current efforts are focusing on developing more potent SP antagonists.

VII. OPIOID RECEPTOR PEPTIDES

Opioid receptors and peptides are present in SCLC cell lines (Roth and Barchas, 1986). SCLC cells have immunoreactive β-endorphin and bind ^{3}H-etorphin with high affinity (Kd = 0.9 nM) to a single class of sites (B_{max} = 52 fmol/mg protein). Because ^{3}H-(D-Ala^2, D-Leu^5)enkephalin (DADLE) and ^{3}H-U-50488H also bound to NCI-H187 cells with high affinity, mu, delta, and kappa opioid receptors may be present on SCLC cells (Maneckjee and Minna, 1992). Morphine and DADLE inhibited SCLC growth, and the growth inhibition was reversed by naloxone. These data suggest that mu and delta receptors inhibit the growth of SCLC in vitro. In contrast, β-endorphin stimulates the clonal growth of some SCLC cell lines

(Davis et al., 1989). These data suggest that epsilon opioid receptors stimulate the growth of SCLC in vitro.

VIII. BRADYKININ

Bradykinin (BK) elevates the cytosolic Ca^{2+} in SCLC cells (Bunn et al., 1990). The increase in cytosolic calcium caused by 10 nM BK was reversed by the B2 antagonist (D-Arg^0, Hyp^3, Thy^{5,8}, D-Phe^7)BK. Also, BK stimulated colony formation of SCLC cells (Sethi and Rozengurt, 1991).

IX. SOMATOSTATIN

Somatostatin (SRIF), a 14-amino-acid peptide, inhibits the increase in SCLC cAMP caused by VIP (Taylor et al., 1988). Also, SRIF inhibits the increased secretion rate of GRP caused by VIP (Taylor et al., 1988). SRIF agonists such as somatuline and octreotide bind with high affinity to SCLC SRIF receptors and inhibit SCLC growth (Taylor et al., 1991; Reubi et al., 1990). SRIF receptors may interact with an inhibitory guanine-nucleotide-binding protein. Therefore, it is possible to alter SCLC growth by using agents that alter neuropeptide secretion.

X. INSULIN-LIKE GROWTH FACTOR

It is generally accepted that agents that stimulate tyrosine kinase activity, such as insulin-like growth factor-I (IGF-I), regulate cancer proliferation. By Western blot analysis, a 16-kDa protein was detected in SCLC cells. ^{125}I-IGF-I bound with high affinity to SCLC cells (Nakanishi et al., 1988). The IGF-I receptor has multiple subunits (α_2,β_2) and the α subunit (130 kDal) binds IGF-I whereas the β subunit (90 kDal) has tyrosine kinase activity. IGF-I (10 ng/ml) stimulated the growth of SCLC cells, and IGF-I was more potent than insulin or IGF-II. These data suggest that IGF-I receptors, but not insulin or IGF-II receptors are present on SCLC cells. An IGF-I receptor MAb (αIR-3, 1 μg/ml) inhibited the growth of SCLC cells. IGF-I is a 70-amino-acid growth factor with essential disulfide bonds. When IGF-I binds to its receptor, protein substrates get phosphorylated on tyrosine amino acid residues, increasing SCLC growth.

XI. SUMMARY

We demonstrated that neuropeptides such as GRP, NT, and β-endorphin are synthesized in and secreted from SCLC cells (Fig. 6). The peptides bind to G-protein-coupled receptors that contain approximately 400 amino acids and alter second-messenger production. GRP, NMB, PACAP, BK, AVP, NT, and CCK receptors stimulate phosphatidyl inositol turnover and elevate SCLC cytosolic

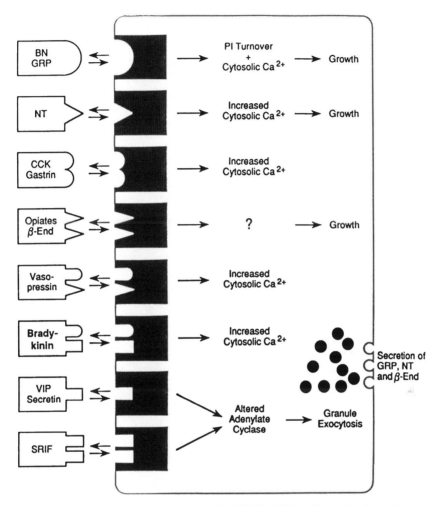

Figure 6 Schematic illustration of classic SCLC. SCLC cells synthesize and secrete GRP, NT, and β-endorphin. Cell surface receptors are also present for GRP, NT, CCK, AVP, BK, opiates, VIP, and SRIF.

Ca^{2+}. Also, GRP, NMB, PACAP, BK, AVP, NT, and CCK elevate DAG and stimulate PKC. VIP and PACAP elevate cAMP and stimulate protein kinase A. Both PKC and protein kinase A will phosphorylate protein substrates on serine and threonine amino acid residues and ultimately stimulate SCLC proliferation. Thus the mechanism by which neuropeptides stimulate growth is very different from that of growth factors such as IGF-I.

SCLC growth can be inhibited by MAb against growth factor receptors. The αIR-3 MAb inhibits tyrosine phosphorylation caused by IGF-I. MAbs against neuropeptide receptors do not currently exist, but numerous synthetic peptide antagonists are available. SP antagonists such as (APTTL)SP block GRP, NMB, and AVP receptors. In addition, high-affinity antagonists for GRP, CCK, AVP, opioid, and BK receptors are available. It remains to be determined whether peptide receptor antagonists will be useful in the treatment of SCLC.

ACKNOWLEDGMENTS

The author thanks Drs. A. Makheja and M. Draoui for helpful discussions. This work was supported in part by NCI Grants CA-48071 and 53477.

REFERENCES

Aguayo, S.M., King, T.E., Kane, M.A., Sherrit, K.M., Silvers, W., Nett, L.M., Petty, T.L., and Miller, Y. (1992). Urinary levels of bombesin-like peptides in asymptomatic cigarette smokers: a potential risk marker for smoking-related diseases. *Cancer Res. 52*: 2727s–2731s.

Alexander, R.W., Upp, J.R., Jr., Poston, G.J., Gupta, V., Townsend, C.M., Jr., and Thompson, J.C. (1988). Effects of bombesin on growth of human small cell lung carcinoma in vivo. *Cancer Res. 48*:1439–1441.

Allen, A.E., Carney, D.N., and Moody, T.W. (1988). Neurotensin binds with high affinity to small cell lung cancer cells. *Peptides 9*(1):57–61.

Battey, J.F., Way, J., Corjay, M.H., Shapira, H., Kusano, K., Harkins, R., Wu, J.M., Slattery, T., Mann, E., and Feldman, R. (1991). Molecular cloning of the bombesin/GRP receptor from Swiss 3T3 cells. *Proc. Natl. Acad. Sci. USA 88*:395–399.

Bepler, G., Zeymer, U., Mahmoud, S., Fiskum, G., Palaszynski, E., Rotsch, M., Willey, J., Koros, A., Cuttitta, F., and Moody, T.W. (1988). Substance P analogues function as bombesin receptor antagonists and inhibit small cell lung cancer clonal growth. *Peptides 9*:1367–1372.

Bunn, P.A., Jr., Dienhart, D.G., Chan, D., Puck, T.T., Tagawa, M., Jewett, P.B., and Braunschweiger, E. (1990). Neuropeptide stimulation of calcium flux in human lung cancer cells: delineation of alternative pathways. *Proc. Natl. Acad. Sci. USA 87*:2162–2166.

Cardona, C., Rabbitts, P.H., Spindel, E.R., Ghatei, M.A., Bleehen, N.M., Bloom, S.R., and Reeve, J.G. (1991). Production of neuromedin B and neuromedin B gene expression in human lung tumor cell lines. *Cancer Res. 51*:5205–5211.

Carney, D.N., Gazdar, A.F., Bepler, G., Guccion, J.G., Marangos, P.J., Moody, T.W., Zweig, M.H., and Minna, J.D. (1985). Establishment and identification of small cell lung cancer cell lines having classic and variant features. *Cancer Res. 45*:2913–2923.

Carney, D.N., Cuttitta, F., Moody, T.W., and Minna, J.D. (1987). Selective stimulation of small cell lung cancer clonal growth by bombesin and gastrin releasing peptide. *Cancer Res. 47*:821–825.

Cuttitta, F., Carney, D.N., Mulshine, J., Moody, T.W., Fedorko, J., Fischler, A., and Minna, J.D. (1985). Bombesin-like peptides can function as autocrine growth factors in human small cell lung cancer. *Nature 316*:823–825.

Davis, T.P., Burgess, H.S., Crowell, S., Moody, T.W., Culling-Bergland, A., and Liu, R.H. (1989). B-endorphin and neurotensin stimulate in vitro clonal growth of human small cell lung cancer cells. *Eur. J. Pharmacol. 161*:283–285.

Davis, T.P., Crowell, S., Taylor, J., Clark, D.L., Coy, D., Staley, J., and Moody, T.W. (1992). Metabolic stability and tumor inhibition of bombesin/GRP receptor antagonists. *Peptides 13*:401–407.

Draoui, M., Battey, J., Birrer, M., and Moody, T.W. (1993). Regulation of GRP gene expression. In: *Growth Factors, Peptides and Receptors* (Moody, T.W., ed.) Plenum Press, New York, pp. 347–352.

Giaccone, G., Battey, J., Gazdar, A.F., Oie, H., Draoui, M., and Moody, T.W. (1991). Neuromedin B is present in lung cancer cell lines. *Cancer Res. 52*:2732s–2736s.

Gozes, I., Davidson, M., Draoui, M., and Moody, T.W. (1993). The VIP gene is expressed in non-small cell lung cancer cell lines. *Biomed. Res. 13*(2):37–40.

Hong, M., and Moody, T.W. (1991). Vasopressin elevates cytosolic calcium in small cell lung cancer cells. *Peptides 12*:1315–1319.

Korman, L.Y., Carney, D.N., Citron, M.L., and Moody, T.W. (1986). Secretin/VIP stimulated secretion of bombesin-like peptides from human small cell lung cancer. *Cancer Res. 46*:1214–1218.

Lee, M., Jensen, R.T., Huang, S.C., Bepler, G., Korman, L., and Moody, T.W. (1990). Vasoactive intestinal polypeptide binds with high affinity to non-small cell lung cancer cells and elevates cyclic AMP levels. *Peptides 11*:1205–1209.

Mahmoud, S., Staley, J., Taylor, J., Bogden, A., Moreau, J.P., Coy, D., Avis, I., Cuttitta, F., Mulshine, J.L., and Moody, T.W. (1991). (Psi-13,14)Bombesin analogues inhibit growth of small cell lung cancer in vitro and in vivo. *Cancer Res. 51*:1298–1802.

Maneckjee, R., and Minna, J.D. (1992). Opioid and nicotine receptors affect the growth regulation of human lung cancer cell lines. *Proc. Natl. Acad. Sci. USA 87*:3294–3298.

Mervic, M., Moody, T.W., and Komoriya, A. (1991). A structure-function study of bombesin's C-terminal domain. *Peptides 12*:1315–1319.

Minna, J.D., Pass, H., Glatstein, E., and Ihde, D.C. (1989). Cancer of the lung. In: *Cancer: Principles and Practice of Oncology*. (DeVita, V.T., Jr., Hellman, S., and Rosenberg, S.A., eds.) J.B. Lippincott, Philadelphia, pp. 591–705.

Moody, T.W., Pert, C.B., Gazdar, A.F., Carney, D.N., and Minna, J.D. (1981). High levels of intracellular bombesin characterize human small cell lung carcinoma. *Science 214*: 1246–1248.

Moody, T.W., Carney, D.N., Cuttitta, F., Quattrocchi, K., and Minna, J.D. (1985a). High affinity receptors for bombesin/GRP-like peptides on human small cell lung cancer. *Life Sci. 37*:105–113.

Moody, T.W., Carney, D.N., Korman, L.Y., Gazdar, A.F., and Minna, J.D. (1985b). Neurotensin is produced by and secreted from classic small cell lung cancer cells. *Life Sci. 37*:105–113.

Moody, T.W., Staley, J., Zia, F., Coy, D.H., and Jensen, R.T. (1992). Neuromedin B receptors are present on small cell lung cancer cells. *J. Pharmacol. Exp. Ther. 263*:311–317.

Moody, T.W., Koros, A.M.C., Reubi, J.C., Naylor, P.H., and Goldstein, A.L. (1993a). VIP analogues inhibit small cell lung cancer growth. *Biomed. Res. 13*(2):131–136.

Moody, T.W., Zia, F., and Makheja, A. (1993b). PACAP elevates cytosolic calcium in small cell lung cancer cell lines. *Peptides 14*:241–246.

Nakanishi, Y., Mulshine, J.L., Kasprzyk, P., Natale, R., Maneckjee, R., Avis, I., Treston, A.M., Gazdar, A.F., Minna, J.D., and Cuttitta, F. (1988). Insulin-like growth factor-1 can mediate autocrine proliferation of human small cell lung cancer cell lines in vitro. *J. Clin. Invest. 82*:354–359.

Reubi, J.C., Waser, B., Sheppard, M., and Macaulay, V. (1990). Somatostatin receptors are present in small cell but not in non-small cell primary lung carcinomas: relationship to EGF receptors. *Int. J. Cancer 45*:269–274.

Roth, K.A., and Barchas, J.D. (1986). Small cell carcinoma cell lines contain opioid peptides and receptors. *Cancer 57*:769–773.

Sethi, T., and Rozengurt, E. (1991). Multiple neuropeptides stimulate clonal growth of small cell lung cancer: effects of bradykinin, vasopressin, cholecystokinin, galanin and neurotensin. *Cancer Res. 51*:3621–3623.

Sethi, T., Langdon, S., Smyth, J., and Rozengurt, E. (1992). Growth of small cell lung cancer cells: stimulation by multiple neuropeptides and inhibition by broad spectrum antagonists in vitro and in vivo. *Cancer Res. 52*:2737s–2742s.

Shaffer, M.M., Carney, D.N., Korman, L.Y., Lebovic, G.S., and Moody, T.W. (1987). High affinity binding of VIP to human lung cancer cell lines. *Peptides 8*:1101–1106.

Spindel, E.R., Sunday, M.E., Hofler, H., Wolfe, J.J., Habener, J.F., and Chin, W.W. (1987). Transient elevation of messenger RNA encoding gastrin-releasing peptide, a putative pulmonary growth factor in human fetal lung. *J. Clin. Invest. 80*:1172–1179.

Spindel, E.R., Giladi, E., Brehm, T.P., Goodman, R.H., and Segerson, T.P. (1990). Cloning and functional characterization of a cDNA encoding the murine fibroblast bombesin/GRP receptor. *Mol. Endocrinol. 4*:1956–1963.

Staley, J., Fiskum, G., and Moody, T.W. (1989). Cholecystokinin elevates cytosolic calcium in small cell lung cancer cells. *Biochem. Biophys. Res. Commun. 163*:605–610.

Staley, J., Jensen, R.T., and Moody, T.W. (1990). CCK antagonists interact with CCK-B receptors on human small cell lung cancer cells. *Peptides 11*:1033–1036.

Staley, J., Coy, D.H., Taylor, J.E., Kim, S., and Moody, T.W. (1991). (Des-met[14])Bombesin analogues function as small cell lung cancer bombesin receptor antagonists. *Peptides 12*:145–149.

Taylor, J.E., Coy, D.H., and Moreau, J.P. (1988). High affinity binding of (^{125}I-Tyr[11])somatostatin-14 to human small cell lung carcinoma (NCI-H69). *Life Sci. 43*:421–427.

Taylor, J.E., Moreau, J.P., Baptiste, L., and Moody, T.W. (1991). Octapeptide analogues of somatostatin inhibit the clonal growth and vasoactive intestinal peptide stimulated cyclic AMP formation in human small cell lung cancer cells. *Peptides 12*:839–844.

Wada, E., Way, J., Shapira, H., Kusano, K., Lebacq-Verheyden, A.M., Coy, D., Jensen, R., and Battey, J. (1991). cDNA Cloning, characterization and brain region-specific expression of a neuromedin-B preferring bombesin receptor. *Neuron 6*:421–430.

Yoder, D., and Moody, T.W. (1987). High affinity binding of cholecystokinin to small cell lung cancer cells. *Peptides 8*:103–107.

DISCUSSION

Polak: Could you comment further on VIP as a protective growth factor and the apparently discrepant findings that VIP may have antiproliferative actions in the normal lung?

Moody: VIP is biologically active in the small cell lung cancer (SCLC) cells. We have found that almost all SCLC cells have VIP receptors that elevate cyclic AMP. This will cause two effects on the SCLC cells: (1) the elevated cyclic AMP may cause the cells to differentiate, which may inhibit proliferation; (2) the cyclic AMP will stimulate protein kinase A, resulting in the phosphorylation of proteins. This will cause exocytosis of granules that contain GRP and stimulate growth. The net result we observe is that VIP stimulates growth and a VIP-receptor antagonist inhibits growth.

Said: We have recently reported that VIP *inhibits* the proliferation of two SCLC cell lines and decreases thymidine incorporation, while stimulating cyclic AMP production. You describe a pregrowth and proliferation action for VIP. Are our results different because different cell lines have been used?

Moody: The growth assays performed in your laboratory and mine differ. Your assays may reflect differentiation due to elevated cyclic AMP. This will result in decreased proliferation. Our assays may reflect protein kinase A stimulation. We have found that VIP elevates c-*fos* mRNA and this will stimulate growth. Our assays are consistent in that the VIP-receptor antagonist inhibits SCLC growth in vitro and in vivo.

27

Neuroregulation of Pulmonary Immunity: The Roles of Substance P and Vasoactive Intestinal Peptide

Edward J. Goetzl, Sunil P. Sreedharan, and H. Benfer Kaltreider
University of California Medical Center, San Francisco, California

I. INTRODUCTION

Neuropeptides and some other neuromediators have diverse potent effects on nonneural cells, including alterations in growth and differentiation, direct elicitation of primary functions, and modifications in responsiveness to other mediators (1–3). Neuropeptides contribute to both regulation of normal physiological processes and the pathogenesis of a range of diseases, including immunological and inflammatory disorders. The effective potency and biospecificity of a neuropeptide mediator are determined by many factors characteristic of each tissue compartment and target cell. The most important of such determinants are patterns of peptidergic innervation and neural responsiveness, which assure appropriate delivery of neuropeptides to the site of a reaction, the availability and properties of cells bearing receptors for neuropeptides, and the activities of peptidases capable of degradatively inactivating neuropeptides. The peptidergic neuroanatomy and potential effects of neuropeptides so far delineated in lung tissues suggest that further studies will elucidate distinctive roles for many neuropeptides in the mediation and regulation of pulmonary immune and inflammatory responses.

Increases and, less commonly, decreases in the number of neuropeptide immunoreactive nerve fibers have been observed in chronic inflammatory lesions of human skin and respiratory tissues (4,5). The consistently greater number of peptidergic nerve fibers in the affected skin of patients with atopic dermatitis than in normal skin contrasted with the lesser than normal number in skin lesions of patients with urticaria or systemic lupus erythematosus (4). Immunoreactive neuropeptides also were found free in fluids of cutaneous inflammatory lesions and in nasal and bronchoalveolar lavage (BAL) fluids from allergic and asthmatic patients after challenge with antigens implicated in their allergies, but not in fluids from nonallergic control subjects (1,5). Although the specific contributions of the neuropeptides to hypersensitivity reactions have not been defined fully, the possibilities include regulation of macrophage and lymphocyte accumulation and functions in respiratory and cutaneous compartmental immune responses (1).

Nucleic acid probes and antibodies for respective detection of the genetic messages and antigens of neuropeptides and neuropeptide receptors have been developed recently. The application of such sensitive and specific reagents in immune and inflammatory reactions should permit more precise elucidation of the mechanisms of neuroendocrine regulation of immunity and hypersensitivity. Immune recruitment of neural mediators and lymphocyte recognition and responses to neuropeptides will be discussed first and then new findings of neuromodulation of pulmonary immune responses in a mouse model.

II.   IMMUNE STIMULATION OF PEPTIDERGIC NEURAL PATHWAYS

Cells that mediate each of the major responses of the immune system, with different time courses and predominant effects, are capable of communicating to neighboring neurons the initiation of the response, through paracrine activity of one or more immune cytokines and low-molecular-weight factors (1) (Fig. 1). Immediate hypersensitivity reactions are characterized by the IgE-dependent release from mast cells of the phospholipid platelet-activating factor (PAF), which enhances neural proliferation, and histamine, which facilitates neural transmission. The eosinophils typical of immediate hypersensitivity reactions, as well as some other types of leukocytes and epithelial cells, secrete a distinct neuroactive isomer of 8,15-dihydroxy-eicosatetraenoic acid (8,15-di-HETE). Nanogram quantities of this specific 8,15-di-HETE transform the fundamental state of reactivity of nociceptive neurons to other stimuli, as exemplified by the prolonged cutaneous hyperalgesia of inflammation that is antagonized by equimolar concentrations of another isomer of 8,15-di-HETE (reviewed in 2).

The macrophages found in subacute and chronic inflammatory reactions and in cellular immune responses generate interleukin 1 (IL-1) and tumor necrosis factor (TNF), which increase proliferation and other cellular activities of both astroglial cells and neurons (reviewed in 2). IL-1 and, to a lesser extent, TNF have two other stimulatory effects on neural cells. The first is augmentation of production of nerve growth factor (NGF) (Fig. 1), which has many potent neurotrophic and immuno-

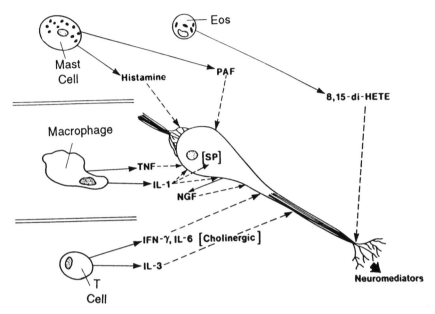

Figure 1   Immune system signaling of neurons. Eos = eosinophils; PAF = phospholipid platelet-activating factor; NGF = nerve growth factor; SP = substance P; TNF = tumor necrosis factor; IL = interleukin; IFN = interferon; HETE = hydroxy-eicosatetraenoic acid.

enhancing effects. The second is a greater expression of preprotachykinin mRNA and the consequent increase in intracellular content of SP, by a mechanism that permits specific antagonism by interferon-gamma (IFN-γ).

The IFN-γ and IL-6 produced by lymphocytes stimulate astrocytes and some other neural cells to proliferate, without apparent specificity (Fig. 1). In contrast, IL-3 is the most selective of the neurally active lymphokines as it preferentially stimulates neurite outgrowth and associated biochemical activities of cholinergic neurons (6). Positive immunoneural feedback loops may operate through such pathways as the stimulation by IL-1 and IFN-γ of neural cell production of TNF and IL-6, which further augments neural cellular activities (reviewed in 2).

III. LYMPHOCYTE RECEPTORS FOR NEUROPEPTIDES

Lymphocytes recognize and respond specifically to diverse neuropeptides by expressing distinct sets of cell membrane receptors (reviewed in 2). Of these structures, the best characterized are the guanine nucleotide-binding (G) protein-associated receptors for substance P (SP) and vasoactive intestinal peptide (VIP)

(Table 1). Three different classes of cellular receptors, designated NK-1, NK-2, and NK-3, bind SP and other tachykinins with a distinct rank-order of affinities for SP, neurokinin (NK) A, and NKB (7). The NK-1 type of tachykinin receptor, which binds SP with greatest affinity, is localized preferentially in the central and peripheral nervous system, smooth muscle, endothelial cells, fibroblasts, and both T and B lymphocytes (reviewed in 2). The cloning of rat and human NK-1 receptors revealed a single 46-kDa protein with the characteristic seven trans-membrane domains and signature sequences of other G-protein-associated receptors. Expression in mammalian cells proved that recombinant NK-1 receptors bound SP with high affinity and the expected specificity (Table 1).

Two different lymphocyte receptors for VIP have been cloned, both of which are typical structurally of other G-protein-associated receptors (Table 1). The high-affinity neuroendocrine type of VIP receptor was cloned initially from rat lung tissue (8) and a line of human colonic adenocarcinoma cells (9). This 52-kDa protein of 457 amino acids is representative of the secretin-parathyroid hormone receptor family, binds VIP with a kD of 0.1–1 nM, and transduces VIP-evoked increases in the intracellular concentration of cyclic AMP with an EC_{50} of 1 nM. Genetic message encoding the high -affinity receptor for VIP in humans is found predominantly in lung tissue, gastrointestinal epithelial cells, and subsets of lymphocytes, with weaker expression in brain, heart, kidney, liver, and placenta (9). The human low-affinity receptor for VIP, which was isolated from a line of human pre-B lymphocytes (10), is a 42-kDa protein of 362 amino acids that has the greatest structural homologies with receptors for peptide mediators of inflammation, such as the intercrines and other chemotactic factors. This receptor binds VIP with a Kd of 2.5–59 nM, as is typical of most lymphocyte native VIP receptors, and is found widely in many human tissues with the greatest prominence in immune cells.

Table 1 G-Protein-Associated Lymphocyte Receptors for Neuropeptides

Receptor	Structure	Receptor family	Ligand specificity	Affinity Kd (nM)	Tissue distribution
SPR (NK-1)	46 kDa 407 aa	Tachykinin	SP $\gg$ NKA > NKB	0.16	Lungs CNS Immune system
VIP Rh	52 kDa 457 aa	Secretin	VIP = PACAP > PHM > secretin	0.1–1	Lungs CNS Immune system
VIP R1	42 kDa 362 aa	Peptide mediators of inflammation	VIP > PHM = secretin > glucagon	2.5–59	Immune system Many other tissues

CNS = central nervous system.

IV. LYMPHOCYTE RESPONSES TO NEUROPEPTIDES

The specific recognition of neuropeptides by lymphocyte receptors elicits responses that range from modulation of proliferation and cytokine production to substantial alterations in expression of adherence proteins and immunoglobulin synthesis (reviewed in 2) (Table 2). The effects of SP and VIP on cytokine production are typical of the former quantitative regulatory activities of neuropeptides. At 10^{-13} M to 10^{-10} M, SP enhanced the expression of IL-2 mRNA in human peripheral blood T lymphocytes and T cells of the Jurkat and HUT-78 lines, which had been incubated concurrently with phorbol myristate acetate (PMA) and phytohemagglutinin (PHA) (11). The maximal increases in secretion of IL-2 evoked by SP and SP(4-11) were two- to fourfold. Similarly, hepatic granuloma cells obtained from mice infected with schistosomiasis and enriched for T lymphocytes, but not spleen cells from the same mice, spontaneously secreted IL-5 and responded to nM or higher concentrations of VIP with increased production of IL-5 (12). The maximal increases in IL-5 secretion by granuloma T cells were twofold. The expression of IL-2 receptors and the activity of NK cells are modified to a similar extent.

VIP alters lymphocyte levels of surface adherence proteins and immunoglobulin secretion to a more pronounced extent, which often includes qualitative changes. In initial studies, human B lymphoblastoid cells of the Raji line, which express a mean of 27,950 receptors for VIP with a mean Kd of 0.8 nM, were aggregated homotypically by nanomolar concentrations of VIP and, to a greater extent, by PMA (13). The homotypic aggregation of human B cells by VIP was cyclic AMP-dependent and mediated by conversion of type 1 lymphocyte function-associated (LFA-1) adhesive protein to the active configurational state

Table 2 Lymphocyte Functional Responses to SP and VIP

Function	SP	VIP
Homing/localization	Lymph nodes	Intestines, lymph nodes
Adherence	ND	Homotypic, heterotypic
Proliferation	Enhances responses to mitogens	Inhibits T-cell MLRs and responses to mitogens
Ig production	Increases IgA	Suppresses IgG and IgM and increases IgE
Cytokine production	Increases IL-1, IL-2, IL-6, TNF-γ, IFN-γ	Increases IL-5 and decreases IL-2
Expression of cytokine receptors	Increases IL-2 Rs	ND
NK cell activity	Augments	Diminishes

R = receptor; MLR = mixed lymphocyte reaction; ND = not demonstrated.

capable of binding to intercellular adherence molecule-1 (ICAM-1). The effects of VIP on lymphocyte homing and tissue localization (2,3) presumably also are mediated by activation of adhesive proteins that determine lymphocyte interactions with endothelial cells.

As short-term exposure of human lymphocytes to VIP led to only minimal alterations in the secretion of IgM and none for IgG, the effects of persistently low concentrations of VIP expected in peptidergically innervated lymphoid tissue were studied in vitro. Human blood mixed lymphocytes were incubated with twice-daily additions of 10^{-10} M to 10^{-8} M VIP for up to 18 days to maintain the concentration of VIP at 0.2–0.4 times that present immediately after the introduction of the VIP (14). The production of IgG and IgM elicited by a marginally stimulatory concentration of pokeweed mitogen (PWM) was suppressed by up to 90%, with a maximal effect at day 9 by 10^{-9} M VIP, which was not associated with changes in the composition of T or B cells. No changes in IgA production were observed with VIP, but the secretion of IgE induced by IL-4 in the same system was increased three- to 10-fold by 10^{-9} M VIP (14). The PMA-stimulated production of IgM by the SKW 6.4 line of EBV-transformed human B cells, which express the low-affinity type of receptors for VIP, was suppressed a mean maximum of 40% by 10^{-12} M to 10^{-9} M VIP after 7 days (15). There was no suppression of unstimulated secretion of IgM or of the proliferation of the B cells.

SP and VIP both affect substantially the homing and tissue localization of lymphocytes, as delineated in sheep lymph nodes and rodent gastrointestinal tissues (reviewed in 2) (Table 2). As for their contributions to other immune functions, VIP and SP have opposite effects on lymphocyte traffic in most models. For example, VIP enhances the trapping of lymphocytes in sheep lymph nodes, with a rank-order of retention of B cells > CD8+ T cells > CD4+ T cells. In contrast, SP increases the release of lymphocytes, with the greatest effect on CD4+ T cells.

V. PULMONARY NEUROIMMUNE RESPONSES

Optimal expression of pulmonary immunity to inhaled antigens requires responses of lymphocytes in the airway mucosa, pulmonary parenchyma, and hilar lymph nodes, concertive communication among these compartments, and the integrated activities of nonimmune systems in the lungs. Native lung parenchyma, consisting of alveoli, interstitium, and associated vasculature, normally has few lymphocytes. Intrapulmonary antigenic challenge of sensitized subjects results in a temporally-ordered influx of lymphocytes and other immunocompetent mononuclear leukocytes, with distinct phenotypic and functional characteristics. The fluid and cells harvested by BAL reflect accurately the constituents and activities of pulmonary parenchymal immune responses, as defined by studies of mechanically and enzymatically disrupted lung tissue. A mouse model has been developed for investigations of lung parenchymal immune responses to intratracheally administered antigens, which permits analyses of helper and suppressor T lymphocytes, cytotoxic lymphocytes, and B lymphocytes competent to produce antibodies of several isotypes (16). The importance of CD4+ T lymphocytes for effective expression of pulmonary immune responses to inhaled antigens has been documented by the results of studies of primed mice challenged with sheep erythrocyte antigen (16) and mice exposed to *Pneumocystis carinii* after total depletion of CD4+ T lymphocytes (17).

Although the T lymphocytes obtained from lungs responding to antigenic challenges bear surface markers typical of sensitized memory cells, little is known of the molecular and cellular determinants of lymphocyte adhesion to endothelium, transvascular migration, and selective retention in compartments of the pulmonary immune system. The results of preliminary studies suggest that some neuropeptides, as well as antigen and immune cytokines, have a central role in the recruitment and activation of lymphocytes in the parenchyma of lungs manifesting immune responses.

Intense influx of immune cells into the lung parenchyma occurs rapidly after the intratracheal challenge of primed mice with sheep erythrocytes, attains a peak at 4–6 days and is characterized by a 300- to 1000-fold increase in the number of lymphocytes in BAL fluid (Table 3). At the time of maximal response, the lymphocytes are 65% CD4+ T cells, 15% CD8+ T cells, and 15% B cells, with a maximum of 300–900 antibody-forming B cells per 10^6 lymphocytes. The CD4+ T cells increase in number earliest, predominate at the time of maximal response,

Table 3 Pulmonary Neuroimmune Responses

	Days after antigen challenge				
	0	1	2	4	6
Total leukocytes, BAL fluid ($\times$ 10^6/mouse)	1.7	1.8	2.5	20	14
Lymphocytes, BAL fluid ($\times$ 10^6/mouse)	0.1	0.2	0.4	8.2	5.8
IgM+ plus IgG+ B cells (#/10^6 lymphocytes, tissue)	0	0	30	175	390
SP, pM	198	542	569	572	441
VIP, pM	197	472	1191	1665	1546

BAL fluid was collected in final concentrations of 10^{-4} M phenylmethylsulfonylfluoride and 5 μg/ml of DL-thiorphan to minimize peptidolysis. The values for SP and VIP are the mean of two radioimmunoassay determinations for samples extracted by solid phase (Sep-Pak, Waters Associates, Milford, MA). The levels of leukocytes and lymphocytes are representative of those found in 5–12 experiments.

and bear surface markers characteristic of memory T cells. The sheep erythrocyte antigen elicits antibodies of the IgG and IgM class, without a clear response of IgE (Table 3). SP and VIP in BAL fluid and lymphocyte receptors (Rs) for these neuropeptides have been quantified in the same mouse model.

The mean concentrations of immunoreactive SP and VIP were similar prior to intratracheal antigen challenge of sensitized mice and increased significantly within 24 hr of the intratracheal introduction of sheep erythrocytes (Table 3). The maximal concentrations of SP and VIP were attained on days 1 and 4 after antigen challenge, respectively, and each was maintained for up to 6 days. High-performance liquid chromatographic resolution of SP and VIP in some BAL fluids prior to radioimmunoassays confirmed the identification of the neuropeptides. The capacity of lymphocytes to respond to SP and VIP is dependent on their expression of the corresponding receptors, which were quantified by a polymerase chain reaction–based technique for quantitative amplification of mRNA encoding SPRs and VIPRs. SPR message was detected in lymphocytes from BAL fluid obtained on days 4 and 6 after sheep erythrocyte challenge, but not in an equal number of lymphocytes from days 1 or 2 after antigen or from day 0 prior to antigen. Similar analyses of the mRNA encoding VIP Rh and VIP R1 (Table 1) showed that the messages for both were readily detectable on days 0–6, with apparent maxima on day 6 and days 2–4, respectively.

VI. SUMMARY

Peptidergic nerves in lung tissues end on or in close proximity to lymphocytes and other immune cells, as well as smooth muscle, epithelial, secretory, and vascular elements. A role for neuropeptides, especially SP and VIP, in the mediation and regulation of pulmonary compartmental immune responses is suggested by their capacity to increase lymphocyte adherence to other lymphocytes and endothelial cells, alter cytokine production, and modify antibody secretion with isotypic specificity. The results of preliminary studies in a mouse model of pulmonary parenchymal immune responses to intratracheal antigen have demonstrated the release of SP and VIP into pulmonary airway secretions in functionally relevant concentrations and an increase in expression of receptors for SP and VIP on the infiltrating lymphocytes. Experimental systems and reagents are now available for comprehensive studies of the possible contributions of peptidergic neurons to pulmonary immune responses and inflammatory diseases.

ACKNOWLEDGMENT

This research was supported by National Institutes of Health Grants R01 AI29912 and U01 AI34570 (EJG), R01 HL34298 (HBK), and R29 DK44876 (SPS).

REFERENCES

1. Ader, R., Felten, D.L., and Cohen, N., eds. (1991). *Psychoneuroimmunology*, 2nd ed. Academic Press, New York.
2. Goetzl, E.J., and Sreedharan, S.P. (1992). Mediators of communication and adaptation in the neuroendocrine and immune systems. *FASEB J. 6*:2646.
3. Ottaway, C.A., and Husband, A.J. (1992). Central nervous system influences on lymphocyte migration. *Brain Behav. Immun. 6*:97.
4. Tobin, D., Nabarro, G., Baart de la Faille, H., van Vloten, W.A., van der Putte, S.C.J., and Schuurman, H-J. (1992). Increased number of immunoreactive nerve fibers in atopic dermatitis. *J. Allergy Clin. Immunol. 90*:613.
5. Nieber, K., Baumgarten, C.R., Rathsack, R., Furkert, J., Oehme, P., and Kunkel, G. (1992). Substance P and beta-endorphin-like immunoreactivity in lavage fluid of subjects with and without allergic asthma. *J. Allergy Clin. Immunol. 90*:646.
6. Kamegai, M., Niijima, K., Kunishita, T., Nishizawa, M., Ogawa, M., Araki, M., Ueki, A., Konishi, Y., and Tabira, T. (1990). Interleukin 3 as a trophic factor for central cholinergic neurons in vitro and in vivo. *Neuron 2*:429.
7. Regoli, D. (1987). Pharmacological receptors for substance P and neurokinins. *Life Sci. 403*:66.
8. Ishihara, T., Shigemoto, R., Mori, K., Takahashi, K., and Nagata, S. (1992). Functional expression and tissue distribution of a novel receptor for vasoactive intestinal peptide. *Neuron 8*:811.
9. Sreedharan, S.P., Patel, D.R., Huang, J-X., and Goetzl, E.J. (1993). Cloning and functional expression of a human neuroendocrine vasoactive intestinal peptide receptor. *Biochem. Biophys. Res. Commun. 193*:546.
10. Sreedharan, S.P., Robichon, A., Peterson, K.E., and Goetzl, E.J. (1991). Cloning and expression of the human vasoactive intestinal peptide receptor. *Proc. Natl. Acad. Sci. USA 88*:4986.
11. Calvo, C-F., Chavanel, G., and Senik, A. (1992). Substance P enhances IL-2 expression in activated human T cells. *J. Immunol. 148*:3498.
12. Mathew, R.C., Cook, G.A., Blum, A.M., Metwali, A., Felman, R., and Weinstock, J.V. (1992). Vasoactive intestinal peptide stimulates T lymphocytes to release Il-5 in murine Schistosomiasis mansoni infection. *J. Immunol. 148*:3572.
13. Robichon, A., Sreedharan, S.P., Yang, J., Shames, R.S., Gronroos, E.C., Cheng, P.P-J., and Goetzl, E.J. (1993). Induction of aggregation of Raji human B lymphoblastic cells by vasoactive intestinal peptide. *Immunology 79*:574.
14. Hassner, A., Lau, M.S., Goetzl, E.J., and Adelman, D.C. (1993). Isotype-specific regulation of human lymphocyte production of immunoglobulins by sustained exposure to vasoactive intestinal peptide (VIP). *J. Allergy Clin. Immunol. 92*:891.
15. Cheng, P.P-J., Sreedharan, S.P., and Goetzl, E.J. (1993). The SKW 6.4 line of human B lymphocytes specifically binds and responds to vasoactive intestinal peptide. *Immunology 79*:64.
16. Curtis, J.L., Byrd, P.K., Warnock, M.L., and Kaltreider, H.B. (1991). Requirement of CD4-positive T cells for cellular recruitment to the lungs of mice in response to a particulate intratracheal antigen. *J. Clin. Invest. 88*:1244.

17. Beck, J.M., Warnock, M.L., Curtis, J.L., Sniezek, M.J., Arraj-Peffer, S.M., Kaltrei-der, H.B., and Shellito, J.E. (1991). Inflammatory responses to pneumocystis carinii in mice selectively depleted of helper T lymphocytes. *Am. J. Respir. Cell Mol. Biol.* 5:186.

DISCUSSION

Said: What are the effects of SP and VIP on immune cytokine production by lymphocytes?

Goetzl: SP increases secretion of IL-1, IL-2, IL-6, TNF-α, and IFN-γ, whereas to date the effects of VIP are limited to a decrease in IL-2 and an increase in IL-5.

Manzini: Do SP or VIP agonists and antagonists affect the lymphocyte responses in the mouse/lung immune model?

Goetzl: These experiments are now in progress, as are studies designed to examine the role of different cytokines.

Linnoila: Are the lymphoid submucosal follicles in the lung tissue innervated by peptidergic neurons and do these follicles contribute to immune responses in your model?

Goetzl: Neuropeptides have been detected in lymphoid follicles of GALT and BALT, but the studies of BAL lymphocytes in the mouse model do not accurately reflect events in the submucosal follicles alone.

28

Sensory Neuropeptides in an Animal Model of Hyperpnea-Induced Bronchoconstriction

Julian Solway and Daniel W. Ray
University of Chicago, Chicago, and Northwestern University and Evanston Hospital, Evanston, Illinois

Allan Garland
UMDNJ–Robert Wood Johnson Medical School, New Brunswick, New Jersey

I. INTRODUCTION

People with asthma commonly experience an asthma attack following exercise. The pathogenesis of exercise-induced bronchoconstriction (EIB) has been the subject of much interest. Most studies of this phenomenon have been conducted in human asthmatic subjects, for until recently there were no animal models that exhibit airway responses that parallel EIB. In 1985, however, Chapman and Danko reported that mechanically hyperventilated guinea pigs develop hyperpnea-induced airflow obstruction, which mimics the hyperpnea-induced broncho-constriction (HIB) routinely seen in human subjects with exercise-induced asthma. This new animal model has now been extensively studied and has been used to generate novel hypotheses about the biochemical mechanisms that operate in human exercise-induced asthma.

In this chapter, we shall review the essential characteristics of hyperpnea-induced airway responses in exercise-induced asthma and in mechanically venti-lated guinea pigs. Then, we shall examine evidence that endogenously released sensory neuropeptides play a critical role in guinea pig HIB. Based on the striking phenomenological similarities we shall see between HIB in guinea pigs and HIB in human exercise-induced asthma, we shall conclude by formulating the hypothesis that endogenous sensory neuropeptide release plays a key role in the pathogenesis of human HIB as well.

II.   EXERCISE-INDUCED ASTHMA

A.   Overview

Exercise is a common precipitant of airflow obstruction in subjects with asthma. That exercise can provoke difficulty in breathing was reported in antiquity by Aretaeus the Cappadocian (Adams, 1856, cited in McFadden and Ingram, 1983). Almost 2 millennia later, Sir John Floyer again observed that "All violent Exercise makes the Asthmatic to breathe short . . . and if the Exercise be continued it ocassions a Fit. . ." (Sakula, 1984). Although physicians wondered how exercise provokes airflow obstruction in asthma, progress in elucidating this process was made slowly. Herxheimer (1946) was the first to suggest that hyperventilation was the feature of exercise that constitutes the stimulus to bronchoconstriction, but he and others (Ferguson et al., 1969; Fisher et al., 1970) held that hyperventilation exerted its effect through generation of airway hypocapnia, a notion that has since been proved wrong (McFadden et al., 1977). Other consequences of exercise or hyperventilation were subsequently considered as potential pathogenetic mechanisms in EIB, including stimulation of airway mechanoreceptors during hyperpnea (Rebuck and Read, 1968; Stanescu and Teculescu, 1970) or lactic acidosis and blood pH changes (Silverman et al., 1972; Vassallo et al., 1972), but further scrutiny eventually disproved these hypotheses as well (Strauss et al., 1977).

A breakthrough in understanding how exercise provokes bronchoconstriction in asthma came in the late 1970s, when several investigative teams demonstrated that the severity of postexercise bronchoconstriction depends on the amount of heat and water lost from the lower airways during the period of exercise (Anderson et al., 1982; Chen and Horton, 1977; Deal et al., 1979a,b; Strauss et al., 1978; Zeballos et al., 1978). Conditions that increase heat and water loss from the intrapulmonary airways (such as increased minute ventilation caused by greater exercise intensity, mouth breathing, and breathing of cool, dry inspired gas) provoke greater postexercise bronchoconstriction, and conditions that minimize lower airway heat and water losses (smaller minute ventilation, nose breathing, and breathing of warm, humid inspired gas) also minimize the bronchoconstrictor response to exercise. Further studies indicate that exercise per se may not be an absolute requirement for EIB, for many features of EIB can be reproduced in the laboratory simply by having subjects perform voluntary isocapnic hyperventilation of cold, dry gas (Deal et al., 1979b). [EIB does differ from HIB in several important ways, though; see below (Ingram and McFadden, 1983).] This breakthrough in understanding explains many prior observations and reconciles earlier apparent inconsistencies. For example, as early as 1864, Salter noted that exercise in a cold environment could worsen the severity of exercise-induced airflow obstruction. In addition, many asthmatic subjects find that running, especially in cold weather, routinely precipitates EIB, whereas swimming presents no such difficulty, even at the same degree of exertion. Reduced respiratory heat and water loss, attributable to elevated inspired air temperature and humidity, can account for most, if not all, of this apparent paradox.

Few now question the importance of respiratory heat and/or water loss in the initiation of exercise-induced airflow obstruction, but there is little certainty about the cellular and biochemical mediator pathways that are subsequently activated. Excess water loss from the central airways during dry gas hyperventilation has been thought by some to cause mucosal surface hypertonicity (Anderson, 1984), and inhalation of hypertonic saline aerosol does cause bronchoconstriction in asthmatic individuals (Anderson and Smith, 1991). Mast cells exposed to hyperosmolar environment can release bronchoactive mediators (Eggleston et al., 1987), and it has been suggested that mast-cell-derived mediators participate in the biochemical pathway leading to EIB/HIB (Kay, 1987). Whereas some studies have found elevated levels of mast-cell-derived mediators in the circulation during EIB (Lee et al., 1982), others have not confirmed these findings (Deal et al., 1980). Furthermore, direct evidence for the participation of mast cell mediators in EIB/HIB is scanty, though recent studies demonstrating inhibition of EIB with potent antihistamines (Badier et al., 1988) increase the likelihood that histamine participates in some fashion. Recent studies demonstrate that inhibition of leukotriene synthesis (Israel et al., 1990) or blockade of the cysteinyl leukotriene-1 receptor (Manning et al., 1990; Robuschi et al., 1992) substantially blunts EIB/HIB. Thus, the role of sulfidopeptide leukotrienes in this response appears more certain, though their cellular origin remains to be identified. In contrast, whereas cholinergic vagal efferent activity was previously held to contribute to airway narrowing after exercise (McNally et al., 1979; Sheppard et al., 1982), further evaluation has demonstrated that neither blockade of upper airway afferent activation with local anesthetics (Fanta et al., 1980) nor blockade of airway muscarinic receptors with atropine (Griffin et al., 1982) prevents EIB or HIB. Thus, cholinergic vagal efferent nerves do not appear to play the major role in effecting airway narrowing following exercise or dry gas hyperpnea.

Just as the cellular and biochemical pathways activated by exercise or dry gas hyperpnea have not been fully established, there is also controversy over which airway structure responds to these mediators to effect airway lumenal narrowing. The long-standing view is that airway smooth muscle, perhaps made hyperresponsive by the asthmatic state, contracts in response to constrictor mediators released during exercise. However, an alternative view suggests that vascular engorgement and edema contribute importantly to lumenal narrowing in exercise-induced asthma attacks (McFadden, 1990); it is held that these vascular changes occur only after the cessation of exercise because they are provoked by the airway rewarming that occurs following, but not during, the period of exercise (see below). Supporting this possibility are the following observations: (1) administration of the vasoconstrictor norepinephrine can blunt (Gilbert and McFadden, 1992a) or reverse HIB (Pichurko et al., 1986); (2) intra-airway airstream temperatures rise faster in asthmatic subjects following cessation of frigid gas hyperpnea than in normals, a finding that might reflect increased bronchial blood flow (Gilbert et al., 1987); and (3) systemic volume loading by intravenous fluid infusion during hyperpnea-induced airflow obstruction worsens the obstruction (Gilbert et al., 1992b). Direct measurements of airway blood flow during EIB or HIB have yet to be made in asthmatic individuals. It may be that both airway smooth muscle and the bronchial vasculature participate in exercise-induced airflow obstruction.

B. Particular Features of Exercise- and Hyperpnea-Induced Bronchoconstriction in Asthma

Now let us consider some of the stereotypical behaviors of EIB or HIB in human asthma. Later, we shall compare these behaviors with those of mechanically hyperventilated guinea pigs. As noted above, exercise and voluntary isocapnic hyperventilation appear to provoke EIB and HIB, respectively, through similar mechanisms involving respiratory heat and water loss in the initial pathogenetic sequence (Deal et al., 1979a,b). Important behavioral differences between the airway responses to exercise or voluntary isocapnic hyperventilation do exist; however, these are likely attributable to additional events present during exercise, but absent during voluntary isocapnic hyperventilation. As such, it seems reasonable to view both provocations as variations of a single entity.

EIB follows a characteristic time course in relation to the period of exercise. Typically, there is bronchodilatation, rather than bronchoconstriction, while exercise is continued at a steady level (Fig. 1a). This exercise-induced bronchodilatation is probably caused in part by endogenous catecholamines released into the circulation during exercise (Pichurko et al., 1986). It is possible, though unproved, that a decrease in nonspecific airway responsiveness observed during exercise (Stirling et al., 1983; Freedman et al., 1988) also contributes to exercise-induced bronchodilatation, or at least to the lack of bronchoconstriction usually found during maintained exercise. Upon cessation of exercise, bronchodilatation quickly abates, and airflow obstruction develops. The airflow obstruction increases in magnitude over several minutes and typically peaks 8–10 min after cessation of exercise. The basis for this time course is not precisely understood but may reflect in part the rapid disappearance of elevated circulating catecholamines following exercise (Pichurko et al., 1986). HIB follows a similar time course as EIB, though the bronchodilatation observed during exercise is absent during hyperpnea, and airflow obstruction peaks earlier than after exercise (typically

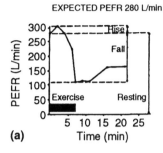
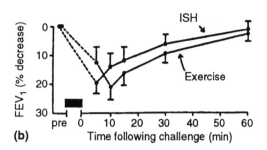

(a) **(b)**

Figure 1 Time course of airway caliber changes during bronchoconstriction induced by exercise or voluntary isocapnic hyperventilation. (a) Bronchodilatation is evident during a period of exercise, and bronchoconstriction follows cessation of exercise (exercise period is shown as the bar), as reflected by changes in peak expiratory flow rate (PEFR) in an 8-year-old child. (Adapted from Godfrey et al., 1973.) (b) Bronchoconstriction induced by isocapnic voluntary hyperventilation (ISH) peaks slightly earlier than does broncho-constriction induced by exercise (as reflected in the nadir of FEV$_1$, the forced expiratory volume in 1 sec), though each follows cessation of hyperpnea or exercise challenge, respectively (the challenge period is shown as the bar). (Adapted from Nagakura et al., 1983.)

at 3–5 min posthyperpnea) (Fig. 1b). These minor differences may be due in part to the absence of an endogenous catecholamine surge during voluntary isocapnic hyperpnea (Pichurko et al., 1986). These is no late-phase bronchoconstriction after either exercise or voluntary isocapnic hyperventilation (Rubenstein et al., 1987), in contrast to the delayed airflow obstruction that characterizes allergen challenge in atopic asthmatic subjects.

The magnitude of peak postexercise or posthyperpnea airflow obstruction in asthma has several well-characterized determinants. Reflecting early clinical observations, laboratory investigations (Deal et al., 1979a,b) have established that the temperature and humidity of inspired air greatly affect the magnitude of EIB and HIB. Asthmatic subjects breathing cold, dry air during exercise or hyperpnea of a given duration and intensity experience substantial postchallenge bronchoconstriction, whereas identical exercise or hyperpnea performed while breathing inspired air warmed to 37°C and fully saturated with water vapor is without effect on airway caliber (Fig. 2). Thus, respiratory heat and/or water loss has emerged as a sine qua non of exercise-induced bronchoconstriction. Additional factors that determine the total heat and/or water lost from the lower airways during a period of exercise or hyperpnea include the minute ventilation during the challenge period and the duration of exercise or hyperpnea. Thus, it is perhaps not surprising that the magnitude of postchallenge airflow obstruction also increases progressively with minute ventilation during exercise or voluntary isocapnic

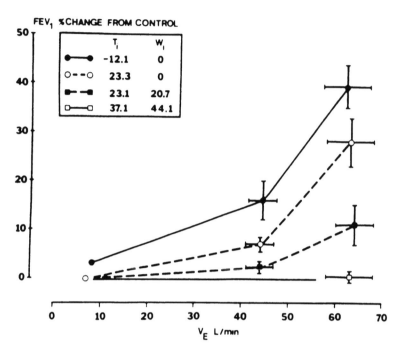

Figure 2 Stimulus-response curves demonstrating the dependence of the magnitude of hyperpnea-induced bronchoconstriction on both inspired gas conditions and minute ventilation during a fixed period of hyperpnea by asthmatic subjects. Four inspired gas conditions are shown: T_i is inspired gas temperature in °C; W_i is inspired gas water content in mg/L. Note that increasing minute ventilation of cool, dry air increases the severity of HIB, but hyperventilation of gas warmed and humidified to alveolar conditions does not provoke bronchoconstriction, even at high minute ventilation. (Adapted from Deal et al., 1980. Reproduced from the *Journal of Clinical Investigation* by copyright permission of the American Society for Clinical Investigation.)

hyperventilation of a given duration (Fig. 2), and that prolonging the duration of isocapnic hyperpnea at fixed minute ventilation worsens airflow obstruction (Blackie et al., 1990) (Fig. 3).

One feature of EIB that has received much attention is the "refractory period" that follows an episode of exercise. Even though an asthmatic individual may develop substantial bronchoconstriction following a first bout of exercise, the magnitude of bronchoconstriction induced by additional bouts performed in rapid sequence may be smaller than the first. This "refractoriness" to EIB wanes with the duration separating consecutive exercise periods and is completely absent by 2 hr after an exercise challenge. Inhibitors of cyclooxygenase can abolish the refractory period after exercise (O'Byrne and Jones, 1986), suggesting that a

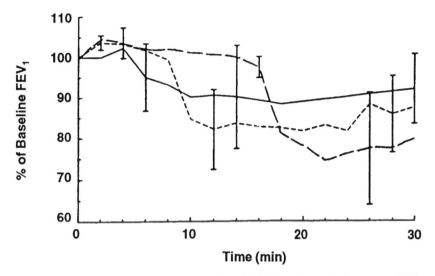

Figure 3 Effect of prolonging the duration of frigid gas isocapnic hyperpnea (ISH) at fixed minute ventilation on the time course and magnitude of HIB in human asthmatic subjects. (Solid line) 4 min ISH; (short dashed line) 8 min ISH; (long dashed line) 16 min ISH. Note that substantial bronchoconstriction is always delayed until the cessation of hyperpnea, and that prolonging the period of hyperpnea increases the severity of post-hyperpnea bronchoconstriction. (Adapted from Blackie et al., 1990.)

bronchodilator prostaglandin may be involved in this phenomenon. In contrast, it appears that there is little or no refractoriness (Stearns et al., 1981) following voluntary isocapnic hyperpnea, though reports to the contrary do exist (Margol-skee et al., 1988; Rosenthal et al., 1990).

A number of pharmacological interventions have been performed in an effort to discern the mediator pathways involved in EIB and HIB. Initial reports suggested that atropine administration can block EIB (Sheppard et al., 1982). However, subsequent studies have clarified that muscarinic blockade results primarily in bronchodilatation and has only minor influence on the position of the stimulus response curve between minute ventilation during isocapnic hyperventilation and airflow obstruction (Griffin et al., 1982). Thus, cholinergic neurotransmission does not appear to play a principal role in EIB/HIB in human asthma. In contrast, a number of compounds clearly do blunt EIB or HIB. Inhalation of β_2-adrenergic agonists prior to exercise or hyperpnea challenge blunts the post-challenge bronchoconstriction (Rossing et al., 1982) and so does pretreatment with cromolyn sodium (Eggleston et al., 1972). Administration of newly available inhibitors of 5-lipoxygenase (Israel et al., 1990) or of the cysteinyl leukotriene-1 receptor (Manning et al., 1990; Robuschi et al., 1992) has demonstrated a clear-cut

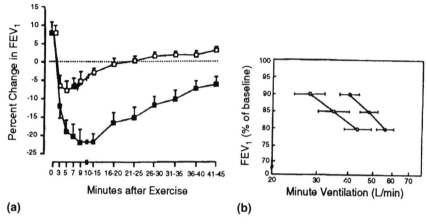

(a)   **(b)**

Figure 4 Role of sulfidopeptide leukotrienes in EIB or HIB in asthma. (a) Pretreatment with a cysteinyl leukotriene-1 receptor antagonist (MK-571) (open squares) blunts postexercise bronchoconstriction, compared with paired trials after placebo treatment (solid squares). (From Manning et al., 1990.) (b) Pretreatment with a 5-lipoxygenase inhibitor (A-64077, solid circles) shifts the minute ventilation versus posthyperpnea bronchoconstriction (expressed as % of baseline FEV_1) stimulus-response curve to the right, compared with paired trials after placebo pretreatment (open circles). (From Israel et al., 1990.)

role for sulfidopeptide leukotrienes in these responses, as reflected in the abilities of these drugs to reduce hyperpnea- or exercise-induced airflow obstruction (Fig. 4).

III. HYPERPNEA-INDUCED AIRWAY RESPONSES IN GUINEA PIGS

A. Hyperpnea-Induced Bronchoconstriction

In 1985, Chapman and Danko first reported that anesthetized guinea pigs develop mild airflow obstruction (less than twofold increase in airway resistance) after 10 min of mechanical hyperventilation with room air. Bronchoconstriction arose after the cessation of hyperpnea, in a time course typical for HIB in human asthma. This initial study provided the first clear-cut demonstration of an airway response in animals that directly parallels HIB in human asthma and revealed preliminary insights into the pathways involved: The severity of guinea pig HIB appeared to be unaffected by elevation of inspired carbon dioxide fraction to 10%, by cervical vagotomy, or by administration of chlorpheniramine, methysergide, atropine, indomethacin, FPL 55712, nifedipine, or verapamil. However, both salbutamol and aminophylline could prevent or reverse this response. Thus, although airway smooth muscle appeared to effect airway narrowing, there was no clear-cut evidence for the participation of any other cell type or mediator pathway.

Shortly thereafter, our group in Chicago also discovered that guinea pigs' airways respond to dry gas hyperpnea (Ray et al., 1988). We sought to compare and contrast these responses with HIB in human asthma, so we designed our initial experiments specifically to that end. In our studies, each pentobarbital-anesthetized male Hartley guinea pig was intubated through high cervical tracheostomy and then mechanically ventilated using a circuit that incorporates a short common segment between inspiratory and expiratory paths. This configuration prevents inspired gas from acquiring heat or water vapor during traversal of a long common segment that might have acquired heat or condensate during each previous exhalation. (One can view this approach as analogous to having human asthmatic subjects breathe through the mouth, rather than through the nose, to increase heat and water losses from the lower airways during exercise.) Respiratory system resistance was monitored in a whole-body plethysmograph. All animals were pretreated with propranolol to achieve β-adrenergic blockade, thereby obviating any potential modulating influence of changing levels of circulating catecholamines on airway responses; even so, we have also shown that propranolol administration does not influence the magnitude of these hyperpnea-induced airway responses (Ray et al., 1991). In some experiments, Evans blue dye was given to label circulating albumin; excess extravasation of Evans blue dye was taken as evidence for bronchovascular hyperpermeability. During mechanically imposed hyperpnea, inspired carbon dioxide was enriched to 5% to maintain eucapnia. Although inspired oxygen fraction during and after hyperpnea was also increased (to avoid severe bronchoconstriction-induced arterial hypoxemia), such FiO_2 elevation does not affect the magnitude of guinea pig HIB (Ray et al., 1991).

Confirming Chapman's earlier observation, our mechanically hyperventilated guinea pigs exhibited dry gas HIB, which follows the characteristic time course shown in Figure 5 (Ray et al., 1988). These data were gathered using dry gas at room temperature as the inspirate, but similar observations have also been made using frigid dry gas (Ingenito et al., 1990). During 5 min of mechanically imposed isocapnic hyperpnea, there is little change in respiratory system resistance (Rrs) over baseline prehyperpnea values. However, after cessation of hyperpnea and return to quiet breathing of humidified gas, progressive airflow obstruction ensues peaking at 3 min posthyperpnea and then abating spontaneously. No late-phase bronchoconstriction is observed for at least 4 hr after the initial constrictor response. The magnitude of the peak bronchoconstrictor response to dry gas hyperpnea is reproducible upon consecutive identical challenges of fixed minute ventilation, duration, and inspired gas conditions. This finding allows the performance of up to four hyperpnea challenges in each animal and thereby allows direct comparison of the effects of diverse hyperpnea conditions or pretreatments.

Whereas animals subjected to hyperventilation with dry gas develop substan-

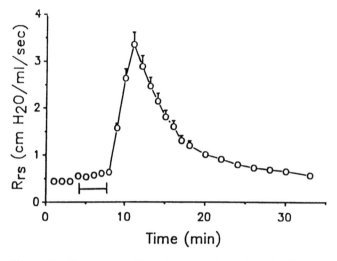

Figure 5 Time course of bronchoconstriction induced by 5 min room temperature, dry gas hyperpnea in guinea pigs. Hyperpnea period is shown by the bar. Rrs, respiratory system resistance. There is little bronchoconstriction during dry gas hyperpnea, but substantial bronchoconstriction ensues following cessation of hyperpnea during return to quiet breathing. (Adapted from Ray et al., 1988.)

tial posthyperpnea airflow obstruction, guinea pigs hyperventilated with fully humidified inspired gas at body temperature have no discernible response (Ray et al., 1988; Ingenito et al., 1990; Garland et al., 1991). Thus, heat and water loss from the lower airways during hyperpnea is a sine qua non of guinea pig HIB, just as it is in human HIB. Furthermore, as in human asthma, increasing the minute ventilation of dry gas during a fixed period of hyperpnea increases the severity of HIB in a stimulus-response relationship (Fig. 6), and prolonging the period of dry gas hyperpnea, at fixed minute ventilation, causes progressively greater posthyperpnea obstruction (Fig. 7). An explanation for why the major portion of bronchoconstriction is always delayed until the cessation of hyperpnea is found in the results shown in Figure 8 (Ray et al., 1991). Guinea pigs were subjected to 10 min dry gas hyperpnea and then returned to quiet breathing of warm humid gas or to isocapnic hyperpnea with warm humid gas. As shown in Figure 8, continued hyperventilation after switching to warm humid inspirate completely suppressed the bronchoconstriction that was marked when animals were quietly ventilated after dry gas hyperpnea. Thus, it appears that hyperpnea per se suppresses the appearance of bronchoconstriction that otherwise would have occurred in the absence of continued hyperventilation. It seems possible that repeated deep inhalations during hyperpnea stretch the airway wall and inhibit its

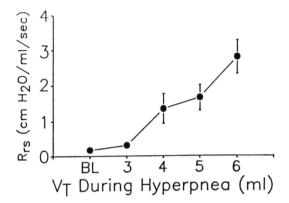

Figure 6 Stimulus-response relationship between the severity of HIB and minute ventilation during a fixed 5-min period of dry gas hyperpnea in guinea pigs. In this experiment, minute ventilation was increased by increasing tidal volume (V_T) at fixed breathing frequency of 150 breaths/min; similar results have also been obtained by increasing breathing frequency at fixed tidal volume. The severity of HIB [shown as peak post-hyperpnea respiratory system resistance (Rrs)] increases with minute ventilation during 5 min isocapnic dry gas hyperpnea. Baseline (BL) Rrs is also shown. (Adapted from Ray et al., 1988.)

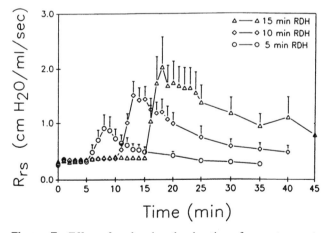

Figure 7 Effect of prolonging the duration of room temperature, dry gas hyperpnea (RDH) at fixed minute ventilation on the time course and magnitude of HIB in guinea pigs. Prolonging the period of hyperpnea causes progressively greater posthyperpnea obstruction in guinea pigs. (Adapted from Ray et al., 1991.)

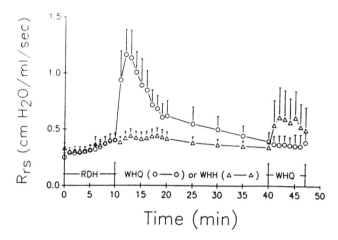

Figure 8 A nonthermal effect of hyperpnea per se suppresses dry gas HIB in guinea pigs. Animals subjected to 10 min isocapnic room temperature, dry gas hyperpnea (RDH) followed by return to 30 min quiet breathing of warm, humidified gas (WHQ) exhibit posthyperpnea bronchoconstriction, whereas the same animals subjected to RDH followed by return to 30 min hyperpnea of warm, humidified gas (WHH) do not. Thus, continued hyperpnea per se suppresses HIB. Some animals whose HIB was suppressed by WHH manifest HIB when finally returned to WHQ at 40 min. (Adapted from Ray et al., 1991.)

narrowing (Ray et al., 1991). Recent evidence from investigators in Vancouver suggests that a similar mechanism may explain the corresponding time course of HIB in human asthma (Blackie et al., 1990) (Fig. 3).

Tantalum bronchograms have localized hyperpnea-induced airway narrowing in guinea pigs to the central conducting bronchi (Fig. 9) (Ray et al., 1990). In contrast to the relatively uniform bronchoconstriction induced by intravenous methacholine, dry gas hyperpnea caused a unique proximal-greater-than-distal axial distribution of lumenal narrowing in these airways. As the minute ventilation during a fixed period of dry gas hyperpnea was increased, the most central airways narrowed further, and bronchoconstriction also appeared in more distal airways that did not narrow at lower minute ventilation. These local dependencies of bronchoconstriction on minute ventilation correspond closely with the expected changes in local airway heat and water loss as minute ventilation is increased (Ingenito et al., 1986; Ray et al., 1989b). Thus, it seems most likely that each airway narrows in response to the local thermal stimulus—i.e., the local airway heat and water loss. Corresponding axial distributions of airway narrowing have yet to be determined in human asthma.

If the local airway heat and water losses determine the severity of local airway narrowing, what are the cellular and biochemical pathways that transduce and

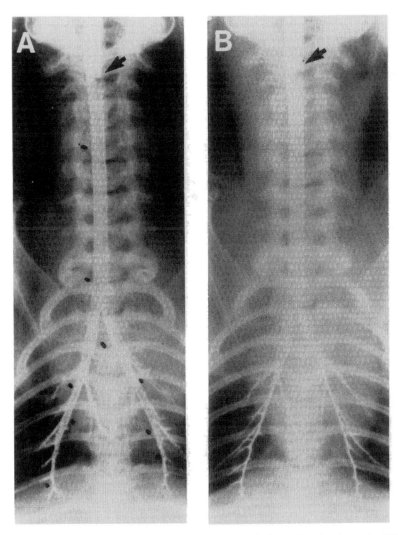

Figure 9 Tantalum bronchograms of a guinea pig before (A) and at the peak of (B) HIB. The arrow points to the distal end of the tracheostomy tube. Black marks in A are registration marks made for comparison of identical airways on serial radiographs. Note the substantial hyperpnea-induced narrowing of airways beyond the proximal main bronchi. (From Ray et al., 1990.)

mediate this phenomenon? Like Chapman and Danko (1985), we found that cholinergic neurotransmission does not play the major role, for neither cervical vagotomy nor additional atropine administration blunts guinea pig HIB (Ray et al., 1988). In contrast to prior studies, though, we have recently documented that eicosanoid mediators *do* play a modulatory role in guinea pig HIB (Garland et al., 1993). We gave piroxicam to block cyclooxygenase activity; A63162 to block 5-lipoxygenase function; BW755c to inhibit both enzyme pathways; or ICI198,615 to block cysteinyl leukotriene-1 receptors. Each of these eicosanoid-modulating interventions blunted guinea pig HIB by 50–80%. It is noteworthy that the inhibitory effect on HIB of sulfidopeptide synthesis or receptor blockade parallels similar effects of these interventions in human HIB/EIB (Fig. 4). Interestingly, dry gas hyperpnea elevates bronchoalveolar lavage fluid levels of prostaglandins D_2 and $F_{2\alpha}$, thromoxane A_2, and leukotriene B_4 (Ingenito et al., 1990). The precise cellular sources of these mediators have not yet been identified.

B. The Role of Sensory Neuropeptides in Guinea Pig HIB

Soon after we demonstrated that heat and water losses from the airway mucosa were necessary precipitants of guinea pig HIB (Ray et al., 1988), we wondered how these thermal events are transduced into the biological response described above. We reasoned that the mucosal cooling and/or drying that could result from local respiratory heat and water losses might act as a local irritant, perhaps leading to stimulation of the irritant-sensitive C-fiber nerves that supply the central airways (Solway and Leff, 1991). If so, then C-fiber tachykinin release could possibly occur and might lead to bronchoconstriction through activation of specific neurokinin receptors on airway smooth muscle.

To test the potential role of C-fibers and endogenous sensory neuropeptide release in guinea pig HIB, we compared stimulus-response (i.e., minute ventilation vs. severity of HIB) curves obtained in control animals with stimulus-response curves obtained in two experimental groups, in which sensorineural function was intentionally manipulated (Ray et al., 1989a). Animals in the first experimental group were depleted of C-fiber neuropeptides by prior systemic pretreatment with capsaicin, a known C-fiber toxin. It had already been established that systemic capsaicin pretreatment reduces physiological responses that depend on endogenous C-fiber neuropeptide release. Animals in the second group received phosphoramidon, an agent that inhibits neutral endopeptidase (NEP); this enzyme is chiefly responsible for the cleavage and inactivation of C-fiber tachykinins released within the airway wall. As such, inhibition of NEP prolongs the longevity of endogenously released tachykinins and increases the magnitude of tachykinin-mediated physiological responses. Figure 10 shows the results of this study. Depletion of endogenous sensory neuropeptides with systemic capsaicin pretreatment virtually ablated the prominent bronchoconstrictor response to dry gas hyperpnea seen in control animals. In marked contrast,

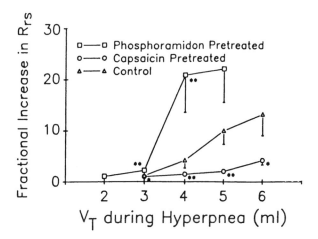

Figure 10 Evidence that tachykinins participate in guinea pig HIB. Shown are minute ventilation versus HIB stimulus-response curves in three groups of guinea pigs. Broncho-constriction is quantified as fractional increase in Rrs over baseline values. (Triangles) Control animals; (Circles) systemic capsaicin-pretreated (tachykinin-depleted) animals; (squares) phosphoramidon-treated (tachykinin-potentiated) animals. Tachykinin depletion markedly blunts guinea pig HIB, whereas inhibition of neutral endopeptidase with phos-phoramidon markedly potentiates HIB. V_T-tidal volume. (Adapted from Ray et al., 1989a.)

potentiation of the activity of endogenously released tachykinins by phosphor-amidon administration also markedly enhanced the bronchoconstrictor response to dry gas hyperpnea. In separate studies, we found that neither intervention altered airway constrictor responsiveness to cholinergic stimulation (Ray et al., 1989a). Together, these findings strongly implicated endogenous tachykinin release in the pathogenesis of guinea pig HIB. But more direct pharmacological evidence for the participation of tachykinins in guinea pig HIB has now been gathered using recently available specific NK-1 and NK-2 receptor antagonists (Solway et al., 1993). Administration of SR-48968 (a potent and specific NK-2 receptor antago-nist) substantially blunts guinea pig HIB (Fig. 11), proving the role of NK-2 receptors in this response. Administration of ($\pm$)CP-96345 (an NK-1 receptor antagonist) also blunts guinea pig HIB, but the meaning of this result is blurred by the fact that this agent can also block membrane calcium channels. Thus, it is conceivable that ($\pm$)CP-96345 exerted its effect on guinea pig HIB through an action other than NK-1 receptor blockade. Further studies will be necessary to clarify the precise role of NK-1 receptors in this response.

If endogenous sensory neuropeptide release plays a critical role in guinea pig HIB, then how do eicosanoids also exert an effect? Because capsaicin pretreat-ment can fully ablate the guinea pig's bronchoconstrictor response to even substantial stimulation by dry gas hyperpnea (Ray et al., 1989a; Garland et al.,

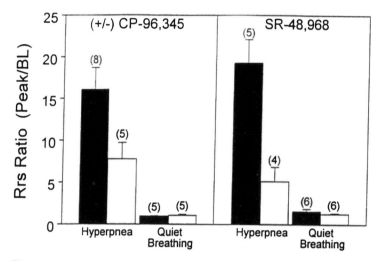

Figure 11 Pharmacological evidence for the participation of tachykinin receptors in guinea pig HIB. Administration of NK-1 [(±)CP-96345] or NK-2 (SR-48968) receptor antagonists blunts HIB induced by dry gas hyperpnea in guinea pigs. Neither drug had any effect on airway tone during quiet breathing. (Filled bars) Vehicle-treated animals; (open bars) drug-treated animals. Numbers of animals in each group are shown in parentheses above each bar. Rrs Ratio—ratio of respiratory system resistance at the peak of HIB normalized to prehyperpnea baseline (BL) Rrs values. (From Solway et al., 1993. Reproduced from the *Journal of Clinical Investigation* by copyright permission of the American Society for Clinical Investigation.)

1991), it seems likely that eicosanoids do not act directly on airway smooth muscle to cause bronchoconstriction under these conditions. (If eicosanoid mediators did act directly on airway smooth muscle during HIB, one might expect persistent partial bronchoconstriction despite obliteration of C-fiber function with capsaicin pretreatment.) Instead, it appears more probable that eicosanoid mediators modulate the magnitude of endogenous sensory neuropeptide release, through an action on sensory C-fibers themselves. In support of this possibility is a recent report by Ellis and Undem (1991), who showed that administration of a cysteinyl leukotriene-1 receptor antagonist blunts bronchoconstriction induced by endogenously released sensory neuropeptides, but has no effect on bronchoconstriction induced by exogenously administered tachykinins.

C. Hyperpnea-Induced Bronchovascular Hyperpermeability (HIBVH)

Since tachykinins are released during dry gas hyperpnea, and since endogenous tachykinin release can itself lead to bronchovascular hyperpermeabiity, we wondered whether dry gas hyperpnea might also provoke microvascular leak within the airways, in addition to the bronchoconstriction reviewed above. Using Evans blue extravasation as an index of microvascular leak, we found that dry gas hyperpnea (but *not* hyperpnea with warm, humidified gas) indeed does increase bronchovascular permeability to macromolecules, and that this effect is most marked within the trachea and main bronchi (Garland et al., 1991, 1993). However, to our surprise, it appears that this phenomenon is not exclusively mediated by endogenously released tachykinins. Systemic capsaicin pretreatment did significantly blunt hyperpnea-induced bronchovascular leak within the intrapulmonary airways, but it had no effect on HIBVH in the extrapulmonary airways (Fig. 12). Confirming this conclusion, neither NK-1 receptor blockade nor NK-2 receptor blockade with (±)CP-96345 or SR-48968, respectively, affected HIBVH in this airway region (Solway et al., 1993). Further studies suggest that eicosanoid mediators may play a contributory role, as combined cyclooxygenase plus 5-lipoxygenase inhibition with BW755c did blunt this response slightly (Garland et al., 1993). The mediators principally responsible for HIBVH in the extrapulmonary

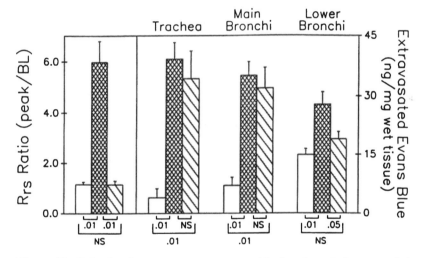

Figure 12 Role of endogenous sensory neuropeptides in guinea pig hyperpnea-induced bronchovascular hyperpermeability. Systemic capsaicin pretreatment blunts guinea pig HIB (quantified as Rrs Ratio as in Fig. 11) and HIBVH (quantified as extravasated Evans blue dye) within the intrapulmonary airways, but has no effect on hyperpnea-induced bronchovascular dye extravasation in the trachea or main bronchi, where this effect is most marked. (Open bars) Normal animals subjected to 10 min quiet ventilation; (cross-hatched bars) normal animals subjected to 10 min dry gas hyperpnea; (diagonal hatched bars) capsaicin-pretreated animals subjected to 10 min dry gas hyperpnea. Rrs ratio—ratio of peak respiratory system resistance after hyperpnea to baseline (BL) prehyperpnea value. (From Garland et al., 1991.)

airways remain to be established. In any case, it seems unlikely that edema that might have accumulated during the period of hyperpnea-induced bronchovascular hyperpermeability can account for the airflow obstruction induced by dry gas hyperpnea, for the magnitude of HIBVH within the intrapulmonary airways [the site of airway narrowing as shown bronchographically (Ray et al., 1990)] correlates extremely poorly with the magnitude of HIB. It is unknown whether bronchovascular dilatation occurs during or after dry gas hyperpnea.

IV. IS GUINEA PIG HIB A "MODEL" OF HIB IN HUMAN ASTHMA?

Physiological events that transpire in animals in response to various stimuli can truly be considered to "model" analogous events in human disease only to the extent that they behave identically phenomenologically and have identical biological mechanisms. We have seen that guinea pig HIB exhibits striking phenomenological similarities with HIB in human exercise-induced asthma: (1) The onset of airflow obstruction follows the cessation of dry gas hyperpnea and then spontaneously resolves without a late phase of obstruction; (2) the bronchoconstrictor response to hyperpnea is ablated by heating and humidification of inspired gas to alveolar conditions; (3) HIB is reproducible on consecutive identical dry gas hyperpnea challenges; (4) the severity of HIB follows a stimulus-response relationship with minute ventilation during a fixed period of hyperpnea and with the duration of hyperpnea at fixed minute ventilation; (5) the mechanism of HIB is largely independent of cholinergic neurotransmission; and (6) leukotriene synthesis or receptor blockade blunts the bronchoconstrictor response. Furthermore, there is evidence in both species that the bronchial vasculature, as well as airway smooth muscle, responds to dry gas hyperpnea, though the relationship of this response to airflow obstruction remains unproved in both species.

These similarities span a broad range of behavioral and mechanistic (i.e., cellular and mediator) features and therefore suggest that the precise mechanisms involved in guinea pig HIB may also operate during HIB in human asthma. Based on this rationale, we think it reasonable to hypothesize that endogenous sensory neuropeptide release may play a critical role in human HIB, as it does in guinea pigs. To date, there has been no information gathered in human asthmatic subjects that directly supports or disproves the potential role of sensory neuropeptides in human exercise-induced asthma, or indeed in any pathological airway response in human asthma. It is noteworthy that substance P has been recovered from the bronchoalveolar lavage fluid of atopic subjects after allergen challenge (Nieber et al., 1991), but such an association between mediator recovery and the occurrence of pathophysiological airway responses (allergen-induced airway narrowing) by no means proves causal participation of the mediator recovered. Instead, direct evidence for the importance of a mediator pathway usually comes from pharmacological studies in which synthetic or receptor blockade in the pathway supposed important is shown to reduce the airway response under study. Unfortunately, the lack of mechanistic pharmacological tools suitable for use in humans has hampered this approach to evaluate the participation of tachykinins or other sensory neuropeptides in asthma. Recently, though, novel pharmacological agents have made it possible to block the human NK-1 receptor (CP-99994) or both NK-1 and NK-2 receptors simultaneously (FK-224). Although the latter agent has been shown to inhibit bradykinin-induced bronchoconstriction in asthmatic subjects (Ichinose et al., 1992), the role of tachykinin receptors in EIB/HIB remains to be explored.

ACKNOWLEDGMENTS

This work was supported by NHLBI Grants HL41009, HL02205, HL-08519, and HL-02376, and by the Allen and Hanburys Foundation.

REFERENCES

Adams, F. (1856). *The Extant Works of Arataeus the Cappadocian.* Syndenham Soc., London, p. 316.

Anderson, S.D. (1984). Is there a unifying hypothesis for exercise-induced asthma? *J. Allergy Clin. Immunol. 73*:660.

Anderson, S.D., and Smith, C.M. (1991). Osmotic challenges in the assessment of bronchial hyperresponsiveness. *Am. Rev. Respir. Dis. 143*(3, Part 2):S43–S46.

Anderson, S.D., Schoeffel, R.E., Follet, R., Perry, C.P., Daviskas, E., and Kendall, M. (1982). Sensitivity to heat and water loss at rest and during exercise in asthmatic patients. *Eur. J. Respir. Dis. 63*:459–471.

Badier, M., Beaumont, D., and Orehek, J. (1988). Attenuation of hyperventilation-induced bronchospasm by terfenadine: a new antihistamine. *J. Allergy Clin. Immunol. 81*:437–440.

Blackie, S.P., Hilliam, C., Village, R., and Pare, P.D. (1990). The time course of bronchoconstriction in asthmatics during and after isocapnic hyperventilation. *Am. Rev. Respir. Dis. 142*:1133–1136.

Chapman, R.W., and Danko, G. (1985). Hyperventilation-induced bronchoconstriction in guinea pigs. *Int. Arch. Allergy Appl. Immunol. 78*:190–196.

Chen, W.Y., and Horton, D.J. (1977). Heat and water loss from the airways and exercise-induced asthma. *Respiration 34*:305–313.

Deal, E.C., McFadden, E.R., Jr., Ingram, R.H., Jr., Strauss, R.H., and Jaeger, J.J. (1979a). Role of respiratory heat exchange in production of exercise-induced asthma. *J. Appl. Physiol. 46*(3):467–475.

Deal, E.C., McFadden, E.R., Jr., Ingram, R.H., Jr., and Jaeger, J.J. (1979b). Hyperpnea and heat flux: initial reaction sequence in exercise-induced asthma. *J. Appl. Physiol. 46*(3):476–483.

Deal, E.C., Wasserman, S.I., Soter, N.A., Ingram, R.H., Jr., and McFadden, E.R. Jr. (1980). Evaluation of the role played by mediators of immediate hypersensitivity in exercise-induced asthma. *J. Clin. Invest. 65*:659–665.

Eggleston, P.A., Bierman, C.W., Pierson, W.E., et al. (1972). A double blind trial of the effect of cromolyn sodium on exercise-induced bronchospasm. *J. Allergy Clin. Immunol. 50*:57.

Eggleston, P.A., Kagey-Sobotka, A., and Lichtenstein, L.M. (1987). A comparison of the osmotic activation of basophils and human lung mast cells. *Am. Rev. Respir. Dis. 135*: 1043–1048.

Ellis, J.L., and Undem, B.J. (1991). Role of peptidoleukotrienes in capsaicin-sensitive sensory fibre–mediated responses in guinea pig airways. *J. Physiol. 436*:469–484.

Fanta, C.H., Ingram, R.H., Jr., and McFadden, E.R., Jr. (1980). A reassessment of the effects of oropharyngeal anesthesia in exercise-induce asthma. *Am. Rev. Respir. Dis. 122*:381.

Ferguson, A., Addington, W., and Gaensler, E.A. (1969). Dyspnea and bronchospasm from inappropriate post-exercise hyperventilation. *Am. Intern. Med. 71*:1063–1072.

Fisher, H.K., Holton, P., Buxton, R.St.J., and Nadel, J.A. (1970). Resistance to breathing during exercise-induced asthma attacks. *Am. Rev. Respir. Dis. 101*:885–896.

Freedman, S., Lane, R., Gillett, M.K., and Guz, A. (1988). Abolition of methacholine induced bronchoconstriction by the hyperventilation of exercise or volition. *Thorax 43*: 631–636.

Garland, A., Ray, D.W., Doerschuk, C.M., Jackson, M., Eappen, S., Alger, L., and Solway, Jr. (1991). The role of tachykinins in hyperpnea-induced bronchovascular hyperpermeability in guinea pigs. *J. Appl. Physiol. 70*(1):27–35.

Garland, A., Jordan, J.E., Ray, D.W., Spaethe, S., Alger, L., and Solway, J. (1993). Role of eicosanoids in hyperpnea-induced airway responses in guinea pigs. *J. Appl. Physiol. 75*(6):2797–2804.

Gilbert, I.A., and McFadden, E.R., Jr. (1992). Airway cooing and rewarming. The second reaction sequence in exercise-induced asthma. *J. Clin. Invest. 90*:699–704.

Gilbert, I.A., Fouke, J.M., and McFadden, E.R., Jr. (1987). Heat and water flux in the intrathoracic airways and exercise-induced asthma. *J. Appl. Physiol. 63*(4):1681–1691.

Gilbert, I.A., Winslow, C.J., Lenner, K.A., Nelson, J.A., and McFadden, E.R., Jr. (1992). Vascular volume expansion and thermally induced asthma. *Eur. Respir. J.* (in press).

Godfrey, S., Silverman, M., and Anderson, S.D. (1973). Problems of interpreting exercise induced asthma. *J. Allergy Clin. Immunol. 52*:199.

Griffin, M.P., Fung, K.F., Ingram, R.H., Jr., and McFadden, E.R., Jr. (1982). Dose-response effects of atropine on thermal stimulus-response relationships in asthma. *J. Appl. Physiol. 53*:1576.

Herxheimer, H. (1946). Hyperventilation asthma. *Lancet 1*:82–87.

Ichinose, M., Nakajima, N., Takahashi, T., Yamauchi, H., Inoue, H., and Takishima, T. (1992). Protection against bradykinin-induced bronchoconstriction in asthmatic patients by neurokinin receptor antagonist. *Lancet 340*(8830):1248–1251.

Ingenito, E.P., Solway, J., McFadden, E.R., Jr., Pichurko, B.M., Cravalho, E.G., and Drazen, J.M. (1986). Finite difference analysis of respiratory heat transfer. *J. Appl. Physiol. 61*(6):2252–2259.

Ingenito, E.P., Pliss, L.B., Ingram, R.H., Jr., and Pichurko, B.M. (1990). Bronchoalveolar lavage cell and mediator responses to hyperpnea-induced bronchoconstriction in the guinea pig. *Am. Rev. Respir. Dis. 141*:1162–1166.

Ingram, R.H., Jr., and McFadden, E.R., Jr. (1983). Are exercise and isocapnic voluntary hyperventilation identical bronchial provocations? *Eur. J. Respir. Dis. 64*(Suppl 128): 242–245.

Israel, E., Dermarkarian, R., Rosenberg, M., Sperling, R., Taylor, G., Rubin, P., and Drazen, J.M. (1990). The effects of 5-lipoxygenase inhibitor on asthma induced by cold, dry air. *N. Engl. J. Med. 323*:1740–1744.

Kay, A.B. (1987). Provoked asthma and mast cells. *Am. Rev. Respir. Dis. 135*:1200–1203.

Lee, T.H., Nagy, L., Nagakura, T., Walport, M.J., and Kay, A.B. (1982). Identification and partial characterization of an exercise-induced neutrophil chemotactic factor in bronchial asthma. *J. Clin. Invest. 69*:889–899.

Manning, P.J., Watson, R.M., Margolskee, D.J., Williams, V.C., Schwartz, J.I., and O'Byrne, P.M. (1990). Inhibition of exercise-induced bronchoconstriction by MK-571, a potent leukotriene D_4-receptor antagonist. *N. Engl. J. Med. 323*:1736–1739.

Margolskee, D.J., Bigby, B.G., and Boushey, H.A. (1988). Indomethacin blocks airway tolerance to repetitive exercise but not to eucapnic hyperpnea in asthmatic subjects. *Am. Rev. Respir. Dis. 137*:842–846.

McFadden, E.R., Jr. (1990). Hypothesis: exercise-induced asthma as a vascular phenomenon. *Lancet 335*(8694):880–883.

McFadden, E.R., Jr., and Ingram, R.H., Jr. (1983). Exercise-induced airway obstruction. *Annu. Rev. Physiol. 45*:453–463.

McFadden, E.R., Jr., Stearns, D.R., Ingram, R.H., Jr., and Leith, D.E. (1977). Relative contributions of hypcoapnia and hyperpnea as mechanisms in post-exercise asthma. *J. Appl. Physiol. Respir. Envir. Exer. Physiol. 42*:22–27.

McNally, J.F., Enright, P., Hirsch, J.E., and Souhrada, J.F. (1979). The attenuation of exercise-induced bronchoconstriction by oropharyngeal anesthesia. *Am. Rev. Respir. Dis. 119*:247–252.

Nagakura, T., Lee, T.H., Assoufi, B.K., Newman-Taylor, A.J., Denison, D.M., and Kay, A.B. (1983). Neutrophil chemotactic factor in exercise- and hyperventilation-induced asthma. *Am. Rev. Respir. Dis. 128*:294–296.

Nieber, K., Baumgarten, C., Witzel, A., Rathsack, R., Oehme, P., Brunnee, T., Kleine-Tebbe, J., and Kunkel, G. (1991). The possible role of substance P in the allergic reaction, based on two different provocation models. *Int. Arch. Allergy Appl. Immunol. 94*:334–338.

O'Byrne, P.M., and Jones, G.L. (1986). The effect of indomethacin on exercise-induced bronchoconstriction and refractoriness after exercise. *Am. Rev. Respir. Dis. 134*:69–72.

Pichurko, B.M., Sullivan, B., Porcelli, R.J., and McFadden, E.R., Jr. (1986). Endogenous adrenergic modification of exercise-induced asthma. *J. Allergy Clin. Immunol. 77*: 796–801.

Ray, D.W., Hernandez, C., Munoz, N., Leff, A.R., and Solway, J. (1988). Bronchoconstriction elicited by dry gas hyperpnea in guinea pigs. *J. Appl. Physiol. 65*(2):934–939.

Ray, D.W., Hernandez, C., Leff, A.R., and Solway, J. (1989a). Tachykinins mediate bronchoconstriction elicited by isocapnic hyperpnea in guinea pigs. *J. Appl. Physiol. 66*(3):1108–1112.

Ray, D.W., Ingenito, E.P., Strek, M., Schumacker, P., and Solway, J. (1989b). Longitudinal distribution of canine respiratory heat and water exchanges. *J. Appl. Physiol. 66*(6): 2788–2798.

Ray, D.W., Eappen, S., Hernandez, C., Jackson, M., Leff, A.R., and Solway, J. (1990). Distribution of airway narrowing during hyperpnea-induced bronchoconstriction in guinea pigs. *J. Appl. Physiol. 69*(4):1323–1329.

Ray, D.W., Hernandez, C., Eappen, S., and Solway, J. (1991). Time course of broncho-constriction induced by dry gas hyperpnea in guinea pigs. *J. Appl. Physiol. 70*(2):504–510.

Rebuck, A.S., and Read, J. (1968). Exercise-induced asthma. *Lancet 1*:429–431.

Robuschi, M., Riva, E., Fuccella, L.M., Vida, E., Barnabe, R., Rossi, M., Gambaro, G., Spagnotto, S., and Bianco, S. (1992). Prevention of exercise-induced bronchoconstric-tion by a new leukotriene antagonist (SK&F 104353). A double-blind study versus disodium cromoglycate and placebo. *Am. Rev. Respir. Dis. 145*:1285–1288.

Rosenthal, R.R., Laube, B.L., Hood, D.B., and Norman, P.S. (1990). Analysis of refractory period after exercise and eucapnic voluntary hyperventilation challenge. *Am. Rev. Respir. Dis. 141*:368–372.

Rossing, T.H., Weiss, J.W., Breslin, F.J., Ingram, R.H., Jr., and McFadden, E.R., Jr. (1982). Effects of inhaled sympathomimetics on obstructive response to respiratory heat loss. *J. Appl. Physiol. 52*:1119–1123.

Rubenstein, I., Levison, H., Slutsky, A.S., Hak, H., Wells, J., Zamel, N., and Rebuck, A.S. (1987). Immediate and delayed bronchoconstriction after exercise in patients with asthma. *N. Engl. J. Med. 317*:482–485.

Sakula, A. (1984). Sir John Floyer's *A Treatise of the Asthma* (1698). *Thorax 39*:248–254.

Salter, H.H. (1864). *On Asthma: Its Pathology and Treatment.* Blanchard and Lea, Philadelphia, pp. 132–153.

Sheppard, D., Epstein, J., Holtzman, M.J., Nadel, J.A., and Boushey, H.A. (1982). Dose-dependent inhibition of cold-air-induced bronchoconstriction by atropine. *J. Appl. Physiol. 53*:169.

Silverman, M., Anderson, S.D., and Walker, S.R. (1972). Metabolic changes preceding exercise-induced bronchoconstriction. *Br. Med. J. 1*:207–209.

Solway, J., and Leff, A.R. (1991). Sensory neuropeptides and airway function. *J. Appl. Physiol. 71*(6):2077–2087.

Solway, J., Kao, B.M., Jordan, J.E., Gitter, B., Rodger, I.W., Howbert, J.J., Alger, L.E., Necheles, J., Leff, A.R., and Garland, A. (1993). Tachykinin receptor antagonists inhibit hyperpnea-induced bronchoconstriction in guinea pigs. *J. Clin. Invest. 92*:315–323.

Stanescu, D.C., and Teculescu, D.B. (1970). Exercise and cough-induced asthma. *Respiration 27*:377–379.

Stearns, D., McFadden, E.R., Jr., Breslin, F., and Ingram, R.H., Jr. (1981). Reanalysis of the refractory period in exertional asthma. *J. Appl. Physiol. 50*:503–508.

Stirling, D.R., Cotton, D.J., Graham, B.L., Hodgson, W.C., Cockroft, D.W., and Dosman, J.A. (1983). Characteristics of airway tone during exercise in patients with asthma. *J. Appl. Physiol. 54*:934–942.

Strauss, R.H., Ingram, R.H., Jr., and McFadden, E.R., Jr. (1977). A critical assessment of the roles of circulating hydrogen ion and lactate in the production of exercise-induced asthma *J. Clin. Invest. 60*:658–664.

Strauss, R.H., McFadden, E.R., Jr., Ingram, R.H., Jr., Deal, E.C., Jr., and Jaeger, J.J.

(1978). Influence of heat and humidity on the airway obstruction induced by exercise in asthma. *J. Clin. Invest. 61*:433–440.

Vassallo, C.L., Gee, J.B.L., and Domm, B.M. (1972). Exercise-induced asthma: observations regarding hypocapnia and acidosis. *Am. Rev. Respir. Dis. 105*:42–49.

Zeballos, R.J., Shturman-Ellstein, R., McNally, J.F., Jr., Hirsch, J.E., and Souhrada, J.F. (1978). The role of hyperventilation in exercise-induced bronchoconstriction. *Am. Rev. Respir. Dis. 118*:877–884.

DISCUSSION

Widdicombe: In the guinea pig, what is the relative contribution of smooth muscle contraction versus mucosal swelling that causes airflow obstruction? Have you tried beta agonists?

Solway: In pilot experiments, we found that albuterol administration before or during HIB prevents or reverses the response. It appears unlikely that HIB-induced vascular hyperpermeability accounts for the airflow obstruction, based on a study comparing the amount of vascular permeability with the amount of airway obstruction; there was poor correlation between the measurements. It is possible that vasodilatation participates, but this remains to be studied.

It is also important to recognize that we must apply great caution in extrapolating animal observations to humans. Once we have neuropeptide antagonists that can be used clinically, we can begin to assess the contributions of neuropeptides to human disease.

29

Nitric Oxide and Guanylyl Cyclases: Correlation with Neuropeptides

Wolfgang Kummer, Axel Fischer, Rudolf E. Lang, and Xiongbin Lin
Philipps-University Marburg, Marburg, Germany

Doris Koesling
Free University of Berlin, Berlin, Germany

Bernd Mayer
Karl-Franzens-University, Graz, Austria

Regis Olry
University of Quebec at Trois-Rivières, Trois-Rivières, Quebec, Canada

I. INTRODUCTION

Guanosine 3′,5′-cyclic monophosphate (cGMP) acts as a second messenger mediating relaxation of smooth muscle cells including those of the airways and pulmonary vasculature (see review by Said, 1992). Its formation from GTP is catalyzed by guanylyl cyclases (GCs), which can be subdivided into at least two families:

Membrane-bound GCs belong to the group of receptor-linked enzymes. They consist of an N-terminal extracellular receptor domain, a single transmembrane domain, and an intracellular C-terminal domain carrying a protein kinase motif and the catalytically active GC domain sharing sequence homologies with adenylate cyclases. Three isoforms (GC-A, GC-B, GC-C) have currently been cloned. Endogenous ligands for GC-A and GC-B are natriuretic peptides; atrial natriuretic peptide (ANP) preferentially binds to GC-A and, with lower affinity, to GC-B; C-type natriuretic peptide (CNP) binds to GC-B; brain natriuretic peptide (BNP) prefers GC-B to GC-A, but in neither case is the preferred ligand compared with

ANP (GC-A) or CNP (GC-B (Chang et al., 1989; Koesling et al., 1991; Ohyama et al., 1992). The endogenous activator of the epithelial intestinal GC (GC-C) has recently been identified as a peptide composed of 15 amino acid residues, termed guanylin (Currie et al., 1992). This GC-C is different from the natriuretic peptide C receptor, which does not contain catalytic GC activity (Porter et al., 1990).

Activation of soluble GCs occurs in response to nitric oxide (NO), but other modes of activation, e.g., interaction with catalase and activation by hydroxyl radicals, also have been suggested (Böhme et al., 1978; Murad et al., 1978; Cherry et al., 1990). Soluble GC was originally identified as a heterodimer containing a prosthetic heme group and consisting of an α_1- (77 kDa) and β_1-subunit (70 kDa), each carrying a catalytic GC domain (Kamisaki et al., 1986; Humbert et al., 1990; Koesling et al., 1990; Nakane et al., 1990). Later, an isoform of GC subunits was cloned from rat kidney, which, based on sequence homology to the β_1-subunit, was termed β_2 (Yuen et al., 1990). So far, there are no reports on a functionally active expression of this isoform. Another isoform was isolated from a human fetal brain library and termed α_2-subunit on the basis of sequence homologies (Harteneck et al., 1991). This α_2-subunit is able to form a catalytically active heterodimer with the β_1-subunit, which, however, is less active than the α_1/β_1-heterodimer (Harteneck et al., 1991). From adult human brain, cDNA clones corresponding to other 70-kDa and 82-kDa subunits have been sequenced (Giuili et al., 1992).

Both membrane-bound and soluble GCs as well as endogenous activators are abundant in the lower airways (Gardner et al., 1986; Kamisaki et al., 1986; Kuno et al., 1986; Sakamoto et al., 1986; Mayer and Böhme, 1989; Koesling et al., 1990; Secca et al., 1991; Schmidt et al., 1992; Fischer et al., 1993a), suggesting a prominent role of cGMP-mediated intracellular effects in these tissues. At least partly, these events may be triggered by neuroeffector mechanisms: Besides their nonneuronal production sites, natriuretic peptides are synthesized and utilized by central and peripheral neurons (Jirikowski et al., 1986; Kawata et al., 1989; Nohr et al., 1989; Saper et al., 1989; Ueda et al., 1991), which is highlighted by the fact that BNP and CNP were originally isolated from nervous tissue (Sudoh et al., 1988, 1990), and the enzyme generating NO from L-arginine, NO synthase (NOS), is contained in a subset of nerve fibers innervating the lower airways (Schmidt et al., 1992; Fischer et al., 1993a,b). Based on this background, the present chapter will review recent advances on the distribution of natriuretic peptide receptors and soluble GC in the lower airways, with particular emphasis on the sensory innervation and the coexistence of other neuropeptides with NOS in the airway innervation.

II. NITRIC OXIDE SYNTHASE IN THE INNERVATION OF THE LOWER AIRWAYS: ORIGIN OF NERVE FIBERS AND COEXISTENCE WITH NEUROPEPTIDES

Three different isoforms of NOS (neuronal, macrophage, endothelial) have been cloned, showing about 50–60% sequence homologies (Bredt et al., 1991; Janssens et al., 1992; Xie et al., 1992). Neuronal NOS is constitutively expressed, although neuronal injury increases immunoreactive NOS and enzyme activity (Fiallos-Estrada et al., 1993) and is activated by an intracellular rise of $[Ca^{2+}]$ via calmodulin (Bredt and Snyder, 1990; Klatt et al., 1992). This enzyme has been demonstrated by means of either immunohistochemistry or NADPH-diaphorase histochemistry, which in neuronal tissue is assumed to specifically demonstrate catalytic NOS activity (Hope et al., 1991; Dawson et al., 1991), in nerve fibers innervating the lower airways of every mammalian species studied so far (Schmidt et al., 1992; Dey et al., 1993; Fischer et al., 1993a,b; Kobzik et al., 1993). There is general agreement that neuronal "nitroxidergic" mechanisms play a crucial role in the inhibitory noncholinergic, nonadrenergic (iNANC) innervation of airway smooth muscle, although gradual differences appear to exist between species as well as between different parts of the tracheobronchial tree (Tucker et al., 1990; Li and Rand, 1991; Belvisi et al., 1991, 1992; Stuart-Smith and Hirshman, 1993). Smooth muscle relaxation is also mediated by neuronally released vasoactive intestinal peptide (VIP) via activation of adenylate cyclase, and the combined action of NO and VIP enhances the effectiveness of relaxation (Said, 1992). Colocalization of NOS and VIP has been observed in autonomic nerve fibers innervating systemic arterial vascular beds in the guinea pig (Kummer et al., 1992a), and, thus, particular attention was paid to the possible coexistence of these two independently acting relaxants in the airway innervation.

In the guinea pig, NOS-containing axons are distributed in the smooth muscle of the trachea and major bronchi, in the lamina propria, and around bronchial and pulmonary blood vessels (Fischer et al., 1993a). Double-labeling histochemistry revealed a high degree of coexistence of VIP and NOS in perivascular axons and in a considerable number of nerve terminals in tracheal and major bronchial smooth muscle. Ultrastructurally, NOS-immunoreactive axon terminals in the tracheal muscle contain abundant small, clear vesicles, so in addition to the neuropeptide VIP, a classical neurotransmitter such as acetylcholine may be coutilized by this type of axon. However, NOS and VIP did not colocalize in all of the axons innervating guinea pig airway smooth muscle: VIP-immunoreactive terminals were more numerous than NOS-containing endings, and some axons contained NOS without VIP immunoreactivity.

In contrast, NOS-containing nerve fibers in the lamina propria were nonreactive to VIP. They were also distinct from substance P (SP)-immunoreactive axons, although both kinds of nerve fibers were often closely associated (Kummer et al., 1992b).

The origin of the majority of these fibers might be extrinsic to the airways, because only a few local ganglion cells situated along the pulmonary veins close to the atria contained NOS, whereas tracheal and peribronchial ganglia were devoid of NOS-containing neurons (Fischer et al., 1993a). Likewise, VIP-immunoreactive tracheal and peribronchial ganglion cells have not been observed in untreated guinea pigs by Bowden and Gibbins (1992) and Kummer et al.

(1992c), although such neurons have been detected after systemic (Triepel et al., 1986) or local colchicine treatment (Shimosegawa et al., 1990). In the peripheral ganglia projecting to the guinea pig lower airways, NOS-containing neurons have been identified in superior cervical and stellate sympathetic ganglia (Kummer, 1992; Fischer et al., 1993a) and in the vagal sensory jugular and nodose ganglia (Kummer et al., 1992b; Fischer et al., 1993a). In sympathetic ganglia, more than 95% of NOS-containing neurons are noncatecholaminergic. Surprisingly, there was only little coexistence with neuropeptides known to be present in subpopulations of noncatecholaminergic sympathetic neurons: Less than 10% of NOS-containing sympathetic neurons exhibited immunoreactivity to VIP (Fig. 1) or neuropeptide Y (NPY), and coexistence with calcitonin gene-related peptide (CGRP) was not observed (Olry et al., 1993). In sensory vagal ganglia, some neurons were simultaneously immunoreactive to NOS and SP, but perikarya exhibiting only one of these immunoreactivities were more numerous.

Which, if any, of these NOS-containing neuronal populations in sympathetic and sensory vagal ganglia contributes to the innervation of the lower airways was

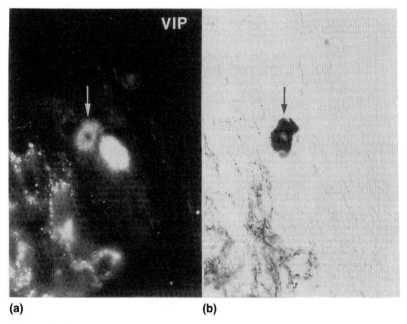

(a)   (b)

Figure 1 Guinea pig sympathetic ganglion showing two postganglionic VIP-immunoreactive neurons (a). The same section was subsequently reacted for NADPH-diaphorase histochemistry to reveal catalytic activity of NOS (b). One of the VIP-immunoreactive neurons also contains NOS (arrows).

investigated by retrograde neuronal labeling, injecting the fluorescent tracer Fast blue into the posterior wall of the lower cervical trachea. Consistent with a previous study utilizing the same experimental setup (Kummer et al., 1992c), approximately 5% of postganglionic sympathetic neurons labeled by the tracer were noncatecholaminergic and did not display VIP immunoreactivity. Most of these neurons were NOS immunoreactive. Confirming previous results (Kummer et al., 1992c), none of sensory nodose neurons and about half of the sensory jugular neurons projecting to the trachea exhibited immunoreactivity to SP. In contrast to this clear separation of SP-immunoreactive airway neurons between these two ganglia, retrogradely labeled NOS-immunoreactive perikarya were localized in both ganglia. They accounted for 10–20% of Fast blue–labeled neurons in the jugular and for 5–10% in the nodose ganglion. An exact quantification of their numbers was difficult because of the occurrence of variable intensities of NOS immunolabeling in vagal sensory ganglia, which were not seen in sympathetic ganglia. An unequivocal colocalization of SP and NOS immunoreactivities in retrogradely labeled neurons has so far not been established.

These findings show a dual, vagal sensory and postganglionic sympathetic, nitroxidergic nerve supply to the guinea pig lower airways, which is mainly distinct from the previously characterized neuropeptide-containing pathways. The question remains, however, where the NOS/VIP-containing perivascular and intramuscular fibers originate. The perivascular axons may arise from the local ganglia situated along the pulmonary vasculature. As for the NOS/VIP-immunoreactive axons innervating the airway smooth muscle, it should be noted that, in contrast to our results, Bowden and Gibbins (1992) observed VIP immunoreactivity in about 20% of sympathetic neurons retrogradely labeled from the guinea pig trachea. The reason for this discrepancy is at present unclear, and it is not known whether these neurons also contained NOS. According to recent in vitro experiments, a hitherto neglected source of extrinsic innervation must be considered: Canning and Undem (1993) observed excitatory and inhibitory NANC effects in a preparation of guinea pig trachea attached to the esophagus. Removal of the esophagus left the constrictory responses to vagal stimulation intact but abolished the inhibitory response, suggesting that the cell bodies mediating iNANC effects are attached to or located even within the esophagus (Canning and Undem, 1993).

The distribution of NOS-containing nerve fibers in the lower airways of humans is similar to that in guinea pig, except that additional nerve fibers are observed around tracheal and bronchial glands (Fischer et al., 1993b). Colocalization of VIP and NOS immunoreactivities was frequent at all these locations. In contrast to the guinea pig, about 50% of local airway neurons exhibited NOS. Many of them were innervated by VIP-immunoreactive nerve terminals (preganglionic parasympathetic?) that did not contain NOS. In most cases, NOS and VIP immunoreactivities were colocalized in intrinsic neurons, but neurons with only one of these immunoreactivities were also present. Thus, it can be assumed that NOS/VIP-immunoreactive fibers in human trachea and bronchi to a large extent are contributed by intrinsic parasympathetic neurons. However, there might be an additional supply from extrinsic ganglia: NOS-immunoreactive neurons in the human stellate ganglion are more numerous than in any other mammalian species studied so far, and several of them do have additional VIP immunoreactivity. Furthermore, NOS-immunoreactive neurons are located in human sensory vagal ganglia, but in these cases no colocalization with VIP has been observed.

Similar distributions of NOS/VIP-immunoreactive axons have been observed in the ferret (Dey et al., 1993). In this species, however, VIP-immunoreactive nerve fibers penetrate the respiratory epithelium, and these VIP-immunoreactive axons do not have NOS immunoreactivity. Neuronal perikarya showing colocalization of VIP and NOS immunoreactivities were consistently found in local ganglia situated in the superficial and glandular plexus, but were rarely located in ganglia of the longitudinal nerve trunks (Dey et al., 1993). A detailed analysis of NOS immunoreactivities and NOS colocalizations with neuropeptides in sympathetic and sensory ganglia supplying the airways in the ferret is yet missing.

In summary, the nitroxidergic innervation of the lower airways does not represent a single entity, but appears to include parasympathetic, sympathetic, and sensory pathways. A major coexisting neuropeptide in the nitroxidergic innervation of airway and vascular smooth muscle is VIP, but additional separate populations of nerve fibers utilizing only one of these mediators also exist.

III. SOLUBLE GUANYLYL CYCLASE IN THE INNERVATION OF THE LOWER AIRWAYS

The lung is a major source of soluble GC (Gerzer et al., 1981; Koesling et al., 1988, 1990; Nakane et al., 1988, 1990), and histochemistry and immunohistochemistry revealed an almost ubiquitous distribution of soluble GC in pulmonary cells (Secca et al., 1991; Schmidt et al., 1992), suggesting a widespread range of NO-mediated functions. Our own immunohistochemical investigations of rat and guinea pig lung and trachea largely confirmed these earlier reports with respect to the α_1- and β_1-subunits of soluble GC. Most prominent immunoreactivities to the α_2- and β_2-subunits, however, were found in axons within larger nerve fiber bundles. Accordingly, sensory ganglia (nodose ganglion and dorsal root ganglia) were investigated for the occurrence of the different isoforms of soluble GC. The α_1-isoform was immunohistochemically localized to the satellite cells of sensory ganglia (Fig. 2), which is in full accordance with the observation of cGMP increase in this cell type in response to stimulation with NO (Morris et al., 1992). From this finding, it has been suggested that NO is utilized as a mediator in neuronal-glial signaling (Morris et al., 1992). The α_2 and β_2-isoforms, however, were localized to sensory neuronal somata and axons, primarily of large diameter (Fig. 2). These immunohistochemical findings could be supported by Western blotting. The preferential distribution of α_2- and β_2-GC immunoreac-

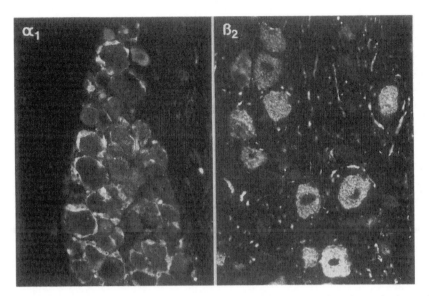

Figure 2 Rat cervical dorsal root ganglion. An antiserum against the α_1-isoform of soluble GC labels primarily the satellite cells, whereas predominantly large-diameter sensory neurons and their axons are immunoreactive to β_2-GC antiserum.

tivity to sensory neurons of large diameter suggests only little colocalization with neuropeptide immunoreactivities, but this topic has to be investigated in more detail. Indeed, patch-clamp studies performed on cultured rat sensory neurons revealed an involvement of the NO-cGMP pathway in the desensitization of small diameter neurons to stimulation with the algesic peptide bradykinin (McGehee et al., 1992).

Altogether, the present findings may point to an involvement of NO/soluble GC-mediated mechanisms in the local effector function of sensory neurons in the airways, but precise data on this topic are lacking.

IV. MEMBRANE-BOUND GUANYLYL CYCLASE IN THE INNERVATION OF THE LOWER AIRWAYS

Although the predominant natriuretic receptor type identified in the lung is not coupled to guanylyl cyclase activity (C-type receptor; Redmond et al., 1990), rat lung membranes contain a membrane-bound GC that can be activated by ANP (Chang et al., 1990). Polymerase chain reaction has helped to identify GC-B mRNA in rat lung (Ohyama et al., 1992), which is supported by our findings obtained by Northern blotting. Binding studies with various analogs have suggested the presence of GC-B at pulmonary endothelial cells (Redmond et al.,

1990). In addition, we obtained evidence for GC-A mRNA in the rat lung. At present, it is not clear whether these pulmonary receptors are physiologically addressed by circulating, paracrine, or neurally released natriuretic peptides. However, the generally neuronal localization of the preferred endogenous GC-B ligand, C-type natriuretic peptide (Ueda et al., 1991), raises the possibility of a contribution of the airway innervation to an intrapulmonary CNP/GC-B messenger pathway.

On the other hand, our in situ hybridization studies on rat nodose ganglion neurons utilizing ^{35}S-labeled GC-A and ANP C-receptor cRNAs demonstrated GC-A mRNA in almost all sensory neurons of this ganglion. This implies that those neurons projecting to the airways, whether or not they contain neuropeptides, are also included. Therefore, the possibility has to be considered that the sensory and local effector functions of sensory airway neurons are under some sort of control by natriuretic peptides. This assumption is supported by the observation that ANP induces a sympathetic inhibition by acting on vagal C-fibers (Schultz et al., 1988).

V. CONCLUSIONS

A. Neuropeptides and NO

Neuropeptides and NO interact in several ways in the airways: (1) NO and VIP are coutilized by a subset of airway nerve fibers. (2) In humans, intrinsic NOS-containing neurons are controlled by VIP-immunoreactive nerve endings (preganglionic parasympathetic?). (3) The presence of isoforms of soluble GC in sensory neurons suggest that they are influenced by NO-mediated mechanisms.

B. Neuropeptides and Membrane-Bound GC

The presence of GC-B mRNA in lung tissue suggests a physiological action of its endogenous ligand, i.e., the neuropeptide CNP. Circulating or paracrine acting ANP might influence the local effector function of peptide-containing sensory neurons that express GC-A mRNA.

ACKNOWLEDGMENT

This study was supported by the Deutsche Forschungsgemeinschaft.

REFERENCES

Belvisi, M.G., Stretton, D., and Barnes, P.J. (1991). Nitric oxide as an endogenous modulator of cholinergic neurotransmission in guinea-pig airways. *Eur. J. Pharmacol.* *198*:219–221.

Belvisi, M.G., Stretton, C.D., and Barnes, P.J. (1992). Nitric oxide is the endogenous neurotransmitter of bronchodilator nerves in human airways. *Eur. J. Pharmacol. 210*: 221–222.

Böhme, E., Graf, H., and Schultz, G. (1978). Effects of sodium nitroprusside and other smooth muscle relaxants on cyclic GMP formation in smooth muscle and platelets. *Adv. Cyclic Nucleotide Res. 9*:131–143.

Bowden, J.J., and Gibbins, I.L. (1992). Vasoactive intestinal peptide and neuropeptide Y coexist in non-noradrenergic sympathetic neurons to guinea pig trachea. *J. Autonom. Nerv. Syst. 38*:1–20.

Bredt, D.S., and Snyder, S.H. (1990). Isolation of nitric oxide synthase, a calmodulin-requiring enzyme. *Proc. Natl. Acad. Sci. USA 87*:682–685.

Bredt, D.S., Hwang, P.M., Glatt, C.E., Lowenstein, C., Reed, R.R., and Snyder, S.H. (1991). Cloned and expressed nitric oxide synthase structurally resembles cytochrome P-450 reductase. *Nature 351*:714–718.

Canning, B.J., and Undem, B.J. (1993). The parasympathetic ganglia mediating NANC relaxations of guinea pig trachealis are extrinsic to the trachea. *Am. Rev. Respir. Dis. 147*:A284.

Chang, C-H., Kohse, K.P., Chang, B., Hirata, M., Jiang, B., Douglas, J.E., and Murad, F. (1990). Characterisation of ANP-stimulated guanylate cyclase activation in rat lung membranes. *Biochim. Biophys. Acta 1052*:159–165.

Chang, M., Lowe, D.G., Lewis, M., Hellmiss, R., Chen, E., and Goeddel, D.V. (1989). Differential activation by atrial and brain natriuretic peptides of two different receptor guanylate cyclases. *Nature 341*:68–71.

Cherry, P.D., Omar, H.A., Farrell, K.A., Stuart, J.S., and Wolin, M.S. (1990). Superoxide anion inhibits cGMP-associated bovine pulmonary arterial relaxation. *Am. J. Physiol. 259*:H1056–H1062.

Currie, M.G., Fok, K.F., Kato, J., Moore, R.J., Hamra, F.K., Duffin, K.L., and Smith, C.E. (1992). Guanylin: an endogenous activator of intestinal guanylate cyclase. *Proc. Natl. Acad. Sci. USA 89*:947–951.

Dawson, T.D., Bredt, D.S., Fotuhi, M., Hwang, P.M., and Snyder, S.H. (1991). Nitric oxide synthase and neuronal NADPH diaphorase are identical in brain and peripheral tissues. *Proc. Natl. Acad. Sci. USA 88*:7797–7801.

Dey, R.D., Dalal, G., Pinkstaff, C.A., Mayer, B., Kummer, W., and Said, S.I. (1993). Nitric oxide synthase and vasoactive intestinal peptide are colocalized in neurons of the ferret tracheal plexus. *Am. Rev. Respir. Dis. 147*:A288.

Fiallos-Estrada, C.E., Kummer, W., Mayer, B., Bravo, R., Zimmermann, M., and Herdegen, T. (1993). Long-lasting increase of nitric oxide synthase immunoreactivity, NADPH-diaphorase reaction and c-JUN co-expression in rat dorsal root ganglion neurons following sciatic nerve transection. *Neurosci. Lett. 150*:169–173.

Fischer, A., Mundel, P., Preissler, U., Philippin, B., and Kummer, W. (1993a). Nitric oxide synthase in guinea pig lower airway innervation. *Neurosci. Lett 149*:157–160.

Fischer, A., Hoffmann, B., Hauser-Kronberger, C., Mayer, B., and Kummer, W. (1993b). Nitric oxide synthase in the innervation of the human respiratory tract. *Am. Rev. Respir. Dis. 147*:A662.

Gardner, D.G., Deschepper, C.F., Ganong, W.F., Hane, S., Fiddes, J., Baxter, J.D., and

Lewicki, J. (1986). Extra-atrial expression of the gene for atrial natriuretic factor. *Proc. Natl. Acad. Sci. USA 83*:6697–6701.

Gerzer, R., Böhme, E., Hofmann, F., and Schultz, G. (1981). Soluble guanylate cyclase purified from bovine lung contains heme and copper. *FEBS Lett. 132*:71–74.

Giuili, G., Scholl, U., Bulle, F., and Guellaen, G. (1992). Molecular cloning of the cDNAs coding for the two subunits of soluble guanylyl cyclase from human brain. *FEBS Lett. 304*:83–88.

Harteneck, C., Wedel, B., Koesling, D., Malkewitz, J., Böhme, E., and Schultz, G. (1991). Molecular cloning and expression of a new α-subunit of soluble guanylyl cylase. Interchangeability of the α-subunits of the enzyme. *FEBS Lett. 292*:217–222.

Hope, B.T., Michael, G.J., Knigge, K.M., and Vincent, S.R. (1991). Neuronal NADPH diaphorase is a nitric oxide synthase. *Proc. Natl. Acad. Sci. USA 88*:2811–2814.

Humbert, P., Niroomand, F., Fischer, G., Mayer, B., Koesling, D., Hinsch, K.-H., Gausepohl, H., Frank, R., Schultz, G., and Böhme, E. (1990). Purification of soluble guanylate cyclase from bovine lung by a new immunoaffinity chromatographic method. *Eur. J. Biochem. 190*:273–278.

Janssens, S.P., Schimouchi, A., Quertermous, T., Bloch, D.B., and Bloch, K.D. (1992). Cloning and expression of a cDNA encoding human endothelium-derived relaxing factor/nitric oxide synthase. *J. Biol. Chem. 267*:14519–14522.

Jirikowski, G.F., Back, H., Forssmann, W.G., Stumpf, W.E. (1986). Coexistence of atrial natriuretic factor (ANF) and oxytocin in neurons of the rat hypothalamus. *Neuropeptides 8*:243–249.

Kamisaki, Y., Saheki, S., Nakane, M., Palmieri, J., Kuno, T., Chang, B., Waldman, S.A., and Murad, F. (1986). Soluble guanylate cyclase from rat lung exists as a heterodimer. *J. Biol. Chem. 261*:7236–7241.

Kawata, M., Hirakawa, M., Kumamoto, K., Minamino, N., Kangawa, K., Matsuo, H., and Sano, Y. (1989). Brain natriuretic peptide in the porcine spinal cord: an immuno-histochemical investigation of its localization and the comparison with atrial natriuretic peptide, substance P, calcitonin gene-related peptide, and enkephalin. *Neuroscience 33*: 401–410.

Klatt, P., Heinzel, B., John, M., Kastner, M., Böhme, E., and Mayer, B. (1992). Ca^{2+}/calmodulin-dependent cytochrome c reductase activity of brain nitric oxide synthase. *J. Biol. Chem. 267*:11374–11378.

Kobzik, L., Drazen, J., Bredt, D., Lowenstein, C., Snyder, S., Gaston, B., Sugarbaker, D., Loszcalzo, J., and Stamler, J. (1993). Nitric oxide synthase (NOS) in the lung: immuno-logic and histochemical localization in human and rat tissue. *Am. Rev. Respir. Dis. 147*:A515.

Koesling, D., Herz, J., Gausepohl, H., Niroomand, F., Hinsch, K-D., Mülsch, H., Böhme, E., Schultz, G., and Frank, R. (1988). The primary structure of the 70 kDa subunit of bovine soluble guanylate cyclase. *FEBS Lett. 239*:29–34.

Koesling, D., Harteneck, C., Humbert, P., Bosserhoff, A., Frank, R., Schultz, G., and Böhme, E. (1990). The primary structure of the large subunit of soluble guanylyl cyclase from bovine lung. *FEBS. Lett. 266*:128–132.

Koesling, D., Böhme, E., and Schultz, G. (1991). Guanylyl cyclases, a growing family of signal-transducing enzymes. *FASEB J. 5*:2785–2791.

Kummer, W. (1992). Neuronal specificity and plasticity in the autonomic nervous system. *Ann. Anat. 174*:409–417.

Kummer, W., Fischer, A., Mundel, P., Mayer, B., Hoba, B., Philippin, B., and Preissler, U. (1992a). Nitric oxide synthase in VIP-containing vasodilator nerve fibres in the guinea pig. *NeuroReport 3*:653–655.

Kummer, W., Fischer, A., and Mayer, B. (1992b). Substance P and nitric oxide: participation in airway innervation. *Regul. Peptides* (Suppl. 1):S92.

Kummer, W., Fischer, A., Kurkowski, R., and Heym, C. (1992c). The sensory and sympathetic innervation of guinea-pig lung and trachea as studied by retrograde neuronal tracing and double-labelling immunohistochemistry. *Neuroscience 49*:715–737.

Kuno, T., Andresen, J.W., Kamisaki, Y., Waldman, S.A., Chang, L.Y., Saheki, S., Leitman, D.C., Nakane, M., and Nurad, F. (1986). Copurification of an atrial natriuretic factor receptor and particulate guanylate cyclase from rat lung. *J. Biol. Chem. 261*:5817–5823.

Li, C.G., and Rand, M.J. (1991). Evidence that part of the NANC relaxant response of guinea-pig trachea to electrical field stimulation is mediated by nitric oxide. *Br. J. Pharmacol. 102*:91–94.

Mayer, B., and Böhme, E. (1989). Ca^{2+}-dependent formation of an L-arginine-derived activator of soluble guanylyl cyclase in bovine lung. *FEBS Lett. 256*:211–214.

McGehee, D.S., Goy, M.F., and Oxford, G.S. (1992). Involvement of the nitric oxide–cyclic GMP pathway in the desensitization of bradykinin responses of cultured rat sensory neurons. *Neuron 9*:315–324.

Morris, S., Southam, E., Braid, D.J., and Garthwaite, J. (1992) Nitric oxide may act as a messenger between dorsal root ganglion neurons and their satellite cells. *Neurosci. Lett. 137*:29–32.

Murad, F., Mittal, C.K., Arnold, W.P., Katsuki, S., and Kimura, H. (1978). Guanylate cyclase: activation by azide, nitrocompounds, nitric oxide, and hydroxyl radical and inhibition by hemoglobin and myoglobin. *Adv. Cyclic Nucleotide Res. 9*:145–158.

Nakane, M., Saheki, S., Kuno, T., Ishii, K., and Murad, F. (1988). Molecular cloning of a cDNA coding for 70 kilodalton subunit of soluble guanylate cyclase from rat lung. *Biochem. Biophys. Res. Commun. 157*:1139–1147.

Nakane, M., Arai, K., Saheki, S., Kuno, T., Buechler, W., and Murad, F. (1990). Molecular cloning and expression of cDNAs coding for soluble guanylate cyclase from rat lung. *J. Biol. Chem. 265*:16841–16845.

Nohr, D., Weihe, E., Zentel, H.J., and Arendt, R.M. (1989). Atrial natriuretic factor-like immunoreactivity in spinal cord and in primary sensory neurons of spinal and trigeminal ganglia of guinea-pig: correlation with tachykinin immunoreactivity. *Cell Tissue Res. 258*:387–392.

Ohyama, Y., Miyamoto, K., Saito, Y., Minamino, N., Kangawa, K., and Matsuo, H. (1992). Cloning and characterization of two forms of C-type natriuretic peptide receptor in rat brain. *Biochem. Biophys. Res. Commun. 183*:743–749.

Olry, R., Mayer, B., and Kummer, W. (1993). Co-localization of nitric oxide synthase with tyrosine hydroxylase and neuropeptides in guinea pig sympathetic ganglia (submitted).

Porter, J.G., Arfsten, A., Fuller, F., Miller, J.A., Gregory, L.C., and Lewicki, J.A. (1990). Isolation and functional expression of the human atrial natriuretic peptide clearance receptor cDNA. *Biochem. Biophys. Res. Commun. 171*:796–803.

Redmond, E.M., Cahill, P.A., and Keenan, A.K. (1990). Atrial natriuretic factor recognized two receptor subtypes in endothelial cells cultured from bovine pulmonary artery. *FEBS Lett. 269*:157–162.

Said, S.I. (1992). Nitric oxide and vasoactive intestinal peptide: cotransmitters of smooth muscle relaxation. *NIPS 7*:181–183.

Sakamoto, M., Nakao, K., and Morii, N. (1986). The lung as a possible target organ for atrial natriuretic polypeptide secreted from the heart. *Biochem. Biophys. Res. Commun. 135*:515–520.

Saper, C.B., Hurley, K.M., Moga, M.M., Holmes, H.R., Adams, S.A., Leahy, K.M., and Needleman, P. (1989). Brain natriuretic peptide: differential localization of a new family of neuropeptides. *Neurosci. Lett. 96*:29–34.

Schmidt, H.H.H.W., Gagne, G.D., Nakane, M., Pollock, J.S., Miller, M.F., and Murad, F. (1992). Mapping of neural nitric oxide synthase in the rat suggests frequent colocalization with NADPH diaphorase but not with soluble guanylyl cyclase, and novel paraneural functions for nitrinergic signal transduction. *J. Histochem. Cytochem. 40*: 1439–1456.

Schultz, H.D., Gardner, D.G., Deschapper, C.F., Coleridge, H.M., and Coleridge, J.C.G. (1988). Vagal C-fiber blockade abolishes sympathetic inhibition by atrial natriuretic factor. *Am. J. Physiol. 255*:R6–13.

Secca, T., Vagnetti, D., Dolcini, B.M., and Di Rosa, I. (1991). Cytochemical and biochemical observations on the alveolar guanylate cyclase of golden hamster lung. *Tissue Cell 23*:67–74.

Shimosegawa, T., Foda, H.D., and Said, S. (1990). [Met]Enkephalin-arg[6]-gly[7]-leu[8]-immunoreactive nerves in guinea-pig and rat lungs: distribution, origin, and coexistence with vasoactive intestinal polypeptide immunoreactivity. *Neuroscience 36*: 737–750.

Stuart-Smith, K., and Hirshman, C.A. (1993). Regional variation in the response to nitric oxide (NO) in porcine airway smooth muscle. *Am. Rev. Respir. Dis. 147*:A514.

Sudoh, T., Kangawa, K, Minamino, M., and Matsuo, H. (1988). A new natriuretic peptide in porcine brain. *Nature 332*:78–80.

Sudoh, T., Minamino, N., Kangawa, K., and Matsuo, H. (1990). C-type natriuretic peptide (CNP): a new member of natriuretic peptide family identified in porcine brain. *Biochem. Biophys. Res. Commun. 168*:863–870.

Triepel, J., Heym, C., and Reinecke, M. (1986). VIP- and PHI-immunoreactive neurons in the guinea-pig autonomic nervous system. *J. Autonom. Nerv. Syst.* (Suppl.):331–338.

Tucker, J.F., Brave, S.R., Charalambous, Lista, Hobbs, A.J., and Gibson, A. (1990). L-N^G-nitro arginine inhibits non-adrenergic, non-cholinergic relaxations of guinea-pig isolated tracheal smooth muscle. *Br. J. Pharmacol. 100*:663–664.

Ueda, S., Minamino, N., Abuyara, M., Kangawa, K., Matsukara, S., and Matsuo, H. (1991). Distribution and characterization of immunoreactive porcine C-type natriuretic peptide. *Biochem. Biophys. Res. Commun. 175*:759–767.

Xie, Q., Cho, H.J., Calaycay, J., Mumford, R.A., Swiderek, K.M., Lee, T.D., Ding, A., Troso, T., and Nathan, C. (1992). Cloning and characterization of inducible nitric oxide synthase from mouse macrophages. *Science 256*:225–228.

Yuen, P.S.T., Potter, L.R., and Garbers, D.L. (1990). A new form of guanylyl cyclase is preferentially expressed in rat kidney. *Biochemistry 29*:10872–10878.

30

Neuropeptides in Bronchoalveolar Lavage in Allergic Asthma

Gert H. H. Kunkel, H. Ratti, M. Zhang, and J. Niehus
Free University of Berlin, Berlin, Germany

Karen Nieber and Jens Furkert
Research Institute of Molecular Pharmacology, Berlin, Germany

I. INTRODUCTION

From experimental and clinical studies, it has been recognized in recent years that allergic bronchial asthma is a complex process not only limited to the release of immediate mediators such as histamine, bradykinin, platelet-activating factor, and metabolites of the arachidonic acid, and metabolism of mast cells and basophils. A number of other factors are involved, including genetic factors, cell-cell inter-actions by interleukins, as well as nonallergic pharmacological or environmental stimuli (1). Since bronchial hyperreactivity (BH) is one of the main diagnostic features of bronchial asthma, it is important to evaluate its underlying causes. There is no doubt today that one of the requirements for the development of BH is airway inflammation (2). Histologically, epithelial damage is associated with inflammatory cell infiltration as a characteristic feature of asthma and therefore may be causally related to the accompanying airway hyperreactivity (3,4). In recent years, a new, but still controversial, hypothesis integrates neuropeptides such as vasointestinal peptide, neurokinin A, calcitonin gene-related peptide, and substance P into the pathophysiological considerations of the development of bronchial-inflammatory diseases; here we discuss allergic bronchial asthma.

Many neuropeptides have been identified in the human airways (5). Their precise physiological and pathophysiological role in the airways still remains

653

uncertain. However, their potent effects on various aspects of airway function suggest that they are involved in airway function. Therefore, it is not unreasonable to assume that neuropeptide control mechanisms might be affected by inflammatory processes, or vice versa, i.e., neurogenic inflammation.

The effects seen after the release of tachykinins by several stimuli, such as electrical, mechanical, chemical, or pharmacological agents, are consistent with neurogenic inflammation. Antidromic electrical stimulation leads to increased vascular permeability, neutrophil adhesion to the vascular endothelium, submucosal gland secretion, and smooth muscle contraction (6). In accordance with the axon reflex hypothesis (7), it has been suggested that mechanical and chemical stimulation of airway sensory nerve endings causes antidromic release of substance P from axon collaterals, leading to the inflammatory effects previously discussed. The best-evaluated neuropeptide so far is substance P, but few studies have examined the effects on airway function, predominantly the upper airways, in humans. These studies indicate that neuropeptides such as substance P make an important contribution to airway function in allergic disease (8–10).

Another neuropeptide in the opioid family (Leu-encephalin, dynorphin) is β-endorphin, for which only some animal experimental data exist, indicating that opioids may also be involved in airway control (11).

Another important factor in context with neurogenic inflammation is the main degrading enzyme neutral endopeptidase (NEP 3.4.24.11). It has been shown that NEP normally limits the effects of endogenously released tachykinins. Experimental data exist that indicate that NEP activity is altered after exposure to various stimuli and that this finding is associated with an increased neurogenic inflammatory response. Toluene diisocyanate–induced asthma, experimental influenza infections, and cigarette smoke exposure in animals (12–14) lead to decreased NEP activity in airway tissue. NEP was localized in the airway by immunocytochemical methods in animals and humans with polyclonal and monoclonal antibodies. It is mainly seen in the basal cells of the airway epithelium and in airway smooth muscle, nerves, submucosal glands, and postcapilliary venous endothelium (15,16). There are no experimental data available so far with respect to bronchial asthma and bronchial hyperreactivity in humans. To obtain more information about the role of substance P and β-endorphin, we measured the concentration of substance P in bronchoalveolar lavage (BAL) fluid in grass pollen–allergic patients before and after allergen provocation. The results were compared with the findings in nonallergic subjects.

II. METHODS

The study protocol was approved by the local ethics committee. After giving written informed consent, six healthy nonallergic subjects (three women, three men; age range 19–41 years) and six patients with grass pollen asthma (two women, four men; age range 18–43 years) were enrolled in the study out of season.

The diagnosis was confirmed by history, skin prick test, and radioallergosorbent test. The history of seasonal asthma was from 2 to 8 years. The following criteria were used for selection of allergic patients:

1. Age range 18–45 years.
2. History of mild seasonal grass pollen asthma.
3. Functional evidence of reversible obstructive airway disease by nonspecific (histamine) and specific (grass pollen allergen) bronchial provocation test.
4. Normal pulmonary function for 4 weeks before the examination without medication.
5. FEV_1 immediately before lavage $>80\%$ of predicted value.
6. No drugs were allowed during this study.

The bronchial provocation was performed locally under bronchoscopic control in the right middle lobe bronchus. The individual allergen provocation dose was determined prior to this examination by inhalative provocation and considered positive with a drop of FEV_1 by 20% and a rise of airway resistance of 100%. The BAL was repeated 8 and 24 hr later. Commercially available grass pollen extracts from ALK-Scherax were used. The local provocation dose varied between 10 and 2500 quality units (SQU).

Substance P and β-endorphin were measured directly in the lavage fluid in triple sets and calculated to pg/ml lavage fluid with highly specific radioimmunoassays (RIA). The minimal detectable concentration was 15 fmol/ml = 20 pg/ml lavage fluid for substance P and 5 fmol/ml = 17 pg/ml for β-endorphin, respectively. For statistical analysis, independent samples were compared with the t-test, and statistical analysis of the multiple time points was performed by analysis of variance (ANOVA) (details described in Ref. 18). Substance P and β-endorphin are expressed as immuno-like reactivity (ILR).

III. RESULTS

Comparing the baseline concentrations for substance P, there was a significant difference ($p < 0.01$) between patients and volunteers. The baseline concentration of SP was 6.8 times higher in allergic persons. The same finding was observed for β-endorphin, where the baseline was 14.3 times higher in patients than in volunteers (Table 1). After segmental provocation with grass pollen allergen (10–2500 SQU ALK Grass extract), SP in BAL fluid increased by 37.9%, $p < 0.05$) (Fig. 1) and β-endorphin by 110% (Fig. 2). Both values returned to baseline levels within 24 hr.

To determine whether there is any relationship between neuropeptides and NEP in lavage fluids and/or serum of patients, we developed a new ELISA method for the detection of NEP in serum and lavage fluid (19). We have evaluated the correlation between NEP in serum and bronchial reactivity to acetylcholine (ACh)

Table 1   Baseline Concentration of SP and β-Endorphin in BAL
Fluid of Healthy, Nonatopic Subjects (Volunteers) and Patients
Allergic to Grass (Allergics)

	Volunteers	Allergic persons
Substance P (ILR) BAL (pg/ml)	37 ± 9	253 ± 24*
β-Endorphin (ILR) BAL (pg/ml)	3.5 ± 0.6	50 ± 16.3*

*$p < 0.01$.
Mean ± SEM of six subjects in each group.

in 75 patients. We found a highly significant correlation between serum NEP and
FEV_1 ($p < 0.003$), a significant correlation between serum NEP and the decrease of PEF after ACh provocation (%) ($p < 0.05$), and a highly significant
correlation between serum NEP and FEV_1 in the group of patients with any
response to ACh ($p < 0.008$).

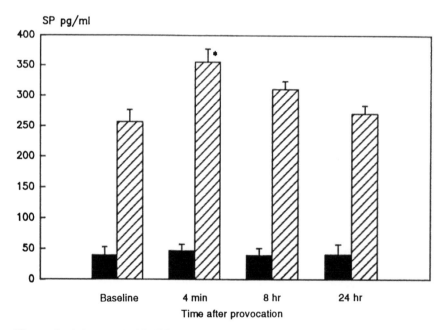

Figure 1   Substance P (SP) ILR concentration in BAL fluids before and after allergen
provocation in nonallergic subjects (volunteers, $n = 6$) and patients allergic to grass pollen
($n = 6$). *$p < 0.05$ after versus before provocation.

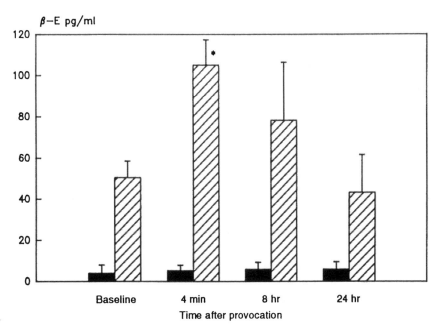

Figure 2 β-Endorphin (β-E) ILR concentration in BAL fluids before and after intrasegmental allergen provocation in nonallergic subjects (volunteers, $n = 6$) and patients allergic to grass pollen ($n = 6$). *$p < 0.05$ after versus before provocation.

IV. DISCUSSION

It is well accepted today that airway narrowing associated with asthma is mainly due to an inflammatory response of the airways. The mechanism of inflammation is combined with different inflammatory cells and several mediators. To study these different mechanisms, BAL techniques have been developed. In the attempt to understand the pathophysiology, considerable interest has also been focused on airway neuropeptides, especially substance P. The present results provide evidence that baseline concentrations of substance P and β-endorphin in BAL fluid of allergic asthmatics was significantly elevated. Moreover, allergen provocation induces an additional immediate rise of those neuropeptides, but only in allergic patients and not in volunteers. This finding is probably due to an increased activity of sensory nerves and/or epithelium alterations. In agreement with the hypothesis that neurogenic inflammation is caused by an axon reflex (7), our results are indicative of an abnormally elevated activity of nonadrenergic, noncholinergic nerves with liberation of substance P. The role of β-endorphin as a member of the opioid family still remains unknown since opioids are mainly confined to diffuse granule-containing cells. Bartho et al. (11) showed that the noncholinergic contraction of the guinea pig–isolated main bronchi as a result of a direct electrical stimulation was inhibited by enkephalin. This effect was abolished by naloxone; since naloxone alone induced no change in the evoked contraction, it is reasonable to assume that there is probably no endogenous opioid control in the bronchi of untreated guinea pigs. However, it is also possible that opioids have a completely different site of action.

The metalloenzyme neutral endopeptidase has been described in a wide variety of tissue and is present in the airway epithelium and airway smooth muscle, nerves, postcapillary venous endothelium, submucosal glands, and perineurium of the vagus nerve. The correlation found between different parameters connected with ACh response (FEV_1, PEF) indicates that NEP is also involved in the neural control of airway reactivity. The fact that there was no correlation between FEV_1 and serum NEP in patients with bronchial hyperreactivity defined by World Health Organization standards (decrease of FEV_1 more than 20% and/or increase of RAW more than 100%) implies that the observed correlations are an expression of an early state of inflammation, whereas in patients with bronchial hyperreactivity as an expression of chronic bronchial inflammation, the neural aspect of inflammation may be less important than the cellular aspect. However, this finding might also be explained by the low number of patients in this group ($n = 21$), compared to the total group of patients responding to acetylcholine with a drop of FEV_1 in any range (20).

REFERENCES

1. Barnes, P.J. (1987). Neuropeptides in the lung: localisation, function and pathophysiological implications. *J. Allergy Clin. Immunol. 79*:285–295.
2. Chung, K.F. (1986). Role of inflammation in the hyperreactivity of the airway in asthma. *Thorax 41*:657–662.
3. Laitinen, L.A., Heino, M., Laitinen, A., Kava, T., and Haahtela, T. (1985). Damage of the airway epithelium in bronchial reactivity in patients with asthma. *Am. Rev. Respir. Dis. 131*:599–606.
4. Kirby, J.G., Hargreave, F.G., Gleich, G.J., and O'Byrne, P.M. (1987). Bronchioalveolar cell profiles of asthmatic and non-asthmatic subjects. *Am. Rev. Respir. Dis. 136*:379–383.
5. Polak, J.M., and Bloom, S.R. (1986). Regulatory peptides of the gastrointestinal and the respiratory tracts. *Arch. Int. Pharmodyn. 280*(Suppl.):16–49.
6. Nadel, J.A. (1989). Neutral endopeptidase modulates neurogenic inflammation in airways. *Allergologie 12*:136–138.
7. Barnes, P.J. (1986). Asthma as an axon reflex. *Lancet 1*:242–245.
8. Pujol, J.L., Bousquet, J., and Grenier, J. (1990). Substance P activation of bronchoalveolar machrophages from asthmatic patients and normal subjects. *Clin. Exp. Allergy 19*:625–628.

9. Nieber, K., Baumgarten, C., Witzel, A., and Kunkel, G. (1991). The possible role of substance P in the allergic reaction, based on two different provocation models. *Int. Arch. Allergy Appl. Immunol. 94*:334–338.

10. Nieber, K., Baumgarten, C., Rathsack, R., Furkert, J., Laake, E., Müller, S., and Kunkel, G. (1991). Effect of azelastine on substance P content in bronchoalveolar and nasal lavage fluids of patients with allergic asthma. *Clin. Exp. Allergy 23*:69–71.

11. Bartho, L., Amann, R., Saria, A., Szolcsanyi, J., and Lembeck, F. (1987). Peripheral effects of opioid drugs on capsaicine-sensitive neurons of the guinea pig bronchus and rabbit ear. *Naunyn-Schmiedeberg's Arch. Pharmacol. 336*:316–320.

12. Sheppard, D., Thompson, J.E., Scypinski, L., Dusser, D., Nadel, J.A., and Borson, D.B. (1988). Toluene diisocynate increases airway responsiveness to substance P and decreases airway responsiveness to substance P and decreases airway enkephalinase. *J. Clin. Invest. 81*:1111–1115.

13. Jakoby, D.B., Tamaoki, J., Borson, D.B., and Nadel, J.A. (1988). Influenza infection increased airway smooth muscle responsiveness to substance P in ferrets by decreasing enkephaline. *J. Appl. Physiol. 64*:2653–2658.

14. Dusser, D.J., Djokic, T.D., Borson, D.B., and Nadel, J.A. (1989). Free radicals from cigarette smoke induced bronchoconstrictor hyperresponsiveness to substance P by inactivating airway neutral endopeptidase. *FASEB J. 3A*:534 (abstract).

15. Sekizawa, K., Tamaoki, J., Graf, T.D., Basbaum, C.B., Borsun, D.B., and Nadel, J.A. (1987). Enkephalinase inhibitor potentiates mammalian tachykinin-induced contraction in ferret trachea. *J. Pharmcol. Exp. Ther. 243*:1211–1217.

16. Sekizawa, K., Tamaoki, J., Nadel, J.A., and Borson, D.B. (1987). Enkephalinase inhibitor potentiates substance P– and electrically induced contraction in ferret trachea. *J. Appl. Physiol. 63*:1401–1405.

17. Hennig, R., Laake, E., Müller, S., Schilling, W.J., and Slapke, J. (1991). A method of intrasegmental allergen provocation followed by bronchoalveolar lavage in patients with pollen asthma. *J. Asthma 28*(4):251–254.

18. Nieber, K., Baumgarten, C.R., Rathsack, R., Furkert, J., Oehme, G., and Kunkel, G. (1992). Substance P and β-endophorin-like immuno-reactivity in lavage fluid of patients with and without allergic asthma. *J. Allergy Clin. Immunol. 90*:646–652.

19. Zhang, M., Niehus, J., Schnellbacher, T., Müller, S., Graf, K., Schultz, K.D., Baumgarten, C., Lucas, C., and Kunkel, G. A new ELISA for the detection of human neutral endopeptidase NEP EC 3.4.24.11 in human serum and blood leucocytes (submitted).

20. Ratti, H., Niehus, J., Zhang, M., and Kunkel, G. (1993). Correlation between neutral endopeptidase (NEP) EC 3.4.24.11 in serum and bronchial hyperreactivity to acetylcholine. *Allergy J. 2*:44 (abstract).

31

Morphology of Neurogenic Inflammation in the Airways: Plasma Protein Movement Across the Airway Mucosa and Epithelium

Marek L. Kowalski
Faculty of Medicine, Medical Academy, Łódź, Poland

Maria Śliwinska-Kowalska
Institute of Occupational Medicine, Łódź, Poland

James N. Baraniuk
Georgetown University Medical Center, Washington, D.C.

Michael A. Kaliner
Institute for Asthma and Allergy, Washington, D.C.

I. INTRODUCTION

Irritation of the rodent airway mucosa initiates local reflexes that induce vasodilatation and increased vascular permeability. This phenomenon is known as neurogenic inflammation (1,2). Neurogenic inflammation in the airway mucosa may also be induced by antidromic electrical stimulation of the vagal nerve with the release of vasoactive neuropeptides from nonadrenergic, noncholinergic, sensory nerve endings (1–3). Electrical stimulation of the vagal nerves, intratracheal capsaicin, and intravenous substance P induce mucosal edema formation and movement of plasma proteins into the lumens of tracheobronchial airways (4–8), indicating that neurogenic inflammation involves increased vascular and epithelial permeability for plasma proteins (7,9). This hypothesis was tested by assessing the distribution and identities of leaking vessels and morphological changes in epithelial cells generated following electrical stimulation of the vagal nerves in rat airways. The movement of plasma proteins within the airway mucosa and lumen was traced using autoradiography of the trachea following intravenous injection of ^{125}I-bovine serum albumin (^{125}I-BSA).

II. MATERIALS AND METHODS

A. Induction of Neurogenic Inflammation in the Airways

Sprague-Dawley rats were anesthetized with methohexital sodium (Brevital; Eli Lilly and Co., Indianapolis, IN), 60 mg/kg intraperitoneally. Neurogenic inflammation was induced as previously described (6). In brief, both cervical vagal nerves were exposed and transected. The distal ends were placed on bipolar, shielded electrodes (Harvard subminiature electrode, No. 50-1650; Harvard Apparatus Co., Boston, MA) and stimulated with square wave pulses (20 V, 10 Hz, 5 msec) using a Grass 88 stimulator (Grass Instruments, Quincy, MA). After this electrical vagal stimulation (EVS), animals were sacrificed by exsanguination, the airways were dissected, and the tissue was processed for light or electron microscopy.

B. Histological Assessment of Vascular and Epithelial Permeability

To localize leaking vessels, animals were injected with Monastral blue B (Sigma Chemical Co., St. Louis, MO) (0.1 ml of 3% suspension/100 g of body weight) 3 min before EVS (6,10). To remove intravascular tracer, animals were perfused with 500 ml of saline after EVS. For whole mount preparations, the lungs were fixed in 10% formalin for 48 hr, cleared in glycerin for 72 hr, and mounted in Permount (Fisher Scientific Co., Fair Lawn, NJ). For histological assessment, samples were fixed in Carnoy's solution for 16 hr at 20°C and embedded in paraffin. The 4-μm sections were cut at different levels of the tracheobronchial tree and stained with toluidine blue. Each experiment included at least three control (sham stimulated) and three vagally stimulated animals. Multiple sections were analyzed in each animal.

C. Electron Microscopy

The tissue samples were fixed in 3% glutaraldehyde and 2% paraformaldehyde in 0.1 cacodylate buffer for 2 hr, rinsed, placed in 1% buffered osmium tetroxide/ferrocyanide for 1 hr at 4°C, dehydrated, and embedded in plastic. Areas containing Monastral blue B dye were trimmed and sectioned on an ultramicrotome using a diamond knife. The sections were stained with Reynold's lead citrate and examined by transmission electron microscopy (11).

D. Autoradiography

Immediately prior to EVS animals received 3 μCi/100 g of ^{125}I-BSA intravenously. After EVS, the tracheas were excised and fixed in 4% formaldehyde/PBS. Tissues were embedded in paraffin and 6-μm-thick sections were cut. Slides were dewaxed in xylene followed by ethanol and then dried in air. In a darkroom, slides were dipped in melted nuclear track emulsion (NTB-2, Eastman Kodak, Rochester, NY). After 24 hr, they were developed and fixed in Kodak reagents, and then counterstained with alcian blue.

III. RESULTS

A. Permeable Vessels in the Airways

There was a dramatic increase in the amount of visually apparent Evans blue extravasated in the trachea and main bronchi. However, precise location of permeable vessels with Evans blue dye was impossible because the dye quickly diffuses away from the site of leakage. Since Evans blue is not electron dense, electron microscopy cannot be used to identify the site of leakage. For this purpose the macromolecular tracer Monastral blue B was used since its large particles (50–300 nm) do not cross intact endothelium, but instead mark only permeable vessels where the endothelial cells have been retracted to leave bare their basement membranes that bind the dye. After EVS and other neurogenic stimulation (6,12), Monastral blue B identifies leaking vessels in the trachea and in the first four to five divisions of the bronchi. The dense network of permeable venules was located primarily in the dorsal wall and between the cartilaginous rings of the trachea and bronchi (Figs. 1 and 2). In paraffin sections of the airways, the leaking vessels were localized to the subepithelial region and were often immediately adjacent to the epithelium (Fig. 3). No Monastral blue B–stained vessels were seen in control, sham-stimulated animals.

Electron microscopy (Figs. 4, 5, and 6) demonstrated that the leaking vessels had the characteristics of postcapillary venules and were 10–60 μm in diameter. Endothelial cells of permeable vessels were contracted and separated to form gaps of 0.25–2 μm diameter. Monastral blue B particles were trapped in the endothelial basal lamina, which is exposed to the vascular lumen when endothelial cells contract. Occasionally red blood cells were observed crossing the endothelium of permeable vessels (Fig. 4). Venules from control rats showed normal junctions between endothelial cells, and no Monastral blue particles were seen within vascular walls.

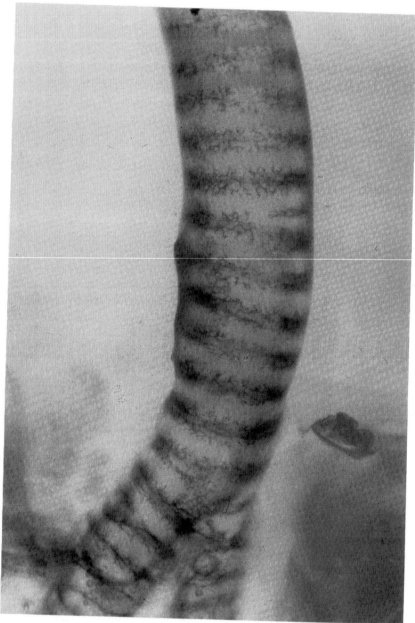

Figure 1 Rat trachea after EVS. Whole mount of the rat trachea and bronchi after 90 sec of electrical stimulation of both vagal nerves (20 V, 10 Hz, 5 msec). Note the dense network of leaking vessels between intercartilaginous rings of the airways (original magnification: ×16).

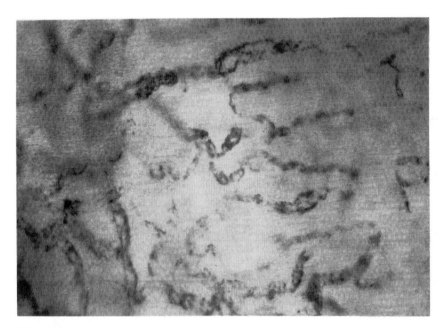

Figure 2 Rat tracheal whole mount. Higher magnification (original magnification ×25) of the whole mount of the rat trachea after electrical vagal nerve stimulation demonstrates a network of leaking vessels in the mucosa.

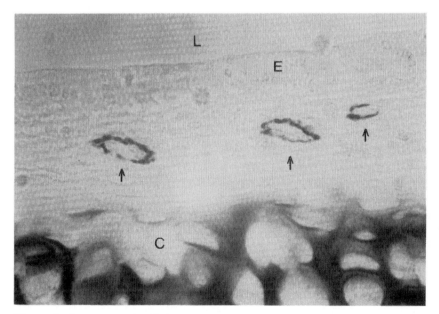

Figure 3 Rat trachea after EVS. Rat tracheal mucosa was counterstained with Toluidine blue after EVS. Arrows indicate Monastral blue B–labeled leaking vessels close to the airway epithelium. E = epithelium; C = cartilage; L = lumen of the airways (original magnification: ×800).

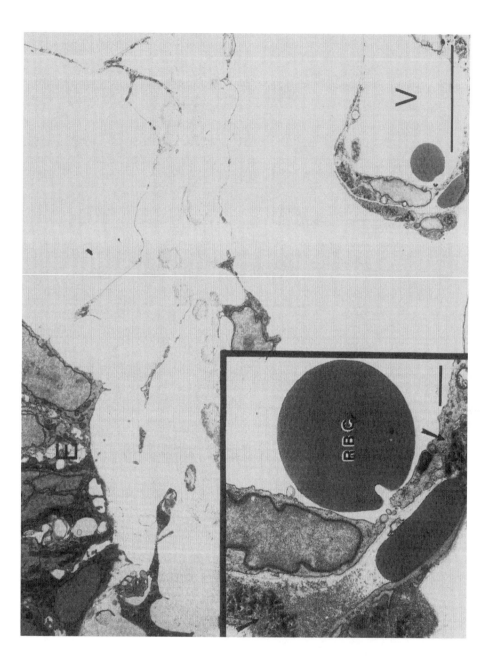

B. Epithelial Changes

Normal rat tracheal epithelium consisted of a single layer of ciliated and nonciliated cells. Blood vessels were located in the lamina propria beneath the airway epithelium (Fig. 6a). In the absence of vagal stimulation, epithelial cells were closely opposed. After 1 min of electrical stimulation, postcapillary venules demonstrated leakage of Monastral blue B, and the interstitial space of the lamina propria became widened as extracellular fluid accumulated (Figs. 4 and 6b).

EVS induced morphological changes in the airway epithelium with widening of the basolateral spaces between ciliated, secretory, and basal cells. The widened spaces were between 1 and 3 μm in size and were most frequently seen between the basal portions of epithelial cells. Widened spaces were only occasionally found in the more superior, luminal half of the basolateral spaces of these cells. The area of tight junctions between apical portion of epithelial cells seemed to be intact.

The epithelial secretory cells were contracted and degranulated with few granules remaining within the cytoplasm.

C. Movement of Plasma Proteins

The movement of plasma proteins from vessels, through the interstitial spaces, and across the epithelium was traced by following the extravasation of intravenously injected ^{125}I-BSA by autoradiography. Distribution of the tracheobronchial ^{125}I-BSA was detected as silver grains on autoradiograms (Figs. 7 and 8). EVS induced extravasation of intravenously injected ^{125}I-BSA from the walls of permeable vessels identified by Monastral blue B staining and into the interstitial space. ^{125}I-BSA rapidly became distributed throughout the mucosa from the adventitial connective tissue to the epithelium, and into the glycocalyx in the airway lumen.

IV. DISCUSSION

In 1976 in Hungary, Szolcsanyi first demonstrated that electrical stimulation of the vagus nerve could induce vascular permeability and edema formation in the skin and large airways mucosa of rodents (1). Jancso suggested that the reaction resulted from activation of nonadrenergic, noncholinergic, capsaicin-sensitive,

---

Figure 4 Subepithelial region of rat trachea after EVS. Electron micrograph of the tracheal subepithelial layer following EVS, demonstrating a leaking vessel (V) with contracted endothelial cells. Monastral blue B dye particles are seen trapped below the basement membrane. The wide "empty" space between the epithelium (E) and the vessel reflects subepithelial edema. (Bar = 5 μm.) Inset micrograph shows the magnified wall of the leaking vessel with a red blood cell partially crossing the endothelium and basement membrane. RBC = red blood cell; arrows indicate Monastral blue particles; bar = 1 μm.

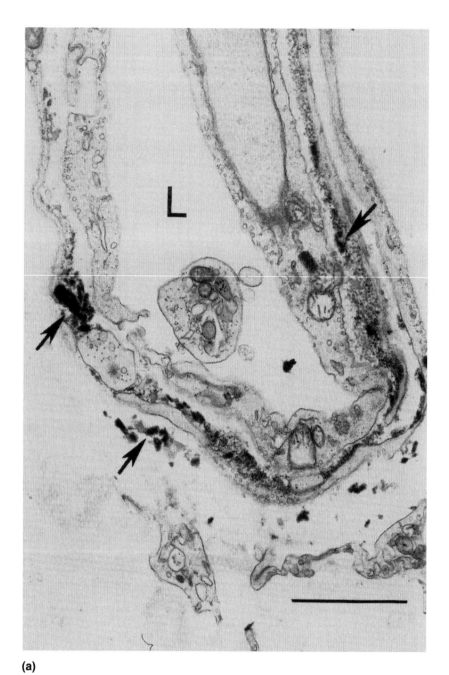

(a)

Figure 5 Leaking vessel in rat airway mucosa after EVS. (a) Monastral blue B particles

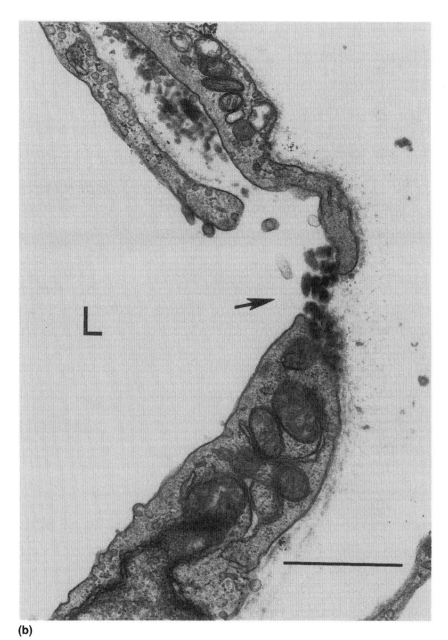

(b)

(arrows) are trapped in the endothelial basal lamina, which is exposed to the vascular lumen by a gap in the endothelium. L = vessel lumen; bar = 2 μm. (b) Leaking vessel in rat airway mucosa after EVS. Endothelial cells of permeable vessels are contracted and separated to form a gap (arrow). L = lumen; bar = 0.5 μm.

(a)

Figure 6 Rat tracheal epithelium after EVS. (a) Transmission electron micrograph of normal rat tracheal mucosa demonstrating the single layer of tightly apposed ciliated and nonciliated cells. Blood vessels are seen in the lamina propria beneath the airway epithelium (bar = 5 μm). (b) Bronchial mucosa after 1 min of electrical stimulation. Ciliated cells are compressed and intercellular gaps between cells have formed (stars). Secretory cell (right) are empty and compressed, indicating exocytosis has occurred. In the lamina propria, a leaking postcapillary venule with Monastral blue B deposit below the endothelial basement membrane (arrow) is seen. The wide spacing of lamina propria structures indicates the presence of edema fluid. (Bar = 5 μm.) (c) After 1 min of EVS, spaces are identified between the basal portion of ciliated cells (C) and between ciliated and basal cells (B). Junctions between upper portions of ciliated cells are intact. Secretory cells (S) are empty and vacuolized, indicating exocytosis. (Bar = 5 μm.)

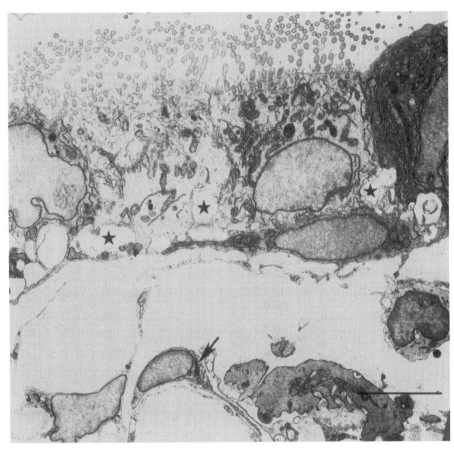

(b)

sensory nerve fibers (2,3,13) since EVS-induced reactions in the airway mucosa resembled neurogenic responses to capsaicin and the neuropeptide substance P (2, 7,14,15).

The current study demonstrates the morphological changes induced during neurogenic inflammation in the rodent airways. Based on the ultrastructural appearance of Monstral blue B–labeled vessels we confirmed that EVS-induced vascular permeability occurs exclusively within postcapillary venules in the airway mucosa. Leakage of intravascular tracers occurs as a result of contraction of endothelial cells and the transient formation of gaps between endothelial cells that lays bare the underlying basement membrane. These gaps permit the

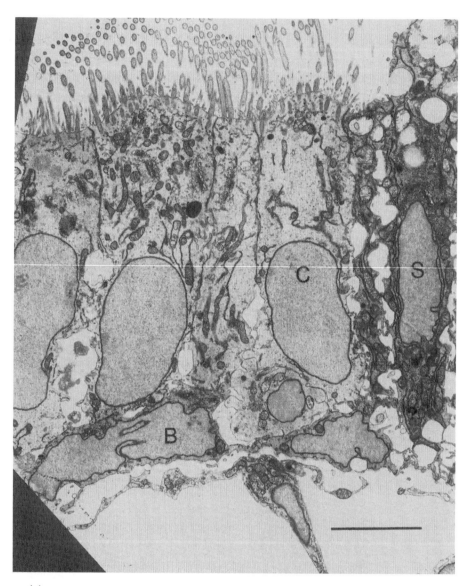

(c)

Figure 6   (Continued)

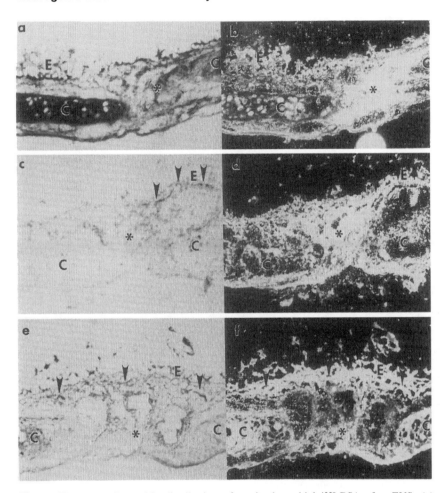

Figure 7 Autoradiographic distribution of tracheobronchial ^{125}I-BSA after EVS. (a) Bright-field view of the posterior membranous portion of the rat trachea showing epithelium (E), monastral blue—stained vessels (*), and cartilage (C). (b) In the dark-field view of the same field, dense white silver grains are seen in the interstitial region near a Monastral blue B—stained vessel (*). (c) Bright-field view showing large vessels (*) and small, Monastral blue B—stained postcapillary venules beneath the epithelial basement membrane (arrowheads). (d) Dark-field view of (c) showing silver grains throughout the interstitial region (*), epithelium (E), and the Monastral blue—stained venules (arrowheads). (e) Bright-field image showing Monastral blue B—stained vessels (arrowheads) and epithelium. (f) Dark-field view of (e) showing ^{125}I-BSA in Monastral blue B—stained venules and the epithelium. The series of autoradiographs indicates passage of radiolabeled albumin from vessels and through the interstitium to the epithelium. (Bar = 25 μm.)

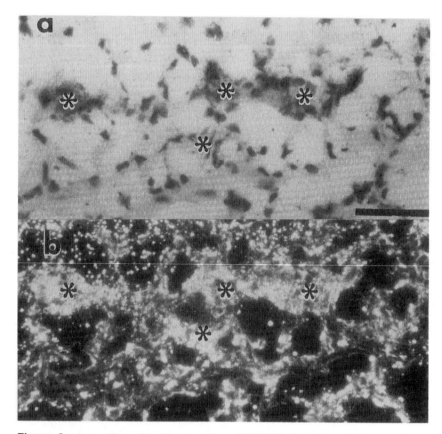

Figure 8 Autoradiographic demonstration of ^{125}I-BSA in Monastral blue B–stained venules. (a) Monastral blue B (*) was identified in postcapillary venules. (b) Dark-field image of the same field demonstrates ^{125}I-BSA-produced silver grains in the walls of venules (*). (Bar = 25 μm.)

rapid extravasation of plasma, which becomes interstitital edema fluid, and the limited extravasation of formed blood elements including erythrocytes and platelets. No leakage was seen within true capillaries. This response is similar to dermal neurogenic inflammation (11,16) and to the effects of histamine, bradykinin, and substance P (17). This neurogenic inflammatory reaction occurring after 1 min of electrical nerve stimulation appears to be the direct effect of neuromediators released from activated nerve endings on vascular endothelium, since previous studies have not revealed evidence of concomitant activation of inflammatory cells (8,18,19).

Neurogenic inflammation generated by electrical stimulation of the vagal nerves also induces changes in the airway epithelium with widening of intracellular spaces between epithelial cells in the basal half of the epithelium and discharge of granules from secretory cells. The morphology of EVS-induced gaps between epithelial cells depends on the cell type involved since larger gaps can appear between secretory and basal cells than between ciliated cells (12). This may result from a specialized distribution of sensory nerves between basal cells in the epithelium or different types of tight junctions between cells. There appear to be fewer fibrils in the junctions between secretory compared to ciliated cells (20). Secretory cells became compressed and contained few granules after EVS compared with cells in control, nonstimulated epithelium, indicating that exocytosis had been stimulated by EVS.

Rapid movement of plasma proteins from vessels, through the interstitium and into the airway lumen, has been demonstrated in neurogenic inflammation (4–6). This phenomenon has been observed after other neurogenic stimuli including substance P, capsaicin, and during allergic reactions in the airways (4,6,7). Extravasated proteins track rapidly through the airway mucosa and may pass across the epithelium by pericellular and intracellular pathways. Airway epithelial intracellular junctions are effective barriers that may regulate the flux of fluid and protein into the airway lumen and limit the access of inhaled material into the interstitial space (21–24). Electron microscopic photomicrographs indicate that after vagal stimulation basolateral intercellular spaces in the epithelium become dilated compared to unstimulated controls, supporting the concept that the interstitial fluid passes across the epithelium via the lateral intercellular spaces. Although intercellular spaces between basal portions of epithelial cells were clearly widened, the apical portions of epithelial cells seemed to be closely apposed. However, the resolution of the current microphotographs did not allow either precise localization of the site of leakage into the airway lumen or assessment of morphological and functional integrity of tight junctions. The extravasation of ^{125}I-BSA was tracked by autoradiography and confirmed that plasma proteins can rapidly pass from the vessels, fill the interstitial spaces of the lamina propria and peribronchial tissue as edema fluid, and pass across the epithelium into the airway lumen.

The mechanisms driving plasma out of vessels and through the epithelium are not clear. Hydrostatic pressures may contribute to this transendothelial flux and edema formation. Transmission of hydrostatic pressures may contribute to the flux of interstitial fluid across the epithelium since at sufficiently high hydrostatic pressures, the epithelial cells may be compressed laterally and allow extravasated plasma to flow into the airway lumen (25). This suggests a passive or permissive role of epithelial cells in the transepithelial fluid flux. However, the current electron microscopic investigations suggest that "packets" of fluid are generated between epithelial cells rather than a general widening of interepithelial spaces, suggesting that other factors such as active transport, modulation of intercellular junctions, or other active mechanisms also contribute to the transepithelial flux in vivo. Neurotransmitters released by electrical vagal nerve stimulation appear to regulate this epithelial phenomenon, induce secretory cell exocytosis, and post-capillary venule permeability, indicating that these neuropeptides regulate the integrity and function of epithelial and endothelial cells in vivo.

REFERENCES

1. Szolcsanyi, J. On the specificity of pain producing and sensory neuron blocking effect of capsaicin. In: *Symposium on Analgesics*. Knoll, J., and Vizzi, E. (eds.). Budapest: Akademiai Kiado, 1976:167–172.

2. Lundberg, J.M., and Saria, A. (1982). Capsaicin-sensitive vagal neurons involved in control of vascular permeability in rat trachea. *Acta Physiol. Scand. 115*:521–523.

3. Lundberg, J.M., Lundblad, L., Anggard, A., Martling, C-R., Thodersson-Norheim, E., Stjarne, P., Hokfelt, T., Saria, A.. Bioactive peptides in capsaicin-sensitive C-fiber afferents of the airways. Functional and pathophysiological implications. In: *The Airways. Neural Control in Health and Disease*. Kaliner, M., and Barnes, P. (eds.). New York: Marcel Dekker, 1988:417–445.

4. Persson, C.G.A., Erjefalt, I., and Andersson, P. (1986). Leakage of macromolecules from guinea-pig tracheobronchial microcirculation. Effects of allergen, leukotrienes, tachykinins and anti-asthma drugs. *Acta Physiol. Scand. 127*:95–105.

5. Erjefalt, I., and Persson, C.G.A. (1989). Inflammatory passage of plasma macromolecules into airway wall and lumen. *Pulm. Pharmacol. 2*:93–102.

6. Kowalski, M.L., Didier, A., and Kaliner, M.A. (1989). Neurogenic inflammation in the airways. I. Neurogenic stimulation induces plasma protein extravasation into the rat airway lumen. *Am. Rev. Respir. Dis. 140*:101–109.

7. Didier, A,. Kowalski, M.L., Jay, J., and Kaliner, M.A. (1990). Neurogenic inflammation, vascular permeability and mast cells: capsaicin densensitization fails to influence IgE-anti-DNP induced vascular permeability in rat airways. *Am. Rev. Respir. Dis. 141*:398–406.

8. Kowalski, M.L., Sertl, K., Didier, A., and Kaliner, M.A. Substance P induces mast cell–independent vascular permeability and plasma protein exudation into the airway lumen. *Int. Arch. Allergy Immunol.* (in press).

9. Sertl, K., Kowalski, M.L., Slater, J., and Kaliner, M.A. (1988). Passive sensitization and antigen challenge increases vascular permeability in rat airways. *Am. Rev. Respir. Dis. 138*:1295–1299.

10. Joris, I., De Girolami, U., Wortham, D., and Majno, G. (1982). Vascular labelling with Monastral blue B. *Stain Technol. 57*:177–183.

11. Kowalski, M.L., Sliwinska-Kowalska, M., and Kaliner, M.A. (1990). Neurogenic inflammation, vascular permeability, and mast cells. II. Further evidence indicating that mast cells are not involved in neurogenic inflammation. *J. Immunol. 145*:1214–1221.

12. McDonald, D.M. (1988). Neurogenic inflammation in the rat trachea. I. Changes in venules, leucocytes and epithelial cells. *J. Neurocytol. 17*:583–603.

13. McDonald, D.M. (1988). Neurogenic inflammation in the rat trachea. Identity and

distribution of nerves mediating the increase in vascular permeability. *J Neurocytol* *17*:605–628.

14. Saria, A., Lundberg, J.M., Skotfisch, G., and Lembeck, F. (1983). Vascular protein leakage in various tissues induced by substance P, capsaicin, bradykinin, serotonin, histamine and by antigen challenge. *Naunyn Schmiedeberg's Arch. Pharmacol.* *324*:212–213.

15. Sertl, K., Wiedermann, C.J., Kowalski, M., Hurtado, S., Plutchok, J., Linnoila, I., Pert, C.B., and Kaliner, M.A. (1988). Substance P: the relationship between receptor distribution in rat lung and the capacity of substance P to stimulate vascular permeability. *Am. Rev. Respir. Dis.* *138*:151–159.

16. Baraniuk, J.N., Kowalski, M.L., and Kaliner, M.A. (1990). Relationship between permeable vessels, nerves and mast cells in rat cutaneous neurogenic inflammation. *J. Appl. Physiol.* *68*:2305–2311.

17. Pietra, G.G., and Magno, M. (1978). Pharmacological factors influencing permeability of the bronchial circulation. *Fed. Proc.* *37*:2466–2470.

18. Kowalski, M.L., Didier, A., Lundgren, J.D., Igarashi, Y., and Kaliner, M.A. The role of sensory innervation and mast cells in neurogenic plasma protein exudation into the airway lumen. *Eur. Respir. J.* (in press).

19. Bethel, R.A., Brokaw, J.J., Evans, T.W., Nadel, J.A., McDonald, D.M. (1988). Substance P–induced increase in vascular permeability in the rat trachea does not depend on netrophils or other components of circulating blood. *Exp. Lung Res.* *14*:769–779.

20. Scheneeberger, E.E. (1980). Heterogeneity of tight junction morphology in extrapulmonary and intrapulmonary airways of the rat. *Anat. Rec.* *198*:193–208.

21. Inoue, S., and Hogg, J.C. (1974). Intracellular junctions in the tracheal epithelium in guinea pigs. *Lab. Invest.* *31*:68–71.

22. Simani, A.S., Inoue, S., and Hogg, J.C. (1974). Penetration of respiratory epithelium of guinea-pigs following exposure to cigarette smoke. *Lab. Invest.* *31*:75–80.

23. Boucher, R.C., Pare, P.D., Gilmore, N.J., Moroz, L.A., and Hogg, J.C. (1977). Airway mucosal permeability in the *Ascaris suum* sensitive rhesus monkey. *J. Allergy Clin. Immunol.* *60*:134–140.

24. Nathanson, I., and Nadel, J.A. (1984). Movement of electrolytes and fluid across airways. *Lung* *162*:125–137.

25. Persson, G.C.A., Erjefalt, I., Gustafsson, B., and Luts, A. (1990). Subepithelial hydrostatic pressure may regulate plasma exudation across the mucosa. *Int. Arch. Allergy Appl. Immunol.* *92*:148–153.

Index

About the Editors

MICHAEL A. KALINER is Medical Director of the Institute for Asthma and Allergy, Washington, D.C., and Section Chief, Asthma and Immunology, Washington Hospital Center. He is the editor or coeditor of nine books, including *The Mast Cell in Health and Disease*, *The Airways: Neural Control in Health and Disease*, and *Asthma: Its Pathology and Treatment* (all titles, Marcel Dekker, Inc.) and the author or coauthor of some 350 book chapters, professional papers, and reviews. A former Chairman of the American Board of Allergy and Immunology and of the American Academy of Allergy and Immunology's Education Council, he serves as editor of the Clinical Allergy and Immunology series (Marcel Dekker, Inc.), is an editor of four journals, and is a member of the American Society for Clinical Investigation and the Association of American Physicians. Dr. Kaliner, a recipient of the U.S. Public Health Service Commendation Award for Meritorious Service and Outstanding Service Award Medal, received the M.D. degree (1967) from the University of Maryland at Baltimore.

PETER J. BARNES is Director of and Professor in the Department of Thoracic Medicine at the National Heart and Lung Institute, University of London, England, and Honorary Consultant Physician at Royal Brompton National Heart and Lung Hospital, London. The author, coauthor, editor, or coeditor of more than 15 books, including *Asthma: Its Pathology and Treatment*, *The Airways: Neural Control in Health and Disease*, and *Pharmacology of the Respiratory Tract* (all titles, Marcel Dekker, Inc.) and the author or coauthor of over 500 articles on respiratory pharmacology and asthma, he is a member of Royal College of Physicians, the British Thoracic Society, and the American Thoracic Society, among other organizations. He is Associate Editor of three journals and on the editorial boards of several journals. Dr. Barnes received the B.A. degree (1969) in natural sciences and medical sciences from the University of Cambridge, England, the B.M., B.Ch. degree (1972) from the University of Oxford, England, the M.A. degree (1973) from the University of Cambridge, and the D.M. (1982) and D.Sc. (1987) degrees from the University of Oxford.

GERT H. H. KUNKEL is Head of the Department for Clinical Immunology and Asthma Outpatient Clinic at the Rudolf Virchow Klinikum, Free University, Berlin, Germany. A past President of the German Society for Allergy and Clinical Immunology (1990–1993) and a recipient of the Karl Heyer Allergy Award (1992), he is the author or coauthor of over 350 professional papers on respiratory medicine and a member of the European Academy of Allergy and Clinical Immunology, the American Academy of Allergy and Clinical Immunology, and the European Society of Pneumology and Clinical and Respiratory Physiology, among other organizations, and he serves on the editorial boards of several journals. Dr. Kunkel received the M.D. degree (1961) from the Free University, Berlin, and received the venia legendi (1971) for internal medicine.

JAMES N. BARANIUK is Assistant Professor of Medicine in the Division of Rheumatology, Immunology, and Allergy at Georgetown University, Washington, D.C. A Fellow of the Royal College of Physicians of Canada and a member of the American Thoracic Society and the American Academy of Allergy and Immunology, among other organizations, he is the author or coauthor of over 100 professional papers and abstracts. Dr. Baraniuk received the B.Sc. degree (1976) in chemistry and microbiology, the B.Sc. (Hons.) degree (1979) in medicine, and the M.D. degree (1981) from the University of Manitoba, Winnipeg, Canada.

Printed and bound by CPI Group (UK) Ltd, Croydon, CR0 4YY

23/10/2024

01777917-0001